Pollution Risks, Impacts & Management

Social, Economic and Environmental Perspectives

Pollution Risks, Impacts and Management

Social, Economic and Environmental Perspectives

Golam Kibria, Ph.D

Global Scientist of Artificial Mussel (AM) Pollution Watch Program, Australia; Principal Researcher Fellow, RMIT University, Australia; Senior Environmental Scientist, Goulburn Murray Rural Water Corporation, Australia and Adjunct Professor, City University of Hong Kong

A. K. Yousuf Haroon, Dr Sci

Consultant (Fisheries), Food and Agriculture Organisation (FAO) of the United Nations (UN); National Technical Advisor, Support to the Bay of Bengal Large Marine Ecosystem (BOBLME) Project (GEF-FAO) and Chief Scientific Officer, Bangladesh Fisheries Research Institute (BFRI)

Gavin Rose, B.Sc (Hons)

Principal Scientist and Director, Kinvara Scientific P/L, Australia; Senior Research Scientist and Group Leader Organic Chemistry, DPIE Victoria, Australia; Research Fellow, Centre for Aquatic Pollution Identification and Management (University of Melbourne, Australia) and Technical Assessor, National Association of Testing Authorities Australia (NATA)

Md. M. Maruf Hossain, Ph.D

Professor, Institute of Marine Sciences, University of Chittagong, Bangladesh; UN/IMO Fellow (Global Protection of Marine Environment) and Pool of Experts on "Global Reporting and Assessments of the State of the Marine Environment"

Dayanthi Nugegoda, Ph.D

Professor of Ecotoxicology, RMIT University, Australia; Lead Researcher, Aquatic Environmental Stress Research Group (AQUEST), Australia; Independent Expert Scientific Panel on Coal Seam Gas and Large Coal Mining and the Forest Stewardship Council (FSC) of Australia

2021

Published by

SCIENTIFIC PUBLISHERS (INDIA)
5 A, New Pali Road
P.O. Box 91
Jodhpur - 342 001 INDIA
E-mail: info@scientificpub.com
Website: http://www.scientificpubonline.com

ISBN: 978-93-89184-96-9
e-ISBN: 978-93-8918-498-3

***Citation :** Kibria G, Haroon AKY, Rose G, Hossain MM and Nugegoda D 2021 Pollution Risks, Impacts and Management. Social, Economic and Environmental Perspectives. Scientific Publishers, Jodhpur, India. ISBN:* 9789389184969

Conflict of interests

There are no conflicts of interest. Golam Kibria has written the entire book (all the 13 chapters). AKY Haroon, Gavin Rose, Md M. Maruf Hossain and Dayanthi Nugegoda peer-reviewed the chapters and agreed to act as co-authors of the book.

Disclaimer and declarations

Declaration 1

The authors' current official title and official address used in the book is in good faith and intention only to reflect the authors' current status and location. However, the works depicted here does not relate to authors office work and was carried out during after hours, weekends and on leaves. This is a voluntary work dedicated for third world and least developed countries, the Asia-Pacific and Africa and Australia for which the authors did not receive any financial benefits or payment or royalty. The views expressed in the book is of the authors only and doesn't reflect the views of the organisation/ countries etc. they presently belong to or belonged in the past.

Declaration 2

All figures/ diagrams/ tables included in the book were acknowledged in the reference section of each chapter as well as at the bottom of each figure/ diagram/ table and were used in good faith and the intention of promoting science and as part of knowledge sharing and promotion of awareness of pollution (existing and emerging) and its risks and impacts, in particular in the third world countries such as the Asia-Pacific and Africa regions where information is not readily available or lacking in pollution.

Declaration 3

This publication may be of assistance to you but the authors or the publisher do not guarantee that the publication is without a flaw of any kind or wholly appropriate for your particular purposes and therefore disclaims all liability for any error, loss or other consequences which may arise from relying on any information in this book.

Printed in India

Author's Profile

Dr. Golam Kibria
(Ph.D. in Aquatic Pollution/ Environmental Management, Australia)

Dr Golam Kibria is currently a *Global Scientist of Artificial Mussel (AM) Pollution Watch Program*, Australia. Golam planned/led pollution risk assessment (heavy metals) in Australia, Bangladesh, Greece, Myanmar, Russia, South Korea, Sri Lanka (as a Lead Scientist of the Global AM Pollution Watch Program). As a *Senior Environmental Scientist,* he planned and managed the collaborative pesticide research in Australia (in collaboration with CSIRO (Federal, Commonwealth Scientific and Industrial Research Organisation), DPI (Department of Primary Industries, Victoria) and the Centre for Ecotoxicology, New South Wales) during 2001-2012 periods.

Dr Golam gained significant professional experience as i. Freshwater Pollution Specialist, ii. Environmental Impact Assessment Specialist, iii. Environment and Waste Officer, iv. Environmental Researcher, v. Climate Change Scientist, vi. Senior Environmental Scientist, vii. Lecturer and viii. Senior Scientific Officer. He supervised or trained up postgraduate research students/academic staffs, examined postgraduate thesis and taught/ trained up technical people, professionals, scientists, and grassroots people in Australia, Asia and Africa. Golam published three books on chemicals, climate change and pollution, four book chapters, 40 peer-reviewed journal papers, 36 peer-reviewed technical reports, nine technical presentations and 22 community capacity building and behaviour changed environmental and sustainability articles.

He has been a *peer-reviewer of 15 International journal papers* (Elsevier's, Springer), *Adjunct Professor*, City University of Hong Kong. Dr Kibria managed programmes & projects related to waste (plastics, e-waste, and food waste), pollution, low-carbon development, climate change, greenhouse gases, energy, water, agriculture, fisheries, pesticides, metals, nutrients, pathogens, public health, and community education. Apart from his Ph.D. training, he received training on wastes management (plastics, organics/ food); sustainable development goals; corporate social responsibility; environmental community engagement; climate science; pollution risks and impacts assessment; cities & climate change; energy security and pollution monitoring using innovative technology. He is in the Editorial Advisory Committee (APN Science Bulletin) and Reviewer of Research proposals of the Asia-Pacific Network of Global Change (APN); Leader/creator of an open *online Environmental Education Program* for developing countries

(*https://www.researchgate.net/project/Online-Environmental-Education-to-Help-Developing-Countries*)

Weblink: *https://www.researchgate.net/profile/Golam_Kibria7*

e-mail: *kibriagolam0@gmail.com*

Dr. A. K. Yousuf Haroon
(Dr. Sci. in Fisheries & Aquaculture, Norway)

Dr A.K. Yousuf Haroon is now a free-lancing Consultant (Fisheries), often works with the *Food and Agriculture Organization (FAO) of the UN*, Bangladesh. His field of interests are marine and inland fisheries, wetlands, climate change implications and adaptation management. Dr. Yousuf co-authored *six books* and four book chapters on fisheries, wetlands, pollution and climate change implications. Besides, he has to his credits more than *55 peer-reviewed research papers* published in international and national journals, 40 peer-reviewed technical reports and proceedings, 13 training modules/ manuals, 10 popular scientific articles, 22 outreach extension materials and one video clip on river shad (*Tenualosa ilisha*) fisheries management.

Dr. Yousuf served the Department of Fisheries (DoF), Bangladesh as a *Scientific Officer*; Department of Biology, University of Basrah, Iraq as a *Technical Teacher*; National Water Plan Project of the Bangladesh Water Development Board (BWDB) as a *Fisheries Consultant*; Bangladesh Fisheries Research Institute (BFRI) as *Senior*, *Principal* and *Chief Scientific Officer*; as *National Technical Advisor,* DANIDA-funded Strengthening Institutional Capacity of DoF (SICD) Project, ASPS II; as *National Advisor*, GEF-funded Support to the Bay of Bengal Large Marine Ecosystem (BOBLME) Project of BFRI; as *Fisheries/ Aquaculture* and *Climate Change Specialist* in FAO, Bangladesh; and as *Environmental Specialist* in the WB Group-funded Sustainable Coastal and Marine Fisheries of Bangladesh (SCMFPB) project, the preparatory phase of the DoF, Bangladesh.

Apart from his Doctor of Science (Dr. Sci.) degree, he got higher educational training from the Centre for Environment and Estuarine Studies (CEES), Solomons, University of Maryland, Maryland, USA; Asian Institute of Technology (AIT), Bangkok, Thailand and Tietgen Business College, Odens, Denmark. He is a member of the Network of Tropical Aquaculture and Fisheries Professionals (NTAF), Penang, Malaysia; Asian Fisheries Society and WorldFish (defunct ICLARM), Malaysia.

Weblink: *http://www.boblme.org/bangladesh.html*
e-mail: *dharoon@ymail.com*

Mr Gavin Rose *(BSc in Chemistry, Australia)*

Mr Gavin Rose has extensive experience in pesticide and organic contaminant residue testing. Previous roles include *Research Fellow*, Centre for Aquatic Pollution Identification and Management (University of Melbourne) and *Senior Research Scientist and Group Leader Organic Chemistry*, DPIE Victoria (now Agriculture Victoria) and State Chemistry Laboratory. Gavin has written or co-authored 20 scientific publications (peer-reviewed) and multiple technical reports since 2009. He has been a *Technical Assessor* for the National Association of Testing Authorities Australia (NATA) since 2002, specialising in ISO 17025 compliance for organic residues in foods and the environment. He was cited as a *NATA Technical Assessor of the Year* (one of three), NATA 2017 Annual Report. October 2017; National Association of Testing Authorities Australia (NATA) 70th Annual Report, page 38-39 *https://www.nata.com.au/phocadownload/annual_reports/NATA_Annual_Report_2017.pdf*

Weblink: *https://www.researchgate.net/profile/Gavin_Rose*

e-mail: *gavinrose51@yahoo.com*

Dr. Md. M. Maruf Hossain *(Ph.D. in Environmental Science, Malaysia)*

Dr. Md M. Maruf Hossain is the *Professor of Institute of Marine Sciences*, University of Chittagong, Bangladesh. His major achievements over the years was *Director* (Institute of Marine Sciences, University of Chittagong, Bangladesh); *UN/IMO fellow* (Global Protection of Marine Environment); *Principal Investigator & Project Coordinator* (impacts of shipbreaking on fisheries and fishers); *National Consultant* (Oil & Chemical Pollution) (NOSCOP) (NORAD/IMO); *International consultant* (Capacity building and enhancement of Fishery, Aquaculture and Biodiversity Conservation, Pollution Control in Bangladesh) (DFID, UK, DelPHE-UK, BC-INSPIRE), higher study linkages programs with UK Universities; National consultant, BOPA/SIDA (SWEDMAR); Member, UNEP team for the regional report publication; *Member of UN/DOLAS program* and the *pool of Experts* on "Global Reporting and Assessments of the State of the Marine Environment, including Socio-economic Aspects"; Expert member for the 2nd Regular Process (2016-2020) by the UN.

He has key research interests in several environmental areas including marine and coastal pollution, ecotoxicology, environmental impact assessment, water quality assessments, marine biodiversity and biosecurity. He has published *two books* (related to ship breaking impacts, EIA, Tannery Wastes) and *65 peer-reviewed journal papers* and technical reports. In addition to his Ph.D. training, Prof Hossain has completed a number of other training courses on environment and related issues such as postgraduate training on 'EIA & its Management' from the University of Aberdeen, Scotland & AIT, Thailand and on 'Environmental Dynamics' from the University Di-Venezia, Italy. Prof Hossain is actively involved with different Environment and Biodiversity group, editorial board and organisations.

Weblink: *https://www.researchgate.net/profile/Mdm_Hossain*

e-mail: *marufimsfcu@gmail.com*

Dr. Dayanthi Nugegoda *(Ph.D. in Ecotoxicology, UK)*

Dr. Dayanthi Nugegoda is the *Professor of Ecotoxicology* at RMIT University, Australia and a *lead researcher* in the Aquatic Environmental Stress (AQUEST) Research Group. Her group has developed novel methods to assess, monitor and evaluate the effect of toxicants and environmental stressors on aquatic organisms, and ecosystems. She has published *>160 peer-reviewed journal articles*, three books, five book chapters and >30 reports to industry. She has graduated 24 PhD and five MSc research students and currently supervises seven PhD candidates.

She served (2012-2014) on the *Independent Expert Scientific Panel* on Coal Seam Gas and Large Coal Mining; and the Forest Stewardship Council (FSC) of Australia. She advises The New South Wales Environment Trust, the Australian Shipowners Association; the Research Councils of Hongkong, South Africa, Norway, France, Croatia and the Netherlands; and the OECD Validation Management Group for Ecotoxicity Tests (VMG-eco); In 2015, she served on the *Scientific Reference Panel* for Onshore Natural Gas and Water, Victoria, and the Science Panel for the Victorian Coastal Council. In 2017 she was invited by the *Lead Scientist of Victoria* to the Scientific Reference Group for Onshore Conventional Gas until 2020. In 2018 she established the Aquatic Pollution Prevention Partnership at RMIT University with Melbourne Water for research in the evaluation and management of pollution in the Melbourne Waterways

Professor Nugegoda was elected *President of the Society for Environmental Toxicology and Chemistry* (SETAC) Australasia 2011–2013. In 2014 she was awarded a visiting Professorship at Tianjin, Chengxian University, China; in 2009 a Visiting Professorship at the Helmholtz Research Centre for Environment and Health, Munich, Germany; and a visiting Research Fellowship at the Flinders University Research Centre for Coastal and Catchment Environments. In 2019 she was awarded the RMIT *Vice-Chancellors Award* for Research Impact.

Weblink: *https://www.researchgate.net/profile/Dayanthi_Nugegoda*

e-mail: *dayanthi.nugegoda@rmit.edu.au*

Synopsis of the book

This book "*Pollution Risks, Impacts & Management. Social, Economic and Environmental Perspectives*" covers the following pollution aspects: **i.** Pollution and pollutants; Arsenic; Heavy metals; Plastic waste and plastic pollution; Persistent organic pollutants; Oil spill and oil pollution; Pesticides; Nutrient pollution, algal toxins and hypoxia; Pharmaceuticals; Surfactants pollution; Synthetic dyes and textile effluents; Sewage pollution; Climate change, pollutants and ecotoxicity; **ii.** Chemistry, toxicity and ecotoxicity of pollutants; **iii**. Pollution impacts on social, economic and environmental sectors; **iv.** Nexus between pollution and poverty, pollution and sustainable development goals (SDGs) and pollution and climate change; **v.** Management of pollutants and **vi.** Guidelines/ regulatory limits.

Preface

Pollution is a global problem. Globally, hundreds of millions of people are consistently at risk of exposure to harmful pollutants via contaminated air, water and food. Harmful pollutants are chemicals which are persistent (P) (stay long in the environment), bioaccumulative (B) (accumulate in the fatty tissues of biota) and toxic (T) (poisonous) (**PBT**)) as well as endocrine-disrupting and carcinogenic (cancer-causing). Although urban areas are usually more polluted than the countryside, however, pollution can spread to remote places. For example, *persistent organic pollutants* (POPs) (organochlorine pesticides, dioxins, PCBs) have been found even in the Arctic and Antarctic areas. Volatile inorganic chemicals such as mercury and POPs can move across vast distances from the southern hemisphere to the northern hemisphere through a repeated cycle of evaporation and condensation, which is called the "*grasshopper effect*". Some chemicals are regarded as "*contaminants of emerging concern*" or *emerging contaminants* which are not commonly monitored in the environment but have the potential to enter the environment and cause known or suspected adverse ecological and/or human health effects. They consist of pharmaceuticals, surfactants/detergents, plasticisers/plastic additives, synthetic dyes, fire retardants and personal care products that are consistently being found in groundwater, surface water, municipal wastewater, drinking water, and food sources.

Arsenic (a metalloid), is one of the most important global environmental pollutants and is a PBT, endocrine-disrupting and carcinogenic. It is potentially toxic to humans, animals and plants and contaminated groundwater, drinking water, rice and vegetables worldwide (reported over 70 countries affecting 200 million people). Peoples of some Asian countries are severely affected due to groundwater arsenic contamination including Bangladesh, West Bengal of India, China and Vietnam. Humans can be exposed to arsenic via drinking water and food. The Agency for Toxic Substances and Disease Registry of the USA (ATSDR) has ranked arsenic as number 1 among the top 275 priority hazardous substances.

Heavy metals (HMs) such as cadmium, chromium, copper, lead, mercury, nickel, zinc is also PBT but some of them are also endocrine-disrupting as well as carcinogenic. The residues of HMs have been detected in water, crops, vegetables and seafood. Four of the top 10 chemicals of major public concern (out of 275 priority list of hazardous substances assessed by the ATSDR) are HMs (Pb, Hg and Cd). Recent research carried out found elevated mercury levels in child-bearing women in 25 countries (all of them in developing countries, where women were exposed to mercury via seafood, gold dust or from localised industrial pollution).

Plastic waste and plastic pollution is ubiquitous and is reported from across the globe (from the Arctic to the Antarctic, from the surface to the sediments, and in every waterway (lakes, rivers, estuaries, coastal areas and oceans). Plastics have been detected throughout the water column (surface, middle, and bottom) and in all the environmental compartments (biota, water, sediments). Plastic debris accounts for more than 75% of marine litter. At some places, plastic litter outnumbered fish larvae and plankton (e.g. in the Danube River, Portuguese coastal waters, North Pacific central gyre). Plastics in the environment as waste pose five major problems and impacts: **i**. ingestion and entanglement of plastics by aquatic biota and wildlife; **ii**. adsorption of POPs and heavy metals by plastics (thus plastic litters can act either as a source or sink or a vehicle to transport toxic chemicals in the aquatic environment); **iii**. colonisation and formation of biofilms (with bacteria, cyanobacteria) on the surface of plastics; **iv**. leaching of additives [(*Bis/di (2-ethylhexyl)phthalate* (DEHP), *Bisphenol A* (BPA), *Dibutyl phthalate* (DBP), *polybrominated diphenyl ethers* (PBDEs), *nonylphenol* (NP)] from plastics and **v**. contamination of seafood and other foods by microplastics (microplastics reported to have contaminated fish, molluscs, canned and dried fish, honey, sugar, beer, salt and drinking water).

Persistent organic pollutants [(*polychlorinated dibenzofurans* (PCDFs); *polychlorinated dibenzo-p-dioxins* (PCDDs), *polychlorinated biphenyls* (PCBs); *perfluorooctane sulfonic acid* (PFOS)] are potentially hazardous to the living organism, persist in the environment for years or decades, bioaccumulate in the fatty tissues of living organisms (including humans) and biomagnify through the food chain and are highly toxic to humans and wildlife. Food is the main source of exposure to PCDDs/PCDFs, PCBs, PBDEs and PFOS to humans. A global survey (in 52 countries) reveals that PCDDs, PCDFs concentrations in women breast milk was very high (>10 pg TEQs/g lipid) in India, and in some European and African countries. Combustion and industrial processes may have contributed to the exposure of human to PCDDs and PCDFs whereas clay consumption (geophagia) by pregnant women could be the likely source of PCDD and PCDF exposure in some African countries.

Oil spills and the resulting **oil pollution** can have a significant impact on recreational and professional fishing, tourism, mariculture, wildlife, ecosystems, human's health, industrial water intakes, salt production and coastal agriculture. Wildlife can also be harmed via exposure to oil through ingestion/inhalation or through the skin and eye irritation.

Though **Pesticides** (herbicides, insecticides, fungicides) use has greatly helped to increase and improve world food production, however, there remain concerns about the adverse effects of pesticides on non-target organisms including humans, fish and wildlife. About 109 pesticides have been identified as endocrine disruptors. Intensive agriculture is contributing to increased pesticide inputs into surface waters (lakes, rivers, estuaries). Pesticide residues may enter the environment via spray drift, vaporisation, surface run-off, unlawful acts, spills and drainage discharges and through leaching or soil dust.

Nutrient pollution (nitrogen, phosphorous) is now considered as the major pollution problem causing algal blooms, algal toxins and hypoxia (hypoxia= dissolved oxygen (DO) <2 mg/L) in freshwater and coastal areas in many countries. Globally hypoxic (DO= <2 mg/L) and anoxic (DO= 0 mg/L) areas are expanding rapidly in coastal areas and there are now more than 500 hypoxic sites across the globe. Hypoxia can cause mass mortality of demersal fishes (fish living on or near the bottom) and benthos (benthos= organisms that live on, in, or near the seabed, river, lake, or stream bottom) resulting in the loss of fisheries resources and impacting livelihoods of fishers. Water with blue-green algal/ cyanobacterial bloom could be unsuitable for agriculture, drinking, livestock drinking, irrigation, wildlife and recreation. Algal toxins can accumulate in crops and seafood and can be transferred to humans via the food and water.

Pharmaceuticals (human and veterinary) [(*non-steroidal anti-inflammatory drugs* (NSAID)/analgesics, antibiotics, *17α-ethinylestradiol*-EE2 (estrogen/ contraceptive pill), diclofenac)] in the environment is of increasing global concern due to their effects on ecological receptors (bacteria, algae, daphnia, fish, frogs, birds, plants/crops). The decline of vultures in the Indian sub-continent (Bangladesh, Pakistan and India) are one of the worst environmental effects of diclofenac toxicity.

Surfactants (e.g. detergents, shampoos, hair care products) discharged into the environment can cause water pollution problems in the form of persistent foams (bright white in colour). Some surfactants (nonylphenol ethoxylates), is persistent, most toxic to aquatic organisms (algae, invertebrates and fish), bioaccumulate in seafood and agricultural foods and cause endocrine disruption in fish.

Synthetic dyes/ Textile dye (azo, anthraquinone, indigo) can enter the aquatic environments via spills, accidental and intentional discharge from industries and by treated or untreated effluents from sewage treatment plants. The discharge of dyes/textile effluents into the environment cause major environmental pollution in the receiving waterways/ aquatic ecosystems (e.g. undesirable colour, impediment of light penetration, reduce photosynthesis, high chemical and biological oxygen demands, elevated heavy metals, toxicity to mammals and aquatic biota). Azo dyes (e.g. Congo red, which is the most commonly used azo dyes) is toxic to many organisms and can be metabolised to benzidine, a known human carcinogen).

Sewage pollution can cause by surface run-off from non-point and point sources, outfalls of wastewater treatment plants (WWTPs) and stormwater drain, leaking of sewer pipes and flooding etc. Sewage water contains biological pollutants/microbial pathogens (bacteria, viruses, protozoa and helminths) and chemical pollutants (organic and inorganic chemicals, pharmaceuticals and personal care products, heavy metals, nutrients, and solids). These pollutants (pathogenic and chemical hazards) can pose serious risks to aquatic biota, water quality, sensitive aquatic ecosystems and public health.

Climate change (the rise in temperatures; an increase in the frequency and intensity of extreme weather events; sea-level rise (alternation of salinity); ocean acidification (decrease in ocean pH) and increase in hypoxia (decrease in dissolved oxygen levels in aquatic environments) are referred to herein in this book as "*Climate change stressors*". The climate change stressors enhanced release, degradation, transportation and mobilisation of both hydrophobic and hydrophilic pollutants from environmental reservoirs. The interactive effects of two stressors [climate change and pollutants] increased the toxicity of several high-risk pollutants (POPs, pesticides, heavy metals, antibiotics, PAHs and PBDEs) to aquatic biota (algae, crustaceans, fish and molluscs).

This book entitled "**Pollution Risks, Impacts and Management. Social, Economic and Environmental Perspectives**" is an attempt to address those above-mentioned important facts and figures relating to harmful pollutants (existing and emerging), its chemistry, toxicity and ecotoxicity and impacts on social, economic and environment sectors. The book has thirteen chapters including **i**. Pollution and pollutants; **ii.** Arsenic; **iii.** Heavy metals; **iv.** Plastic waste and plastic pollution; **v**. Persistent organic pollutants (POPs); **vi**. Oil spill and oil pollution; **vii.** Pesticides; **viii**. Nutrient pollution, algal toxins and hypoxia; **ix**. Pharmaceuticals; **x**. Surfactants pollution; **xi.** Synthetic dyes/textile effluents pollution; **xii**. Sewage pollution; **xiii**. Climate change, pollutants and ecotoxicity.

The book provided a global and regional overview of each of the targeted pollutant. Each pollutant has been critically analysed to assess the potential risk to biota and humans based on the degree/ rating of *persistence* (half-life), *bioaccumulation* (octanol-water partition coefficient (log K_{ow}), bioconcentration factors, *toxicity* [(LD_{50} (the lethal dose that kills 50% of the test animals such as rats, birds, bees)]; endocrine-disruption and carcinogenicity and *ecotoxicity* [(EC_{50}/LC_{50} (EC_{50} is the effective concentration to produce a certain effect in 50% of test organisms such as algae, daphnia; and LC_{50} is the lethal concentration that kills 50% of the test animals such as fish)]. Social, economic and environmental impacts (SEEI) of each pollutant were evaluated in the context of detection in various environmental matrices/ compartments (surface water, drinking water, groundwater, effluents, sediments) and risks on water quality, aquatic biota, wildlife, sensitive ecosystems, residues in seafood/ crops/ fruits/vegetables, marine infrastructure, shipping, recreation, tourism, livelihoods and human health.

The book identified that there is a nexus between pollution and poverty, pollution and sustainable development goals (SDGs) and pollution and climate change. Management of pollutants including guidelines/ regulatory limits for the protection of aquatic ecosystems, water quality (irrigation, drinking, recreation, and recycled water), food quality, aquaculture and sediments are included. It (book) has been structured as a combination of research, text, reference and extension book and written in simple and plain English to benefit a wide range of both technical and non-technical readers. This pollution book (containing latest, authentic and evidence-based data and information) would be beneficial to academic and research institutes, university research students, environmental, climate change, water pollution, fisheries and agricultural scientists/ researchers, economists, social scientists, human health professionals, government planners, regulators, environmental campaigners and trade and business people.

Contents

1 Pollution and Pollutants

Summary

Pollution is a global problem. Although urban areas are usually more polluted than the countryside, however, pollution can spread to remote places. For example, persistent organic pollutants (POPs) have been found in the Arctic and Antarctic areas. In recent time, microplastics have been a major marine pollution problem. Pollution is a major challenge for economic, social and environmental development of a country. It is also a major obstacle and barrier to achieve sustainable development goals (SDGs). Pollution would enhance poverty and increase burdens on the poor since the livelihoods of poor people (both women and men) associated with agriculture, livestock, fishing, aquaculture, wetlands, forestry, seafood and food export business etc. would be affected due to pollution, deterioration of water quality and contamination of food. It is, therefore, essential that chemicals are managed (produce, transport, market, use and dispose of) in a safe and secure manner to protect both the environment and human health. This chapter reviews: **i.** POPs; **ii**. pollution types; **iii**. world's most polluted places; **iv.** harmful pollutants and their characteristics; **v.** examples of persistent, bioaccumulative, toxic (PBT), endocrine-disrupting and carcinogenic pollutants; **vi**. priority pollutants; **vii.** linkages between pollution and poverty, pollution and climate change and pollution and sustainable development goals; **viii.** control and management of pollutants.

Contents

List of Tables

List of Figures

List of Box

1.1. Introduction

Pollution is a global problem because of the introduction of harmful chemicals into the natural environment. Harmful chemicals are those which are persistent, bioaccumulative and toxic (PBT). These harmful chemicals are called pollutants. A pollutant is a substance introduced into the environment that may cause long- or short-term damage to receptors (biota, human) by changing the growth rate of plant or fish species, or by interfering with human amenities, comfort, health or property values. Although urban areas are usually more polluted than the countryside, pollution can spread to remote places. For example, heavy metals (mercury) and persistent organic pollutants (**POPs**) (see Table 1.1 and Box 1.1 for POPs) have been found in the Arctic and the Antarctic areas as some volatile compounds can move across vast distances through a repeated cycle of evaporation and condensation called the **"grasshopper effect/global**

distillation". In recent time, microplastics are becoming a major marine pollution problem which is transported by rivers, wind and tides to the ocean. To date, there are over 5.25 trillion pieces of plastic weighing over 268,940 tonnes floating on the world's oceans. There are five large patches of marine plastic pollution "hot spots", of which, the North/ Great Pacific Garbage Patch is the most polluted with plastic's (with nearly 1.8 trillion pieces of plastic) (Kibria 2017). This chapter is an introduction to pollution and pollutants.

Table 1.1. Twelve (12) initial persistent organic pollutants (POPs) (UNEP 2005) **with brief properties** (Kibria et al 2010).

*[**POPs** are organic pollutants persistent in the environment (lasts for years or even decades before degrading), evaporate and travel long distances through air and water and accumulate in fatty tissues of biota and humans and toxic to both wildlife and humans. The Stockholm Convention[1] on POPs, was adopted in 2001 and entered into force in 2004, is a global treaty to safeguard human health and the environment from highly harmful chemicals that persist in the environment and affect the well-being of humans as well as wildlife; **Half-life** is the time required for a quantity to reduce to half of its initial value; **log K_{ow}** is the partition coefficient between octanol and water that relates to the solubility of the chemical in fats (see also section 1.4).*

file:///C:/Users/User1/Downloads/UNEP-POPS-PUB-Brochure-16NewPOPs-201706.English%20(1).pdf

POPs	Uses	Remarks
1. Aldrin [Insecticide-Cyclodiene organochlorine; Half-life in soil (DT_{50}): 0.5-3 yr; Water solubility – almost insoluble in water; log K_{ow}: 4.32]	Applied to the soil to kill termites, grasshoppers, corn rootworms and other insect pests.	Pesticides Be eliminated
2. Chlordane [Insecticide-Cyclodiene organochlorine; Half-life in soil (DT_{50}): 1 yr; Water solubility: 0.1 mg/L; log K_{ow}: 6]	Used extensively to control termites and as a broad-spectrum insecticide on a range of agricultural crops. Chlordane formulations were usually mixture with heptachlor.	Pesticides Be eliminated
3. DDT [Insecticide- Organochlorine; Half-	Used during World War II to protect soldiers and civilians from	Pesticides Be restricted

[1]***Stockholm Convention** = In 1995, the Governing Council of the UNEP called for taking global action on POPs. The Conventions banned 12 worst Organochlorine chemicals (often nick named as dirty dozen) which are extremely toxic for human health and environment. The Convention on POPs was adopted in 2001 and entered into force on 17 May 2004 with ratification by an initial 128 parties and 151 signatories. For details on Stockholm Convention see link:*

http://en.wikipedia.org/wiki/Stockholm_Convention_on_Persistent_Organic_Pollutants).

POPs	Uses	Remarks
life in soil (DT_{50}): 2-15.6 yr; Water solubility - practically insoluble, 0.012-0.0055 mg/L; log K_{ow}: 4.89-6.914]	malaria, typhus, and other diseases spread by insects. It continues to be applied to control mosquitoes in several developing countries to reduce the incidence of malaria.	
4. Dieldrin [Insecticide- Chlorinated hydrocarbon; Half-life in soil (DT_{50}): 0.5-3 yr; Water solubility: 0.17 mg/L; log K_{ow}: 4.32]	Used principally to control termites and textile pests; has also been used to control insect-borne diseases and insects living in agricultural soils.	Pesticides Be eliminated
5, 6: Dioxins (5. PCDDs, 6. PCDFs) [**PCDDs** (Polychlorinated dibenzo-*p*-dioxins): chlorinated organic chemicals, most dangerous toxic chemical so far known on the earth; Half-life in soil (DT_{50}): 9-15 yr; Water solubility 0.00012-0.00074 mg/L; log K_{ow}: 6.4-8.2 (Srogi 2008); [**PCDFs** (Polychlorinated dibenzofurans): 2nd of the closely related Dioxin families, Half-life in soil (DT_{50}): 7 yr; Water solubility: 0.00013- 0.00042 mg/L; log K_{ow}: 6.22-8.8 (Srogi 2008)]	These chemicals are not directly used, rather produced instantly and unintentionally during incomplete combustion, as well as during manufacture/ production of certain pesticides and chemicals. In addition, certain kinds of metal recycling and pulp and paper bleaching releases dioxins. Dioxins are also been found in automobile exhaust, tobacco, wood and coal smoke.	Unintentional product Be eliminated
7. Endrin [Insecticide- Cyclodiene organochlorine; Half-life in soil (DT_{50}) > 10 yr; Water solubility: insoluble; log K_{ow}: 5.6]	Sprayed on leaves of crops such as cotton and grains; also used to control mice, voles and other rodents.	Pesticides Be eliminated
8. Heptachlor [Insecticide-Cyclodiene organochlorine; Half-life in soil (DT_{50}) 0.75-2 yr; Water solubility: 0.056 mg/L; log K_{ow}: 5.27]	Primarily employed to kill soil insects and termites; also used more widely to kill cotton insects, grasshoppers, other crop pests, and malaria-carrying mosquitoes. Heptachlor formulations regularly contained chlordane.	Pesticides Be eliminated
9. Hexachlorobenzene (HCB) [Fungicide; Chlorinated hydrocarbon; Water solubility: practically insoluble, 0.006 mg/L; log K_{ow}: 5.86]	HCBs are used to kill and control fungi that affect food crops. It is also released as a by-product during the manufacture of certain chemicals and as a result of the processes that give rise to dioxins and furans.	Pesticides/Industrial chemicals/ Unintentional product Be eliminated

POPs	Uses	Remarks
10. Mirex [Insecticide; Organochlorine; Half-life in soil (DT_{50}): 10 yr; Water solubility: not soluble, 0.07 mg/L; log K_{ow}: 5.28]	Is applied mainly to combat fire ants and other types of ants and termites; also used as a fire retardant in plastics, rubber, and electrical goods.	Pesticides Be eliminated
11. Polychlorinated Biphenyls (PCBs) [Often called as 'Dioxin like', Polychlorinated biphenyls (PCBs); Half-life in soil (DT_{50}) – 10.5 yr; Water solubility: 0.00023-0.00086 mg/L; log K_{ow}: 6.04-8.35] https://www.atsdr.cdc.gov/toxprofiles/tp17-c4.pdf	These compounds are used in industry as heat exchange fluids, in electric transformers and capacitors, and as additives in paint, carbonless copy paper, sealants and plastics.	Industrial chemicals Be eliminated
12. Toxaphene [Insecticide; Chlorinated camphene, camphechlor, a mixture of > 175-200 components produced by chlorination of camphene; Half-life in soil (DT_{50}): 1-14 yr; Water solubility: negligible; log K_{ow}: 3.33-6.44]	Also called as Camphechlor, is applied to cotton, cereal grains, fruits, nuts, and vegetables; also used to control ticks and mites in livestock.	Pesticides Be eliminated

1.2. Pollution types

Pollution can be of different types (air, litter, light, noise, plastic, radioactive, sewage, soil, thermal, visual and water), depending on the part of the environment that is getting polluted (Oishimaya 2018). However, major forms of pollution are air, soil and water pollution. This book put an emphasis on different aspects of **water pollution**. Water pollution is the contamination of surface waterways/water bodies (drains, channels, rice fields, swamps, lakes, rivers, billabongs, estuaries, oceans) as well as groundwater/aquifer. Common **pollutants** include arsenic, heavy metals, plastics, dioxins/PCBs/PFOA, petrochemicals/oil, pesticides, nutrients/fertilisers, pharmaceuticals, synthetic dyes, surfactants/detergents, sewage and greenhouse gases (CO_2-ocean acidification) (Kibria et al 2010). Pollution in waterways can be from non-point/diffuse (urban stormwater run-off, agriculture croplands, livestock feedlot) and point sources (factory, sewage treatment plant). Pollutants can enter waterways via the following paths:

- Directly via drains, channels, rivers, estuaries;
- Legal and illegal discharges of waste from factories;
- Discharge of treated or untreated effluents from sewage/ wastewater treatment plants;

- Spills and leaks of chemicals (pesticides, oil);
- Stormwater;
- Air transport of volatile pollutants (mercury, DDTs), and
- Wind, storms, and littering (plastic, cigarette butts, cans, papers).

1.3. World's most polluted places

Hundreds of millions of people are at risk of exposure to toxic pollution worldwide. According to the Black Smith Institute and Green Cross Switzerland, there are currently ten "Worst Polluted Places in the world" (*https://www.worstpolluted.org/projects_reports/display/111*). These ten most polluted places are in developing countries: Argentina (*Matanza-Riachuelo*), Bangladesh (*Hazaribagh*), Ghana (*Agbogbloshie*), Indonesia (*Kalimantan, Citarum River*), Nigeria (*Niger River delta*), Russia (*Dzerzhinsk, Norilsk*), Ukraine (*Chernobyl*) and Zambia (*Kabwe*) (Fig. 1.1; Table 1.2). The main pollutants found in those ten polluted places are mainly inorganic pollutants of heavy metals (cadmium, chromium, copper, lead, mercury, nickel, uranium, zinc) and organic pollutants (dioxins, phenol, petroleum products and toluene) (Table 1.2).

Box 1.1. The Stockholm Convention on Persistent Organic Pollutants and 16 new POPs.

http://en.wikipedia.org/wiki/Stockholm_Convention_on_Persistent_Organic_Pollutants

file:///C:/Users/User1/Downloads/UNEP-POPS-PUB-Brochure-16NewPOPs-201706.English%20(1).pdf

The Stockholm Convention on Persistent Organic Pollutants is an international environmental treaty, signed in 2001 and effective from May 2004, that aims to eliminate or restrict the production and use of persistent organic pollutants (POPs). The Convention is structured to address:

- Intentionally produced POPs, which include pesticides and industrial POPs (e.g., polychlorinated biphenyls (PCBs);
- Intentionally produced POPs for which the only exempted use is restricted to acceptable purposes, e.g. DDT for disease vector control; and
- Unintentionally generated POPs that are produced and released as a result of human activity, e.g., Dioxins and Furans as by-products of combustion.

There are **12 POPs** (Table 1.1) originally listed in the Convention (PCBs and hexachlorobenzene appear in both Annexes A and C because they are both intentionally and unintentionally produced):

- **Annex A** – Intentionally produced chemicals the production of which is to **be prohibited**: Comprises eight pesticide POPs namely aldrin, chlordane, dieldrin, endrin, heptachlor, hexachlorobenzene, mirex, and toxaphene and the group of compounds known as polychlorinated biphenyls (PCBs)

- **Annex B** – Intentionally produced chemicals the production of which is to **be restricted**: DDT
- **Annex C** – Unintentionally produced chemicals the production of which is to be **prevented**: Dioxins, furans, hexachlorobenzene (HCB) and PCBs

Sixteen (16) new chemicals were declared as POPs by the Stockholm Convention in 2017. Among them, seven are pesticides as listed below:

1. Alpha hexachlorocyclohexane	To be eliminated	Pesticides
2. Beta hexachlorocyclohexane	To be eliminated	Pesticides
3. Chlordecone	To be eliminated	Pesticides
4. Decabromodiphenyl ether (Commercial mixture, cDecaBDE)	To be eliminated	Industrial chemical
5. Hexabromobiphenyl	To be eliminated	Industrial chemicals
6. Hexabromocyclododecane	To be eliminated	Industrial chemicals
7. Hexabromodiphenyl ether and heptabromodiphenyl ether (commercial octabromodiphenyl ether	To be eliminated	Industrial chemicals
8. Hexachlorobutadiene	To be eliminated	Industrial chemicals, unintentional product
9. Lindane	To be eliminated	Pesticides
10. Pentachlorobenzene	To be eliminated	Pesticides/Industrial chemicals, unintentional product
11. Pentachlorophenols and its salts and esters	To be eliminated	Pesticides
12. Perfluoroctane sulfonic acid (PFOS), its salts and perfluorooctane sulfonylfluoride	To be restricted	Industrial chemicals
13. Polychlorinated naphthalenes	To be eliminated	Industrial chemicals, unintentional product
14. Short-chain chlorinated paraffins (SCCPs)	To be eliminated	Industrial chemicals
15. Technical endosulfan and its related isomers (organochlorine pesticides)	To be eliminated	Pesticides
16. Tetrabromodiphenyl ether and pentabromodiphenyl ether (commercial pentabromodiphenyl ether)	To be eliminated	Industrial chemicals

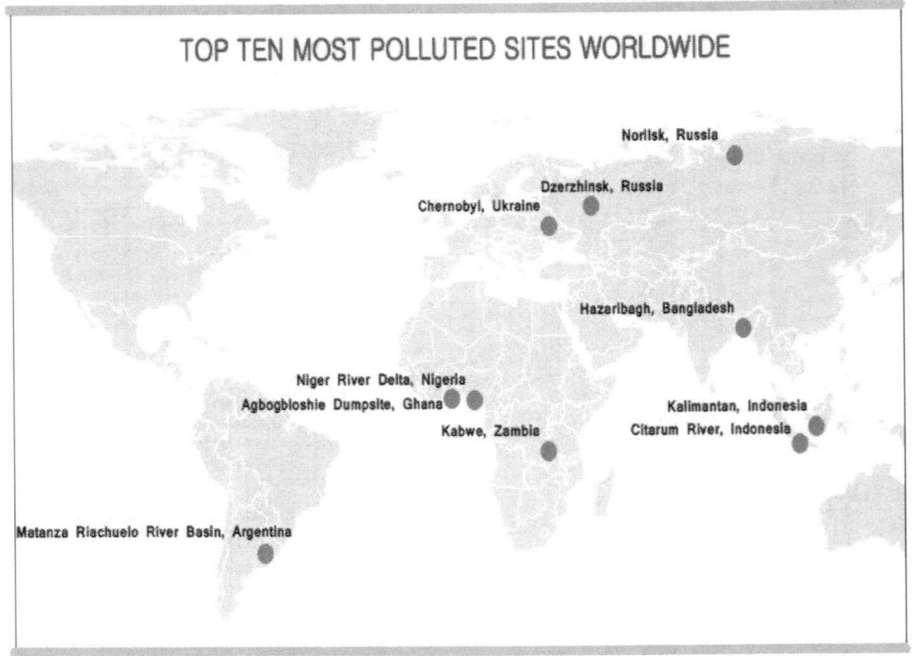

Fig. 1.1. Location of the top ten most polluted sites in the world
http://sciencythoughts.blogspot.com/2013/11/the-ten-most-polluted-places-on-earth.html

Table 1.2. Top ten most polluted sites in the world, identified pollutants and their impacts.

Country & main pollutant	Impacts
Argentina (*Matanza-Riachuelo*) **Pollutant:** Toluene, chromium (Cr) **Affected population**: 20,000⁺	The *Matanza-Riachuelo* River Basin houses several chemical manufacturers. It is estimated that 15,000 industries are actively releasing effluent into the adjacent river. The soil on the banks of the river contained **zinc (Zn), lead (Pb), copper (Cu), nickel (Ni)** and total **chromium (Cr)** that were all above recommended levels. Chromium in soil, for example, had a mean value of 1,141 ppm[2], which is significantly higher than the recommended level of 220 ppm. The basin water is not safe for drinking. *https://www.worstpolluted.org/projects_reports/display/114*
Bangladesh (*Hazaribagh*[3])	There were 270 registered tanneries in Bangladesh and around 90-95 percent is located at Hazaribagh. Every day, these tanneries

[2] ***ppm*** *= parts per million (mg/L).*

[3] ***Hazaribagh, Bangladesh tanneries:*** *There are 270 registered tanneries in Bangladesh and around 90-95 per cent are located at Hazaribagh on about 25 hectares of land. Most of these*

Country & main pollutant	Impacts
Pollutant: Chromium (Cr) **Affected population**: 160,000+	collectively dump 22,000 cubic litres of toxic wastes, including cancer-causing **hexavalent chromium**, into the adjacent Buriganga River (Dhaka's main river and a key water supply, water transport spot and iconic water sports place). Skin and respiratory diseases can result from repeated exposure to hazardous chemicals of the hide tanning processes. A recent research carried out found that the Buriganga River near Hazaribagh is the most polluted river in Bangladesh *https://www.worstpolluted.org/projects_reports/display/111*; Kibria et al 2016a).
Ghana (*Agbogbloshie*) **Pollutant**: Lead (Pb) **Affected population**: 40,000+	Agbogbloshie, in Accra Ghana, is the second largest e-waste processing area in West Africa. Ghana annually imports around 215,000 tons of second-hand consumer electronics from abroad, primarily from Western Europe, and generates 129,000 tons of e-waste every year. Samples taken around the perimeter of Agbogbloshie (Ghana), revealed the presence of Pb levels as high as 18,125 ppm in soil. The US EPA standard for **Pb** in the soil is 400 ppm. Samples taken from five workers on the site found **aluminium (Al), copper (Cu), iron (Fe),** and **Pb** levels above guidelines. *https://www.worstpolluted.org/projects_reports/display/107*
Indonesia (*Citarum River*) **Pollutant:** Cadmium (Cd), chromium (Cr), lead (Pb), and pesticides **Affected population**: 500,000+	The Citarum River provides 80% of surface water to Jakarta's water supply authority, irrigates farms that supply 5% of Indonesia's rice, and is a source of water for around 2,000 factories. A range of contaminants are present in the river, both from industrial and domestic sources. The level of Pb is at more than 1,000 times the USEPA standard in drinking water. A 2013 APN Science bulletin reported that **aluminium (Al), manganese (Mn)**, and **iron (Fe)** concentrations in the river water were 97 ppb[4], 195 ppb, and 194 ppb, respectively. *https://www.worstpolluted.org/projects_reports/display/109*
Indonesia (*Kalimantan*) **Pollutant:** Mercury (Hg) **Affected population**:	Kalimantan is the Indonesian portion of the island of Borneo and is composed of five provinces. In two of those provinces, Central and South, Artisanal Small-scale Gold Mining (ASGM) forms the primary source of income for 43,000 people. It is estimated that more than 1,000 tons of **Hg** are released into the environment each year through this ASGM process. The concentration of Hg in the

use old, outdated, and inefficient processing methods. Together, the tanneries employ anywhere between 8,000 and 12,000 people. In 2019, the government of Bangladesh has ordered to transfer all tanneries to Savar from current place of Hazaribagh; the new shifted place is adjacent to the Bangshi river and seems again will be polluted as was the fate of the Buriganga river.

[4] ***ppb** = parts per billion (0.001 mg/L)*

Country & main pollutant	Impacts
225,000⁺	Kahayan River of Central Kalimantan was 2,260 ng/L. This is more than twice Indonesia's standard for total Hg in drinking water (1,000 ng/L). *https://www.worstpolluted.org/projects_reports/display/113*
Nigeria *(Niger River Delta)* **Pollutant:** Petroleum (BTEX, PAH) **Affected population**: NA	The Niger River Delta is a densely populated region (8% of Nigeria's land mass). It has been the site of major petroleum operations since the late 1950s. Averages of 240,000 barrels of crude oil are spilled in the Niger delta every year. These spills have not only contaminated the surface and ground water of the delta but also the ambient air and locally grown crops with hydrocarbons. *https://www.worstpolluted.org/projects_reports/display/115*
Russia (Dzerzhinsk) **Pollutant:** Dioxins, phenol **Affected population**: 300,000	Dzershinsk was one of Russia's principal sites of chemical manufacturing, including chemical weapons. Today, it is still a significant centre of Russian chemical industry. Between 1930 and 1998, an estimated 300,000 tons of chemical wastes were improperly land filled in Dzershinsk and the surrounding areas. From this waste, around 190 identified chemicals were released into the groundwater. In 2007, water samples taken within the city showed levels of **dioxins and phenol** thousands of times greater than recommended levels. *https://www.worstpolluted.org/projects_reports/display/110*
Russia *(Norilsk)* **Pollutant:** Copper (Cu), nickel oxide, other heavy metals **Affected population**: 135,000	Norilsk is an industrial city founded in 1935. Mining and smelting operations began in the 1930s and Norilsk contained the world's largest **heavy metals** smelting complex as recently as the early 2000s. Nearly 500 tons each of **Cu** and **Ni** oxides and 2 million tons of Sulphur dioxide (SO_2) are released annually into the air. It is estimated that over 130,000 local residents are being exposed to particulates, sulfur dioxide, heavy metals, and phenols each day from air pollution. *https://www.worstpolluted.org/projects_reports/display/116*
Ukraine *(Chernobyl)* **Pollutant:** Uranium, plutonium, cesium-137,	Chernobyl is internationally recognized as one of the worst nuclear disaster (on April 25, 1986). Around 150,000 km^2 of area was affected in the accident. Radioactive contamination from the Chernobyl Power Plant spread over 40% of Europe and parts of Asia, North Africa, and North America immediately following the nuclear disaster. Nearly 400 million people resided in territories that were contaminated with radiation at a level higher than 4

Country & main pollutant	Impacts
strontium-90 and other metals **Affected population**: Up to 10 million	kBq^5/m^2 (Bq=becquerel is the SI derived unit of radioactivity). Today, there are over a dozen artificial radionuclides such as **cesium-137** that can be detected in the surface soil around the power plant. *https://www.worstpolluted.org/projects_reports/display/108*
Zambia *(Kabwe)* **Pollutant:** Lead (Pb) **Affected population**: 300,000$^+$	Kabwe, is the second-largest city in Zambia. A 2006 health study discovered that, on average, children's blood lead levels in Kabwe exceeded the recommended levels by five to ten times. This was the result of contamination from **lead** mining in the area, which is situated around the Copperbelt. The current CDC (Centers for Disease Control and Prevention, USA) recommended level of lead in children's blood is 5 µg/dL (µg/dL =micrograms of lead per decilitre of blood). In some neighbourhoods in Kabwe, blood concentrations of 200 ug/dL or more were recorded in children, and records show average blood levels of children tested ranged between 50 and 100 µg/dL *https://www.worstpolluted.org/projects_reports/display/112*

1.4. Harmful pollutants

Harmful pollutants are chemicals that are persistent, bioaccumulative and toxic (**PBT**) as well as endocrine-disrupting **(ED)** and carcinogenic (cancer-causing) to both biota and human (Table 1.3), highlighted below as bullet points.

- A substance is **persistent** if it does not decay to half of its original quantity within two months (if in water) or six months (if in soil or sediment). Persistent chemicals remain in the environment and they are hydrophobic (insoluble in water) and bioaccumulate in organisms/ biota. They have a log K_{ow} values more than 3.
- The **log K_{ow}** is the ratio of the concentration of a chemical in *n*-octanol (a surrogate for animal lipid) to the concentration in water both at equilibrium and at a constant temperature (also called (*Octanol/ Water Partition Coefficient*). A log K_{ow} of 3 denotes the compound is 1,000 times more soluble in octanol than water.
- **Bioaccumulative** chemicals are those that have a strong tendency to dissolve in fats and remain in the fatty tissue of organisms. A substance is classed as bioaccumulative if it accumulates in the fatty tissue relative to water in a ratio (quantity-wise) of 10,000 to 1 or more.

[5] ***The Becquerel*** *(symbol: Bq) is the SI (Système international) derived unit of radioactivity. One becquerel is defined as the activity of a quantity of radioactive material in which one nucleus decays per second.*

- A chemical is **toxic** if it causes direct damage to organisms that are exposed to them. A substance can be classed as being of concern if it is lethal to at least 50% of test aquatic organisms at a concentration of <1 mg/L or less (see Table 1.4 for toxicity rating). Chemicals such as arsenic, heavy metals, pesticides, petrochemicals/crude oil (BTEX-benzene, toluene, ethylbenzene and xylene and PAHs-Polycyclic aromatic hydrocarbons, microplastics, dioxins, Perfluorooctanoic acid (PFOA) and synthetic dyes fall within the PBT category.
- **Endocrine-disrupting (ED) chemicals** are compounds that alter the normal functioning of the endocrine system, potentially causing disease or deformity in organisms and their offspring (such as hormones (17β-estradiol (E2), 17α ethynylestradiol (EE2)), industrial chemicals (nonylphenols, octylphenols), pesticides (2,4-D, DDT) and heavy metals (cadmium, lead).
- **Carcinogenic** chemicals have the potential to cause cancer. Some examples are metalloid (arsenic), heavy metals (cadmium, chromium, lead, nickel), pesticides (aldrin, dieldrin, DDT), petrochemicals compound (Benzo[*a*]pyrene, Benzene).

Therefore, the risk of pollutants/ harmful chemicals can simply be assessed based on the degree/ rating of persistence, bioaccumulation, toxicity and potential for endocrine-disrupting and carcinogenicity of each harmful pollutant/ chemical. Such a concept (Table 1.5) has been applied throughout this book (Chapter 2 through Chapter 13).

Table 1.3. First tier criteria for chemicals of greatest concern.
(*http://www.defra.gov.au/environment/chemicals/csf/concern/index.htm*, Kibria et al 2010, IARC 2019).

https://www.chemsafetypro.com/Topics/CRA/Toxicology_Dose_Descriptors.html

Criterion	First tier thresholds
Persistence (P)	Half-life[6] (t½) water > 2 months. Half-life (t½) soil/ sediment > 6 months. *http://www.defra.gov.au/environment/chemicals/csf/concern/index.htm*
Bio accumu-	Log K_{ow}[7] > 5 (greatest concern) (unless BCF < 50,000).

[6] ***Half-life (t½)*** *= defined as the time it takes for an amount of a compound to be reduced by half through degradation in an environmental compartment (water, soil or sediment, or air, etc). It is used to measure the persistence of a substance. https://www.chemsafetypro.com/ Topics/CRA/Toxicology_Dose_Descriptors.html.*

[7] ***Log*** K_{ow} *= Is the partition coefficient between octanol and water that relates to the solubility of the chemical in fats. A log* K_{ow} *value of 3 denotes the compound is 1,000 times more soluble in octanol than water.*

Criterion	First tier thresholds
lative (B)	Log K_{ow} > 3 (high concern) (unless BCF < 500). *http://www.defra.gov.au/environment/chemicals/csf/concern/index.htm*
Toxicity or Toxic (T)	LC_{50}[8]/ EC_{50}[9] < 1 mg/L or NOEC[10] < 0.11 mg/L *http://www.defra.gov.au/environment/chemicals/csf/concern/index.htm*
Endocrine Disrupting Effects (EDC)	Evidence of EDC (Kibria et al 2010).
Carcinogenicity	Evidence of carcinogens (group 1, 2A, 2B, 3) (IARC 2019).

Table 1.4. The relationship between LC_{50}, LD_{50} and toxicity rating (Bae and Freeman 2007).

LC_{50} (mg/L)	LD_{50}[11] (mg/kg)	Toxicity rating
< 1 mg/L	< 50 mg/kg	Extremely toxic
1-10 mg/L	50-500 mg/kg	Very toxic
10-100 mg/L	500-5,000 mg/kg	Moderately toxic
>100 mg/L	> 5,000 mg/kg	Relatively non-toxic

[8] ***LC_{50}*** *(median lethal concentration): The concentration of material in the water that is estimated to be lethal to 50% of the test organisms (fish, daphnia). The LC_{50} is usually expressed as a time-dependent value, (24 h or 96 h).*

[9] ***EC_{50}*** *= In ecotoxicity, EC_{50} (median effective concentration) is the concentration of test substance which results in a 50 percent reduction in either algae growth or algae growth rate or Daphina immobilization. They are often obtained from acute aquatic toxicity studies. EC_{50} values are often used for acute environmental hazard classification and calculation of predicted non-effect concentration (PNEC) (https://www.chemsafetypro.com/Topics/CRA/Toxicology_Dose_Descriptors.html).*

[10] ***NOEC*** *= .no observed effect concentration; the concentration in an environmental compartment (water, soil, etc) below which an unacceptable effect is unlikely to be observed. It is typically obtained from chronic aquatic toxicity studies and terrestrial toxicity studies (https://www.chemsafetypro.com/Topics/CRA/Toxicology_Dose_Descriptors.html).*

[11] ***LD_{50}*** *= (median lethal dose), the dose of material that is estimated to be lethal to 50% of the test organisms (rats, mice).*

Table 1.5. Examples of pollutants that are persistent, bioaccumulative, toxic, endocrine-disrupting and carcinogenic [*IARC = International Agency for Research on cancer,* ***na****= data is not available].*

Pollutants (covered in this book)	**Risks of Pollutants** √ = identified pollutant falls within that risk category				
	Persi-stent	**Bioaccu-mulative**	**Toxicity and ecotoxi-city***	**Endocrine disrupting (ED)**	**Carcinoge-nicity**[12] **(IARC groups 1, 2A, 2B, 3)**
Chapter 1- Pollution and pollutants	√	√	√	√	√
Chapter 2- Metalloid-Arsenic *(e.g.* $\boldsymbol{As_2O_3}$ *(arsenic trioxide;* $\boldsymbol{As_2O_5}$ *(arsenic penta-oxide))*	√	√	√	√	√ (As- 1)
Chapter 3- Heavy metals *(e.g.* ***cadmium****(Cd);* ***mercury*** *(Hg),* ***lead*** *(Pb);* ***zinc*** *(Zn)*	√	√	√	√	√ (Cd, Cr, Ni-1; Pb- 2A; Co, Pb, MeHg-2B)
Chapter 4- Microplastics *(e.g.* ***HDPE*** *(high-density polyethylene);* ***DEHP*** *(di (2-thylhexyl)phthalate (plastic additives))*	√	√	√	√	√ (DEHP- 2B)
Chapter 5- Persistent organic pollutants (POPs) *(e.g.* ***TCDD*** *(2, 3, 7, 8 tetrachlorodibenzo-p-dioxin);* ***PCB*** *(polychlorinated biphenyls);* ***PBDE*** *(Polybrominated diphenyl ethers),* ***PFOS*** *(perfluorooctanesulfonate/ perfluorooctanesulfonic acid)*	√	√	√	√	√ (PCDD/F, PCB-1; PBDE-2A; PFOA-2B)
Chapter 6- Oil pollution *(e.g.* ***BTEX*** *(benzene,*	√	√	√	√	√

[12] ***Carcinogenicity*** *is based on* ***IARC classification:*** *group 1 = Carcinogenic to humans; group 2A = Probably carcinogenic to humans; group 2B = Possibly carcinogenic to humans; group 3 = Not classifiable as to carcinogenicity in humans; *** atmospheric life-time of* CO_2 *is 100-1000 years;* CH_4 *is 12 years;* N_2O *is 114 years*

Pollutants (covered in this book)	Risks of Pollutants √ = identified pollutant falls within that risk category				
	Persistent	Bioaccumulative	Toxicity and ecotoxicity*	Endocrine disrupting (ED)	Carcinogenicity[12] (IARC groups 1, 2A, 2B, 3)
*toluene, ethylbenzene and xylenes), **PAHs** (polycyclic aromatic hydrocarbons)*					(Benzo[a]pyrene, Benzene-1; Napthalene, Chrysene, ethylbenzene-2B)
Chapter 7- Pesticides *(e.g. **DDT** (Dichlorodiphenyl-trichloroethane),*	√	√	√	√	√ (Lindane-1; DDT, glyphosate-2A; 2,4-D, mirex, chlorothalonil, toxaphene-2B)
Chapter 8- Nutrient pollution, algal toxins & hypoxia *(e.g. **Microcystins** (toxins producing cyanobacteria/blue-green algae); **PSP** (paralytic shellfish poisoning caused by dinoflagellates), **Hypoxia** (low-dissolved oxygen conditions))*	√	√	√	√	√ (Microcystins-2B)
Chapter 9- Pharmaceuticals *(e.g. **antibiotics** (tetracycline); **NSAID** (nonsteroidal anti-inflammatory drug-diclofenac)*	√	√	√	?	√ (17α-ethinylestradiol (EE2), 17β-estradiol (E2)-1)
Chapter 10- Surfactants pollution *(e.g. **LAS** (Linear alkylbenzene sulfonates-anionic surfactants); **NP** (nonylphenol- non-ionic surfactants)*	Most are non-persistent except for nonylphenol	√	√	√	**na**
Chapter 11- Synthetic dyes & Textile effluents	√	√	√	**na**	√ (all benzidine

Pollutants (covered in this book)	**Risks of Pollutants √ = identified pollutant falls within that risk category**				
	Persistent	**Bioaccumulative**	**Toxicity and ecotoxicity***	**Endocrine disrupting (ED)**	**Carcinogenicity[12] (IARC groups 1, 2A, 2B, 3)**
*(e.g. **azo dyes** (dyes used to treat textiles, leather articles, and some foods); **malachite green** (used in aquaculture in fungicides and ectoparasiticides)*					based azo dyes- 1, others recognised to be carcinogens: Acid dye, Acid red 26, Solvent yellow 1 and Solvent yellow 2) (SCCNFP 2002)
Chapter 12- Sewage pollution *(e.g. **EE2** (17α-ethynylestradiol-hormone); **Illicit drugs** (cocaine, amphetamines, ecstasy); **4-MBC**) (4-Methylbenzylidene camphor (UV filter)*	√	√	√	√	√ (Several chemicals found in sewage water are carcinogenic (heavy metals, hormones, POPs)
Chapter 13- Climate change, pollutants & Ecotoxicity (CO_2, CH_4, N_2O)	√**	?	√ (related to temperature, salinity, ocean acidification, hypoxia)	?	√ (UV radiation and cancer relating to global warming effects)

* ***Toxicity*** *is based on LD_{50} values (rat);* ***ecotoxicity*** *is based on LC_{50} or EC_{50} values (fish or daphnia or algae); ** atmospheric life-time of CO_2 is 100-1000 years; CH_4 is 12 years; N_2O is 114 years*

1.5. Priority pollutants

1.5.1. US EPA's priority pollutants

Priority pollutants refer to a list of 126 specific pollutants that include metalloids, heavy metals, plastics/plastic additives, polychlorinated biphenyls, petroleum/ oil compounds, pesticides and other specific organic chemicals. The priority pollutants are a sub-set of "toxic pollutants". These 126 pollutants were assigned by the United States Environment Protection Agency (US EPA) (*https://19january2017snapshot.epa.gov/sites/production/files/2015-09/documents/priority-pollutant-list-epa.pdf*) as a high priority for development of water quality criteria and effluent limitation guidelines because they are frequently found in wastewater. The priority pollutants fall within the following main categories:

- **Metalloid**: Arsenic; Antimony.
- **Heavy Metals**: Cadmium; Chromium; Copper; Lead; Mercury; Nickel; Selenium; Silver, Thallium; Zinc.
- **Plastics/plastics additives**: Bis(2-ethylhexyl) phthalate; Butyl benzyl phthalate; Di-N-Butyl Phthalate; Di-n-octyl phthalate; Diethyl Phthalate; Dimethyl phthalate; Di-N-Butyl Phthalate; Vinyl chloride.
- **Dioxins:** ***Polychlorinated biphenyls (PCBs):*** PCB-1016 (Arochlor 1016); PCB-1221 (Arochlor 1221); PCB-1232 (Arochlor 1232); PCB-1248 (Arochlor 1248); PCB-1242 (Arochlor 1242); PCB-1254 (Arochlor 1254); PCB-1260 (Arochlor 1260); ***Dioxin:*** 2,3,7,8-TCDD, Polychlorinated dibenzofurans.
- **Crude oil compounds:** ***Polycyclic Aromatic Hydrocarbons (PAHs)*** (Anthracene; Acenaphthene; Benzo(b)fluoranthene; Benzo(a)pyrene; Benzo(ghi)perylene; Benzo(a)anthracene; Benzo(k)fluoranthene; Chrysene; Dibenzo(a,h)anthracene; Fluoranthene; Fluorene; Indeno (1,2,3-cd)pyrene; Naphthalene; Pyrene; Phenanthrene; ***BTEX*** (Benzene; Ethylbenzene; Toluene)
- **Pesticides:** Acrolein; Aldrin; Benzene hexachloride (BHC) (including Alpha-BHC, Beta-BHC, Delta-BHC, Gamma-BHC); Chlordane; DDT (including 4,4-DDT, 4,4-DDE, 4,4-DDD); Diazinon; Dieldrin; Endosulfan (including Alpha-endosulfan, Beta-endosulfan, Endosulfan sulfate); Endrin; Endrin aldehyde; Heptachlor; Heptachlor epoxide; Hexachlorobenzene; Toxaphene.
- **Dyes:** Benzidine.
- **Others:** 1,1-dichloroethane; 1,1-dichloroethylene; 1,1,1-trichloreothane; 1,1,2-trichloroethane; 1,1,2,2-tetrachloroethane; 1,2-dichlorobenzene; 1,2-dichloropropane; 1,2-dichloroethane; 1,2-diphenylhydrazine; 1,2-trans-dichloroethylene; 1,2,4-trichlorobenzene; 1,3-dichlorobenzene; 1,3-dichloropropylene; 1,4-dichlorobenzene; 2-chlorophenol; 2-chloroethyl vinyl ethers; 2-chloronaphthalene; 2-nitrophenol; 2,4-dichlorophenol; 2,4-dimethylphenol; 2,4-dinitrotoluene; 2,4-dinitrophenol; 2,6-dinitrotoluene; 2,4,6-trichlorophenol; 3,3-dichlorobenzidine; 4-bromophenyl phenyl ether; 4-chlorophenyl phenyl ether; 4-nitrophenol; 4,6-dinitro-o-cresol; Acrylonitrile; Asbestos; Beryllium; Bis(2-chloroethyl) ether; Bis(2-chloroisopropyl) ether; Bis(2-chloroethoxy) methane; Bromoform; Carbon tetrachloride; Chlorobenzene;

Chloroethane; Chlorodibro-momethane; Chloroform; Cyanide; Dichloro-bromomethane; Hexachlorobutadiene; Hexachloro-cyclopentadiene; Hexachloroethane; Isophorone; Methylene chloride; Methyl bromide; Nitrobenzene; N-nitrosodimethylamine; N-nitrosodiphenylamine; N-nitrosodi-n-propylamine; 4-chloro-3-cresol, Pentachlorophenol; Phenol; Tetrachloroethylene; Toluene; Trichloroethylene.

1.5.2. *ATSDR's substance priority list*

The Agency for Toxic Substances and Disease Registry (ATSRD), USA listed hazardous substances priority list (Table 1.6). This priority list is based on risk assessment of pollutants that pose the most significant threat to human health (due to their known or suspected toxicity and potential for human exposure). It includes 275 priority hazardous substances - a prioritization of substances based on a combination of their frequency, toxicity, and potential for human exposure. The 275 substances include: **i.** *metalloid/heavy metals* (arsenic, chromium, cobalt, lead, mercury); **ii.** *plastic* additives (Di-n-butyl phthalate, Di(2-ethylhexyl) phthalate, vinyl chloride); **iii.** *crude oil compounds* (Benzene, Benzo(A)anthracene, Benzo(b) fluoranthene, dibenzo(a)anthracene, naphthalene, toluene, xylenes); **iv.** *POPs* (PCBs- aroclor 1232, aroclor 1254, aroclor 1260, DDT, endrin, dieldrin, dioxin, heptachlor, toxaphene); **v.** *pesticides* (acrolein, chlorpyrifos, diazinon, disulfoton, endosulfan, HCH); **vi**. *Dyes* (Benzidine); **vii**. Nutrients (phosphorus); **viii**. *Greenhouse gases* (methane).

Table 1.6. The ATSDR 2017 substances priority list

(out of 275 priority hazardous substances assessed, top 100 priority pollutants are listed below (ATSDR 2017; *https://www.atsdr.cdc.gov/SPL/*); *CAS RN = Chemical Abstracts Service Registry Number; https://www.atsdr.cdc.gov/SPL/#2017spl).*

2017 rank	Substance name	Total points	CAS RN	Remarks
1	**Arsenic** (arsenic trioxide is used as a pesticide to control termites, and to produce clear glass, hide preservatives in tanneries and wood preservatives).	1,674	7,440-38-2	Metalloid
2	**Lead** (used in batteries, pipes, petrol).	1,531	7.439-92-1	Heavy metals
3	**Mercury**	1,458	7.439-97-6	Heavy metals
4	**Vinyl chloride**	1,358	75-01-4	Plastic additives
5	**Polychlorinated biphenyls (PCBs)**	1,345	1.336-36-3	POPs/PCB
6	**Benzene**	1,329	71-43-2	Crude oil compound
7	**Cadmium**	1,320	7.440-43-9	Heavy metals
8	**Benzo(a)pyrene**	1,306	50-32-8	Crude oil compound

2017 rank	Substance name	Total points	CAS RN	Remarks
9	**Polycyclic aromatic hydrocarbons**	1,279	130,498-29-2	Crude oil compound
10	**Benzo(b) fluoranthene**	1,251	205-99-2	Crude oil compound
11	**Chloroform** (used as a solvent for lacquers, floor polishes, resins, adhesives, alkaloids, fats, oils and rubber as well as an anaesthetic).	1,203	67-66-3	Industrial chemicals
12	**Aroclor 1260**	1,191	11,096-82-5	POPs/ PCB
13	**DDT**	1,183	50-29-3	POPs/ Pesticides
14	**Aroclor 1254**	1,172	11,097-69-1	POPs/ PCB
15	**Dibenzo(a)anthracene**	1,156	53-70-3	Crude oil compound
16	**Trichloroethylene**	1,155	79-01-6	Industrial chemical
17	**Chromium hexavalent (CrVI)**	1,148	18,540-29-9	Heavy metals
18	**Dieldrin**	1,144	60-57-1	POPs/ Pesticides
19	**Phosphorus white**	1,141	7,723-14-0	Nutrients (P as nutrient)
20	**Hexachlorobutadiene**	1,130	87-68-3	Industrial chemical
21	**DDE, P,P'-**	1,127	72-55-9	POPs/ Pesticides
22	**Chlordane**	1,126	57-74-9	POPs/ pesticides
23	**Aroclor 1242**	1,126	53,469-21-9	POPs/ PCB
24	**Coal tar creosote** (used as a wood preservative)	1,124	8,001-58-9	Industrial chemical
25	**Aldrin**	1,116	309-00-2	POPs/ Pesticides
26	**DDD, P,P'-**	1,114	72-54-8	POPs/ Pesticides
27	**Aroclor 1248**	1,105	12,672-29-6	POPs/ PCB
28	**Heptachlor**	1,102	76-44-8	POPs/ Pesticides
29	**Aroclor** (Aroclor is a PCB mixture)	1,101	12,767-79-2	POPs/ PCB
30	**Benzidine** (dyes used for the production of textiles, paints, printing inks, paper, and pharmaceutical products).	1,093	92-87-5	Dyes
31	**Acrolein**	1,090	107-02-8	Pesticides
32	**Toxaphene**	1,089	8,001-35-2	POPs/ Pesticides
33	**Tetrachloroethylene**	1,078	127-18-4	Dry-cleaning fluid
34	**Hexachlorocyclohexane, Gamma-**	1,076	58-89-9	Pesticides
35	**Cyanide**	1,071	57-12-5	Used in mining
36	**Hexachlorocyclohexane, Beta-**	1,054	319-85-7	Pesticides

2017 rank	Substance name	Total points	CAS RN	Remarks
37	**Disulfoton** (an organophosphate acetylcholinesterase inhibitor used as an insecticide).	1,049	298-04-4	Pesticides
38	**Benzo(A)anthracene**	1,047	56-55-3	Crude oil compound
39	**1,2-Dibromoethane/ ethylene dibromide** (widely used to control nematodes, wireworms and other soil pests).	1,043	106-93-4	Pesticides
40	**Endrin** (used primarily as an agricultural insecticide in tobacco, apple trees, cotton, sugar cane, rice, cereals, and grains).	1,039	72-20-8	POPs/ Pesticides
41	**Diazinon**	1,038	333-41-5	Pesticides
42	**Hexachlorocyclohexane, Delta-**	1,036	319-86-8	Pesticides
43	**Beryllium**	1,031	7.440-41-7	Heavy metals
44	**Endosulfan**	1,029	115-29-7	Pesticides
45	**Aroclor 1221**	1,028	11,104-28-2	POPs/PCB
46	**1,2-Dibromo-3-chloopropane (dibromochloropropane)** (is the active ingredient in the nematicide Nemagon; it is a soil fumigant formerly used in American agriculture).	1,027	96-12-8	Pesticides
47	**Heptachlor epoxide**	1,022	1,024-57-3	Pesticides
48	**Endosulfan, Alpha**	1,019	959-98-8	Pesticides
49	**Cis-Chlordane**	1,017	5,103-71-9	POPs/ Pesticides
50	**Carbon tetrachloride** (used in the production of refrigeration as a cleaning fluid and degreasing agent in fire extinguishers, and in spot removers).	1,014	56-23-5	Industrial chemicals
51	**Cobalt**	1,013	7,440-48-4	Heavy metals
52	**Aroclor 1016**	1,012	12,674-11-2	POPs/ PCB
53	2,4′-DDD **2,4′-Dichlorodiphenyl)dichloroethane**	1,009	789-02-6	POPs/ Pesticides
54	**Pentachlorophenol** (used as a wood preservative for power line poles, cross arms, fence posts).	1,008	87-86-5	Industrial chemicals
55	**Methoxychlor** (used to control a wide range of insect pests in crops, fruits, vines, flowers,	1,007	72-43-5	Pesticides

2017 rank	Substance name	Total points	CAS RN	Remarks
	vegetables and in forestry).			
56	**Endosulfan sulfate**	1,005	1,031-07-8	Pesticides
57	**Nickel**	996	7,440-02-0	Heavy metals
58	**Di-n-butyl phthalate**	995	84-74-2	Plastic additives
59	**Endrin ketone** (endrin exists in the environment as either endrin aldehyde or endrin ketone).	993	53,494-70-5	POPs/ Pesticides
60	**Dibromochloropropane** (a highly persistent, lipophilic, brominated organochlorine which is effectively used against nematodes).	984	6,7708-83-2	Pesticides
61	**Benzo(k)fluoranthene**	970	207-08-9	Crude oil
62	**Trans-chlordane**	969	5,103-74-2	Pesticides
63	**Endosulfan, beta**	968	33,213-65-9	Pesticides
64	**Chlorpyrifos**	965	2,921-88-2	Pesticides
65	**Xylenes, total**	964	1330-20-7	Crude oil compound
66	**Chromium trioxide**	961	1,333-82-0	Heavy metals
67	**Aroclor 1232**	959	11,141-16-5	POPs/PCB
68	**Endrin aldehyde**	959	7,421-93-4	POPs/Pesticides
69	**Methane**	952	74-82-8	Greenhouse gases
70	**3,3-dichlorobenzidine** (widely used in the production of diarylide yellow pigments which is used in the production of printing inks).	942	91-94-1	Dyes ingredient
71	**2-Hexanone** (a ketone, used as a general solvent and in paints).	941	591-78-6	Industrial chemical
72	**2,3,7,8-Tetrachlordibeno-p-diozin (TCDD)**	941	1,746-01-6	POPs/ Dioxin
73	**Benzofluoranthene**	937	56832-73-6	Crude oil compound
74	**Toluene**	917	108-88-3	Crude oil
75	**Zinc**	915	7,440-66-6	Heavy metals
76	**Pentachlorobenzene** (used as an intermediate in the manufacture of pesticides, particularly the fungicide pentachloronitrobenzene).	907	608-93-5	Industrial chemicals
77	**Di(2-ethylhexyl) phthalate**	906	117-81-7	Plastic additives
78	**Chromium**	895	7,440-47-3	Heavy metals
79	**Aroclor 1240**	889	71,328-89-7	POPs/PCB
80	**2,4,6-Trinitrotoluene**	879	118-96-7	Industrial chemicals

2017 rank	Substance name	Total points	CAS RN	Remarks
	(one of the most commonly used explosives for military, industrial, and mining applications).			
81	**Naphthalene** (used mainly as a precursor to other chemicals, industrial production of phthalic anhydride, azo dyes, insecticide 1-naphthyl-*N*-methylcarbamate (*carbaryl*).	877	91-20-3	Crude oil compound
82	**1,1-Dichloroethane** (used as a feedstock in chemical synthesis, chiefly of 1,1,1-trichloroethane, also as a solvent for plastics, oils and fats, as a degreaser, as a fumigant in insecticide sprays etc.).	876	75-35-4	Industrial chemicals
83	**Bromodichloroethane**	868	683-53-4	Industrial chemicals
84	**DDD, O,P'-**	867	53-19-0	POPs/ Pesticides
85	**2,4,6-Trichlorophenol**	867	88-06-2	Industrial chemicals
86	**Bis(2-Chloroethyl) ether**	867	111-44-4	Industrial chemicals
87	**Hydrazine** **(used as gas producers and propellants, a precursor to pesticides and pharmaceuticals, rocket fuels).**	862	302-01-2	Industrial chemicals
88	**Methylene chloride** (used in many different industries: paint stripping, paint remover manufacturing, pharmaceutical manufacturing, metal cleaning and degreasing, adhesives manufacturing and used in polyurethane foam production).	860	75-09-2	Industrial chemicals
89	**2,4-Dinitrophenol** (used as an antiseptic and as a non-selective bioaccumulating pesticide).	859	51-28-5	Industrial chemicals
90	**4,4′-Methylenenbis(2-chloroaniline)** (used as a curing agent in polyurethane foam production).	859	101-14-4	Industrial chemicals
91	**1,2-Dichloroetahne** (95% of the world's production of 1,2-dichloroethane is used in the production of vinyl chloride monomer).	852	107-06-2	Industrial chemicals
92	**Thiocyanate**	847	302-04-5	Industrial chemicals
93	**Hexachlorobenzene** (a fungicide, formerly used in seed treatment, especially on wheat to control	844	118-74-1	Pesticides

2017 rank	Substance name	Total points	CAS RN	Remarks
	the fungal diseases).			
94	**Asbestos** (used for insulation in buildings, as roofing materials, water supply lines and clutches, brake linings, gaskets and pads for automobiles).	841	1,332-21-4	Industrial chemicals
95	**RDX (cyclonite)**	833	121-82-4	Industrial chemicals
96	**Radium-226** (used in medicine, producing radon gas from radium chloride for use in radiotherapy in cancer treatment).	833	13,982-63-3	Industrial chemicals
97	**Uranium** (used by the military to power nuclear submarines and in manufacturing nuclear weapons).	832	7,440-61-1	Heavy metals
98	**2,4-Dinitrotoluene** (used in the production of toluene diisocyanate, which is used to produce flexible polyurethane foams).	832	121-14-2	Industrial chemicals
99	**Ethion** (organophosphate insecticide).	831	563-12-2	Pesticides
100	**4,6-Dinitro-o-cresol**	828	534-52-1	Industrial chemicals

1.6. Linkages between pollution and poverty

Pollution would enhance poverty and increase burdens on the poor (poor may become poorer) since the livelihoods of poor people and farmers (in particular ultra-poor living in developing nations) associated with or depending on available natural resources (agriculture, livestock, fishing, aquaculture, wetlands, forestry, seafood and food export business etc) would be affected due to pollution, deterioration of water quality and contamination of soil, water and food. **Firstly**, water contaminated with chemical pollutants [like arsenic, heavy metals, microplastics, POPs, oil/petrochemical compounds, pesticides, nutrients/ fertilisers, pharmaceuticals, surfactants/detergents, synthetic dyes and sewage effluents, etc.] could be unsuitable for crop irrigation, and vegetables growing, human and livestock drinking, recreation, stock and domestic water supply, aquaculture/fisheries, industries, hence, would impact on food production, food yield and businesses associated with water, food, and water trading. **Secondly**, some chemical pollutants are hydrophobic/non-polar (As, Pb, Hg, DDT, PAHs, microplastics, synthetic dyes), those can bioaccumulate in crops, vegetables and seafood and can be transferred to humans via the food chain. Foods (crops, vegetables, fish) contaminated with hazardous chemicals and exceeds the guideline threshold values (such as maximum residue level or MRL) can be rejected locally and internationally or have to be sold at a lower price. **Thirdly**, exposure of chemical pollutants (via water or food) may cause lethal (mortality) and sub-lethal (reduced growth, reproduction) effects on biota (fish, prawns, mussels), crops, vegetables, therefore, may cause a reduction or even loss of crop's yield and thereby cause an impact on poor farmers income. **Fourthly**, as a consequence of pollution, sensitive native species may be eliminated or reduced, and aquatic organisms such as fish, lobsters, shrimps, sharks would move away from the highly polluted estuaries and coasts to a new territory/ or a country resulting in the loss to the local economy and livelihoods of local people (Kibria et al 2010, Kibria 2016a,b). **Fifthly**, sea-level rise (caused by greenhouse gases (GHGs)/ carbon pollution) would cause significant losses of the world's coastal wetlands and its biodiversity (e.g. Royal Bengal tiger habitat in Bangladesh and India, island's biodiversity in biodiversity 'hot spots" in the world, coral reefs, river dolphins, fisheries), contamination of surface and groundwater with salt, salinisation of rice lands and freshwater aquaculture facilities, flooding of low-lying coastal areas, and forced migration, therefore, would impact on poor's who mostly depends on the ecosystem's goods and services (fish, fisheries, fishing and aquaculture, wetlands and forest biodiversity) as a source of food, animal protein, revenues, and livelihoods) but lacks resources, capacity and capabilities to adapt to climate change (Kibria 2014, Kibria et al 2017). **Lastly**, both chemical and climate change/GHGs pollutants would have a profound and pervasive threat to human health and wellbeing via water and food contamination, water and food-borne diseases (cholera and malaria), endocrine disruption, diabetes, cancer and other diseases, thus, predisposing poor/ vulnerable people to diseases and increased risks to their lives and livelihoods (as they can hardly afford expensive medicines or don't have access to medical treatments). The risks and impacts of specific pollutants on social, economic and environmental sectors have been discussed in detail from chapter 2 through chapter 13.

1.7. Linkages between pollution and climate change

Climate change/global warming would facilitate the transport and fate of pollutants in the environment. For example, the projected rise in temperatures would cause increased volatilisation of chemicals (Hg, POPs, pesticides) and increased wash off chemicals (pesticides, pharmaceuticals, metals, POPs) during flooding (briefly discussed in chapters 2, 5, 7, 13). During wild/ forest fires substantial amounts of Hg and dioxins could be released into the atmosphere (briefly discussed in chapters 2, 13). The melting/thawing of snows and permafrost would release a very significant amount of Hg, POPs trapped in snows and permafrost (briefly discussed in chapters 2, 7, 13). The uptake and toxicity of metals in seafood organisms may be enhanced with increasing temperatures/global warming (briefly discussed in chapters 2, 7, 8, 9). Sea-ice represent a sink for microplastic particles (MPs), however, melting of ice due to climate change/global warming could release them back into the environment (briefly discussed in chapter 4, 13). The warmer climates and climate extreme events could be more favourable to the proliferation of insect pests and plant diseases resulting in increased use of chemicals pesticides (briefly discussed in chapter 7). The expected rise in surface water temperatures could accelerate the growth of harmful blue-green algae/cyanobacteria blooms (BGA) which produce toxins (microcystins). In addition, global warming would worsen hypoxia/ deoxygenation in the open ocean and coastal waters impacting marine life (benthic organisms- fish and shellfish) (briefly discussed in chapter 8). Climate change could increase the incidence of sewer overflows due to increased rainfall intensity during storms. This could increase biological micropollutants, chemical micropollutants and nutrients and solids into the environment (briefly discussed in chapter 12).

1.8. Linkages between pollution and sustainable development goals (SDGs)

On 25 September 2015, the 193 countries of the UN General Assembly adopted the 2030 Development Agenda, a new Sustainable Development Goals (**SDGs**) (which has 17 specific goals and 169 associated targets). The SDGs are the blueprint to achieve a better and more sustainable future for including those related to poverty, food security, water security, good health, inequality, climate change, environmental degradation, peace and justice. However, pollutants like arsenic, heavy metals, pesticides, pharmaceuticals, microplastics, POPs (PCDDs, PCBs, PBDEs, PFOA), oil compounds (BTEX, PAHs), nutrients, surfactants/ detergents and synthetic dyes would undermine achieving SDGs since it (pollutants) can contaminate both waters (drinking, irrigation and recreational) and food (crops, vegetables, fish) and affect public health and social life. Therefore, it will prevent/impede achieving several SDGs including SDG6 (ensure clean water supply), SDG3 (ensure healthy lives for all), SDG1 (end poverty in all its forms everywhere), SDG2 (end hunger, achieve food security and improved nutrition and promote sustainable agriculture), SDG8 (jobs and sustainable economic growth), and SDG 14 (conserve marine resources and pollution in natural waterbodies).

1.9. Control and management of pollutants

1.9.1. *Monitoring of chemical residues in the environment*

Evidence of potential adverse effects or the risk of pollutants can be identified by chemical measurements of the pollutant concentrations in the environment. Monitoring helps to predict the level of risks (whether the risk is high, moderate or low) to various receptors such as human, wildlife or aquatic biota and whether the measured concentrations are within the recommended guideline threshold values and standards for the protection of human and environmental health. The actual concentrations of hydrophobic and hydrophilic pollutants can be measured by grab sampling or passive sampling or use of biomonitors (Kibria et al 2010).

1.9.2. *Awareness, education and capacity development*

Communities and industries should be educated about the threats and risks posed by pollutants so that they can have a responsibility to take care of their local environment: **i.** use of green and sustainable chemical products (e.g. green detergents); **ii.** disposing chemicals and wastes properly in council bins or resource recovery centres and not to flush out in the toilet (pesticides, medicines, paints, oil, solvents, plastics); **iii.** using chemicals based on the materials safety data sheet.

1.9.3. *Treatment of effluents*

Industries should be regulated to treat wastes/effluents appropriately effluent treatment plants (ETP) to reduce pollutants entering the waterways. Physical treatment (nano- and ultra-filtration, sorption), biological (microbes, macrophytes) and chemicals (coagulation and flocculation) treatments must be followed before discharging wastes/effluents in the environment to prevent pollution. All industries be forced to follow all legislated standards, guidelines, laws, acts and environmental codes of practices (ECOPs).

1.9.4. *Recycling and reuse of water*

Other ways to prevent water pollution is to recycle and reuse the wastewater for beneficial uses. Reclaimed water or recycled water that is treated to remove solids and impurities can be used in sustainable landscape irrigation or to recharge groundwater aquifers, to meet commercial and industrial water needs. This process helps to achieve sustainability and water conservation goals and prevents discharging the treated water to rivers and oceans. Recycled water can also be used for streamflow augmentation to benefit ecosystems and improve aesthetic uses (Kibria et al 2016b). However, it would be necessary to ensure that reuse does not ultimately discharge the pollutant or assist in the uptake to the food chain where it will be used for agriculture or aquaculture.

1.9.5. *Preventing soil erosion*

Preventing soil erosion would help reduce nutrients and pesticides load entering waterways and thus control pollution and algal blooms.

1.9.6. *Enforcing laws, acts and regulations to prevent water pollution*

Water pollution can be prevented by implementing a number of legislated standards, guidelines, practice codes and acts aiming to improve water quality and water usages. The rule of thumb should be 'Polluter's to Pay (PtoP)' for any sort of breaching of legislated standards, guidelines, laws, acts and ECOPs and remedial measures of environmental pollution. For example, in Victoria, Australia the State Environment Protection Policies (SEPP, Waters) address water quality requirements across Victoria by setting out the beneficial uses of water and the environmental quality indicators and objectives to protect them. SEPPs are established under the *Environment Protection Act 1970 as* a legal framework to protect and improve Victoria's water environments, as well as the social and economic values they support.

https://www.water.vic.gov.au/waterways-and-catchments/rivers-estuaries-and-waterways/state-environment-protection-policy

1.9.7. *Containment of pollutants*

Landfill containment systems (lining and capping systems such as compacted soil liners, synthetic membrane liners, and amended soil liners) serve as barriers to contain wastes and their derivatives (leachates and gases) within the waste disposal facilities so that they do not pollute groundwater resources, soil, air, and the environment (Akinwumi et al 2019). In addition, the litter trap can be used to trap floating plastics and other debris (Fig. 1.2A), retention basin/ artificial wetlands (Fig. 1.2B) to filter water through physical and biological processes to treat and remove pollutants from stormwater before it enters creeks, rivers and oceans (Fig. 1.2C). Use of floating turbidity curtains would help contain/ control the migration of suspended silt and sediment into the waterways.

1.2A. Litter trap catches floating waste including plastics, cigarettes butts, vegetation etc.

1.2B. Artificial wetlands to filter, treat and remove pollutants from stormwater.

1.2C. Floating turbidity curtains to contain suspended silt and sediments.

Fig. 1.2. Different techniques to contain and prevent pollution into waterways (all images from *https://www.google.com/search*).

1.10. Conclusion

The use of chemicals may have many benefits and are even essential for agriculture, horticulture, fisheries, aquaculture, animal husbandry and human health. In every case chemical should be used judiciously and with great care, following all legislated standards, guidelines, laws, acts and ECOPs (Environmental Community of Practices). Because pollution is now considered as a major challenge for economic, social and environmental development of a country. It is also a major obstacle and barrier to achieve globally agreed sustainable development goals (SDGs). For example, contamination of surface water and groundwater by high-risk pollutants would undermine achieving several SDGs such as SDG 6 (ensure clean water and sanitation for all), SDG 3 (ensure healthy lives for all), SDG 1 (end poverty in all its forms everywhere), SDG 2 (end hunger, achieve food security and improved nutrition, and promote sustainable agriculture), SDG 14 (life below water-conserve and reduce pollution of waterways). Several pollutants are persistent, bioaccumulative and toxic as well as endocrine-disrupting and carcinogenic posing significant threats and risks to water quality, fisheries, agriculture, food production, biota and human health. Such pollutants are arsenic, cadmium, chromium, lead, mercury, nickel, DDT, endosulfan, PCBs, Benzo[a]pyrene, benzene, nonylphenols, di(2-ethylhexyl) phthalate (DEHP), Bisphenol A (BPA), microcystins and malachite green. Water pollution has become a major concern worldwide, especially in developing countries. Pollution would enhance poverty and increase burdens on the poor since the livelihoods of poor people (both women and men) associated with agriculture, fishing, aquaculture, seafood, food export business etc. would be affected due to pollution, deterioration of water quality and contamination of food. Currently, most of the world top most polluted places are in poor developing countries (Argentina, Bangladesh, Ghana, Indonesia, Nigeria, Zambia

(Figure 1.2) where acts, laws, regulations pertinent to pollution prevention are not strongly enforced or lacking. It is, therefore, essential for regular monitoring to ascertain whether the measured concentrations of pollutants in all environmental matrixes (water, food, soil, biota) are within the recommended guideline threshold values and standards for the protection of human and environmental health. And where guidelines threshold values are exceeded effective regulatory action is implemented.

1.11. References

Akinwumi II, Booth CA, Ojuri OO et al 2019 Containment of pollution from urban waste disposal sites. Chapter 17 (pages 223-234). Urban Pollution: Science and Management, First Edition. Edited by Susanne M. Charlesworth and Colin A. Booth. John Wiley & Sons Ltd. Published 2019 by John Wiley & Sons Ltd. *https://doi.org/10.1002/9781119260493.ch17*

ATSDR (Agency for Toxic Substances and Disease Registry) 2017 The ATSDR 2017 substances priority list. *https://www.atsdr.cdc.gov/SPL/*

Bae J-S and Freeman HS 2007 Aquatic toxicity evaluation of new direct dyes to the *Daphnia magna. Dyes and Pigments* **73(1)** 81-88

IARC (International Agency for Research on Cancer) 2019 Agents Classified by the IARC Monographs, Volumes 1–123 *https://monographs.iarc.fr/wp-content/uploads/2018/09/ClassificationsAlphaOrder.pdf*

Kibria G 2014 Sea-level rise and its impact on wetlands, water, agriculture, fisheries, aquaculture, public health, displacement, infrastructure and adaptation. DOI: 10.13140/RG.2.1.1267.2487/5. *file:///C:/Users/User1/Downloads/Kibriaetal.2017_Climatechange-Fisheries_Areview_31Dec16%20(7).pdf*

Kibria G 2017 Plastic Waste, Plastic Pollution- A Threat to All Nations. DOI: 10.13140/RG.2.2.11169.51048 *https://www.researchgate.net/publication/319391174_Plastic_Waste_Plastic_Pollution-_A_Threat_to_All_Nations*

Kibria G, Haroon AKY, Nugegoda D and Rose G 2010 *Climate change and chemicals: Environmental Biological aspects.* New India Publishing Agency. New Delhi. 470 p. ISBN: 9789380235301. *https://www.researchgate.net/publication/261216635_Climate_Change_and_Chemicals_Environmental_and_Biological_Aspects*

Kibria G, Hossain MM, Mallick D et al 2016a Monitoring of metal pollution in waterways across Bangladesh and ecological and public health implications of pollution. *Chemosphere* **165** 1-9 *DOI: 10.1016/j.chemosphere.2016.08.121*

Kibria G, Haroon AKY and Nugegoda D 2016b Climate Change and Water security. Impacts, future projections, adaptations and mitigations. 330 p, ISBN: 978-93-85516-26-9.

https://www.researchgate.net/publication/303218799_Climate_Change_Water_Security

Kibria G, Haroon AKY and Nugegoda D 2017 Climate change impacts on tropical and temperate fisheries, aquaculture, and seafood security and implications - A review. *Livestock Research for Rural Development.* 29, Article #022 *http://www.lrrd.org/lrrd29/1/kibr29022.htm*

Oishimaya SN. 2018 "How Many Types of Pollution Are There?" *World Atlas*, Oct. 9, 2018, worldatlas.com/articles/how-many-types-of-pollution-are-there.html. *https://www.worldatlas.com/articles/how-many-types-of-pollution-are-there.html*

SCCNFP 2002 Opinion of the scientific committee on cosmetic products and on-food products intended for consumers concerning the safety review of the use of certain azo-dyes in cosmetic in cosmetic products.

SCCNFP/0495/01, final. adopted by the SCCNFP during the 19th plenary meeting of 27 February 2002 *https://ec.europa.eu/health/archive/ph_risk/committees/sccp/documents/out155_en.pdf*

Srogi K 2008 Levels and congener distributions of PCDDs, PCDFs and dioxin-like PCBs in environmental and human samples: a review. *Environmental Chemistry Letter* **6** 1-28

UNEP (United Nations Environment Programme) 2005 Ridding the world of POPs: A guide to the Stockholm Convention on persistent organic pollutants *http://chm.pops.int/Portals/0/Repository/CHM-general/UNEP-POPS-CHM-GUID-RIDDING.English.PDF*

Web sites

http://en.wikipedia.org/wiki/Stockholm_Convention_on_Persistent_Organic_Pollutants

https://www.worstpolluted.org/projects_reports/display/11

http://sciencythoughts.blogspot.com/2013/11/the-ten-most-polluted-places-on-earth.html

https://19january2017snapshot.epa.gov/sites/production/files/2015-09/documents/priority-pollutant-list-epa.pdf

■■■■

2 Arsenic Risks, Impacts and Management

Summary

Arsenic (a metalloid) is one of the most important global environmental pollutants and is a persistent, bioaccumulative, toxic, endocrine-disrupting and carcinogenic The Agency for Toxic Substances and Disease Registry of the USA has ranked arsenic as number 1 among the top 275 priority hazardous substances. Arsenic contamination of drinking water is a worldwide problem (reported over 70 countries affecting 200 million people). This chapter reviews **i**. chemistry, toxicity and ecotoxicity of arsenic; **ii**. impacts of arsenic on social, economic and environmental sectors including drinking water, soil, rice and vegetables, cow's milk, freshwater and marine fish, humans, social life and SDGs; **iii.** bioaccumulation of arsenic and arsenobetaine in fish/ seafood organisms; **iv**. management of arsenic risks.

Contents

List of Tables

List of Figures

List of Box

2.1. Introduction

Arsenic (As) have properties intermediate between those of a metal and a non-metal. It is often referred to as a metalloid or semi-metal (IARC 2004). Arsenic constitutes approximately 0.00005% of the Earth's crust and is the 20th most abundant element in the Earth's crust (Smedley and Kinniburgh 2002). As is present in soils and it also reported in drinking water obtained from wells drilled into arsenic-rich ground strata (Fig. 2.1) in many parts of the world. It occurs both in inorganic and organic forms, of which inorganic As is generally more toxic. The **arsenite** is more soluble, mobile and is absorbed faster than the arsenate, whereas **arsenate** is less soluble. Among the inorganic As, arsenite (As^{III}) is 60 times more toxic than arsenate (As^{V}) (Ferguson and Gavis 1972, WHO 1981). Inorganic arsenic compounds are about 100 times more toxic than organic arsenic compounds (dimethyl-arsinic acid, DMAA and monomethylarsonic acid, MMAA)) (Nagy and Korom 1983). Inorganic As^{III} and As^{V} are the major As species present in groundwater; other species of arsenic are DMAA, MMAA; methylated arsenic species are also present in natural water in some particular circumstances. As ranks No. 1 in the Agency for Toxic Substances and Disease Registry's (ATSDR) Priority List of Hazardous Substances, and is a known human carcinogen. Arsenic poisoning can impact the nervous system, gastrointestinal system, and renal system, and can create skin lesions, diabetes, cardiovascular diseases, and cancer (WHO 2010, ATSDR 2011). Its Chemical Abstracts Service Registry Number (CAS RN) is 7440-38-2 (ATSDR 2017; https://www.atsdr.cdc.gov/SPL/).

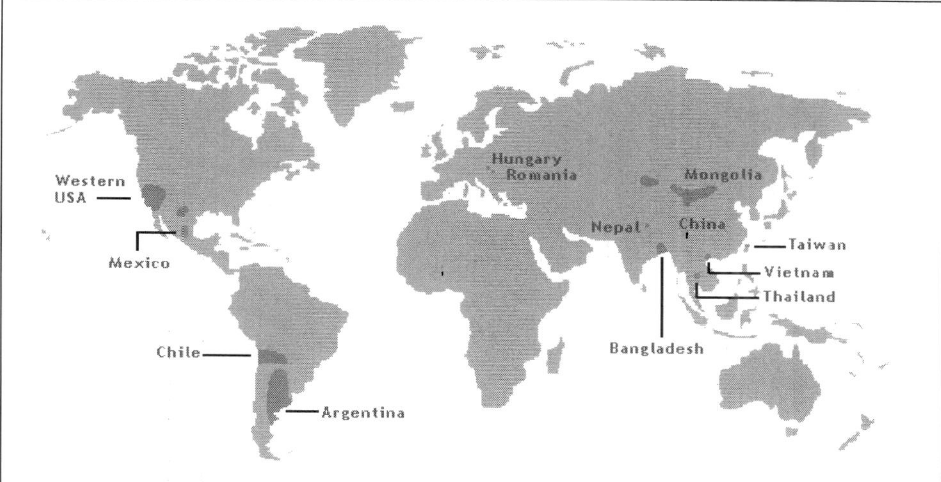

Fig. 2.1. Groundwater arsenic contamination areas in the world
http://en.wikipedia.org/wiki/Arsenic_contamination_of_groundwater

[Original from Arsenic risk areas worldwide Data source: "Arsenic contamination in south-east Asia region: Technologies for arsenic mitigation" by M. Feroze Ahmed of the World Bank Group's "Water Week 2004" forum].

As is commonly found in association with coal, gold, nickel, cobalt, antimony and iron ores or as arsenides or sulphides in copper, lead and zinc sulphide ores. As is relatively concentrated in gold and sulfide-bearing ore deposits. Pyrite (FeS) and arsenopyrite[1] (FeAsS) are typical examples of sulfide-bearing minerals with high arsenic contents. Arsenopyrite is reported to be the most abundant As-containing mineral (up to 4.5 wt. % As) and usually occurs under anaerobic sub-surface or marine environments (Henke 2009, Shahid et al 2018). Human activities (mining and manufacturing industries) may lead to an increase in the amount of arsenic in the environment. Industrial effluents also contribute arsenic to water in some areas (see Table 2.1). This chapter reviews **i**. Chemistry, toxicity and ecotoxicity of arsenic; **ii**. Impacts of arsenic on social, economic and environmental sectors including drinking water, soil, rice and vegetables, cow's milk, freshwater and marine fish, humans, social life and SDGs; **iii.** Bioaccumulation of arsenic and arsenobetaine in fish/ seafood organisms; **iv**. Management of arsenic risks.

Table 2.1. Anthropogenic sources of arsenic
(Commonwealth of Australia 2005, Naidu et al 2006, IARC 2012)

[As = Arsenic].

Sources	Examples
Agriculture (pesticides and herbicides)	As compounds were widely used in pesticides and herbicides. Lead arsenates was used in horticulture; arsenic trioxide in termite and cattle tick control; and monosodium methyl arsenate, disodium methyl arsenate and sodium arsenate in agriculture as herbicides. Old orchards can have high As concentrations due to accumulations of As in the soil.
Animal farms	Organic As species such as, arsanilic acid and carbarsone are used as a growth promoter (in feed) for poultry and pigs; manure used from these farms could be a source of arsenic (*note:* poultry litter found to contain 15-25 mg/kg total As).
Atmospheric deposition	Emissions of arsenic into the atmosphere can come from smelting and refining of non-ferrous metals, iron, and steel production and coal burning. Other sources are wood burning, agricultural chemicals, glass and cement manufacturing and waste incineration. These activities can lead to atmospheric arsenic emissions that are later deposited onto the land.
Coal (fossil fuel)	Coal combustion process produces fly and bottom ashes which may contain substantial concentrations of As. Deposition of fly ash onto land could increase soil arsenic level. Natural coal may contain an average of 1.05-3.4.2 mg/kg of total As. As level in coal from Czechoslovakia varies around 900-1,500 mg/kg. In Guizhou province, China some coals often contain As level as high as 100-9,600 mg/kg.

[1] ***Arsenopyrite** is an iron arsenic sulfide.*

Sources	Examples
	https://www.sourcewatch.org/index.php/ Heavy_metals_and_coal#cite_note-sludge-2
Glass making	Arsenic trioxide and pentoxide are used to produce clear glass.
Irrigation	Waste effluents of irrigation water can lead to the accumulation of arsenic in soils. Use of arsenic-contaminated groundwater (e.g. in Bangladesh and West Bengal, India) resulted in the gradual accumulation of As in topsoils and in the food chain (vegetables, rice).
Laboratory procedures	Arsenic trichloride is used in organo-arsenic chemistry and in chemical analysis procedures.
Mining and smelting	As is a natural constituent of many alloys, including copper, lead, zinc, silver and germanium. Soils can be contaminated with As during mining and smelting processes. As in streams, shallow wells and bore holes in gold mining areas of Ghana ranges from <2 to 175 µg/L and As in Topolnitza River (copper smelter area), Bulgaria was reported to vary from 750 to 1,500 µg/L.
Tannery (hide preservation)	Arsenic trioxide and sodium arsenate are used as hide preservatives. Tannery wastes are, therefore, contaminated with As.
Wastes/ sewage sludge (Urban and industrial)	Sewage sludge (bio-solids) can contain As concentrations up to 2-50 mg/kg of dry sludge. Wastes from pulp and paper industries may contain small amounts of As.

2.2. Sources of arsenic contamination of drinking water/ groundwater, shallow tube wells

- **Natural contamination** of groundwater aquifers by As is reported from various parts of the world (over 70 countries). As contamination generally occurs with fluctuating water tables and flooding cycles, in particular, in acid sulphide/ sulphate soils or where iron and/ or manganese-enriched layers or saline-layer aquifers occur. The most noteworthy occurrences are from Asia and America, where millions of people are at risk due to natural contamination of drinking water including Bangladesh, West Bengal (India), inner Mongolia (China), Mexico, Taiwan, USA and Vietnam (Table 2.2, Fig. 2.1).
- **Mining related** arsenic problems have been reported from Australia (gold mining), Brazil (gold mining), Canada (uranium mining), Ghana

(gold mining), Greece (lead and silver mining), Mexico and Thailand (tin mining), UK, parts of the USA (gold mining and Pb, Zn and Ag mining) (Table 2.2).

- Arsenic associated with **geothermal** water includes hot springs and have been reported from parts of Argentina, Chile, Italy, Mexico, New Zealand, the USA (Table 2.2).

Table 2.2. Concentration of arsenic in ground waters reported from some vulnerable regions/countries of the world [(Ng 2005, Naidu et al 2006, Garelick et al 2008, Smedley et al 2008, Sharma and Sohn 2009, Kibria et al 2010, Chakraborti et al 2013, Herath et al 2016)

[As = Arsenic; As guidelines for drinking water is 10 µg/ L; na = data is not available; NSW = New South Wales, Australia].

Country	Sources of arsenic exposure	As concentration range	Estimated population exposed	Remarks
Afghanistan	Wells (15-65 m depth)	10-500 µg/L*	na	• Location: Ghazni metropolitan area • As source: aquifer sediments (Smedley et al 2008).
Argentina	Drinking water, groundwater	100-2,000 µg/L	200,000	• Location(s): The Chaco-Pampean Plain, Cordoba province, North-western Argentina, Carcarana River • As source: Natural, volcanic ash
Australia	Ground-water, shallow wells	7,000 µg/L (Perth) (Appleyard et al 2006); 52-337 µg/L (NSW) (Smith et al 2003); < 1-300,000 µg/L (Victoria) (Mukherjee et al 2006).	na	• Locations(s): Perth, Northern new South Wales and Victoria • As sources: Pyritic sediments (Appleyard et al 2006); Al hydroxides and Fe oxyhydroxides (Smith et al 2003); Gold mining (Mukherjee et al 2006).

Country	Sources of arsenic exposure	As concent-ration range	Estimated population exposed	Remarks
Bangladesh	Natural – deriving from geological strata; Drinking water (shallow tube wells/ groundwater) (<100 m depth)	< 1-4,730 μg/L	57 million exposed to As > 10 μg/L and 35 million exposed to As > 50 μg/L (Chakraborti et al 2013)	• **Out of 64 districts**, As level in **60 districts** have exceeded WHO recommended guidelines of 10 μg/L and in 51 districts it exceeded Bangladesh recommended guidelines of above 50 μg/L (Chakraborti et al 2013). • Affected aquifers are shallow (< 100-150 m deep) • Deeper groundwater's (> 100-150 m depth) have generally low As concentrations (< 10 μg/L); • As problems in groundwater is related to flood plain, deltaic sediments (Herath et al 2016). • High As ground waters are also high in iron (> 0.2 mg/L), manganese (> 0.5 mg/L, bicarbonate (> 500 mg/L), and often phosphorus (> 0.5 mg/L), low chloride (< 60 mg/L), sulphate (< 1 mg/L), nitrate and fluoride (<1 mg/L), and pH (~ > 7) • As^{3+} (arsenite) is the dominant species found in the groundwater's of Bangladesh.
Brazil	Surface water	0.4–350 μg/L	na	• Location: Minas Gerais • As source: Gold mining • (Mukherjee et al 2006).

Country	Sources of arsenic exposure	As concentration range	Estimated population exposed	Remarks
Cambodia	Ground water	1-1,610 μg/L	0.5-1 million	• Location: Mekong River • As source: Iron phases in sediment aquifers (Berg et al 2007).
Canada	Ground-water, shallow water	1.5-738 μg/L	na	• Locations: Nova Scotia; Northern Saskatchewan • As source: Uranium mines (Donahue and Hendry 2003, Nriagu et al 2007).
Chile	Natural – associated with quaternary volcanism; Groundwater	900-1,040 μg/L	437,000	• Locations: High As concentrations are reported from the arid region (Northern Chile-Antofagasta, Calama, Tocopilla) • As source: Volcanic sediments and high salinity, thermal springs, mining (Nordstrom 2002, Nriagu et al 2007).
China (inner Mongolia)	Drinking water (bores)	1-2,400 μg/L	600,000	• na
Croatia	Shallow and deep groundwater	27-240 μg/L	na	• Location: Osijek area • As source: Aquifer sediments (Romić et al 2011).
Germany	Groundwater	10-150 μg/L	na	• As source: Aquifer sediments (Nordstrom 2002).
Ghana	Mining activities; Groundwater	1-175 μg/ L	100,000	• As source: Gold mining (Nordstrom 2002).

Country	Sources of arsenic exposure	As concentration range	Estimated population exposed	Remarks
Greece	Groundwater	10,000 μg/L	na	• As source: Mine tailings (Nriagu et al 2007).
Hungary and Romania	Natural; Drinking water (groundwater wells)	0.7-240 μg/L	400,000	• Location: Pannonian Basin • As source: Aquifer sediments (Appleyard et al 2006).
India (West Bengal)	Natural – deriving from geological strata; Shallow tube wells	< 10-3,700 μg/L	100,000 Over 5 million exposed to As > 50 μg/L; 300,000 suffering from arsenicosis.	• Location(s): West Bengal and Uttar Pradesh • As source: Alluvial/ deltaic sediments • Affected aquifers are shallow (< 100-150 m deep) • Deeper groundwater's (> 100-150 m depth) have generally low As concentrations (< 10 μg/L) (Chakraborti et al 2013).
Italy	Groundwater	0.1–6940 μg/L	na	• Locations: Vulcano Island and Phlegrean Fields • As source: Volcanic aquifers (Aiuppa et al 2003).
Japan	Groundwater wells	1-293 μg/L	na	• Location(s): Fukuoka, Sendai and Takatsuki • As sources: Factory wastes containing arsenic sulphide (Mukherjee et al 2006).
Laos	Tubewells	277.8 μg/L	na	• Location: Laos • As source: Alluvial and deltaic aquifers (Nordstrom 2002).

Country	Sources of arsenic exposure	As concent-ration range	Estimated population exposed	Remarks
Mexico	Natural; Groundwater	8-620 µg/L	400,000	• Locations: Zimapan and la Lagunéra region • As sources: Volcanic sediments, mining (Armienta et al 1997; Boochs et al 2014).
Nepal	Natural; Drinking water	<10-34 µg/L	550,000 exposed to As > 50 µg/L and 3.19 million exposed to As >10 µg/L	• 90% people in Terai regions rely on groundwater for drinking water (47% of Nepal's population) • out of 15,000 tube wells analysed, 23% contained As above 10 µg/L and 5% above 50 µg/L.
New Zealand	Shallow water	21 µg/L	na	• Location: Taupo Volcanic Zone • As source: Volcanic activities.
Pakistan	Surface and groundwater	3-106 µg/L	50-60 million people	• As hot spots are around Lahore and Hyderabad (Podgorski et al 2017).
Slovakia	Groundwater	37-39 µg/L	na	• Locations: Banska Bystrica and Nitra counties. • As source: Aquifer sediments (Leonardi et al 2012).
Spain	Groundwater	<1-100 µg/L	na	• Alluvial sediments (Nordstrom 2002).
Taiwan	Natural; Groundwater	1.3-1,820 µg/L	10,000 (?)	• Chianan plains, Lanyang plains and Coastal zones (Nordstrom 2002; Maity et al 2011).

Country	Sources of arsenic exposure	As concentration range	Estimated population exposed	Remarks
Thailand	Shallow groundwaters, shallow wells; mining activities	1-5,000 μg/L	15,000	• Ron Phibun district • As problems are related to mining and dredged alluvium (Kim et al 2011).
UK	Groundwater	<1–80 μg/L	na	• Location: Southwest England • As source: Mining (Nriagu et al 2007).
USA	Natural, geothermal and mining related sources; Drinking water	10-48,000 μg/L	13 million exposed to As = 10 μg/L.	• Location: Western USA • As sources: Natural, geothermal and the mining related sources (Welch et al 1988).
Vietnam	Natural; Groundwater	1-3,100 μg/L	10 million	• Location: Red River delta • As source: iron phases in sediment aquifers (Berg et al 2007)

**μg/L is also often termed as parts per billion (ppb)*

2.3. Chemistry, toxicity and ecotoxicity

2.3.1. Chemistry and speciation

Arsenic (As, atomic number = 33, relative atomic mass, 74.92) is a ubiquitous element, which occurs naturally in the earth's crust. As exists in four oxidation/valance states, **i.** $^{+}V(5)$ (pentavalent arsenate, AsV; As^{5+}; As_2O_5), **ii.** $^{+}III(3)$ (trivalent arsenite, As^{3+}; As_2O_3; AsIII), **iii.** $^{0}As(0)$ - elemental arsenic), and **iv.** $^{-}III(3)$ (arsine, AsH_3-III) (Akter et al 2005, Sharma and Sohn 2009, IARC 2012, Chowdhury et al 2018; Table 2.3, Table 2.4). There are three major groups of As compounds as follows: **a.** inorganic arsenic compounds, **b.** organic arsenic compounds and **c.** arsine gas (IARC 2012). As can enter into the environment and biological system in various inorganic and organic forms (species) (Table 2.3). These are arsenous acid (As^{III}), arsenic acid (As^{V}), monomethyl arsonic acid (MMAA), dimethyl arsinic acid (DMAA), arsenocholine, arsenobetaine and arsenosugars. Arsenobetaine and arsenocholine are collectively called as **'fish**

arsenic'. The physical and chemical properties of some selected arsenic compounds are given in Table 2.4.

Table 2.3. Some arsenic compounds of environmental importance (IPCS 2001, Ng 2005).

Type of arsenic	Name	Formula or structure
Inorganic (trivalent)	Arsenous acid	$As(OH)_3$
	Arsenenous acid	$HAsO_2$
	Arsenic trioxide	As_2O_3 (or As_4O_6)
	Arsenic chloride	As_2Cl_3
	Arsenic sulphide	As_2S_3
Inorganic (pentavalent)	Arsenic pentoxide	As_2O_5
	Arsenic acid	$AsO(OH)_3$
	Arsenenic acid	$HAsO_}$
	Arsenates, salts of *ortho*-arsenic acid	$H_2AsO^-_4$, $HAsO^{2-}_4$, AsO^{3-}_4
Organic	Methylarsine	CH_3AsH_2
	Dimethylarsine	$(CH_3)_2AsH$
	Trimethylarsine	$(CH_3)_3As$
	Monomethylarsonic acid	$CH_3AsO(OH)_2$,MMA^V
	Monomethylarsenous acid	$CH_3As(OH)_2$, MMA^{III}
	Dimethylarsinic acid	$(CH_3)_2AsO(OH)$, DMA^V
	Dimethylarsenous acid	$(CH_3)_2AsOH$, DMA^{III}
	Trimethylarsinic oxide	$(CH_3)_3AsO$, TMAO
	Tetramethylarsonium ion	$(CH_3)_4As^+$, TMA
	Arsenobetaine (fish arsenic)	$(CH_3)_3As^+CH_2COO^-$, AB
	Arsenocholine (fish arsenic)	$(CH_3)As^+CH_2$ CH_2OH, AC
	Dimethylarsinoylribosides	1-11
	Triaklylarsonioribosides	12, 13
	Dimethylarsonoulribitol sulfate	14
	Glycerophosphorarsenochloine	15
	Glycerophosphatidylarsenochloine	16

Table 2.4. Physical and chemical properties of arsenic compounds
(WHO 2001a, Kumari et al 2016).

[na = data is not available].

Chemical name	Synonyms	Molecular weight	Melting point (°C)	Oxidation state[2]	Water solubility
Arsenic acid $AsO(OH)_3$ (inorganic)	na	141.95	na	+5	Freely soluble
Arsenic As (inorganic)	Metallic arsenic Arsenic-75, As black, Colloidal As	74.92	613	0	Insoluble
Arsenic penta-oxide As_2O_5 (inorganic)	Arsenic oxide Arsenic acid anhydride	229.84	315	+5	Freely soluble 1,500 g/L at 16°C
Arsenic tri-oxide As_2O_3 (inorganic)	Arsenic oxide Arsenous acid anhydride, white arsenic, arsenolite	197.82	312.3	+3	Slightly soluble 37 g/L at 20°C
Arsine AsH_3 (inorganic)	Arsenic hydride, arsenic trihydride	85.91	na	+3	Freely soluble
Dimethylarsinic acid $(CH_3)_2HAS0_2$ (organic)	Cacodylic acid Dimethylarsinic acid	138.01	200	+5	Soluble 829 g/L at 22°C
Methanearsonic acid $CH_3H_2AsO_3$ (organic)	Methyl arsenic acid, Monomethylarsonic acid	139.98	na	+5	Freely soluble
Sodium arsenate Na_2HAsO_4 (inorganic)	Disodium arsenate Sodium biarsenate	85.91	na	+5	Very soluble

[2]***Oxidation state*** *= the degree of oxidation (loss of electrons) of an atom in a chemical compound.*

Chemical name	Synonyms	Molecular weight	Melting point (°C)	Oxidation state[2]	Water solubility
Sodium dimethyl arsenate $(CH_3)_2NaAsO_2$ (organic)	Sodium cacodylate	159.98	na	+5	Readily soluble

Speciation of inorganic As in soil is largely controlled by reduction and oxidation processes (redox). Under oxidizing and aerobic (oxygen-rich) conditions, the predominant form of As in soil and water is arsenate (As^V), whereas under anaerobic (absence of oxygen) and reducing/ or waterlogged (< 200 m^V; mv=megavolt) conditions, arsenite (As^{III}) predominates (Akter et al 2005, Ng 2005). The rate of conversion is dependent on the Eh (Redox potential) and pH of the soil as well as other physical, chemical and biological factors. In short, at moderate or high Eh, inorganic As occurs as a series of pentavalent arsenate oxyanions (As^V) such as $H_3AO_4^-$, H_2AO_4, $HAsO_4^{2-}$ and AsO_4^{3-}. However, under most reducing (acid and mildly alkaline conditions (low Eh values) arsenite (As^{III}) predominates. Microbial activity can influence As speciation via various mechanisms such as redox[3] reactions with Fe and As and via (de)methylation of As species. Speciation determines how arsenic compounds interact with their environment. For example, the behaviour (mobility) of arsenite (As^{III}) and arsenate (As^{IV}) in soil differs considerably depending on soil pH and soil type. At a lower pH (e.g. pH 5.8), mobility of arsenate is slightly more than arsenite. When pH changes from acidic to a neutral or basic level, mobility of both arsenite and arsenate increases with increasing pH. In strongly adsorbing soils, transport rate and speciation are influenced by organic carbon content and microbial population. Both arsenite and arsenate are transported at a slower rate in strongly adsorbing soils than in sandy soils (IPCS 200, Ng 2005, Sharma and Sohn 2009). In **groundwater**, inorganic As commonly exists as either arsenate (As^V) or arsenite (As^{III}), the latter being considered to be more mobile and more toxic to living organisms (Korte and Fernando 1991). Redox potential (Eh)[4],

---

[3]***Redox*** *is an oxidation – reduction reaction, it is a type of chemical reaction that involves a transfer of electrons between two species. For example, if one species gains electrons (reduction), another species (or more than one species) must lose electrons (oxidation). For example, the redox reaction in which zinc displaces copper in an aqueous solution of copper sulfate:* $CuSO_4(aq) + Zn(s) \rightarrow ZnSO_4(aq) + Cu(s)$

(https://www.chemicool.com/definition/redox-reactions-electrochemistry.html).

[4]***Eh*** *= Eh is a measure of the redox (oxidation-reduction) state of a solution or, its solutes. Eh is a measurement of electrical potential and thus commonly expressed in volts. Values of Eh in nature range from -0.6 to +0.9V, with 0.0 characterizing a solution with no drive to either oxidize or reduce. Accurate measurement of Eh is much more difficult than that of pH, because measurement of Eh requires equilibration with many solutes and even solids (http://railsback.org/Fundamentals/8150RedoxPlot07LS.pdf).*

pH[5], and dissolved oxygen (DO) are all important factors controlling As speciation and chemistry in groundwater (Akter et al 2005).

2.3.2. Toxicity of arsenic

Arsenic is one of the most important global environmental pollutants and is a persistent, bioaccumulative, toxic (Kaur et al 2011, Shahid et al 2018), endocrine disrupting[6] (Davey et al 2008) and carcinogenic[7] (IARC 2012). It is potentially toxic to humans, animals and plants. Inorganic arsenicals are proven carcinogens in humans (Table 2.5).

Table 2.5. IARC (International Agency for Research on Cancer) classification of inorganic and organic arsenic (IARC 2012, IARC 2019).

Substances	IARC carcinogen group	Remarks
Arsenic and inorganic arsenic compounds	Group 1	Carcinogenic to humans
Arsenobetaine and other organic arsenic compounds that are not metabolized in humans	Group 3	Not classifiable as to their carcinogenicity to humans
Dimethylarsinic acid	Group 2B	Possibly carcinogenic to humans
Methylarsonic acid	Group 2B	Possibly carcinogenic to humans
Monomethylarsonic acid	Group 2B	Possibly carcinogenic to humans

Ingestion of contaminated water and food (crops and vegetables) are the two major pathways of arsenic exposure and poisoning to humans, animals and biota (Jain and Ali 2000, Akter et al 2005). The toxicity of As in the environment is affected by temperature, pH, Eh, organic content, phosphate concentration, adsorption to solid matrices, presence of other substances and toxicants, duration of exposure and the arsenic species present. A wide range of As toxicity has been determined that depends on As speciation (Ng 2005, Sharma and Sohn 2009). In general, inorganic arsenicals are more **toxic** than organo-arsenicals to biota and trivalent arsenic (arsenite, As^{III}) is more acutely toxic than pentavalent (arsenate,

[5]***pH*** *(the negative logarithm of the hydrogen ion concentration [H+]; a pH of 0–7 indicates acidity, a pH of 7–14 indicates basicity, and neutral solutions have a pH of 7)*

[6] ***Endocrine disruptors*** *are chemicals that may interfere with the body's endocrine system and produce adverse developmental, reproductive, neurological, and immune effects in both humans and wildlife https://www.niehs.nih.gov/health/topics/agents/endocrine/index.cfm*

[7] ***Carcinogenic*** *= cancer-causing*

As^{V}) form (UN 2001, Akter et al 2005). The toxicity of different arsenic species varies in the following orders:

Arsenite (As^{III}) > Arsenate (As^{V}) > Monomethylarsonate (MMAA) > Dimethylarsinate (DMAA) > Arsenocholine > Arsenobetaine (Stugeron et al 1989, Jain and Ali 2000).

The LD_{50}[8] (median lethal dose) values of different arsenicals are given in Table 2.6 which shows that arsenite (As^{III}) is most toxic (LD_{50} range 15-42 mg/kg) and arsenobetaine is least toxic (LD_{50} > 10,000 mg/kg).

Table 2.6. LD_{50} of different arsenic species to rats

(*http://www.epa.gov/fishadvisories/forum/2004/presentations/wednesday/abernathy.pdf*; MSDS-Thermo Fisher 2009-Arsenic; MSDS-Labchem 2000- Arsenic Trioxide; Akter et al 2005).

Arsenic compound	Arsenic species	LD_{50} (mg/kg)
Inorganic As	Arsenite [As(III)]	10-42
	Arsenate [As(V)]	20-800
Organic As	MMA-monomethylarsonic acid	700-1,800
	DMA-dimethyarsinic acid	1,200 -2,600
	AsC-arsenocholine (marine)	6,500
	AsB-arsenobetaine (marine)	≥10,000

2.3.3. Ecotoxicity of arsenic

As is toxic to algae, invertebrates (*Daphnia*) and fish (Table 2.7). Based on LC_{50}[9] (median lethal concentration) results (Table 2.7), As is very toxic to moderately toxic to fish. Lab-based toxicity experiments conducted reveals that arsenite (As^{III}) is generally more toxic than arsenate to biota. For example, the 96 h LC_{50} for freshwater fish range from 10.8 to 91 mg/L for trivalent arsenic (As^{III}) and from 4.8 to > 360 mg/L for pentavalent arsenic (Table 2.7). Various studies reported LC_{50} values of As for salmonids fish to be between 3 mg/L and 167 µg/L (USEPA 1985). Sub-lethal effects of As on fish include reduced growth, anaemia, gallbladder inflammation, and alterations in liver and kidney (Rankin and Dixon 1994, Woodward et al 1994).

[8] ***LD_{50}*** *= the lethal dose of material that is estimated to be lethal to 50% of the test organisms (rat)*

[9] ***LC_{50}*** *= The lethal concentration of material in the water that is estimated to be lethal to 50% of the test organisms (fish)*

Table 2.7. Laboratory-based acute toxicity of various arsenic species
(based on various sources and compiled in WHO 2001b; Tisler and Zagorc-Konean 2003; Akter et al 2008; *http://www.who.int/mediacentre/factsheets/fs210/en/print.htm*)

[Ecotoxicity classification: extremely toxic: if LC_{50} or EC_{50} are below 1 mg/L; very toxic: if LC_{50} or EC_{50} is between 1 and 10 mg/L; moderately toxic: if LC_{50} or EC_{50} are between 10 and 100 mg/L (Tisler and Zagorc-Konean 2003, Bae and Freeman 2007)].

Biota	Value range	References
Algae	48 µg to 202 mg/L (As^V)	WHO 2001a
	Scenedesmus subspicatus, LC_{50}: 34.7 mg/L for 72 h (Arsenic)	Tisler and Zagorc-Konean 2003
Aquatic invertebrates	LC_{50}/ EC_{50}: 0.68 to 73.5 mg/L for 48 h (As^{III}) LC_{50}/ EC_{50}: 3.6 to 49.6 mg/L (As^V)	WHO 2001b
Aquatic invertebrates/ *Daphnia*	*Daphnia magna*, EC_{50}: 2.5 (2.4-2.7) mg/L for 48 h (Arsenic)	Tisler and Zagorc-Konean 2003
Aquatic vertebrates (96 h LC_{50}/ EC_{50})	Freshwater fish: 10.8-91 mg/L for 96 h (As^{III}) Freshwater fish: 4.8 - > 360 mg/L for 96 h (As^V) Freshwater fish: 1.9-1,412 mg/L for 96 h (Monomethylarsonate) Marine fish: 12.7-28.5 mg/L (As^{III}) Marine fish: 21.4-157 mg/L (As^V)	WHO 2001b
Fish	*Anabus testudinus,* adult, LC_{50}: 18.21 mg/L for 96 h (Sodium arsenate)	Akter et al 2008
	Catla, fingerlings, LC_{50}: 20.41 mg/L for 96 h (As^{III} - Arsenite)	Lavanya et al 2011
	Channa punctatus, fingerlings, LC_{50}: 10.8 mg/L for 96 h (As^{III})	Shukla et al 1987
	Chanos, fingerlings, LC_{50}: 7.29 mg/L for 96 h (Arsenic)	Chou et al 2006
	Cirrhosus mrigala, adult, LC_{50}: 24.5 mg/L for 96 h (As^{III})	Kousar and Javed 2014
	Clarias batrachus, adult, LC_{50}: 84 mg/L for 48 h (As^{III})	Bhattacharya and Bhattacharya 2007

Biota	Value range	References
	Cyprinus carpio, adult, LC_{50}: 32 mg/L for 96 h (As^{III})	Kovendan et al 2013
	Danio rerio, juvenile, LC_{50}: 43 mg/L for 96 h (As^{V} - Arsenate)	Liu et al 2008
	Labeo rohita, juvenile, LC_{50}: 28.3 mg/L for 96 h (As^{III})	Vutukuru et al 2007
	Oncorhynchus mykiss, LC_{50}: 15.3 (12.9-17.3) mg/L for 96 h (Arsenic)	Tisler and Zagorc-Konean 2003; MSDS-Thermo Fisher 2009; MSDS-Labchem 2000
	Oncorhynchus shawytscha, juvenile, LC_{50}: 21.4 mg/L for 96 h (As^{V})	Hamilton and Buhl 1990
	Oreochromis mossambicus, adult, LC_{50}: 28.68 mg/L for 96 h (Arsenic)	Akter et al 2008
	Oryzias latipes, fingerlings, LC_{50}: 14.6 mg/L for 7d (As^{III})	Suhendrayatna et al 2002

2.4. Impacts of arsenic on social, economic and environmental sectors

2.4.1. Arsenic contamination of drinking water

Globally an estimated 200 million people, mostly in Asian countries, are most severely affected by groundwater As contamination (WHO 2012), Bangladesh is the most serious of all. Prolonged human exposure to As-contaminated groundwater has occurred in many countries (Table 2.2). The groundwater As poisoning in Bangladesh is regarded as the largest disaster/ mass poisoning in the history of human civilization (where more than 70 million people are believed to be drinking As-poisoned water or exposed to As-contaminated water). In Bangladesh, most of the shallow tube-wells (20-75 m deep) used for drinking water and/ or irrigation is contaminated with high levels of As, whereas tube-wells deeper than 150 m is relatively free from arsenic (Kibria et al 2010, chapter 7). WHO 2010 report estimated 9,100 deaths and 125,000 disability-adjusted life years in Bangladesh due to the consumption of As-contaminated drinking water in 2001. The other countries affected by groundwater As contamination are India

(West Bengal), China (inner Mongolia), Nepal, Taiwan, Thailand, Vietnam, Argentina, Chile, USA, Mexico) (see Table 2.2).

2.4.2. Arsenic contamination of soil

Elevated concentration of As is reported in soils in Bangladesh where irrigation was carried out with As-contaminated groundwater (up to 81 mg/kg in contaminated soils compared to 0.10-2.75 mg/kg in uncontaminated soils) (Huq et al 2006 a, b).

2.4.3. Arsenic contamination of rice and vegetables

The elevated level of As in water and soil can result in elevated concentrations of As in food since crops receiving As-contaminated irrigated water can uptake As during the phyto-extraction process and can bioaccumulate in different degrees in different parts of plants (e.g. roots, stems, and grain). The highest quantities of As residues are found in plant roots and moderate As accumulation has been found in the vegetative parts such as leaves and stem and the smallest quantities in fruits and seeds (Table 2.8). Because of the highest As accumulation or concentrations in roots, humans can be exposed to As from eating of root vegetables such as carrot, beet, potato or arum. The effects of elevated levels of soil As on plants include a reduction of crop yield or growth of plants. inhibition of seed germination and seedling establishment. According to a review by Brammer and Ravenscroft 2009, the effects of As toxicity on rice includes the following symptoms:

- delayed seedling emergence
- reduced plant growth
- yellowing and wilting of leaves
- brown necrotic spots on older leaves
- reduced grain yields (e.g. rice yields in conventional paddy fields in Bangladesh decreased from 8.92 to 2.99 t/ha with an increase of soil As levels from 26.3 to 575 mg/kg).

Table 2.8. Distribution of arsenic (mg/kg) in rice plant, grain, leaf, stems and roots (Huq et al 2006b).

Plant part	Minimum-Maximum*	Average
Root	9.06-73.4	29.1
Stem	0.35-25	4.29
Leaf	0.06-3.32	1.21

Plant part	Minimum-Maximum*	Average
Grain	0.02-2.81	0.29
Husk	0.02-0.15	0.07

** Recommended guideline threshold levels in food is 1 mg/kg*

Several research studies have found high concentrations of As in vegetables and rice in areas where concentrations of As in soil and water is also high. Higher concentrations of As were also reported in rice plants (*boro* rice/ winter rice cultivated in Bangladesh and Assam, Bihar and West Bengal of India) in the following orders:

Rice roots > Rice stem > Rice leaf > Rice grain > Rice husk (Table 2.8).

It is, therefore, evident that the food chain could be a significant pathway of ingestion of As by the people of Bangladesh (Kibria et al 2010). Several research studies have found high concentrations of As in vegetables and rice in areas where concentrations of arsenic in soil and water are also high for example, the greatest concentration of arsenic was found in leafy vegetables[10] (in particular in arum or *kochu;* up to 153.2 mg/kg) followed by other vegetables (Table 2.9). High concentrations of As in leafy vegetables and rice have been found where concentrations of arsenic in soil and water are also high.

Table 2.9. Arsenic in common food plants of Bangladesh from uncontaminated and contaminated areas (Huq et al 2006 a, b).

Common English name *(local Bangla/Bengali name)*	**Botanical name**	**Uncontaminated areas (mg/kg)**	**Contaminated areas (mg/kg) ***
Arum (kochu)	*Colocassia antiqourum*	0.077-0.387	0.13-153.2
Aubergine/Egg plant (begun)	*Solanum melongena*	0.23	2.3
Bean (shim)	*Dolicos lablab*	0.092	0.3-1.16
Bitter gourd (karola)	*Momordicum charantia*	1.56	2.12

[10] ***Leaf vegetables*** *are plant leaves eaten as a vegetable*

Common English name *(local Bangla/Bengali name)*	**Botanical name**	**Uncontaminated areas (mg/kg)**	**Contaminated areas (mg/kg) ***
Chilli (morich)	*Capsicum* spp.	0.41	1.52
Green papaya (pepe)	*Carica papaya*	0.212-0.46	0.04-2.22
Indian spinach (pui shak)	*Brasilia alba*	0.102-0.146	0.07-1.00
Long bean (barboti)	*Vicia laba*	0.3	0.37-2.83
Potato (alu)	*Solanum tuberosum*	0.62	0.71-2.43

** Recommended guideline threshold levels of arsenic in food is 1 mg/kg*

2.4.4. Arsenic contamination of cows' milk and dung

A new research study (Ghosh et al 2013) revealed some alarming news that As contamination is not only limited to crops and vegetables (Table 2.9) in Bangladesh but also to cow's milk (Table 2.10). This is a new threat to the cattle farming industry since paddy straw accounts for the main source of dairy ration in Bangladesh. Additionally, small-scale dairy farms provide an unprecedented range of relatively low-cost milk and milk products for consumers. The research found that paddy straw fed to cows is the main source of arsenic contamination in dairy rations. They reported more arsenic in Boro rice (winter crop; October-March crops) straws compared to *Aus* and *Aman* rice (*kharif I* and *kharif II*; kharif crops = monsoon crops; July–October period) straws. Their results showed that a wide range of As in milk of cows, cow urine and dung in Bangladesh (Table 2.10). This creates a potential dietary risk to human. It is important to mention here that, in Bangladesh and India, cow dung is generally used as manure and cooking fuel. Burning As-rich cow dung cakes in unventilated ovens release As loads to the environment and thereby can affect people through direct inhalation (Pal et al 2007). It is also possible that recontamination of the food chain with As can occur in case the cow dung is used as fertilizer in conventional or organic farming.

Table 2.10. Arsenic in cow milk, urine and dung in some selected areas of Bangladesh (Ghosh et al 2013).

Districts (Bangladesh)	**Milk (µg/L)**	**Urine (µg/L)**	**Cow dung (µg/kg) ***
Chandpur	18.8	150.3	1,590.0

Districts (Bangladesh)	Milk (µg/L)	Urine (µg/L)	Cow dung (µg/kg) *
Faridpur	24.2	92.3	1,767.1
Jashore	49.3	103.0	1,804.5
Madaripur	24.7	171.2	1,495.5
Satkhira	16.1	93.2	1,841.1

** Dry weight*

2.4.5. Arsenic contamination of freshwater and marine fish

A compilation of global data reveals that several marine fish (hake, sole, mackerel, whiting, mullet) consumed by humans have exceeded the respective maximum allowable limits of arsenic (Table 2.11). However, it should be noted that up to 90% of As in fish muscle is present in the non-toxic arsenobetaine form (Zorob et al 1998, Goyer and Clarkson 2001, Bosch et al 2016) (see the next section 2.4.6 for bioaccumulation of arsenobetaine in zooplankton, phytoplankton and in seafood/ fish).

Table 2.11. Arsenic (As) concentrations in some commonly consumed marine fish in some selected countries where it exceeded the respective maximum allowable limits

[maximum allowable limits of As for seafood (fish, crustaceans) in Australia and New Zealand is 2 mg/kg (FSA 2016), the regulatory limits in South Africa is set at 3.0 mg/kg (DOH 2004, Bosch et al 2016)].

Country/ region	Fish species	mg/kg	References
Croatia, Croatian coasts	Hake, *Merluccius merluccius*	23.03	Juresa and Blanusa 2003
Croatia, Croatian coasts	Horse mackerel, *Trachurus trachurus*	6.85	Juresa and Blanusa 2003
Croatia, Croatian coasts	Sardine, *Sardina pilchardus*	8.08	Juresa and Blanusa 2003
Croatia, Croatian coasts	Shell-fish, *Pitaria chione*	8.40	Juresa and Blanusa 2003

Country/ region	Fish species	mg/kg	References
Italy, Central Adriatic Sea	Atlantic mackerel, *Scomber scombrus*	30.76	Perugini et al 2014
Italy, Central Adriatic Sea	Blue whiting, *Micromesistius poutassou*	35.30	Perugini et al 2014
Italy, Central Adriatic Sea	European hake, *Merluccius*	38.70	Perugini et al 2014
Italy, Central Adriatic Sea	Red mullet, *Mullus barbatus*	59.91	Perugini et al 2014
USA, Aleutian Islands, Alaska	Flathead sole, *Hippoglossoides elassodon*	19.5	Burger et al 2014
USA, Aleutian Islands, Alaska	Rock Sole, *Lepidopsetta bilineata*	4.34	Burger et al 2014

2.4.6. Arsenic bioaccumulation in aquatic/ seafood organisms (fish, shrimp, crabs, mussels)

Many investigations were carried out to assess bioaccumulation potential of As in different parts of fish's/ seafood organs (Table 2.12). It reveals that high accumulation of As generally occurs in liver and gills and less in muscles (the part that is eaten by a human) (Table 2.12A). In addition, **arsenobetaine**[11] (AsB) constituted the dominant arsenic species accumulated in marine fish and seafood organisms (Table 2.12B). Arsenobetaine is described as the only non-toxic arsenic compound (Caumette et al 2012). It seems therefore that, the consumption of seafood is a low health risk to human considering organic As is much less toxic. The transfer of arsenobetaine to fish/ seafood organism seems to occur via the marine food chain (Phytoplankton → Zooplankton → Fish) (Box 2.1).

[11] ***Arsenobetaine*** *is an organoarsenic compound that is the main source of As found in fish and is relatively non-toxic (it is often called ''seafood/ fish arsenic'')*

Table 2.12. Bioaccumulation of arsenic and arsenobetaine in fish/ seafood organisms.

[As = Arsenic; AsB = arsenobetaine; ASC = arsenocholine; DMA = dimethylarsenic acid; dimethylarsinate; - = data not available].

Sequence	Fish and other aquatic species studied	The sequence of accumulation of As	Remarks	References
2.12A The sequence of bioacc-umulation of arsenic in **different parts/ organs** of fish	Common carp, *Cyrinus carpio* (China).	Intestine → bone → gill → liver → muscle → brain	Fish exposed to arsenite (As^{III})	Han et al 2012
	Herbivorous fish *(Catla catla, Labeo rohita, Cirrhosus mrigala) and* carnivorous fish (*Rita rita, Mystus sperata* and *Wallago attu*) (Pakistan).	Liver → kidney → gills → intestine → reproductive organs → scales → fins → bones → muscle → fats	The mean As concentrations in the body organs of carnivorous fish were significantly higher than that observed in herbivorous fish species.	Jabeen et al 2012
	Walking catfish, *Clarias batrachus* (India).	Liver → gills → blood → muscles → skin → brain	Fish exposed to sodium arsenite (As^{III})	Kumar and Benerjee 2012
	Rainbow trout, *Oncorhynchus mykiss* (Czech Republic).	Muscle → liver → spleen → bile	The total As in muscle bonded in the form of non-toxic **arsenobetaine**.	Celechovska et al 2011
	Rohu, *L. rohita,* Mrigal, *C. mrigala,* Katal, *C. catla,* Grass carp, *Ctenopharyngodon idella* (Pakistan).	Kidney → liver → gills → bones → gut → skin → muscles → fins → scales	-	Kousar and Javed 2014
	Stinging catfish, *Heteropneustis fossilis* (Bangladesh).	Liver → intestine → muscles	Fish exposed to arsenite (As^{III})	Begum et al 2013

Sequence	Fish and other aquatic species studied	The sequence of accumulation of As	Remarks	References
	Tilapia, *Tilapia mossambicus* (India).	Gills → skin → muscles		Thakur and Mhatre 2015
	Brown trout (*Salmo trutta).*	Operculum → liver → gills → muscles	The operculum, liver, and gills showed the highest As concentrations, whereas levels in muscle were lower. Bioaccumulation occurred preferentially in the gills, operculum, and liver.	Culioli et al 2009
2.12B. Bioacc-umulation of arsenobetaine **in fish and marine biota**	Fish (36 species) (Korea)	AsB → DMA → ASC → As(V) → As(III) → MMA For all the fish samples, AsB (mean fraction: 90.6%) was dominant among the six arsenicals. As(III) and As(V) were detected at low concentrations.	Salinity is believed to play a key role in increased As concentrations in marine biota as well as in water and sediment (Hong et al 2014)	Choi et al 2015
	Fishes, bivalves, crabs, and shrimps, snail (South Korea).	As accumulation in biota: Crab → bivalves → shrimp → gastropods → fishes Arsenicals were mainly localized in the intestine of fish and bivalve. Arsenobetaine was the most dominant form of As found in	Concentrations of AsB in fishes were greater in samples where concentrations of As were greater in water and sediment. Concentrations of As in biota were dependent on corresponding waterborne	Hong et al 2014

Sequence	Fish and other aquatic species studied	The sequence of accumulation of As	Remarks	References
		fishes, bivalves, crabs, and shrimps, while As^{III} was predominant in freshwater snail.	concentrations of As, particularly for fish and filter-feeder organisms such as bivalves and shrimps Distribution and bioaccumulation of As was strongly correlated with water salinity.	
	Gastropods (three marine gastropods, *Thais bitubercularis, T. distinguenda* and *Morula musiva)* (from Phuket, Thailand).	All gastropod samples had arsenobetaine (AsB) as the predominant arsenic compound (93-95% of the total As in the extracts) followed by arsenocholine (3.1–4.6%) and tetramethylarsonium ion (0.21–2.2%).	-	Francesconi et al 1998
	Mangrove leaves, corals, algae, molluscs, fish and crustaceans (Thailand).	AsB was the predominant form of As in the fishes, crustaceans and gastropods. The coral samples contained mainly arsenosugars and **arsenobetaine**.	Mangrove leaves contained only low concentrations of total As (methylarsonate and dimethylarsinate).	Khokiattiwong et al 2009
	Marine fishes (38 species) (Spratly islands of China).	AsB was the dominant As species in 38 species of tropical marine fishes from the Spratly islands.	As(III) and As(V) were detected at low concentrations.	Li et al 2017

Sequence	Fish and other aquatic species studied	The sequence of accumulation of As	Remarks	References
	Marine ffishes (pelagic and benthic), shrimps, and crabs (China).	As *accumulation in marine biota*: Crabs → Shrimps → benthic fishes → pelagic fishes. As *species accumulation in marine biota:* **AsB** → DMA → MMA ≈ As(V) and As(III). As speciation data showed that AsB was the dominant form of As in wild marine organisms.	The highest level of As was found in crabs (13.4–35.1 μg/g), followed by shrimps (8.52–27.6 μg/g), benthic fish (3.45–28.6 μg/g), and pelagic fish (1.22–5.23 μg/g). **Arsenobetaine (AsB)** (87.3–99.8%) was the most dominant form of As found in all marine organisms. As species concentrations in benthic fish and shrimps were correlated with food sources (sediment) and water salinity.	Zhang et al 2018
	Marine molluscs (whelk, clam, blood clam, oyster), and crustaceans (barnacle, crab) (South China).	The less-toxic **arsenobetaine** (AsB) constituted 80.6–98.8% of all As compounds in crustaceans and mollusks, and dimethylarsinic acid (DMA) constituted 0.47–3.44 %.	The high percentage of **AsB** might reflects a particular ability to selectively bioaccumulate AsB over other arsenicals.	Zhang et al 2013
	Mussels (*Mytilus edulis*) (Australia, Perth).	**Arsenobetaine** was dominant As species found in mussels exposed to various arsenic	The results show that **arsenobetaine** and arsenocholine are efficiently accumulated from	Gailer et al 1995

Sequence	Fish and other aquatic species studied	The sequence of accumulation of As	Remarks	References
		compounds. **Arsenobetaine** → Arsenocholine → Tetramethylarsonium ion → Trimethylarsine oxide → Dimethylarsinate → Methylarsonate → Arsenate → Arsenite	seawater by blue mussels.	
	Snakehead (*Channa striata*) and non-carnivorous fish (Thailand).	Gills → Liver → Muscles *C. Striata* collected from the contaminated sites contained markedly higher concentrations of As.	Dimethylarsinate (DMA) was the dominant As species in fish from the contaminated sites. Non-carnivorous fish species (*Danio regina*, *Rasbora heteromorpha* and *Puntius orphoides*) had lower As concentrations in muscle compared to carnivorous *C. Striata*.	Jankong et al 2007
	Tilapia (*Oreochromis* spp.) (China).	Intestine → liver → → muscle	More than **90% of As** was converted into organic form in liver, gill, and muscle. The percentage of an organic form of As in muscle was the highest.	Pei et al 2019

Box 2.1. Facts about arsenobetaine in marine organisms.

- Arsenobetaine is often called ''seafood/ fish arsenic'' since it is the main As compound found in marine organisms that are consumed by humans as seafood (Cullen and Reimer 1989).
- Arsenobetaine is an organoarsenic compound that is the main source of As found in fish and is relatively non-toxic (Caumette et al 2012, Bosch et al 2016).
- Arsenobetaine concentrations in marine organisms are much higher than those in freshwater organisms.
- Arsenobetaine is generally not ubiquitous in freshwater organisms as it is in marine organisms, while arsenosugars dominate in freshwater environment (Caumette et al 2012).
- Phytoplankton contains arsenosugars as organoarsenic compounds but arsenobetaine is not identified in phytoplankton (Caumette et al 2012). For example, arsenobetaine was not detected in phytoplankton *Skeletonema costatum* and *Heterosigma akashiwo* (collected from the Sea of Japan), whereas arsenosugars contained 65% and 84% respectively (Shibata et al 1996).
- Marine zooplankton always contains arsenobetaine (as a minor compound in herbivorous zooplankton e.g. *Calanus* sp.) (Shibata et al 1996) and as a major compound in carnivorous zooplankton (e.g. *Themisto* sp. and *Sagitta* sp.) (Shibata et al 1996, Edmonds et al 1997, Takeuchi et al 2005). For example, zooplankton *Themisto* sp. and *Euphausia* sp.) (collected from the Sea of Japan) contained 77% and 55% arsenobetaine respectively (Shibata et al 1996).
- Nearly all higher trophic marine animals contain arsenobetaine such as marine mammals, fish, sponges, sea anemones, jellyfish, crayfish, lobsters, crabs, shrimps, gastropods, clams, squids, octopus, mussels, and turtles. Though the proportions of arsenobetaine found in marine organisms can differ on their trophic position and feeding habit (Cullen and Reimer 1989, Kuinto et al 2008, Reimer et al 2010).
- Arsenobetaine in the marine environment is formed in organisms of low trophic levels and accumulated through the food chain, e.g. by the degradation of dimethylated arsenosugars from algae (Francesconi 2010). Once phytoplankton is ingested by zooplankton, arsenosugars in phytoplankton are degraded by bacterial communities associated with zooplankton digestive systems, leading to the formation of arsenobetaine, which is then accumulated by the organism (Caumette et al 2012).
- Arsenobetaine is selectively accumulated by the organism to serve as an osmolyte (to regulate the osmotic pressure in response to water stress and mitigate potentially damaging effects of salinity fluctuations) (Burg and Ferraris 2008).

2.4.7. Arsenic effects on human health

As is not essential for humans. The International Agency for Research on Cancer (IARC) of the World Health Organization (WHO) and the US Environmental Protection Agency (EPA) has recognized As and inorganic arsenic compounds as Group 1 human carcinogen (Table 2.5). As species, such as monomethylarsonic acid (MMA) and dimethylarsenic acid (DMA) is slightly toxic to humans, whereas inorganic species, such as arsenite [As(III)] and arsenate [As (V)], is the most toxic form of As. The Agency for Toxic Substances and Disease Registry (ATSDR) has ranked arsenic as number 1 among the top 275 priority hazardous substances (ATSDR 2017) (see Chapter 1, Table 1.6).

Humans can be exposed to As via drinking water and food (crops and vegetables) (as mentioned in sections 2.4.1 through 2.4.5). Ingesting inorganic As greater than 50 μg/L through drinking water is a cancer risk (Morales et al 2000). The long-term ingestion or exposure (10-15 years) of As can lead to a disease called 'arsenicosis', 'arseniasis, and arsenicism' (Ng 2005). The acute and chronic health effects of inorganic As to humans are highlighted below:

- Acute exposure of As causes abdominal pain, vomiting, diarrhoea, muscular pain, weakness, with flushing of the skin rashes, muscular cramping (Ng et al 2003, Rahman et al 2009).
- Chronic intoxication causes hyperpigmentation and hypopigmentation, hyperkeratosis, skin, bladder and lung cancer and diabetes, cardiovascular diseases, black foot disease and developmental neurotoxicity (Smith et al 2000, Ng 2005, Rahman et al 2009).

2.4.8. Arsenic effects on social life

The As problem (due to arsenicosis) can cause a significant social problem, for example, marriage, employment, and social interaction can be difficult or impossible (Thakur et al 2013). It is reported that a child with arsenicosis (arsenic-related disease) are not allowed to attend a school or social or religious functions and even they are not allowed to take baths in their village ponds in Bangladesh. Moreover, arsenicosis enhances the economic burden of the poor since poor arsenicosis patients have to spend a big proportion of their income on treatment, which finally diminishes the household income. Separation and marital breakdown are also reported. In some circumstance, affected people are sent back to their homes by their employer believing that the disease may spread from one to another (in principle, arsenic-related diseases are not a spreadable disease) (Abedin and Shaw 2013).

2.5. Managing arsenic problems

Millions of people in Bangladesh and other poor developing countries continue to drink well water containing elevated levels of arsenic (Table 2.2). The most common As mitigation option in poor countries would be well switching, i.e. switching from an As-unsafe well to an arsenic-safe drinking water source, use of deep tube wells (deep aquifers are generally lower in As such as > 150-200 m in case of Bangladesh (Kinniburg and Smedley 2001) and use of As filters and pond sand filter (Table 2.13). There is also a need for As-education program for an increasing As awareness about safe uses of As-contaminated water and the health implications of As exposure (George et al 2013). There are at least eight low-cost safe alternative options that can be implemented to provide arsenic-safe and pathogen-free water as listed below:

- deep tube wells that tap deeper and arsenic-free aquifers;
- rainwater harvesting;
- household arsenic removal filters;
- community As removal filters;
- rural piped water supply that provides safe water;
- pond sand filters, which remove pathogens from As-free surface water;
- dug wells or install tube wells (hand or shallow or deep), that is, As-safe; and
- switching to neighbouring uncontaminated safe tube wells (green painted) from unsafe tube wells (red painted).

Table 2.13. Possible arsenic mitigation methods (Brammer 2009).

Type	Method
1.Water treatment	a. Filtration/chemical treatment
	b. Co-precipitation with iron
2. Alternative irrigation supply	a. Deep aquifer
	b. River/lake
	c. Reservoir/pond (tank)
3. Agronomic methods	a. Grow dryland crops
	b. Grow crops on raised bed
	c. Use arsenic-tolerant varieties
	d. Provide drainage
	e. Grow rainfed crops
4. Soil amendment	a. Add ferric iron
	b. Add phosphorus

Type	Method
5. Soil rehabilitation	a. Grow hyperaccumulator plants b. Remove topsoil
6. Food preparation	a. Cook in excess water b. Remove tuber/vegetable skin

Other measures to reduce the risks of As exposure to humans would be (see also Table 2.13):

- Use of As tolerant crop varieties.
- Breeding of rice plants that are tolerant of As and have a limited As uptake. Scientists have identified an aromatic rice from Bangladesh (Sylhet district) with very low As content and higher concentrations of essential nutrients, selenium and zinc (Al-Rmalli et al 2012).
- Use of hyper-accumulating plants (ferns), *Pteris vittata, Pityrogramma calomelanos* and *Chrysopogon zizanioides* are reported to be extremely efficient in extracting As from soils and water (Chang et al 2009; Chowdhury et al 2018).
- Clean up or remove As-contaminated top soils.
- The Cooking of rice in excess water or removal of tuber/ vegetable skins before consumption.
- Parboiling and polishing of rice (such practices would reduce inorganic As in rice since it is mainly present in the outer husk and bran layer (Rahman et al 2019).
- Monitoring of As residue concentrations in water (human drinking, irrigation, recreation, and livestock drinking), biota (fish), food, aquatic ecosystems and comparing with guideline threshold values of Australia is given in Table 2.14.

Table 2.14. Water, food and health quality guideline threshold values for arsenic (irrigation, livestock, recycled water, sediments, aquaculture, aquatic ecosystems protection, drinking water, seafood and crops) [Column 1= EPAA 2018; Column 2, 3, 5, 6, 7 = based on ANZECC 2000 a, b; Column 4 = based on NWQMS 2007; Column 8= based on NHMRC and NRMMC 2011; Column 9 = based on Food Standards Australia & New Zealand 2001, 2016, NRS, 2012; Column 10 =based on Abedin et al 2002]

dw = dry weight; NA = not available. TV = trigger values which are the concentrations of the key performance indicators measured for the ecosystem, below which there exists a low risk and adverse biological (ecological) effects.

1	2	3	4	5	6	7	8	9	10
Agricultural soil (Regulatory guidelines)	**Irrigation** (long term TV-short term TV) mg/L	**Livestock drinking** (mg/L)	**Recycled water** (mg/L)	**Aquaculture** (freshwater = F; marine = M)	**Sediment** (low TV-high TV)	**Aquatic ecosystems protection** (freshwater = F; marine = M)	**Human health** (drinking water)	**Human health** (seafood, meat, cereals, vegetables) (maximum level)	**Maximum residue limit (MRL) for food**
20 mg/kg	0.1-2 mg/L	0.5 mg/L	0.007 mg/L	<50 µg/L (F) < 30 µg/L (M)	20-70 mg/kg dw	- As^{III} 95% protection in freshwater: 24 µg/L - As^{V} 95% protection in freshwater: 13 µg/L	10 µg/L	***Inorganic arsenic*** Crustaceans, fish: 2 mg/kg; Molluscs:1 mg/kg Seaweed: 1mg/kg Cereal grains and milled cereal products:1 mg/kg Meat of cattle. Pigs, sheep: 0.005 mg/kg ***Organic arsenic*** Cereal grains and milled *cereal products:*1 mg/kg Salt: 0.5	1-2 mg/kg

2.6. Conclusion

Arsenic is one of the most important global environmental pollutants and is a persistent, bioaccumulative, toxic, endocrine-disrupting and carcinogenic metalloid. Inorganic arsenicals are more toxic than organo-arsenicals to biota and trivalent arsenic (arsenite, As^{III}) is more acutely toxic than pentavalent (arsenate, As^{V}) form. The Agency for Toxic Substances and Disease Registry, the USA has ranked As as number 1 among the top 275 priority hazardous substances. As contamination of drinking water is a worldwide problem (reported over 70 countries affecting 200 million people). Peoples of some Asian countries are severely affected including Bangladesh, West Bengal of India, China and Vietnam. Millions of people are now at health risk from contamination of drinking water and food (crops, vegetables, livestock) with As. The elevated level of As in water and soil can result in elevated concentrations of As in food since crops receiving As-contaminated irrigated water can uptake As during the phyto-extraction process and can bioaccumulate in different degrees in different parts of plants (e.g. roots, stems, and grain). However, the risk to fish and seafood appears to be low since non-toxic arsenobetaine (AsB) is mainly accumulated in marine fish and seafood organisms. As contamination would undermine achieving global sustainable development goals (SDGs) (SDGs 1, 2, 3, 8 and 14) since As can contaminate both waters (drinking, irrigation and recreational) and food (crops, vegetables, fish) and affect public health, food, water and livelihoods security and social life (as discussed in chapter 1 section 1.8). The most common As mitigation options in poor countries are **i.** well switching, i.e. switching from an arsenic-unsafe (red painted) well to an arsenic-safe (green painted) well, **ii.** use of deep tube wells (deep aquifers are generally lower in As such as > 150-200 m deep in case of Bangladesh), **iii.** use of As-tolerant crop varieties, **iv**. use of arsenic filters and pond sand filter and **v.** rainwater harvesting. There is also a need of As awareness improvement education program for the rural and coastal community about safe uses of arsenic-contaminated water and the health implications of As exposure.

2.7. References

Abedin MJ, Cotter-Howells J and Meharg AA 2002 Arsenic uptake and accumulation in rice (*Oryza sativa* L.) irrigated with contaminated water. *Plant and Soil* **240** 311-319

Abedin MA and Shaw R 2013 *Arsenic contamination in Bangladesh: contemporary alarm and future strategy*. Chapter 9. p. 191- 216. *In*: R. Shaw et al. (eds.), Disaster Risk Reduction Approaches in Bangladesh, 191. Disaster Risk Reduction *DOI 10.1007/978-4-431-54252-0_9* Springer Japan

Aiuppa A, D'Alessandro W, Federico C, Palumbo B and Valenza M 2003 The aquatic geochemistry of arsenic in volcanic ground waters from southern Italy. *Applied Geochemistry* **18** 1283–96

Akter KF, Owens G, Davey DE, and Naidu R 2005 Arsenic speciation and toxicity in biological systems. *Reviews* of *Environmental Contamination* and *Toxicology* **184** 97–149

Akter MS, Ahmed MK, Akhand MAA, Islam MM 2008 Acute toxicity of arsenic and mercury to fresh water climbing perch, *Anabas testudineus* (Bloch). *World Journal of Zoology* **3(1)** 13–18

Al-Rmalli SW, Jenkins RO, Michael J. Watts MJ and Haris PI 2012 Reducing human exposure to arsenic, and simultaneously increasing selenium and zinc intake, by substituting non-aromatic rice with aromatic rice in the diet. *Biomedical Spectroscopy and Imaging* **1(4)** 365-381 *DOI: 10.3233/BSI-12002841.*

ANZECC 2001a Australian and New Zealand Environment and Conservation Council (ANZECC) and Agriculture and Resource Management Council of Australia and New Zealand (ARMCANZ). Australian and New Zealand Guidelines for Fresh and Marine Water Quality, Volume 1, The Guidelines (Chapters 1-7). *http://www.mincos.gov.au/publications/Australian_and_New_Zealand_guidelines_fo r_fresh and_marine_water_quality.*

ANZECC 2000b Australian and New Zealand Environment and Conservation Council (ANZECC) and Agriculture and Resource Management Council of Australia and New Zealand (ARMCANZ). Australian and New Zealand Guidelines for Fresh and Marine Water Quality, Volume 2, Aquatic ecosystems – rationale and background information. National Water Quality Management Strategy. (Chapter 8). *http://www.mincos.gov.au/publications/australian_and_new_zealand_guidelines_ for_fresh_and_marine_ water_quality/volume_2*

Appleyard SJ, Angeloni J and Watkins R 2006 Arsenic-rich groundwater in an urban area experiencing drought and increasing population density, Perth, Australia. *Applied Geochemistry* **21** 83–97

Armienta MA, Rodriguez R, Aguayo A, Ceniceros N, Villaseñor G and Cruz O 1997 Arsenic contamination of groundwater at Zimapán, Mexico. *Hydrogeology Journal* **5** 39–46

ATSDR 2017 Agency for Toxic Substances and Disease Registry ATSDR). The ATSDR 2017 substances priority list. *https://www.atsdr.cdc.gov/SPL/*

ATSDR 2011 Agency for Toxic Substances Disease Registry (ATSDR). U.S. Department of Health and Human Services. Toxic substances portal.

Bae J-S and Freeman HS 2007 Aquatic toxicity evaluation of copper-complexed direct dyes to the *Daphnia magna. Dyes and Pigments* **73** 126-132

Begum A, Mustafa AI, Amin MN, Banu N, Chowdhury TR 2013 Accumulation and histopathological effects of arsenic in tissues of shingi fish (Stinging Catfish), *Heteropneustesfossilis* (Bloch, 1794). *Journal of the Asiatic Society of Bangladesh Sci* **39(2)** 221–230

Berg M, Stengel C, Pham TK et al 2007 Magnitude of arsenic pollution in the Mekong and Red River Deltas-Cambodia and Vietnam. *Science of the Total Environment* 2007 **372(2-3)** 413-425

Bhattacharya T, Ray AK and Bhattacharya S 1987 Blood glucose and hepatic glycogen interrelationship in *Channa punctatus:* a parameter of non-lethal toxicity bioassay with intestinal pollutants. *Indian Journal of Experimental Biology* **25** 539–541

Boochs PW, Billib M, Gutiérrez C, and Aparicio J 2014 Groundwater contamination with arsenic, Regiốn Lagunera, México. Proceedings of the 5th International Congress on Arsenic in the Environment, May 11-16, 2014, Buenos Aires, Argentina. *file:///C:/Users/User1/Downloads/BuenosAires_2014Boochs-Billibe.a._ArsenicLagunera-MEX.pdf*

Bosch AC, O'Neill B, Sigge GO, Kerwath SE and Louwrens C Hoffman LC 2016 Heavy metals in marine fish meat and consumer health: a review. *Journal of the Science of Food and Agriculture* **96** 32–48 *DOI 10.1002/jsfa.7360*

Brammer H and Ravenscroft P 2009 Review article - Arsenic in groundwater: A threat to sustainable agriculture in south and south-east Asia. *Environment International* **35** 647-654.

Burg MB and Ferraris JD 2008 Intracellular organic osmolytes: function and regulation. *Journal of Biological Chemistry* **283(12)** 7309–7313

Burger J, Gochfeld M, Jeitner C, Pittfield T and Donio M 2014 Heavy metals in fish from the Aleutians: Inter-specific and locational differences. *Environmental Research* **131** 119–130

Caumette G, Koch I, Reimer KJ 2012 Arsenobetaine formation in plankton: A review of studies at the base of the aquatic food chain. *Journal of Environmental Monitoring* **14** 2841–2853

Celechovska O, Harkabusova V, Macharackova B et al 2011 Accumulation of arsenic during the growing period of rainbow trout (*Oncorhynchus mykiss*). *Acta Veterinaria Brno* **80** 219–225 *doi:10.2754/avb201180020219*

Chakraborti D, Rahman MM, Das B et al 2013 Groundwater arsenic contamination in Ganga–Meghna–Brahmaputra plain, its health effects and an approach for mitigation. *Environmental Earth Sciences* Special issue. August 2013 *DOI 10.1007/s12665-013-2699-y*

Chang J-S., Yoon I-H and Kim K-W 2009 Short communication -Heavy metal and arsenic accumulating fern species as potential ecological indicators in As-contaminated abandoned mines. *Ecological Indicators* **9 (6)** 1275-1279 *doi:10.10.16/j.ecoind.2009.03.011*

Choi SD, Son HS, Choi M, and Park MK 2015 Accumulation features of arsenic species in various fishes collected from coastal cities in Korea. *Ocean Science Journal* **50** 741–750

Chou BYH, Liao CM, Lin MC and Cheng HH 2006 Toxicokinetics/ toxicodynamics of arsenic for farmed juvenile milkfish *Chanos chanos* and human consumption risk in BFD-endemic area of Taiwan. *Environment International* **32** 545–553

Chowdhury A, Datta R and Sarkar D 2018 Heavy metal pollution and remediation. Chapter 3.10.in bookOROK-9780128092705. Chapter in *Green Chemistry* January 2018. *https://doi.org/10.1016/B978-0-12-809270-5.00015-73.*

Commonwealth of Australia 2005 *Managing arsenic in water supplies.* Interim AusAID guide and operating procedures. Commonwealth of Australia. 22 p *http://www.dcita.gov.au/cca*

Culioli J-L, Calendini S, Mori C and Orsini A 2009 Arsenic accumulation in a freshwater fish living in a contaminated river of Corsica, France *Ecotoxicology and Environmental Safety* **72** (2009) 1440–1445

Cullen WR and Reimer KJ 1989 Arsenic speciation in the environment. *Chemical Reviews* **89** 64

Davey JC, Nomikos AP, Wungjiranirun M et al. 2008 Arsenic as an endocrine disruptor: Arsenic disrupts retinoic acid receptor– and thyroid hormone receptor–edited gene regulation and thyroid hormone–mediated amphibian tail metamorphosis. *Environmental Health Perspectives* **16(2)** 165-172

DOH (Department of Health) 2004 South African DOH, Foodstuffs, cosmetics and disinfectants act, 1972 (Act no. 54 of 1972). *Government Gazette* No. R. **500** 4–6 (2004).

Donahue R and Hendry M 2003 Geochemistry of arsenic in uranium mine mill tailings, Saskatchewan, Canada. *Applied Geochemistry* **18** 1733–1750

Edmonds JS, Shibata Y, Francesconi KA et al 1997 Arsenic Transformations in Short Marine Food Chains studied by HPLC-ICP MS. *Applied Organometalic Chemistry* **11(4)** 281–287

EPAA 2018 Environment Protection Authority of Australia 2018 Classification and management of contaminated soil for disposal. Information Bulletin 105. Hobart, TAS 7001 Australia. *https://epa.tas.gov.au/Documents/Information%20 Bulletin%20105%20-%20Classification%20of%20Contaminated% 20Soils%20%28IB105%29%20V3_2018.pdf*

Ferguson JF and Gavis J 1972 A review of the arsenic cycle in natural water. *Water Research* **6** 1259-1274

Food Standards Australia New Zealand (FSA) 2016 *Schedule 19. Maximum levels of contaminants and natural toxicants*. Food Standards Australia New Zealand (FSA). *http://www.foodstandards.gov.au/code/Documents/Sched%2019%20Contaminant%2 0MLs%20v157.pdf*

Food Standards Australia New Zealand (FSA) 2001 Generally expected levels (GELs) for metal contaminants. Additional Guidelines. To Maximum Levels in Standard 1.4.1 - Contaminants and Natural Toxicants *http://www.foodstandards.gov.au/ code/userguide/documents/GELs_0801.pdf*

Francesconi KA, Goessler W, Panutrakul S, Irgolic KJ 1998 A novel arsenic containing riboside (arsenosugar) in three species of gastropod. *Science of the Total Environment* **221** 139–148

Francesconi KA 2010 Arsenic species in seafood: Origin and human health implications *Pure and Applied Chemistry* **82** 373–381

Garelick H, Jones H, Dybowska A and Valsami-Jones E 2008 Arsenic pollution sources. In: D.M. Whitacre (ed.), *Reviews of Environmental Contamination*. 197: 17-60. *doi: 10.1007/978-0-387-79284-2_2, # Springer Sciencez+Business Media, LLC 2008.*

Gailer JK, Francesconi A, Edmonds JS and K. Irgolic J 1995. Metabolism of arsenic Compounds by the blue mussel, *Mytilus edulis.* After accumulation from seawater spiked with arsenic Compounds. *Applied Organometallic Chem*istry **9** 34-355

George CM, Factor-Litvak P, Khan K et al. 2013 Approaches to increase arsenic awareness in Bangladesh: An evaluation of an arsenic education program. *Health Education & Behaviour* **40(3)** 331– 338 *DOI: 10.1177/1090198112454735*

Ghosh A, Majumder S, Awal MA and Rao RD 2013 Arsenic exposure to dairy cows in Bangladesh. *Archives of Environmental Contamination and Toxicology* **64** 151–159

Goyer RA and Clarkson TW 2001 Toxic effects of metals, in Casarett and Doull's *Toxicology: the basic science of poisons*, ed. by Klaassen CD, McGraw-Hill, New York, p. 811–867

Hamilton SJ and Buhl KJ 1990 Safety assessment of selected inorganic elements to fry of chinook salmon (*Oncorhynchus shawytscha). Ecotoxicology and Environmental Safety* **20(3)** 307–324

Han Z, Li J, Zhang M, Chunxia LV 2012 Effect of montmorillonite on arsenic accumulation in common carp. *African Journal of Biotechnology* **11(22)** 6160–6168

Henke K 2009 Arsenic: environmental chemistry, health threats and waste treatment. John Wiley & Sons

Herath I, Vithanage M, Bundschuh J et al 2016 Natural arsenic in global ground waters: Distribution and geochemical triggers for mobilization. *Current Pollution Reports* **2** 68–89 DOI 10.1007/s40726-016-0028-2

Hong S, Jong Seong Khim JS, Park J and Son H-S et al 2014 Species- and tissue-specific bioaccumulation of arsenicals in various aquatic organisms from a highly industrialized area in the Pohang City, Korea *Environmental Pollution* **192** 27-35

Huq SMI, Correl R and Naidu R 2006a *Arsenic accumulation in food sources in Bangladesh.* Chapter 15: 283-293 pp. *In*: Managing arsenic in the environment. From soil to human health. Edited by *Naidu et al.* CSIRO, Victoria, Australia.

Huq SMI, Joarder JC, Parvin S, Correll, R. and Naidu R 2006b Arsenic contamination in food-chain: Transfer of Arsenic into food materials through groundwater irrigation. *J. Health Population and Nutrition*. September **24(3)** 305-316

IARC (International Agency for Research on Cancer) 2019 Agents Classified by the IARC Monographs, Volumes 1–123 *https://monographs.iarc.fr/wp-content/uploads/2018/09/ClassificationsAlphaOrder.pdf*

IARC (International Agency for Research on Cancer) 2012 A review of human carcinogens. Part C: Arsenic, metals, fibres, and dusts/ IARC Working Group on the Evaluation of Carcinogenic Risks to Humans (2009: Lyon, France). ISBN 978 92 832 1320 8.

IARC (International Agency for Research on Cancer) 2004 Some drinking-water disinfectants and contaminants, including arsenic. *IARC Monogr Eval Carcinog Risks Hum*, **84** 1–477 PMID:15645577 *https://www.ncbi.nlm.nih.gov/books/NBK304375/pdf/Bookshelf_NBK304375.pdf*

IPCS (International Program on Chemical Safety) 2001 *Arsenic and arsenic compounds.* Geneva. World Health Organisation. International Program on Chemical Safety - Environmental Health Criteria 224. Published under joint sponsorships of UNEP, ILO, WHO, Geneva. *http://www.inchem.org/documents/ehc/ehc/ehc224.htm*

Jain CK and Ali I 2000 Arsenic: Occurrence, toxicity and speciation techniques. *Water Research* **34(17)** 4304-4312

Jabeen G, Javed M, Azmat H 2012 Assessment of heavy metals in the fish collected from the river Ravi. Pakistan. *Pakistan Veterinary Journal* **32(1)** 107–111

Jankong P, Chalhoub C, Kienzl N et al 2007 Arsenic accumulation and speciation in freshwater living in arsenic contaminated waters. *Environmental Chemistry* **4** 11–17

Juresa D and Blanusa M 2003 Mercury, arsenic, lead and cadmium in fish and shell-fish from the Adriatic Sea. *Food Additives and Contaminants* **20** 241–246

Kaur S, Kamli MR, and Ali A 2011 Role of arsenic and its resistance in nature**.** *Canadian Journal of Microbiology* **57(10)** 769-74; *doi: 10.1139/w11-062. Epub 2011 Sep 21.*

Khokiattiwong S, Kornkanitnan N, Goessler W, Kokarnig S and Francesconi KA 2009 Arsenic compounds in tropical marine ecosystems: similarities between mangrove forest and coral reef. *Environ. Chem.* **6** 226–234

Kibria G, Haroon AKY, Nugegoda D and Rose G 2010 *Climate change and chemicals: Environmental Biological aspects.* New India Publishing Agency. New, Delhi. 470 p. ISBN: 9789380235301. *http://researchbank.rmit.edu.au/view/rmit:12529*

Kinniburgh DG. and Smedley, PL (editors) 2001 Arsenic contamination of groundwater Bangladesh. Volume 1. Summary British Geological Survey Report WC/00/19

Keyworth. BGS and DPHE 2001 ISBN 0 85272 384 9 *file:///C:/Users/User1/Downloads/Vol1Summary%20(1).pdf*

Kim K-W, Chanpiwat P, Hanh H, Phan K and Sthiannopkao S 2011 Arsenic geochemistry of groundwater in Southeast Asia. *Frontiers in Medicine* **5** 420–33

Korte NE and Fernando Q 1991 A review of As(III) in groundwater. *Critical Reviews in Environmental Control* **21(1)** 211–239

Kousar S and Javed M 2014 Heavy metals toxicity and bioaccumulation patterns in the body organs of four fresh water fish species. *Pakistan Veterinary Journal* **34(2)** 161–164

Kovendan K, Vincent S, Janarthanan S and Saravanan M 2013 Expression of metallothionein in liver and kidney of freshwater fish *Cyprinus carpio* var. *communis* (Linn) exposed to arsenic trioxide. *American Journal* of *Scientific and Industrial* Research **4(1)** 1–10

Kumar R and Banerjee TK 2012 Analysis of arsenic bioaccumulation in different organs of the nutritionally important catfish, *Clarias batrachus* (L.) exposed to the trivalent arsenic salt, sodium arsenite. *Bulletin Environmental Contamination and Toxicology* **89(3)** 445–449

Kumari B, Kumar V, Amit K. Sinha AK et al 2016 Toxicology of arsenic in fish and aquatic systems *Environmental Chemistry Letters* **14(55)** 1-22 *DOI 10.1007/s10311-016-0588*-9

Kunito T, Kubota R, Fujihara J, Agusa T and Tanabe S 2008 Arsenic in marine mammals, seabirds, and sea turtles. *Reviews of Environmental Contamination and Toxicology*, ed. D. M. Whitacre, Springer, New York **195** 31–69

Lavanya S, Mathan R, Chokkalingam K, Annamalai M 2011 Hematological, biochemical and ion regulatory responses of Indian major carp, *Catla catla* during chronic sub lethal exposure to inorganic arsenic. *Chemosphere* **82** 977–985

Leonardi G, Vahter M, Clemens F, Goessler W, Gurzau E, and Hemminki K, et al 2012 Inorganic arsenic and basal cell carcinoma in areas of Hungary, Romania, and Slovakia: a case–control study. *Environmental Health Perspectives* **120** 721–6

Li JX, Sun CJ, Zheng et al 2017 Determination of trace metals and analysis of arsenic species in tropical marine fishes from Spratly islands. *Marine Pollution Bulletin* **122** 464–469

Liu F, Gentles A, and Theodorakis CW 2008 Arsenate and perchlorate toxicity, growth effects, and thyroid histopathology in hypothyroid zebra fish, *Danio rerio*. *Chemosphere* **71(7)** 1369–137

Maity JP, Nath B, Chen C-Y, Bhattacharya P, Sracek O, and Bundschuh J et al. 2011 Arsenic-enriched groundwaters of India, Bangladesh and Taiwan—comparison of hydrochemical characteristics and mobility constraints. *Journal of Environmental Science* and *Health* **46** 1163–76

Morales KH, Ryan L, Kuo TL et al. 2000 Risk of internal cancers from arsenic in drinking water. *Environmental Health Perspectives* **108** 655-661

MSDS-Thermo Fisher 2009 (revised 2019) Arsenic Safety data sheet. 9p *https://www.fishersci.com/shop/msdsproxy?productName=SA449500&productDescription=ARSENIC*

MSDS-Labchem 2000 Arsenic Trioxide Safety data sheet. 7p *http://www.labchem.com/tools/msds/msds/75512.pdf*

Mukherjee A, Sengupta MK, Hossain MA, Ahamed S, Das B,Na yak B, et al 2006 Arsenic contamination in groundwater: a global perspective with emphasis on the Asian scenario. *Journal of Health Population and Nutrition* **24** 142–63

Nagy G and Korom I 1983 SpaleHautsymptome der ArsenvergifturgautGrund der Arsenendemic in\BugacĐAlsomonostor. *Z. Hautkr* **58** 961-964

Naidu R, Smith E, Owens G., Bhattacharya P and Nadebaum P (eds.) 2006 *Managing arsenic in the environment. From soil to human health.* CSIRO, Victoria, Australia.

Ng JC 2005 Environmental contamination of arsenic and its toxicological impacts on humans. *Environmental. Chemistry* **2** 146-160

Ng JC, Wang J and Shraim A 2003 A global health problem caused by arsenic from natural sources. *Chemosphere* **52** 1353-1359

NHMRC (National Health and Medical Research Council) and National Resource Management Ministerial Council (NRMMC) 2011 Australian Drinking Water Guidelines Paper 6 National Water Quality Management Strategy. National Health and Medical Research Council and National Resource Management Ministerial Council, Commonwealth of Australia, Canberra.

Nriagu J, Bhattacharya P, Mukherjee A et al 2007 Arsenic in soil and groundwater: an overview. In book- *Trace Metals and other Contaminants in the Environment* 3–60; ISBN: 9780444518200

Nordstrom DK 2002 Worldwide occurrences of arsenic in ground water. *Science* **296** 2143–2145

NRS (National Residue Survey) 2006 National residue Survey. Annual report 2005-2006. Australian Government Department of Agriculture, Fisheries and Forestry, Meat and Livestock Australia. Canberra, ACT 2601. 186 p.

National Residue Survey (NRS) 2012 National Residue Survey Annual Report 2010–2011. Australian Government Department of Agriculture. Fisheries and Forestry, Canberra (218 p).

NHMRC and NRMMC 2011 National Health and Medical Research Council (NHMRC) and National Resource Management Ministerial Council (NRMMC) Australian Drinking Water Guidelines Paper 6 National Water Quality Management Strategy. National Health and Medical Research Council and National Resource Management Ministerial Council, Commonwealth of Australia, Canberra.

NWQMS 2007 National Water Quality Management Strategy (NWQMS) Australian Guidelines for Water Recycling: Managing Health and Environmental Risks (Phase 2) — Augmentation of Drinking Water Supplies. Natural Resource Management Ministerial Council. Environment Protection and Heritage Council. Australian Health Ministries Conference. Biotex Pty Ltd, Canberra (140 p.).

Pal A, Nayak B, Das B, Hossain MA, Ahamed S, Chakraborti D 2007 Additional danger of arsenic exposure through inhalation from burning of cow dung cakes laced with arsenic as a fuel in arsenic affected villages in Ganga–Meghna–Brahmaputra plain. *Journal of Environmental Monitoring* **9** 1067–1070

Pei J, Zuo J, Wang X et al 2019 The Bioaccumulation and tissue distribution of arsenic species in Tilapia. *International Journal of Public Health* **16** 757 *doi:10.3390/ijerph16050757*

Perugini M, Visciano P, Manera M, Zaccaroni A, Olivieri V and Amorena M 2014 Heavy metal (As, Cd, Hg, Pb, Cu, Zn, Se) concentrations in muscle and bone of four commercial fish caught in the central Adriatic Sea, Italy. *Environmental Monitoring and Assessment* **186** 2205–2213

Podgorski, JE, Eqani SAMA, Tasawar Khanam T et al 2017 Extensive arsenic contamination in high-pH unconfined aquifers in the Indus Valley. *Science Advances* **3(8)** e1700935 *DOI: 10.1126/sciadv.1700935*

Rahman H, Carey M, Hossain M, et al 2019 Modifying the parboiling of rice to remove inorganic arsenic, while fortifying with calcium. *Environmental Science & Technology DOI: 10.1021/acs.est.8b06548*

Rahman MM, Ng JC and Naidu R 2009 Chronic exposure of arsenic via drinking water and its adverse health impacts on humans. *Environmental Geochemistry and Health* **31** 189-200; *DOI: 10.1007/s10653-008-9235-0*

Rankin MG and Dixon DG 1994 Acute and chronic toxicity of waterborne arsenite to rainbow trout (*Oncorhynchus mykiss*). *Canadian Journal of Fisheries and Aquatic Sciences* **51(2)** 372–380

Reimer KJ, Koch I and Cullen WR 2010 in Organometallics in Environment and Toxicology. *The Royal Society of Chemistry* **7** 165–229

Romić Ž, Habuda-Stanić M, Kalajdžić B and Kuleš M 2011 Arsenic distribution, once nitration and speciation in groundwater of the Osijek area, eastern Croatia. *Applied Geochemistry* **26** 37–44

Sharma VK and Sohn M 2009 Review article -Aquatic arsenic: toxicity, speciation, transformation and remediation. *Environment International* **35(4)**743-59.

Shibata Y, Sekiguchi M, Otsuki A and Morita M 1996 Arsenic compounds in zoo- and phyto-plankton of marine Origin. *Applied Organometallic Chemistry* **10** 713–719

Shahid M, Dumat C, Niazi K et al 2018 Global scale arsenic pollution: increase the scientific knowledge to reduce human exposure. *DOI: 10.4000/vertigo.21331*

Shukla JP, Shukla KN, Dwivedi UN 1987 Survivality and impaired growth in arsenic treated fingerlings of *Channa punctatus,* a fresh water murrel. *Acta Hydrochimica et Hydrobiologica* **15(3)** 307–311

Smedley PL and Kinniburgh DG 2002 A review of the source, behaviour and distribution of arsenic in natural waters. *Applied Geochemistry* **17** 517-568. *DOI :10.1016/S0883-2927(02)00018-5*

Smedley PL, Nicolli HB, Macdonald DMJ, Kinniburgh DG 2008 Arsenic in groundwater and sediments from La Pampa Province, Argentina. In: Bundschuh J, ArmientaMA, Birkle P, Bhattacharya P, Matschullat J, Mukherjee AB, editors. Natural arsenic in groundwaters of Latin America. Boca Raton: Taylor & Francis; 2008. p. 35–45.

Smith AH, Lingas EO, Rahman, M 2000 Contamination of drinking-water by arsenic in Bangladesh: a public health emergency. *Bulletin of the World Health Organization* **78(9)** 1093-1103 *http://www.who.int/bulletin/archives/78(9)1093.pdf*

Smith JVS, Jankowski J and Sammut J 2003 Vertical distribution of As(III) and As(V) in a coastal sandy aquifer: factors controlling the concentration and speciation of arsenic in the Stuarts Point groundwater system, northern New South Wales, Australia. *Applied Geochemistry* **18** 1479–96

Stugeron RE, Siu KWM, Willie SN and Berman SH 1989 Quantification of arsenic species in a river water reference material for trace metals by graphite furnace atomic absorption spectrometry technique. *Analyst* **114** 1393-1396

Suhendrayatna, Ohki A, Nakajima T, Maeda S 2002 Studies on the accumulation and transformation of arsenic in freshwater organisms I. Accumulation, transformation and toxicity of arsenic compounds on the Japanese Medaka, *Oryzias latipes. Chemosphere* **46** 319–324

Takeuchi M, Terada A, Nanba K et al 2005 Distribution and fate of biologically formed organo-arsenicals in coastal marine sediment. *Applied Organometallic Chemistry* **19** 945–951

Thakur J and Mhatre M 2015 Bioaccumulation of heavy metals in *Tilapia mossambicus* fish from industrially polluted Patal-ganga River, India. *International Journal of Advanced Research* **3(2)** 486–490

Thakur BK, Gupta, V and Chattopadhyay, U 2013 *Arsenic groundwater contamination related socio-economic problems in India*: Issues and Challenges. p. 163-182. *In*: S. Nautiyal et al. (eds.), Knowledge Systems of Societies for Adaptation and Mitigation of Impacts of Climate Change, Springer. *DOI: 10.1007/978-3-642-36143-2_10, _ Springer-Verlag Berlin Heidelberg 2013*

Tisler T and Zagorc-Konean J 2003 Aquatic toxicity of selected aquatic chemicals as a basic criterion for environmental classification. *Journal of Aquatic Toxicity in Environmental Classification. Arh Hig Rada Tokiskol* **54** 207-213

UN (United Nations) 2001 Synthesis report on arsenic in drinking water. In: Source and Behaviour of Arsenic in Natural Waters, p. 1-48. *http://www.who.int/water_sanitaton_health/dwq/en/arsenicun1.pdf.*

USEPA (US Environmental Protection Agency) 1985 Ambient water quality criteria for Arsenic, 1984. US Environmental Protection Agency, Office of Water Regulations and Standards, Criteria and Standards Division, Washington, DC, p. 66.

Vutukuru SS, Prabhath NA, Raghavender M and Yerramilli A 2007 Effect of arsenic and Cr on the serum amino-transferases activity in Indian major carp, *Labeo rohita. International Journal of Environmental Research and Public Health* **4** 224–227

Welch AH, Lico MS and Hughes JL 1988 Arsenic in ground water of the Western United States. *Grounndwater* 26 333-347 10.1111/j.1745-6584.1988.tb00397.x

WHO (World Health Organization) 2012 Arsenic, fact sheet No 372 [Internet]. Geneva: World Health Organization; 2012. Available from: *http://www.who.int/mediacentre/factsheets/fs372/en/*

WHO (World Health Organization) 2010 Preventing diseases through healthy environment: action needed on chemicals of major public Health concern. World Health Organization (WHO) 2010 Available from: *http://www.who.int/ipcs/assessment/public_health/chemicals_phc/en/.*

WHO (World Health Organization) 2001a Arsenic in Drinking-water. WHO/SDE/WSH/ 03.04/75/Rev/1 *https://www.who.int/water_sanitation_health/dwq/chemicals/ arsenic.pdf*

WHO (World Health Organization of the United Nations) 2001b Arsenic in drinking water. Fact sheet 210. *http://www.who.int/mediacentre/factsheets/fs210/en/print.htm*

WHO (World Health Organization) 1981 Arsenic. *Environmental health criteria* 18. WHO, Geneva.

Woodward DF, Brumbaugh WG, Delonay AJ, Little EE and Smith CE 1994 Effects on rainbow trout fry of a metals-contaminated diet of benthic invertebrates from the

Clark Fork River, Montana. *Transaction of the American Fisheries Society* **123(1)** 51–62

Zhang W, Guo, Z, Song, D, Du S, Zhang L 2018 Arsenic speciation in wild marine organisms and a health risk assessment in a subtropical bay of China. *Science of the Total Environment* **626** 621–629

Zhang W, Wang W X and Zhang L 2013 Arsenic speciation and spatial and interspecies differences of metal concentrations in mollusks and crustaceans from a South China estuary. *Ecotoxicology* **22** 671–682

Zorob GK, Mckiernan JW and Caruso JA 1998 ICP-MS for elemental speciation studies. *Mikrochimica Acta* **128** 145–168.

■■■■

3 Heavy Metals Risks, Impacts and Management

Summary

Heavy metals (HMs) are commonly defined as a set of metals that possess densities at least 5 times higher than that of water and with an atomic weight greater than 40. It includes Cadmium (Cd), Copper (Cu), Chromium (Cr), Lead (Pb), Mercury (Hg) and Zinc (Zn). HMs are persistent, toxic and bioaccumulative but some of them are also endocrine-disrupting as well as carcinogenic. Globally more than 50% of soil contamination is caused by HMs. The anthropogenic sources of HMs can be from 22 sources/industries including agriculture (fertilizers), cement, coal-fired power plants, e-wastes, manures, tanneries, textiles, ship-breaking yards, oil refineries, mining, pesticides, sewage treatment plants, and urban run-off. Four of the top 10 chemicals of major public concern (out of 275 priority list of hazardous substances assessed by ATSDR, USA) are HMs (Pb, Hg and Cd). HMs would undermine achieving global sustainable development goals (SDGs). This chapter reviews: **i.** chemistry, toxicity and ecotoxicity of HMs; **ii.** impacts of Cd, Cu, Hg, Pb, Zn, Cr, Ni on social, economic and environmental sectors including water, crops, vegetables, freshwater and marine fish and human's health; **iii**. Hg levels in child-bearing women in 25 countries and health risks to women who are frequent users of facial makeup; **iv**. linkages among heavy metals load & bioaccumulation and climate change; **v.** management and remediation of HMs.

Contents

List of Tables

List of Figures

List of Boxes

3.1. Introduction

Heavy metals (HMs) are commonly defined as a set of metals that possess densities at least five times higher than that of water (Vamerali et al 2010, WHO 2010, Alloway 2013). Heavy/ trace metals (cadmium - Cd, copper - Cu, chromium - Cr, lead - Pb, mercury - Hg, zinc - Zn) are elements with metallic properties and atomic weight greater than 40 (Table 3.1) and characterised by similar atomic electron configurations in the outer orbital's. In this chapter, trace/heavy metals (cadmium, copper, lead, mercury, zinc) are referred to as **HMs**. They fall under the category of **PBT** (i.e. persistent, bioaccumulative and toxic) or harmful pollutants but some of them are also **endocrine-disrupting** as well as **carcinogenic** (Kibria et al 2010a, Kibria et al 2016a, Kibria et al 2016b). In general, substances which are toxic, bioaccumulative and carcinogenic are of great concern if they are frequently detected in soil or water or food. Based on their immediate threat to human health and the environment, four of the top 10 chemicals of major public concern reviewed by the World Health Organization (WHO) and ATSDR (Agency for Toxic Substances Disease Registry, US Department of Health and Human services) are HMs including lead, mercury and cadmium (see Chapter 1, Table 1.6).

Table 3.1. Metallic elements with an atomic weight greater than 40
(WHO/FAO/IAEA 1996, Mohammed et al 2011, Tchounwou et al 2012)

[= HMs which are essential elements and often regarded as nutrients for humans].*

Heavy metals	Symbol	Atomic weight	Heavy metals	Symbol	Atomic weight
Cadmium	Cd	112.40	Mercury	Hg	200.59

Heavy metals	Symbol	Atomic weight	Heavy metals	Symbol	Atomic weight
*Chromium	Cr	51.99	*Molybdenum	Mo	95.95
*Cobalt	Co	58.93	*Nickel	Ni	58.71
*Copper	Cu	63.55	Platinum	Pt	195.09
Gold	Au	196.97	*Selenium	Se	78.97
*Iron	Fe	55.85	Silver	Ag	107.87
Lead	Pb	207.2	Tin	Sn	118.69
*Manganese	Mn	54.94	Titanium	Ti	47.90
*Magnesium	Mg	24.305	*Zinc	Zn	65.38

https://en.wikipedia.org/wiki/History_of_the_periodic_table#/media/File:Periodic_table_(18-col,_enwiki),_black_and_white.png

HMs are natural constituents of the earth's crust and created due to natural phenomena such as weathering and volcanic activity contributing significantly to HM pollution. Some HMs are essential elements (Co, Cr, Cu, Fe, Mg, Mn, Ni, Si and Zn) acting as micronutrients for various biochemical and physiological functions (WHO/FAO/IAEA 1996, Mohammed et al 2011, Tchounwou et al 2012) that are required. However, most of HMs, regardless of being micronutrients or not, can be toxic to all forms of life at high concentrations due to the formation of complex compounds within the cell.

HMs are widely used in several industrial, agricultural, medical, and technological applications. They can be toxic to living organisms even at very low levels of exposure. HMs can exist in soils, waters and organisms in a wide variety of chemical forms and in combination with other materials. Globally, more than 10 million sites of soil pollution have been reported, of which, more than 50% of the sites are contaminated with HMs and/or metalloids (including Arsenic - As) (Table 3.2) (He et al 2015). HMs can cause water and food contamination via polluted soil, water and air, and is a threat to water quality, aquatic biodiversity, crops and seafood and animal and human health. **This chapter emphasizes** the risks and impacts of HMs principally, **Cd, Cu, Hg, Pb** and **Zn** on social, economic and environmental sectors.

Table 3.2. Soil pollution in the world and the proportion of heavy metals/metalloid pollution (He et al 2015).

Country	**Number of pollution sites**	**Heavy metal/(metalloid) pollution (%)**
Australia	>50,000	> 60%
China	1.0 million km^2	> 80%
European Union	>80,000	37%
Global	10,000,000	50%
USA	100,000	70%

3.2. Potential environmental sources of HMs

The HMs can come from both natural and anthropogenic sources:

- The natural sources are geological minerals, wind-blown silicate dust, volcanic emissions, sea spray and combustion (Kibria et al 2010a).
- The anthropogenic sources can be from 22 sources/industries (Kibria 2018; Table 3.3) including batteries, cement, coal combustion, detergents, e-wastes, face makeup (ladies) products, fertilizers, manures, mine effluents, paintings, paper and pulp mills, pesticides, plastics, oil refinery, sewage sludge, stormwater/urban run-off, smelting, ship-breaking yards, tanneries, textile effluents, waste incineration and wood preservatives (Table 3.3 and Box 3.1). **Major sources** of HMs are from coal combustion, e-wastes and mine effluents (Table 3.3).

3.3. Chemistry, toxicity and ecotoxicity

3.3.1. Speciation of metals

Metals may exist in a number of physico-chemical states in aqueous systems. These include particulate matter and dissolved forms such as simple inorganic species, organic complexes and elements adsorbed on a variety of colloidal particles (see Table 3.4). The variations in dissolved and particulate concentrations of 14 metals (Al, As, Ba, Ca, Cd, Co, Cr, Cu, Fe, Mn, Mg, Ni, Pb and Zn) was found to be a function of water discharge, solid transport, biological activity and water temperature in the Po River, Italy (Pettine et al 1994). According to Snodgrass1980, the following factors affect the solid-phase speciation of metals:

Table 3.3. Potential environmental sources of selected heavy metals (HMs)

([1]Bobby 2014; [2]Marcotullio 2007; [3]Cipurkovic et al 2014; [4]*https://www.sourcewatch.org/index.php/Heavy_metals_and_coal*; [5]Sani and Shehu 2018; [6]Aonghusa and Gray 2002; [7]Ari 2016; [8]Environmental Defense Canada 2011; [9]Kibria et al 2012a; [10]Kibria et al 2016a; [11]Zhang et al 2012; [12] Kibria 2012b; [13]Grant 2016; [14]Kumar et al 2015; [15]Tariq et al 2016; [16]Whitt et al 2016; [17]Hahladakis et al 2018; [18]Alonso-Hernández et al 2011; [19]Akpoveta and Osakwe 2014; [20]Shamuyarira and Gumbo 2014; [21]Al-Ameri et al 2018; [22]Mohammed et al 2012; [23]Artiola et al 2019; [24]Tewari et al 2001; [25]Islam et al 2013; [26]Yaseen and Scholz 2019; [27]Zhang et al 2008; [28]Bhattacharya et al 2002; [29]Bhardwaj et al 2014; [30]Ogunbileje et al2013; [31]Perkins et al 2014). *[√== sources of metals based on literatures consulted (referenecs 1 to 31 as given above of this Table)].*

Heavy metals sources	[1,2]**Batteries**	[3,30]**Cement**	[4]**Coal combustion**	[2,5,6]**Detergents**	[7,31]**E-wastes**	[8]**Face make up (ladies)**	[2,9,10]**Fertilisers**	[11]**Manures (animal)**	[12]**Mineeffluents**	[13]**Paintings**	[14]**Paper& pulp mills**	[2,9,15]**Pesticides**	[16,17]**Plastics**	[18,19]**Oil refinery**	[2,20]**Sewage sludge**	[21,22]**Storm water**	[23]**Smelting**	[24]**Ship breaking yards**	[25]**Tanneries**	[26,29]**Textile/dyes effluents**	[2,27]**Waste incineration**	[28]**Wood preservatives**
Aluminium (Al)			√		√				√										√			
Antimony (Sb)			√		√					√			√									
Arsenic (As)			√	√	√	√	√	√	√			√		√	√		√		√	√	√	√
Barium (Ba)		√	√		√				√	√												
Beryllium (Be)			√		√	√																

Heavy metals sources	[1,2]Batteries	[3,30]Cement	[4]Coal combustion	[2,5,6]Detergents	[7,31]E-wastes	[8]Face make up (ladies)	[2,9,10]Fertilisers	[11]Manures (animal)	[12]Mineeffluents	[13]Paintings	[14]Paper& pulp mills	[2,9,15]Pesticides	[16,17]Plastics	[18,19]Oil refinery	[2,20]Sewage sludge	[21,22]Storm water	[23]Smelting	[24]Ship breaking yards	[25]Tanneries	[26,29]Textile/dyes effluents	[2,27]Waste incineration	[28]Wood preservatives
Boron (B)																				√		
Cadmium (Cd)	√	√	√	√	√	√	√	√	√	√	√	√	√	√	√	√	√	√	√	√	√	
Chromium (Cr)		√	√	√	√		√	√	√	√	√	√	√	√	√		√		√	√	√	√
Cobalt (Co)		√	√						√	√		√	√				√	√	√			
Copper (Cu)		√	√	√	√		√	√	√	√	√	√		√	√	√	√	√	√	√	√	√
Gold (Au)					√																	
Indium (In)					√																	
Iron (Fe)			√		√				√		√							√	√	√		
Lead (Pb)	√		√	√	√	√	√	√	√	√	√	√	√	√	√	√	√	√	√	√	√	
Lithium (Li)	√				√																	
Magnesium (Mg)			√						√													
Manganese (Mn)	√	√	√				√		√	√				√				√	√	√		

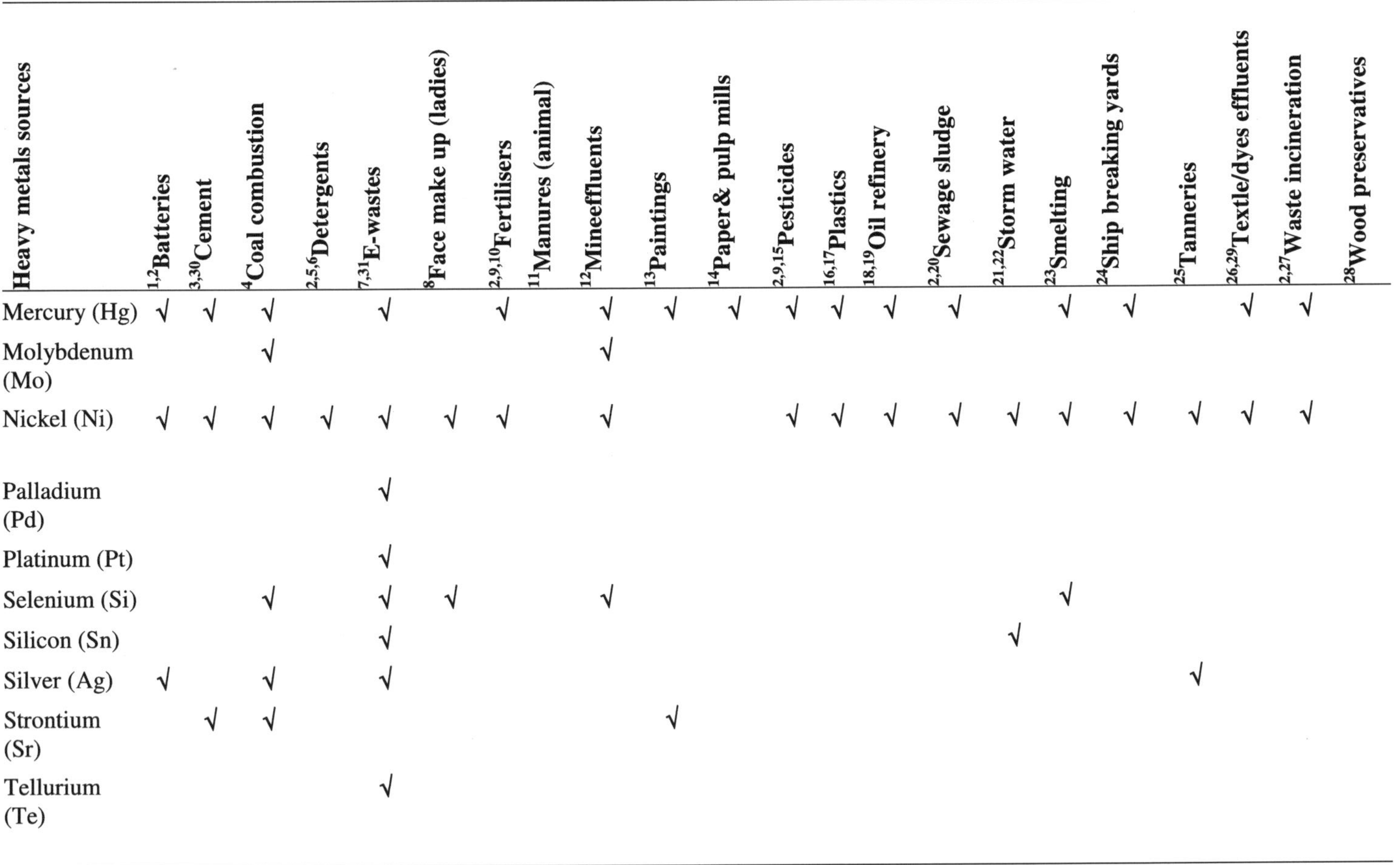

Heavy metals sources	[1,2]Batteries	[3,30]Cement	[4]Coal combustion	[2,5,6]Detergents	[7,31]E-wastes	[8]Face make up (ladies)	[2,9,10]Fertilisers	[11]Manures (animal)	[12]Mineeffluents	[13]Paintings	[14]Paper& pulp mills	[2,9,15]Pesticides	[16,17]Plastics	[18,19]Oil refinery	[2,20]Sewage sludge	[21,22]Storm water	[23]Smelting	[24]Ship breaking yards	[25]Tanneries	[26,29]Textile/dyes effluents	[2,27]Waste incineration	[28]Wood preservatives
Mercury (Hg)	√	√	√		√		√		√	√	√	√	√	√	√		√	√		√	√	
Molybdenum (Mo)			√						√													
Nickel (Ni)	√	√	√	√	√	√	√		√			√	√	√	√	√	√	√	√	√	√	
Palladium (Pd)					√																	
Platinum (Pt)					√																	
Selenium (Si)			√		√	√			√								√					
Silicon (Sn)					√											√						
Silver (Ag)	√		√		√														√			
Strontium (Sr)		√	√							√												
Tellurium (Te)					√																	

Heavy metals sources	[1,2]Batteries	[3,30]Cement	[4]Coal combustion	[2,5,6]Detergents	[7,31]E-wastes	[8]Face make up (ladies)	[2,9,10]Fertilisers	[11]Manures (animal)	[12]Mineeffluents	[13]Paintings	[14]Paper& pulp mills	[2,9,15]Pesticides	[16,17]Plastics	[18,19]Oil refinery	[2,20]Sewage sludge	[21,22]Storm water	[23]Smelting	[24]Ship breaking yards	[25]Tanneries	[26,29]Textile/dyes effluents	[2,27]Waste incineration	[28]Wood preservatives
Thallium (Ti)																						
Tin (Sn)			√		√								√	√								
Uranium (U)							√		√													
Vanadium (V)		√	√											√								
Zinc (Zn)	√		√	√	√		√	√	√	√	√	√	√	√	√	√	√	√	√	√	√	√

- Direct adsorption onto inorganic clays.
- Adsorption/ complexation with organic particulates; detritus or bacteria.
- Association with humic[1] and fulvic acids[2]
- Ionic exchange with clay colloids.
- Precipitation.

Box 3.1. Selected major industrial sources of heavy metals.

- **Cement industries:** Cement kiln dust (CKD)[3] contains several heavy metals (Ca, Cd, Cr, Cu, Fe, Hg, Ni, Mn, Pb, Zn) that can spread along a large area through wind and rain, waste streams/effluents, ground-leaching from cooling tanks, etc. These metals accumulate in plants, animals and soil, and pose a significant hazard and threats to human health (Ogunbileje et al 2013) (see also Table 3.3).
- **Coal industries**: A variety of chemicals (mostly metals) are associated with coal that is either found in the coal directly or in the layers of rock that lie above and between the seams of coal. Burning coal produces airborne compounds, known as fly ash[4] and bottom ash (collectively referred to as coal ash), which can contain large quantities of HMs that settle or wash out of the atmosphere into oceans, streams, and land. Many of the toxic HMs released in the mining and burning of coal are, for example, As, Ba, Cd, Cr, Co, Cu, Pb, Hg, Ni and Zn *https://www.sourcewatch.org/index.php/Heavy_metals_and_coal#cite_note-sludge-2* (see also Table 3.3).
- **E-waste[5] processing plants:** Discarded electrical and electronic equipment and end-of-life electrical appliances could be a significant source of HMs, for example, Cd (from switches, connectors, ink or tonner of photocopying machines/ printers, mobile phones); Hg (from thermostats, sensors, monitors, fluorescence lamps, liquid crystal display backlights); Li, Ni (from batteries); Pb (from light bulbs, TVs, batteries, cathode ray tubes); and Zn (from cathode-ray tubes, metal coatings) (Perkins et al 2014).

---

[1] *Humic = Humic substances are organic compounds that are important components of humus (humus is an organic component of soil formed by the decomposition of leaves and other plant material by soil microorganisms).*

[2] ***Fulvic acids** are similar to humic acids, with differences being the carbon and oxygen contents, acidity, degree of polymerization, molecular weight, and colour*

[3] ***Cement kiln dust (CKD)** is a by-product of the cement manufacturing process (fine-grained, solid, highly alkaline waste).*

[4] ***Fly ash** or flue ash is a coal combustion product that is composed of the particulates that are driven out of coal-fired boilers together with the flue gases. Ash that falls to the bottom of the boiler's combustion chamber is called bottom ash.*

[5] ***E-waste** include computers, TVs, printers/ copiers, mobile phones, refrigerators etc (broken or unwanted)*

- **Mining industries:** Mining and agriculture waste contributes to major heavy metal sources in the developed countries (Araújo et al 2013).

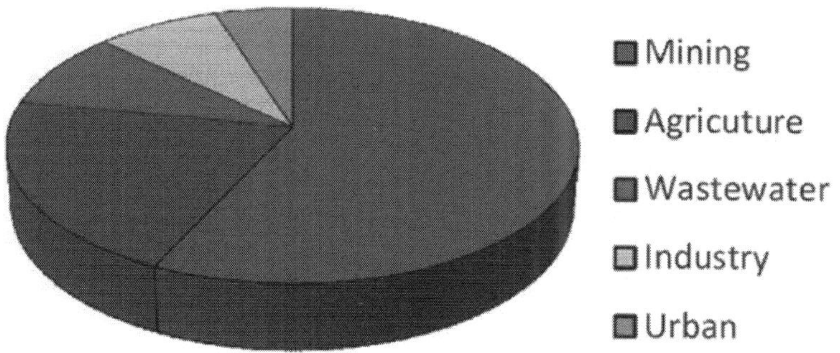

Fig. 3.1. Amount of metals containing waste produced in developed Countries (Araújo et al 2013).

Table 3.4. Possible forms of heavy metals (Florence et al 1992)

[Fulvate = salt or ester of fulvic acid; Fulvic acids = organic acids (natural compounds and components of the humus; soil organic matter)].

Physico-chemical forms	**Possible examples**	**Approximate diameter (nm)**
Simple ionic species	$Zn(H_2O)_6^{2+}$	0.8
Inorganic complexes	$CdCl^+$, $PbCO_3$	1
Differing valency states	Cr(III), Cr(VI)	1
Lipid soluble complexes	CH_3HgCl	1
Organo-metallic species	$CH_3As(OH)_2$	1
Weak complexes	Cu fulvate	2
Adsorbed on colloidal particles	Cu^{2+}/ humic acid/ Fe_2O_3	10-500
Particulate	Metals adsorbed into or contained within clay particles	>450

3.3.2: Factors relating to the speciation of metals

Metal speciation in freshwater (salinity ≤ 2.5 ppt[6]) is strongly influenced by a variety of physico-chemical parameters including hardness (Ca and Mg concentration), alkalinity, pH, natural dissolved organic matter (DOM) and redox potential (oxidizing or reducing power of a solution). Water hardness is reported to be the only physico-chemical variable that is quantitatively related to metal toxicity in freshwater (Bubb and Lester 1991, Markich et al 2001; Table 3.5).

Table 3.5. Metal speciation and factors for controlling it
(Bubb and Lester 1991, Markich et al 2001).

[EDTA = ethylene diamine tetra acetic acid; NTA = Nitrilo tri acetic acid].

Metals	Complexes	Predominant species in water	Factors controlling speciation
Cadmium (Cd)	• Form a wide range of soluble complexes such as with carbonate, sulphate, chloride, hydroxide and with synthetic chelating agents (EDTA and NTA), of which cadmium carbonate and $Cd^{2+}$are the dominant inorganic forms in river water (pH-dependent). • 70-90% of Cd in the freshwater streams can be present as free hydrated ion or other labile complexes. • Unlike Hg, Cd does not form organo-metallic complexes.	Cd^{2+}	pH*, inorganic complexation, organic complexation.
Copper (Cu)	• 90% of Cu in average river water may be complexed by dissolved humic matter.	Cu^{2+} +(most common); copper hydroxide and carbonate complexes (at higher pH).	pH*, inorganic complexation, organic complexation.
Lead (Pb)	• Main Pb complexes are carbonates (pH 7-9) and lead hydroxides (in potable and good quality river water).	Pb^{2+}, lead hydroxides (pH~6); lead hydroxide and	pH*, inorganic complexation, organic complexation.

[6]***PPT** = Parts per thousand*

		carbonate species (at higher pH).	
Mercury (Hg)	• In oxygenated surface water, the mercuric ion (Hg^{2+}) is predominant. • In the absence of organics, mercury exits as hydroxy species in river water [$Hg(OH)0_2$, ($HgOCl_2$) and in seawater as $HgCl_2$.	Hg^0, Hg^2 (mercury II), Organo-mercurial (mono- and di-methyl-mercury).	CH_3Hg content, inorganic complexation, organic complexation.
Zinc (Zn)	• In aerobic water, a significant fraction of Zn exists as simple ionic forms such as carbonate, and hydroxyl complexes, hydrated and stable organic compounds. • Also adsorbed onto organic and inorganic colloids (e.g. humics) in particular at higher pH (above pH 7).	Zn^{2+} (predominant form); $ZnOH^+$, Zn $(OH)_2$ (zinc hydroxides at higher pH).	pH*, inorganic complexation, organic complexation.

Important in freshwaters only.

3.3.3. Bioavailability of metals

The bioavailability (the ability of a metal to bind or traverse/cross the cell membrane) of metal (here HMs) is generally dependent on their physico-chemical form. Speciation of metals are summarised below (Bubb and Lester 1991, Markich et al 2001, Worms et al 2006):

- The bioavailability of metal [Cd, Cr (III), Cu, Ni, Pb, Zn] in freshwater decreases with increasing hardness.
- A decrease in pH may increase the bioavailability of metals to freshwater organisms.
- Natural DOM (humic and fulvic acids) may reduce metal bioavailability in surface waters by binding metals in metal-organic complexes.
- Metal bioavailability decreases with increasing salinity due to the formation of metal-chloride complexes, which are considered to have a relatively low bioavailability.

3.3.4. Bioaccumulation of metals

The bioaccumulation[7] of HMs in natural waters is related to their ability to cross biological barriers, such as plasma membrane (Worms et al 2006). It is a process by which chemical substances are accumulated/ absorbed by aquatic organisms from the water, directly (via gill or through consumption of food/diet (containing the chemical; Fig. 3.2) and eliminated from the organism via the respiratory exchange, faecal egestion, metabolic biotransformation of the parent compound and growth dilution (Fig. 3.2).

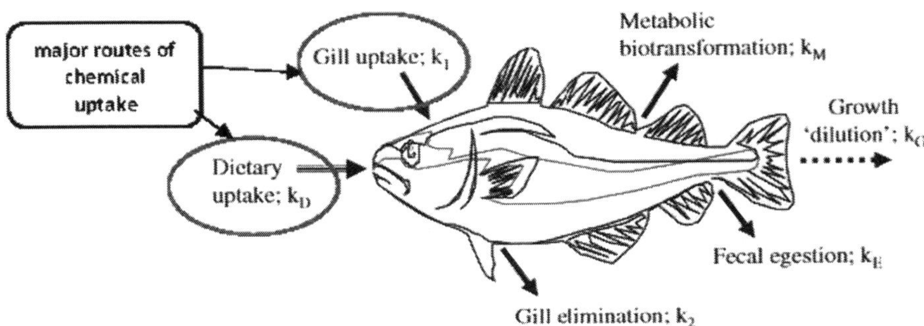

Fig. 3.2. Bioaccumulation of chemicals in fish, showing major routes of chemical uptake (diet, gill) and chemical elimination (gill, faecal, growth and metabolic) [Arnot and Gobas 2006, page 261].

[k_1= the chemical uptake rate constant from the water at the respiratory surface; k_D= is the uptake rate constant for chemical in the diet; k_2, k_E, k_G, k_M are rate constants representing chemical elimination from the organism via the respiratory surface, faecal egestion, growth dilution and metabolic biotransformation, respectively].

Most metals are insoluble in water but soluble in fats and resistant to biological and chemical degradation - properties that result in their accumulation in the tissues of organisms and their biological amplification in the food chain and food webs. Aquatic flora and fauna such as algae, invertebrates (oysters, mussels) and fish can accumulate HMs to several orders of magnitude (thousands to million times) above background levels in the aquatic environment, thus human and other predators at the top of the food chain can be at risks from eating HMs contaminated crops and seafood. The notable example of metal

[7]***Bioaccumulation*** *describes the accumulation and enrichment of contaminants in aquatic organisms (fish, prawns, mussels), relative to that in the environment. Bioaccumulation is the net result of all uptake and loss processes, such as respiratory and dietary uptake, and loss by egestion, passive diffusion, metabolism, transfer to offspring and growth (https://www.sciencedirect.com/topics/pharmacology-toxicology-and-pharmaceutical-science/bioaccumulation).*

accumulation is methyl-mercury[8] (MeHg) in fish and cadmium, copper, lead, zinc, and mercury in oysters and mussels. Apparently, there are three main mechanisms by which bioaccumulation of metals in aquatic organisms can occur:

- via respiration (gills and skin surface).
- by adsorption onto the body surface and
- from ingestion of foods (solids and dissolved matter) (Kibria et al 2010a).

However, the dominant pathway of accumulation appears to be via food. It is well known that metal accumulation in aquatic organisms does not occur uniformly across tissues. This may be due to the complex forms of the metals and due to certain organs and tissues performing specific roles in metal uptake, retention and excretion. For example, in crustaceans, such as in prawns, accumulation of metals is highest in the head (due to high metal storage in hepatopancreas) (Evans-Illidge 1997). For burrowing and detritus-feeding organisms such as eels, worms and bivalves the major pathway of metals accumulation appears to be through ingestion of sediment. For pelagic fish, exposure of metals would be limited since these fish have minimal contact with bottom sediments. On the other hand, molluscs and bivalves (e.g. mussels, oysters) when living in heavily polluted environments can accumulate very high concentrations of HMs such as Cd, Pb and Zn. Because of their ability to concentrate HMs, bivalves have been used widely to monitor HM levels in the aquatic environment.

In fish, most of the metals are accumulated in the kidney and liver and to a lesser extent in the muscles (the part eaten by a human). However, **MeHg** is reported to accumulate in fish tissues (muscles) at higher concentrations, which may be related to high bio-concentration factors of methyl-mercury for fish (see Table 3.6 for bio-concentration factor (BCF)[9] in aquatic organisms. Bioconcentration is the process by which a chemical substance is absorbed by an organism from the ambient environment only through its respiratory and dermal surfaces but chemical exposure in the diet is not included (Arnot and Gobas 2006). Large fish, known for their longevity, such as shark, swordfish, marlin and tuna can accumulate very high concentrations of mercury. In particular,

---

[8]***Methylmercury*** *is an organometallic cation with the formula* $[CH_3Hg]^+$. *It is a bioaccumulative environmental toxicant (a potent neurotoxin). Methylmercury is formed from inorganic mercury by the action of microbes that live in aquatic systems including lakes, rivers, wetlands, sediments, soils and the open ocean (Ullrich et al 2010). This methylmercury production has been primarily attributed to anaerobic bacteria in the sediment.*

[9]***BCF*** *= Bioconcentration factor (BCF) can be expressed as the degree to which a chemical can be concentrated in the tissues of an organism in the aquatic environment or the ratio of the concentration of a chemical in an organism to the concentration of the chemical in the surrounding environment; BCF= CB/CWD. CB is the chemical concentration in the organism (g/kg) and CWD is the freely dissolved chemical concentration in the water (g/L) (Arnot and Gobas 2006).*

sharks have particular characteristics that make them prone to bioaccumulation of high levels of potentially toxic micro-pollutants (because they are long-lived; have comparatively slow rates of growth; are large in size; belong to the higher trophic levels and are top predator/ **apex predator** of the Oceans' and seas' food chain). Kibria and Haroon (2015) provided a global overview of bioaccumulation of inorganic and organic pollutants in shark species. The review reveals that Cd, Cu, DDT, Hg, Pb, PCBs, PCDD/Fs and Zn have bioaccumulated in a number of sharks species (see section 3.4.6 for Hg accumulation in sharks, fish and cephalopods).

Table 3.6. Bioconcentration factors (BCF) of toxic heavy metals to aquatic organisms

[note that highest BCF is with methyl-mercury].

Metals	Aquatic species	Bio-concentration factors	References
Cadmium (Cd)	Aquatic organisms.	10,000-100,000	Reeder et al 1979
Chromium (Cr)	-	100-1,000	CCREM 1987
Copper (Cu)	Various species including phytoplankton, zooplankton, macrophytes, macro-invertebrates and fish.	100-26,000	Spear and Pierce 1979
Lead (Pb)	Invertebrates and fish.	499-1,700	USEPA 1985
Methyl-mercury (MeHg)	Fish.	1,000,000-100,000,000	ANZECC 2000b
Nickel (Ni)	Oysters, mussels.	370	Florence et al 1994
Zinc (Zn)	Freshwater animals.	50-1,130	USEPA 1987

3.3.5. Heavy metal's transportation- grasshopper effect

HMs can enter into an environment via **i.** rainwater run-off from agriculture lands, forests and urban areas, **ii.** effluents discharged from industries (mining, sewage treatments plants, landfill), **iii**. atmospheric fallout of insoluble precipitates; and **iv.** as volatile compounds, for example, certain volatile chemicals such as heavy metals (mercury) and POPs (persistent organic pollutants) can move across vast distances from the southern hemisphere to the northern hemisphere through a repeated cycle of evaporation and condensation, which is called the **"grasshopper effect"** (Fig. 3.3). For example, Hg can evaporate from soils and surface water into the atmosphere and can travel to another location where it condenses in rain or snow. When conditions become warm, Hg can once again evaporate. This repeating cycle of evaporation, transportation and condensation released Hg in one part of the world (low latitude or warm region) can be transported through the atmosphere to regions far away from the original source (high latitude or cold region). It is expected that as global warming increases (due to climate change), chemicals would volatilize/ evaporate more readily into the atmosphere, increasing their presence in the air and other matrices (see section 3.5 on climate change and heavy metals load and bioaccumulation) *https://www.unenvironment.org/news-and-stories/story/grasshopper-effect-serves-pollutants-plates-arctic-peoples*

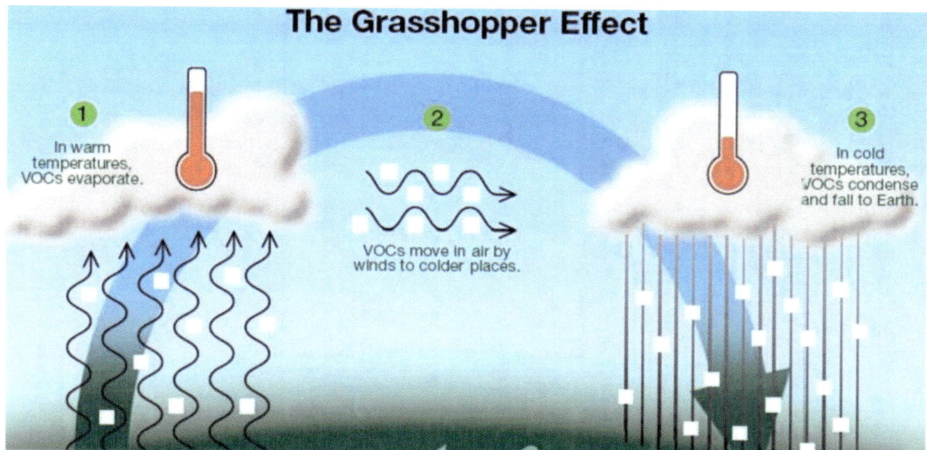

Fig. 3.3. Mercury and POPs (persistent organic pollutants) transportation from hotter regions to colder regions through a repeated cycle of evaporation and condensation, which is called "Grasshopper Effect"
[*https://insideclimatenews.org/news/20090609/arctic-indigenous-peoples-being-poisoned-industry-thousands-miles-away*].

3.3.6. Toxicity and ecotoxicity of heavy metals

Some HMs are essential to aquatic life and some are non-essential. Metals such as Cu and Zn is known to be essential, whereas Cd, Hg and Pb are regarded as non-essential and highly toxic to aquatic life (Table 3.7). However, an abundance of both essential and non-essential heavy metal supplies could be hazardous and highly toxic to living organisms, as summarized below:

- Several HMs are highly toxic to biota such as fish, daphnia, algae (Cd, Cu, Hg, Ni, Pb, U, Zn) (ANZECC 2000a; Kibria et al 2010a; Kibria et al 2016a,b) (see Table 3.7).
- Some of the HMs are highly bioaccumulative such as Cd, Cu, Hg, Pb, Zn (ANZECC 2000a; Kibria et al 2010a; Kibria et al 2016a,b).

Table 3.7. Chemistry, toxicity and ecotoxicity of harmful heavy metals

[na = data is not available; Toxicity classification: Rat LD_{50} < 50 mg/kg=extremely toxic; 50-500 mg/kg= very toxic; 500-5,000 mg/kg=moderately toxic; Fish LC_{50}< 1 mg/L= extremely toxic; 1-10 mg/L= very toxic; 10-100 mg/L= moderately toxic].

Heavy metals	**Solubility**	**Toxicity** **Toxicity to mammal** (LD_{50} (Rat) oral)	**Ecotoxicity** **Toxicity to aquatic organisms** (LC_{50} (fish/ crustaceans, frog, worm; EC_{50} (algae/ daphnia).	**Remarks**
Cadmium	Insoluble in water	Rat: LD_{50}: 225 mg/kg	• Fish, *Oncorhynchus mykiss,* LC_{50}: 0.006 mg/L for 96h flow-through (static). • Fish, *Cyprinus carpio,* LC_{50}: 0.24 mg/L for 96 h static. • Algae, *Selenastrum capricornutum,* EC_{50}: 0.0023 mg/L for 72 h. (MSDS-Thermo Fisher 2018, Cadmium; MSDS-Merck 2006, Cadmium; MSDS-ESPI Metals 2015, Cadmium).	• Carcinogenic to humans (Table 3.8). • Very toxic to aquatic organisms.
Chromium (VI)	Soluble in water	Rat: LD_{50}: 52 mg/kg (MSDS-Merck 2007, Chromium VI)	• Moderately toxic to fish (Fish, *Labeo rohita,* LC_{50}: 39.40-58.76 mg/L. (Bakshia and Panigrahi 2018). • Daphnia, *Daphnia magna* (Water flea), EC_{50}: 0.162 mg/L for 48 h. (MSDS-Merck 2007, Chromium VI)	• CrVI carcinogenic to humans (Table 3.8). • Very toxic to mammals and daphnia and moderately toxic to fish.

Heavy metals	Solubility	Toxicity Toxicity to mammal (LD_{50} (Rat) oral)	Ecotoxicity Toxicity to aquatic organisms (LC_{50} (fish/ crustaceans, frog, worm; EC_{50} (algae/ daphnia).	Remarks
Copper	Insoluble in water	Rat: LD_{50}: 472 mg/kg	• Fish, *Oncorhynchus mykiss,* LC_{50}: 0.052 mg/L for 96 h flow through. • Fish, *Cyprinus carpio,* LC_{50}: 0.08 mg/L for 96 h static. • Daphnia, *Daphnia magna* (Water flea), EC_{50}: 0.03 mg/L for 48 h static (MSDS-Fisher 2015, Copper).	• Very toxic to aquatic organisms (fish, daphnia).
Lead	Insoluble in water	na	• Fish, *Micropterus dolomieu* (smallmouth bass) $LC_{50:}$ 2.2 mg/L for 96 h. • Algae, *Skeletonema costatum,* EC_{50}: 7.94 mg/L for 10 d (MSDS- Jamestown 2015, Lead)	• Pb (inorganic): Probably carcinogenic to humans (Table 3.8). • Toxic to fish.
Mercury	Insoluble in water	Rat: LD_{50}:2,330 mg/kg	• Fish, *Cyprinus carpio,* LC_{50}: 0.16 mg/L for 96 h flow-semi static. • Fish, rainbow trout, LC_{50}: 0.16-0.90 mg/L for 96 h. • Daphnia, *Daphnia magna* (Water flea), LC_{50}: 0.0052 mg/L for 96 h static. • Algae, *Chlorella pyrenoidosa,* EC_{50}: 0.3 mg/L for 5 h. (MSDS-Thermo Fisher 2014, Mercury; MSDS-Chem Supply 2017, Mercury; MSDS-Merck 2006, Mercury).	• Methyl-mercury (MeHg) possibly carcinogenic to humans (Table 3.8). • Very toxic to aquatic life.

Heavy metals	**Solubility**	**Toxicity** **Toxicity to mammal** (LD_{50} (Rat) oral)	**Ecotoxicity** **Toxicity to aquatic organisms** (LC_{50} (fish/ crustaceans, frog, worm; EC_{50} (algae/ daphnia).	**Remarks**
Nickel	Insoluble in water	Rat: LD_{50}: > 9,000 mg/kg (MSDS-Thermo Fisher 2010, Nickel)	• Fish, *Cyprinus carpio*, LC_{50}: 10.4 mg/L for 96 h static. • Fish, *Cyprinus carpio,* LC_{50}: 1.3 mg/L for 9 6h semi-static. • Daphnia, *Daphnia magna* (Water flea), EC_{50}: 1 mg/L for 48 h static. • Algae, *Pseudokirchneriella subcapitata* (green algae), EC_{50}: 0.174 - 0.311 mg/L for 96 h static. (MSDS-Thermo Fisher 2010, Nickel).	• Carcinogenic to humans (Table 3.8). • Very toxic to aquatic life.
Zinc		Rat: LD_{50}: > 2,000 mg/kg (MSDS-Merck 2017, Zinc)	• Fish, *Pimephales promelas* (fathead minnow), LC_{50}: 0.238 - 0.269 mg/L for 96 h. • Daphnia, *Daphnia magna* (Water flea), EC_{50}: 0.356 mg/L; 48 h. • Algae, *Pseudokirchneriella subcapitata* (green algae), EC_{50}: 0.106 mg/L for 2 h. (MSDS-Merck 2017, Zinc)	• Very toxic to aquatic life.

- Some of the HMs are endocrine-disrupting (Cd, Hg, Pb, Zn) (ANZECC 2000a; Kibria et al 2010a; Kibria et al 2016a,b).
- Several HMs are carcinogenic (IARC Class 1, 2A, 2B) to humans including cadmium, chromium, lead, methyl-mercury, nickel (IARC 2019) (see Table 3.8).
- Four of the top 10 chemicals of major public concern [out of 275 priority list of hazardous substances assessed by The Agency for Toxic Substances and Disease Registry (ATSDR), US Department of Health and Human Service are HMs- lead, mercury and cadmium (ATSDR 2017) (see Chapter 1, Table 1.6)].

Table 3.8. IARC classification of metals (IARC 2019)
[IARC = International Agency for Research on Cancer;
https://monographs.iarc.fr/agents-classified-by-the-iarc/].

Substances	IARC carcinogen group	Comments
Cadmium and its cadmium compound	1	Carcinogenic to humans.
Chromium metallic	3	Not classifiable as to carcinogenicity in humans.
Chromium (III) compounds	3	Not classifiable as to carcinogenicity in humans.
Chromium (VI) compounds	1	Carcinogenic to humans.
Cobalt and cobalt compounds	2B	Possibly carcinogenic to humans.
Lead	2B	Possibly carcinogenic to humans.
Lead compounds inorganic	2A	Probably carcinogenic to humans.
Lead compounds (organic)	3	Not classifiable as to carcinogenicity in humans.
Mercury and inorganic mercury compounds (Hg)	3	Not classifiable as to carcinogenicity in humans.
Methyl-mercury compounds (MeHg)	2B	Possibly carcinogenic to humans.
Nickel metallic and alloys (Ni)	2B	Possibly carcinogenic to humans.
Nickel compounds (Ni)	1	Carcinogenic to humans.

3.4. Impacts of heavy metals on social, economic and environmental sectors

3.4.1. Heavy metals residues in Australian waterways

Table 3.9 provides an account of metals detected in environmental waters (irrigation channels, streams, rivers, lakes, estuaries and sediments) in Australia. It reveals that water quality with respect to HMs pollution has exceeded at a number of investigated sites (identified with a star symbol in Table 3.9). In most cases, Cd, Cu, Pb, Zn exceeded the guidelines of threshold values. Agriculture, mining, sewage effluents and industrial discharges appear to be the major sources of HMs in Australian waterways.

Table 3.9. Metals detected in environmental waters (surface water and sediments) in Australia

[dw = dry weight; G-MW =Goulburn Murray water; na = data is not available; NCCMA = North Central Catchment Management Authority; star ( ) = where metal concentrations have exceeded the guidelines values].*

Metals	Range	Water or sediment	Locations
Arsenic (As) in sediment	<5-7 mg/kg	Sediment	Intensive agriculture (potato) and forestry, Latrobe catchment (2011-2012) (Mossop et al 2012).
	As (arsenic) sediment quality guideline trigger values: 20-70 mg/kg dw (ANZECC 2000a)		
Cadmium (Cd) in surface water	Detected	Freshwater	Passive samples, G-MW catchment, Victoria (2007-08) (Kibria et al 2010b).
	< 0.4 μg/L	Freshwater	G-MW catchment, Victoria (Irrigation channels and rivers) (2009-10) (Kibria et al 2010c).
	< 0.4 μg/L	Freshwater	Yarra catchments, Victoria (peri-urban streams) (2009-10) (Kibria et al 2010c).
	< 0.4 μg/L	Freshwater	NCCMA, Victoria (ephemeral streams) (2009-10) (Kibria et al 2010c).

Metals	Range	Water or sediment	Locations
	< 0.4 µg/L	Freshwater	G-MW catchment, Victoria (Irrigation channels and rivers) (2004-06) (Rose and Kibria 2006).
	< 0.05 µg/L	Estuarine water	Central Port Philip Bay, Victoria (1991)
	0.20 µg/L	Estuarine water	Corio Bay, Victoria (1991)
	***1.90 µg/L**	Estuarine water	North Lake Macquarie, NSW (1987)
	0.20 µg/L	Estuarine water	South Lake Macquarie, NSW (1987)
	0.37 µg/L	Estuarine water	Port Augusta, South Australia (1983)
	0.32 µg/L	Estuarine water	Port Pirie, S.A. (off-shore) (1983)
	0.05 µg/L	Estuarine water	Derwent River, Tasmania (na)
	0.07 µg/L	Estuarine water	Ross River, Queensland (1986)
			(all data based on Batley 1995)
	Cd water quality guideline trigger values: Freshwater aquatic ecosystems protection value: 0.2 µg/L; Marine aquatic ecosystems protection value: 0.7 µg/L (ANZECC 2000a).		
Cadmium (Cd) in sediment	< 1 mg/kg	Freshwater sediment	Intensive agriculture (potato) and forestry, Latrobe catchment (2011-2012) (Mossop et al 2012).
	***0.1-13 mg/kg**	Estuarine sediment	Corio Bay, Victoria (1991) (surface sediment) (Batley 1995).
	< 0.5-2 mg/kg	Estuarine sediment	Port Phillip Bay (1976) (surface sediment) (Batley 1995).
	< 5 mg/kg	Estuarine sediment	Port Phillip Bay, near Werribee Treatment (1991) (surface sediment) (Batley 1995).

Metals	Range	Water or sediment	Locations
	***160 mg/kg**	Estuarine sediment	North Lake Macquarie, NSW (1987) (sediment from 55 cm depth) (Batley 1995).
	4 mg/kg	Estuarine sediment	South Lake Macquarie, NSW (1987) (Batley 1995).
	3 mg/kg	Estuarine sediment	Blackwattle Bay, Sydney Harbour (1986) (sediments from 10 cm depth) (Batley 1995).
	Cd sediment quality guideline trigger values: 1.5-10 mg/kg dw (ANZECC 2000a).		
Chromium (Cr) in freshwater and sediment	Detected	Freshwater	Passive samples, G-MW catchment, Victoria (2007-08) (Kibria et al 2010b).
	11-101 mg/kg	Sediment	Intensive agriculture (potato) and forestry, Latrobe catchment (2011-2012) (Mossop et al 2012).
	Cr water quality guideline trigger values: Freshwater aquatic ecosystems protection value: 0.01 µg/L (CrIV); Cr sediment quality guideline trigger values: 80-370 mg/kg dw (ANZECC 2000a).		
Cobalt (Co) in freshwater and sediment	Detected	Freshwater	Passive samples, G-MW catchment, Victoria (2007-08) (Kibria et al 2010b).
	8-48 mg/kg	Sediment	Intensive agriculture (potato) and forestry, Latrobe catchment (2011-2012) (Mossop et al 2012).
	Co water quality guideline trigger values: Freshwater aquatic ecosystems protection value: 1.4 µg/L (Nagpal 2004); Co sediment quality guideline trigger values: na		
Copper (Cu) in surface water	Detected	Freshwater	Passive samples, G-MW catchment, Victoria (2007-08) (Kibria et al 2010b).

Metals	Range	Water or sediment	Locations
	***< 0.9-2.47 µg/L**	Freshwater	G-MW catchment, Victoria (Irrigation channels, and rivers) (2009-10) (Kibria et al 2010c).
	***< 0.9-7.0 µg/L**	Freshwater	Yarra catchments, Victoria (peri-urban streams) (2009-10) (Kibria et al 2010c).
	***1.1-1.9 µg/L**	Freshwater	NCCMA, Victoria (ephemeral streams) (2009-10) (Kibria et al 2010c).
	***< 0.9-3.9 µg/L**	Freshwater	G-MW catchment, Victoria (Irrigation channels, and rivers) (2004-06) (Rose and Kibria 2006).
	0.6 µg/L	Estuarine water	Central Port Philip Bay, Victoria (1991)
	1.1 µg/L	Estuarine water	Corio Bay, Victoria (1991)
	***1.5 µg/L**	Estuarine water	North Lake Macquarie, NSW (1987)
	1.2 µg/L	Estuarine water	South Lake Macquarie, NSW (1987)
	0.45 µg/L	Estuarine water	Port Augusta, South Australia (1983)
	0.25 µg/L	Estuarine water	Port Pirie, S.A. (off-shore) (1983)
	1.2 µg/L	Estuarine water	Derwent River, Tasmania (na)
	0.24 µg/L	Estuarine water	Ross River, Queensland (1986)
			(all data based on Batley 1995)
	Cu water quality guideline trigger values: Freshwater aquatic ecosystems protection value: 1.4 µg/L; Marine aquatic ecosystems protection value: 1.3 µg/L (ANZECC 2000).		
Copper (Cu) in sediments	12-53 mg/kg	Freshwater sediment	Intensive agriculture (potato) and forestry, Latrobe catchment (2011-2012) (Mossop et al 2012).

Metals	Range	Water or sediment	Locations
	2-50 mg/kg	Estuarine sediment	Corio Bay, Victoria (1991) (surface sediment) (Batley 1995).
	1.5-8 mg/kg	Estuarine sediment	Port Phillip Bay (1976) (surface sediment) (Batley 1995).
	< 5-75 mg/kg	Estuarine sediment	Port Phillip Bay, near Werribee Treatment (1991) (surface sediment) (Batley 1995).
	170 mg/kg	Estuarine sediment	North Lake Macquarie, NSW (1987) (sediment from 55 cm depth) (Batley 1995).
	20 mg/kg	Estuarine sediment	South Lake Macquarie, NSW (1987) (Batley 1995).
	180 mg/kg	Estuarine sediment	Blackwattle Bay, Sydney Harbour (1986) (sediment from 10 cm depth) (Batley 1995).
	Cu sediment quality guideline trigger values: 65-270 mg/kg dw (ANZECC 2000a).		
Iron (Fe) in freshwater and sediments	Detected	Freshwater	Passive samples, G-MW catchment, Victoria (2007-08) (Kibria et al 2010b).
	22,100-66,800 mg/kg	Sediment	Intensive agriculture (potato) and forestry, Latrobe catchment (2011-2012) (Mossop et al 2012).
	Fe water quality guideline trigger values: Freshwater aquatic ecosystems protection value: na; Fe sediment quality guideline trigger values: na		
Lead (Pb) in surface water	Detected	Freshwater	Passive samples, G-MW catchment, Victoria (2007-08) (Kibria et al 2010b).

Metals	Range	Water or sediment	Locations
	< 0.21-0.37	Freshwater	G-MW catchment, Victoria (Irrigation channels, and rivers) (2009-10) (Kibria et al 2010c).
	< 0.2-3.1 µg/L	Freshwater	Yarra catchments, Victoria (peri-urban streams) (2009-10) (Kibria et al 2010c).
	< 0.21 µg/L	Freshwater	NCCMA, Victoria (ephemeral streams) (2009-10) (Kibria et al 2010c).
	< 0.21-2.4 µg/L	Freshwater	G-MW catchment, Victoria (Irrigation channels, and rivers) (2004-06) (Rose and Kibria 2006).
	< 0.8 µg/L	Estuarine water	Central Port Philip Bay, Victoria (1991)
	< 0.8 µg/L	Estuarine water	Corio Bay, Victoria (1991)
	1.6 µg/L	Estuarine water	North Lake Macquarie, NSW (1987)
	0.1 µg/L	Estuarine water	South Lake Macquarie, NSW (1987)
	0.54 µg/L	Estuarine water	Port Augusta, South Australia (1983)
	***5.1 µg/L**	Estuarine water	Port Pirie, S.A. (offshore) (1983)
	0.23 µg/L	Estuarine water	Derwent River, Tasmania (na)
	0.11 µg/L	Estuarine water	Ross River, Queensland (1986)
			(all data based on Batley 1995)
	Pb water quality guideline trigger values: Freshwater aquatic ecosystems protection value: 3.4 µg/L; Marine aquatic ecosystems protection value: 4.4 µg/L (ANZECC 2000a).		
Lead (Pb) in sediments	7-19 mg/kg	Sediment	Intensive agriculture (potato) and forestry, Latrobe catchment (2011-2012) (Mossop et al 2012).

Metals	Range	Water or sediment	Locations
	***2-210 mg/kg**	Estuarine sediment	Corio Bay, Victoria (1991) (surface sediment) (Batley 1995).
	8-22 mg/kg	Estuarine sediment	Port Phillip Bay (1976) (surface sediment) (Batley 1995).
	< 20-140 mg/kg	Estuarine sediment	Port Phillip Bay, near Werribee Treatment (1991) (surface sediment) (Batley 1995).
	***1200 mg/kg**	Estuarine sediment	North Lake Macquarie, NSW (1987) (sediment from 55 cm depth) (Batley 1995).
	68 mg/kg	Estuarine sediment	South Lake Macquarie, NSW (1987) (Batley 1995).
	***520 mg/kg**	Estuarine sediment	Blackwattle Bay, Sydney Harbour (1986) (sediment from 10 cm depth) (Batley 1995).
	Pb sediment quality guideline trigger values: 50-200 mg/kg dw (ANZECC 2000a).		
Mercury (Hg) in surface water	Detected	Freshwater	Passive samples, G-MW catchment, Victoria (2007-08) (Kibria et al 2010b).
	< 0.2 µg/L	Freshwater	G-MW catchment, Victoria (Irrigation channels, and rivers) (2009-10) (Kibria et al 2010c).
	< 0.2 µg/L	Freshwater	Yarra catchments, Victoria (peri-urban streams) (2009-10) (Kibria et al 2010c).
	< 0.2 µg/L	Freshwater	NCCMA, Victoria (ephemeral streams) (2009-10) (Kibria et al 2010c).

Metals	Range	Water or sediment	Locations
	< 2 µg/L	Estuarine water	Central Port Philip Bay, Victoria (1991)
	< 2 µg/L	Estuarine water	Corio Bay, Victoria (1991)
	na	Estuarine water	North Lake Macquarie, NSW (1987)
	na	Estuarine water	South Lake Macquarie, NSW (1987)
	na	Estuarine water	Port Augusta, South Australia (1983)
	na	Estuarine water	Port Pirie, S.A. (offshore) (1983)
	***0.034 µg/L**	Estuarine water	Derwent River, Tasmania (na)
	na	Estuarine water	Ross River, Queensland (1986)
			(all data based on Batley 1995)
	Hg water quality guideline trigger values: Freshwater aquatic ecosystems protection value: 0.06 µg/L; Marine aquatic ecosystems protection value: 0.1µg/L (ANZECC 2000a).		
Mercury (Hg) in sediments	< 0.1-0.4 mg/kg	Sediment	Intensive agriculture (potato) and forestry, Latrobe catchment (2011-2012) (Mossop et al 2012)
	Hg sediment quality guideline trigger values: 0.15-1 mg/kg *(ANZECC 2000a).*		
Nickel (Ni) in freshwater and sediment	Detected	Freshwater	Passive samples, G-MW catchment, Victoria (2007-08) (Kibria et al 2010b).
	8-62 mg/kg	Sediment	Intensive agriculture (potato) and forestry, Latrobe catchment (2011-2012) (Mossop et al 2012).
	Ni water quality guideline trigger values: Freshwater aquatic ecosystems protection value: 11 µg/L; Ni sediment quality guideline trigger values: na (ANZECC 2000a).		
Zinc (Zn) in surface water	Detected	Freshwater	Passive samples, G-MW catchment, Victoria (2007-08) (Kibria et al 2010b).

Metals	Range	Water or sediment	Locations
	1.57- 3.75 µg/L	Freshwater	G-MW catchment, Victoria (Irrigation channels, and rivers) (2009-10) (Kibria et al 2010c).
	< 1.2-18 µg/L	Freshwater	Yarra catchments, Victoria (peri-urban streams) (2009-10) (Kibria et al 2010c).
	1.9-5.8 µg/L	Freshwater	NCCMA, Victoria (ephemeral streams) (2009-10) (Kibria et al 2010c).
	***< 0.02-8.6 µg/L**	Freshwater	Passive samples, G-MW catchment, Victoria (2007-08) (Kibria et al 2010b).
	< 0.002 µg/L	Estuarine water	Central Port Philip Bay, Victoria (1991)
	< 0.002 µg/L	Estuarine water	Corio Bay, Victoria (1991)
	5.2 µg/L	Estuarine water	North Lake Macquarie, NSW (1987)
	1.0 µg/L	Estuarine water	South Lake Macquarie, NSW (1987)
	<10 µg/L	Estuarine water	Port Augusta, South Australia (1983)
	47 µg/L	Estuarine water	Port Pirie, S.A. (offshore) (1983)
	3.4 µg/L	Estuarine water	Derwent River, Tasmania (NA)
	NA µg/L	Estuarine water	Ross River, Queensland (1986)
			(all data based on Batley 1995)
	Zn water quality guideline trigger values: Freshwater aquatic ecosystems protection value: 8.0 µg/L; Marine aquatic ecosystems protection value: 15 µg/L (ANZECC 2000a).		
Zinc (Zn) in sediments	38-84 mg/kg	Sediment	Intensive agriculture (potato) and forestry, Latrobe catchment (2011-2012) (Mossop et al 2012).
	4-400 mg/kg	Estuarine sediment	Corio Bay, Victoria (1991) (surface sediment) (Batley 1995).

Metals	Range	Water or sediment	Locations
	21-40 mg/kg	Estuarine sediment	Port Phillip Bay (1976) (surface sediment) (Batley 1995).
	9-300 mg/kg	Estuarine sediment	Port Phillip Bay, near Werribee Treatment (1991) (surface sediment) (Batley 1995).
	***2400 mg/kg**	Estuarine sediment	North Lake Macquarie, NSW (1987) (sediment from 55 cm depth) (Batley 1995).
	150 mg/kg	Estuarine sediment	South Lake Macquarie, NSW (1987) (Batley 1995).
	***1150 mg/kg**	Estuarine sediment	Blackwattle Bay, Sydney Harbour (1986) (sediment from 10 cm depth) (Batley 1995).
	Zn sediment quality guideline trigger values: 200- 410 mg/kg (ANZECC 2000a).		

3.4.2. Identification of heavy metal pollution "hot spots" using innovative artificial mussel (AM) technology

As part of a global program, comprehensive research was carried out to identify HMs pollution "hotspots"[10] in waterways both in Australia (2010-2012) and Bangladesh (2013-2015) using *"artificial mussel"* (AM) technology (see Box 3.2 for an explanation on AM tool). The Australian study identified Cu and Zn "hot spots" in waterways close to intensive orchards (pome and stone), vineyards and areas growing olives and tomatoes (see Kibria et al 2012a). In Bangladesh, five HMs pollution "hot spots" (Cu, Fe, Mn, Ni, Pb) were located at the Buriganga River, Dhaka which received effluents from tanneries, textiles, garments, domestic wastes, dyeing industries (see Kibria et al 2016a, 2016b).

[10]***pollution "hotspots"**- The places/ areas/ sites with high concentrations of a pollutant. For example, sites where elevated pollutant concentration was detected compared to other sites can be called as pollution "hotspots (Kibria et al 2016a) (compare with "**biodiversity "hotspots"**: biodiversity hotspots are areas that support natural ecosystems that are largely intact and where native species and communities associated with these ecosystems are well represented (https://www.environment.gov.au/biodiversity/conservation/hotspots).*

3.4.3. Impacts of cadmium

Cd is naturally present in soils at concentrations ranging between 0.1 and 1 mg/kg. It is readily available for uptake by food crops. Cd is a known human carcinogen (Table 3.8) and ranks No. 7 in ATSDR's Priority List of Hazardous Substances (see Table 1.6. in Chapter 1 for ATSDR substances priority list). Potential environmental sources of Cd are batteries, cement kiln dust, coal combustion, e-wastes, fertilisers, oil refineries, mining, pesticides, paper and pulp mills, paints, ship-breaking yards, smelting, textile effluents (Table 3.3). The chemistry, toxicity and ecotoxicity of Cd can be found in section 3.3 (Table 3.7). The following section discusses the contamination of rice and seafood (freshwater and marine fish) by Cd and effects of Cd on human health.

Box 3.2. What is an artificial mussel (AM) technology?

AM is an innovative pollution monitoring device/tool. The tool allows to continuously monitor micro-pollutants such as toxic HMs in waterways (24h/7d). Research carried out found that AM accumulates several toxic and high-risk metals including As, Cd, Co, Cr, Cu, Fe, Hg, Mn, Ni, Pt, Pb, Se, U, Zn. The tool is a tube (60 mm x 25 cm) with an opening on both sides. AM can be deployed in rivers, estuaries, coasts, ocean, groundwater bores, irrigation channels, toxic and anoxic environment (where there is no life) to monitor, to assess risks and to identify metal's pollution "hot spots". The global AM watch program covered six continents (20 countries) (see Kibria et al 2017, Kibria 2018). The lead author of this book (Dr. Golam Kibria) is a Global Scientist of Artificial Mussel (AM) Pollution Watch Program, who planned or led AM Watch Research Program in seven countries (Australia, Bangladesh, Greece, Myanmar, Russia, South Korea, Sri Lanka) as part of a global program. For example, a research carried out in Australia using AM found that climate variability (dry vs wet year) can influence the transport of heavy metals into waterways. Elevated levels of Cu were found during the dry year (drought) while elevated levels of Zn were found during the wet year (floods) showing climate variability affects outflows from the catchment, thereby affecting the inputs, transport and availability of individual metals (Kibria et al 2012a). Another research carried out across Bangladesh using the same AM tool, identified seven pollution " hot spots" (Kibria et al 2016b).

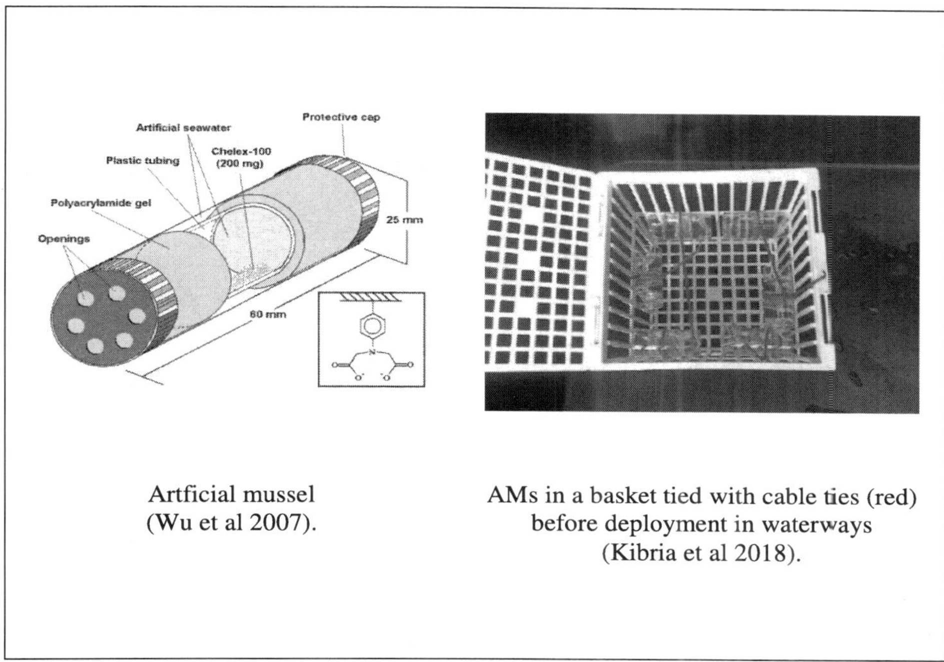

Artficial mussel (Wu et al 2007).

AMs in a basket tied with cable ties (red) before deployment in waterways (Kibria et al 2018).

3.4.3.1. Cadmium contamination of rice

Cultivation of crops for human or animal consumption on contaminated soil or with contaminated irrigation water can potentially lead to uptake and accumulation of metals in edible plant parts (through the roots of plants to edible leaves, fruits and seeds) with a possible risk to human and animal health. Therefore, people can be exposed to Cd upon consumption of Cd containing plants or animals. Very recently, Meharag et al (2013) carried out a global research to assess Cd (mg/kg) levels in rice grain in twelve countries (Bangladesh, Cambodia, France, Ghana, India, Italy, Japan, Nepal, Spain, Sri Lanka, Thailand, USA) in four continents (Asia, Africa, Europe, North America). The study found that Cd levels in rice grain were the highest in Bangladesh, Sri Lanka and India (Fig. 3.4). In Bangladesh and Sri Lanka, per capita rice intake is highest. The elevated levels of Cd found in rice in Bangladesh and Sri Lanka are a real concern from health perspectives since rice intake among poor and ultra-poor could be three times per day, therefore, the likelihood of Cd exposure via food intake is likely to be elevated when compared to food safety guideline values. Human ingestion of rice-derived Cd is dependent on three factors: **i.** the concentration of Cd in grain consumed; **ii.** the rate of rice ingestion; and **iii**. body mass of the consuming individual.

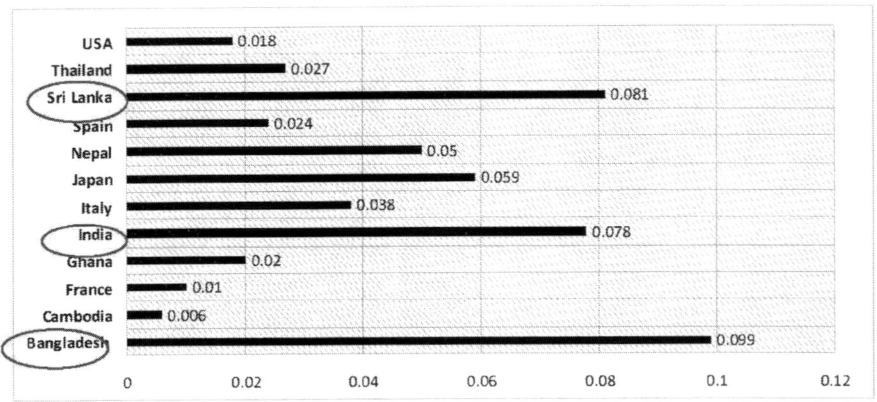

Fig. 3.4. Cadmium (mg/kg) concentration in the rice grain in twelve countries of four continents (Asia, Africa, Europe, North America)

(graph is plotted basing on data from Meharag et al 2013) [Cd levels in rice grain were the highest in Bangladesh, Sri Lanka and India].

3.4.3.2. Cadmium contamination of freshwater and marine fish

Cd is highly toxic to plants and animals. LC_{50} for fish, *Oncorhynchus mykiss* is 0.006 mg/L and EC_{50} for Algae, *Selenastrum capricornutum* is 0.0023 mg/L for 72 h (Table 3.7). Cd can damage the nervous and reproductive systems and reduce the ability of osmoregulation of aquatic organism (ANZECC 2000b). A compilation of global data reveals that several marine fish (Albacore, Atlantic mackerel, Black bellied angler, Blue whiting, Brown ray, European anchovy, European conger eel, European pilchard, European hake, Rose fish, Grey mullet, Horned octopus, Red mullet, Rose fish) consumed by humans have exceeded the respective maximum allowable limits of cadmium (Table 3.10).

Table 3.10. Cadmium concentrations in commonly consumed marine fish in some selected countries where it exceeded the respective maximum allowable limits

[Maximum allowable limit of Cd for seafood are: fish muscle meat: 0.05 mg/kg; swordfish: 0.30 mg/kg; crustaceans: 0.30 mg/kg; cephalopods and molluscs: 1.0 mg/kg) (EC 2008)].

Country/ region	Fish species	Range (mean) mg/kg	References
Italy Mediterranean Sea	*Lophius budegassa* Black bellied angler	0.06 to 0.11 (0.09)	Storelli and Barone 2013
Italy Mediterranean Sea	*Raja miraletus* Brown ray	0.04-0.13 (0.08)	Storelli and Barone 2013

Country/ region	Fish species	Range (mean) mg/kg	References
Italy Mediterranean Sea	*Conger* European conger eel	0.08-0.12 (0.11)	Storelli and Barone 2013
Italy Sicily, Mediterranean Sea (nearby a petro-chemical industrial area)	*Sardina pilchardus* European pilchard	(0.045)	Copat et al 2012
Italy Sicily, Mediterranean Sea (nearby a petroch-emical industrial area)	*Engraulis encrasicolus* European anchovy	(0.366)	Copat et al 2012
Italy Mediterranean Sea	*Mullus barbatus* Red mullet	0.02-0.12 (0.08)	Storelli and Barone 2013
Italy Sicily, Mediterranean Sea	*Mullus barbatus* Red Mullet	(0.084)	Copat et al 2012
Italy Central Adriatic Sea	*Micromesistius poutassou* Blue whiting	(0.10) muscle	Perugini et al 2014
Italy Central Adriatic Sea	*Merluccius* European hake	(0.06) muscle	Perugini et al 2014
Italy Central Adriatic Sea	*Mullus barbatus* Red mullet	(0.07) muscle	Perugini et al 2014
Italy Mediterranean Sea	*Helicolenus dactylopterus* Rosefish	0.07-0.12 (0.10)	Storelli and Barone 2013
Mediterranean Sea	*Thunnus alalonga* Albacore	0.03–0.10 (0.05) (flesh) 7.07–11.67 (9.22) liver	Storelli and Marcotrigiano 2003
Mediterranean Sea	Horned octopus *Eledone moschata*	0.16–0.46 (0.33) flesh 4.20– 11.33 (6.72) hepatopancreas	Storelli and Marcotrigiano 2003

Country/ region	Fish species	Range (mean) mg/kg	References
Nigeria Fish samples from markets in Ogbomoso, Oyo State	Atlantic mackerel *Scomber scombrus*	0.49	Ogundiran et al 2014
Nigeria Fish samples from markets in Ogbomoso, Oyo State	*Sardina pilcladus* European pilchard	0.19	Ogundiran et al 2014
Turkey Black Sea	*Mugil auratus* Grey mullet	0.10-0.40 (flesh) 0.15-0.50 (liver)	Filazi et al 2003

3.4.3.3. *Effects of cadmium on human health*

The Cd has been classified as a category 1 carcinogen and is not regarded as an essential element to human life. It is known that serious and systematic health problems can develop as a result of excessive dietary intake of Cd and Pb in the human body. Cd can cause kidney failure and has been statistically associated with an increased risk of cancer in the lung, endometrium[11], bladder, and breast (European Food Safety Authority 2012).

3.4.4. *Impacts of copper*

Background Cu concentrations in soil depend on geology and typically vary between 2 and 50 mg of Cu/kg. Copper is an important essential element for all living organisms and deficiency in plants and ruminants occur in soils with low levels of available Cu. However, elevated soil Cu concentrations can cause toxic effects in all terrestrial organisms (plants, invertebrates and micro-organisms). Cu falls within the USEPA priority list of pollutants (ranked 120th) (see section 1.5.1 of chapter 1). The widespread use of Cu has resulted in significant anthropogenic inputs to topsoils through atmospheric deposition and agricultural practices (fertilizers, pesticides, sewage sludge etc.). Potential environmental sources of Cu are cement kiln dust, coal combustion, mining effluents, paints, pesticides, sewage sludge, ship-breaking yards, textile effluents (Table 3.3). The chemistry, toxicity and ecotoxicity of Cu can be found in section 3.3 (Table 3.7). The following section discusses the contamination of food (vegetables, rice) by Cu and effects of Cu on human health.

---

[11]***endometrium*** *is the mucous membrane lining the uterus, which thickens during the menstrual cycle in preparation for possible implantation of an embryo*

3.4.4.1. Copper contamination of food

Fruit and vegetables in several countries including Australia, China, India and Egypt are often contaminated with Cu (Table 3.11). Use of inadequately treated sewage water/wastewater in growing crops and vegetables caused most food contamination by Cu.

Table 3.11. Heavy metal (copper) contamination from diverse sources in global food crops

[Recommended maximum levels of Cu for vegetables are 73.3 mg/kg (Chary et al 2008)].

Country where investigated	Sources of heavy metal contaminants affecting food chains	Copper concentrations recorded in food crops (dry weight)	References
Australia French beans, beetroot, kale	Urban stormwater	0.016–0.66 mg/kg	Tom et al 2014
Australia Rice and other paddy crops and vegetables (**imported** from Bangladesh, India, Pakistan, Thailand, Italy, Canada and Egypt)	Arsenic- and metal-contaminated groundwater	Rice: 1.0–9.4 mg/kg Vegetables: 1–29 mg/kg	Rai et al 2019
China Maize, green cabbage, radish, turnip, spinach, cauliflower, and lettuce	Sewage effluent (inadequately treated)	9-15 mg/kg	Khan et al 2008
China *Brassica* sp., food grains, and leafy vegetables	Both sewage and industrial waste (from smelter) drained into river water used for irrigation	3.58-18.59 mg/kg	Liu et al 2005
Egypt Various fruits and vegetables	Inadequately treated wastewater	Fruits:1.22-18.3 mg/kg Vegetables: 0.83-5.69 mg/kg	Radwan and Salama 2006

India *Brassica* sp., *Chenopodium* sp., leafy and root vegetables	Sewage effluent (inadequately treated)	Crops (Rice, wheat, sorghum, maize): 8.71-57.2 mg/kg Vegetables (spinach, rape, cauliflower, cucumber, radish): 5.42-23.1 mg/kg	Rattan et al 2005
India Spinach, Ladies finger, brinjal, Coriander leaves, Amaranthus	Sewage wastewater (in adequately treated)	Amaranthus: 1.4 mg/kg; Brinjal: 0.7 mg/kg; Coriander leaves: 1.2 mg/kg; Ladies finger: 0.6 mg/kg; Spinach: 0.09 mg/kg	Chary et al 2008

3.4.4.2. Copper contamination of freshwater and marine fish

Cu is highly toxic to fish, invertebrates and amphibians. The reported LC_{50} for fish, *Oncorhynchus mykiss* is 0.052 mg/L and EC_{50} for *Daphnia magna* is 0.03 mg/L (Table 3.7). Cu can damage gill and kidney and reduce the growth of aquatic organisms such as fish. Cu can bioaccumulate in fish and molluscs with high potential to bio-concentrate in molluscs (ANZECC 2000b, Kibria et al 2010a).

3.4.4.3. Effects of copper on human health

Cu is an essential trace element for humans, however, elevated levels of Cu or Cu poisoning may result in cirrhosis of the liver (liver diseases) and in extreme cases, the death (Kibria 2014).

3.4.5. Impacts of lead

Pb is among the elements that have been most extensively used by man over time. Pb is the 2nd most hazardous substances according to the ATSDR's Priority List of Hazardous Substances (see Table 1.6. in Chapter 1 for ATSDR substances priority list) and is a "probable human carcinogen" (IARC group 2A) (Table 3.8). Potential environmental sources of Pb are batteries, cement kiln dust, coal combustion, lead oil, mining effluents, paper and pulp mills, paints, smelting, sewage sludge, ship-breaking yards, textile effluents, wood preservatives (Table 3.3). The chemistry, toxicity and ecotoxicity of Pb can be found in section 3.3 (Table 3.7). The following section discusses the contamination of food (rice and vegetables) and seafood (freshwater and marine fish) by Pb and the effects of Pb on human health.

3.4.5.1. Lead contamination of food

Fruit and vegetables in several countries including Australia, China, India and Egypt are often contaminated with Pb (Table 3.12). Use of inadequately treated sewage water/wastewater in growing crops and vegetables caused most food contamination by Pb.

Table 3.12. Heavy metal (lead) contamination from diverse sources in global food crops

[Recommended maximum levels of Pb for vegetables are 0.3 mg/kg (Chary et al 2008)].

Country where investigated	Sources of heavy metal contaminants affecting food chains	Lead concentrations recorded in food crops (dry weight)	References
Australia French beans, beetroot, kale	Urban stormwater.	0.001–0.11 mg/kg	Tom et al 2014
Australia Rice and other paddy crops and vegetables (**imported** from Bangladesh, India, Pakistan, Thailand, Italy, Canada and Egypt)	Arsenic- and metal-contaminated groundwater.	Rice: 0.016–0.248 mg/kg Vegetables: 0.035–0.495 mg/kg	Rai et al 2019
China Maize, green cabbage, radish, turnip, spinach, cauliflower, and lettuce	Sewage effluent (inadequately treated).	3-5 mg/kg	Khan et al 2008
China *Brassica* sp., food grains, and leafy vegetables	Both sewage and industrial wastes (from smelter) drained into river water used for irrigation.	1.9-3.7mg/kg	Liu et al 2005
Egypt Various fruits and vegetables	Inadequately treated wastewater	Fruits:0.15-8.8 mg/kg Vegetables: 0.01-0.58 mg/kg	Radwan and Salama 2006
India *Brassica* sp., *Chenopodium* sp., leafy and root vegetables	Sewage effluent (inadequately treated)	0.02-0.13 mg/kg	Rattan et al 2005

Country where investigated	Sources of heavy metal contaminants affecting food chains	Lead concentrations recorded in food crops (dry weight)	References
India Spinach, Ladies finger, brinjal, Coriander leaves, Amaranthus	Sewage wastewater (inadequately treated)	Amaranthus: 2.9 g/kg; Brinjal: 3 mg/kg; Coriander leaves: 2.7 mg/kg; Spinach: 3.1 mg/kg; Ladies finger: 3.6 mg/kg	Chary et al 2008

3.4.5.2. Lead contamination of freshwater and marine fish

Pb can adversely affect aquatic organisms including algae, invertebrates, and fish. High Pb levels can cause damage to gill epithelium and affected fish may die from suffocation. It may further cause a wide range of effects in fish including muscular and neurological degeneration and destruction, growth inhibition, mortality, reproductive problems. The reported LC_{50} for fish, small-mouth bass is 2.2 mg/L and EC_{50} for algae is 7.94 mg/L (Table 3.7). A compilation of global data reveals that several marine fish (Atlantic mackerel, canned sardines, common Rudd, European anchovy, salmon, Pacific cod, Red nose labeo) consumed by humans have exceeded the respective maximum allowable limits of lead (Table 3.13).

Table 3.13. Lead concentrations in commonly consumed marine fish in some selected countries

[where it exceeded the respective maximum allowable limits; Max Allowable Limit of Pb for seafood are: fish muscle meat: 0.30 mg/kg; crustaceans: 0.50 mg/kg; cephalopods: 1.0 mg/kg; and molluscs: 1.5 mg/kg (EC 2006)].

Country/ region	Fish species	Range (mean) mg/kg	References
Brazil Different brands of sardines purchased from local markets	*Sardinella brasiliensis* Canned sardines	077-2.15 (1.87)	Tarley et al 2001
Bulgaria Biosphere Reserve "Srebarna"	*Scardinius erythrophthalmus* Common rudd	0.76-10.42 (4.31) Muscle	Hristov and Kirin 2014
Italy Sicily, Mediterranean Sea	*Engraulis encrasicolus* European anchovy	(0.32)	Copat et al 2012

Country/ region	Fish species	Range (mean) mg/kg	References
Lithuania	*Salmo salar* Salmon	(0.413) muscle (0.237) gills	Idzelis et al 2012
Nigeria Fish samples purchased from markets in Ogbomoso, Oyo State	*Scomber scombrus* Atlantic mackerel	(0.46)	Ogundiran et al 2014
Nigeria Fish samples purchased from markets in Ogbomoso, Oyo State	*Gadus macrocephalus* Pacific cod	(0.202)	Ogundiran et al 2014
South Africa Polluted Olifants river	*Labeo rosae* Red-nose labeo	(0.8) muscle	Jooste et al 2014

3.4.5.3. Effects of lead on human health

Cultivation of crops and vegetables for human or animal consumption on Pb contaminated soil/water can potentially lead to uptake and accumulation of Pb in edible plant parts with a resulting risk to human and animal health. Pb is considered as a potential carcinogen and can damage cardiovascular, kidney, nervous system, circulatory systems in humans (Kibria 2014).

3.4.6. Impacts of mercury

Hg ranked No. 3 in ATSDR's Priority List of Hazardous Substances (see Table 1.6. in Chapter 1 for ATSDR substances priority list). Hg exists in various forms: elemental (or metallic) and inorganic (to which people may be exposed through their occupation), and organic (e.g., methyl-mercury (MeHg), to which people may be exposed through their diets such as seafood and rice). Once in the environment, Hg can be transformed by bacteria into MeHg. MeHg then bioaccumulates in fish and sharks. Methyl-mercury do also biomagnifies, for example, large predatory fish (marlin, tuna, swordfish) and sharks (black-tip shark, great white shark, little sleeper shark, tiger shark, short-fin mako shark, smooth hammerhead shark) are more likely to have high levels of Hg as a result of eating many smaller fish that have acquired mercury through ingestion of plankton (Table 3.14) (*https://www.who.int/news-room/fact-sheets/detail/mercury-and-health*; Kibria et al 2010a). MeHg is a possible carcinogen to humans (IARC group 2B) (Table 3.8). Potential environmental sources of Hg are cement kiln dust, coal combustion, e-waste processing sites, fertilizers, manometers, mining effluents and pulp and paper production, sewage sludge, ship-breaking yards (Table 3.3). The chemistry, toxicity and ecotoxicity of Hg can be found in section 3.3 (Table 3.7). The following section discusses the contamination of rice and seafood by Hg and the effects of Hg on human health.

3.4.6.1. Mercury contamination of rice

Rice consumption can be the main pathway of MeHg exposure to humans in Hg mining areas, recent research reveals (carried out by Meng et al 2011) that MeHg in rice grown in an abandoned Hg mining area found to contain elevated levels of MeHg in the edible portion of rice than that of other locally grown crop plant. The research also found that newly deposited Hg is more readily transformed into organic MeHg and accumulated in rice plants. MeHg in soil was first absorbed by roots and then translocated to the above-ground parts (leaf and stalk). During the full rice-growing season only a very small amount of MeHg was retained in the root section and most of this MeHg was transferred to seed during the ripening period (Fig. 3.5). Therefore, rice consumption can an important pathway of MeHg exposure to humans in mining areas.

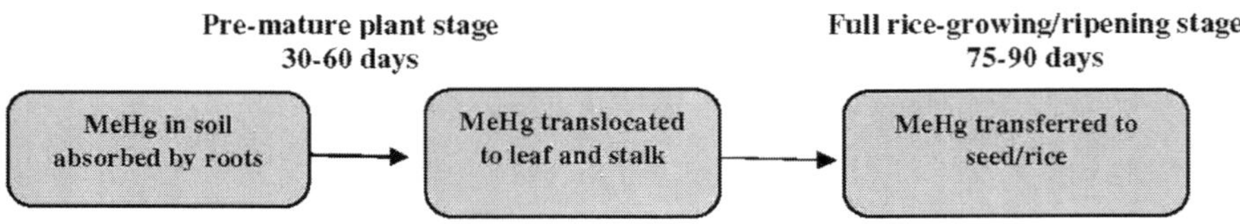

Fig. 3.5. Transfer of methyl-mercury (MeHg) in the soil to rice seed
(Adapted based on Meng et al 2011).

3.4.6.2. Mercury contamination of seafood (sharks, fish, cephalopods, prawns/ shrimps)

Inorganic Hg has low toxicity to aquatic organisms whereas organic Hg complexes (e.g. MeHg) are more toxic and tend to bioaccumulate. Organo Hg compounds are 4 to 31 times more toxic than inorganic Hg (ANZECC 2000b). Bioconcentration factors for MeHg for fish are very high compared to other metals (Table 3.6). Organo-mercurials are readily absorbed into fish, therefore organic Hg in the form of methyl- and ethyl-Hg makes its way up the food chain through fish (Peralta-Videa et al 2009). Hg compounds can damage vital tissues and organs including gill, liver, kidney, brain and skin in fish and may reduce reproductive success (ANZECC 2000b). The primary effect of Hg on fish populations is most likely to be reduced reproductive success resulting from maternally derived Hg to embryonic and larval stages. It may also cause reductions in gonad size, circulating reproductive steroids and gamete production (ANZECC 2000b, Kibria 2014). MeHg has a half-life[12] of approximately 2 years

[12] ***Half-life*** *is the time required for a quantity to reduce to half of its initial value*

in fish, which allows larger and older fish (predatory species) to contain higher levels of MeHg. Therefore, the consumption of fish contaminated with MeHg is the main source of exposure of humans. The notable **'Minamata disease'** is a classic example due to intake of bioaccumulated and biomagnified Hg from marine fish.

A review of global data shows that Hg have bioaccumulated in **cephalopods** (octopus, squid), **fish** (albacore, anchovy, carps, croaker, dolphin fish, eel, freshwater catfish, hake, hilsa, louvar, marlin, mullet, pomfret, rosefish, swordfish, trevally, tuna), **prawns/ shrimps**, rays, dogfish, chimaera and several **sharks** (bigeye thresher shark, black-mouth cat shark, black-tip shark, blue shark, broad-nosed seven gill shark, blunt-nose six gill shark, bonnet head shark, bull shark, carcharhinid shark, common thresher shark, copper shark, gatuzo shark, Greenland shark, great white shark, gulper shark, kite-fin shark/ seal shark, lemon shark, little sleeper shark, night shark, New Zealand lantern-shark, Pacific sleeper shark, pintarrojo shark, plunket's shark, sandbar shark, short-fin mako shark, silky shark, silver trip shark, small-spotted cat shark and smooth hammerhead shark) (Table 3.14). The analysis of global data on Hg bioaccumulation in sharks and fish and other aquatic species reveals the following facts:

- Hg levels in several sharks and fish species have exceeded food standard guideline values of 1.0 μg/g in seafood (Table 3.14).
- Hg accumulation in sharks was highest in muscle and liver (Table 3.14) (shark's meat, soup is used as human food, in industry (skin, shark liver oil) and as medicinal purpose (vitamin and cancer cure).
- Very high levels of Hg (> 2 μg/g) were found in several **sharks species** (black-tip shark, black-mouth cat shark, blunt-nose six gill shark, bull shark, common thresher shark, gulper shark, little sleeper shark, New Zealand lantern-shark, pintarrojo shark, short-fin mako shark, smooth hammerhead shark, spotted shark, tiger shark, gatuzo shark, great white shark, velvet lantern/ velvet belly shark) and **fish species** (albacore tuna, blue marlin) (Table 3.14).
- Hg levels were positively correlated with length and weight of sharks/ fish.
- Sharks those feed at a high trophic level (piscivorous sharks) had higher Hg levels.
- Hg concentrations were comparably much higher in sharks than in teleost fishes.
- Organic Hg (MeHg) accounted for most of Hg accumulated in sharks' muscle.
- In the Bay of Bengal region, the Hg levels in seafood organisms (sharks, fish, prawns/ shrimps, squid) were much lower (below < 1.0 μg/g) (Table 3.14).

- In Belgium, Brazil, Israel, Italy, Japan, the Mediterranean Sea, Portugal, Spain, the USA, the Hg levels in sharks and fish were generally above >1.0 µg/g (Table 3.14).

Table 3.14. Bioaccumulation of inorganic pollutants (Hg in µg/g wet weight) in muscle or liver in sharks, fish, cephalopods and other aquatic species reported from the world

****star** with a bold indicates where guideline values of 1.0 µg/g exceeded*
*[**MRL** = maximum residue limits; **Organic mercury** = methyl-mercury or MeHg;*
***THg** = total mercury; **ww** = wet weight;*
Hg levels with respect to various guidance values: **Australia and New Zealand:** 1.0 mg/kg (1.0 µg/g) for the barramundi, billfish, blue-fin tuna, gemfish, halibut, marlin, mullet, pike, orange roughy, rays, all species of shark, sword fish; 0.5 mg/kg (0.5 µg/g) for other fish, crustaceans and molluscs (ANZFSC 2016); **European Commission:** 1 ppm (European Commission 2008); **UK**: 0.5 ppm in edible fish, with up to 1 ppm allowed for certain predatory fish species ;**USA:** USFDA: 1.0 ppm (1.0 µg/g) (USFDA 2001); **WHO:** 0.5 ppm total Hg (0.5 µg/g) (Marrugo-Negrete et al 2008, Burger and Gochfeld 2011].

Country, location and reference	**Species and common name**	**Hg (total) concentrations** Range (mean) (µ g/g, wet wt); M = muscle; L = liver	**Remarks**
Adriatic and Ionian Seas (Marcotrigiano and Storelli 2003)	*Squalus acanthias* Spiny dogfish	[13]***6.53±2.19**	Exceeded food standard guideline values of 1.0 µg/g in seafood.
Argentina-Bahia Blanca Estuary (Marcovecchio et al 1991)	*Mustelus schmitti* Gatuzo shark	***M: 0.03-3.26 (0.77)** ***L: 0.00-2.31 (0.79)**	Hg concentrations were similar in both muscle and liver in all the shark's species
	Halaelurus bivis Pintarrojo shark	***M: 0.89-3.04 (2.20)** ***L: 0.78-2.99 (2.26)**	Higher Hg concentrations found in *H. bivius* and *Notorhinchus* sp. This could be related to feeding habits/diets (both species fed mainly on fish; fishes are considered as a great source of Hg in aquatic environments)
	Notorhinchus sp Spotted shark	***M: 0.95-3.43 (2.99)** ***L: 0.85-2.85 (2.11)**	The metal concentrations increased proportionally to the total length of the sharks.

13***star** *with a bold (in column 3) indicates where guideline values of 1.0 µg/g exceeded*

Country, location and reference	Species and common name	Hg (total) concentrations Range (mean) (µ g/g, wet wt); M = muscle; L = liver	Remarks
Australia -South East Australia (Pethybridge et al 2010)	*Centroselachus crepidater* Long-nose velvet dogfish	***M: 1.05-1.60** L: 0.24-0.46	Hg accumulation in all species followed the following trends: muscle (59-82%) > liver (1-25%) > kidney (1-5%) > skin (0-0.5%). More than half the total Hg burden was found in the muscle tissue. MeHg accounts for the majority (> 93%) of total Hg in shark muscle tissue (in particular shark's species occupying deeper environments).
	Chimaera lignaria Carpenter's chimaera	***M: 0.86-1.15** ***L: 0.68-1.49**	
	Centrophorus zeehaani Southern dog fish	***M: 2.47** L: 0.18	
	Dalatias licha Kite-fin shark/ seal shark	***M:** 1.06 L: 0.49	
	Etmopterus baxteri New Zealand lantern-shark	***M: 1.55-2.86** ***L: 0.47-1.3**	
	Proscymnodon plunketi Plunket's shark	***M: 2.69-4.50** ***L: 0.47-1.3**	
	Rhinochimaera pacifica Pacific spookfish	M: 0.4 L: 0.5	
	Squalus acanthias Spurdog/picked dogfish	***M: 0.64-1.45** L: 0.61-0.83	
	Squalus megalops (DD) Short-nose spurdog	***M: 0.75-0.79** L: 0.38-0.79	

Country, location and reference	Species and common name	Hg (total) concentrations Range (mean) (μ g/g, wet wt); M = muscle; L = liver	Remarks
	Squalus mitsukurii Short-spine spurdog	***M: 2.83-3.23** L: NA	
Australia-Tasmania (Davenport 1995)	*Prionace glauca* Blue shark	***M: 0.27-1.20** L: NA	-
Bangladesh (fish from rivers, lakes and estuaries) (Holsbeek et al 1997)	*Hilsa ilisha* River shad	M: 0.002-0.060 (0.019) (MeHg 31%)	Based on this study of 1997, Hg contamination of Bangladesh fishes from rivers, lakes and estuaries (ranged 0.002-0.427 μg/g, wet wt) appears to be low compared to other countries. Low Hg levels in Bangladesh fish could be due to the absence of mining or the sweeping/ washing/ flooding of land by annual monsoon.
	Catla Gangetic carp	M: 0.010-0.070 (0.029) MeHg (75%)	
	Labeo rohita Gangetic carp	M: 0.028-0.070 (0.042) MeHg (69%)	
	Notopterus Bronze featherback	M: 0.033-0.154 (0.064) MeHg (79%)	
	Puntinus sarana Olive barb	M: 0.009-0.050 (0.021) MeHg (73%)	
	Heteropneustes fossilis Asian stinging catfish	M: 0.018-0.083 (0.034) MeHg (79%)	
	Mystus seenghala Giant river catfish	M: 0.029-0.427 (0.104) MeHg (87%)	
	Wallago attu Freshwater catfish	M: 0.060-0.302 (0.145) MeHg (85%)	

Country, location and reference	Species and common name	Hg (total) concentrations Range (mean) (μ g/g, wet wt); M = muscle; L = liver	Remarks
Bay of Bengal (covering deep-sea areas of Andaman, Bangladesh, India, Sri Lanka) (Sompongchaiyakul et al 2008)	*Alopias superciliosus* Bigeye thresher shark	M: 0.198-0862 (0.514)	Hg levels were found to be positively correlated with the length and weight of the fish.
	Carcharhinus brachyurus Copper shark	M: 0.108-0.419 (0.251)	Tille trevally and sword fish had high Hg levels. The Hg levels in seafood organisms (sharks, fish,) were much lower (below < 1.0 μg/g).
	Carcharhinus falciformis Silky shark	M: 0.074-0.158 (0.122)	
	Caranx tille Tille trevally	M: 0.782-0.990 (0.886)	
	Katsuwonus pelamis Skipjack tuna	M: 0.005-0.625 (0.110)	
	Xiphias gladius Sword fish	***M:0.023-1.245** (0.478)	
	Thunnus albacares Yellow-fin tuna	M: 0.061-0.124 (0.092)	
Bay of Bengal (eastern part) (Panutrakul et al 2008)	*Sthenoteuthis oualaniensis* Oceanic purple back squid	(0.040) (edible part) (0.034) (visceral mass)	The concentration of Hg in both edible tissue (mantle, arm and tentacle) and visceral mass (ink sac, digestive gland, gills, and gonads) were lower than Australia and New Zealand safety standards of 1.0 μg/g, wet wt (ANZFSC 2016).
Bay of Bengal (Coastal area of Bangladesh) (Hossain	*Penaeus indicus* White shrimp	0.027-0.124 (0.063) (soft body parts)	Based on this study of 1994, Hg contamination of Bangladesh shrimp and

Country, location and reference	Species and common name	Hg (total) concentrations Range (mean) (μ g/g, wet wt); M = muscle; L = liver	Remarks
1994)	*Metapenaeus monoceros* Brown shrimp	0.023-0.113 (0.061) (soft body parts)	fishes (ranged 0.07-0.142 μg/g, wet wt) appears to be low compared to other countries. Hg contamination to seafood (shrimp and fishes) could be due to effluents discharged from the Karnafuli paper and Rayon Mills and Charcoal plant in Chattogram/Chittagong.
	Penaeus monodon Giant tiger shrimp	0.012-0.107 (0.044) (soft body parts)	
	Tenualosa ilisha River shad	0.007-0.046 (0.029) (muscle)	
	Coilia dussumieri Gold-spotted grenadier anchovy	0.018-0.050 (0.032) (soft body parts)	
	Pampus chinensis Silver pomfret	0.025-0.105 (0.057) (soft body parts)	
	Johnius belangeri Belanger's croaker	0.012-0.142 (0.049) (muscle+gill)	
Belgium-Celtic Sea (Domi et al 2005)	*Galeorhinus galeus* Tope shark	***M: 0.5-1.9 (1.1)**	The highest levels of muscle Hg was detected in those sharks (starry smooth hounds, black-mouthed cat sharks and lesser-spotted dogfish) which feed at a high trophic level (fish). Sharks generally absorb the pollutants through the diet (fish) and fish are the most critical source of Hg for marine predators like sharks.
	Galeus melastomus Black-mouth catshark	***M: 1-4.4 (2.1)**	
	Mustelus asterias Starry smooth-hound	***M: 1.7-3.1 (2.2)**	

Country, location and reference	**Species and common name**	**Hg (total) concentrations** Range (mean) (μ g/g, wet wt); M = muscle; L = liver	**Remarks**
	Squalus acanthias Spiny dogfish/ picked dogfish	M: 0.2-0.3 (0.2)	
	Scyliorhinus canicula Lesser-spotted dogfish	***M: 0.8-3.1 (1.9)**	
Brazil-Offshore waters (Penedo de Pinho et al 2002)	*Carcharhinus signatus* Night shark	***M: (1.77)**	The study found that feeding habits influenced total Hg level in sharks. The highest Hg concentrations were detected in piscivorous shark species (*C. signatus, S. megalops, S. mitsukurii*) feeding on higher trophic levels (fish), while species that feed mainly on invertebrates (*M. canis* and *M. norrisi*) had lower Hg concentrations. Piscivorous, *S. mitsukurii* had also higher values for methyl-mercury concentration compared to *M. canis,* which feeds mainly on small invertebrates.
	Mustelus canis Dusky smooth-hound	M: (0.41)	
	Mustelus norrisi Narrow-fin smooth-hound	M: (0.36)	
	Squalus megalops Short-nose spurdog	***M: (1.90)**	
	Squalus mitsukurii Short-spine spurdog	***M: (2.22)**	
Italy-Adriatic Sea (Storelli et al 2001)	*Prionace glauca* Blue shark	M: (0.31)	-
Italy- Mediterranean Sea (Storelli and Barone 2013)	*Lophius budegassa* Black-bellied angler	0.96±0.32	

Country, location and reference	Species and common name	Hg (total) concentrations Range (mean) (µ g/g, wet wt); M = muscle; L = liver	Remarks
	Raja miraletus Brown ray	***1.09±0.39**	
	Conger European conger eel	***1.14±0.46**	
	Mullus barbatus Red mullet	0.43±0.55	
	Helicolenus dactylopterus Rosefish	***1.04±0.56**	
Italy-Ionian Sea (Storelli et al 2003)	*Sphyrna zygaena* Smooth hammerhead shark	*** M: 8.55-21.07 (12.15)** ***L: 32.31-39.48 (35.89)** MeHg: M: 7.45 (14.00) L: 19.10–26.30 (23.77)	Diet (benthic fish) could have caused the high levels of Hg concentrations in Hammerhead shark. High Hg concentrations in liver compared to muscle may be related to the role played by the liver in terms of pollutant biotransformation.
Italy- Adriatic Sea, Italy (Perugini et al 2014)	*Merluccius* European hake	0.59±0.14	-
Israel, Mediterranean Sea (Hornung et al 1993)	*Centrophorous granulosus* Gulper shark	***M: 0.48-8.37** ***L: 0.60-10.5**	Hg accumulation in all species followed the following trends: muscle > liver > kidney > intestines > gonads The level of Hg in sharks was considerably higher which could be related to higher levels of Hg in the water of the region. The mercury content in the
	Etmopterus spinax Velvet lantern/velvet belly shark	***M: 1.83-4.58** ***L: 1.97-2.97**	

Country, location and reference	Species and common name	Hg (total) concentrations Range (mean) (μ g/g, wet wt); M = muscle; L = liver	Remarks
	Galeus melastomus Black-mouth cat shark	***M: 0.99-8.76** ***L: 0.28-5.90**	muscle of *C. granulosus* increased with increasing body weight. increasing body weight until
	Hexanchus griseus Blunt-nose six gill shark	***M: 0.41-4.55** ***L: 0.42-13.3**	
	Somniosus rostratus Little sleeper shark	***M: 0.86-5.60** ***L: 1.14-22.81**.	
Japan-Ishigaki Island (Endo et al 2008)	*Galeocerdo cuvier* Tiger shark	*M: 0.38-1.34 (0.78) ***L: 0.11-20.1 (1.17)**	The highest concentrations of Hg in liver and muscle were found in the bull shark which is a well-known predator. Continuous intake of Hg via food and the slower growth rate of mature sharks may cause higher Hg accumulation.
	Carcharhinus albimarginatus Silver trip shark	***M: 1.19-2.35 (1.80)** ***L: 0.26-1.41 (0.70)**	
	Carcharodon leucas Bull shark	***M: (3.65)** ***L: (28.1)**	
	Carcharhinus plumbetus Sandbar shark	***M: (1.66)** ***L: (3.62)**	
Mediterranean Sea (Storelli and Marcotrigiano 2004)	*Thunnus alalunga* Albacore tuna	**M: *0.88–2.34 (1.56)** ***L: 0.95–4.30 (2.41)**	Hg concentrations exceeded the prescribed legal limit of 1 μg/g wet weight. Higher accumulation of metals in the liver can be related to its physiological functions (liver is an organ
	Eledone moschata Horned octopus	M: 0.22–0.47 (0.36) L: 0.44–2.20 (0.76)	

Country, location and reference	**Species and common name**	**Hg (total) concentrations** Range (mean) (μ g/g, wet wt); M = muscle; L = liver	**Remarks**
			in which metals primarily tend to concentrate).
Mexico-Baja California Sur (Escobar-Sánchez et al 2010)	*Sphyrna zygaena* Smooth hammerhead shark	***M: 0.005-1.93 (0.73)** Fin: 0.005–0.009 (0.007)	The main preys of the smooth hammerhead sharks are cephalopods (squids, octopus), those could be the major route of Hg bioaccumulation in sharks (cephalopods have special capabilities to accumulate HMs at various levels in their tissues).
New England: Northwest Atlantic Ocean (Teffer et al 2014)	*Thunnus alalonga* Albacore tuna	M: 0.294-0.683 (0.46)	Hg concentrations were highest in mako and thresher sharks and significantly lower in teleosts (albacore, yellow-fin tuna, dolphin fish). A linear relationship was found between body size and Hg concentration.
	Alopias vulpinus Common thresher shark	***M: 0.206-3.209 (0.88)**	
	Isurus oxyrinchus Short-fin mako shark	***M: 0.755-4.933 (2.65)**	
	Thunnus albacares Yellow-fin tuna	M: 0.044-0.503 (0.32)	
	Coryphaena hippurus Dolphin fish	M: 0.021-0.648 (0.20)	
Portugal-Atlantic Ocean (Coelho et al 2010)	*Scylioprhinus canicula* Small-spotted cat shark	M: 0.13-0.8; *Organic Hg*: 0.05-0.52	Bioaccumulation: muscle > heart > liver > gills > pancreas
Portugal-Azore and Canary Islands (Branco et al 2004)	*Prionace glauca* Blue shark	***M:** *Total Hg:* 0**.16-1.84** M: *Organic Hg:*0.15-1.23	Organic Hg accounted for most Hg accumulated in sharks' muscle.
Spain-Fish retailers from Madrid	*Luvarus imperialis*	***0.40-1.50 (0.99)**	Around 90% of louvar samples and around 35%

Country, location and reference	Species and common name	Hg (total) concentrations Range (mean) (μ g/g, wet wt); M = muscle; L = liver	Remarks
(Herreros et al 2008)	Louvar (predatory fish)		of sword-fish portions exceeded MRL for Hg (MRL for luvar: 0.5 and MRL for sword fish 1 μg/g).
	Xiphias gladius Swordfish (predatory fish)	***0.20-1.90 (0.93)**	
USA-Florida (Florida, Cape Canaveral) (Nam et al 2010)	*Negaprion brevirostris* Lemon shark	M: 0.18–0.81 (0.311) *(Organic mercury 0.294 or 95% of THg)* L: 0.049–0.396 (0.155) *Organic mercury 0.014 or 9% THg)*	Organic Hg accounts for nearly all of the THg in muscle (> 95%).
USA-California Coast (Suk et al 2009)	*Isurus oxyrinchus* Short-fin mako	***M: 0.15-2.90 (1.13)**	Found strong linear relationships between body size and Hg level.
	Alopias vulpinus Common Thresher shark	M: 0.00-0.70 (0.13)	All of the common thresher's muscle had total Hg levels below the US Food and Drug Administ-ration's established action level of 1.00 μg/g for commercial fish.
	Alopias superciliosus Big-eye thresher shark	M: 0.46-0.47 (0.46)	
	Notorynchus cepedianus Broad-nose seven gill shark	M: 0.48	
	Isurus oxyrinchus Short-fin mako	***M: 0.40-3.10 (1.32)**	

Country, location and reference	Species and common name	Hg (total) concentrations Range (mean) (μ g/g, wet wt); M = muscle; L = liver	Remarks
USA-Hawaii-Pacific Ocean (Boush and Thielke 1983)	*Galeocerdo cuvieri* Tiger shark	***M: 2.09-6.14 (4.02)**	-
	Carcharhinus limbatus Black-tip shark	***M: 7.29-28.22 (12.93)**	
USA-Gulf of Mexico (Cai et al 2007)	*Carcharhinus* spp. Carcharhinid shark	***M:0.46–4.08 (1.61)**	Hg bioaccumulation identified in three species was of high range (> 1.0-10.52 μg/g wet weight). Significant positive relationships were found between Hg concentration and size (length and/ or weight) of all species examined. Hg concentration was positively associated with trophic position.
	Makaira nigricans Blue marlin	***THg:4.95–18.72 (10.52)**	
	Euthynnus alletteratus Little tunny	***THg: 0.24–2.52 (1.08)**	
USA-Atlantic Coast of Florida (Adams and McMichael 1999)	*Carcharhnius leucas* Blue shark	***M: 0.24-1.70** (0.77)	Hg levels were directly related to shark size. Through bioaccumulation, larger, older adults of shark species potentially contained excessive Hg burdens.
	Sphyrna tiburo Bonnet head shark	***M: 0.13-1.50** (0.50)	
USA-Cumberland and Prince William sound (McMeans et al 2007)	*Somniosus microcephalus* Greenland shark	L: (0.492)	-
	Somniosus pacifica Pacific sleeper shark	L: (0.122)	

Country, location and reference	**Species and common name**	**Hg (total) concentrations** Range (mean) (μ g/g, wet wt); M = muscle; L = liver	**Remarks**
USA-California Bight (Mull et al 2012)	*Carchardon carcharias* Great white shark	***M: 0.41-10.3 (2.78)** L: 0.016-0.708 (0.206)	-
USA-New Jersey Coast (Burger and Gochfeld 2011)	*Isurus oxyrinchus* Short-fin mako	***M: (1.83)**	Hg levels showed significant positive correlation.
	Thunnus thynnus Atlantic blue-fin tuna	0.52±0.03	
World (Neff 2002)	Sharks (world average)	M: 0.035-52.5	

3.4.6.3. Effects of mercury on human health

Elevated levels of Hg can damage the human nervous system, brain and kidney (WHO 1996). Most human exposure to Hg occurs from eating fish and shellfish contaminated with MeHg. The documented cases of Hg poisoning in Japan ("Minamata disease") resulted in many deaths (caused by consumption of large quantities of MeHg contaminated fish and shell-fish (5.61-35.7 ppm[14]) from Minamata Bay (Harada 1995). Additionally, the surviving victims and their offspring displayed mental retardation, cerebral palsy, muteness, which was observed in children exposed while in the foetal stage (WHO 1996, Kibria et al 2010a).

A recent research carried out found elevated Hg levels in child-bearing women in 25 countries (all of them in developing countries). In those 25 countries, women were exposed to Hg via **i.** seafood contamination, **ii.** direct exposure to gold dust and **iii.** localized industrial pollution (Box 3.3, Fig. 3.6). Hg level > 1 ppm can cause brain damage, IQ loss, kidney and heart damage, whereas Hg level > 0.58 ppm can cause foetal neurological damage.

[14]***ppm = parts per million*** *or* ***milligrams per liter (mg/L)***

Box 3.3. Elevated levels of mercury in child–bearing women in 25 Countries

A Global research study carried out by IPEN (www.ipen.org) and Biodiversity Research Institute (www.briloon.org) found elevated Hg levels in child-bearing women in 25 countries (Bell et al 2017, Fig. 3.6). The study found three factors resulted in elevated levels of Hg in mothers including a **fish-rich diet**, the practice of **artisanal and small-scale gold mining** (**ASGM**) and **proximity to industrial locations**. The Hg level has exceeded the threshold level of 1 ppm (parts per million) of women in Cook Islands, Indonesia, Kenya, Kiribati, Marshall Islands, Myanmar, Nepal, Nigeria, Solomon Islands, Thailand, Tonga, and Tuvalu. Women from Alaska, Albania, Chile, Kazakhstan, Ukraine, and Vanuatu exceeded the 0.58 ppm Hg level.

The research findings are summarised below:

- The women of Indonesia, Kenya, and Myanmar show significantly elevated Hg levels. These women were exposed during **ASGM** activities (direct exposure to gold dust during panning, sluicing and ball milling) and from consuming local fish, rice contaminated with Hg from ASGM.
- Women from Small Island Developing States (SIDS) in the Pacific (the Cook Islands, Kiribati, Marshall Islands, Solomon Islands, Tonga, and Tuvalu) were found to have very high levels of Hg body burden. These women have a diet rich in seafood (**large predatory fish**). Distant air emissions of Hg from coal-fired power plants, cement kilns and other industries contaminate ocean fish that serve as a primary protein source for Pacific Islanders.
- Women from locations in Albania, Chile, Nepal, Nigeria, Kazakhstan, and Ukraine have Hg levels of concern due to localized **industrial pollution** (Chlor-alkali, cement kiln, waste incineration, pulp and paper plants) of waterways and suspected fish contaminations.

3.4.7. Impacts of zinc

Zinc is an essential mineral. Zn is naturally present in all soils in typical background concentrations 10–100 mg/kg. Zn ranked No. 75th in ATSDR's Priority List of Hazardous Substances (see Table 1.6. in Chapter 1 for ATSDR substances priority list). Potential environmental sources of Zn are coal combustion, e-wastes, mine effluents, paper and pulp mills, sewage sludge, smelting, stormwater, textile effluents and waste incineration (Table 3.3). The chemistry, toxicity and ecotoxicity of Zn can be found in section 3.3 (Table 3.7). The following section discusses the contamination of food and fish by Zn and effects of Zn on human health.

3.4.7.1. Zinc contamination of food

Fruits and vegetables in several countries including Australia, China, India and Egypt are often contaminated with Zn (Table 3.15). Using inadequately treated sewage water/wastewater in growing crops and vegetables caused most food contamination by Zn.

Table 3.15. Heavy metal (zinc) contamination from diverse sources in global food crops

[Recommended maximum levels of Zn for vegetables: 9.4 mg/kg (Chary et al 2008)].

Country where investigated	Sources of heavy metal contaminants affecting food chains	Zinc concentrations recorded in food crops (dry weight)	References
Australia French beans, beetroot, kale	Urban stormwater	0.038–0.145 mg/kg	Tom et al 2014
Australia Rice and other paddy crops and vegetables (imported from Bangladesh, India, Pakistan, Thailand, Italy, Canada and Egypt)	Arsenic- and metal-contaminated groundwater	Rice: 10.9–24.5 mg/kg Vegetables: 17–183 mg/kg	Rai et al 2019
China Maize, green cabbage, radish, turnip, spinach, cauliflower, and lettuce	Sewage effluent (inadequately treated)	30-70 mg/kg	Khan et al 2008
China *Brassica* sp., food grains, and leafy vegetables	Both sewage and industrial waste (from smelter) drained into river water used for irrigation	32.01–69.26 mg/kg	Liu et al 2005
Egypt Various fruits and vegetables	Inadequately treated wastewater	Fruits: 1.36-10.5 mg/kg Vegetables: 7.16-20.9 mg/kg	Radwan and Salama 2006
India *Brassica* sp., *Chenopodium* sp., leafy and root vegetables	Sewage effluent (inadequately treated)	Fruits: 49.7-78.8 mg/kg Vegetables: 46.7-91.9 mg/kg	Rattan et al 2005

Country where investigated	Sources of heavy metal contaminants affecting food chains	Zinc concentrations recorded in food crops (dry weight)	References
India Spinach, Ladies finger, brinjal, Coriander leaves, Amaranthus	Sewage wastewater (inadequately treated)	Amaranthus: 8 mg/kg Brinjal: 4.5 mg/kg Coriander leaves: 5.4 mg/kg Spinach: 10 mg/kg Ladies finger: 3.7 mg/kg	Chary et al 2008

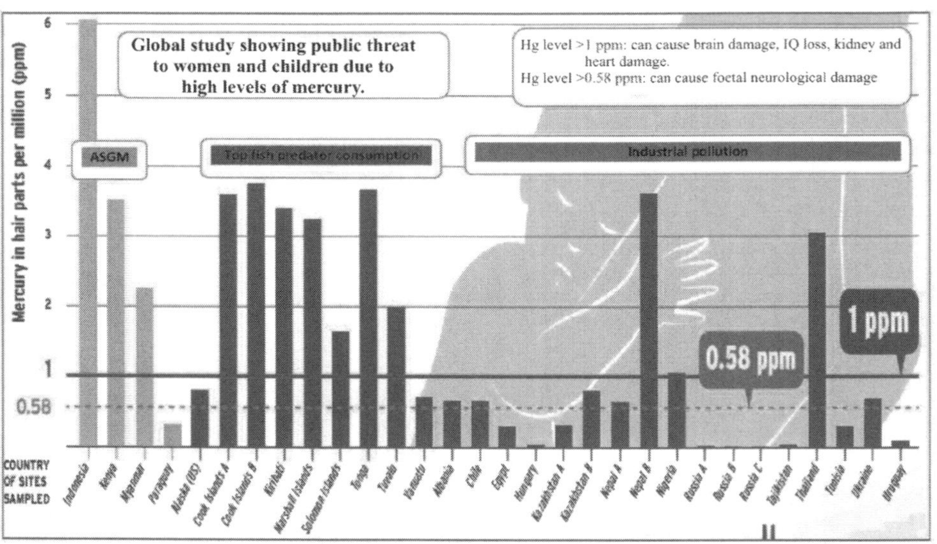

Fig. 3.6. Elevated mercury levels in women in 25 countries (Bell et al 2017); *https://ipen.org/sites/default/files/documents/FINAL%20hg-infographic-Sept-2017_EN.pdf*

3.4.7.2. Zinc contamination of freshwater and marine fish

Growth, survival, and reproduction of aquatic plants and animals can be adversely affected by elevated Zn levels. Zn can cause gill damage, reduce growth and kidney damage of aquatic organisms such as fish (ANZECC 2000b, Kibria et al 2010a).

3.4.7.3. Effects of zinc on human health

Though Zn is an essential element, however, excessive concentration in food and food plants are of great concern because of their toxicity to humans (stomach cramps, skin irritations, respiratory disorder.

3.4.8. Other heavy metals (Cr and Ni)

Chromium (CrVI) is known to be a human carcinogen (IARC group 1) (Table 3.8). CrVI (hexavalent chromium) ranked No. 17th in ATSDR's Priority List of Hazardous Substances (see Table 1.6. in Chapter 1 for ATSDR substances priority list). Potential environmental sources of Cr are coal combustion, electroplating, e-wastes, mining effluents, paints, textile effluents, tannery and wood preservatives (Table 3.3). The chemistry, toxicity and ecotoxicity of Cr can be found in section 3.3 (Table 3.7). Cr is very toxic to mammal and daphnia and moderately toxic to fish (Table 3.7). CrVI is easily absorbed across the gills and bio-concentrates up the food chain (Sanyal et al 2017, Bakshi and Panigrahi 2018). Cr toxicity impacts the immune system, renal and respiratory systems and can cause cancer (Chowdhury et al 2018).

Nickel (Ni) is known to be a human carcinogen (IARC group 1) (Table 3.8). Ni ranked No. 57th in ATSDR's Priority List of Hazardous Substances (see Table 1.6. in Chapter 1 for ATSDR substances priority list). Potential environmental sources of Cr are cement, coal, e-wastes, mine effluents, sewage sludge, smelting, ship-breaking yards, tannery, textile effluents (Table 3.3). Ni is very toxic to aquatic life (Table 3.7). Tilapia growth was significantly affected by Ni poisoning and accumulated principally in the viscera and the head (Vargas et al 2009). Soil usually contains between 4 and 80 parts of nickel in a million parts of soil (ppm; 1 ppm = 1,000 ppb). The highest soil concentrations (up to 9,000 ppm) (as dust) are found near industries that extract nickel from ore. The most serious harmful human health effects from exposure to Ni are chronic bronchitis, reduced lung function and cancer of the lung and nasal sinus (*https://www.atsdr.cdc.gov/phs/phs.asp?id=243&tid=44*).

3.4.9. Case studies

3.4.9. Case study 1: Hidden heavy metals in face makeup- risks to beautiful women who are frequent users of makeup?

Canada based Environmental Defense released a research report on the health risks associated with the toxic HMs found in makeup (Environmental Defense Canada 2011). The research reveals that cosmetics could be a source of exposure to many of the HMs to women. They (Environmental Defense Canada) tested 49 face makeup items, including 05 foundations; 04 concealers; 04 powders; 05 blushes or bronzer; 07 mascaras; 02 eyeliners; 14 eye shadows; 08 lipsticks or glosses. According to the research, women who frequently use makeup such as lipstick throughout each day can be consuming large amounts of Pb in a lifetime. This HM may be ingested and absorbed through the skin (such as broken skin). As (arsenic) a group, HMs can build up in the body over time. Many HMs are suspected as hormone disruptors and respiratory toxins. The research findings are highlighted below:

- Seven of the eight metals of concern (Table 3.16) were found in 49 different face makeup items, of which **arsenic (As), cadmium (Cd)** and **lead (Pb)** are known to be "toxic".

- Of the chemicals of most concern, As was detected in 20 per cent of the products, Cd was detected in 51 per cent, Pb was detected in 96 per cent of the items tested (Table 3.16).

Table 3.16. Percentage of products (face makeup) containing tested heavy metals of concern [Environmental Defense Canada 2011; *http://www.ottawacitizen.com /pdf/HeavyMetalHazard.pdf*]

Heavy metals tested for	% of items with detectable metals	Average level (µg/g)	Health threats
Arsenic, (As)	20	1.8	Carcinogenic, cause arsenicosis[15].
Beryllium (Be)	90	0.8	Impact on immune systems.
Cadmium (Cd)	51	0.3	Carcinogenic, lung damage, kidney, bone and lung disease.
Lead (Pb)	96	4.6	Possibly carcinogenic; damage to the brain and kidneys.
Mercury, (Hg)	0	0	Disruptions to nervous system, damage to kidney and brain functions.
Nickel, (Ni)	100	25.1	Carcinogenic, contact dermatitis (eczema).
Selenium (Se)	14	1.48	Hair loss, neurological damage.
Thallium (Tl)	61	0.2	Suspected human carcinogen, extremely toxic.

3.5. Linkages among heavy metals load, bioaccumulation and climate change

Climate change/global warming will facilitate increased volatilization of chemicals such as Hg, POPs, therefore, would increase their presence in the air and other environmental matrices (soil, water), in the future. Higher ocean temperatures are likely to increase the (re)emissions of volatile Hg species (AMAP 2016). Where intense rainfall and floods are expected to increase, as a

[15]***Arsenicosis*** = *Arsenic poisoning*; it *occurs due to elevated levels of arsenic in the body*

result, run-off of pollutants (heavy metals/metalloids, pesticides, nutrients, pharmaceuticals, endocrine-disrupting chemicals) in water bodies will increase (Kibria et al 2010a, Kibria 2014). The climate change (hotter and drier days, long heatwaves, and drought) is increasing the frequency and severity of wildfires across the globe (Kibria 2017, Ku et al 2018). During wild/ forest fires substantial amounts of Hg is released into the atmosphere (as gaseous Hg) as well as run-off (as ash Hg) (Fig. 3.7). The run-off of wildfire, ash would increase Hg load into aquatic environments, where microbes can convert the element into highly toxic MeHg (Ku et al 2018; *https://phys.org/news/2018-12-wildfire-ash-mercury.html*).

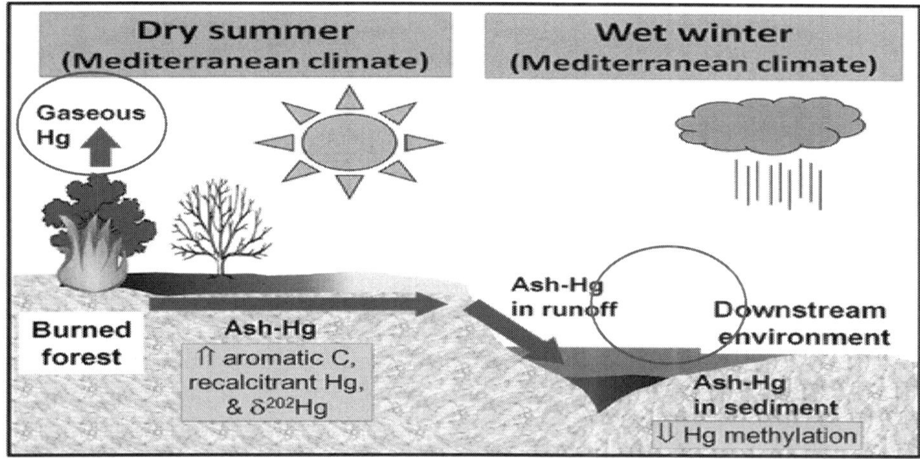

Fig. 3.7. Pathways of mercury entering an environment (gaseous Hg and ash Hg) after wild/ forest fire (Ku et al 2018; page 14149).

Temperatures in the Arctic Circle are increasing at a rate of about double compared with the other parts of the world. The rise in temperatures and consequent melting/thawing of snows and permafrost would release a very significant amount of Hg trapped in snows and permafrost in melt water into Arctic lakes and waterways (Durnford and Dastoor 2011, Obrist et al 2017) and bioaccumulate in wild-life (animals and fish; *https://news.nationalgeographic.com/2018/02/melting-arctic-permafrost-toxic-mercury-environment/*). Furthermore, it is expected that uptake and toxicity of metals in seafood organisms maybe enhanced with increasing temperatures/global warming (Kibria et al 2010a, Kibria et al 2013 and Kibria et al 2016a,b). Temperature related increases in the uptake, bioaccumulation and toxicity of metals (As, Cu, Co, Cd and Pb) have been reported for several marine organisms, including crustaceans, echinoderms and molluscs (Hutchins et al 1996; Khan et al 2006; Wang et al 2005; Mubiana and Blust 2007). At higher temperatures the metabolism of aquatic organisms is increased, and oxygen concentration is reduced in water and therefore, the rate of water in-flow into the animal can increase to extract more oxygen, which can increase the entrance of dissolved chemical pollutant(s) into the body (Marques et al 2010, Anwar 2013). Ocean warming facilitates methylation of mercury and subsequent uptake of MeHg in fish and mammals has been found to increase by 3–5% for each 10℃ rise in water temperature (Booth and Zeller 2005).

3.6. Management and remediation of heavy metals

The main purpose of HM remediation is to minimize the risk of these toxic compounds to human and ecosystem health. The selection of appropriate remediation method depends on the characteristics of the site, extent of the HM contamination, and regulatory limits for the HMs of concern in that regulatory domain. The following are some of the existing tools (physical, chemical, biological and ecological approaches) of remediation of HMs in soil-food crops sub-systems to mitigate human health risks (Chowdhury et al 2018, Rai et al 2019):

3.6.1 Physical methods

Physical remediation methods are very expensive and time-consuming. Commonly used physical remediation methods are as follows:

- **Soil replacement:** Remove the HM-contaminated soil partially or completely or replace contaminated soil with non-contaminated soil, often called the "dig-and-haul" method. It is only suitable if the contamination is of small-scale.
- **Soil washing:** Soil washing helps to minimize the level of HM contamination in severely contaminated soil. This process can be applied *in-situ* by forcing washing solution through the soil matrix, or *ex-situ* by physically excavating the soil and washing it in reactors. The main purpose of this method is to make the HMs soluble in the washing reagents and to separate them out from the soil. Excavate the soil and wash with chemicals in reactors; chemicals are surfactants (detergents, foaming agents, emulsions, and dispersants), chelating agents (binding agents), organic acids. This technique is more effective in sandy and granular soils than in clay-rich soil (Chowdhury et al 2018, Rai et al 2019) as clay creates problems with solid-liquid separation after washing (Maturi and Reddy 2008).
- **Soil isolation:** It is to isolate contaminated soil from uncontaminated soil to prevent off-site transport of HMs (Chowdhury et al 2018, Rai et al 2019).
- **Source reduction:** Avoiding inadequately treated effluent and sewage sludge could significantly reduce the HM accumulation in food crops (see Tables 3.11, 3.12, 3.15 for food contamination by sewage and industrial wastes and effluents). The use of industrial effluents and

sewage sludge as fertilizers are the primary HM contamination sources in the soil–crop systems in developed countries. In developing countries, irrigation with inadequately treated effluent or sludge is the main HM contaminating source for food crops (Rai et al 2019).

- **Vitrification (solidification):** This technology involves heating the soil at high temperature (1,400-2,000°C) to decompose or volatilize organic matter and to form vitreous materials such as oxide solids. This technology can be applied both in *ex-situ* or *in-situ* processes. In *ex-situ* vitrification processes, soils are excavated and subjected to pre-treatment (such as screening, dewatering and blending) and are fed to the reactors for melting. Collection of volatile metal species (such as Hg) is very important during this process. Cooling of melted soil produces vitreous materials, which are even stronger than concrete. For *in-situ* vitrification processes, electrodes are inserted into the contaminated area vertically to heat the soil. The soil needs to be a little wet to provide the necessary conductance. The resistive heating melts the soil and moves both outward and vertically. It is an expensive option and a large amount of energy investment is the main disadvantage of this process (Chowdhury et al 2018, Rai et al 2019).

3.6.2 Chemical methods

Chemical remediation methods involve the immobilization or extraction of HMs from the soil using different chemical agents.

- **Immobilization:** The immobilization process reduces the metal solubility and hence their mobility in the soil. Several immobilizing agents such as clay, organic composts, cement, minerals, phosphates, and zeolites (adsorbents and catalysts) can be used to reduce HM toxicity through precipitation, sorption, and complexation. Unconventional low-cost industrial residues such as red mud and termitaria can also be used to immobilize HMs in soil (Chowdhury et al 2018, Rai et al 2019).
- **Extraction:** This process involves the addition of chemical agents to bring the metals into the aqueous phase (by increasing the metal solubility) to extract/remove the metals from the system. Different types of inorganic effluents, chelating agents, and surfactants are commonly used in this method to bring the HMs into the aqueous phase (than extracting/removing the metals from the system). Synthetic chelating agents such as EDTA (ethylene diamine tetra acetic acid), N-hydroxy ethylene diamine tri acetic acid, nitrile triacetate, and natural chelating agents such as Ethylene diamine-N, N'-di-succinic acid (EDDS) have been successfully used to enhance soil Pb extraction (Chowdhury et al 2018). The role of potassium phosphate for As extraction from soil has also been studied. Chemically induced metal

extraction processes are often coupled with phytoremediation methods (see below) and are found to be more effective.

3.6.3. Biological/ ecological methods

Phytoremediation: Biological organisms such as hyper-accumulator plants[16] can be used to accumulate HMs into their shoots and leaves. A hyper-accumulator plant is capable of growing in water with very high concentrations of metals, absorbing these metals through their roots, and concentrating extremely high levels of metals in their tissues. Hyper-accumulator plants (Brassicaceae and Fabaceae family members) can accumulate various HMs (Table 3.17) at concentrations of 100 to 1,000-fold (higher than the plants belonging to the non-hyper-accumulator group) (Chowdhury et al 2018). Some plants such as the ferns, *Pteris vittata* and *Pityrogramma calomelanos* are capable of accumulating high concentrations of toxic As^{III} (Chang et al 2009). It is also known that naturally grown or cultured green, brown and red seaweeds (macro-algae) uptake HMs from the aquatic environment. Green seaweed (e.g. *Enteromorpha intestinalis*) has more ability to uptake HMs as compared to brown (*Hydroclaththrus clathratus, Padina tetrastromatica, Sargassum oligocystum*) and red seaweeds (Hoq and Islam 2019). Water hyacinth (*Eichhornia crassipes*) has been successfully used for wastewater treatment including removal of pollutants like heavy metals (Cr and Cu) (Lissy and Madhu 2010). In order to select hyper-accumulator plants, there are four basic “hall marks/criteria” according to Rascio and Navari-Izzo 2011

- plants should have a greater capability of taking up HMs from the soil.
- plants should have the capability of faster and effective root-to-shoot translocation of metals.
- plants should be able to sequester/sink huge amounts of HMs in the leaves (be tolerant to leaf tissues).
- plants should be able to detoxify/remove toxic substances.

Table 3.17. Plant species commonly used for phytoremediation of heavy metals (compiled by Chowdhury et al 2018).

Plant species	Metals of concern
Atriplex halimus	Cd, Zn
Brassica spp.	Cd, Cr, Cu, Ni, Pb, Zn

[16]***Hyperaccumulator plants*** - *A hyperaccumulator is a plant capable of growing in water with very high concentrations of metals.*

Plant species	Metals of concern
Cichorium intybus	Pb
Cynodon dactylon	Cd, Cu, Pb, Zn
Cyperus alternifolius	Cd, Cu, Mn, Pb, Zn
Pteris vittata	As, Pb
Salix viminalis	Cd, Zn
Thlaspi caerulescens	Cd, Zn
Triticum aestivum	Cu, Cd, Cr, Zn, Fe, Ni, Mn, Pb
Chrysopogon zizanioides	Al, As, Cd, Cu, Fe, Hg, Mn, Pb, Zn

- **Biochar (charcoal):** Biochar[17] derived from waste could effectively sequester HMs by altering the physico-chemical conditions of soil and reducing the phyto-availability of hazardous elements. As a low-cost, eco-friendly tool, biochar amendment has gained increasing popularity for arsenic (As) removal from soils related to food crops and drinking water (reviewed by Rai et al 2019). Recent research found that the biochars produced from long roots of *E. crassipes* (water hyacinth) at 200, 300, 400 and 500°C were able to remove Pb, Zn, Cu and Cd efficiently (Li et al 2018). These are suitable and less expensive techniques/ options for developing countries where water hyacinth is regarded as a menace.

3.6.4. Monitoring

As part of remediation and reducing risks of HMs to the environment, food, water and public health, on-going national monitoring of HMs of soil, water, food, ecosystems and comparing with the guideline threshold values should be prioritized (see Table for 3.18 for HMs guidelines for water quality and food quality).

[17]***Biochar** is charcoal used as a soil amendment for both carbon sequestration and soil health benefits.*

Table 3.18. Water, food and health quality guideline threshold trigger values (TV) values for heavy metals [irrigation, livestock, recycled water, sediments, aquaculture, aquatic ecosystems protection, drinking water, seafood and crops) [Column 1 = EPAA 2018; 2, 3, 5, 6, 7 = based on ANZECC 2000 a, b; Column 4 = based on NWQMS 2007; Column 8 = based on NHMRC and NRMMC 2011; Column 9 = based on Food Standards Australia NZ 2001, 2016, NRS, 2012;

dw *= dry weight;* ***na*** *= not available;* ***TV*** *= trigger values which are the concentrations of the key performance indicators measured for the ecosystem, below which there exists a low risk and adverse biological (ecological) effects].*

	1	2	3	4	5	6	7	8	9
Metals	**Agricultural soil** (Regulatory guidelines)	**Irrigation** (long term TV-short term TV) mg/L	**Livestock drinking** (mg/L)	**Recycled water** (mg/L)	**Aquaculture** (fresh water = Fw; marine = Mr)	**Sediment** (low TV-high TV)	**Aquatic ecosystems protection** (fresh water = Fw; marine = Mr)	**Human health** (drinking water)	**Human health** (seafood, meat, cereals, vegetables) (maximum level)
Cd	3 mg/kg	0.01-0.005 mg/L	0.01 mg/L	0.002 mg/L	< 0.2-1.8 µg/L (Fw) < 0.5-5 µg/L (Mr)	1.5-10 mg/kg dw	0.2 µg/L (Fw) 0.7 µg/L (Mr)	0.002 mg/L	Molluscs (excluding dredge/bluff oysters and queen scallops): 2 mg/kg; Rice, wheat: 1 mg/kg Root and tuber: 1 mg/kg Leafy vegetable: 1 mg/kg Meat of cattle, sheep and pig (excluding offal): 0.05 mg/kg Liver of cattle, sheep and pig: 1.25 mg/kg Salt: 0 .5 mg/kg Peanuts: 0.5 mg/kg Chocolate and cocoa products: 0.5 mg/kg

	1	2	3	4	5	6	7	8	9
Metals	**Agricultural soil** (Regulatory guidelines)	**Irrigation** (long term TV-short term TV) mg/L	**Livestock drinking** (mg/L)	**Recycled water** (mg/L)	**Aquaculture** (fresh water = Fw; marine = Mr)	**Sediment** (low TV-high TV)	**Aquatic ecosystems protection** (fresh water = Fw; marine = Mr)	**Human health** (drinking water)	**Human health** (seafood, meat, cereals, vegetables) (maximum level)
Co	100 mg/kg	0.05-0.1 mg/L	1 mg/L	na	na	na	na (Fw) 1 µg/L (Mr)	na	na
Cr	50 mg/kg	0.01-1 mg/L	1 mg/L	0.05 mg/L	< 20 µg/L (Fw) <20 µg/L (Mr)	80-370 mg/kg dw	1 µg/L CrVI (Fw) 4.4 µg/L CrVI (Mr)	0.05 mg/L CrVI	na
Cu	100 mg/kg	0.2-5 mg/L	0.4 mg/L for sheep; 1 mg/L for cattle; 5 mg/L (for pigs, poultry)	2 mg/L	< 5 µg/L (Fw) < 5 µg/L (Mr)	65-270 mg/kg dw	1.4 µg/L (Fw) 1.3 µg/L (Mr)	2 mg/L	Crustacean: 10 mg/kg Fish: 0.5 mg/kg Molluscs: 5 mg/kg Wheat: 5 mg/kg Nuts: 15 mg/kg Meat of cattle, sheep, pigs: 1 mg/kg Edible offal of cattle, pigs and sheep: 5 mg/kg
Fe	NA	0.2-10 mg/L	na	na	< 10 µg/L (Fw) < 10 µg/L (Mr)	na	na	na	na
Hg	1 mg/kg	0.002-0.002 mg/L	0.002 mg/L	0.01 mg/L	< 1 µg/L (Fw) < 1 µg/L (Mr)	0.15-1 mg/kg dw	0.06 µg/L inorganic (Fw) 0.1 µg/L	0.001 mg/L	Gemfish, billfish (including marlin), southern blue-fin tuna,

	1	2	3	4	5	6	7	8	9
Metals	**Agricultural soil** (Regulatory guidelines)	**Irrigation** (long term TV-short term TV) mg/L	**Livestock drinking** (mg/L)	**Recycled water** (mg/L)	**Aquaculture** (fresh water = Fw; marine = Mr)	**Sediment** (low TV-high TV)	**Aquatic ecosystems protection** (fresh water = Fw; marine = Mr)	**Human health** (drinking water)	**Human health** (seafood, meat, cereals, vegetables) (maximum level)
							inorganic (Mr)		barramundi, ling, orange roughy, rays and all species of shark: 1.5 mg/kg Other fish, fish products, crustaceans and molluscs: 1.5 mg/kg Salt: 0.1 mg/kg
Mn	500 mg/kg	0.2-10 mg/L	na	0.5 mg/L	< 10 μg/L (Fw) < 10 μg/L (Mr)	na	1900 μg/L (Fw) NA (Mr)	0.5 mg/L	na
Ni	60 mg/kg	0.2-2 mg/L	1 mg/L	0.02 mg/L	< 100 μg/L (Fw) <100 μg/L (Mr)	21-52 mg/kg dw	11 μg/L (Fw) 7 μg/L (Mr)	0.02 mg/L	na
Pb	300 mg/kg	2-5 mg/L	0.1 mg/L	0.01 mg/L	1-7 μg/L (Fw) 1-7 μg/L (Mr)	50-220 mg/kg dw	3.4 μg/L (Fw) 4.4 μg/L (Mr)	0.01 mg/L	Fish: 0.5 mg/kg Molluscs: 2 mg/kg Cereals, pulses and legumes: 0.2 mg/kg Meat of cattle, sheep, pig and poultry (excluding offal): 0.1 mg/kg Edible offal of cattle, sheep, pig and poultry:0.5

	1	2	3	4	5	6	7	8	9
Metals	**Agricultural soil** (Regulatory guidelines)	**Irrigation** (long term TV-short term TV) mg/L	**Livestock drinking** (mg/L)	**Recycled water** (mg/L)	**Aquaculture** (fresh water = Fw; marine = Mr)	**Sediment** (low TV-high TV)	**Aquatic ecosystems protection** (fresh water = Fw; marine = Mr)	**Human health** (drinking water)	**Human health** (seafood, meat, cereals, vegetables) (maximum level)
									mg/kg Fruit: 0.1 mg/kg Infant formula products: 0.02 mg/kg
U	NA	0.01-0.1 mg/L	0.2 mg/L	na	na	na	na	0.017 mg/L	na
Zn	200 mg/kg	2-5 mg/L	20 mg/L	3 mg/L	5 µg/L (Fw) 5 µg/L (Mr)	200-410 mg/kg dw	8 µg/L (Fw) 15 µg/L (Mr)	na	Crustaceans: 25 mg/kg Fish: 5 mg/kg Oysters: 130 mg/kg Meat of cattle, sheep, pigs: 60 mg/kg Edible offal of cattle, pigs and sheep: 35 mg/kg

3.7. Conclusion

HMs (Cd, Cu, Hg, Pb, Zn) are persistent, toxic, bioaccumulative but some of them are also endocrine-disrupting as well as carcinogenic. HMs can enter the environment via rainwater run-off, effluents discharged from industries, atmospheric fallout and long-range transport of volatile compound "Hg" (warmer to colder region) via 'grasshopper effect'. Several HMs are carcinogenic (IARC Class 1, 2A, 2B) to humans including Cd, Cr, Pb, MeHg, Ni, etc. Cd, Cu, Hg, Pb, Zn contaminate crop, vegetable, and marine fish across the globe. HM contamination of soil, water and food would undermine achieving SDGs 1, 2, 3, 8 and 14 (as discussed in chapter 1 section 1.8). A Global research revealed elevated Hg levels in child-bearing women in 25 countries who were exposed via **i.** eating contaminated seafood (the Pacific Island nations), **ii.** the practice of artisanal and small-scale gold mining and **iii.** living in close proximity to industries (Chlor-alkali, cement kiln, waste incineration, pulp and paper plants). A Canada based organization's research reveals that cosmetics could be as well a source of exposure to many of the toxic and carcinogenic HMs to women (Pb, Ni, As, Cd) who are frequent users of facial makeup. Climate change and global warming would enhance HMs load into the environment (e.g. via an increase in forest fires, run-off, floods, melting/thawing of snows and permafrost) and would enhance bioaccumulation of toxic HMs in seafood organisms (e.g. via an increase in the metabolism of fish in a warming world). Some of the best measures and remediation techniques to reduce risk of HMs in soil-food crops sub-systems would be avoiding the use of inadequately treated effluent and sewage sludge in growing food crops and monitoring of HMs in soil, food, water and ecosystems to safeguard the environment and human health.

3.8. References

Adams DH and McMichael RH 1999 Mercury levels in four species of sharks from the Atlantic Coast of Florida. *Fisheries Bulletin* **97** 372–379

Akpoveta OV and Osakwe SA 2014 Determination of heavy metal contents in refined petroleum products. *IOSR Journal of Applied Chemistry* (*IOSR-JAC)* 2278-5736. **7(6)** 01-02

Al-Ameri M, Hatt B, Coustumer, SL et al 2018 Accumulation of heavy metals in stormwater bioretention media: A field study of temporal and spatial variation. *Journal of Hydrology* **567** 721-731

Alonso-Hernández CM, Bernal-Castillo J, Bolanos-Alvarez Y, Gómez-Batista M and Diaz-Asencio M 2011 Short communication Heavy metal content of bottom ashes from a fuel oil power plant and oil refinery in Cuba. *Fuel* **90** 2820–2823

Alloway BJ 2013 Sources of heavy metals and metalloids in soils. *In*: Alloway B. (eds) Heavy Metals in Soils. Environmental Pollution, vol **22** Springer, Dordrecht, Germany.

AMAP 2016 Arctic Monitoring and Assessment Programme (AMAP) *Influence of Climate change on transport, levels, and effects of contaminants in Northern Areas –*

Part 2. *In*: P. Carlsson, J.H. Christensen, K. Borgå, R. Kallenborn, K. Aspmo Pfaffhuber, J.Ø. Odland, L.-O. Reiersen, and J.F. Pawlak (eds.). Arctic Monitoring and Assessment Programme (AMAP), Oslo. 52 p. ISBN: 978-82-7971-099-8 ©; Arctic Monitoring and Assessment Programme (AMAP), 2016

Anawar HM 2013 Impact of climate change on acid mine drainage generation and contaminant transport in water ecosystems of semiarid and arid mining areas. *Physics and Chemistry of the Earth* **58–60** 13–21

ANZECC 2000a Australian and New Zealand Environment and Conservation Council (ANZECC) and Agriculture and Resource Management Council of Australia and New Zealand (ARMCANZ). Australian and New Zealand Guidelines for Fresh and Marine Water Quality, Volume 1, The Guidelines (Chapters 1-7).

http://www.mincos.gov.au/publications/Australian_and_New_Zealand_guidelines_for_fresh and_marine_water_quality.

ANZECC 2000b Australian and New Zealand Environment and Conservation Council (ANZECC) and Agriculture and Resource Management Council of Australia and New Zealand (ARMCANZ). *Australian and New Zealand Guidelines for Fresh and Marine Water Quality*, Volume 2, Aquatic ecosystems – rationale and background information. National Water Quality Management Strategy. (Chapter 8)

http://www.mincos.gov.au/publications/australian_and_new_zealand_guidelines_for_fresh_and_marine_ water_quality/volume_2.

ANZFSC 2016 Maximum levels of contaminants and natural toxins. Australia New Zealand Food Standards Code. Schedule 19.

https://www.foodstandards.gov.au/code/Documents/Sched%2019%20Contaminant%20MLs%20v157.pdf

Aonghusa CN and Gray NF 2002 Laundry detergents as a source of heavy metals in Irish domestic wastewater of *Environmental Science and Health*, Part A, **37(1)** 1-6 *DOI: 10.1081/ESE-100108477*

Ari V 2016 A review of technology of metal recovery from electronic waste; *http://dx.doi.org/10.5772/61569file:///C:/Users/User1/Downloads/49287.pdf*

Araújo CST, Carvalho DC, Rezende HC, Almeida ILS et al 2013 *Bioremediation of Waters Contaminated with heavy metals using Moringa oleifera Seeds as Biosorbent,* Chapter **10** 225-254 p. INTECH. *http://dx.doi.org/10.5772/56157*

Arnot JA and Gobas FAPC 2006 A review of bioconcentration factor (BCF) and bioaccumulation factor (BAF) assessments for organic chemicals in aquatic organisms. *Environmental Reviews* **14** 257–297 *doi: 10.1139/A06-005*

Artiola JF, Walworth JL, Musil SA and Crimmins MA 2019 Chapter 14 - Soil and Land Pollution. *Environmental and Pollution Science* (Third Edition). 219-235 *https://doi.org/10.1016/B978-0-12-814719-1.00014-8*

ATSDR 2017 Agency for Toxic Substances Disease Registry, U.S. Department of Health and Human services (ATSDR) 2017 *ATSDR's Substance Priority List. https://www.atsdr.cdc.gov/spl/index.html#2017spl*

Bakshia A and Panigrahi AK 2018 "A comprehensive review on chromium induced alterations in fresh water fishes". *Toxicology Reports* **5** 440–447 *doi:10.1016/j.toxrep.2018.03.007*

Bately GE 1995 Heavy metals and tributyltin in Australian coastal and estuarine waters. The State of the Marine Environment Report for Australia: Pollution – Technical

Annex 2. Dept. of the Environment, Water, Heritage and the Arts. Ocean Rescue 2000 Program.

http://www.environment.gov.au/coasts/publications/somer/annex2/batley.html.

Bhardwaj V, Kumar P and Singhal G 2014 Toxicity of heavy metals pollutants in textile mills effluents. *International Journal of Scientific & Engineering Research* **5(7)** 664

https://www.ijser.org/paper/Toxicity-of-Heavy-Metals-Pollutants-in-Textile-Mills-Effluents.html

Bell L, Evers D, Johnson J, Regan K, DiGangi J, Federico J, Samanek J 2017 *Mercury in women of child-bearing age in 25 countries.* IPEN and BRI.69 p; *http://ipen.org/mercury-and-women*; *https://ipen.org/sites/default/files/documents/updateNov14_mercury-women-report-v1_6.pdf*

Bhattacharya P, Mukherjee AB, Jacksa G and Nordqvistc S 2002 Metal contamination at a wood preservation site: characterisation and experimental studies on remediation. *The Science of the Total Environment* **290** 165–180

Bobby 2014 What metals are used in batteries?

https://www.upsbatterycenter.com/blog/metals-used-batteries/

Booth S and Zeller D 2005 Mercury, food webs, and marine mammals: Implications of diet and climate change for human health. *Environmental Health Perspectives* **113(5)** 521–526

Boush GM and Thieleke JR 1983 Mercury Content in sharks. *Bulletin of Environmental Contamination and Toxicology* **30** 284-290

Branco V, Canário J, Vale C, J Raimundo J and Reis C 2004 Total and organic mercury concentrations in muscle tissue of the blue shark (*Prionace glauca* L. 1758) from the Northeast Atlantic. *Marine Pollution Bulletin* **49** 854-874

Bubb JM and Lester JN 1991 The impact of heavy metals on low land rivers and the implications for man and the environment. *The Science of the Total Environment* **100** 207-233

Burger J and Gochfeld M 2011 Mercury and selenium levels in 19 species of saltwater fish from New Jersey as a function of species, size, and season. *Science of the Total Environment* **409 (8)** 1418-29 *doi: 10.1016/j.scitotenv.2010.12.034. Epub 2011Feb 2*

Cai Y, Rooker JR, Gill, GA et al 2007 Bioaccumulation of mercury in pelagic fishes in NW Gulf of Mexico and its relationship with length, location, collection year, and trophic level. *Canadian Journal of Fisheries and Aquatic Sciences* **64(3):** *http://www.freepatentsonline.com/article/Canadian-Journal-Fisheries-Aquatic-Sciences/163465849.html*

CCREM (Canadian Council of Resource and Environment Ministers)1987 Canadian water quality guidelines. Canadian Council of Resource and Environment Ministers, Ontario, Canada.

Chang JS, Yoon IH and Kim KW 2009 Short Communication-Heavy metal and arsenic accumulating fern species as potential ecological indicators in As-contaminated abandoned mines. *Ecological Indicators Ecological Indicators* **9(6)** 1275-1279 *doi:10.10.16/j.ecoind.2009.03.011.*

Chary NS, Kamala, CT, and Raj, DSS 2008 Assessing risk of heavy metals from consuming food grown on sewage irrigated soils and food chain transfer. *Ecotoxicology and Environmental Safety* **69** 513–524

Chowdhury A, Datta R and Sarkar D 2018 Heavy metal pollution and remediation. Chapter 3.10. 2018, Pages 359-373. In book. *Green Chemistry in Practice DOI: 10.1016/B978-0-12-809270-5.00015-7*

Cipurkovic A, Trumic I, Hodžic Z et al 2014 Distribution of heavy metals in Portland cement production process. *Advances in Applied Science Research* **5(6)** 252-259

Coelho JP, Santos H, Reis AT et al 2010 Mercury bioaccumulation in the spotted dogfish (*Scyliorhinus canicula*) from the Atlantic Ocean. *Marine Pollution Bulletin* **60(8)** 1372-1375

Copat C, Bella F, Castaing M, Fallico R, Sciacca S and Ferrante M 2012 Heavy metals concentrations in fish from Sicily (Mediterranean Sea) and evaluation of possible health risks to consumers. *Bulletin of Environmental Contamination and Toxicology* **88(1)** 78–83

Davenport SR 1995 Mercury in blue sharks and deepwater dogfish from around Tasmania. *Australian fisheries* **54(3)** 20-22

http://hdl.handle.net/102.100.100/234782?index=1

Domi N, Bouquegneau JM, Das K 2005 Feeding ecology of five commercial shark species of the Celtic Sea through stable isotope and trace metal analysis. *Marine Environmental Research* **60** 551–569

Durnford D and Dastoor A 2011 The behaviour of mercury in the cryosphere: A review of what we know from observations. *Journal of Geophysical Research* **116(D6)** D06305.

EC 2006 Commission of the European Communities (EC) 2006 Commission regulation (EC) No 1881/2006. *Official Journal of the European Union* **364** L 364/5-24 *https://eur-lex.europa.eu/LexUriServ/LexUriServ.do?uri=OJ:L:2006:364:0005: 0024:EN:PDF*

Endo T, Hisamichi Y, Haraguchi K, Kato Y, Ohta C and Koga N 2008 Hg, Zn and Cu levels in the muscle and liver of tiger sharks (*Galeocerdo cuvier*) from the coast of Ishigaki Island, Japan: Relationship between metal concentrations and body length. *Marine Pollution Bulletin* **56** 1774–1780

Environmental Defense Canada 2011 *Heavy metals hazard. The Health risks of hidden heavy metals in face makeup. Environmental Defence* **116** Spadina Avenue, Suite 300. Toronto, Ontario M5V 2K6;

http://www.ottawacitizen.com/pdf/HeavyMetalHazard.pdf

EPAA 2018 Environment Protection Authority of Australia 2018 Classification and management of contaminated soil for disposal. Information Bulletin 105. Hobart, TAS 7001 Australia.

https://epa.tas.gov.au/Documents/Information%20Bulletin%20105%20-%20Classification%20of%20Contaminated%20Soils%20%28IB105%29%20V3_2018.pdf

Escobar-Sánchez O, Galván-Magana F and Roslies-Martinez R 2010 Mercury and selenium bioaccumulation in the smooth hammerhead shark, *Sphyrna zygaena* Linneaus, from the Mexican Pacific Ocean. *Bulletin* of *Environmental Contamination* and *Toxicology* **84(4)** 488-491

European Commission 2008 Commission European. Commission regulation No. 629/2008/setting maximum levels for contaminants in foodstuffs. *Official Journal of the European Union* **51** 4-10

https://www.fsai.ie/uploadedFiles/Commission_Regulation_EC_No_629_2008.pdf

European Food Safety Authority 2012 Cadmium dietary exposure in the European population. *European Food Safety Authority Journal* **10**(1) 25-51 *doi:10.2903/j.efsa.2012.2551*

Florence TM, Morrison GM and Stauber JL 1992 Determination of trace element speciation and the role of speciation in aquatic toxicity. *The Science of the Total Environment* **125** 1-13

Evans-Illidge E 1997.*Heavy metals in commercial prawns and crayfish species in the Torres Strait*. Commonwealth Coastal Action Program. Great Barrier Reef Marine Park Authority; 111 p.

Filazi A, Baskaya R, Kum C and Hismiogullari SE 2003 Metal concentrations in tissues of the Black Sea fish *Mugil auratus* from Sinop-Icliman, Turkey. *Human and Experimental Toxicology* **22(2)** 85-87

Florence TM, Stauber JL and Ahsanullah M 1994 Toxicity of nickel ores to marine organisms. *Science of the Total Environment* **148** 139–155

Florence TM, Morrison GM and Stauber JL 1992 Determination of trace element speciation and the role of speciation in aquatic toxicity. *The Science of the Total Environment* **125** 1-13

Food Standards Australia New Zealand (FSA) 2016 *Schedule 19. Maximum levels of contaminants and natural toxicants*. Food Standards Australia New Zealand (FSA). *http://www.foodstandards.gov.au/code/Documents/Sched%2019%20Contaminant%20MLs%20v157.pdf*

Food Standards Australia New Zealand (FSA) 2001 Generally expected levels (GELs) for metal contaminants. Additional Guidelines. To Maximum Levels in Standard 1.4.1 - Contaminants and Natural Toxicants (*http://www.foodstandards.gov.au/code/userguide/documents/GELs_0801.pdf*).

Grant D 2016 Toxic Art: Is Anyone Sure What's In A Tube of Paint? *https://observer.com/2016/10/toxic-art-is-anyone-sure-whats-in-a-tube-of-paint/##targetText=Many%20of%20the%20traditional%20pigments,%2C%20manganese%2C%20strontium%20and%20zinc*.

Harada M 1995 Minamata Disease: Methyl-mercury Poisoning in Japan Caused by Environmental Pollution *Critical Reviews in Toxicology* **25(1)** 1-24

Hahladakis JN, Velis CA, Weber R, Lacovidoua E, and Purnell P 2018 An overview of chemical additives present in plastics: Migration, release, fate and environmental impact during their use, disposal and recycling. *Journal of Hazardous Materials* **344** 179–199

He Z, Shentu J, Yang X, Baligar VC, Zhang T and Stoffella PJ 2015 Heavy Metal Contamination of Soils: Sources, Indicators, and Assessment. *Journal of Environmental Indicators* **9** 17-18

Herreros MA, Iñigo-Nuñez S, Sanchez-Perez E, Encinas T and Gonzalez-Bulnes A 2008 Contribution of fish consumption to heavy metals exposure in women of childbearing age from a Mediterranean country (Spain). *Food and Chemical Toxicology* **46** 1591–1595

Hristov S and Kirin D 2014 Accumulation of lead (Pb) in *Scardinius erythrophthalmus* and *Ceratophyllum demersum* from freshwater ecosystem biosphere reserve Srebarna, Bulgaria. *Scientific Papers Series D Animal Science* **LVII** 290–297

Hoq ME and Islam MM 2019 Heavy metals in naturally grown and cultured seaweeds from the Bay of Bengal coast, Bangladesh. Abstract published in Book of Abstracts, 1st International Conference on Sustainable Fisheries (ICSF)-2019, Faculty of Fisheries, Sylhet Agriculture University, Bangladesh.

Holsbeek L, Das HK and Joiris CR 1997 Mercury speciation and accumulation in Bangladesh freshwater and anadromous fish. *The Science of the Total Environment* **198** 201-210

Hornung H, Krom MD Y, Cohen Y and Bernhard M 1993 Trace metal content in deep-water sharks from the eastern Mediterranean-Sea. *Marine Biology* **115** 331–338

Hossain MM 1994 Concentration of total mercury in shrimps and fish species from the Bay of Bengal, Bangladesh. *Marine Research* **32** 27-32

Hutchins DA, Teyssii, JL, Boisson, F, Fowler SW and Fisher NS 1996Temperature effects on uptake and retention of contaminant radionuclides and trace metals by the brittle star *Ophiothrix fragilis*. *Marine Environmental Research* **41(4)** 363–378

IARC 2019 IARC monographs on the identification of carcinogenic hazards to humans.

https://monographs.iarc.fr/agents-classified-by-the-iarc/

Idzelis RL, Sauliute G, Grigeleviciute J and Svecevicius G 2012 Bioaccumulation of lead in body tissues of Atlantic salmon (*Salmo salar* L.): Experimental investigation and comparative analysis. *Science Future Lithuanian* **4** 423–429

Islam S, Islam F, Bakar MA, Das S and Bhuiyan HR 2013 Heavy metals concentration at different tannery wastewater canal of Chittagong city in Bangladesh. *International Journal of Agriculture, Environment and Biotechnology* **6(3)** 355-361 *DOI: 10.5958/j.2230-732X.6.3.003*

Jooste A, Marr S, Addo-Bediako A and Luus-Powell W 2014 Metal bioaccumulation in the fish of the Olifants River, Limpopo province, South Africa, and the associated human health risk: A case study of red nose labeo, *Labeo rosae* from two impoundments. *African Journal of Aquatic Science* **39** 271–277

Khan S, Cao Q, Zheng YM, Huang YZ, Zhu, YG 2008 Health risks of heavy metals in contaminated soils and food crops irrigated with wastewater in Beijing, China. *Environmental Pollution* **152** 686–692 *10.1016/j.chemosphere.2016.08.121*

Khan MA, Ahmed QSA, Catalin B, Khodadoust A, Ajayi O and M. Vaughn 2006 Effect of temperature on heavy metal toxicity to juvenile crayfish, *Orconectes immunis* (Hagen). *Environmental Toxicology* **21** 513–520

Kibria G, Lau TC, Wu R 2012a Innovative 'Artificial Mussels' technology for assessing spatial and temporal distribution of metals in Goulburn–Murray catchments waterways, 461 Victoria, Australia: Effects of climate variability (dry vs. wet years). *Environment International* **50** 38-46 *https://doi.org/10.1016/j.envint.2012.09.006*

Kibria G 2012b Mining: Friend or Foe? Economic, Environmental & Social Impacts- An Overview. *DOI: 10.13140/RG.2.1.1685.0403*

https://www.researchgate.net/publication/265379518_Mining_Friend_or_Foe_Economic_Environmental_Social_Impacts-_An_Overview

Kibria G 2014 Climate Change and Chemical Contaminants. *DOI: 10.13140/RG.2.1.2733.6169*

Kibria G 2017 Bush fire/Forest fire/Wildfire Risk Increases Globally but Why? *DOI: 10.13140/RG.2.2.35126.60481*

Kibria G 2018 Presentation: Heavy metals-Its Risk on the Environment, Water, Food and Public Health. *DOI: 10.13140/RG.2.2.19560.93445*

Kibria G and Haroon AKY 2015 Pollutants bioaccumulation in sharks and shark seafood security. 13 p. *DOI: 10.13140/RG.2.1.1529.3926*

Kibria G, Haroon AKY, Nugegoda D and Rose G 2010a *Climate change and chemicals: Environmental Biological aspects.* New India Publishing Agency. New Delhi. 470 p. ISBN: 9789380235301

https://www.researchgate.net/publication/261216635_Climate_Change_and_Chemicals_Environmental_and_Biological_Aspects

Kibria G, Rose G, Lau TC and Lung YK and Chan AKY and Wu R 2010b *Monitoring heavy Metals in Goulburn-Murray waterways using passive sampling with artificial mussels (AM) – Pilot Study* (trial of AM technology). Report prepared under a research collaboration agreement between Goulburn Murray Rural Water Corporation, Tatura, Australia, the City University of Hong Kong, the University of Hong Kong, and the Department of Primary Industries, Werribee, Victoria, Australia. 29 p.

Kibria G, Allinson G, Pettigrove V, Slessar P, Lau TC and Wu R 2010c *Monitoring trace metals in North and Central Victorian Waterways, Australia, using Artificial Mussel (AM) Technology* (2009-2010). Report prepared under a research collaboration agreement between Goulburn Murray Rural Water Corporation, Tatura, Australia, the City University of Hong Kong, the University of Hong Kong, and the Department of Primary Industries, Werribee, Victoria, Australia. GMW docs: 2972226: 34p *DOI: 10.13140/RG.2.1.1534.3047/1*

Kibria G, Lau TC and Wu R 2012a Innovative 'Artificial Mussels' technology for assessing spatial and temporal distribution of metals in Goulburn-Murray catchments waterways, Victoria, Australia: Effects of climate variability (dry vs. wet years). *Environment International* **50** 38-46

http://dx.doi.org/10.1016/j.envint.2012.09.006

Kibria G, Haroon AKY, Nugegoda D 2013 *Climate change and Agricultural food production: Impacts, vulnerabilities and remedies.* New India Publishing Agency. New Delhi. ISBN: 9789381450512;

https://www.researchgate.net/publication/261178054_Climate_Change_and_Agricultural_Food_Production_Impacts_Vulnerabilities_and_Remedies.

Kibria G, Hossian MM, Mallick D. Lau TC and Wu R 2016a Trace/Heavy Metal Pollution Monitoring in Estuary and Coastal Area of Bay of Bengal, Bangladesh and Implicated Impacts. *Marine Pollution Bulletin* **105(1)** 393-402 *DOI: 10.1016/j.marpolbul.2016.02.021*

Kibria G, Hossian MM, Mallick D. Lau TC and Wu R 2016b Monitoring of metal pollution in waterways across Bangladesh and ecological and public health implications of pollution. *Chemosphere* **165** 1-9

https://doi.org/10.1016/j.chemosphere.2016.08.121

Kibria G, Hossain MM, Wu R and Lau TC 2017 Artificial Mussel Monitoring Programme - the way to go both in Australia and Bangladesh *DOI: 10.13140/RG.2.2.34243.14889*

Kibria G, Hossain MM and Wu R 2018 Presentation: Application of innovative artificial mussel (AM) technology for heavy metals monitoring in rivers, irrigation channels, creeks, bores, estuaries, and oceans-a global AM watch program. *DOI:10.13140/RG.2.2.24109.79842/1*

https://www.researchgate.net/publication/264859047_Monitoring_trace_metals_in_North_and_Central_Victorian_Waterways_Australia_using_Artificial_Mussel_%28AM%29_Technology_%28Research_collaboration_between_GM_Water_Australia_the_University_of_Hong_Kong_and_Department_of_Primary_Industries_Victoria_Australia.%29

Kumar V, Chopra AK, Kumar S, Singh J and Thakur RK 2015 Effects of pulp and paper mill effluent disposal on soil characteristics in the vicinity of Uttaranchal Pulp and Paper Mill, Haridwar (Uttarakhand), India. *International Journal of Agricultural Science Research* **4(6)** 117-125

Ku P, Tsz-Ki Tsui M, Nie X, Chen H et al 2018 Origin, reactivity, and bioavailability of Mercury in wildfire ash. *Environmental Science and Technology* **52(24)** 14149-14157

https://pubs.acs.org/doi/abs/10.1021/acs.est.8b03729

Lissy AMP N and Madhu BG 2010 Removal of Heavy Metals from Waste Water Using Water Hyacinth. Proc. of Int. Conf. on Advances in Civil Engineering 2010 © 2010 ACEE *DOI: 02.ACE.2010.01.39 42*

https://pdfs.semanticscholar.org/ad6f/47f23ce9c021c15491a30bcca8204f661043.pdf

Li Q, Tang L, Hu J, Jiang M, Shi X, Zhang T, Li Y, Pan X 2018 Removal of toxic metals from aqueous solutionby biochars derived from long-root *Eichhornia crassipes. R. Soc. Open Sci.* **5(10)** 180966 *http://dx.doi.org/10.1098/rsos.180966*

Liu W, Zhao JZ, Ouyang ZY, Söderlund L and Liu GH 2005 Impacts of sewage irrigation on heavy metal distribution and contamination in Beijing, China. *Environment International* **31(6)** 805–812

Marcotrigiano GO and Storelli MM 2003 Heavy metal, polychlorinated biphenyl and organochlorin pesticide residues in marine organisms: Risk evaluation for consumers. *Veterinary Research Communications* **27(Suppl)** 183–195

Marcotullio PJ 2007 Urban water-related environmental transitions in Southeast Asia. *Sustainability Science* **2 (1)** 27-54

Marcovecchio JE, Moreno, and VJ, Pėrez A 1991 Metal Accumulation in Tissues of Sharks from the Bahia Blanca Estuary, Argentina. *Marine Environmental Research* **31** 263-274

Markich SJ, Brown PL, Batley GE, Apte SC and Stauber JL 2001 Incorporating metal speciation and bio-availability into water quality guidelines for protecting aquatic ecosystems. *Australasian Journal of Ecotoxicology* **7**109-122

Marrugo-Negrete J, Verbel JO, Ceballos EL and Benitez LN 2008 Total mercury and methyl-mercury concentrations in fish from the Mojana region of Colombia. *Environ. Geochem. Health* **30** 21–30

Marques A, Nunes ML, Moore SK and Strom MS 2010 Climate change and seafood safety: Human health implications. *Food Research International* **43** 1766–1779

Maturi K and Reddy KR 2008 Extractants for the removal of mixed contaminants from soils. *Soil and Sediment Contamination* **17(6)** 586-608

McMeans BC, Borga K, Bechtol WR, Higginbotham D and Fisk AT 2007 Essential and non-essential element concentrations in two sleeper shark species collected in arctic waters. *Environmental Pollution* **148** 281-290

Meharag AA, Norton G, Deacon C, Williams P et al 2013 Variation in rice cadmium related to human exposure. *Environmental Science and Technology* **47** 5613-5618

Meng B, Feng X, Qiu G et al 2011 The process of methyl-mercury accumulation in rice (*Oryza sativa* L.) *Environ. Sci. Technol.* **45(7)** 2711-2717 *doi: 10.1021/es103384v. Epub 2011 Mar 2.*

Mohammed T, Aryal R, Vigneswaran S et al 2012 Removal of heavy metals in stormwater by hydrous ferric oxide. *Water Management* **165** 171–178 *http://dx.doi.org/10.1680/wama.2012.165.3.171*

Mohammed AS, Kapri A, and Goel R 2011 Heavy metal pollution: source, impact, and remedies. *In*: Khan M., Zaidi A., Goel R., Musarrat J. (eds). Biomanagement of Metal-Contaminated Soils. *Environmental Pollution* **20** Springer, Dordrecht, Germany. *https://doi.org/10.1007/978-94-007-1914-9_1*

Mossop D, Kellar C, Jeppe K, Myers J, Rose G, Weatherman K et al 2012 Impacts of intensive agriculture and plantation forestry on water quality in the Latrobe catchment, Victoria. Publication number 1528 April 2013 Authorised and published by EPA Victoria, 200 Victoria Street, Carlton EPA Victoria, Monitoring and Assessment Unit Centre for Aquatic Pollution Investigation and Management

MSDS-Chem Supply 2017 *Mercury Safety data sheet;*

https://www.chemsupply.com.au/documents/MA0591CH4B.pdf

MSDS-ESPI Metals 2015 *Cadmium Safety data sheet;*

https://www.espimetals.com/index.php/msds/476-Cadmium

MSDS-Fisher 2015 *Copper Safety data sheet;*

https://www.fishersci.com/store/msds?partNumber=S25267&productDescription=COPPER+METAL+FOIL+100G+.005&vendorId=VN00115888&countryCode=US&language=en

MSDS- Jamestown 2015 *Lead Safety data sheet.*

https://www.biodex.com/sites/default/files/documents/sds-lead-bricks-002-246-248.pdf

MSDS-Merck 2006 (revised 2017) *Cadmium Safety data sheet.* 17 p; *http://www.merckmillipore.com/AU/en/product/msds/MDA_CHEM-102004*

MSDS-Merck 2006 (revised 2017) *Chromium VI Safety data sheet.* 21 p *http://www.merckmillipore.com/AU/en/product/msds/MDA_CHEM-100229?Origin=PDP&ReferrerURL=https%3A%2F%2Fwww.google.com%2F file:///C:/Users/User1/Downloads/100229_SDS_AU_EN%20(3).PDF*

MSDS-Thermo Fisher 2018 *Cadmium Safety data sheet.* 9 p *https://www.fishersci.com/store/msds?partNumber=C3500&productDescription=CADMIUM+METAL+GRAN+PURIF+500G&vendorId=VN00033897&countryCode=US&language=en*

MSDS-Merck 2006 (revised 2017) *Mercury Safety data sheet.* 8 p; *http://dept.harpercollege.edu/chemistry/msds/Lead%20metal%20foil%20sheets%20 Fisher.pdf*

MSDS-Thermo Fisher 2014 (revised 2018) *Mercury Safety data sheet* 8 p *https://www.fishersci.com/store/msds?partNumber=M1416LB&productDescription =MERCURY+MTL+INST+GRD+REAG+6LB&vendorId=VN00033897&countryC ode=US&language=en*

MSDS-Thermo Fisher 2010 (revised 2018) *Nickel powder Safety data sheet.* 8 p; *https://www.fishersci.com/shop/msdsproxy?productName=AC193610250&productD escription=NICKEL+POWDER+99.9%25+-+25GR&catNo=AC19361-0250&vendorId=VN00032119&storeId=10652*

MSDS-Merck 2017 *Zinc Safety data sheet.* 16 p; *http://www.merckmillipore.com/AU/en/product/msds/MDA_CHEM-108774*

Mubiana VK and Blust R 2007 Effects of temperature on scope for growth and accumulation of Cd, Co, Cu and Pb by the marine bivalve *Mytilus edulis. Marine and Environmental Research* **63(3)** 219–235

Mull CG, Blasius ME, O'Sullivan JB and Lowe CG 2012. Chapter 5: Heavy Metals, Trace Elements, and Organochlorine Contaminants in Muscle and Liver Tissue of Juvenile White Sharks, *Carcharodon carcharias*, from the Southern California Bight. p. 59-75. DOI: 10.1201/b11532-7. *In:* Domeier, ML (ed.) Global perspectives on the biology and life history of the white shark. CRC Press, Taylor & Francis Group.
https://www.researchgate.net/publication/281381586_Heavy_Metals_Trace_Elemen ts_and_Organochlorine_Contaminants_in_Muscle_and_Liver_Tissue_of_Juvenile_ White_Sharks_Carcharodon_carcharias_from_the_Southern_California_Bight

Nagpal NK 2004 Technical report, water quality guidelines for cobalt. Water, Air and Climate Change Branch. Ministry of Water, Land and Air Protection, Victoria BC V8W 9M1, Canada

http://www.env.gov.bc.ca/wat/wq/BCguidelines/cobalt/cobalt_tech.pdf

Nam DH, Adams DH, Reyier EA, Basu N 2010 Mercury and selenium levels in lemon sharks (*Negaprion brevirostris*) in relation to a harmful red tide event. *Environmental Monitoring and Assessment* **176** 549–559

National Residue Survey (NRS) 2012 National Residue Survey Annual Report 2010–2011. Australian Government Department of Agriculture. Fisheries and Forestry, Canberra (218 p)

NHMRC and NRMMC 2011 National Health and Medical Research Council (NHMRC) and National Resource Management Ministerial Council (NRMMC) Australian Drinking Water Guidelines Paper 6 National Water Quality Management Strategy. National Health and Medical Research Council and National Resource Management Ministerial Council, Commonwealth of Australia, Canberra.

NWQMS 2007 National Water Quality Management Strategy (NWQMS) Australian Guidelines for Water Recycling: Managing Health and Environmental Risks (Phase 2) — Augmentation of Drinking Water Supplies. Natural Resource Management Ministerial Council. Environment Protection and Heritage Council. Australian Health Ministries Conference. Biotex Pty Ltd, Canberra (140 p.).

Neff JM 2002 Bioaccumulation in Marine Organisms: Effect of Contaminants from Oil well produced water. Elsevier Science. 468 p.

Obrist D, Agnan Y, Jiskra M, Olson CL et al 2017 Tundra uptake of atmospheric elemental mercury drives Arctic mercury pollution. *Nature* **547** 20-204 *doi:10.1038/nature22997*

Ogunbileje Jo, Sadagoparamanujam VM, Anetor JI, Farombi EO et al 2013 Lead, mercury, cadmium, chromium, nickel, copper, zinc, calcium, iron, manganese and chromium (VI) levels in Nigeria and United States of America cement dust. *Chemosphere* **90(11)** 2743-2749

Ogundiran MA, Adewoye SO, Ayandiran TA and Dahunsi SO 2014 Heavy metal, proximate and microbial profile of some selected commercial marine fish collected from two markets in south western Nigeria. *African Journal of Biotechnology* **13** 1147–1153

Panutrakul S, Rankiri PP, Krishantha J and Chirdsak C 2008 Heavy Metal Contents in Purpleback Squid (*Sthenoteuthis oualaniensis)* from the Bay of Bengal. The Ecosystem-Based Fishery Management in the Bay of Bengal. Department of Fisheries (DoF), Ministry of Agriculture and Cooperatives, Thailand. p 233-244

Pethybridge H, Cossa D and Butler ECV 2010 Mercury in 16 demersal sharks from southeast Australia: Biotic and abiotic sources of variation and consumer health implications. *Marine Environmental Research* **69(1)** 18-26

Pettine M, Camusso M, Martinotti W et al 1994 Soluble and particulate metals in the Po River: Factors affecting concentrations and partitioning. *Science of the Total Environment* **145(3)** 243-265

Penedo de Pinho P, Guimaraes JRD, Martins AS et al 2002 Total mercury in muscle tissue of five shark species from Brazilian Offshore waters: Effects of feeding habit, sex, and length. *Environmental Research* Section A **89** 250-258

Peralta-Videa, JR, Lopez ML, Narayan M, Saupe G and Gardea-Torresdey J 2009 Review-The biochemistry of environmental heavy metal uptake by plants: implications for the food chain. *The International Journal of Biochemistry and Cell Biology* **41(8-9)** 1665-1677; doi:10.1016/j.biocel.2009.03.005

Perkins DN, Drisse BMN, Nxele T and Sly PD 2014 E-Waste: A Global Hazard. Icahn School of Medicine at Mount Sinai. *Annals of Global Health* **80** 286-295

Perugini M, Visciano P, Manera M, Zaccaroni A, Olivieri V and Amorena M 2014 Heavy metal (As, Cd, Hg, Pb, Cu, Zn, Se) concentrations in muscle and bone of four commercial fish caught in the central Adriatic Sea, Italy. *Environmental Monitoring and Assessment* **186** 2205–2213

Radwan MA and Salama AK 2006 Market basket survey for some heavy metals in Egyptian fruits and vegetables. *Food and Chemical Toxicology* **44** 1273–1278

Rai, PK, Lee SS, Zhang M, Tsang YF, Kim KH 2019 Heavy metals in food crops: Health risks, fate, mechanisms and management. *Environment International* **125** 365–38

Rascio N, Navari-Izzo F 2011 Heavy metal hyper-accumulating plants: how and why do they do it? And what makes them so interesting? *Plant Sci.* **180** 169-81

Rattan RK, Datta SP, Chhonkar PK, Suribabu K, Singh, AK 2005 Long-term impact of irrigation with sewage effluents on heavy metal content in soils, crops and groundwater-a case study. *Agriculture Ecosystems and Environment* **109** 310–322

Reeder SW, Demayo A and Taylor MC 1979 *Cadmium*. *In*: Guidelines for surface water quality. Vol. 1: Inorganic chemical substances, Water Quality Branch, Inland Waters Directorate, Environment Canada, Ottawa.

Rose G and Kibria G 2006 Pesticide Monitoring in Goulburn-Murray Water's Irrigation Supply Channels covering the Six Irrigation Areas [2004-2006 Irrigation Season Study Report]. Report Prepared under a research collaboration agreement between G-MW and PIRVic. Goulburn Murray Rural Water Authority (G-MW), Tatura and Primary Industries Research, Vic, Werribee. *https://www.researchgate.net/publication/264899276_Pesticide_Monitoring_in_Irrigation_Water_Using_Innovative_Passive_Sampling_Technology_Victoria_Australia_%28research_collaboration_between_GM_Water_Australia_and_Department_of_Primary_Industries_Victoria_Australia%29*

Sanyal T, Kaviraj A and Saha S 2017 Toxicity and bioaccumulation of chromium in some freshwater fish, *Human and Ecological Risk Assessment* **23** *DOI: 10.1080/10807039.2017.1336425*

Sani A and Shehu A 2018 Determination of some heavy metals concentration in selected detergents used in Kano metropolis, Nigeria. *Environmental and Toxicology Studies Journal* **2(1)** 4

Shamuyarira KK and Gumbo JR 2014 Assessment of heavy metals in municipal sewage sludge: A case Study of Limpopo province, South Africa. *International Journal of Environmental Research and Public Health* **11** 2569-2579 *doi:10.3390/ijerph110302569*

Sompongchaiyakul P, Hantow J, Sornkrut S, Sumontha M and Jayasinghe RPPK 2008 An Assessment of mercury concentration in fish tissues caught from three compartments of the Bay of Bengal. The Ecosystem-Based Fishery Management in the Bay of Bengal. Department of Fisheries (DoF), Ministry of Agriculture and Cooperatives, Thailand. p 221-232

https://www.researchgate.net/publication/233903687_An_Assessment_of_Mercury_Concentration_in_Fish_Tissues_Caught_from_Three_Compartments_of_the_Bay_of_Bengal

Sondgrass WJ 1980 *Distribution and behaviour of nickel in the aquatic environment. In*: J.O. Nriagu (ed.). Nickel in the Environment. John Wiley and Sons, New York, Chapter **9** 203-274

Spear PA and Pierce RC 1979 *Copper in the aquatic environment*: Chemistry, distribution and toxicology. NRCC No 16454. Associate Committee on Scientific Criteria for Environmental Quality, National Research Council of Canada, Ottawa

Storelli MM and Barone G 2013 Toxic metals (Hg, Pb and Cd) in commercially important demersal fish from Mediterranean Sea: Contamination levels and dietary exposure assessment. *Journal of Food Science* **78(2)** T362–T366

Storelli MM and Marcotrigiano GO 2004 Content of mercury and cadmium in fish (*Thunnus alalunga*) and cephalopods (*Eledone moschata*) from the south-eastern Mediterranean Sea. *Food Additives & Contaminants* **21** 1051–1056 *DOI: 10.1080/02652030400023127*

Storelli MM, Ceci E, Storelli A and Marcotrigiano GO 2003 Polychlorinated biphenyl, heavy metal and methylmercury residues in hammerhead sharks: contaminant status and assessment. *Marine Pollution Bulletin* **46(8)** 1035-1039

Storelli MM, Giacominelli-Stuffer R and Marcotrigiano GO 2001 Total mercury and methyl-mercury in tuna fish and sharks from the South Adriatic Sea. *Italian Journal of Food Sciences* **13(1)** 101–10

Suk SH, Smith SW and Ramon DA 2009. Bioaccumulation of mercury in pelagic sharks from the Northeast Pacific Ocean. *California Cooperative Oceanic Fisheries Investigations Reports* **20** 172-177

Tariq SR, Shafiq M and Chotana GA 2016 Distribution of heavy metals in the soils associated with the commonly used pesticides in cotton fields. *Scientifica* **2016** 11 *http://dx.doi.org/10.1155/2016/7575239*

Tarley CRT, Coltro WKT, Matsushita M and De Souza NE 2001 Characteristic levels of some heavy metals from Brazilian canned sardines (*Sardinella brasiliensis*). *Journal of Food Composition and Analysis* **14** 611–617

Tchounwou PB, Yedjou CG, Patlolla AK and Sutton DJ 2012 Heavy metals toxicity and the environment. EXS **101** 133–164 *doi: 10.1007/978-3-7643-8340-4_6*

Teffer AK, Staudinger MD, Taylor DL and Juanes F 2014 Trophic influences on mercury accumulation in top-pelagic predators from offshore New England waters of the northwest Atlantic Ocean. *Marine Environmental Research* **101** 124–134

Tewari A, Joshi H, Trivedi R et al 2001 The Effect of Ship Scrapping Industry and its Associated Wastes on the Biomass Production and Biodiversity of Biota in in situ Condition at Alang. *Marine Pollution Bulletin* **42(6)** 461-468 *DOI: 10.1016/S0025-326X(00)00185-5*

Tom M, Fletcher TD, McCarthy, DT 2014 Heavy metal contamination of vegetables irrigated by urban stormwater: a matter of time? *PLoS One* **9(11)** 112-441 *doi:10. 1371/journal.pone.0112441*

Ullrich Susanne, Tanton Trevor and Abdrashitova Svetlana 2001 "Mercury in the aquatic environment: A review of factors affecting methylation". *Critical Reviews in Environmental Science and Technology* **31(3)** 241–293 *doi:10.1080/20016491089226*

USEPA 1987 *Ambient water quality criteria for zinc-1987*. EPA-440/5-87-003. Criteria and standard division, US Environment Protection Agency, Washington, DC

USEPA 1985 *Ambient water quality criteria for lead — 1984*. United States Environmental Protection Agency, Washington DC. EPA-440/5-84-027

USFDA 2001 US Food and Drug Administration (USFDA). Fish and Fisheries Products Hazards and Controls Guidance. 3rd edition. Appendix 5. FDA & EPA Safety Levels in Regulations and Guidance. Available Source:

http://vm.cfsan.fda.gov/~comm/haccp4x5.html

Vamerali T, Bandiera M and Mosca G 2010 Field crops for phytoremediation of metal-contaminated land. A review. *Environmental Chemistry Letters* **8(1)**117

Vargas AMM, and Nozaki J 2009 Effects of daily nickel intake on the bio-accumulation, body weight and length in tilapia (*Oreochromis niloticus*). *Toxicological & Environmental Chemistry* **91(4)** 751-759

Wang J, Chuang CY and Wang WX 2005 Metal and oxygen uptake in the green mussel *Perna viridis* under different metabolic conditions. *Environmental Toxicology and Chemistry* **24(10)** 2657–2664

Whitt M, Brown W, Danes JE and Keith Vorst KL 2016 Migration of heavy metals from recycled polyethylene terephthalate during storage and microwave heating. *Journal of Plastic Film and Sheeting* 32(**2**) 189–207

WHO 2010 Preventing diseases through healthy environment: action needed on chemicals of major public health concern. World Health Organization (WHO) 2010 Available from: *http://www.who.int/ipcs/assessment/public_health/chemicals_phc/en/.*

WHO 1996 Guidelines for drinking water quality, 2nd edition. Vol. 2. *Health criteria and other supporting information* (p. 940-949) and Addendum to Vol. **2**. 1998 (p. 281-283), Geneva, World Health Organisation

http://www.who.int/water_sanitation_health/GDWQ/summary_tables/sumtab.htm.

WHO/FAO/IAEA 1996 World Health Organization. Switzerland: Geneva; 1996. Trace Elements in Human Nutrition and Health ISBN: 92 4 156173 4. 361p

https://www.who.int/nutrition/publications/micronutrients/9241561734/en/

Worms I, Simon DF, Hassler CS and Wilkinson KJ 2006 Bioavailability of trace metals to aquatic organisms: importance of chemical, biological and physical processes on bio-uptake. *Biochimie* **88** 1721-1731

Wu RSS, Lau TC, Fung WKM, Ko PH, and Leung KMY 2007 An artificial mussel for monitoring heavy metals in a marine environments. *Environmental Pollution* **145** 104–10

Yaseen DA and Scholz M 2019 Textile dye wastewater characteristics and constituents of synthetic effluents: A critical review. *International Journal of Environmental Science and Technology* **16** 1193–1226 *https://doi.org/10.1007/s13762-018-2130-z*

Zhang F, Li Y, Yang M and Li W 2012 Content of heavy metals in animal feeds and manures from farms of different scales in northeast China. *International Journal of Environmental Research and Public Health* **9(8)** 2658–2668

Zhang H, Pin-Jing He P-J and Shao L-M 2008 Fate of heavy metals during municipal solid waste incineration in Shanghai. *Journal of Hazardous Materials* **156** 365–373

■■■■

4 Plastic Waste and Plastic Pollution Risks, Impacts and Management

Summary

Plastic waste and plastic pollution are ubiquitous and is reported from across the globe (from the Arctic to the Antarctic, from the surface to sediment, and in every waterway (lakes, rivers, estuaries, coastal and oceans). Plastics have been detected throughout the water column (surface, middle, and bottom) and in all the environmental compartments (biota, water, sediment). Plastic debris accounts for more than 75% of marine litter. At some places, plastic litter outnumbered fish larvae and plankton (the Danube River, Portuguese coastal waters, North Pacific central gyre). Plastic litters can act either as a source or sink or a vehicle to transport hydrophobic toxic chemicals/ contaminants in the aquatic environment. Plastics in the environment as waste pose five major problems and impacts- ingestion and entanglement of plastics by aquatic biota and wildlife; adsorption of persistent organic pollutants and heavy metals by plastics; colonisation and formation of biofilms (bacteria, cyanobacteria) on the surface of plastics; leaching of additives of the plastics and contamination of seafood and other foods by plastic particles. This chapter reviews: **i.** Chemistry, toxicity and ecotoxicity of plastics including chemical additives used; **ii**. Impacts of plastics on social, economic and environmental sectors including plastic residues in global waterways; adsorption of contaminants/pollutants by plastics; plastics in seafood (fish, mollusc, canned and dried fish), honey, sugar, beer, salt and drinking water; impacts of plastics on marine life, tourism, shipping and human's health; **iii.** Plastic pollution and climate change; **iv**. Management and mitigation measures to reduce impacts on the environment, biota, wildlife, seafood and human health.

Contents

List of Tables

List of Figures

List of Box

4.1. Introduction

The term plastic is derived from the Greek word "*plastikos*" (meaning able to be moulded). The first completely synthetic plastic, **bakelite**, was made in 1907 by Leo Baekeland. Plastics are persistent contaminants (Thompson 2017). Plastic waste/ plastic pollution is ubiquitous and is reported from across the globe (from the Arctic to the Antarctic, from the surface to sediment (see section 4.3) (Cressy 2016, Kibria 2017). Around 80% of marine litter is made of plastic. Recent estimates suggest that 4.8 to 12.7 million metric tons of plastic wastes entered the global marine environment in 2010 (Jambeck et al 2015). Rough estimates suggest that just 1% of the plastic in the ocean's floats on the surface, 5% drifts ashore onto beaches and the rest falls to the sea bottom. Plastics are a synthetic material/synthetic **polymer**[1] and made from petrochemicals (see section 4.2 for chemistry, toxicity and ecotoxicity of plastics). The common types of plastic are polyethylene, polyester, polyethylene terephthalate, polyvinyl chloride, polypropylene, polystyrene, polyamide and polycarbon (Table 4.1). In this chapter, the term plastic and polymer has been interchangeably used.

The plastics in most demand worldwide are polyethylene and polypropylene (Lackner 2015, Crawford and Quinn 2017). Plastics are ideally suited for a wide variety of applications, including packaging (accounts for 40% use), building and construction (accounts for 20% use), automotive and transport (accounts for 9% use), electrical and electronics (accounts for 6% use), agriculture (accounts for 3% use) and others (medical, sports, etc. accounts for 22% use) (Lackner 2015, Sebillie et al 2016). It is cheap, light-weight, strong, malleable and resistant to external shocks. They can be moulded or shaped (by heat and pressure). In 2014, the top three global producers of plastics were China, Europe and North America at 26%, 20% and 19%, respectively. The global plastic production was 0.35 million metric t (mmt) in 1950 and in 2016 it stood at 339.0 mmt (*https://www.statista.com/statistics/282732/global-production-of-plastics-since-1950/*).

---

[1]***A polymer*** *(poly-, "many" + -mer, "part") is a large molecule, or macromolecule, composed of many repeated subunits; Polymers, both natural and synthetic, are created via polymerization of many small molecules, known as monomers. The terms polymer and resin are often synonymous with plastic (https://en.wikipedia.org/wiki/Polymer)*

Table 4.1. Some common plastics, their application and recycling potential

[Hansen et al 2013, USEPA 2016, Crawford and Quinn 2017, Kibria 2018, *https://www.ryedale.gov.uk/attachments/article/690/Different_plastic_polymer_types.pdf; https://www.thebalancesmb.com/what-recycling-symbols-mean-4126251; https://www.bpf.co.uk/plastipedia/polymers/LDPE.aspx*; *Note: phthalates are reported to be present at up to 50% of the mass of finished PVC plastics (Oehlmann et al 2009); *Recycling symbols based on the American Society for Testing and Materials (ASTM) designator code]*.

***Recycling symbols (ASTM code)**	**Plastic types Density**	**Abbreviation & Recyclability**	**Uses/ Applications**
01 PET	Polyethylene terephthalate Density: 1.33-1.4 g/cm^3	PET [**easy to recycle**]	Beverage/drink bottles, fibers, filling material fleece, food containers, plastic film and sheeting, strapping, recording tapes.
02 PE-HD	High-density polyethylene (HDPE) Density 0.940–0.965 g/cm^3	HDPE [**can be recycled**]	Beverage bottles, buckets, chemical containers, corrosion protection cover (on steel cables), crates, food containers, gasoline tanks, helmets, kitchen utensils transport boxes, pipes (gas, sewage and water), tubing, and waste containers.
03 PVC	Polyvinyl chloride (rigid) Density: 1.38-1.53 g/cm^3	PVC [**commonly not acceptable**] (*presence of **phthalates** in PVC)	Cladding/covering, containers, electrical conduit, flame retardant properties (cables, bottles, and containers), gutters, panels for buildings, pipes, sandwich panels, window frames.
03 PVC	Polyvinyl chloride (plasticised)	PVC [**commonly not acceptable**]	Balls, cable insulation, dolls, flooring, footwear, garden hose, mats and sheeting's, membranes for swimming pools, packaging, plastic covers, wall panels.
04 PE-LD	Low-density polyethylene Density: 0.917–0.930 g/cm^3	LDPE [**not often recycled**]	Bags (grocery/plastic), cable isolation, carton linings, food packaging films/foils, kitchen tools, laboratory equipment, mulch film, pallet sheets, shrink wrap, tubes, waste bins, outdoor furniture, squeezable bottles.

***Recycling symbols (ASTM code)**	**Plastic types Density**	**Abbreviation & Recyclability**	**Uses/ Applications**
05 PP	Polypropylene (PP) Density: 0.925–0.959 g/cm^3	PP [**occasionally recycled**]	Automotive parts, battery cases, bottle caps and container lids, boxes of all kinds, cable insulation, dishware's, fibres and filaments, foils, garden/outdoor furniture, kitchen utensils, medical containers, packaging tape, toys, parts of washing machines, pipes, plates, rope, suction tubes, toys, suitcases.
06 PS	Polystyrene (PS) Density 1.04-1.07 g/cm^3	PS [**difficult to recycle**]	Boxes for CD and DVD, disposable cutlery and tableware, foamed food containers, single-use packaging for food, pharmaceuticals and cosmetics and disposable tableware.
06 PS	Polystyrene foams (expanded/ extruded)	EPS, XPS [**difficult to recycle**]	Disposable food containers and cups, egg cartoons and cups, foam boxes, meat trays, packaging foam, thermal insulation materials (buildings, shock-absorbing packaging, life belts).
07 O	Polycarbonate Density: 1.2 g/cm^3	PC [**difficult to recycle/never recycled**]	Cycling helmets, eye protection, food containers, hair dryers, information storage discs, protective glass for rear lights of cars, pump wheels, propellers, security windows.
07 O	Polymethylme-thacrylate Density: 1.17-1.2 g/cm^3	PMA/PMMA [**difficult to recycle/never recycled**]	Aircraft canopies and windows, contact lenses, fluorescent light diffusers, protection glass for instruments, vehicle lighting covers and lenses, window glass/glazing.
07 O	Polytetrafluo-roethylene Density: 2.15 g/cm^3	PTEF [**difficult to recycle/never recycled**]	Artificial blood vessels and other implants, automotive applications, breathable fabrics, electrical applications in airplanes and

*Recycling symbols (ASTM code)	Plastic types Density	Abbreviation & Recyclability	Uses/ Applications
			robots, non-stick coatings for cookware and bake ware, coating in chemical apparatuses, plumbing tape.
07 O	Acrylonitrile butadiene Styrene	ABS **[difficult to recycle/never recycled]**	Automotive applications, door handles, cabinets for office machines, television and radios, loudspeakers, telephones, computers, tools, toys (building bricks), electrical and electronic equipment, inner liners for refrigerators and frost boxes, pipes, suitcases.
07 O	Polyamide/Nylon Density: 1.12-1.16 g/cm^3	PA **[difficult to recycle/never recycled]**	Automotive applications, buttons for electronic devices, cabinets, fibers, gears, fishing nets and lines, kitchen machines and utensils, monofilament bearings.
07 O	Polychloroprene (neoprene)	CR **[difficult to recycle/never recycled]**	Automotive gaskets, footwear, gloves, diving garments, medical bandages and supports, weather balloons.

4.1.1. Pros and cons of plastics

Plastics do have several advantages and disadvantages. Some are listed below (Lackner 2015, Kibria 2017, Rogers 2019).

- **Advantages** (they are low-cost, has good and excellent technical properties; has easy processability; can save energy and resources compared with other materials and recycling is possible).
- **Disadvantages** (they are based on petrochemicals i.e. fossil fuels; some plastics are difficult to recycle; microplastics are not biodegradable and persist in the environment for a very long period (see section 4.2); uncontrolled combustion of plastics can release toxic substances (CO_2) that causes climate change (see section 4.4 on plastics and climate change); microplastics can harm biota (fish, birds, turtles) as biota often mistake microplastics as food; microplastics do a**d**sorb toxic and harmful pollutants that can be transferred to biota and humans via the food chain; microplastics

are a source of air pollution in the form of dust and airborne fibrous particles and can contaminate food (seafood, salt, honey, beer) and drinking water (see section 4.3).

4.1.2. Classification of plastics

Based on thickness/size, plastics are classified as **nanoplastic** (<1 μm), **microplastic** (1 μm-5 mm), **mesoplastic** (5-25 mm) and **macroplastic** (> 25 mm). Microplastics are further classified into primary microplastics and secondary microplastics (based on morphology/appearance) (Alimi et al 2018; Crawford and Quinn 2017, Germanov et al 2018).

- Primary microplastics include **microbeads** (microbeads are commonly made from PE, PP, PET, PA and polymethyl methacrylate (PMA) and found in cosmetics, personal care products (soap, toothpaste, facial scrub), nail polish, lipsticks, microfibres used in textiles, virgin pellets and cleaning products) (Potter 2017, Germanov et al 2018) and **pellets** (nurdles) (nurdles are a very small pellet of plastic which serves as raw material in the manufacture of plastic products).
- Secondary microplastics (resulting from environmental degradation of larger plastic items as a result of weathering, wave action, wind abrasion, biodegradation and ultraviolet photo-degradation) include **synthetic fibres** (fishing gears, textiles) and **fragments** (plastic bags, bottles, car tyres), and **films** (Table 4.2).

Table 4.2. Classification of plastics based on size and morphology
(Crawford and Quinn 2017, Alimi et al 2018)
[terms mentioned in this Table 4.2 are also used in research and investigations to describe and identify microplastics].

Type and Abbreviation	Size	Definition
Fibre (FB)	<5 mm-1 mm	A strand or filament of plastic less than 5 mm to 1 mm in size along its longest dimension.
Film (FI)	<5 mm-1 mm	A thin sheet or membrane-like piece of plastic less than 5 mm to 1 mm in size along its longest dimension.
Foam (FM)	<5 mm-1 mm	A piece of sponge, foam, or foam-like plastic material less than 5 mm to 1 mm in size along its longest dimension.
Fragment (FR)	<5 mm-1 mm	An irregular shaped piece of plastic less than 5 mm to 1 mm in size along its longest dimension.

Type and Abbreviation	Size	Definition
Macroplastic (MAP)	≥25 mm	Any piece of plastic equal to or larger than 25 mm in size along its longest dimension.
Mesoplastic (MEP)	<25 mm-5 mm	Any piece of plastic less than 25 mm to 5 mm in size along its longest dimension.
Microbead (MBD)	< 1mm-1 μm	A small spherical piece of plastic less than 1 mm to 1 μm in diameter.
Microfibre (MFB)	< 1 mm-1 μm	A strand or filament of plastic less than 1 mm to 1 μm in size along its longest dimension.
Microfilm (MFI)	<1 mm-1 μm	A thin sheet or membrane-like piece of plastic less than 1 mm to 1 μm in size along its longest dimension.
Microfoam (MFM)	<1 mm-1 μm	A piece of sponge, foam, or foam-like plastic material less than 1 mm to 1 μm in size along its longest dimension.
Microfragment (MFR)	<1 mm-1 μm	An irregular-shaped piece of plastic less than 1 mm to 1 μm in size along its longest dimension.
Microplastic (MP)	<5 mm-1 μm	Any piece of plastic less than 5 mm to 1 mm in size along its longest dimension.
Mini-microplastic (MMP)	<1 mm-1 μm	Any piece of plastic less than 1 mm to 1 μm in size along its longest dimension.
Nanoplastic (NP)	<1 μm	Any piece of plastic less than 1 μm in size along its longest dimension.
Pellet (PT)	<5 mm-1 mm	A small spherical piece of plastic less than 5 mm to 1 mm in diameter.

4.1.3. Plastic transportation and sources in the aquatic environment

Plastic pollution in freshwater and marine environments has been identified as a global problem. It is estimated that plastic debris accounts for 60–80% of marine litter, reaching to as high as 90–95% in some areas (Xanthos and Walker 2017). Around 80% of plastic pollution in the marine environment originates from land-based sources while the remainder comes from ocean-based sources (fishing nets, fishing ropes) (Sebillie et al 2016). Plastics can enter or be transported to the aquatic environment through many pathways including the following (Lambert et al 2014, IUCN 2018, USEPA 2018).

- intentional littering/ open dumping of bags, bottles, and other plastic items (during large public gatherings/ events, or from coastal tourism);
- urban and stormwater run-off;
- extreme weather events (floods and cyclones);
- wind-blown debris;
- wastewater/sewage treatment discharges (as most wastewater treatment facilities don't filter out microplastics/ microbeads);
- sewage overflows;
- landfill[2] waste (mismanaged plastics)
- accidental spillage during transport and handling of plastic pellets/ virgin pellets;
- release of microfibres during washing/laundering of synthetic clothes/ textiles;
- discharge of microbeads from use of personal care products;
- fibres from fishing nets and lines (plastic debris from fishing activities may be a key source in some areas (Galgani et al 2000, Andrady 2011, Lambert et al 2014);
- improper or ineffective solid waste management.

The Ocean plastics originate from synthetic textiles (35%), tyres (28%), urban dusts (24%), road marking (7%), ship hull anti-rot paints (3.7%); personal hygiene products (2%), plastic pellets (0.2%) (*https://www.dw.com/en/river-of-plastic-trash-is-flooding-our-oceans/a-39817471*); See also section 4.3.1.

4.2. Chemistry, toxicity and ecotoxicity of plastics

4.2.1. Chemistry

Plastics are a synthetic material/ synthetic organic **polymer** (*poly* = many; *mer* = part; polymers are composed of many repeated subunits of monomers (*mono* = one; *mer* = part). In fact, plastics are simply a chain of molecules linked together.

[2]*A **landfill** site (also known as a tip, dump, rubbish dump, garbage dump or dumping ground) is a site for the disposal of waste materials by burials. Landfill is the oldest and most common form of waste disposal.*

These chains are called polymers (Fig. 4.1). Ethylene and propylene are important sources of plastic products. Since plastic is made from non-renewable petrochemicals, so, consequently, they are principally based on the carbon (up to 87.6% C) and hydrogen (up to 14.5% H) atom (Table 4.3). It usually, consists of additives, fillers, colours (Lackner 2015) (section 4.2.3).

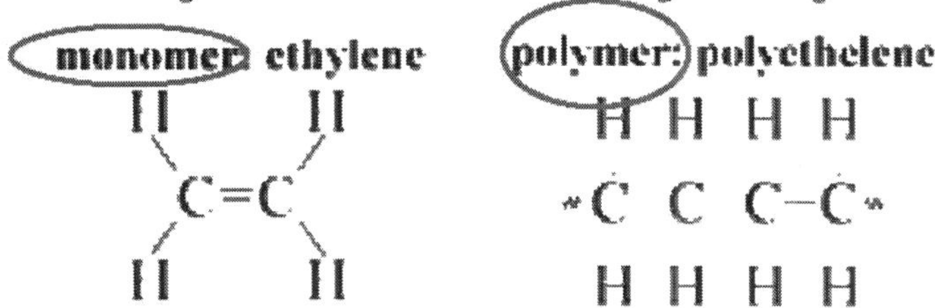

Fig. 4.1. Structure of monomer (ethylene) and polymer (polyethylene)
[*https://www.ck12.org/c/physical-science/carbon-monomers-and-polymers/lesson/Carbon-Monomers-and-Polymers-MS-PS/*]

Table 4.3. Chemical composition (% by weight) of representative plastics in solid wastes (Saquing et al 2010).

Polymer types	Main monomer	C	H	Cl	N	O	Ash
i. Pure polymers							
HDPE (high density polyethylene)	PE (polyethylene)	87.6%	14.5%	-	0.1%	-	0.0%
LDPE (low density polyethylene)	PE (polyethylene)	84.6%	6.1%	-	<0.002 %	-	0.0%
MDPE (Medium-density polyethylene)	PE (polyethylene)	85.2%	5.8%	-	0.004%	-	0.0%
PVC (polyvinyl-chloride)	PVC (polyvinylchloride)	38.9%	4.7%	56.7%	0.1%	-	0.0%

Polymer types	Main monomer	C	H	Cl	N	O	Ash
ii. Single resin polymer							
Disposable cold cup	PS (polystyrene)	84.0%	7.1%	-	0.0%	-	7.1%
Drinking water container	PE (polyethylene)	84.4%	7.6%	-	0.0%	-	
Furniture foam	PU (Polyurethane)	58.7%	5.2%		5.8%	29.1%	5.2%
Prescription drug bottle	PP (Polypropylene)	84.3%	8.2%	-	<0.1%	-	
PVC pipe	PVC (polyvinylchloride)	37.6%	5.1%	53.6%	<0.02%	-	5.8%
iii. Composite materials							
Nylon carpet	PA6 (polyamide 6 nylon)	35.1%	4.6%	-	4/0%	14.1%	36.4%
Vinyl flooring	PVC (polyvinylchloride)	19.4%	4.5%	-	0.1%	-	57.7%

4.2.2. Glass transition temperature (Tg) of plastics

The Glass Transition Temperature (Tg)[3] is one of the most important properties of amorphous (non-crystalline) polymers. Tg is the range of temperatures where the polymer transition occurs from hard, glassy material to a soft, rubbery material. Rubbery or "soft: plastics have a low Tg (e.g. PE, LDPE, HDPE, PP) whereas glassy or "hard" plastics have a high Tg (e.g. PVC, PET) (Table 4.4). Tg property is related to the adsorption/ sorption capacity of pollutants by plastics. For example, the rubbery polymers have higher sorption capacities, a greater diffusion, and greater affinity for contaminants whereas glassy polymers have lower sorption (see section 4.3 adsorption of pollutants by plastics). In general, rubbery polymers (PE) are expected to allow greater diffusion of contaminants into the polymer than glassy polymers (PVC, PET) (Alimi et al 2018).

*[3]A **glass transition temperature** (Tg) is the temperature at which a polymer turns from a ductile material (solid materials that can be deformed) to a hard, brittle material.*

Table 4.4. Glass transition temperature (Tg) of some selected polymers

[Saquing et al 2010, Alimi et al 2018; *https://www.misumi-techcentral.com/tt/en/mold/2011/12/106-glass-transition-temperature-tg-of-plastics.html;http://www.appstate.edu/~clementsjs/polymerproperties/ plastics_low_temp.pdf*;
note: at room temperature, plastics that have a low Tg (e.g. PE has a Tg -20°C) are classified as rubbery or "soft" while those that have a high Tg (e.g. PVC has a Tg of 80-90°C) are considered glassy or "hard" (Gauthier 1995)].

Rubbery plastics		**Glassy plastics**	
Plastic-type	**Tg (°C)**	**Plastic-type**	**Tg (°C)**
PE (Polyethylene)	-125	**PVC** (Polyvinyl chloride)	+87
HDPE (High-density polyethylene)	-90	**PP** (Polypropylene)	0
LDPE (Low-density polyethylene)	-110	**PS** (Polystyrene)	+100
POM (Polyoxymethylene)	-30	**PET** (Polyethylene terephthalate)	69
PUR (Polyurethane)	-20	**PA** (nylon)/Polyamide	+47
Drinking water container	-20	**Soda bottle**	+79

4.2.3. Chemical additives in plastic materials

In plastic materials, a number of **additives** (plasticisers, flame retardants, heat stabilisers, antioxidants, biocides, colourants, fillers etc.) are incorporated to enhance polymer properties, improve the performance, functionality and ageing properties (prolonging polymer life) of the polymer. The most commonly used chemical additives are listed below (based on OECD 2009, Hansen et al 2013, Hahladakis et al 2018):

- **Plasticizer**s (for improving the flexibility, durability and stretchability of polymeric film) [e.g. Benzyl butyl phthalate (BBP), Bis/di (2-ethylhexyl)phthalate (DEHP), Dibutyl phthalate (DBP)]; typical **amou**nt range is 10-70% w/w (weight weight);
- **Flame retardants** (to reduce the risk of ignition and retard combustion) [e.g. polybrominated diphenyl ethers (PBDEs]; typical amount range is 3-25% w/w (weight weight);

- **Heat stabilisers and antioxidants** (heat stabilisers to prevent thermal degradation of polymers; antioxidants to inhibit degradation of polymers [e.g. (Bisphenol A (BPA); Cd, Pb compounds; nonylphenol; octyl phenol]; typical amount range is 0.05-3% w/w (weight weight);
- **Biocides** (to protect plastics from microorganisms like bacteria, algae, fungi and mould) [e.g. arsenic (As), tin (Sn), triclosan]; typical amount range is 0.001-1% w/w (weight/weight);
- **Colourants** (used in highly transparent plastics (PS, PMMA and cellulose plastics) to give a bright transparent colour) [e.g. azo colorants]; typical amount range is 0.25-5% w/w (weight weight);
- **Fillers** (inert materials to reduce polymer costs) [e.g. calcium carbonate, clay, zinc oxide, metal powder, wood powder, siliceous earth]; typical amount range is up to 50% w/w (weight/weight);
- **Reinforcements** [e.g. glass fibres, carbon fibres]; typical amount range is 15-30% w/w (weight/ weight).

One of the risks of plastic additives is that they can leak and potentially migrate from various plastic products. Research carried out reveals that a number of toxic substances were released from plastic products including **phthalates[4], brominated flame retardants, bisphenol-A, lead, tin and cadmium, nonylphenol[5]** and many other volatile organic compounds (Hahladakis et al 2018). Some of the plastic additives are very harmful to biota and humans being toxic, bioaccumulative, endocrine disrupting and carcinogenic (see section 4.2.4 and Table 4.5). Fasano et al (2012) reported migration of plastic additives (phthalates, alkylphenols, 2,2-bis(4-hydroxyphenyl) propane, bisphenol A and di(2-ethylhexyl) adipate (DEHA) from **packaging materials** (tuna cans, marmalade cap, yogurt packaging, polystyrene dish, teat, bread bag, film, baby's bottle, aseptic plastic laminate paperboard carton and plastic wine top) into food. Li et al (2016a) studied the migration of five phthalates from disposable table wares (feeding bottles, paper bowls, plastic cups, Dixie cup/ disposable paper cups) to drinking water, where concentrations of DBP and DEHP in the drinking water samples exceeded the guideline limit levels for drinking water (8 ng/ mL and 3 ng/mL respectively). BPA can leach from polycarbonate plastic that is

---

[4]***Phthalates*** *are a group of chemicals (related to the organic acid, phthalic acid) used to make plastics more flexible and harder to break. They are often called plasticisers. Phthalates are used widely in polyvinyl chloride (PVC) plastics, which are used to make products such as plastic packaging film and sheets, garden hoses, inflatable toys, blood-storage containers, medical tubing, pipes, vinyl wall and floor coverings, roofing materials etc.*

[5]***Nonylphenol*** *are non-ionic surfactants/detergent-like substances. Nonylphenols fall into the general chemical category of alkylphenols. They are used in manufacturing antioxidants, lubricating oil additives, laundry and dish detergents, emulsifiers, and solubilizers. These compounds are also precursors to the commercially important non-ionic surfactants alkylphenol ethoxylates and nonylphenol ethoxylates, which are used in detergents, paints, pesticides, personal care products, and plastics. Nonylphenol is an endocrine disruptor.*

commonly used as can linings. It has been detected in food cans. Studies carried out by Ye et al 2009 found that canned foods contributed at least 60% to the total daily intake of BPA among adults in Japan (Miyamoto and Kotake 2006) and the EU (European Commission 2008).

4.2.4. Toxicity and ecotoxicity of plastic additive compounds

Toxicity and ecotoxicity of most detected plastic additive chemicals including phthalates, PBDEs, Bisphenol-A, nonylphenols and heavy metals are summarised in Table 4.5. Plastic additive chemicals fall within the category of persistence, bioaccumulative, toxic, endocrine disrupting and some are also carcinogenic as highlighted below:

i. Plastics are persistent: The lifespan of plastic is estimated to be hundreds to thousands of years (Wang et al 2016). Some disposable plastic products (water bottle, tooth brush, fishing line) can take over 400 years to biodegrade (Fig. 4.2).

ii. Plastic additive compounds are bioaccumulative: Phthalates (DBP, DEHP), Bisphenol-A (BPA), PBDEs, nonylphenol are hydrophobic and highly bioaccumulative having a log K_{ow} or Octanol-water partition coefficient values > 3.4 (BPA) to 8.05 (BDE-153) (Table 4.5).

iii. Plastic additive compounds are very toxic: Phthalates (DBP, DEHP), Bisphenol-A (BPA), PBDEs, nonylphenol and heavy metals are highly toxic to aquatic biota (algae, daphnia and fish) with EC_{50}/LC_{50} values < 1 mg/L to 10 mg/L (Table 4.5).

iv. Plastic additive compounds are endocrine-disrupting: Phthalates (DBP, DEHP), Bisphenol-A (BPA), PBDEs, nonylphenol and heavy metals can act as an endocrine disruptor on animals and humans (Kibria et al 2010, Table 4.5 (see Box 4.1 for endocrine disputing chemicals). Nonylphenol has endocrine disruption effects as it induced vitellogenin[6] (an egg yolk precursor[7]) in male rainbow trout (*Oncorhynchus mykiss*) at 20.3 mg/L (Jobling et al 1996), and medaka (*Oryzias latipes*) at 0.1 mg/L (Tabata et al 2001). Salmon fry (*Salmo salar*) exposed to 1.0 mg/L of BPA for 6 days had yolk-sac edema (buildup of fluid in fish) and haemorrhage (Crain et al 2007).

v. Plastic additive compounds are carcinogenic: Phthalates (DEHP) has been classified as IARC group 2B (possibly carcinogenic to humans), some heavy metals (Cd) are carcinogenic to humans. PBDEs are a potential carcinogen (IARC group 2A- probably carcinogenic) and can increase the risk of thyroid cancer in humans (Table 4.5).

---

[6]***Vitellogenin*** *a protein present in the blood, from which the substance of egg yolk is derived.*

[7]***A precursor*** *is a compound that participates in a chemical reaction that produces another compound, or a chemical substance that gives rise to another more important chemical product*

Plastic items, years to biodegrade

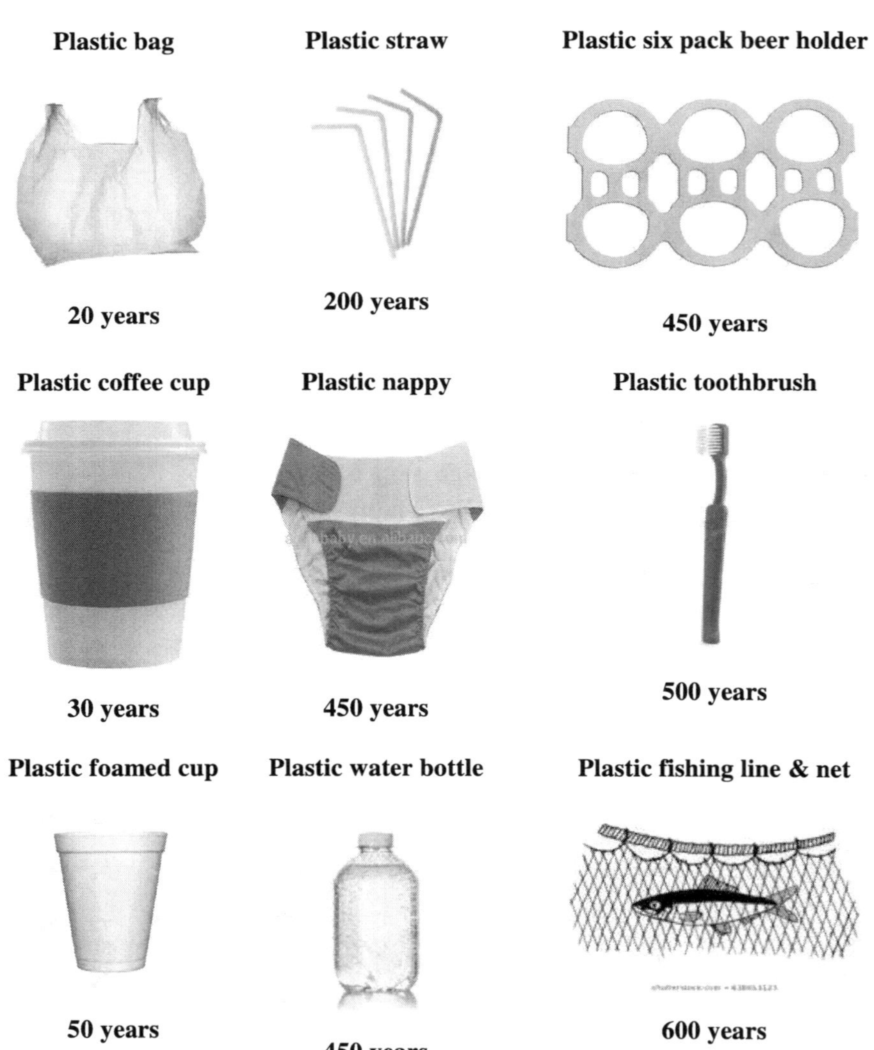

Fig. 4.2. The lifecycle of some plastic products

https://www.telegraph.co.uk/news/2018/01/10/stark-truth-long-plastic-footprint-will-last-planet/

https://www.wwf.org.au/news/blogs/the-lifecycle-of-plastics#gs.iywjrd

Box 4.1. Endocrine disputing chemicals (EDCs)

EDCs comprise a broad-class of exogenous substances including many man-made chemicals that are widely dispersed and detected in the environment. EDCs interfere with the hormonal system altering endocrine regulation of essential physiological function. Endocrine disruption can result in adverse effects on the development of the brain and nervous system, growth and function of the reproductive system in fish. Endocrine disruptors affect many aspects of transcription and transcriptional regulation that influences gene expression. Some molecules identified as endocrine disruptors are listed below:

- natural and synthetic steroid estrogens [17β-estradiol (**E2**), estrone (**E1**), and 17 α-ethinylestradiol (**EE2**)];
- plastic additives [Bisphenol-A (**BPA**), plasticizers (**phthalates-BBP**, **DBP**, **DEHP**)];
- synthetic chemicals [polychlorinated biphenyls (**PCBs**), polybro-minated biphenyls (**PBBs**), dioxins];
- alkyl phenol [nonylphenols (**NP**), octylphenols (**OP**)];
- pesticides [methoxychlor, chlorpyrifos, dichlorodiphenyl-trichloroethane (**DDT**), vinclozolin, dieldrin, heptachlor epoxide, and hexachlorobenzene, endosulfan];
- pharmaceutical agents [diethylstilbestrol (**DES**)];
- metals [**As, Cd**, **Hg**, **Pb**, **Zn**];
- polybrominated diphenyl ethers (**PBDEs**), dioxins.
- triclosan (an antibacterial and antifungal agent)
- polycyclic aromatic hydrocarbons/ petroleum hydrocarbons [**BTEX** and **PAHs**] (BTEX = benzene, toluene, ethylbenzene, xylene; PAH = Polycyclic aromatic hydrocarbons).

(Kibria et al 2010, Caballero-Gallardoa et al 2016).

Table 4.5. Selected plastic additive substances and their chemistry, toxicity and ecotoxicity

*[**BDE-15** = 4,4'-Dibrominated diphenyl ether; **BDE-47** = 2,2',4,4'-Tetrabrominated diphenyl ether; **BDE-99** = 2,2',4,4',5-Pentabrominated diphenyl ether; **BDE-153** = 2,2',4,4',5,5'-Hexabrominated diphenyl ether; **DeBDE** = Decabromodiphenyl ether; **PBDEs** = polybrominated diphenyl ethers; **PeBDE** = pentabromodiphenyl ether; note: **PBDEs** have 209 congeners (variants) arranged in ascending numerical order using a numbering system: **Penta BDE** = BDE 81-127; **Octa BDE** = BDE 194-205 and **Deca BDE** = BDE 209; Toxicity classification: **Rat** LD_{50} < 50 mg/kg = extremely toxic; 50-500 mg/kg = very toxic; 500-5,000 mg/kg = moderately toxic; Ecotoxicity classification: **Fish** LC_{50} < 1 mg/L = extremely toxic; 1-10 mg/L = very toxic; 10-100 mg/L = moderately toxic].*

	Chemistry		**Toxicity**	**Ecotoxicity**	
Chemical	**Water solubility**	**Log K_{ow}**	**Acute toxicity to mammal LD_{50} (Rat) oral**	**Acute aquatic toxicity** (algae, daphnia and fish) (EC_{50}/LC_{50})	**Remarks**
Phthalate (Di-n-Butylphthalate (DnBP)	11.2 mg/L (Staples et al 1997)	4.45 (Staples et al 1997)	Rat: LD_{50}: 7499 mg/kg (MSDS-PubChem-Dibutyl phthalate 2005)	• Algae, *Pseudokirchneriella subspicata* EC_{50}: 0.4 mg/L for 96 h (Adams et al 1995) • Algae, *Scenedesmus subspicatus* EC_{50}: 9 mg/L for 48 h (Kuhn and Pattard 1990) • Water flea, *Daphnia magna* EC_{50}: 3.0 mg/L for 48 h (Adams et al 1995) • Water flea, *D. magna* EC_{50}: 5.2 mg/L for 48 h (McCarthy and Whitmore 1985) • Fish, rainbow trout, *Oncorhynchus mykiss* LC_{50}: 1.5-6.5 mg/L for 96 h (Mayer and Ellersieck 1986) • Fish, *Lepomis macrochirus* LC_{50}: 0.73-6.5 mg/L for 96 h (Mayer and Ellersieck 1986).	Endocrine disruptors (van Wezel et al 2000); bioaccumulative (Oehlmann et al 2009); ATSDR substance priority list rank #56 (ATSDR 2017).
Phthalate (Di-2-ethylhexyl phthalate (DEHP) (also called Bis(2-	0.003 mg/L (Staples et al 1997, EU 2008)	7.5 (Staples et al 1997, EU 2008)	Rat: LD_{50}: 30,000 mg/kg (MSDS- CDH-DEHP 2006), IARC- gr 2B (IARC 2019)	• Algae, *P. subspicata* EC_{50}: 0.1 mg/L for 96 h (Adams et al 1995) • Algae, *S. subspicatus* EC_{50}: 130 mg/L for 72 h (Scholz 1995)	DEHP has been classified as IARC group: 2B (possibly carcinogenic to humans) (IARC 2019); endocrine

	Chemistry		Toxicity	Ecotoxicity	
Chemical	**Water solubility**	**Log K_{ow}**	**Acute toxicity to mammal LD_{50} (Rat) oral**	**Acute aquatic toxicity** (algae, daphnia and fish) (EC_{50}/LC_{50})	**Remarks**
ethylhexyl phthalate)) (DEHP)				• Water flea, *D. magna* EC_{50}: 0.1-2 mg/L (data compiled by van Wezel et al 000) • Fish, *O. mykiss* LC_{50}: >0.32 mg/L for 96 h (Adams et al 1995) • Fish, *O. mykiss* LC_{50}: >20 mg/L for 96 h (Defoe et al 1990) • Fish, sheepshead minnow, *Cyprinodon variegatus* LC_{50}: 0.17 mg/L for 96 h (Adams et al 1995) • Fish, sheepshead minnow, *C. variegatus* LC_{50}: 550 mg/L for 96 h (Heitmuller et al 1981).	disruptos (van Wezel et al 2000); Bio accumulative (Oehlmann et al 2009); ATSDR substance priority list rank number#77 (ATSDR 2017).
Phenols Bisph-enol-A	120–300 mg/L (Staples et al 1998)	3.4 (Staples et al 1998)	Rat: LD_{50}: 2,000-5,000 mg/kg (MSDS-Kumho-Bisphenol A 2015)	• Algae, *S. capricornutum* EC_{50}: 2.7-3.1 mg/L for 96 h (cell count/ cell volume) (Alexander et al 1988) • Algae, *S. capricornutum* EC_{50}: 2.5 mg/L for 96 h (cell growth) (Stephenson et al 1983) • Water flea, *D. magna* EC_{50}: 3.9 mg/L for 48 h (immobilisation) (Stephenson 1983) • Water flea, *D. magna* EC_{50}: 20 mg/L for 48 h (immobilisation) (Hendriks et al 1994) • Fish, fathead minnow *Pimephales promelas* LC_{50}: 4.6-4.7 mg/L for 96 h (mortality) (Alexander et al 1988) • Fish, *Salmo gairdneri* LC_{50}: 3.0-3.5 mg/L for 96 h (mortality) (Reiff and Phil 1979).	Bioaccumulative ,endocrine disrupting (Oehlmann et al 2009), high to moderately toxic to aquatic life.

	Chemistry		Toxicity	Ecotoxicity	
Chemical	**Water solubility**	**Log K_{ow}**	**Acute toxicity to mammal LD_{50} (Rat) oral**	**Acute aquatic toxicity** (algae, daphnia and fish) (EC_{50}/LC_{50})	**Remarks**
PBDEs Polybrominated diphenyl ethers (Flame retardants)	13.3 μg/L, < 1 μg/L and < 1 μg/L for penta BDE, Octa BDE and Deca BDE respectively (USEPA 2017, ATSDR 2017)	5.9-6.2 (Darnerud et al 2001) 5.86, 6.78, 7.39, 8.05 for BDE-15, BDE-47, BDE-99, BDE-153 respectively) (Kuramochi et al 2007)	Rat LD_{50}: 500-5000 mg/kg (PentaBDE) (Gill et al 2004), IARC- gr 2A (IARC 2019)	• Algae, EC_{50}:>1 mg/L for 96 h (growth) (DeBDE) (Walsh et al 1987, USEPA 2018) • Water flea, *D.magna* EC_{50}: 0.058 mg/L for 48 h (PeBDE) (Scanlan et al 2015) • Water flea, *D. magna* EC_{50}: 0.165 mg/L for 48 h (PeBDE) (Tang et al 2010) • Water flea, *D. magna* EC_{50}: 0.02 mg/L for 48 h (DeBDE) (immobilization) (Nakari and Huhtala 2010, USEPA 2018) • Fish, Turbot, *Psetta maxima* LC_{50}: 0.014-0.027 mg/L for 48 h (embryos and larvae) (BDE-47) (Mhadhbi et al 2012) • Fish, *P. maxima* LC_{50}: 0.029-0.038 mg/L for 48 h (embryos and larvae) (BDE-99) (Mhadhbi et al 2012).	Extreme persistence, toxic (USEPA 2016), potential carcinogens, neurotoxins and endocrine disrupting (Akortia et al 2016),bioaccumulative, highly toxic to aquatic life and human health (Lu et al 2018), PBDEs can increases the risk of thyroid cancer in humans (Zhang et al 2008).
Nonylphenol	3 mg/L (MSDS-Sasol-Nonylphenol 2015)	4.36-4.60 (De Araujo et al 2018)	Rat: LD_{50}: 300-2000 mg/kg (MSDS-Sasol-Nonylphenol 2015)	• All Algae, EC_{50}: 0.025-0.75 mg/L (Naylor 1995) • Water flea, *D. magna* LC_{50}: 0.19 mg/L for 48 h (Naylor 1995) • Water flea, *D. magna* EC_{50}: 0.3 mg/L for 48 h (Comber et al 1993) • Water flea, *D. magna* EC_{50}: >0.01-1 mg/L for 48 h (MSDS-Sasol-Nonylphenol 2015) • All fish, LC_{50}: 0.13-1.4 m/L (Nylor 1995) • Killifish, *Fundulus heteroclitus* EC_{50}: 0.21-5.4 mg/L (newly hatched	Persistent, bioaccumulative, endocrine disrupting (Soares et al 2008), highly toxic to aquatic life (algae, daphnia and fish) having EC_{50}/LC_{50} values<1 mg/L.

	Chemistry		Toxicity	Ecotoxicity	
Chemical	**Water solubility**	**Log K_{ow}**	**Acute toxicity to mammal LD_{50} (Rat) oral**	**Acute aquatic toxicity** (algae, daphnia and fish) (EC_{50}/LC_{50})	**Remarks**
				larvae to embryos) (Kelly and Di Giulio 2000) • Fish, *P. promelas* > 0.1-1 mg/L for 96 h (MSDS-Sasol-Nonylphenol 2015).	
Heavy metals Cd, Cr, Pb	See Table 3.7 in Chapter 3			See Table 3.7 in Chpater 3	Persistent, bioaccum-ulative, toxic, endocrine disrupting and carcinogenic.

4.3. Impacts of plastics on social, economic and environmental sectors

4.3.1. Plastic residues in global waterways

Table 4.6 provides an account of synthesised global data and information related to plastics pollution in waterways. The **summary of Table 4.6** is highlighted below:

- **Global plastic pollution:** Microplastics (MPs) have been detected in waterways (lakes, rivers, estuaries, coastal and oceans) across the globe including the Antarctic sea, Arctic Ocean, Argentina, Australia, Austria, the Bay of Bengal, Belgium, Brazil, Canada, China, European coasts, France, Germany, Great Pacific Garbage Patch (between Hawaii and California), Greece, Greenland, Hong Kong, India, Indian Ocean, Indonesia, Iran, Italy, Japan, the Maldives, Mediterranean Sea, Mongolia, the Netherlands, New Zealand, North Pacific Central Gyre, Norway, Oceania, Papua New Guinea, Portugal, Singapore, South Africa, Sub-Antarctica, Sweden, Switzerland, Taiwan, Turkey, UK and USA (Table 4.6).
- **Plastic pollution "Hotspots"[8] in the world:** The plastic pollution 'hot spots' so far been identified includes: East Asia Seas (Japan); Rhine River, Germany; Great Pacific Garbage Patch (between Hawaii and

[8]***Pollution "hotspots"-*** *The places/ areas/ sites with high concentrations of a pollutant. For example, sites where elevated pollutant concentration was detected compared to other sites can be called as pollution "hotspots".*

California); North Pacific Gyre; North Pacific Central Gyre; North West Pacific; Lake Winnipeg (Canada), North Atlantic Gyre; Mediterranean Sea (Fig. 4.3 and Table 4.6)

- **Microplastics (MPs) detection:** MPs (1-5 mm) were detected throughout the water column (surface, middle, bottom) (Lattin et al 2004) and in all the environmental compartments (biota, water, sediment) (Van Cauwenberghe et al 2015) (Table 4.6).
- **MP types and shapes:** MPs detected in the environmental waters include: **i.** Acrylic, EPS, HDE, LDPE, PA/Nylon, PE, PES, PET, PE, PMA, PP, PS, PVC; **ii.** fragments, fibres, film/filaments, line, foam, pellet, spheres/spherules (Table 4.6).
- **Plastic debris:** Plastic debris accounted for 75% in New Zealand (Galgani et al 2000) to 96% in the Maldives (Imhof et al 2017) and 99.9% in the Great Pacific Garbage Patch (Leberton et al 2018). In New Zealand, fishing contributed 40% of plastic debris (Gregory and Ryan 1997), whereas it is 46% in the Great Pacific Garbage Patch (Leberton et al 2018) (Table 4.6).

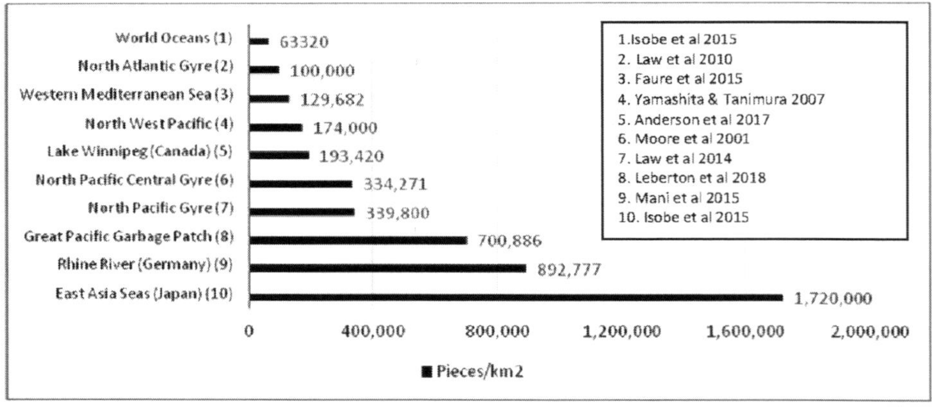

Fig. 4.3. Plastic pollution "Hot spots" in the world
(plotted based on references 1-10 given in the box of the figure).

- **Plastic sources:** MPs entered the freshwater and marine environment via i. synthetic ***textile fibres*** (rayon, polyester and acrylic) released during washing of clothes (Andrady 2011, Browne et al., 2011, Cole et al 2014, Anderson et al 2017, Peng et al 2017 (a single garment can produce >1,900 fibres per wash; Browne et al 2011); ***ii. fishing and aquaculture*** activities**:** fishing gears, handlines and ropes (Andrady 2011, Browne et al 2011, Cole et al 2014, Frias et al 2014), foams used in fishing boats and seafood markets (Fok et al 2017) and styrofoam buoys for aquaculture (oyster, mussel) (Kim et al 2015); **iii.** *WWTP* influents/effluents (Magnusson and Norén 2014, Leslie et al 2017) and wastewater discharge (Dris et al 2015); **iv.** ***biosolids*** containing plastics applied in agriculture farming and forestry (Leslie et al 2017); **v.** ***atmospheric*** fallout of microfibres (Dris et al 2015); **vi.** ***cosmetics and personal care products*** containing microbeads (released during showering/bathing) (Liebezeit and Dubaish 2012); **vii.** ***ship breaking yards*** (insulating, fabrics, packaging materials used in ships) (Reddy et al 2006); **viii.** ***urban and industrial outflows*** containing plastic fibres (Naji et al 2017); **ix.** ***seasonal outputs*:** wet season/ high run-off can have high plastic inputs in waterways compared to the dry season (Cheung et al 2016); **x.** ***other sources*** (beach visitors and recreational activities causing plastic pollution in beaches (Jayasiri et al 2013), densely populated coasts and metropolitan cities, etc. could be a significant source of plastics pollution (Faure et al 2015, Naidoo et al 2015).

- **More plastics than fish/ plankton:** Plastic litters outnumbered fish larvae in the Danube River, Austria (Lechner et al 2014); MPs were present in 61% of zooplankton samples in Portuguese coastal waters (Frias et al 2014); the mass of plastic was approximately six times that of plankton in the North Pacific central gyre (Moore et al 2001).
- **Plastic sinks: i.** polar sea ice could act as a plastic sink (though melting of ice due to global warming could release them back to the environment) (Obbard et al 2014), **ii.** deep-sea sediments (at 1,100-5,000 m depths) (Van Cauwenberghe et al 2013) and **iii**. freshwater sediments (Castañeda et al 2014) are sinks for plastics.

Table 4.6. Selected examples of microplastics (MPs) concentrations reported from lakes, rivers, estuaries, coasts, oceans, and gyres across the globe

*[**ABS** = acrylonitrile butadiene styrene; **ALKD** = alkyd resin; **AN** = acrylonitrile; **HDPE** = high-density polyethylene; **LDPE** = low-density polyethylene; **EPP** = ethylene propylene; **EPR** = ethylene propylene rubber; **EPS** = expanded polystyrene; **na** = data is not available: **Nylon** = polyamide; **PA** = polyamide/ nylon; **PAN** = polyacrylonitrile; **PE** = polyethylene; **PEP** = poly(ethylene-propylene); **PES** = polyester; **PET** = polyethylene terephthalate; **PEVA** = polyethylene vinyl acetate; **PMA** = polymethylacrylate; **PP** = polypropylene; **PS** = polystyrene; **PSS** = polystyrene sulfonate; **PTFE** = polytetrafluoroethylene; **PU** = polyurethane; **PVC** = polyvinyl chloride; **PVOH** = polyvinyl alcohol; **MPs** = microplastics, **WWTP** = waste water treatment plants; **CP** = cellophane; **dw.**= dry weight; **Fragments** = break down of larger plastic items; see Table 4.2 for definition of microplastics, mesoplastics, macroplastics, microbead, fibre, film, fragment, pellet].*

Country/ Region, ecosystems **References** SW = surface water	**Microplastics (MP)** Types/shapes	**MPs concentration** (items/particles/pieces) *(different units used by different authors)*	**Remarks**
Antarctica Ross Sea,	PE, PP; fragments (71.9%) and	0.0032 to 1.18 items/m^3, (mean:	na

Country/ Region, ecosystems **References** SW = surface water	**Microplastics (MP)** Types/shapes	**MPs concentration** (items/particles/pieces) *(different units used by different authors)*	**Remarks**
SW-Marine; (Cincinelli et al 2017)	fibres (12.7%)	0.17±0.34 items/m^3)	
Arctic Ocean Beaufort, Chukchi Seas, Ice core (Obbard et al 2014)	Rayon (54%), PES (21%), PA (16%), PP (3%), PS (2%), acrylic (2%) and PE (2%)	38–234 items/m^3 (38 to 234 particles per cubic meter of ice)	Polar sea ice could be a major sink of plastics; melting of ice could release them back.
Argentina SW-Marine (Gregory and Ryan 1997)	na	124-224 items/km (mean, estimated)	Microplastics: 37-72% of marine debris.
Australia Tasmania, SW-Marine (Gregory and Ryan 1997)	na	213 items/ km (mean, estimated)	71% plastic debris (40% fisheries related) 177 km surveyed.
Australia Victoria, SW- Marine (Gregory and Ryan 1997)	na	47.12 items/km (mean, estimated)	62% plastic debris; 179 km surveyed.
Austria Danube River (Water, Fish) (Lechner et al 2014)	na	Plastic: 316.8 items per 1,000 m^3 Fish larval abundance: 275.3/individuals/ 1,000 m^3	Plastic litter outnumbered fish larvae in the river.
Bay of Bengal South West, Marine (Eriksen et al 2017)	Film (49.5%), fragments (32.5%), line (13%), foam (4.5%) and pellet (1%)	Few hundreds to 20,000 items/km^2	Plastic abundances were much higher near Nicobar Islands (>100,000 microplastics/ km^2).
Belgium Marine beach biota, water, sediments (Van Cauwenberghe et al 2015)	Seawater: 0.4 particles/litre; *sediments:* 6.0 particles/kg dry sediment; mussels, *Mytilus edulis*: 0.2 particles/g tissue; lugworms, *Arenicola marina:* 1.2 particles/g/tissue; mussel faeces: 1 particle/g tissue.		MP was present in all environmental compartments, tissues, faeces, water, sediment samples.

Country/ Region, ecosystems **References** SW = surface water	**Microplastics (MP)** Types/shapes	**MPs concentration** (items/particles/pieces) *(different units used by different authors)*	**Remarks**
Black Sea SW- Marine (Aytan et al 2016)	Fibres (49.4%), films (30.6%) and fragments (20%)	600-1,200 particles/m^3	Black Sea is regarded as a "hotspot" for MP pollution (high MP concentrations).
Brazil Goiana Estuary, SW (Frias et al 2014)	41.08% soft plastic, 29.11% paint chips, 28.42% hard plastic and 1.4% threads	26.04 particles per 100 m^3	Main sources of MP: Fishing gears, ropes, handlines.
Brazil Sediments, Marine Beach (Costa et al 2010)	Fragments (96.7%) and Virgin plastic pellets (3.3%).	310 particles/kg dry sediment	The presence of virgin plastic pellets implies long-range marine transport.
Canada Lake Winnipeg, SW Freshwater (Anderson et al 2017)	Fibres, films and foam; Fibres (major): 90%	193,420 items per km^2	Majority of microplastics were synthetic textiles (fibres).
Canada St. Lawrence River, Sediments (Castañeda et al 2014)	Microbeads (Polyethylene)	13,832 microbeads per m^2 (mean)	Freshwater sediments can act as a sink for plastic pollutants.
China Shanghai; Sediments-Rivers (7 sites) (Peng et al 2018)	[PP (57.1%), PES (17.1%), Rayon (11.4%)]; [spheres (88.98%) and fibre (7.55%)]	802 items/kg dw (80.2 items/ 100 g dw)	Clothes' washing constitutes a large portion of MPs (polyester, rayon sourced from clothes).
China Changjiang Estuary, Sediments (Peng et al 2017)	Fibres (93%), fragments (6%), pellets (1%); Rayon, polyester, acrylic	121 items per 100 g dw	Detection of rayon, polyester and acrylic indicate MPs in the Estuary was from washing of clothes.

Country/ Region, ecosystems **References** SW = surface water	**Microplastics (MP)** Types/shapes	**MPs concentration** (items/particles/pieces) *(different units used by different authors)*	**Remarks**
China Guangdong Province, Sediments Beach (Fok et al 2017)	Foams (96%), resin pellets (2%) and fragments (2%)	6,675 items m^2	Foams are heavily used by fishing boats and seafood markets.
China Yangtze Estuary, SW (Zhao et al 2014)	*Yangtze estuary:* Fibres (79.1%) > Granules> Films *Coastal waters*: Fibres (83.2%)> Granules> Films	*Yangtze estuary*: 4,137.3 particles/m^3; *Coastal waters:* 0.167 particles/m^3	Transparent and coloured plastics could be related to packaging, clothing and fishing line (Cole et al., 2014).
Deep-Sea (1,100-5,000 m depths) Sediments; Gulf of Guinea, Mediterranean Sea, Atlantic Ocean, Southern Ocean (Van Cauwenberghe et al 2013)	na	Microplastics were recovered from the top 1 cm of the sediment samples (0.5 MPs/25 cm^2)	Microplastics have so far reached the most remote of marine environments: the deep sea.
European Coasts Baltic, North, Celtic Sea, Bay of Biscay, Mediterranean Sea, Adriatic Sea (Galgani et al 2000)	0 to 101 000 pieces of debris per km^2 (70% was plastic); Plastic items/ha were: 2.63 (Adriatic Sea), 1.56 (Celtic Sea),1.128 (Bay of Biscay), 1.05 (East-Corsica), 0.92 (Gulf of Lion), 0.754 (North Sea), 0.45 (Baltic Sea), 0.149 (Channel East) and 0.064 (Bay of Seine).		Plastic (mainly bags and bottles) accounted for > 70% of total number of marine debris,
France Greater Paris (Urban), SW, atmospheric fallout, wastewater, River-seine (Dris et al 2015)	Fibres (>90 %) *Atmospheric MP fallout:* 118 particles per m^2per day (range: 29 to 280); *Raw wastewater:* 260–320 x 10^3 particles m^3;*Rivers*: 4–108 particles m^3		Atmospheric fallout reveals a significant source of fibres in the freshwater ecosystem.
Germany The Rhine, Sediments-freshwater 11 locations (820 km)	[PA, PE, PP, PS); opaque spherules-45%, fragments - 37%, transparent	892,777 particles/km^2	Microplastic particles in the Rhine are much higher when compared to other Rivers and

Country/ Region, ecosystems **References** SW = surface water	**Microplastics (MP)** Types/shapes	**MPs concentration** (items/particles/pieces) *(different units used by different authors)*	**Remarks**
(Mani et al 2015)	spherules-13% and fibres-2.5%		lakes.
Germany The Rhine and Main Rivers; Sediments (Klein et al 2015)	[PE (49%), PP (26%), PS (10%); (spheres-50%, fibres- 13%]	*Rhine River*: 228–3, 763 particles/kg; *Main River*: 786–1,368 particles/kg	All sediments analysed contained microplastic particles.
Germany East Frisian Islands, Sediments Beach (Liebezeit and Dubaish 2012)	Granular and fibrous types	496/10 g sediment	Most of granular materials could be cosmetics derived MP which passed through sewage treatment plants.
Great Pacific Garbage Patch (GPGB) (between California and Hawaii) SW (1.6 million km^2) (Leberton et al 2018)	PE, PP; Hard plastics, sheets and films (47%); nets, ropes lines (52%); 700,886 pieces/km^2; 69.58 kg/m^2; *Microplastics*: 1.7 trillion pieces; *Mesoplastics:* 56 billion pieces; *Macroplastics:* 821 million pieces; *Megaplastics*: 3.2 million pieces.		Of all samples collected, 99.9% were plastics; estimated 1.8 trillion pieces (80,000 tons) of plastic in the GPGB; 46% of the total plastic mass (fishing nets).
Greece Kea Island, Beaches (Kaberi et al 2013)	*Pellets* (82% PE, 11% PP, 7% PET; *Fragments* (71% PE, 24% PP, 5% PS)	275 items/m^2	MP pollution spread to remote Kea Island far from land-based sources.
Greenland (arctic) Microplastic (water column) (Amelineau et al 2016)	PES (polyester) (53%), HDPE, LDPE (23%); filaments (97.2%)	2.38 items per m^3 0.99 and 2.3 pieces per m^3 (2005, 2014 respectively)	Sea-ice represent a sink for MP particles (can be released upon melting (due to climate change impacts).
Hong Kong 25 beaches, Sediments (Fok and Cheung 2015)	EPS: 92%; fragments 5%; pellets- 3%.	5,595 items/ m^2	5,595 items/m^2 is higher than international averages; Hong Kong is a hotspot of marine plastic pollution.

Country/ Region, ecosystems **References** SW = surface water	**Microplastics (MP)** Types/shapes	**MPs concentration** (items/particles/pieces) *(different units used by different authors)*	**Remarks**
Hong Kong **Pear River Estuary,** Sandy beaches (Cheung et al 2016)	MPs: dry season: 87%; wet season: 92%	Dry season (January to March): 889 items/m^2; Wet season (June to August) 5,595 items/m^2	Seasonal variations: wet season (high plastic inputs), dry season (decreased plastic inputs).
India Four sandy beaches, Arabian Sea (Jayasiri et al 2013)	na	items/m^2	Beach visitors, coastal inhabitants and recreational activities can cause plastic pollution in beaches.
India Ship-breaking yards, sediments (Reddy et al 2006)	Polyurethane, nylon, polystyrene, polyester, and glass wool	81.43 mg/kg of sediment.	Polyurethane, nylon, polystyrene, polyester, glass wool is used in constructing ships (insulating materials, fabrics, packaging).
Indian Ocean The Chagos Archipelago, Sediments Beach (Readman et al 2013)	Nylon, PE, PES, PP and rayon	4.55 pieces per 50 mL of sediment	Indicate that microplastics can accumulate in remote locations.
Indonesia Surface sediment (Yonaet al 2019)	Plastic fragments (54.34%), fibres (41.45%), films (4.21%)	206.04–896.96 particle/kg; highest in mangrove area (896.96 particles/kg)	Domestic wastes and fisheries activities were the main causes of the high microplastic particles.
Iran Persian Gulf, The Strait of Hormuz, Sediments Beach (Naji et al 2017)	PE, PA/nylon, PET Fibres (83%), film (11%), fragments (6%)	Particles per kg: Bostanu (Industrial area) (1258); Gorsozan (Urban area) (122)	Likely plastic sources: beach debris, discarded fishing gear, urban and industrial outflows that contain fibres from clothes.
Italy Venice (10 sites), Sediments	PE (48.4%), PP (34.1%), poly(ethylene-propylene)	672–2,175 particles/kg	Higher concentrations observed in landward sites.

Country/ Region, ecosystems **References** SW = surface water	**Microplastics (MP)** Types/shapes	**MPs concentration** (items/particles/pieces) *(different units used by different authors)*	**Remarks**
(Vianello et al 2013)	(5.2%), PEST (3.6%)		
Japan East Asia Seas, Surface water (Isobe et al 2015)	PE (68%), PP (19%)	1.72 million items/km^2	"Hot spots" of pelagic[9] microplastics of 16 times were greater than in the North Pacific and 27 times greater than in the world oceans.
The Maldives Lhaviyani Atoll, Beach Sediments (Imhof et al 2017)	PE, PP, PS (major), PU, PA, PVC; EPS (61%), fragments of foil remnants (20%), larger plastic products (15%); 1029±1,134 plastic particles/m^2		96% of the particles were identified as plastic debris.
Mediterranean Sea Surface water (41 samples) (Faure et al 2015)	Fragments (73%), films (14%), foams (5 %), lines (2%)	130,000 items/km^2 (microplastics); 62% plastic debris	The densely populated countries in the Mediterranean sea generated MP concentrations similar to the North Atlantic gyre (about 100,000 particles/km^2).
Mongolia Lake Hovsgol (Free et al 2014)	Fragments and films; plastic bottles (37%), fishing gear (25%), plastic bags (16%) and plastic fragments (18%)	20,264 particles/km^2	Polluted with household microplastics.
Netherlands WWTPs effluent (Leslie et al 2017)	*WWTP effluents*: 68–910 particles/L (influents); 51–81 particles/L (effluents); 510–760 particles/kg (sewage sludge) (weight weight)		Biosolids application in farming and forestry could be a source of plastic pollution.

[9]***Pelagic fish*** *live and feed away from the bottom in the open water column and* ***demersal fish*** *live and feed on or near the bottom of seas or lakes (the demersal zone).*

Country/ Region, ecosystems **References** SW = surface water	**Microplastics (MP)** Types/shapes	**MPs concentration** (items/particles/pieces) *(different units used by different authors)*	**Remarks**
New Zealand Surface water Marine (Gregory and Ryan 1997)	na	748 items/km (mean, estimated)	About 75% plastic debris (40 % fisheries related); 467 km surveyed.
North Atlantic Sub-tropical Gyre (Law et al 2010)	na	100,000 particles/km^2	-
North Pacific Central Gyre SW, Marine (Moore et al 2001)	Fragments, thin plastic films, polypropylene (98% plastic pieces)	Mean 334,270 pieces/km^2 (Range 31,982-969,777 pieces)	The ratio of plastic to plankton (mass) was 6.1:1.
North Pacific gyre Eastern Pacific Ocean (Law et al 2014)	na	339,800 particles/km^2	na
Norway Arctic polar waters (Lusher et al 2015)	Fibres (95%), fragments (4.9%), films (<0.1%)	0.34±0.31 particles/m^3	MPs Black (45%) and blue (29%).
Oceania Ducie Atoll, Pitcairn Islands, South Pacific Oceania (25°S) (Gregory and Ryan 1997)	na	184.7 items/km (mean, estimated)	About 77% plastic debris (23% fisheries related).
Oceania Raoul Island, Oceania (30°S) (Gregory and Ryan 1997)	na	360 items/km (mean, estimated)	>80% plastic debris (25% fisheries related); <8 km surveyed.
Papua New Guinea Motupore Island, Bootless Bay (Smith 2012)	na	13.74 items m^2	89.7% of items were plastic.

Country/ Region, ecosystems **References** SW = surface water	**Microplastics (MP)** Types/shapes	**MPs concentration** (items/particles/pieces) *(different units used by different authors)*	**Remarks**
Portugal **Coastal waters** (Frias et al 2014)	PE, PP, PA	0.002- 0.036 items per m^3	Microplastics were present in 61% of zooplankton samples.
Singapore **Coast,** Sediments (Nur Hazimah and Obbard 2014)	PE, PP, PVC, PA/nylon	37 particles/kg	-
South Africa Sediments, five urban estuaries (Naidoo et al 2015)	Fragments, fibres dominated	Highest average plastic concentrations (745.4 particles per 500 mL)	Pollution levels were highest around metropolitan areas and low levels were away from the city centre.
South Korea Surface water, Coastal Soya Island (Kim et al 2015)	EPS (>99% of MP), PP (0.13%), PE (0.10%)	EPS: 46,334 items/m^2 (the highest level globally)	Styrofoam buoys used for aquaculture (oyster, mussel, sea squirt) are the major sources of plastics.
Sub Antarctic Surface water, Marine (Indian Ocean, South Africa, Prince Edward Island (Gregory and Ryan 1997)	na	28.16 items/km (mean, estimated)	About 88% plastic debris (22% fisheries related), 4.9 km surveyed.
Sub-Antarctic Surface water, Marine (Atlantic Ocean), (Gregory and Ryan 1997)	na	677.6 items/km (mean, estimated)	About 77% plastic debris (25% fisheries related), 13.6 km surveyed.
Sweden WWTP effluent (Magnusson and Norén 2014)	PE, PP Influent water: 15.1 particles per m^3; Effluent water: 8.25 particles per m^3; Sludge: 16.7±1.96·10 particles per dw		The total amount of microplastics was higher in influent than in effluent.

Country/ Region, ecosystems **References** SW = surface water	**Microplastics (MP)** Types/shapes	**MPs concentration** (items/particles/pieces) (*different units used by different authors*)	**Remarks**
Switzerland Lake Geneva, Surface water (Faure et al 2012, Faure et al., 2013)	Fragments/films	51,556 particles/km^2 (average)	-
Taiwan Sediments, four beaches (Kunz et al. 2016)	PE (44%), PP (43%), PS (12%) and ABS (1%)	Plastic particles per 0.0125 m^3; Range: 10-532	Microplastic particles were present in all the beaches studied.
Turkey Sediments, Iskenderun, Mersin Bays (Gündoğdu and Çevik 2017)	Fragments (60.1%)	Iskenderun Bay = 0.2254 item/m^2; Mersin Bay: 0.6827 item/m^2	Microplastics comprised 87% of the debris.
UK Sediment from beaches, from estuarine, sub-tidal sediments around Plymouth (Thompson et al 2004)	Nine polymers: acrylic, alkyd, poly (ethylene: propylene), PA(nylon), PES, PE, PMA, PP, PVC	na	Polymers identified were used in clothing, packaging and rope.
UK Tamar estuary, Surface water (Sadri and Thompson 2014)	PE (40%), PS (25%) and PP (19%)	0.028 items m^3 (mean)	Plastics comprised 82% of the debris.
UK Southampton, Surface water (Gallagher et al 2016)	PE, PP, and cellophane Fibres (majority of the plastics sampled)	960 items/m^3	PE, PP fibres from polyester rich clothes and ropes used in fishing and shipping (Andrady 2011, Browne et al 2011).
USA Sediments- Beach, Hawaiian Islands (20 samples) (McDermid and McMullen 2004)	Fragments (87%), pellets (11%), line (1.34%), foam (0.15%), film (0.06%)	2333 items/m^2	Of the debris,72% by weight was plastic; small-plastic debris can occur at the remote beaches.

Country/ Region, ecosystems **References** SW = surface water	**Microplastics (MP)** Types/shapes	**MPs concentration** (items/particles/pieces) (*different units used by different authors*)	**Remarks**
USA Southern California's coastal waters (Moore et al 2002)	Fragments (92-100%), thin plastic films (5%), Styrofoam, fishing line and plastic pellets (2%)	7.25 pieces/m^3 (0.002 g/m^3)	Plastic and plankton (plastic 2.5: plankton 1), large plastic material associated with the fishing and shipping industry.
USA Northeast Pacific Surface water (595 samples) (Doyle et al 2011)	Fragments, fishing net and line fibres	0.004-0.19 particles/ m^3, 0.014-0.209 mg dry mass/m^3	Plastic particles: surface samples (8.75%-84.0%).
USA Southern California offshore Santa Monica Bay sites (Lattin et al 2004)	Plastics were present throughout the water column (surface, middle, bottom)	3.92 pieces/m^3 (0.03 g/m^3)	The density of plastics on the surface was considerably higher after the storm (and after a rain event).
World Wide 12 countries (sediments, beach) 18 sites, six continents (Browne et al 2011)	PES (56%), acrylic (23%), PP (7%), PE (6%), PA fibres (3%)	Fibres/250 mL sediment (2 (Australia) to 31 (Portugal, U.K.)	Microplastic fibres derived from sewage via washing clothes (a single garment can produce >1,900 fibres per wash).
World Wide (deep sea sediments- 3500 m), Atlantic, Indian Oceans, Mediterranean Sea (Woodall et al 2014)	Rayon (non-plastic polymer) > PES> PA, acetate, acrylic	Plastic microfibre abundance mean 13.4 pieces/50 mL (range: 1.4 to 40 pieces)	Deep-sea sediments are a sink for microplastics.
Worldwide Open ocean- five subtropical gyres (3,070 samples, 442 sites) (Cózar et al 2014)	Global load: plastic debris (SW open ocean) in kilotons (kt); North Pacific Ocean: 4.8 kt (2.3-12.4); North Atlantic Ocean: 2.7 kt (1.0-6.7); South Atlantic Ocean: 2.6 kt (1.7-5.4); Indian Ocean: 2.2 kt (0.8-5.1); South Pacific Ocean: 2.1 kt (0.8-5.6); Total: 14.4 (6.6-35.2)		The high plastic load in the North Pacific Ocean could be related to the most densely populated coast in the world.

4.3.2. Adsorption of contaminants by plastics

Adsorption is the adhesion of atoms, ions, molecules, or pollutants from gas, water or dissolved solids to a surface of an adsorbent (such as plastic). In adsorption, molecules adhere to the surface of a plastic/polymer (compare **a<u>d</u>sorption** vs. **a<u>b</u>sorption** in Fig. 4.4). Adsorption plays an important role in the transport of contaminants. Because of their **hydrophobic surface**[10], microplastics can adsorb and concentrate hydrophobic organic contaminants (PAHs, DDTs, PCBs) (Mato et al 2001, Ogata et al 2009, Wright and Kelly 2017) and inorganic heavy metals (Cd, Ni, Pb, Zn) (Holmes et al 2012, Rochman et al 2014, Wright and Kelly 2017) Currently an **International Plastic Pellet Watch Program** has been established for global monitoring of POPs in coastal areas and the oceans across the globe using beached plastic resin pellets (*http://www.pelletwatch.org/; shige@cc.tuat.ac.jp*).

A<u>d</u>sorption

[a surface pehenemenon where a substance (e.g. POP) is adhere or stick to the surface of a substance, such as, the plastic]

Adsorption of persistent organic pollutants (POPs) to microplastic where water borne chemical pollutants (POPs) form a superficial monomolecular layer on a substrate (the microplastic).

A<u>b</u>sorprtion

[a bulk phenemenon where a substance (e.g. POP) is absorbed into the substance, such as, the plastic]

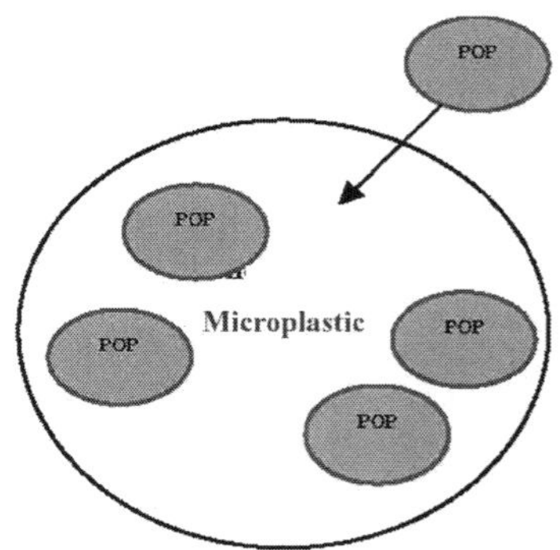

Absorption of persistent organic pollutants (POPs) to microplastic where waterborne chemical pollutants (POPs) diffuse into the bulk of the plastic material.

Fig. 4.4. Adsorption and absorption of chemicals
(Crawford and Quinn 2017, *https://biodifferences.com/difference-between-absorption-and-adsorption.html#ComparisonChart*)

[10] ***Hydrophobic surface*** *is a surface that has the ability to repel water. Hydrophobic molecules include the alkanes, oils, fats, and greasy substances.*

The physical (size, shape and surface area) and the chemical properties (hydrophobicity, binding affinity and capacity) of specific plastic/polymer allow adsorbing a number of hazardous chemical contaminants from the surrounding aquatic environment to the surface of microplastics. Table 4.7 **synthesize research results** related to adsorption of contaminants/chemicals/pollutants by various types of plastics, the main findings of this research are highlighted below:

- **Plastics as a source and sink of pollutants:** Microplastics can act as either a source or sink or a vehicle for transport of toxic organic chemicals/ contaminants and heavy metals in the aquatic environment and aquatic biota (Mato et al 2001, Lee et al 2014, Wu et al 2016, Teuten et al 2007, Wang and Wang 2018).
- **Plastic types:** Research carried out found that 10 types of plastic can adsorb a variety of pollutants/chemicals from waterways. These are HDPE, LDPE, MDPE, PA, PE, PEP, POM, PP, PS, PVC (Table 4.7). Most contaminants were adsorbed in polyethylene (PE), which could be related to the rubbery nature of PE. The rubbery plastic is reported to have higher sorption capacity, greater diffusion and greater affinity for contaminants (Alimi et al 2018) (see Table 4.3 for the glass transition temperature for various rubbery and glassy plastics).
- **Contaminants adsorbed in plastics:** Contaminants adsorbed in plastics fall within the following groups/categories: **i.** *heavy metals* (Al, Cd, Co, Cr, Fe, Mn, Ni, Pb); **ii.** *Hydrocarbons* (benzene, toluene, naphthalene, o-xylene, PAHs, phenanthrene); **iii**. *industrial chemicals* (chlorobenzene, n-hexane); **iv.** *lubricating oil*; **v.** *pesticides* (4 hexachlorocyclohexanes, DDT, DDE), **vi.** *Perfluoro chemicals* (PFOS, FOSA); **vii**. *pharmaceuticals and personal care products* (4-methylbenzylidene camphor, carbamazepine, 17α-ethinyl estradiol); **viii.** *persistent organic pollutants* (e.g. PCBs); **ix**. *surfactants/ detergents* (alkylbenzene, nonylphenol, octyl-phenol, triclosan) (Table 4.7).
- **Factors related to adsorption of contaminants:** Several factors can influence adsorption of contaminants into plastics: **i**. ***temperature and salinity***: greater adsorption of PAHs found at 21°C compared to 10°C in PE (Crawford and Quinn 2017); an increase of salinity enhanced the sorption of triclosan in PE (Wu et al 2016) and the sorption of lubricating oil in PE and PS increases with increasing concentration of salinity (Hu et al 2007); **ii.** ***particle size and surface area:*** the sorption capacity of PCBs in PP increased as particle size decreased (Zhan et al 2016), the uptake rate of PAHs were four times higher with HDPE than LDPE (with four times higher surface areas with HDPE) (Muller et al 2001), while the sorption of lubricating oil on nano-PE (having more surface area) was significantly higher than that of micro-PS (less surface area) (Hu et al 2017) (surface area becomes larger with decreasing particle size (Zhan et al 2016)); **iii. *rough vs virgin plastic pellets:*** rough microbeads adsorbed more contaminants (phenanthrene, DDT) than smooth pellets in PE (Napper et al 2015); **iv.**

aged vs pristine plastics: heavy metal contaminants were adsorbed more in aged pellets compared to pristine pellets in PE plastic (Holmes et al 2012); higher levels Cu and Zn were sorbed onto aged PVC compared to pristine polystyrene particles (Brennecke et al 2016); ***v. discoloured vs non-discoloured plastics:*** discolored/fouled PE resin pellets had sorbed higher concentrations of PCBs than non-discoloured one (Endo et al 2005); ***vi. rubbery vs glassy polymers:*** rubbery polymers (HDPE, LDPE) have greater adsorption of contaminants than glassy polymers (PET and PVC). This is related to greater diffusion (permeability and greater free volume) with rubbery polymers than the glassy polymers (Rochman et al 2013).

- **Plastics as a passive sampler**: a number of high risk pollutants/ contaminants were adsorbed in microplastic PE, such as benzene, DDT, DDE, 17α-ethinyl estradiol, FOSA, heavy metals, NP, OP, PAHs, PCBs, PFOS, toluene, naphthalene, phenanthrene, toluene, xylene (Muller et al 2001, Fries and Zarfl 2012). Therefore, PE can be used as a passive sampler for monitoring pollution/organic contaminants in waterways. Beached plastic pellets have already been used globally to monitor hydrophobic organic contaminants (PAHs, PCBs, DDTs) as part of International Plastic Pellet Watch Program (*http://www.pelletwatch.org/maps/map-5.html*).

Table 4.7. Contaminants/pollutants adsorbed in various types of plastic polymers

*[**DDE** = Dichloro-diphenyl-dichloro-ethylene; **DDT** = Dichloro-diphenyl-trichloro-ethane; **PFOSA** = perfluoro-octane-sulfonamide; **HDPE** = high density polyethylene; **Kd** = The adsorption-desorption distribution coefficient, it is an important parameter for understanding the mobility of a compound in the environment (partitioning) and its distribution between water, sludge, soil and sediment compartments (Ecetoc 2019); **LDPE** = low density polyethylene; **MDPE** = medium-density polyethylene; **NP** = nonyphenol; **OP** = octylphenol; **PA** = polyamide; **PE** = polyethylene; **PAHs** = polycyclic aromatic hydrocarbons; **PCBs** = polychlorinated biphenyl; **PEP** = plastic eroded pellets; **PFOS** = perfluorooctanesulfonate; **POM** = polyoxymethylene; **PP** = polypropylene; **PS** = polystyrene; **PVC** = polyvinyl chloride]*

Country References	**Type of plastics and Size (diameter)**	**Contaminant(s) adsorbed Sorption capacity**	**Remarks**
Australia (Muller et al 2001)	**HDPE, LDPE**	**PAHs** Sorption capacity: HDPE> LDPE	-The uptake rate was four times higher with HDPE than LDPE (with four times higher the surface area of HDPE). -HDPE can be used to sample trace organic contaminants.
Austria (Huffer and	**PA** (109 mm), **PE** (152 mm), **PS** (168 mm),	**Hydrocarbons** (n-Hexane, cyclohexane,	Sorption by the PS, PVC was higher (related to strong sorption coefficients/ hydrophobicity of

Country References	**Type of plastics and Size (diameter)**	**Contaminant(s) adsorbed Sorption capacity**	**Remarks**
Hofman 2016)	**PVC** (58 mm)	benzene, toluene, chlorobenzene, ethylbenzoate naphthalene) Sorption capacity: PS> PVC> PE> PA	the two plastics).
China (Wang and Wang 2018)	**PE** (100-150 μm), **PS**, **PVC**, sediment	**Phenanthrene (Phe)** Sorption capacity: PE> PS> PVC> natural sediment	-MPs have the capacity of sorbing chemicals from water (potential as sinks and carriers of contaminants). -MPs can act as transport and bioaccumulation of toxic organic chemicals.
China (Wu et al 2016)	**PE** debris (250-280μm) Dissolved organic matter	**Pharmaceuticals and personal care products (PPCS)** (carbamazepine-CBZ , 4-methylben-zylidene camphor (4MBC), triclosan (TCS), 17α -ethinyl estradiol (EE2)) Sorption capacity: 4MBC> TCS> EE2> CBZ	-Sorption capacity of PPCPs was related to their hydrophobicity (log K_{ow}). -Increase of dissolved organic matter (DOM) content reduced the sorption of PPCS. -Increase of salinity enhanced the sorption of TCS. -Microplastics can act as carriers for the transport of contaminants.
China (Zhan et al 2016)	**PP** (0.425-0.85 mm)	**-PCB**-3,3′,4,4′-tetrachlorobiphenyl (PCB77) (simulated seawater) -Sorption capacity: simulated seawater> ultrapure water> n-hexane	Sorption capacity increases as particle size and temperature decreases but not pH.
China (Hu et al	**PS** (20-40 mm), **PE** (50 mm)	**Lubricating oil**	-Sorption of lubricating oil on nano-PE was significantly higher than that of micro-PS.

Country References	Type of plastics and Size (diameter)	Contaminant(s) adsorbed Sorption capacity	Remarks
2017)		Sorption capacity: PE> PS	-Sorption capacity increases with increasing the concentration of salinity in PE and PS.
Germany (Fries and Zarfl 2012)	**HDPE, LDPE**	**PAHs** Sorption capacity: LDPE> HDPE	-LPDE showed greater sorption than HDPE. -PE can be used as passive samplers for monitoring of organic contaminants (PAHs).
Greece (Karapangioti and Klontza 2008)	PE (2-3 mm), **PEP** (-), **PP** (2-3 mm), **POM** (<2mm)	**Phenanthrene (Phe)** (synthetic freshwater) Sorption capacity: PEP> PE> POM> PP	Diffusion into the polymer was not affected by salinity.
Hong Kong (Wang et al 2015)	**PE** (150 µm), **PS** (230 µm), **PVC** (250 µm)	**PFOS, PFOSA** Sorption capacity: PE> PVC >PS	-Sorption of PFOS and FOSA on PE and PVC were stronger than that of PS. -PFOS were found to be readily sorbed by PE and PS under a high salt concentration/ seawater.
Japan (Endo et al 2005)	**PE, PS**	**PCBs** Sorption capacity: PE> PS	-Discoloured or fouled PE resin pellets had higher concentrations of PCBs than non-discoloured one. -Plastic resin pellets are a dominant route of transport of contaminants at remote places. -The resin pellets can be used to monitor pollution in seawater.
Japan (Mato et al 2001)	**PP** virgin pellets	**DDE, NP, PCBs** (seawater)	-Contaminants (PCBs, DDE) sorbed about 1 million times more contaminants than immediate seawater. -The plastic resin pellets serve both as a transport medium and a

Country References	Type of plastics and Size (diameter)	Contaminant(s) adsorbed Sorption capacity	Remarks
			potential source of toxic chemicals in the marine environment.
Japan, Malaysia (Mato et al., 2002)	**PE, PP** pellets	**DDE, NP, OP, PAHs, PCBs** (seawater) Sorption capacity: PE> PP	Greater adsorption of PCB, DDE, and PAHs by PE than by PP.
Netherlands (Velzeboer et al 2014)	**PE** (10-18 μm), **PS** (70 nm)	**PCBs** Sorption capacity: PS> PE	Salinity increased sorption for both polymer's (nano-PS and micro-PE).
Netherlands (Liu et al 2016)	**PS** (70 mm)	**PAHs**	-Sorption unaffected by aggregate size. -Nanoplastic (high surface area) binds hydrophobic chemicals very effectively and increases their hazard when taken up by biota.
No country names (Crawford and Quinn 2017)	**PE** (3.8 mm)	**PAHs** Sorption capacity: 21°C> 10°C	Higher adsorption of PAHs at higher temperature.
Portugal (Brennecke et al 2016)	**PS** (0.7-9 mm), **PVC** (1.06 x 0.8 mm)	**Cu, Zn** Sorption capacity: PVC> PS	-Adsorption of Cu was significantly greater in PVC fragments due to higher surface area and polarity of PVC. -Plastics can act as vectors to transport heavy metal and aquatic biota can ingest microplastics contaminated heavy metals causing negative effects.
South Korea (Lee et al 2014)	**PE, PP, PS** (< 250 μm)	**PAHs,** 4 **hexachloro-cyclohexanes** (HCHs) and 2	Since microplastics undergo slow decomposition, they may play an important role in the fate and transport of hydrophobic

Country References	Type of plastics and Size (diameter)	Contaminant(s) adsorbed Sorption capacity	Remarks
		chlorinated benzenes (CBs) (seawater) -Sorption capacity: HCHs: PP> PE CBs: PE> PP; PAHs: PS> PE> PP	chemicals in the marine environment.
UK (Holmes et al 2012)	**PE** (beached/ aged), **PE** (virgin/new) (1-5 mm)	**Cd, Co, Cr, Cu, Ni, Pb, Zn** (filtered seawater) Sorption capacity: Beached PE> Virgin PE	-Heavy metal contaminants were adsorbed more in aged-pellets compared to pristine pellets. -Plastics are an important vehicle for the transport of metals in the marine environment. -Invertebrates, fish, birds and mammals can mistake plastics as food (Teuten et al 2009).
UK (Napper et al 2015)	**PE** microbeads (164-327 μm)	3-H **phenanthrene**, 14-C-**DDT** (seawater)	-Microbeads adsorbed greater amounts of POPs (DDT) than Phe. -Rough microbeads adsorb higher contaminants than smooth ones. -Microbeads (used in cosmetics) sorb hydrophobic contaminants onto their surfaces and may transfer to biota upon ingestion. -Research shows that hydrophobic organic compounds (PAHs, PCBs and PBDEs) can be transferred from PE to the fish causing liver toxicity and pathology (Rochman et al 2013).
UK (Bakir et al 2014)	**PE, PVC** (200-250 μm)	**Phenanthrene**, 4,4'-**DDT** Sorption capacity: Phe: PE> PVC; DDT: PVC> PE	-Increases in salinity had no significant effects on the sorption capacity of PVC and PE for Phe. -Sorption capacity for DDT was significantly higher in freshwater. -Microplastics have the potential to sorb substantially more contaminants in estuaries (due to higher concentrations of

Country References	Type of plastics and Size (diameter)	Contaminant(s) adsorbed Sorption capacity	Remarks
			contaminants than in riverine and marine waters). -Estuarine systems could be an important source (due to flushing, dredging) and sink of contaminated plastics.
UK (Teuten et al 2007)	Plastic particles (**PE, PP, PVC**) (200-250 μm)	**Phenanthrene** (Phe) Sorption capacity: PE>PP>PVC	-PE sorbed more phenanthrene than either PP or PVC (the K_d value with PE, PP and PVC were 38,100, 2,190 and 1,650 respectively). The plastics may be an important vehicle for transporting contaminants to organisms.
USA (Saquing et al 2010)	**HDPE, LDPE, MDPE, PVC** (0.25 mm)	**Toluene** (landfill) Rubbery polymers (PE)> Glassy polymers (PVC)	-The affinity of hydrophobic organic contaminants (HOC (toluene) > biopolymer. -The diffusivity of toluene is rapid in rubbery plastics but slow in glassy plastics. -Plastics in landfills are potential sinks of hydrophobic organic contaminants.
USA (Rochman et al 2013b)	**HDPE, LDPE, PET, PP, PS, PVC** (2 mm)	**PAHs** Sorption capacity: PS=HDPE=LDPE> PP>PET>PVC	-PS sorbed PAHs greater concentrations than PP, PET, and PVC, similar to HDPE and LDPE. -Plastic debris (PS) can act as a source and sink for PAHs in the marine environment.
USA (Rochman et al 2013a)	**HDPE, LDPE, PET, PP, PVC**	**PAHs, PCBs** Sorption capacity: HDPE> LDPE> PP> PET> PVC	Concentrations of PAHs and PCBs sorbed to HDPE, LDPE, and PP were consistently much greater than concentrations sorbed to PET and PVC.
USA (Rochman et al 2014)	**HDPE, LDPE, PP, PVC** (3 mm), **PET** (3	**Al, Cd, Co, Cr, Fe, Mn, Ni, Pb**	-The study shows that a complex mixture of metals (including USEPA priority metals-Cd, Ni,

Country References	Type of plastics and Size (diameter)	Contaminant(s) adsorbed Sorption capacity	Remarks
	mm)	Sorption capacity: HDPE<LDPE, PVC, PET, PP, PS	Zn) accumulated in most commonly produced plastics. -HDPE accumulated lesser concentrations of metals (Cd, Cr, Ni, Pb, Zn) than the other four plastic-types.
USA (Wu et al 2001)	**HDPE, PVC**	**Alkylbenzene, Toluene, o-xylene** Sorption capacity: PVC> HDPE> biopolymers	Alkylbenzene sorption to plastics was greater than to biopolymer composites.
USA (Pascall et al 2005)	**PE, PS, PVC**	**PCBs** Sorption capacity: PE> PS> PCB	PE showed the highest uptake of PCBs and partition coefficients' (which could be related to the rubbery nature of PE; rubbery polymers have higher sorption capacities (Alimi et al 2018; also see Table 4.3).

4.3.3. Plastics in seafood, honey, sugar, beer, salt and drinking water

Globally, fish/seafood provides approximately 4.3 billion people with 15% of their animal protein intake (FAO 2014). Apart from other contaminants found in seafood (PCBs, DDTs, heavy metals), plastics have been a new contaminant in fish and seafood. Recent studies on microplastics have confirmed that commercially important fish, bivalves, shrimps are contaminated with microplastics. Microplastics have also contaminated other foods (honey, sugar, beer, salt) and drinking water. Table 4.8 provides an account of synthesised global data and information related to plastics contamination of food and drinking water. The **summary of Table 4.8** is highlighted below:

- **Plastics in seafood (fish, shrimps, bivalves/molluscs, dried fish, canned fish):** In China, all the sea fish and freshwater fish examined (21 species) found ingested micro/ mesoplastics where bentho-pelagic, demersal (bottom dwelling) fishes had more plastics than in pelagic (surface to mid depths) fishes (Jabeen et al 2017). Similarly, all the pelagic and demersal species caught from the English Channel ingested plastics (Lusher et al 2013, Collard et al 2017). Fish sold in the Indonesian and USA markets were found contaminated with plastic debris particles (100% and 20% in Indonesia and the USA respectively) (Rochman et al 2015). About 64-100% fish species ingested microplastics caught form the Adriatic Sea, Italy (Avio et al 2015). Plastics were also found in shrimps (Devriese et al 2015), bivalves/ mussels (Mathalon and Hill 2014 Li et al 2015), canned fish (Karami et al 2018), dried fish (Karami et al 2017a). The plastic polymers identified in seafood were **types** (CA, PA/nylon, PE, PET, PES, PP, PS, PVC); **shapes** (fibres, films, filaments, fragments, pellet) (Table 4.8).

- **Plastics in honey:** Honey from Germany, France, Italy, Spain and Mexico were 100% contaminated with plastic fibres and fragments (Liebezeit and Liebezeit 2013) (Table 4.8).
- **Plastics in sugar:** Five commercial sugars analysed from Germany were with fibres and fragments (Liebezeit and Liebezeit 2013) (Table 4.8).
- **Plastics in beer:** Around 24 beer brands of Germany and 12 beer brands of USA tested had plastics fragments and fibres (Liebezeit and Liebezeit 2014; Kosuth et al., 2018) (Table 4.8).
- **Plastics in sea salt:** Different brands of sea salt from Australia, China, France, Iran, Japan, Malaysia, NZ, Portugal, South Africa, Spain, Turkey, the USA were found contaminated with plastic fragments, filaments, film (Yang et al 2015, Karami et al 2017b, Iñiguez et al 2017, Gündoğdu 2018, Kosuth et al 2018).
- **Plastics in drinking water:** Nearly 81% of global tap water samples found contaminated with plastic fibres (Kosuth et al 2018).

Table 4.8. Examples of the occurrence of microplastics in seafood (fish, molluscs, shrimp), honey, sugar, beer, salt and drinking water (tap water)

[See Table 4.2 for definitions of plastic fibre, film, fragment, and pellet].

Country or Location	Species/parts/items examined or analysed	Research findings	References
1. Seafood			
Australian and Malaysian markets (from 13 countries)	20 brands of **canned sardines** and **sprats**	20% of canned seafood contaminated with **plastic fragments; fragments** (46.6%)> films (26.6%) > filaments (26.6%); the most abundant plastic polymers were PP and PET.	Karami et al 2018
Belgium, France	**Shrimp** (*Crangon crangon*); whole shrimp and peeled	63% of the shrimp ingested microplastics mainly **synthetic fibre;** 0.68 plastic particles/g	Devriese et al 2015

Country or Location	Species/parts/items examined or analysed	Research findings	References
(Southern North Sea and Channel area)	shrimp.	individual	
Canada (Nova Scotia)	Farmed and wild **mussels**	Microfibers in farmed mussels were higher than in wild mussels; fibres: 170/5 wild mussels; 375/5 farmed mussels; farmed mussels were grown on plastic PP lines.	Mathalon and Hill 2014
China (Shanghai market)	9 commercial **molluscs** (bivalves) - soft tissue	Fibres (65%) followed by fragments; plastic particles: 4.3 to 57.2 items/individual; plastic polymers: PE, PET, PA; MPs in shell-fish could represent a risk for human health.	Li et al 2015
China (coastal, freshwater)	Fish: 27 species (**21 sea fish** and **6 freshwater fish)**- stomachs and intestines	All the fish species ingested micro/ mesoplastics; **benthopelagic, demersal fishes had more plastics** than **pelagic fishes**; fibres, fragment, pellet, cellophane, PE, PET, PES; transparent, black and blue plastics.	Jabeen et al 2017
China (Coast)	**Mussels (wild, farmed)** (*Mytilus edulis)* from 22 sites	Wild *M. edulis* contained more microplastics (2.7 items/g) compared to farmed groups (1.6 items/g); microplastics: fibres and fragments: CP, PET, PES.	Li et al 2016b
English Channel	Fish- 10 species (five pelagic and five demersal)- gastrointestinal tracts	All the **pelagic** and **demersal species ingested plastic;** 36.5% fish ingested plastics; fibres (68.3%), fragments, (6.1%), beads (11.5%); PA (35.6%), rayon (57.8%).	Lusher et al 2013
English Channel, Medit-erranean Sea, North-eastern	**Wild-caught planktivorous fish** (sardines, herrings, anchovies) - liver	Nearly 80-100% fish contaminated with plastic particles (fibres); **sardines ingested more fibres than herrings** and **anchovies**; polymers: PE (37%), PP (26%),	Collard et al 2017

Country or Location	Species/parts/items examined or analysed	Research findings	References
Atlantic (Bay of Biscay)		PET (16%), polyacrylonitrile (PAN 7%), PS (5%), PA, 5%.	
Indonesia, USA (fish markets)	Indonesia (75 fish/ 11 species), USA (64 fish/12 species) - digestive tracts of fish and whole shell-fish	About 55% and 67% of all species in Indonesia, the USA were contaminated with plastic particles; Indonesia (100% debris in fish is plastic, majority fragments); USA (20% debris in fish is plastic-majority fibres); **chemicals adsorbed in fish can** be **transferred to humans** via **diets (seafood).**	Rochman et al 2015
Iran (Persian Gulf)	Five species of molluscs (three bivalves and two gastropods)-soft tissue	MPs were detected in all five **gastropods and bivalve species; 3.7 to 17.7** particles/individual (soft tissue); fibres (>50%), fragments (26%), film (14%); PE, PET, PA; the trophic transfer of MPs in the food web is possible and the consumption of seafood may be a source of human microplastic intake.	Naji et al 2018
Italy (North Adriatic Sea)	125 fish examined - gastrointestinal tract	Around 64-100% species ingested microplastics; fragments (57%), line (23%), film (11%); PE, PET, PS, PVC, PA/nylon, PP; benthic species contained highest microplastics and pelagic species were lowest.	Avio et al 2015
Italy, Mediter-ranean Sea (central)	Pelagic fish-stomachs	18.2% of fish ingested plastic fragments (micro-, meso-, macro plastics); mesoplastics were abundant in swordfish and microplastics in albacore and blue-fin tunas; transparent and white plastic fragments were commonly ingested.	Romeo et al 2015

Country or Location	Species/parts/items examined or analysed	Research findings	References
Malaysia (from local market)	**Dried fish species** (whole fish excluding the viscera and gills)	**Dried fish** contaminated with plastic **fragments, filaments, films**; PP **(47.2%)>PE (41.6%), PS (5.56%)> PA/nylon** (2.77%); dried fish consumed as a whole, can translocate a significant number of MPs to consumers.	Karami et al 2017a
North Pacific Ocean (North Pacific Central Gyre)	Six fish species (five mesopelagic and one epipelagic species); 670 fish examined	35% of the fish had ingested plastic, averaging 2.1 pieces per fish; fragments (94%), film (3%), fishing line (2%); the plastic fragments consisted of white (58.2%), clear (16.7%), blue (11.9%), black (3.6%), and yellow (1%).	Boerger et al, 2010
North Sea	Seven common North Sea fish species; 1203 fish examined.	2.6% of fish contaminated with plastics; the highest frequency (>13%) was in cod and the lowest in herring (2%); PE, PP, PET, SA (styrene acrylate); ingested plastic: 5.4% of the fish from the southern North Sea, 1.2% in the northern North Sea.	Foekemam et al 2013
Norway (Norwegian coast)	Fish: Atlantic cod (*Gadus morhua*); 302 fish stomachs examined	Micro- (18.8%), meso- (75%) and macro-plastics found in **3% of cod** stomachs; fibres, fragments, granule, film; 9 different with polyester (polycyclohexylene-dimethylene terephthalate-PCT) was the most frequently detected in stomachs.	Brate et al 2016
Turkey (The Medit-erranean coast)	Fish were collected from 10 locations; 1337 specimens of fish examined/28 species stomachs, intestines	34% of fish had microplastic; **pelagic** fish **ingested more microplastic** than **demersal** species; fibres (70%), hard plastics (20.8%), nylon, rubber; blue colour plastic was the most dominant; fish that ingested a higher number of microplastic particles originated from the sites that also had a higher	Güven et al 2017

Country or Location	Species/parts/items examined or analysed	Research findings	References
		particle count in the seawater and sediment.	
2. Honey			
Germany, France, Italy, Spain and Mexico	Honey (from local supermarkets or producers); mostly from Germany	100% contaminated with fibres and fragments; both coloured and transparent fibres and fragments were recorded; *possible contamination routes* (airborne, by bees into the hive or introduced during honey collection/ processing).	Liebezeit and Liebezeit 2013
3. Sugar			
Germany	Five commercial sugars from a local supermarket	Transparent and coloured fibres (mean 217 ± 123/kg of sugar) and fragments (32 ± 7/kg of sugar) were found; unrefined cane sugar had **560 fibres and 540 fragments per kilogra**m of honey.	Liebezeit and Liebezeit 2013
4. Beer			
Germany	24 beer brands (bought from local supermarkets) analysed	100% beer brands tested had plastics fragments, fibres and granules; fragments (most abundant); *possible contamination routes* (atmospheric deposition, clothing worn by workers, materials used in the production process).	Liebezeit and Liebezeit 2014
USA	Beer, 12 brands	Fibers and fragments; **98.4% were fibres** (4.05 particles/L)	Kosuth et al 2018
5. Sea-salt			
Australia, France, Iran, Japan,	Sea-salt (bought from local supermarkets)	100% samples with fragments, filaments, film; 46.7–806 particles/kg of salt; polymers were:	Karami et al 2017b

Country or Location	Species/parts/items examined or analysed	Research findings	References
Malaysia, NZ, Portugal, South Africa		fragments (63.8%), filaments (25.6%), (10.6%); PP (40.0%), PE (33.3%); *possible contamination routes* (atmospheric deposition, local coastal water).	
China (Coastal water)	Sea-salt, 15 brands of sea salts, lake salts, and rock/well salt	**100% contaminated with fragments**, filaments, film; 681 microplastics/kg sea salt; PET (most common in sea salts); cellophane (dominant microplastic in lake salts, rock/well salts); based on the study, the adults will ingest approximately **1000 microplastic particles** /year just from table salts.	Yang et al 2015
Spain	Sea-salt (21 different commercial table salt)	**100% contaminated with fibres;** 50–280 MP particles/kg of salt; PET (most common polymer), followed by PP and PEPE.	Iñiguez et al 2017
Turkey	Sea-salt (16 brands of table salts from the Turkish market)	**100% contaminated with** fibres, fragments, films; MP particle content: 16**–84 item/kg in sea** salt; ***common plastic polymers***: PE (22.9%), PP (19.2%); sea salt contamination may demonstrate the extent of **MPs pollution** in the environment.	Gündoğdu 2018
USA	Sea-salt (12 brands from grocery stores)	**100% contaminated with fibres (major), fragments**; 46.7–806 particles/kg of salt.	Kosuth et al 2018
6. Drinking water (tap), bottled water and mineral water and groundwater			
Tap water World (14 countries)	Tap water; 159 samples	Nearly 81% of global tap water samples were contaminated with plastic fibres; [particles/L mean: 5.45; Germany 0.91, France 1.82, India 6.24, USA 9.24]; out of 14 countries, the highest particles per litre were in the USA); people may	Kosuth et al 2018

Country or Location	Species/parts/items examined or analysed	Research findings	References
		be ingesting between 3,000 and 4,000 MPs from tap water every year.	
Mineral water 22 different returnable, single-use plastic bottles	Grocery stores from Germany	100% of samples contaminated with plastic fragments; 2–241 particles/L; most of the particles in returnable plastic bottles were PET, 84% and PP; 7%; the study indicate that **plastic packaging** can also **emit microparticle**s since a wide range of food is packed or wrapped in plastic.	Schymanski et al 2018

4.3.4. Impacts on marine life

Plastic debris can affect marine organisms through entanglement, ingestion, suffocation (Gall and Thompson 2015), harm and even death. Lost or discarded fishing gear (sometimes called "**ghost nets**"[11]) is one of the main categories of debris resulting in the entanglement of aquatic animals on a global scale (Macfadyen et al 2009; Wilcox et al 2013). Plastics are mistaken as food and ingested by various marine organisms including invertebrates (Murray Cowie 2011), fishes (Carpenter and Smith 1972; Davison and Asch 2011), sea turtles (Lazar and Gracan 2011), seabirds (Spear et al 1995) and whales (Tarpley and Marwitz 1993).

- **Examples of plastics ingestion:** At least 208 species are affected (Gall and Thompson 2015); a sharp increase in plastic ingestion in seabirds and marine turtles was reported (with a rate of 1.7% increase per year for seabirds (Wilcox et al 2015) and 0.7% for turtles (Schuyler et al 2014).
- **Examples of plastics entanglement:** Around 250 species are known to have become entangled in plastic debris including over 50 species of marine mammals, over 100 species of seabirds, many species of fish and seven species of marine turtles (Thompson 2017); marine turtles mistake floating plastics (bags or sheeting) as food (jellyfish) causing gastrointestinal blockage (Schuyler et al 2014), injury and reproductive impediments (several entanglements were due to discarded **fishing gear/ "ghost nets"**) (Wilcox et al 2013).

[11]***Ghost nets*** *are fishing nets that have been left or lost in the ocean by fishermen. They can entangle fish, dolphins, sea turtles, sharks, dugongs, crocodiles, seabirds, crabs, and other creatures, including the occasional human diver (http://www.eurocbc.org/page54.html).*

- Around 49 commercially important species such as sardines, herring, hake, whiting and red mullet are known to ingest microplastics (Kuhn et al 2015, GESAMP 2016) (see also Table 4.8).
- Shell-fish including mussels, clams, oysters and scallops also ingest microplastics (Kuhn et al 2015) (see also Table 4.8).

4.3.5. Impacts on aesthetic value, recreation and tourism

Plastic waste damages the aesthetic value of tourist destinations, leading to decreased tourism-related incomes and major economic costs related to the cleaning and maintenance of the sites (IUCN 2018). Coastal areas, beaches and oceans are used by recreational users for swimming, diving and a number of water sports. Plastic pollution could discourage such users from visiting affected areas. A coast littered with plastic does not look as pretty and welcoming as a pristine beach (Sebille et al 2016).

4.3.6. Impacts on maritime infrastructure

Plastics can entangle propellers, anchors, rudders. It can block water intake pipes and valves (Thompson 2017).

4.3.7. Impacts on shipping

Entanglement of anchors in abandoned plastic fishing gear and fouling of a vessel's propeller with plastics can cause vessel breakdowns and in extreme cases led to the loss of human lives (Sebille et al 2016). The propeller of life-boat can be entangled with plastics (the sinking of the Ferry M/V Soe-Hae in 1993 was, in part, caused by rope around the propellers and resulted in 292 deaths) (Cho 2005).

4.3.8. Impacts on the local economy

The plastic litter can have several economic implications (Sebille et al 2016):

- Losses of revenues associated with a decline in tourism;
- Losses of revenues due to losses of fisheries and aquaculture products and other foods (due to food contamination);
- The clean-up costs associated with removing the plastic litter from beaches; and
- Damage to vessels and equipment (propeller)

4.3.9. Impacts on human health

Humans can be exposed to microplastics via contaminated seafood (fish, shell-fish), honey, sugar, beer, salt and tap water (as discussed in section 4.3.3). In particular, high-risk pollutants/chemicals adsorbed on microplastics (as discussed in section 4.3.2) can be transferred to humans through the food chain pathways. In addition, plastic debris can support a diverse microbial community (the hard surface of plastics provides an ideal environment for opportunistic microbial colonisers to form biofilms). Hence the colonisation, survival and transport of pathogens on polymers present a potential risk to human health (Keswani et al 2016). Pathogenic *Vibrio parahaemolyticus* were reported on a number of microplastic particles (polyethylene, polypropylene and polystyrene) from North/Baltic Sea (Kirstein et al 2016). Bacterial counts including the cyanobacteria group were consistently found abundant on plastic marine debris in the Western Mediterranean basin (Dussud et al 2018).

4.4. Linkages between plastic pollution and climate change

Plastics are generally made from fossil fuels (ethylene and propylene). If the plastic is incinerated, it increases carbon emissions (which cause climate change) (because plastic contains majority of carbon) (IUCN 2018) (see Table 4.3). The USEPA estimates that, as many as five ounces of carbon dioxide are emitted for each ounce of polyethylene (PET) produced (one ounce = 0.029 Liter) - the type of plastic most commonly used for beverage bottle (Glazner 2015). Sea-ice (polar sea ice, the Antarctic sea ice) represent a sink for microplastic particles (MPs), however, melting of ice due to climate change/global warming could release them (MPs) back into the environment (Obbard et al 2014). Royer et al (2018) investigated the production of hydrocarbon gases from different polymer plastics at ambient temperature. The study reveals that, many plastics as waste give off harmful greenhouse gasses (GHGs) including methane (CH_4). CH_4 is the most potent atmospheric GHG. The CH_4 production with Low-density Polyethylene (LDPE) is the highest compared to other plastics as follows: LDPE> PS> PET> PP> HDPE> Acrylic> PA (Table 4.9). Plastic waste would be another unrecognized source of climate-relevant trace GHGs that could increase as more plastic is produced to meet demands and accumulated in the environment.

Table 4.9. Mean production rates of CH_4 from a variety of plastics incubated in water under ambient solar radiation (light) and dark conditions (Royer et al 2018)

*[**pmol**=picomoles per litre; **1 mol**=1000,000,000,000 picomoles; a picomole is one-trillionth of a mole; **ns** = final concentrations not significantly different from those in the control treatment]*

Plastics	**CH_4 (pmol $g^{-1}d^{-1}$)**	
	Light	**Dark**
Polycarbonate (PA)	10±2	ns
Acrylic	30±3	ns

Plastics	**CH_4 (pmol $g^{-1}d^{-1}$)**	
	Light	**Dark**
High-density Polyethylene (HDPE)	90±10	ns
Polypropylene (PP)	170±10	ns
Polyethylene Terephthalate (PET)	500± 20	50 ± 10
Polystyrene (PS)	730± 110	120 ± 10
Low-density Polyethylene (LDPE)	4100± 200	ns

4.5. Management and mitigation measures

To reduce impacts from plastic waste and plastic pollution on the environment, biota, wildlife, seafood and human health, the following strategies can be applied:

- Implement and enforce 3 Rs (reduce, reuse and recycle plastics) strategy to reduce plastic waste generation;
- Use alternatives to plastic (jute bags, environmentally friendly bags, reuse bags);
- Prevent or penalize illegal plastic waste dumping;
- Encourage industries to use biodegradable or bioplastic made from renewable raw materials such as starch or cellulose or bio-synthesized materials;
- Ban and levy tax on using plastic bags (to curb the growth of single-use plastics);
- Collect and remove old or abandoned fishing nets from marine and freshwater environments;
- Research and innovate efficient wastewater treatment plant that can remove nanoplastics, microplastics and microbeads;
- Raise public awareness education on harm caused by plastic pollution (latest, evidence and science-based data and information);
- Avoid cosmetics that contain microbeads or ban the use of microplastic in cosmetics (many countries including UK, USA already banned inclusion of microbeads in cosmetics);
- Implement microbeads legislation;

- Equip washing machines with fiber filters since clothes are a major source of synthetic fibers;
- Educate the community to dispose of personal plastic wastes appropriately (do not encourage to burn as burning of plastics would emit GHGs that cause climate change!);
- Use technology and mechanical interventions to capture plastic litters in streams and rivers before it gets to the oceans (see Fig. 1.2A in chapter 1);
- Monitor plastic pollution in lakes, rivers, coasts and oceans to assess the risk posed by plastic litters; and
- Monitor seafood and other foods and drinking water to assess risks and reduce risks to biota, wildlife and humans.

4.6. Conclusion

Plastic waste and plastic pollution are ubiquitous and is reported from across the globe (from the Arctic to the Antarctic, from the surface to sediment, and in every waterway (lakes, rivers, estuaries, coastal and oceans). Plastics have been detected throughout the water column (surface, middle, and bottom) and in all the environmental compartments (biota, water, sediment). Plastic debris accounted for more than 75% of marine litter. At some places, plastic litter outnumbered fish larvae and plankton (the Danube River, Portuguese coastal waters, North Pacific central gyre). Plastics litter can act either as a source or sink or a vehicle to transport hydrophobic toxic chemicals/contaminants in the aquatic environment. Plastics in the environment as waste pose five major problems and impacts: **i**. ingestion and entanglement of plastics by aquatic biota and wildlife; **ii**. adsorption of persistent organic pollutants and heavy metals by plastics; **iii.** colonisation and formation of biofilms (bacteria, cyanobacteria) on the surface of plastics; **iv.** leaching of additives from plastics and **v.** contamination of seafood and other foods by plastic particles. Research carried out reveals that, a number of contaminants/ pollutants are adsorbed in microplastics (heavy metals, petroleum hydrocarbons, pesticides, perfluoro chemicals, pharmaceuticals, personal care products, persistent organic pollutants and surfactants/detergents). Microplastic particles contaminate seafood (fish, shrimps, bivalves/molluscs, dried fish, and canned fish), honey, sugar, beer, sea salt and drinking tap water. Plastic debris affects many marine organisms through entanglement, ingestion, and suffocation causing harm and even death. Lost or discarded fishing gear (sometimes called as "ghost nets") is one of the main categories of debris resulting in the entanglement of animals on a global scale. Plastics are mistaken as food/plankton and ingested by various marine organisms including invertebrates, fishes, sea turtles' seabirds and whales. Plastic waste damages the aesthetic value of tourist destinations, leading to decreased tourism-related incomes and major economic costs related to cleaning and maintenance of the sites. It can block water intake pipes and valves. Humans can be exposed to microplastics via contaminated seafood, honey, sugar, beer, salt and tap water. The high-risk pollutants/ chemicals adsorbed on microplastics can be transferred to humans through the food chain pathway. Due to their incredibly small size, nanoplastics present a series of complicated toxicological and analytical challenges. Nanoplastics are a particular challenge since these particles can enter the cells of the organisms (Bank and Hansson 2019). To reduce impacts from plastic wastes and plastic pollution a number of measures can be undertaken including: **i.** 3Rs (reduce, reuse and recycle plastics) strategy; **ii.** use of alternatives to plastic (jute bags, environmentally friendly bags, reuse bags); **iii.** Use of biodegradable or bioplastic; **iv.** imposing ban, levy tax on using single-use of plastic bags; **v.** use of efficient wastewater treatment plant that can remove nanoplastics, microplastics and microbeads; **vi.** awareness education on harm caused by plastic pollution; **vii.** and monitoring of plastic in waterways, seafoods, other foods and drinking water.

4.7. References

Adams WJ, Biddinger G.R, Robillard KA and Gorsuch JW 1995 A summary of the acute toxicity of 14 phthalate esters to representative organisms. *Environmental Toxicology and Chemistry* **14** 1569-1574

Akortia E, Okonkwo JJO, Lupankwa M et al 2016 A review of sources, levels and toxicity of polybrominated diphenyl ethers (PBDEs) and their transformation and transport in various environmental compartments. *Environmental Reviews* **24(3)** 253-273 *doi 10.1139/er-2015-0081*

Alexander HC, Dill DC, Smith PD, Guiney PD and Dorn PB 1988 Bisphenol A: Acute aquatic toxicity. *Environmental Toxicology and Chemistry* **7** 19-26

Alimi QS, Budarz JF, Hernandez LM and Tufenkji N 2018 Microplastics and Nanoplastics in aquatic environments: Aggregation, Deposition, and Enhanced Contaminant Transport. *Environmental Science and Technology* **52** 1704–1724

Amelineau F, Bonnet D, Heitz O, et al., 2016 Microplastic pollution in the Greenland Sea: Background levels and selective contamination of planktivorous diving seabirds. *Environmental Pollution* **219** 1131–1139

Andrady LA 2011 Microplastics in the marine environment. *Marine Pollution Bulletin* **62** 1596–1605

Anderson PJ, Warrack S, Langen V et al 2017 Microplastic contamination in Lake Winnipeg, Canada. *Environmental Pollution* **225** 223–232

ATSDR 2017 Toxicological profile for polybromide diphenyl ethers (PBDEs). US Department of Health and Human services. Public Health Service Agency for Toxic Substances and Disease Registry; *https://www.atsdr.cdc.gov/ToxProfiles/tp.asp?id=901&tid=183*

Avio CG, Gorbi S and Regoli F 2015 Experimental development of a new protocol for extraction and characterization of microplastics in fish tissues: First observations in commercial species from Adriatic Sea, *Marine Environmental Research* **111** 18-26 *http:// dx.doi.org/10.1016/j.marenvres.2015.06.014*

Aytan U, Valente A, Senturk Y et al., 2016 First evaluation of neustonic microplastics in Black Sea waters. *Marine Environmental Research* **119** 22-30

Bakir A, Rowland SJ and Thompson RC 2014 Transport of persistent organic pollutants by microplastics in estuarine conditions. *Estuarine, Coastal and Shelf Science* **140** 14-21

Bank, M and Hansson SV 2019 *The Plastic Cycle:* A Novel and Holistic Paradigm for the Anthropocene. *Environmental Science & Technology* **53(13)** 7177-7179 *https://doi.org/10.1021/acs.est.9b02942*

Boerger CM, Lattin GL, Moore SL and Moore CJ 2010 Plastic ingestion by planktivorous fishes in the North Pacific Central Gyre. *Marine Pollution Bulletin* **60(12)** 2275-2278

Bråte ILN, Eidsvoll DP, Steindal CC and Thomas KV 2016 Plastic ingestion by Atlantic cod (*Gadus morhua*) from the Norwegian coast. *Marine Pollution Bulletin* **112** 105–110; *https://doi.org/ 10.1016/j.marpolbul.2016.08.034*

Brennecke D, Duarte B, Paiva F et al 2016 Microplastics as vector for heavy metal contamination from the marine environment. *Estuarine*, *Coastal* and *Shelf Science* **178** 189-195

Browne MA, Crump P, Niven SJ et al 2011 Accumulation of microplastic on shorelines worldwide: sources and sinks. *Environmental Science and Technology* **45** 9175–9179 *doi:10.1021/es201811s*

Carpenter E and Smith K 1972 Plastics on the Sargasso Sea surface. *Science* **175** 1240–1241

Castañeda RA, Avlijas S, Simard MA et al 2014 Microplastic pollution in St. Lawrence river sediments. *Canadian Journal of Fisheries and Aquatic Sciences* **71** 1767–1771

Cheung PK, Cheung LTO, Fok L 2016 Seasonal variation in the abundance of marine plastic debris in the estuary of a subtropical macro-scale drainage basin in South China. *Science of the Total Environment* **562** 658–665

Cho DO 2005 Challenges to marine debris management in Korea. *Coastal Management* **33** 389-409

Cincinelli A, Scopetani C, Chelazzi D et al 2017 Microplastic in the surface waters of the Ross Sea (Antarctica): Occurrence, distribution and characterization by FTIR. *Chemosphere* **175** 391–400

Cole M, Webb H, Lindeque PK et al 2014 Isolation of microplastics in biota-rich seawater samples and marine organisms. *Scientific Reports* **4** *http://dx.doi.org/10.1038/srep04528.*

Collard F, Gilbert B, Eppe G et al 2017 Morphology of the filtration apparatus of three planktivorous fishes and relation with ingested anthropogenic particles. *Marine Pollution Bulletin* **116** 182–191

Comber MHI, Williams TD and Stewart KM 1993 The effects of nonylphenol on *Daphnia magna*. *Water Research* **27** 273–276

Costa MF, Ivar do Sul J, Silva-Cavalcanti JS et al 2010 On the importance of size of plastic fragments and pellets on the strandline: a snapshot of a Brazilian beach. *Environmental Monitoring and Assessment* **168** 299–304

Cózara A, Echevarríaa F, González-Gordilloa, JI et al 2014 Plastic debris in the open ocean. *PNAS* **111(28)** 10239–10244

Crain DA, Eriksen M, Iguchi T et al., 2007 An ecological assessment of bisphenol-A: evidence from comparative biology. *Reproductive Biology* **24** 225-239

Crawford CB and Quinn B 2017 Microplastic pollutants. Chapter 6. The interactions of microplastics and chemical pollutants. Persistent organic pollutants (POPs). Page 131-156. Elsevier Inc.

Cressey D 2016 The plastic Ocean. *Nature* **536** 263-265

Darnerud PO, Eriksen GS, Johannesson T et al 2001. Polybrominated diphenyl ethers: occurrence, dietary exposure, and toxicology. *Environmental Health Perspectives* **109(1)** 49-68

Davison P and Asch R 2011 Plastic ingestion by mesopelagic fishes in the North Pacific Subtropical Gyre. *Marine Ecological Progress Series* **432** 173–180

De Araujo FG, Bauerfeldt F and CID YP 2018 Nonylphenol: Properties, legislation, toxicity and determination. *Anais da Academia Brasileira de Ciências* **90(2 Suppl. 1)** 1903-1918

Defoe DL, Holcombe GW, Hammermeister DE and Biesinger KE 1990 Solubility and toxicity of eight phthalate esters to four aquatic organisms. *Environmental Toxicology and Chemistry* **9** 623-636

Devriese LI, Van de Meulen MD, Maes T et al 2015 Microplastic contamination in brown shrimp (*Crangon crangon*, Linnaeus 1758) from coastal waters of the southern north sea and channel area. *Marine Pollution Bulletin* **98** 179–187

Doyle MJ, Watson W, Bowlin NM and Sheavly SB 2011 Plastic particles in coastal pelagic ecosystems of the Northeast Pacific ocean. *Marine Environmental research* **71** 41–52 *doi:10.1016/j.marenvres.2010.10.001*

Dris R, Gasperi J, Rocher V et al 2015 Microplastic contamination in an urban area: a case study in Greater Paris. *Environmental Chemistry* **12** 592

Dussud C, Meistertzheim AL, Conan P et al 2018 Evidence of niche partitioning among bacteria living on plastics, organic particles and surrounding seawaters. *Environmental Pollution* **236** 807-816

Ecetoc 2019 Technical report 123. Adsorption-desorption distribution (Kd) and organic carbon-water partition (KoC) coefficients *http://www.ecetoc.org/report/measured-partitioning-property-data/adsorption-desorption-distribution-kd-and-organic-carbon-water-partition-koc-coefficients/* (accessed 11 Dec2019)

Endo S, Takizawa R, Okuda K et al 2005 Concentration of polychlorinated biphenyls (PCBs) in beached resin pellets: Variability among individual particles and regional differences. *Marine Pollution Bulletin* **50(10)** 1103-1114

European Commission 2008 Updated European Risk Assessment Report: 4,4′-isopropylidenediphenol (Bisphenol-A). Ispra, Italy.

Eriksen M, Liboiron M, Kiessling T et al 2017 Microplastic sampling with the AVANI trawl compared to two neuston trawls in the Bay of Bengal and South Pacific. *Environmental Pollution* **232** 430–439

EU 2008 BIS (2-ethylhexyl) phthalate (DEHP). Summary Risk Assessment Report. EUR 23384 EN/2; *https://echa.europa.eu/documents/10162/060d4981-4dfb-4e40-8c69-6320c9debb01*

Fasano E, Bono-Blay F, Cirillo T, Montuori, P Lacorte S 2012 Migration of phthalates alkylphenols, bisphenol A and di(2-ethylhexyl)adipate from food packaging, *Food Control* **27(1)** 132–138

Faure F, Caini C, Potter G et al 2015 An evaluation of surface micro- and mesoplastic pollution in pelagic ecosystems of the Western Mediterranean Sea. *Environmental Science and Pollution Research* **22** 12190–12197

Faure F, Corbaz M, Baecher H et al 2013 Pollution due to plastics and microplastics in Lake Geneva. Koblenz, Germany. In: 6th International Conference on Water Resources and Environment Research;

https://infoscience.epfl.ch/record/187023?ln=en

Faure F, Corbaz M, Baecher H and de Alencastro LF 2012 Pollution due to plastics and microplastics in Lake Geneva and in the Mediterranean Sea. *Archives des Sciences* **65** 157–164

Foekemam EM, Gruijter C, Mergia MT et al 2013 Plastic in North Sea fish. *Environmental Science and Technology* **47(15)** 8818-8824

Fok L and Cheung PK 2015 Hong Kong at the Pearl River Estuary: a hotspot of microplastic pollution. *Marine Pollution Bulletin* **99** 112-118

http://dx.doi.org/10.1016/j.marpolbul.2015.07.050.

Fok L, Cheung PK, Tang G et al 2017 Size distribution of stranded small plastic debris on the coast of Guangdong, South China. *Environmental Pollution* **220** 407–412

FAO 2014 FAO Yearbook; Fishery and Aquaculture Statistics 2012. ISBN 978-92-5-008293-6 *http://www.fao.org/3/a-i3740t.pdf*

Frias JPGL, Otero V and Sobral P 2014 Evidence of microplastics in samples of zooplankton from Portuguese coastal waters. *Marine Environmental Research* **95** 89–95

Free CM, Jensen OP, Mason SA et al 2014 High-levels of microplastic pollution in a large, remote, mountain lake. *Marine Pollution Bulletin* **85(1)** 156-163

Fries E and Zarfl C 2012 Sorption of polycyclic aromatic hydrocarbons (PAHs) to low- and high-density polyethylene (PE). *Environmental Science and Pollution Research* **19(4)** 1296-1304

Galgani F, Leaute F, Moguedet JP et al 2000 Litter on the sea floor along European coasts. *Marine Pollution Bulletin* **40** 516-527 *DOI: 10.1016/S0025-326X(99)00234-9*

Gall SC and Thompson RC 2015 The impact of debris on marine life. *Marine Pollution Bulletin* **92** 170–179

Gallagher A, Rees A, Rowe R et al 2016 Microplastics in the Solent estuarine complex, UK: An initial assessment. *Marine Pollution Bulletin* **102** 243–249

Gauthier MM 1995 Engineered materials handbook Edition: Desk. ASM. ISBN: 978-0-87170-283-8; *https://www.asminternational.org/handbooks/-/journal_content/56/10192/06560G/PUBLICATION*

Germanov, E.S, Marshall, A.D, Bejder, L, Fossi, M. C and Loneragan, NR 2018 Microplastics: No Small Problem for Filter- Feeding Megafauna. *Trends in Ecology and Evolution* **33(4)** 227-232

GESAMP 2016 Sources, Fate and Effects of Microplastics in the Marine Environment – Part Two of a Global Assessment. In: IMO/FAO/UNESCO-IOC/UNIDO/WO/IAEA/UN/UNEP/UNDP Joint Group of Experts on the Scientific Aspects of Marine Environmental Protection (ed. P. J. Kershaw and C. M. Rochman), 220 p.

Gill U, Chu I, Ryan JJ, and Feeley M 2004 Polybrominated diphenyl ethers: human tissue levels and toxicology. Reviews of *Environmental Contamination* and *Toxicology* **183** 55-97

Glazner E 2015 Plastic pollution and climate change; *https://www.plasticpollutioncoalition.org/blog/2015/11/17/plastic-pollution-and-climate-change*

Gregory MR and Ryan PG 1997 Pelagic plastics and other seaborne persistent synthetic debris: A review of southern hemisphere perspectives. Chapter 5, p 49. *In*: Marine Debris: Sources, Impacts, and Solutions (1997) J.M. Coe and D.B. Rogers (eds.). 49

Gündoğdu S 2018 Contamination of table salts from Turkey with microplastics. Food Additives and Contaminants: Part A: Chemistry, Analysis, Control, Exposure & Risk Assessment **12** 1–9

Gundoğdu S and Cevik C 2017 Micro- and mesoplastics in Northeast Levantine coast of Turkey: The preliminary results from surface samples. *Marine Pollution Bulletin* **118** 341–347

Güven O, Gökdag K, Jovanovic B and Kõdeys AE 2017 Microplastic litter composition of the Turkish territorial waters of the Mediterranean Sea, and its occurrence in the gastrointestinal tract offish. *Environmental Pollution* **223** 286–294 *https://doi.org/10.1016/j.envpol.2017.01.025*.

Hahladakis JN, Velis CA, Weber R, Iacovidou E and Purnell P 2018 An overview of chemical additives presents in plastics: Migration, release, fate and environmental impact during their use, disposal and recycling.*Journal of Hazardous Materials* **344** 179-199. *doi: 10.1016/j.jhazmat.2017.10.014. Epub 2017 Oct 9.*

Hansen E, Nilsson NH, Lithner D, Lassen C 2013 Hazardous substances in plastic materials. COWI in cooperation with Danish Technological Institute. *http://www.byggemiljo.no/wp-content/uploads/2014/10/72_ta3017.pdf*

Heitmuller PT, Hollister TA and Parrish PR 1981 Acute toxicity of 54 industrial chemicals to sheepshead minnows (*Cyprinodon variegatus*). *Bulletin Environmental Contamination and Toxicology* **27** 596-604

Hendriks AJ, Maas-Diepeveen JL, Noordsij A and Van der Gaag MA 1994 Monitoring response of XAD- concentrated water in the Rhine Delta: A major part of the toxic compounds remains unidentified. *Water Research* **28(3)** 581-598

Holmes LA, Turner A and Thompson, RC 2012 Adsorption of trace metals to plastic resin pellets in the marine environment. *Environmental Pollution* **160(1)** 42-48

Hu J.-Q, Yang S.-Z, Guo L et al 2017 Microscopic investigation on the adsorption of lubrication oil on microplastics. *Journal of Molecular Liquids* **227** 351-355

Hüffer T and Hofmann T 2016 Sorption of non-polar organic compounds by micro-sized plastic particles in aqueous solution. *Environmental Pollution* **214(17)** 194-201

IARC 2019 IARC monographs on the identification of carcinogenic hazards to humans. *https://monographs.iarc.fr/agents-classified-by-the-iarc/*

Imhof HK, Sigl R, Brauer E et al 2017 Spatial and temporal variation of macro-, meso- and microplastic abundance on a remote coral island of the Maldives, Indian Ocean. *Marine Pollution Bulletin* **116** 340–347

Iñiguez ME, Conesa JA and Fullana A 2017 Microplastics in Spanish table salt. *Scientific Reports* **7** 8620

Isobe A, Uchida K, Tokai T and Iwasaki S 2015 East Asian seas: a hot spot of pelagic microplastics. *Marine Pollution Bulletin* **101** 618-623 *DOI: 10.1016/j.marpolbul.2015.10.042*

IUCN 2018 Marine plastics. International Union for Conservation of Nature. IUCN I issues briefs. Switzerland.

https://www.iucn.org/sites/dev/files/marine_plastics_issues_brief_final_0.pdf

Jabeen K, Su L, Li J et al 2017 Microplastics and mesoplastics in fish from coastal and fresh waters of China. *Environmental Pollution* **221** 141–149

https://doi.org/10.1016/ j.envpol.2016.11.055

Jambeck JR, Geyer R, Wilcox C et al 2015 Plastic waste inputs from land into the ocean. *Science* **347** (6223) 768-771

Jayasiri HB, Purushothaman CS and Vennila A 2013 Quantitative analysis of plastic debris on recreational beaches in Mumbai, India. *Marine Pollution Bulletin* **77** 107–112

Jobling S, Sheahan D, Osborne A et al 1996 Inhibition of testicular growth in rainbow trout (*Oncorhynchus mykiss*) exposed to estrogenic alkylphenolic chemicals. *Environmental Toxicology and Chemistry* **15** 194-202

Kaberi H, Tsangaris C, Zeri C et al 2013 Microplastics along the shoreline of a Greek island (Kea isl., Aegean Sea): types and densities in relation to beach orientation, characteristics and proximity to sources. 4th international conference on environmental management, engineering, planning and economics (CEMEPE) and SECOTOX conference, Mykonos Island, Greece

Karapanagioti HK and Klontza I 2008 Testing phenanthrene distribution properties of virgin plastic pellets and plastic eroded pellets found on Lesvos island beaches (Greece). *Marine Environmental Research* **65(4)** 283-290

Karami A, Golieskardi A, Ho YB et al 2017a Microplastics in eviscerated flesh and excised organs of dried fish. *Scientific Reports* **7** 547

Karami A, Golieskardi A, Keong Choo C et al 2017b The presence of microplastics in commercial salts from different countries. *Scientific Reports* **7** 46173

Karami A, Golieskardi A, Choo CK et al 2018 Microplastic and mesoplastic contamination in canned sardines and sprats. *Science of the Total Environ.* **612** 1380-1386

Kelly SA and Di Giulio RT 2000 Developmental toxicity of estrogenic alkylphenols in killifish (*Fundulus heteroclitus*). *Environmental Toxicology and Chemistry* **9** 2564–2570

Keswani A, Oliver DM, Gutierrez T and Quilliam RS 2016 Microbial Hitchhikers on Marine plastic debris: human exposure risks at bathing waters and beach environments. *Marine Environmental Research* **118** 10–19

Kibria G 2017 Plastic waste and plastic pollution- A Threat to All Nations. *DOI: 10.13140/RG.2.2.11169.51048*

Kibria G 2018 Presentation: Plastic pollution- sources, global production, global "hotspots", impacts on biodiversity & seafood; Adsorption of organic & inorganic chemicals and mitigation. *DOI: 10.13140/RG.2.2.35028.24967/3*

Kibria G, Haroon AKY, Nugegoda D and Rose G 2010 *Climate change and chemicals: Environmental Biological aspects.* New India Publishing Agency. New Delhi. 470 p.

ISBN: 9789380235301; *https://www.researchgate.net/publication/261216635_Climate_Change_and_Chemicals_Environmental_and_Biological_Aspects*

Kim I-S, Chae D-H, Kim S-K et al 2015 Factors influencing the spatial variation of microplastics on high-tidal coastal beaches in Korea. *Archives of Environmental Contamination and Toxicology* **69** 299–309

Klein S, Worch E and Knepper TP 2015 Occurrence and spatial distribution of microplastics in river shore sediments of the rhine-main area in Germany. *Environmental Science and Technology* **49(10)** 6070–6076

Kosuth M, Mason SA, Wattenberg EV 2018 Anthropogenic contamination of tap water, beer, and sea salt. *PLoS ONE* **13(4)** e0194970; *https://doi.org/10.1371/journal.pone.0194970*

Kirstein IV, Kirmizi S, Wichels A et al 2016 Dangerous hitchhikers? evidence for potentially pathogenic *Vibrio* spp. on microplastic particles. *Marine Environmental Research* **120** 1–8

Kuhn R and Pattard M 1990 Results of the harmful effects of water pollutants to green algae (*Scenedesmus subspicatus*) in the cell multiplication inhibition test. *Water Research* **24** 31-38

Kuhn S, Bravo Rebolledo EL and van Franeker JA 2015 Deleterious Effects of Litter on Marine Life. *In*: Marine Anthropogenic Litter. M. Bergman, L. Gutow & M. Klages (eds.), 75–116. Heidelberg: Springer.

Kunz A, Walther BA, Lowemark L et al 2016 Distribution and quantity of microplastic on sandy beaches along northern coast of Taiwan. *Marine Pollution Bulletin* **111** 126–135

Kuramochi H, Maeda K and Kawamoto K 2007 Physicochemical properties of selected polybrominated diphenyl ethers and extension of the UNIFAC model to brominated aromatic compounds. *Chemosphere* **67** 1858–1865

Lackner M 2015 Bioplastics. Biobased plastics as renewable and/or biodegradable alternatives to petro plastics. In book: Kirk-Othmer Encyclopedia of Chemical Technology. Edition: 6 Chapter: Bioplastics - Biobased plastics as renewable and/or biodegradable alternatives to petroplastics. Publisher. John Wiley & Sons, Inc. *DOI: 10.1002/0471238961.koe00006*

Lambert SC, Sinclair CJ and Boxall A 2014 Occurrence, degradation, and effect of polymer-based materials in the environment. *Reviews of Environmental Contamination and Toxicology* **227** 1-53

Lattin GL, Moore CJ, Zellers AF et al 2004 A comparison of neustonic plastic and zooplankton at different depths near the southern California shore. *Marine Pollution Bulletin* **9** 291-294

Law KL, Morét-Ferguson SE, Goodwin DS et al 2014 Distribution of surface plastic debris in the Eastern Pacific Ocean from an 11-year data set. *Environmental Science and Technology* **48** 4732–4738 *doi:10.1021/es4053076*

Law KL, Morét-Ferguson S, Maximenko NA et al 2010 Plastic accumulation in the North Atlantic subtropical gyre. *Science* **329** 1185-1188 *doi:10.1126/science.1192321*

Lazar B and Gracan R 2011 Ingestion of marine debris by loggerhead sea turtles, *Caretta caretta*, in the Adriatic Sea. *Marine Pollution Bulletin* **62** 43–47

Lebreton L, Slat B, Ferrari F et al 2018 Evidence that the Great Pacific Garbage Patch is rapidly accumulating plastic *Nature Scientific Reports* **8** 4666 *DOI:10.1038/s41598-018-22939-w 22 March 2018*

Lechner A, Keckeis H, Lumesberger-Loisl F et al 2014 The Danube so colourful: A potpourri of plastic litter outnumbers fish larvae in Europe's second largest river. *Environmental Pollution* **188** 177–181

Lee H, Shim WJ and Kwon JH 2014 Sorption capacity of plastic debris for hydrophobic organic chemicals. *Science of the Total Environment* **470(471)** 1545-1552

Leslie HA, Brandsma SH, van Velzen MJM et al 2017 Microplastics en route: Field measurements in the Dutch river delta and Amsterdam canals, wastewater treatment plants, North Sea sediments and biota. *Environment International* **101** 133–142

Li J, Yang D, Li L et al 2015 Microplastics in commercial bivalves from China. *Environmental Pollution* **207** 190–195

Li C, Xu J, Chen D, and Xiao Y 2016a Detection of phthalates migration from disposable table wares to drinking water using hexafluoroisopropanol-induced catanionic surfactant coacervate extraction. *Journal of Pharmaceuticals Analysis* **6** 292–299

Li J, Qu X, Su L et al 2016b Microplastics in mussels along the coastal waters of China. *Environmental Pollution* **214** 177-184

Liebezeit G and Dubaish F 2012 Microplastics in beaches of the East Frisian Islands Spiekeroog and Kachelotplate. *Bulletin of Environmental Contamination* and *Toxicology* **89** 213–217

Liebezeit G and Liebezeit E 2013 Non-pollen particulates in honey and sugar. *Food Additives and Contaminants* Part A Chem. *Anal. Control Expo. Risk Assess* **30(12)** 2136–2140

Liebezeit G and Liebezeit E 2014 Synthetic particles as contaminants in German beers. *Food Additives and Contaminants* Part A **31(9)** 1574–1578

Liu L, Fokkink R, Koelmans AA 2016 Sorption of polycyclic aromatic hydrocarbons to polystyrene nanoplastic. *Environmental Toxicology and Chemistry* **35(7)** 1650-1655

Lu C, Yang S, Yan Z et al 2018 Deriving aquatic life criteria for PBDEs in China and comparison of species sensitivity distribution with TBBPA and HBCD. *Science of the Total Environment* **640–641** 1279–1285

Lusher AL, McHugh M, and Thompson RC 2013 Occurrence of microplastics in the gastrointestinal tract of pelagic and demersal fish from the English Channel. *Marine Pollution Bulletin* **67(1)** 94-99

Lusher A, Tirelli V, O'Connor L and Officer R 2015 Microplastics in Arctic polar waters: the first reported values of particles in surface and sub-surface samples. *Scientific Reports* **5** 14947 *DOI: 10.1038/srep14947*

Macfadyen G, Huntington T and Cappell R 2009 Abandoned, Lost or Otherwise Discarded Fishing Gear. *In*: UNEP Regional Seas Reports and Studies, 115. Rome: UNEP/FAO

Magnusson K and Noren F 2014 Screening of microplastic particles in and down-stream a wastewater treatment plant. *https://www.diva-portal.org/smash/get/diva2:773505/FULLTEXT01.pdf*

Mani T, Hauk A, Walter U et al 2015 Microplastics profile along the Rhine River. *Scientific Reports* **5** 17988

Mato Y, Isobe T, Takada H et al 2001 Plastic resin pellets as a transport medium for toxic chemicals in the marine environment. *Environmental Science and Technology* **35(2)** 318-324

Mato Y, Takada H, Zakaria MP et al 2002 Toxic chemicals contained in plastic resin pellets in the marine environment: Spatial difference in pollutant concentrations and the effects of resin type. *Kankyo Kagakukaishi* **(15)** 415–423

Mayer FL and Ellersieck MR 1986 Manual of Acute [1]oxicity: Interpretation and data base for 410 chemicals and 66 species of freshwater animals. *Resource Publication* **160** U.S. Department of the Interior, Fish and Wildlife Service, Washington, DC

McCarthy JF and Whitmore DK 1985 Chronic toxicity of di-n-butyl and di-n-octyl phthalate to *Daphnia magna* and the fathead minnow. *Environmental Toxicology and Chemistry* **4** 167-179

McDermid KJ and McMullen TL 2004 Quantitative analysis of small-plastic debris on beaches in the Hawaiian archipelago. *Marine Pollution Bulletin* **48** 790-794 *http://dx.doi.org/10.1016/j.marpolbul.2003.10.017*

Mathalon A and Hill P 2014 Microplastic fibres in the intertidal ecosystem surrounding Halifax Harbor, Nova Scotia. *Marine Pollution Bulletin* **81** 69–79

Mhadhbi L, Fumega J, Boumaiza M and Beiras R 2012 Acute toxicity of polybrominated diphenyl ethers (PBDEs) for turbot (*Psetta maxima*) early life stages (ELS) *Environmental Science and Pollution Research* **19(3)** 708-17 *708–717 DOI 10.1007/s11356-011-0602-5*

Miyamoto K and Kotake M 2006 Estimation of daily bisphenol-a intake of Japanese individuals with emphasis on uncertainty and variability. *Environmental Science* **13** 15–29

Moore CJ, Moore SL,Weisberg SB, Lattin GL, Zellers AF 2002 A comparison of neustonic plastic and zooplankton abundance in southern California's coastal waters. *Marine Pollution Bulletin* **44** 1035–1038

Moore CJ, Moore SL, Leecaster MK and Weisberg SB 2001 A comparison of plastic and plankton in the North Pacific central gyre. *Marine Pollution Bulletin* **2** 1297–1300

MSDS-CDH-DEHP 2006 Safety data Sheet. Bis-(2-Ethyl Hexyl) Phthalate. *https://www.cdhfinechemical.com/images/product/msds/37_2020417248_Bis-(2-EthylHexyl)Phthalate-CASNO-117-81-7-MSDS.pdf*

MSDS-Kumho-Bisphenol A 2015 Safety data Sheet Bisphenol A. *http://www.kpb.co.kr /resources/download/user/eng/catalog/200/BISPHENOL-A%20SDS.pdf*

MSDS-PubChem-Dibutyl phthalate 2005 (modified 2019). Safety Data Sheet Dibutyl phthalate. *https://pubchem.ncbi.nlm.nih.gov/compound/Dibutyl-phthalate*

MSDS-Sasol-Nonylphenol 2015 Safety data sheet. 10 p;

http://www.sasoltechdata.com/MSDS/NONYLPHENOL.pdf

Müller JF, Manomanii K, Mortimer MR et al 2001 Partitioning of polycyclic aromatic hydrocarbons in the polyethylene/water system. *Fresenius Journal of Analytical Chemistry* **371(6)** 816-822

Murray F and Cowie PR 2011 Plastic contamination in the decapod crustacean *Nephrops norvegicus* (Linnaeus, 1758). *Marine Pollution Bulletin* **62** 1207–1217

Naidoo T, Glassom D and Smit, A J 2015 Plastic pollution in five urban estuaries of KwaZulu-Natal, South Africa. *Marine Pollution Bulletin* **101(1)** 473–480

Naji A, Esmaili Z and Khan FR 2017 Plastic debris and microplastics along the beaches of the Strait of Hormuz, Persian Gulf. *Marine Pollution Bulletin* **114** 1057–1062

Naji A, Nuri M and Vethaak AD 2018 Microplastics contamination in molluscs from the northern part of the Persian Gulf. *Environmental Pollution* **2351** 13–120

Nakari T and Huhtala S 2010 In vivo and in vitro toxicity of decabromodiphenyl ethane, a flame retardant. *Environmental Toxicology* **25(4)** 333-338

Napper IE, Bakir A, Rowland SJ and Thompson RC 2015 Characterization, quantity and sorptive properties of microplastics extracted from cosmetics. *Marine Pollution Bulletin* **99(1-2)** 178-185

Naylor CG 1995 Environmental fate and safety of nonylphenol ethoxylates. *Text. Chem. Color.* **27** 29–33

Nur Hazimah MN and Obbard JP 2014 Microplastics in Singapore's coastal mangrove ecosystems. *Marine Pollution Bulletin* **79** 278–283

Obbard RW, Sadri S, Wong YQ et al 2014 Global warming releases microplastic legacy frozen in Arctic Sea ice. *Earth's Future* **2** 315–320

OECD 2009 OECD series on emission scenario documents. Number 3 emission scenario document on plastic additives ENV/JM/MONO(2004)8/REV1. Organisation for Economic Co-operation and Development

Oehlmann J, Schulte-Oehlmann U, Kloas W et al 2009 A critical analysis of the biological impacts of plasticizers on wildlife. *Philosophical Transactions of the Royal Society B* **364(1526)** 2047-2062

Ogata Y, Takada H, Mizukawa K et al 2009 International pellet watch: Global monitoring of persistent organic pollutants (POPs) in coastal waters. 1. Initial phase data on PCBs, DDTs, and HCHs. *Marine Pollution Bulletin* **58 (10)** 1437–1446

Pascall MA, Zabik ME, Zabik MJ and Hernandez RJ 2005 Uptake of polychlorinated biphenyls (PCBs) from an aqueous medium by polyethylene, polyvinyl chloride, and polystyrene films. *Journal of* Agricultural *and* Food Chemistry **53(1)** 164-169

Peng G, Zhu B, Yang D et al 2017 Microplastics in sediments of the Chingjiang Estuary, China. *Environmental Pollution* **225** 283–290

Peng G, Xu P, Zhu B, et al., 2018 Microplastics in freshwater river sediments in Shanghai, China: A case study of risk assessment in mega-cities. *Environmental Pollution* **234** 448–456

Potter A 2017 Microbeads in personal care products

https://www.choice.com.au/health-and-body/beauty-and-personal-care/skin-care-and-cosmetics/articles/microplastics-and-microbeads-in-toothpaste-facial-body-scrubs

Readman JW, Deluna F, Ebinghaus R, Sheppard C et al 2013 Contaminants, pollution and potential anthropogenic impacts in Chagos/British Indian Ocean Territories. Chapter 21 (p 283-298). In *Coral reefs of the United Kingdom overseas territories, coral reefs of the world* (ed. CRC Sheppard). Coral Reefs of the United Kingdom Overseas Territories, Coral Reefs of the World 4, 283 DOI 10.1007/978-94-007-5965-7_21, © Springer Science+Business Media Dordrecht 2013 Amsterdam, The Netherlands: Springer. *https://www.researchgate.net/publication/272477948_Contaminants_Pollution_and_Potential_Anthropogenic_Impacts_in_ChagosBIOT*

Reddy MS, Basha S, Adimurthy S and Ramachandraiah G 2006 Description of the small plastics fragments in marine sediments along the Alang-Sosiya ship-breaking yard, India. *Estuarine, Coastal and Shelf Science* **68** 656660; *DOI: 10.1016/j.ecss.2006.03.018*

Reiff B and Phil M 1979 The acute toxicity of diphenylolpropane to rainbow Trout (*Salmo gairdheri*). *Shell* Toxicology Laboratory (Tunstall), Sittingboume Res. Centre, Kent, England.

Rochman CM, Hoh E, Kurobe T et al 2013 Ingested plastic transfers hazardous chemicals to fish and induces hepatic stress. *Science Report* **3** 3263. *http://dx.doi.org/10.1038/srep03263*

Rochman CM, Hoh E, Hentschel BT et al 2013a Long-term field measurement of sorption of organic contaminants to five types of plastic pellets: implications for plastic marine debris. *Environmental Science and Technology* **47(3)** 1646-1654

Rochman CM, Manzano C, Hentschel BT et al 2013b Polystyrene plastic: a source and sink for polycyclic aromatic hydrocarbons in the marine environment. *Environmental Science and Technology* 2013b, **47(24)** 13976-13984

Rochman CM, Hentschel BT, Swee JT 2014 Long-term sorption of metals is similar among plastic types: implications for plastic debris in aquatic environments. *PLoS One* **9(1)** e85433

Rochman CM, Tahir A, Williams SL et al., 2015 Anthropogenic debris in seafood: plastic debris and fibres from textiles in fish and bivalves sold for human consumption. *Scientific Reports* **5** 14340 *DOI: 10.1038/srep14340*

Romeo T, Pietro B, Peda C et al 2015 First evidence of plastic debris in stomach of large pelagic fish in the Mediterranean Sea. *Marine Pollution Bulletin* **95(1)** 358-631

Rogers K 2019 Microplastics-plastic particulate.

https://www.britannica.com/ technology/microplastic

Royer S-J, FerroÂn S, Wilson ST, Karl DM 2018 Production of methane and ethylene from plastic in the environment. *PLoS* ONE **13(8)** e0200574.

https://doi.org/10.1371/journal.pone.0200574

Saquing JM, Saquing CD, Knappe DR and Barlaz MA 2010 Impact of plastics on fate and transport of organic contaminants in landfills. *Environmental Science and Technology* **44(16)** 6396-6402

Sadri SS and Thompson RC 2014 On the quantity and composition of floating plastic debris entering and leaving the Tamar Estuary, Southwest England. *Marine Pollution Bulletin* **81(1)** 55-60

Scanlan, LD, Loguinov AV, Teng Q et al 2015 Gene transcription, metabolite and lipid profiling in eco-indicator *Daphnia magna* indicate diverse mechanisms of toxicity by legacy and emerging flame retardants. *Environmental Science and Technology* **49** 7400–7410

Scholz N 1995 Determination of the effect of vestinol AH (DEHP) on the reproduction rate of *Daphnia magna*. OECD guideline 202 part II. Huls A. G. Final Report DL 160, Marl, Germany

Schymanski D, Goldbeck C, Humpf H.-U and Fürst P 2018 Analysis of microplastics in water by micro-Raman spectroscopy: release of plastic particles from different packaging into mineral water. *Water Research* **129** 154–162

Sebille EV, Spathi C and Gilbert A 2016 The ocean plastic pollution challenge: towards solutions in the UK. Grantham Institute Briefing paper No 19. Imperial College, London. 16 p

https://www.imperial.ac.uk/media/imperial-college/grantham-institute/public/publications/briefing-papers/The-ocean-plastic-pollution-challenge-Grantham-BP-19_web.pdf

Schuyler QA, Hardesty BD, Wilcox C and Townsend K 2014 Global analysis of anthropogenic debris ingestion by sea turtles. *Conservation Biology* **28** 128-139

Smith S 2012 Marine debris: A proximate threat to marine sustainability in Bootless Bay, Papua New Guinea. *Marine Pollution Bulletin* **64** 1880–1883

Soares A, Guieysse B, Jefferson B et al 2008 Nonylphenol in the environment: A critical review and treatment in wastewaters. *Environmental International* **34** 1033-1049

Spear LB, Ainley DG and Ribic CA 1995 Incidence of plastic in seabirds from the tropical pacific, 1984–91 - Relation with distribution of species, sex, age, season, year and body-weight. *Marine Environmental Research* **40** 123–146

Staples CA, Peterson DR, Parkerton TF and Adams WJ 1997 The environmental fate of phthalate esters: a literature review. *Chemosphere* **35(4)** 667-749

Staples CA, Dome PB, Klecka GM, Oblock ST and Harris LR 1998 A review of the environmental fate, effects, and exposures of bisphenol A. *Chemosphere* **36(10)** 2149-2173

Stephenson RR 1983 Diphenylol propane: Acute Toxicity to *Daphnia magna* and *Selanastrum capricornutum.* Group Research Report. Shell Research Ltd. Sittingboume Research Centre, Kent, England

Tabata A, Kashiwada S, Ohnishi Y et al 2001 Estrogenic influences of estradiol-17 beta, *p*-nonylphenol and bis-phenol-A on Japanese medaka (*Oryzias latipes*) at detected environmental concentrations. *Water Science and Technology* **43** 109-116

Tang BH, Zhu LY and Zhou QX 2010 Joint effect of penta-BDE and heavy metals on aquatic invertebrate *Daphnia magna* survival and its reproduction. *ActaScientiarum Naturalium* Univ. Sunyatseni **49** 93–99

Tarpley RJ and Marwitz S 1993 Plastic debris ingestion by cetaceans along the Texas coast: Two case reports. *Aquatic Mammals* **19** 93–98

Teuten EL, Rowland SJ, Galloway TS and Thompson RC 2007 Potential for plastics to transport hydrophobic contaminants. *Environ. Sci. Technol* **41(22)** 7759-64

Thompson RC, Olsen Y, Mitchell RP et al 2004 Lost at sea: where is all the plastic? *Science* **304** 838

Thompson RC 2017 Future of the Sea: Plastic Pollution; *https://assets.publishing.service.gov.uk/government/uploads/system/uploads/attachment_data/file/634433/Future_of_the_sea_-_plastic_pollution_final.pdf*

USEPA 2016 A summary of literature on the chemical toxicity of plastics pollution to aquatic life and aquatic-dependent wildlife. EPA-822-R-16-009. Environmental Protection December 2016; *https://www.epa.gov/sites/production/files/2016-12/documents/plastics-aquatic-life-report.pdf*

USEPA 2017 Technical fact sheet – Polybrominated Diphenyl Ethers (PBDEs). EPA 505-F-17-015. US United States; Environmental Protection Agency.

https://www.epa.gov/sites/production/files/2014-03/documents/ffrrofactsheet_contaminant_perchlorate_january2014_final_0.pdf

USEPA 2018 Environmental and human health hazards of five persistent, bioaccumulative and toxic chemicals. EPA Document# EPA-740-R1-8001. June 2018. 39 p.

https://www.epa.gov/sites/production/files/2018-06/documents/environmental_human_health_hazard_summary_five_pbts.pdf

Van Cauwenberghe L, Vanreusel A, Mees J et al 2013 Microplastic pollution in deep-sea sediments. *Environmental Pollution* **182** 495–499

Van Cauwenberghe L, Claessens M, Vandegehuchte MB and Janssen CR 2015 Microplastics are taken up by mussels (*Mytilus edulis*) and lugworms (*Arenicola marina*) living in natural habitats. *Environmental Pollution* **19** 10–17 *http://dx.doi.org/10.1016/j. envpol.2015.01.008*

van Wezel AP, Van Vlaardingen P, Posthumus R et al 2000 Environmental risk limits for two phthalates, with special emphasis on endocrine disruptive properties. *Ecotoxicology and Environmental Safety* **46** 305-321

Velzeboer I, Kwadijk CJAF, Koelmans AA 2014 Strong sorption of PCBs to nanoplastics, microplastics, carbon nanotubes, and fullerenes. *Environmental Science and Technology* **48(9)** 4869-4876

Vianello A, Boldrin A, Guerriero P et al 2013 Microplastic particles in sediments of Lagoon of Venice, Italy: First observations on occurrence, spatial patterns and identification. *Estuarine, Coastal and Shelf Science* **130** 54–61

Walsh GE, Yoder MJ, Mclaughlin LL and Lores EM 1987 Responses of marine unicellular algae to brominated organic compounds in six growth media. *Ecotoxicology and Environmental Safety* **14(3)** 215-222

Wang W and Wang J 2018 Different partition of polycyclic aromatic hydrocarbon on environmental particulates in freshwater: Microplastics in comparison to natural sediment. *Ecotoxicology and Environmental Safety* **147** 648-655

Wang F, Shih KM and Li X Y 2015 The partition behavior of perfluorooctanesulfonate (PFOS) and perfluorooctanesulfonamide (FOSA) on microplastics. *Chemosphere* **119** 841-847

Wang J, Tan Z, Peng J, Qiu Q and Li M 2016 The behaviors of microplastics in the marine environment. *Marine Environmental Research* **113** 7–17

http://dx.doi.org/10.1016/j.marenvres. 2015.10.014.

Wilcox C, Hardesty BD, Sharples R et al 2013 Ghostnet Impacts on globally threatened turtles, a spatial risk analysis for Northern Australia. *Conservation Letters* **6** 247–254

Wilcox C, Sebille EV and Hardesty BD 2015 Threat of plastic pollution to seabirds is global, pervasive, and increasing *PNAS* **112 (38)**11899-11904

Woodall LC, Sanchez-Vidal A, Canals M et al 2014 The deep sea is a major sink for microplastic debris. *Royal Society* Open Science **1** 140317

http://dx.doi.org/10.1098/rsos.140317

Wright SL and Kelly FJ 2017 Plastic and human health: A micro issue? *Environmental Science and Technology* **51** 6634-6647

Wu B, Taylor CM, Knappe DR et al 2001 Factors controlling alkylbenzene sorption to municipal solid waste. *Environmental Science and Technology* **35(22)** 4569-4576

Wu C, Zhang K, Huang X and Liu J 2016 Sorption of pharmaceuticals and personal care products to polyethylene debris. *Environmental Science and Pollution Research* **23(9)** 8819-8826

Xanthos D and Walker TR 2017 International policies to reduce plastic marine pollution from single-use plastics (plastic bags and microbeads): A review. *Marine Pollution Bulletin* **118(1–2)** 17-26

Yang D, Shi H, Li L et al., 2015 Microplastic pollution in table salts from China. *Environmental Science and Technology* **49(22)** 13622–13627

Yamashita R and Tanimura A 2007 Floating plastic in the Kuroshio Current area, western North Pacific Ocean. *Marine Pollution Bulletin* **54** 485–488 *doi: 10.1016/j.marpolbul.2006.11.012*

Yona D, Sari SHJ, Iranawati F et al 2019 Microplastics in the surface sediments from the eastern waters of Java Sea, Indonesia F1000 Research 2019 **8 98** *doi: 10.12688/f1000research.17103.1*

Ye X, Pierik FH, Angerer J et al 2009 Levels of metabolites of organophosphates pesticides, phthalates, and bisphenol A in pooled urine specimens from pregnant women participating in the Norwegian mother and child cohort study (MoBA). *International Journal of Hygiene and Environmental Health* **212(5)** 481-91

Zhan Z, Wang J, Peng J et al 2016 Sorption of 3,3,4,4-tetrachlorobiphenyl by microplastics: A case study of polypropylene. *Marine Pollution Bulletin* **110(1)** 559-563

Zhang Y, Guo GL, Han X et al 2008 Do Polybrominated Diphenyl Ethers (PBDEs) Increase the Risk of Thyroid Cancer? *Biosci Hypotheses* **1(4)** 195–199 *doi:10.1016/j.bihy.2008.06.003*

Zhao S, Zhu L, Wang T et al 2014 Suspended microplastics in the surface water of the Yangtze Estuary System, China: First observations on occurrence, distribution. *Marine Pollution Bulletin* **86** 562–568

Websites

http://www.climateactionprogramme.org/news/plastic-microbeads-are-now-banned-in-the-uk

http://wastemanagementreview.com.au/eu-ban-single-use-plastics/

http://www.climateactionprogramme.org/news/starbucks-to-save-1-billion-plastic-straws-with-global-ban-by-2020?utm_source=ActiveCampaign&utm_medium=email&utm_content=Starbucks+to+save+1+billion+plastic+straws+with+global+ban+by+2020+-+Climate+Action+News&utm_campaign=CA+Newsletter+10+July+2018

http://www.basel.int/Implementation/Plasticwastes/Overview/tabid/6068/Default.aspx

■■■■

5 Persistent Organic Pollutants (POPs) Risks, Impacts and Management

Summary

Persistent organic pollutants (POPs) are a group of man-made organic chemicals that are extremely hazardous compounds ever synthesised (being persistent, toxic, bioaccumulative, endocrine disrupting and carcinogenic). There are 28 POPs under Annexures A, B, and C of the Stockholm Convention including their uses [i. aldrin; ii. alpha (α) hexachlorocyclohexane; iii. beta (β)hexachlorocyclohexane; iv. chlordane; v. chlordecone; vi DDT; vii. decabromodiphenyl ether (decaBDE); viii. dieldrin; ix. endosulfan and its isomers; x. endrin; xi. heptachlor; xii. hexabromobiphenyl; xiii. hexachlorobutadiene; xiv. Hexabromocyclododecane; xv. hexachlorobenzene; xvi. hepta -bromodiphenyl ethers; xvii. lindane; xviii. mirex; xxi. pentachlorobenzene; xx. pentachlorophenol, its salts and esters; xxi. perfluorooctane sulfonic acid and its salts (PFOS) & perfluorooctane sulfonyl fluoride (PFOSF); xxii. polychlorinated biphenyls (PCBs); xxiii. polychlorinated dibenzo-p-dioxins (PCDDs); xxiv polychlorinated dibenzofurans (PCDFs); xxv. polychlorinated naphthalenes; xxvi. penta and tetrabromodiphenyl ethers; xxvii. short-chain chlorinated paraffins; xxviii. Toxaphene (comprising 16 pesticides, 7 industrial chemicals, 5 by-products)]. The most common POPs are pesticides (covered in Chapter 7, hence not repeated here), polychlorinated dibenzo-p-dioxins (PCDDs/Dioxins), Polychlorinated dibenzofurans (PCDFs/Furans), polychlorinated biphenyls (PCBs), polybrominated diphenyl ethers (PBDEs), as well as fluorinated compounds (Perfluorooctane sulfonic acid (PFOS) and perfluorooctanoic acid (PFOA). PCDDs/PCDFs, PCBs, PBDEs and PFOS, PFOA have been detected in all the environmental matrices (air, soil, sediments, water, biota/wildlife, terrestrial food (crops, vegetables, meat, eggs) and seafood (fish, prawns, mussels) and humans (humans blood and human breast milk)). Food is the main source of exposure to PCDDs/ PCDFs, PCBs, PBDEs and PFOS, PFOA to humans. This chapter reviews: **i**. chemistry, toxicity and ecotoxicity of POPs including dioxins/furans (PCDDs, PCDFs,) PCBs, Polybrominated diphenyl ethers (PBDEs), Perfluorooctane sulfonate (PFOS) and perfluorooctanoic acid (PFOA); **ii**. impacts of PCDDs, PCDFs, PCBs, PBDEs and PFOS, PFOA on social, economic and environmental sectors including environmental matrices, aquatic biota, food, and human health (including breast milk); **iii**. the linkage between climate change and POPs; and **iv.** management and mitigation measures to reduce impacts of PCDDs, PCDFs, PCBs, PBDEs and PFOS, PFOA on the environment and human health (including guidelines threshold values).

Contents

List of Tables

List of Figures

5.1. Introduction

Persistent organic pollutants (POPs) are a group of manmade organic chemicals that are most hazardous compounds ever synthesised. POPs are polyhalogenated hydrocarbons containing chlorine, bromine or fluorine. POPs are priority pollutants consisting of pesticides, industrial chemicals, chemicals used in consumer products and by-products of certain manufacturing and combustion (includes waste incineration or natural bush/ forest fires) processes (WHO 2010, Alharbi et al 2018). POPs include [i. aldrin; ii. alpha (α) hexachlorocyclohexane; iii. beta (β)hexachlorocyclohexane; iv. chlordane; v. chlordecone; vi DDT; vii. decabromodiphenyl ether (decaBDE); viii. dieldrin; ix. endosulfan and its isomers; x. endrin; xi. heptachlor; xii. hexabromobiphenyl; xiii. hexachlorobutadiene; xiv. Hexabromocyclododecane; xv. hexachlorobenzene; xvi. hepta -bromodiphenyl ethers; xvii. lindane; xviii. mirex; xxi. pentachlorobenzene; xx. pentachlorophenol, its salts and esters; xxi. perfluorooctane sulfonic acid and its salts (PFOS) & perfluorooctane sulfonyl fluoride (PFOSF); xxii. polychlorinated biphenyls (PCBs); xxiii. polychlorinated dibenzo-p-dioxins (PCDDs); xxiv polychlorinated dibenzofurans (PCDFs); xxv. polychlorinated naphthalenes; xxvi. penta and tetrabromodiphenyl ethers; xxvii. short-chain chlorinated paraffins; xxviii. toxaphene)]. Table 5.1 lists 28 POPs under Annexures A, B and C of the Stockholm Convention including their uses. The 28 POPs include 16 pesticides, 7 industrial chemicals, 5 by-products (*note*: some POPs fall within two categories such as both pesticide and industrial chemical or pesticide and by-product).

5.1.1. Characteristics of POPs

POPs possess the following special and unique characteristics:

- have long half-lives and **persist** in the environment for years or decades;
- are hydrophobic and **bioaccumulate** in the fatty tissues of living organisms (including humans) and biomagnify **through the food** chain (e.g. plankton → invertebrates → fish → birds/ mammals → human) (except perfluorinated compounds/ Perfluoroalkylated substances (PFAS), which do not absorb/ accumulate in fatty tissues instead, they bind to proteins in animals) (see section 5.3.3.2);
- are highly **toxic** to humans and wildlife;
- transported over large distances through soil, water, air and can be found worldwide including the Arctic and the Antarctic environment (the long-range transport to the Arctic and the Antarctic occurs via the repeated cycle of evaporation and condensation called the "grass hopper effect") (see Chapter 3, section 3.3.5 for an explanation of "**grasshopper effect**");

- Some POPs are also considered to be **endocrine disrupters** (i.e. alter the **hormonal system,** can damage the reproductive and immune systems of exposed individuals as well as their offspring). An endocrine disruptor is an exogenous substance or mixture that alters function(s) of the endocrine system and consequently causes adverse health effects in an intact organism, or its progeny, or (sub) populations (WHO 2012).
- Some POPs are **carcinogenic** (cancer causing) as well.

Table 5.1. POP chemicals currently controlled under the Stockholm Convention

[WHO 2010, Wikoff et al 2012, Crawford and Quinn 2017,UNEP 2017, Alharbi et al 2018; *http://chm.pops.int/Convention/The%20POPs/tabid/673/language/en-US/Default.aspx;file:///C:/Users/User1/Downloads/UNEP-POPS-PUB-Brochure-16NewPOPs-201706.English%20(3).pdf;* *http://www.pops.int/TheConvention/ThePOPs/AllPOPs/tabid/2509/Default.aspx*].

*[*Chemicals listed under **Annex A (Elimination)** in the Stockholm Convention. Parties must take measures to eliminate the production and use of the chemicals; **Chemicals listed under **Annex B (Restriction)** in the Stockholm Convention. Parties must take measures to restrict the production and use of the chemicals; ***Chemicals listed under **Annex C** of the Stockholm Convention- POPs (**unintentionally produced** chemicals).Parties must take measures to reduce the unintentional releases of chemicals listed under Annex C].*

POPs	Uses/produced	Remarks (chemical class, Stockholm convention guidance)
Aldrin	Control of soil pests (termites, grasshoppers, corn rootworm, and other insect pests).	Pesticide *Elimination
Alpha (α)-Hexachlorocyclohexane	By-product of lindane production.	Pesticide, by-product *Elimination
Beta (β)-Hexachlorocy-clohexane	Formerly used in agriculture.	Pesticide, by-product *Elimination
Chlordane	Used extensively to control termites and as a broad-spectrum insecticide on a wide range of agricultural crops.	Pesticide *Elimination
Chlordecone	Agricultural insecticide, miticide, fungicide.	Pesticide *Elimination

POPs	Uses/produced	Remarks (chemical class, Stockholm convention guidance)
DDT (dichlorodiphenyltri-chloroethane)	Control of insects causing malaria, typhoid and other vector-borne diseases and agricultural use.	Pesticide **Restriction
Decabromodiphenyl ether (DecaBDE) (**PBDEs**)	DecaBDE is one of the congeners of brominated flame retardant, belongs to the group of polybrominated diphenyl ethers(**PBDEs**), DecaBDE containing plastics are used in housings of computers and TVs, wires and cables, pipes and carpets.	Industrial chemical *Elimination
Dieldrin	Control of soil pests; tropical disease vector control (tse-tse flies).	Pesticide *Elimination
Endosulfan and its isomers	Used to control crop pests, tse-tse flies and ectoparasites of cattle and as wood preservative.	Pesticide *Elimination
Endrin	Used in insect and rodent control.	Pesticide *Elimination
Heptachlor	Kills insects, termites.	Pesticide *Elimination
Hexabromobiphenyl	Hexabromobiphenyl belongs to a wider group of polybrominated biphenyls (PBBs), used as flame retardants in synthetic fibres and plastics.	Industrial chemical *Elimination
Heptabromodiphenyl ether (**PBDEs**)	Hexabromodiphenyl ether and heptabromodiphenyl ether are the main components of commercial octabromodiphenyl ether	Industrial chemical *Elimination

POPs	Uses/produced	Remarks (chemical class, Stockholm convention guidance)
	(OctaBDE) of **PBDEs,** used as flame retardants.	
Hexachlorobenzene (HCB)	Wood preservative, fungicide; used as an intermediate in manufacturing of synthetic rubber, dyes, pesticides.	Pesticide/ by-product *Elimination/***Unintentional product
Hexachlorobutadiene *(under investigation)*	Used as a solvent for other chlorine-containing compounds.	By-product *Elimination/ ***Unintentional product
Hexabromocyclododecane (HBCDD) *(under investigation)*	Used as flame-retardant additive, the main uses are in expanded and extruded polystyrene foam insulation.	Industrial chemical *Elimination
Lindane (gamma hexachlorocyclohexane, γ-HCH)	Broad-spectrum insecticide in agriculture; used against ticks and fleas; head lice.	Pesticide *Elimination
Mirex	Control of fire ants and other ants; used as fire retardant in plastics, rubber, and electrical goods.	Pesticide *Elimination
Penta chlorobenzene (PeCB)	Used as pesticide (fungicide), flame retardant, and combined with PCBs in dielectric fluids	Pesticide/Industrial chemical/by-product *Elimination/ ***Unintentional product
Pentachlorophenol, its salts and esters *(under investigation)*	Used as herbicide, insecticide, fungicide, algaecide, disinfectant and as an ingredient in antifouling paint.	Pesticide *Elimination
Perfluorooctane sulfonic acid (**PFOS**), its salts and Perfluorooctane sulfonyl	The current use of PFOS includes: electric and electronic parts, firefighting	Industrial chemical ** Restriction

POPs	**Uses/produced**	**Remarks (chemical class, Stockholm convention guidance)**
fluoride (PFOS-F)	foam, surface treatments of fabric for soil/stain resistance, coating of paper, photo imaging, hydraulic fluids and textiles.	
Polychlorinated biphenyls (**PCBs**)	Primarily used in electrical industry, also as additives in paint, carbo-nless copy paper, sealants and plastics. Still exist in transformers, capacitors.	Industrial chemical *Elimination/ ***Unintentional product.
Polychlorinated dibenzofurans (**PCDFs**) (**Furans**)	Produced unintentionally from many of the same processes that produce dioxins, and also, during the production of PCBs.	By product ***Unintentional product.
Polychlorinated dibenzo-*p*-dioxins) (**PCDDs**) (**Dioxins**)	Unintentionally formed during manufacture of certain pesticides and other chemicals and as product of incineration and open burning of municipal and hazardous wastes as well as automobile emissions, peat, coal, wood and forest fires/bush fires.	By product ***Unintentional product.
Polychlorinated naphthalenes *(under investigation)*	Insulating coatings for electrical wires. Others have been used as wood preservatives, as rubber and plastic additives, for capacitor dielectrics and in lubricants.	By product ***Unintentional product.
Short-chain chlorinated paraffins (SCCPs) *(under investigation)*	Used as a plasticiser in rubber, paints, adhesives, flame retardants for plastics as well as an extreme pressure lubricant in metal working fluids.	By product *Elimination

POPs	Uses/produced	Remarks (chemical class, Stockholm convention guidance)
Tetrabromodiphenyl ether (**TetraBDE**), Pentabromodiphenylether (**PentaBDE**)	TetraBDE and PentaBDE are the main components of **PBDE,** the most common use has been as fire retardant in polyurethane foam.	Industrial chemical *Elimination
Toxaphene (camphechlor)	Agricultural insecticide, tick and mite control in livestock.	Pesticide *Elimination

This chapter emphasizes the risks and impacts of common POPs, principally, polychlorinated dibenzo-p-dioxins (PCDDs/Dioxins), Polychlorinated dibenzofurans (PCDFs/Furans), polychlorinated biphenyls (PCBs), polybrominated diphenyl ethers (PBDEs), as well as fluorinated compounds (Perfluorooctane sulfonic acid (PFOS) and perfluorooctanoic acid (PFOA). The risks and impacts of pesticides including DDTs are covered in chapter 7 and will not be repeated here.

5.2. Chemistry, toxicity and ecotoxicity of POPs

5.2.1. Dioxins, Furans and PCBs

Dioxins, Furans and PCBs are persistent, bioaccumulative and toxic as well as endocrine-disrupting and carcinogenic. Dioxins are a class of structurally and chemically related chlorinated organic chemicals that mainly include polychlorinated dibenzo-p-dioxins (PCDDs or dioxins), Polychlorinated dibenzo furans(PCDFs or furans) and the 'dioxin-like' polychlorinated biphenyls (PCBs) (Kulkarni et al 2008). A dioxin consists of two benzene rings connected by two oxygen atoms, whereas a furan consists of two benzene rings connected by only one oxygen atom. Dioxins and furans have a very similar chemical structure. Polychlorinated biphenyls (PCBs) are a class of organic compounds with 1 to 10 chlorine atoms attached to biphenyl which is a molecule composed of two benzene rings each containing six carbon atoms (Fig. 5.1).

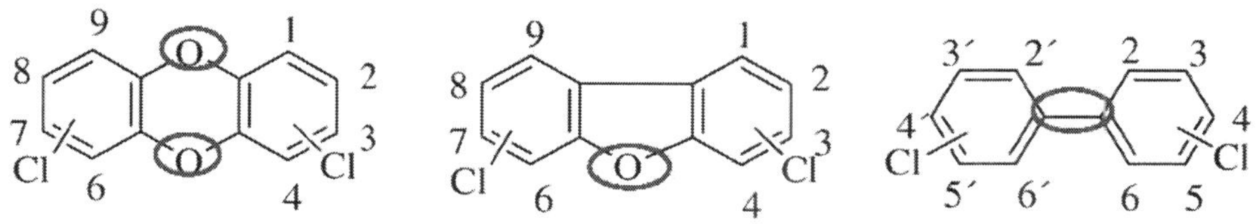

PCDDs	PCDFs	PCBs
Dioxins (in the left)	Furans (in the middle)	PCBs (in the right)
Dioxins have **two oxygen** atoms joining the two benzene rings (e.g. PCDDs)	Furans have **one oxygen** atom joining the two benzene rings (e.g. PCDFs)	PCBs consist of a biphenyl (two benzene rings) with **a carbon to carbon bond** between carbon 1 on one ring and carbon 1' on the second ring, i.e. two phenyl rings attached at one point) with a varying number of chlorines (e.g. PCBs)

Fig. 5.1. Structure of dioxin, furan and PCB (Kulkarni et al 2008).

5.2.1.1. Sources of dioxins, furans and PCBs

Dioxins are unwanted by-products of a wide range of manufacturing processes. Dioxins are released into the environment as a result of combustion activities including power generation, waste incineration (e.g. municipal, medical waste), metal smelting, chlorine bleaching of pulp and paper mill and the manufacture of some chemicals (e.g. pesticides, herbicides). They are also produced from natural processes such as bushfires/forest fires, compost burning, wood stoves and volcanoes. In the USA, forest/wildfires account for a major and a natural source of dioxins (Thrasher 2014; *http://www.dioxinfacts.org/dioxin_health/dioxin_tissues/threshold.html*. In Australia, the major sources of dioxin emissions in the air are bushfires and burning agricultural stubble (the cut stalks of cereal plants) (FSANZ 2012). Electrical and electronic equipment recycling and disassembling (waste electrical and electronic equipment (WEEE) or E-waste) sites and municipal waste dumping sites could be a potential source of PCDD/Fs and PCBs (Gatehouse 2004, Kulkarni et al 2008, Srogi 2008, Wikoff et al 2012, WHO 2016). Although the use of PCBs has been banned all over the world, it is being formed as a by-product during the manufacture of various types of polymeric products such as adhesives, plastics, polythene, pesticides, rubber and related polymeric compounds. Besides, they are also being formed during incomplete incineration as well due to the burning of waste materials in dust bins (*http://www.nio.org/WaterFront/PCB/PCBs.jsp#refer*).Old power transformers are a major source of PCBs. Dioxins, furans and PCBs are semi-volatile and can be transported in either gaseous form or attached to dust and ash particles (Gatehouse 2004).When dioxins are emitted to the atmosphere or land, they can subsequently undergo long-range atmospheric transport due to the **'grasshopper effect'** (Gouin et al 2004) whereby dioxin previously deposited to land surfaces can undergo further transport through successive evaporation cycles (Booth et al 2013). Table 5.2 provides an account of the global deposition of dioxins in/on land and ocean.

Table 5.2. Facts about the global deposition of dioxins in/on land and ocean (Booth et al 2013).

*[TEQ= Toxicity Equivalent Concentration; The **G20 countries**: Argentina, Australia, Brazil, Canada, China, the European Union, France, Germany, India, Indonesia, Italy, Japan, Mexico, Russia, Saudi Arabia, South Africa, South Korea, Turkey, the United Kingdom, and the United States].*

Dioxin production	Dioxin loads	Remarks
i. Annual global dioxin production	287 kg-TEQ[1].	Countries belonging to the **G20** accounted for over **80%** of the estimated annual dioxin emissions (of which, Japan, the USA and China accounts for 30% of the annual global emission).
ii. Dioxin deposition (land and ocean)	**Land:** 163 kg-TEQ (57%). **Ocean:** 115 kg- TEQ (40%).	Approximately 9 kg-TEQ (3%) of the annual dioxin production remains in the atmosphere.
iii. Terrestrial dioxin loads	**Highest load** *East Asia* (China, Japan, South Korea); *South Asia* (much of the Indian sub-continent); *Europe* (most of central, northern and Eastern part); *North America* (large parts of Northern areas).	High local production of dioxin could be related to higher levels of economic activity.

[1] ***TEQ***= *Toxicity Equivalents. Dioxins are generally found in mixtures containing several kinds of dioxins and dioxin-like compounds, each having its own degree of toxicity. To express the overall toxicity of such a mixture as a single number, the concept of "**International Toxic Equivalents" (TEQ**) has been developed. The "Toxic Equivalent" (TEQ) scheme weighs the toxicity of the less toxic compounds as fractions of the toxicity of the most toxic 2,3,7,8-TCDD or TCDD. Each compound is attributed a specific "Toxic Equivalency Factor" (TEF). This factor indicates the degree of toxicity compared to TCDD, which is given a reference value of 1. To calculate the total TCDD toxic equivalent (TEQ) of a dioxin mixture, the amounts of each toxic compound are multiplied with their Toxic Equivalency Factor (TEF) and then added together. The TEQ scheme refers only to **adverse effects (e.g. cancer)** following interactions with the cellular **Ah receptors**. Other toxic effects of dioxins and dioxin-like compounds are not quantified by this method. Toxic Equivalency Factor (TEF) values vary for different animal species https://www.greenfacts.org/en/dioxins/toolboxes/teq-explanations.htm (see also Table 5.5).*

Dioxin production	Dioxin loads	Remarks
	Lowest load Antarctica.	
iv. Ocean dioxin loads	***High concentrations*** Atlantic (northeast and northwest); the Caribbean; Mediterranean; Indian Ocean (northern); Pacific (large parts the north-western Pacific); South China Seas. ***Low concentrations*** North America (Northern parts of the west coast); South America (West coast).	Dioxin deposited to the oceans can act as a sink for dioxin. Dioxin can be accumulated in microplastic particles that are eaten by marine plankton and fishes (see Chapter 4 on plastics), thus human populations with high seafood consumption levels may be exposed to higher levels of dioxin.

5.2.1.2. Congeners[2] in dioxins, furans and PCBs

There are 75 congeners in the dioxin family (PCDD), 135 compounds in the furan family (PCDF) and 209 possible compounds in the PCBs based on the possible number of chlorine atoms (Table 5.3). Only seven PCDD congeners, 10 PCDF congeners and 12 PCB congeners are identified as toxic and have dioxin-like toxicity (USEPA1994,Wikoff et al 2012). Dioxins are the most dangerous toxic chemical so far known on the earth (**'super toxic'**) and considered as a serious threat to public health. The 2, 3, 7, 8 tetrachlorodibenzo-*p*-dioxin (TCDD) is the most toxic dioxin to mammals, birds and fish, which has been the focus of many studies and has received most attention. 2, 3, 7, 8 represent that the four chlorine atoms are attached at position 2, 3, 7, 8 on the dioxin molecule (Fig. 5.1).

Table 5.3. Dioxin families and their known isomer numbers.
(congener = closely related chemicals derived from the same parent compound)

Dioxin families	Congeners	Toxic congeners
Polychlorinated-dibenzo-p (para)-dioxins (PCDDs)	75	7
Polychlorinated dibenzo furans (PCDFs)	135	10
Polychlorinated biphenyls (PCBs)	209	12
Total	**419**	**29**

[2]***Congeners***= *are related chemical substances "related to each other by origin, structure, or function" (IUPAC 1997).*

5.2.1.3. Persistence

PCDDs/PCDFs and PCBs are hydrophobic (low solubility in water and high solubility in lipids) and are very persistent in nature. Once they enter the environment and the living organisms, they will remain for a very long time. PCDDs/PCDFs have a high log K_{ow} values (6-9) whereas PCBs (all congeners) have log K_{ow} values between 4.46 and 8.18 (Hawker and Connell 1988, Gatehouse 2004; Table 5.4). Dioxins have a relatively long atmospheric half-life[3], which enables them to be transported for long distances in the atmosphere. In soils and sediment, dioxins bind strongly to organic matter and persist for long periods (Gatehouse 2004). The **half-life** in soil (DT_{50}) for PCDDs, PCDFs and PCBs are 9-15 years, seven years and 10.5 years respectively (Bacci et al 1990; *https://fortress.wa.gov/ecy/publications/documents/0104010.pdf*). In general, the higher chlorinated congeners (e.g. PCDD/Fs, PCBs) are very resistant to chemical and biological degradation. The half-life of different polychlorinated dibenzo-*p*-dioxins (PCDDs) in adult humans varies between 5.58 and 132 years (Geyer et al 2002), some are cited below:

- 2,3,7,8-Tetrachlorodibenzo-*p*-dioxin or TCDD (range 5.8-9.7 years)
- 1,2,3,7,8-Pentachlorodibenzo-*p*-dioxin or PeCDD (range 12.6-15.7 years)
- 1,2,3,4,7,8-Hexachlorodibenzo-*p*-dioxin or HxCDD (range 8.4-45 years)
- 1,2,3,4,6,7,8-Heptachlorodibenzo-*p*-dioxin or HpCDD (range 25-102 years)
- 1,2,3,4,6,7,8,9-Octachlorodibenzo-p-dioxin or OCDD (range 50-132 years)

5.2.1.4. Bioaccumulation

Dioxins have low solubility in water and high solubility in lipids as represented by its high log K_{ow} values (Hawker and Connell 1988; Table 5.4). Because of their high affinity to lipids, dioxins can accumulate in the bodies of aquatic and terrestrial organisms including humans (e.g. adipose tissue) when they are exposed to these chemicals in water, soil, sediment, or in their diet. Dietary intake is the primary mechanism governing the levels of dioxins and dioxin-like chemicals (PCBs) in biota. More than 90% of human exposure is through food, mainly meat and dairy products, fish and shellfish (WHO 2016).

---

[3]***half-life*** *is the time required for a quantity to reduce to half of its initial value*

5.2.1.5. Toxicity and toxicity equivalency factor (TEF)

It is generally accepted that only 29 out of the 419 congeners of dibenzo-p-dioxins, dibenzofurans and polychlorinated biphenyls are toxic to humans and other organisms (Table 5.3). The 2, 3, 7, 8-tetrachlorodibenzo-*p*-dioxin (TCDD) is most toxic to human and biota (for example, LD_{50} for rat and LC_{50} for fish is 0.022 mg/kg and 0.00000170 mg/L respectively). TCDC is also registered by the WHO as carcinogenic (group 1) for humans (see Table 5.4). PCDDs, PCDFs and PCBs share a common toxic mechanism, which involves binding to a cellular protein, known as the, aryl hydrocarbon (Ah) receptor where they form complexes that translocates to the nucleus and initiate changes in gene expression. Binding to this receptor appears to be the initial step leading to biochemical, cellular, and tissue level changes occurring in organisms exposed to dioxins (Hahn and Stegeman 1992, Gatehouse 2004). The presence of an Ah receptor in an organism also governs whether dioxins are toxic to the organism or not. Though dioxins are highly toxic to fish, mammals, however, they are slightly toxic or non-toxic to amphibians, invertebrates and plants. The lack of sensitivity of dioxins to plants and invertebrates is believed to be related with non-existence of Ah receptors in those organisms (Gatehouse 2004).

Dioxins exist in the environment as complex mixtures rather than as single chemical compound, with the potency of each individual dioxin-like congener in a mixture differing considerably. As a result, the concept of **Toxicity Equivalent Factors** (TEF) or weighing factors for dioxins has been developed (WHO 1998), whereby the toxicity of a mixture is expressed in terms of a single value or TCDD. Toxicity Equivalent Concentration (TEQ) reflects the combined toxicity of all dioxins in the mixture. The TCDD TEQ (or in short TEQ) in an environmental sample (e.g. sediment, soil, organism) is calculated by determining the levels of each congener analytically in the sample, multiplying the concentration of each congener by a congener specific TEF and then adding all the products to give a TEQ. The TEQ can be obtained by the following equation:

$$\mathbf{TEQ = \sum[PCDD_i \times TEFi] + \sum[PCDF_i \times TEFi] + \sum[PCB_i \times TEFi]}$$

(WHO 1998, Reiner et al 2006)

The toxicity of dioxins is expressed as toxic equivalent quantities (TEQs where the most toxic congener TCDD is rated as 1.0 and the less toxic congeners as fractions of this (Table 5.5). Many of the other PCDDs and PCDFs and certain PCBs are less potent than TCDD (Kulkarini et al 2008).

Table 5.4. Chemistry, toxicity and ecotoxicity of PCDDs/Dioxins, PCDFs/Furans, PCBs, PBDEs, PFOS, PFOA

*[(**Toxicity classification**: **Rat** LD_{50}< 50 mg/kg = extremely toxic; 50-500 mg/kg = very toxic; 500-5,000 mg/kg = moderately toxic; **Ecotoxicity classification: Fish** LC_{50}< 1 mg/L= extremely toxic; 1-10 mg/L= very toxic; 10-100 mg/L= moderately toxic; **mM** = milli molar concentration (10^{-3} mol/L), molar concentration or molarity is the number of moles per litre, having the unit symbol mol/L. A solution with a concentration of 1 mol/L is said to be 1 molar, commonly designated as 1 M https://en.wikipedia.org/wiki/Molar_concentration); **na** = data is not available]*

Chemicals	Physico-chemical		Toxicity			Ecotoxicity
	Water solubility	**Log K_{ow}**	**Acute Toxicity to mammal LD_{50} (Rat) oral**	**Carcinogenicity IARC group** (IARC 2019)	**Endocrine disrputor (ED) /ED effects**	**Acute aquatic toxicity to aquatic organisms** (algae, daphnia and fish)
Polychlorinated dibenzo-p-dioxins **(PCDDs/ dioxins)**	0.019 µg/L (Gatehouse 2004)	6-9 (PCDD) (Gatehouse 2004)	Rat LD_{50}: 0.022 mg/kg (TCDD) (Wada 2002)	TCDD: Group 1 (carcinogenic to humans) (WHO 2010, IARC 2019)	Anti-estrogenic and anti-thyroid (WHO 2010)	Fish: Coho salmon, *Oncorhyncus mykiss* sac-fry/newly hatched fry, LC_{50}: 0.00000183 mg/L (1.83 ng/L) for 96 h/21 d (salmon sac-fry) (USEPA 1993, Gatehouse 2004) (TCDD). Fish: Fathead minnows, *Pimephales promelas* juvenile LC_{50}: 0.00000170 mg/L (1.7 ng/L) for 28 d (USEPA 1993a, Gatehouse 2004) (TCDD).
Polychlorinated dibenzofurans **(PCDFs)** (Furans)	0.692 µg/L (Gatehouse 2004)	6-9 (Gatehouse 2004)	Rat LD_{50}: 5.2 mg/kg (Yu et al 2014)	Group 1 (carcinogenic to humans) (IARC 2019)	Endocrine disruptor	na
Polychlorinated biphenyls **(PCBs)**	2.7 - 590 µg/L (Gatehouse 2004)	4.46-8.18 (Gatehouse 2004)	Rat LD_{50}: 2000 to 10,000 mg/kg (low acute toxicity)	Group 1 (carcinogenic to humans) (IARC	Estrogenic, anti-estrogenic and anti-thyroid	Fish: Fathead minnows, LC_{50}: 0.015 mg/L for 96 h Fish, bluegills LC_{50}: 2.74 mg/L for 9 6h

Chemicals	Physico-chemical		Toxicity			Ecotoxicity
	Water solubility	**Log K_{ow}**	**Acute Toxicity to mammal LD_{50} (Rat) oral**	**Carcinogenicity IARC group** (IARC 2019)	**Endocrine disrputor (ED) /ED effects**	**Acute aquatic toxicity to aquatic organisms** (algae, daphnia and fish)
			(http://www.popstoolkit.com/about/chemical/pcbs.aspx)	2019)	(WHO 2010)	*(http://www.popstoolkit.com/about/chemical/pcbs.aspx)*.
Polybrom-inated diphenyl ethers (**PBDEs**)	PentaBDE: 13.3 µg/L, OctaBDE: < 1 µg/L; DecaBDE: < 1 µg/L (USEPA 2017a)	PentaBDE: 6.64-6.97; OctaBDE: 6.29; DecaBDE: 6.285 (USEPA 2017a)	Rat LD_{50}: 500- 5000 mg/kg (PentaBDE) (Gill et al 2004)	2A (probably carcinogenic to humans) (IARC 2019)	Alteration of thyroid hormones (Wikoff et al 2012)	Algae, EC_{50}:>1 mg/L for 96 h (growth) (DeBDE) (Walsh et al 1987). Water flea, *Daphnia magna*, EC_{50}: 0.058 mg/L for 48 h (PeBDE) (Scanlan et al 2015). Fish, Turbot, *Psetta maxima*, LC_{50}: 0.014-0.027 mg/L for 48 h (embryos and larvae) (BDE-47) (Mhadhbi et al 2012).
Perfluoroo-ctane sulfonate (**PFOS**)	498-680 mg/L (pure water) (Van Hoven and Nixon 2000, 3 M Company 2001)	4.49 (Pubchem MSDS 2005) 1.08 (OECD 2002)	Rat LD_{50}: 250 mg/kg (3M 1999, OECD 2002)	na	Endocrine disruptor (ES 2014)	Algae, EC_{50}: 48.2 (45.2-51.1) mg/L (Boudreau et al 2003) to 263 mg/L (Sutherland and Krueger 2001) for 96 h (cell density) (Beach et al 2006). Water flea, *Daphnia* magna, EC_{50}/LC_{50}: 59 (32-87) mg/L for 48 h (survival/immobilization) (Drottar and Krueger 2000). Fish, Rainbow trout, LC_{50}: 7.8 (6.2-9.8) mg/L for 96 h (Robertson 1986).

Chemicals	Physico-chemical		Toxicity			Ecotoxicity
	Water solubility	**Log K_{ow}**	**Acute Toxicity to mammal LD_{50} (Rat) oral**	**Carcinogenicity IARC group** (IARC 2019)	**Endocrine disrputor (ED) /ED effects**	**Acute aquatic toxicity to aquatic organisms** (algae, daphnia and fish)
Perfluoroo-ctanoic acid (**PFOA**)	9.5 X 10^3mg/L (USEPA 2017b)	4.81 (Pubchem 2005 PFOA); 6.08 (CRC CARE 2017)	Rat LD_{50}: 430-680 mg/kg (PHE 2009)	2B (Possibly carcinogenic) (IARC 2019)	Endocrine disruptor (ES 2014)	Algae, *Pseudokirchneriella subcapitata*, EC_{50}: 1.807 (1.76–1.86) mM[3] for 4.5 h (photosynthesis) (Ding et al 2012a). Water flea, *Daphnia* magna, EC_{50}: 0.531 (0.506-0.555) mM for 24 h intoxication/immobilization) (Ding et al 2012b). Zebra fish, LC_{50}: 430 (290-710) mg/L for 6 d (embryos) (Ulhaq et al 2013).

[3] ***mM*** *= milli molar concentration*

Table 5.5. Dioxin, furans and PCBs toxic equivalent factors (TEFs) for mammals, birds and fish.

(WHO 1998, Van den Berg et al 1998)

Sl. No.	Congeners/isomers	Compound name	Mammals (TEF)	Birds (TEF)	Fish (TEF)
Dioxins-Chlorinated dibenzo-p-dioxins					
1	2, 3, 7, 8-TetraCDD	Tetrachlorodibenzo-*p*-dioxin (TCDD)	1	1	1
2	1, 2, 3, 7, 8-PentaCDD	Pentachlorodibenzo-*p*-dioxin (PeCDD)	1	1	1
3	1, 2, 3, 4, 7, 8-HexaCDD	Hexachlorodibenzo-*p*-dioxin (HxCDD)	0.1	0.05	0.5
4	1, 2, 3, 6, 7, 8-HexaCDD	Hexachlorodibenzo-*p*-dioxin (HxCDD)	0.1	0.01	0.01
5	1, 2, 3, 7, 8, 9-HexaCDD	Hexachlorodibenzo-*p*-dioxin (HxCDD)	0.1	0.1	0.01
6	1, 2, 3, 4, 6, 7, 8-HeptaCDD	Heptachlorodibenzo-*p*-dioxin (HpCDD)	0.01	<0.001	0.001
7	1, 2, 3, 4, 6, 7, 8, 9-OctaCDD	Octachlorodibenzo-*p*-dioxin (OCDD)	0.0001	0.0001	<0.0001
Furans: Chlorinated dibenzofurans					
8	2, 3, 7, 8-TetraCDF	Tetrachlorodibenzo-furan (TCDF)	0.1	1	0.05
9	1, 2, 3, 7, 8-PentaCDF	Pentachlorodibenzo-furan (PeCDF)	0.05	0.1	0.05
10	2, 3, 4, 7, 8-PentaCDF	Pentachlorodibenzo-furan (PeCDF)	0.5	1	0.5
11	1, 2, 3, 4, 7, 8-HexaCDF	Hexachlorodibenzo-furan (HxCDF)	0.1	0.1	0.1

Sl. No.	Congeners/isomers	Compound name	Mammals (TEF)	Birds (TEF)	Fish (TEF)
12	1, 2, 3, 6, 7, 8-HexaCDF	Hexachlorodibenzo-furan (HxCDF)	0.1	0.1	0.1
13	1, 2, 3, 7, 8, 9-HexaCDF	Hexachlorodibenzo-furan (HxCDF)	0.1	0.1	0.1
14	2, 3, 4, 6, 7, 8-HexaCDF	Hexachlorodibenzo-furan (HxCDF)	0.1	0.1	0.1
15	1, 2, 3, 4, 6, 7, 8-HeptaCDF	Heptachlorodibenzo-furan (HpCDF)	0.01	0.01	0.01
16	1, 2, 3, 4, 7, 8, 9-HeptaCDF	Heptachlorodibenzo-furan (HpCDF)	0.01	0.01	0.01
17	1, 2, 3, 4, 6, 7, 8, 9-OctaCDF	Octachlorodibenzo-furan (OCDF)	0.0001	0.0001	<0.0001
PCB congeners					
Non-ortho-substituted PCBs (non-coplanar PCBs) (non-dioxin like PCBs)					
18	3, 3' ,4, 4'-tetrachlorobihenyl	Tetrachlorobiphenyl (PCB 77)	0.0001	0.05	0.0001
19	3, 4, 4', 5-tetrachlorobihenyl	Tetrachlorobiphenyl (PCB 81)	0.0001	0.1	0.0005
20	3, 3', 4, 4', 5'-pentachlorobihenyl	Pentachlorobiphenyl (PCB 126)	0.1	0.1	0.005
21	3, 3', 4, 4', 5, 5'-hexachlorobihenyl	Hexachlorobiphenyl (PCB 169)	0.01	0.001	0.00005
Mono-ortho-substituted PCBs (coplanar PCBs) ([4]dioxin-like PCBs)					
22	2, 3, 3' 4, 4'-pentachlorobiphenyl	Pentachlorobiphenyl (PCB 105)	0.0001	0.0001	<0.000005
23	2, 3, 4, 4', 5-pentachlorobiphenyl	Pentachlorobiphenyl (PCB 114)	0.0005	0.0001	<0.000005
24	2, 3', 4, 4', 5-pentachlorobiphenyl	Pentachlorobiphenyl (PCB 118)	0.0001	0.00001	<0.000005

[4] ***coplanar PCBs/dioxin-like PCBs** : The chlorination patterns of the PCBs determine the toxicity of the substance. A number of PCB congeners show **dioxin-like** toxicity. These PCBs have no more than one chlorine atom at the ortho-position (polychlorinated non-ortho and mono-ortho biphenyls). The phenyl rings of these molecules can rotate and adopt a coplanar structure, which leads to the same toxicity as the polychlorinated dibenzo-p-dioxins (PCDDs) and polychlorinated dibenzofurans (PCDFs). A number of PCB congeners, however, have two or more of the ortho-positions in the biphenyl molecules occupied by chlorine molecules. For these, the two phenyl rings are not in the same plane and these PCBs express **non-dioxin-like** toxicity (https://www.atsdr.cdc.gov/csem/csem.asp?csem=30&po=4).*

Sl. No.	Congeners/isomers	Compound name	Mammals (TEF)	Birds (TEF)	Fish (TEF)
25	2', 3, 4, 4', 5-pentachlorobiphenyl	Pentachlorobiphenyl (PCB 123)	0.0001	0.00001	<0.000005
26	2, 3, 3', 4, 4', 5-hexachlorobiphenyl	Hexachlorobiphenyl (PCB 156)	0.0005	0.0001	<0.000005
27	2, 3, 3', 4, 4', 5, 5'-hexachlorobiphenyl	Hexachlorobiphenyl (PCB 157)	0.0005	0.0001	<0.000005
28	2, 3', 4, 4', 5, 5'-hexachlorobiphenyl	Hexachlorobiphenyl (PCB 167)	0.00001	0.00001	<0.000005
29	2, 3, 3', 4, 4', 5, 5'-heptachlorobiphenyl	Heptachlorobiphenyl (PCB 189)	0.0001	0.00001	<0.000005

5.2.2. Polybrominated diphenyl ethers (PBDEs)

PBDEs are organo-bromine compounds that are structurally similar to PCBs (where 2-10 bromine atoms are attached to the diphenyl ether molecule) (Fig. 5.2). PBDEs consists of 209 congeners (Wikoff et al 2012, Crawford and Quinn 2017; *https://www.atsdr.cdc.gov/toxprofiles/tp207-c4.pdf*; Table 5.6). PBDEs have been extensively used as flame retardants[5] in a variety of consumer products such as plastics, textiles, carpets, polyurethane foams, television sets, electronic devices, computers and building materials(Boas et al 2012, ATSDR 2015a). There are three types of commercial PBDE mixtures, including pentabromodiphenyl ether (pentaBDE), octabromodiphenyl ether (octaBDE-BDE183) and decabromodiphenyl ether (decaBDE), which are most widely used across the globe (Muirhead et al 2006, USEPA 2017a). In 1999, decaBDE, octaBDE and pentaBDE comprised approximately 80%, 6%, and 14% of worldwide production of PBDEs, respectively (Ikonomou et al 2002a). DecaBDE consists primarily of the fully brominated congener (BDE 209) (with 10 bromines), OctaBDE's main components are largely heptabrominated (BDE 183) and octabrominated (BDEs 196, 197 and 203) congeners and pentaBDE is composed mostly of tetrabrominated (BDE 47) and pentabrominated (BDEs 99 and 100) congeners (Wikoff et al 2012; Table 5.7). PBDEs can enter soil from discarded products (e.g. in landfills). Bio-solids may also contain PBDEs (ATSDR 2017). The PBDEs are hydrophobic/lipophilic, undergo bioaccumulation and biomagnification in the food chain (Boas et al 2012),

[5]***Flame retardants** are substance that is applied to fabric, wood, or other material in order to make it resistant to catching fire.*

extremely persistence and toxic (USEPA 2016) (see Table 5.4) and endocrine disrupting (Akortia et al 2016, Nugegoda and Kibria 2017). They can be carried to long distances by air (Toms et al 2006).

Table 5.6. Chemical identity of polybrominated diphenyl ether (PBDE) congeners.

[*https://www.atsdr.cdc.gov/toxprofiles/tp207-c4.pdf*; IUPAC=International Union of Pure and Applied Chemistry].

IUPAC[6] number	Compound/substitutes
BDE 1-3	MonoBDE
BDE 4-15	DiBDE (dibromodiphenyl ether)
BDE 16-39	TriBDE (tribromodiphenyl ether)
BDE 40-81	TetraBDE (tetrabromodiphenyl ether)
BDE 82-127	PentaBDE (pentabromodiphenyl ether)
BDE 128-169	HexaBDE (hexabromodiphenyl ether)
BDE 170-193	HeptaBDE (heptabromodiphenyl ether))
BDE 194-205	OctaBDE (octabromodiphenyl ether)
BDE 206-208	NonaBDE (nonabromodiphenyl ether)
BDE-209	DecaBDE (decabromodiphenyl ether)

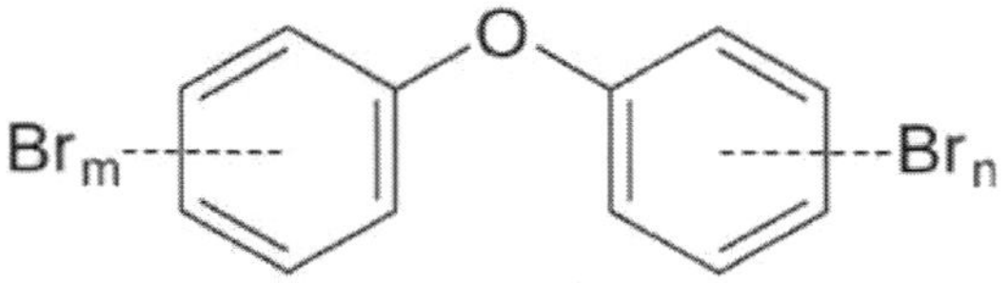

Br_m O Br_n

Fig. 5.2. Chemical structure of polybrominated diphenyl ethers (PBDEs).

[6] ***IUPAC*** *= International Union of Pure and Applied Chemistry*

Table 5.7. General composition of commercial PBDE flame retardants
(JECFA 2006, FSANZ 2007;
h*ttps://www.foodstandards.gov.au/science/surveillance/documents/PBDE_Report_Dec_07.pdf*).

Mixture	Congener composition (% of total)
DecaBDE	97–98% decaBDE < 3% nonaBDEs
OctaBDE	44% heptaBDEs, 31–35% octaBDEs 10–12% hexaBDEs 10–11% nonaBDEs < 1% decaBDEs
PentaBDE	50–60% pentaBDEs 24–38% tetraBDEs 4–8% hexaBDEs

5.2.3. Perfluorooctane sulfonate (PFOS) and perfluorooctanoic acid (PFOA)

Perfluorooctane sulfonate (PFOS) and perfluorooctanoic acid (PFOA) are two kinds of emerging contaminants. Both PFOS and PFOA belong to a large group of compounds called per- and poly-fluoroalkyl substances (PFAS). PFAS substances are fully fluorinated (perfluoro-) alkane (carbon-chain) molecules. Their basic chemical structure is a chain of two or more carbon atoms with a charged functional group end (Fig. 5.3). The functional groups commonly are carboxylic or sulfonic acids, but other forms are also detected in the environment.

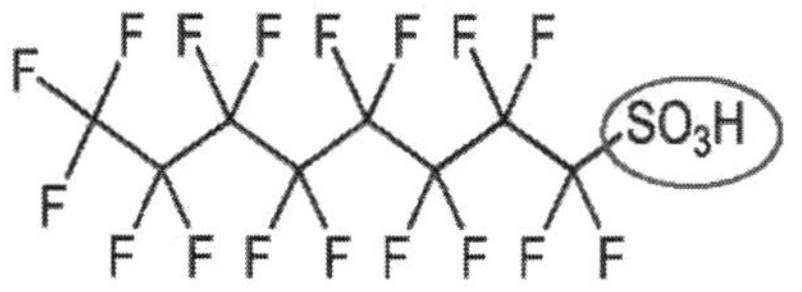

Perflurooctane sulfonate/ Perfluorooctane sulfonic acid (PFOS)

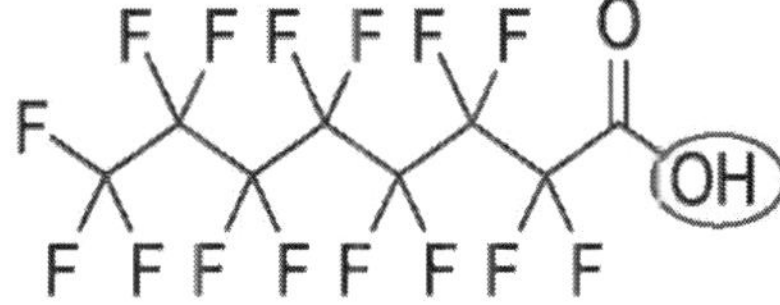

Perflurooctane carboxolate/ perfluorooctanoic acid (PFOA)

Fig. 5.3. Chemical structure of PFOS and PFOA.

PFAS have been used since the 1950s in a range of common household products and speciality applications, including in the manufacture of non-stick cookware; fabric, furniture and carpet stain protection applications; food packaging; some industrial processes; as surfactants, polymers, pesticides, pharmaceuticals, lubricants; and in some types of fire-fighting foams (Giesy et al

2006, Ding et al 2013). Additionally, PFOS and PFOA are the breakdown endpoint of many precursor[7] products (Norström et al 2015). PFOS and PFOA and precursors (Table 5.8) are the most commonly encountered PFAS in the environment and wildlife (Giesy et al 2010). Both old and new landfill sites and areas where firefighting foams have been used can be a potential source of contamination of lands, surface and groundwater with PFOS and PFOA (Gallen et al 2014, ALS 2015, AHPPC 2016, Gallen et al 2017).

Table 5.8. PFAS and their precursor molecules (Giesy et al 2006).

*[**FFAS** = per- and poly-fluoroalkyl substances; **precursor** = a precursor is a compound that participates in a chemical reaction that produces another compound].*

Compound (synonyms)	CAS number	Molecular Formula	Molecular weight
FPAS (per- and poly-fluoroalkyl substances)			
Perfluorobutane sulfonate (C4; PFBS)	29420-49-3	$C_4F_9SO_3$	299
Perfluorohexane sulfonate (C6; PFHxS)	432-50-7	$C_6F_{13}SO_3$	399
Perfluorooctane sulfonate (PFOS)	2795-39-5	$C_8F_{17}SO_3$	499
Perfluorooctane sulfonic Acid	1763-23-1	$C_8F_{17}SO_3H$	500
Pentadecafluorooctanoate (C8; PFOA)	-	$C_7F_{15}COO^-$	413
Pentadecafluorooctanoic acid	335-67-1	$C_7F_{15}COOH$	414
Tridecafluoroheptanoate (C7; PFHpA)	-	C_6F1_3COO	363
Tridecafluoroheptanoic acid	375-85-9	$C_6F_{13}COOH$	364
Heptadecafluorononanoate (C9; PFNA)	-	$C_8F_{17}COO$	463
Heptadecafluoronoic acid	375-95-1	$C_8F_{17}COOH$	464

[7] ***Precursor*** *: In chemistry, a precursor is a compound that participates in a chemical reaction that produces another compound, or a chemical substance that gives rise to another more important chemical product.*

Compound (synonyms)	CAS number	Molecular Formula	Molecular weight
Nonadecafluorodecanoate (C10; PFDA)	-	$C_9F_{19}COO^-$	513
Nonadecafluorodecanoic acid	335-76-2	$C_9F_{19}COO^-$	514
Perfluoroundecanoate (C11; PFUnA)	-	$C_{10}F_{21}COO$	563
Perfluoroundecanoic acid	2058-94-8	$C_{10}F_{21}COO$	564
Perfluorododecanoate (C12; PFDoA)	-	$C_{11}F_{23}COO^-$	613
Perfluorododecanoic acid 3	07-55-1	$C_{11}F_{23}COOH$	614
Perfluorotridecanoate (C13; PFTrA)	-	$C_{12}F_{25}COO$	663
Perfluorotetradecanoate (C14; PFTA)	-	$C_{13}F_{27}COO^-$	713
Perfluorotetradecanoic acid	376-06-7	-	714
Perfluoropentadecanoate (C15, PFPA)	-	$C_{14}F_{29}COO^-$	763
PFAS (Precursors)			
Perfluorooctane sulfonamide (FOSA)	4151-50-2	$C_8F_{17}SO_2NH_2$	499
N-Methyl perfluorooctane	24448-09-7	$C_8F_{17}SO_2N(CH_3)$	557
sulfonamidoethanol (NMeFOSE)	-	C_2H_4OH	-
N-Ethyl perfluorooctane	1691-99-2	$C_8F_{17}SO_2N(C_2H_5)$	571
sulfonamidoethanol (NEtFOSE)	-	C_2H_4OH	-
N-Ethyl perfluorooctane	2991-51-7	$C_8F_{17}SO_2N(C_2H_5)$	585
sulfonamidoacetic acid (PFOSAA)	-	CH_2CO_2H	-

Compound (synonyms)	CAS number	Molecular Formula	Molecular weight
N-Ethyl perfluorooctane sulfonamide (NEtFOSA)	4151-50-2	$C_8F_{17}SO_2NH(C_2H_5)$	-
Perfluorooctane Sulfonylfluoride (POSF)	307-35-7	$C_8F_{17}SO_2F$	502
Perfluorooctanesulfinate (PFO Sulfinate)	-	-	483
6:2 Fluorotelomer Alcohol (6:2 FTOH)	-	$CF_3(CF_2)_5C_2H_4OH$	364
8:2 Fluorotelomer Alcohol (8:2 FTOH)		$CF_3(CF_2)_7C_2H_4OH$	464
10:2 Fluorotelomer Alcohol (10:2 FTOH)	-	$CF_3(CF_2)_9C_2H_4OH$	564

5.2.3.1. Chemistry

All PFAS (per- and poly-fluoroalkyl substances) are highly persistent, bioaccumulative (they do not absorb/ accumulate in fatty tissues instead, they bind to proteins and accumulate in the liver, kidney, and blood (see section 5.3.3.2) and potentially toxic to humans and animals (Concawe 2016). Due to its environmental persistence and ability to bioaccumulate, PFOS has been classified as a persistent organic pollutant (POP) and was added to Annex B (Table 5.1) of the Stockholm Convention in May 2009 (Sharpe et al 2010, Concawe 2016), (hence, their production and use is restricted). PFOS and PFOA have very low volatility because of their ionic nature. Therefore, they will be persistent in water and soil (3M 2000). PFOS and PFOA are moderately soluble in water (Table 5.3) and are stable to hydrolysis[8] in the environment based on their long half-lives in water (the reported half-life of PFOS and PFOA at 25°C is >41 years and 92 years respectively) (DEPA 2015, USEPA 2017b, Deeb et al 2018).

5.2.3.2. Toxicity and ecotoxicity

The oral rat LD_{50}[9] for PFOS is 250 mg/kg body weight (bw) (3M 1999) and for POFA is 430 to1,800 mg/kg (Kennedy et al 2004); therefore, PFOS is little more toxic than PFOA. The International Agency for Research on Cancer (IARC) has classified PFOA as a Group 2B carcinogen (possibly carcinogenic to humans)

[8] ***Hydrolysis:*** *the chemical breakdown of a compound due to reaction with water.*

[9] ***LD_{50}*** *= median lethal* ***dose****, the dose of material that is estimated to be lethal to 50% of the test organisms (rats, mice).*

(Table 5.4).The acute toxicity (LC_{50}[10]) of PFOS to a number of freshwater and saltwater fish species were found to be in the range of 4.7-133 mg/L with *Pimephales promelas* being the most sensitive species (review by Brooke et al 2004). PFOS has been categorized as moderately acute and slightly chronically toxic to aquatic organisms (Table 5.9). PFOA exhibits moderate to low acute toxicity with aquatic species such as fish (LC_{50} ranging from 70 to 2,470 mg/L). In general, PFOS appears to be about 10 times more toxic to aquatic organisms than PFOA (Jensen et al 2012).

Table 5.9. Acute and chronic toxicity data of PFOS (Perfluorooctane sulfonate) (most sensitive endpoints).

*[Brooke et al 2004; **NOEC** =- no observed effect concentration; **End points**: A toxic endpoint is the result of a study (either with algae, daphnia, fish or rat) conducted to determine how dangerous a chemical substance is. The data collected from such studies are generally used to report the relative toxicity of the compound to various regulatory agencies and environmental compliance groups or for research purposes. Toxic endpoints can include mortality, behaviour, reproductive status or physiological and biochemical changes. There are acute (short term) as well as chronic (long-term) end points. https://sciencing.com/definition-toxic-endpoint-6603458.html].*

Acute toxicity	Algae	*Selenastrum capricornutum* (96 h): EC_{50} = 126 mg/L *Skeletonema costatum* (saltwater) (96 h): EC_{50}> 3.2 mg/L
	Invertebrates	*Daphnia magna* (48 h): EC_{50} = 27 mg/L Mysid shrimp, *Mysidopsis bahia* (saltwater) (96 h): LC_{50} = 3.6 mg/L
	Fish	Fathead minnow, *Pimephales promelas* (96 h): LC_{50} = 4.7 mg/L Rainbow trout, *Oncorhynchus mykiss* (saltwater) (96 h): LC_{50} = 13.7 mg/L
Chronic (long-term)	Algae	*S. capricornutum* (96 h): NOEC = 44 mg/L *S. costatum* (saltwater) (96 h): NOEC = >3.2 mg/L Duckweed, *Lemna gibba* (7 d): NOEC = 15.1 mg/L
	Invertebrates	*D. magna*, (28 d): NOEC reproduction = 7 mg/L Mysid shrimp, *M.bahia* (saltwater) (35 d): NOEC reproduction = 0.25 mg/L
	Fish	Fathead minnow, *P.promelas* (42 d): NOEC survival = 0.3 mg/L

[10] ***LC_{50}** = median lethal **concentration**: The concentration of material in the water that is estimated to be lethal to 50% of the test organisms (fish, daphnia). The LC_{50} is usually expressed as a time-dependent value, (24 h or 96 h).*

5.3. Impacts of PCDDs, PCDFs, PCBs, PBDEs and PFOS, PFOA on social, economic and environmental sectors

5.3.1. Impacts of dioxins, furans, and PCBs

5.3.1.1. Impacts on the environmental matrices

Dioxins have been detected in almost all environmental matrices such as air, soil, sediments, vegetation, biota (kangaroo, platypus, birds), seafood, food plants, wild animals, human breast milk) from Austria, Australia, Belgium, Brazil, Bulgaria, Cambodia, Canada, China, Croatia, Czech Republic, Egypt, Finland, Fiji, France, Germany, Greece, Hong Kong, Hungary, India, Indonesia, Ireland, Israel, Italy, Japan, Kazakhstan, Korea, Laos, Latvia, Luxembourg, Malaysia, the Netherlands, New Zealand, Norway, the Philippines, Portugal, Republic of Uzbekistan, Russia, Slovakia, South Africa, Spain, Sweden, Taiwan, UK, Ukraine, USA and Vietnam. The following is a summary of data and information compiled by Kibria et al 2010 on PCDD/Fs and PCBs detected in different environmental matrices across the globe (based on Rappe 1996, Commonwealth of Australia 2004, Reiner et al 2006, Birch et al 2007, LaKind 2007, Tanabe and Kunisue 2007, Mai et al 2007, Srogi 2008, Li et al 2009, Malarvannan et al 2009, Schuhmacher et al 2009, Tsang et al 2009,Wang et al 2009, Zhao et al 2009):

- **Air**: range <1-15,000 fg TEQ/m^3; (TEQ = toxic equivalent; fg[11]= femtogram); (contaminated area had highest PCDD/Fs and the rural area had the lowest PCDD/Fs in the air) (detected/reported countries and regions were Asia, Australia, Europe, South America, Sweden); the highest level of 15,000 fg TEQ/m^3 from Europe.
- **Soils**: range 0.1-98,000 pg TEQ/g; (pg[12] = picogram) (contaminated/ urban soils had the highest PCDD/Fs and rural/remote areas soils had the PCDD/Fs) (detected/reported countries and regions were Australia, Austria, Belgium, Germany, Japan, Ireland, Italy, Luxemburg, Netherlands, New Zealand, Spain, UK, Vietnam); the highest level of 98,000 pg TEQ/g from the Netherlands.
- **Freshwater:** range: 0.001-48 pg TEQ/L (Japan).
- **Sediments**: range: 0.03-101,000 pg TEQ/g lipid (dry weight) (highly contaminated estuarine/marine sediments had the highest PCDD/Fs) (detected/reported countries and regions were Australia, Finland, Germany, Hong Kong, Italy, Japan, Netherlands, New Zealand, North America, North Sea, Norway, Russia, South Africa, Spain, UK); the highest level of 101,000 pg TEQ/g from Europe.

---

[11] *fg = femtogram (1x10^{-15} gram or 0.000,000,000,000,001g)*

[12] *pg = picogram (1x10^{-12} gram or 0.000,000,000,001 g)*

- **Vegetation**: Range: 0.3-1.9 pg TEQ/g (spruce and pine needles) (detected/reported countries and regions were Austria, Germany).
- **Biota, wild animals**: *Kangaroo,* Australia (0.001-25 pg TEQ/g lipid); *Birds of prey,* Australia (maximum 3,900 pg TEQ/g lipid); *Platypus* and *echidna*, Australia (9.3-60 pg/TEQ/g lipid); *Bird-spotted owlet*, India (9.2-2700 pg/g lipid); *Polar bears,* Canadian Arctic (2-23 pg TEQ/g ww; ww = wet weight); *Japan sea*, Whale (0.71-13 pg TEQ/g ww).
- **Food-aquatic: Fish** (range 0.053-329 pg TEQ/g lipid (detected/reported countries and regions were Adriatic coast, Australia, Belgium, China, Finish coast, Hong Kong, Japan, Slovakia, Spain, Taiwan, UK and USA); the highest level of 329 pg TEQ/g detected in herring muscle from Finland coast.
- **Food-animal origin**: dairy products (range 0.0023-2.8 pg TEQ/g fat); pork (range 0.253-1.8 pg/ TEQ/g fat); chickens (range 0.390-2.18 pg TEQ/g fat); chicken eggs (0.248-2.24 pg TEQ/g fat).
- **Food-plant origin**: vegetables (range 0.0086-0.44 pg TEQ/g); cereals (range 0.016-0.94 pg TEQ/g); rice (range 0.0039-0.05 pg TEQ/g).
- **Human breast milk (PCDD/Fs only)**: reported from 43 countries; global range 0.15-48.7 pg WHO-TEQ/g fat; higher level reported from China (48.7 pg TEQ/g fat in 2001), Finland (39 pg TEQ/g fat in 1992-1994), Germany (34 pg TEQ/g fat in 1986-1991), the Netherlands (47 pg TEQ/g fat in 1993), Norway (36 pg TEQ/g fat in 1992-1993) and Russia (42.26 pg TEQ/g fat in 1997); recent global trends show a decline in PCDD/Fs levels in human women milk.
- **Human women milk (PCBs only)**: reported from 24 countries; global range 4.9 pg TEQ/g fat-1,800 ng TEQ/g fat; higher-level reported from Europe (1,800 ng TEQ/g fat from Faroe Island in 2001), Germany (550 ng TEQ/g fat in 1995-97), Russia (520 ng TEQ/g fat in 1996), Kazakhstan (380 ng TEQ/g fat).

5.3.1.2. Impacts on aquatic biota

- **Fish:** Fish is highly acutely toxic to dioxins (TCDD) with LC_{50} for coho salmon, sac-fry is reported to be 0.00000183 mg/L (1.83 ng/L) (USEPA 1993) (Table 5.4). In particular, TCDD caused significant impacts on early life stages of fish (fish fry, yolk-sac, eggs). For example, trout eggs exposed to 0.10 ng TCDD/L for 96 h displayed delayed development and reduced fry growth upon hatching (US EPA 1993). The newly fertilized eggs of brook trout (*Salvelinus fontinalis*) when exposed to TCDD exhibited dose-related toxic effects during and after hatching including sac-fry mortality and associated yolk-sac edema (build-up of fluid), haemorrhages (escape of blood from a ruptured blood vessel), arrested development (stoppage of physical development) (Walker and Peterson 1994). Around 100% mortality occurred when fertilized eggs and sac-fry of rainbow trout was exposed to 10 ng/L TCDD for 96 h (US EPA 1993). Rainbow trout sac-fry exposed to 1ng/L TCDD (ng[13] = nanogram) for 96 h exhibited reduced growth and mortality for up to 160 d after exposure, while sac-fry exposed to 10 ng/L for 96 h had all died by 10 d (US EPA 1993) (see also Table 5.4).

- **Freshwater and marine invertebrates:** No mortalities or reproductive effects were observed when snails (*Physa sp.*), annelid worms (*Paranais* sp.), and mosquito larvae (*Aedes aegypti*) were exposed to 200 ng/L (nominal) TCDD in water under static conditions for 36, 55 and 17 d, respectively. Similarly, no mortalities, growth or reproductive effects were observed when water flea (*Daphnia magna*) and snails (*Physa* sp.) were exposed in a model ecosystem for 33 d to 1,300 ng/L TCDD (US EPA 1993). The lack of sensitivity to dioxin is due to non-existence of Ah[14] receptor in invertebrates (Gatehouse 2004).
- **Plants:** Plants are less sensitive to TCDD than most species of animals. For example, measured water concentrations of up to 1.33 μg/L had no observable effects on algae (*Oedogonium cardiacum*) and duckweed (*Lemna minor*) during a 33-dexposure period (USEPA 1993). The lack of sensitivity to dioxins is also due to non-existence of Ah receptor in plants (Gatehouse 2004).

5.3.1.3. Impacts on food

Dioxin can enter the food chain through a number of different routes including **i.** aerial transport and deposition of dioxins and dioxin-like PCB dust on leafy vegetables, pastures, crops and grazing animals; **ii.** contamination of water, soil, and sediment; **iii**. bioaccumulation in fatty tissues in fish and livestock; **iv.** use of contaminated sewage sludge for growing crops and vegetables. Dioxins can enter the food chain when animals eat plants on which dioxins have fallen. In oceans, rivers and lakes, filter-feeding animals can absorb dioxins when they filter sediments floating in the water (FSANZ 2012). Food is the main source of polychlorinated dibenzo-p-dioxin (PCDD), dibenzofuran (PCDF) and coplanar polychlorinated biphenyl (dioxin like PCBs) exposure for human, accounting for 90% (WHO 2016) to 98% of the total intake (Fries 1995). According to several researchers, **high-fat food** such as seafood/fish, vegetable fats and oils, dairy products, meat and hen's eggs are the main contributors to human exposure to dioxin (Sorgi et al 2008, Kanan and Samara 2018, Guo et al 2019) (Table 5.10). Seafood (fish) could be a major source of dioxin compared to other foods (Table 5.10). The predominant route of dioxin exposure to fish in the wild is through the

---

[13] ***ng*** = *nanogram (10^{-9} gram or 0,000,000,001 gram).*

[14] ***Ah receptor*** = *aryl hydrocarbon receptor (Ahr); Ahr is a protein that in humans is encoded by the AHR gene. The Ahr is a transcription factor that regulates gene expression (https://en.wikipedia.org/wiki/Aryl_hydrocarbon_receptor)*

diet) (Gatehouse 2004). The tissue concentrations of dioxins in some Australian commercial seafood fish species (Australian salmon, Barramundi, Bream, Flathead, Garfish, Golden perch, Green-backed flounder, King George whiting, River cobbler, Sand whiting, Short-finned eel) are given in Table 5.11. It shows that dioxin levels in Australian fish ranged from 0.009 pg TEQ/g wet weight (ww) to about 0.49 pg TEQ/g ww (mean 0.09 pg TEQ/g ww).

5.3.1.4. Impacts on human health (human breast milk and short-/long-term exposure)

PCDDs, PCDFs, PCBs in human breast milk: Recently, a global survey (52 countries) reveals that PCDDs, PCDFs concentrations in human breast milk were very high (>10 pg TEQs/g lipid) in **India**, and in some **European** (Netherlands, Italy, Germany, Spain, Luxemburg, Belgium), and **African** (Egypt, Congo, Côte d'Ivoire) countries (van den Berg et al 2017; Fig. 5.4). Combustion and industrial processes may have contributed to the human exposure of PCDDs and PCDFs (UNEP 2011). Clay consumption (geophagia[15]) by pregnant women could be the likely source of PCDD and PCDF exposure in some African countries that can lead to high levels in human milk (Reeuwijk et al 2013). The global survey further found high PCB levels in human breast milk in East and West Europe (Czech Republic, Slovakia, Italy, Spain, Germany and the Netherlands) and lowest in non-European countries (van den Berg et al 2017; Fig. 5.4). In all cases, the guideline of safe level has exceeded for PCDDs, PCDFs and PCBs compounds for the breastfed infant (Fig. 5.4). PCDDs, PCDFs and PCBs in human milk are a good reflection of the body burden.

Short- and long-term exposure of dioxins: *Short-term exposure* of humans to high levels of dioxins may result in skin lesions, such as chloracne (a rare skin condition) and patchy darkening of the skin and altered liver function. *Long-term exposure* is linked to impairment of the immune system, developing nervous system, endocrine system and reproductive functions. Dioxin exposure has further been linked to birth defects, inability to maintain pregnancy, decreased fertility, reduced sperm counts, **endometriosis** (a disorder in which tissue that normally lines the uterus grows outside the uterus), diabetes, learning disabilities, lung problems; and lowered testosterone levels (Srogi 2008, WHO 2016; *http://www.ejnet.org/dioxin/*). TCDD was classified by IARC as a "known human carcinogen (**IARC group 1**) (IARC 2019) (see also Table 5.12 for health hazards associated with POPs in food).

---

[15] ***Geophagia*** *is defined as deliberate consumption of earth, soil, or clay.*

Table 5.10. Contribution of different food groups to the dioxin exposure (after Srogi 2008).

Country and study period	Dairy (%)	Meat, poultry (%) (where shown with an * means the data includes meat, poultry & eggs)	Eggs (%)	Fish (%)	Cereals, cereal products, vegetables, fruit, vegetable fats and oils (%)	References
Belgium (2001)	30	31	-	39	-	Focant et al 2002
China (2000)	16	35	21	28	-	Wu et al 2002
Finland (1999)	2	*2	-	94	2	Kiviranta et al 2004
Finland (1999)	8	7	2	82	1	Kiviranta et al 2001
Finland (1999)	14	*5	-	72	9	Kiviranta et al 2004
Finland (1999)	16	6	4	63	11	SCOOP 2000
France (1999)	33	13	2	26	26	SCOOP 2000
Germany (1998)	39	30	11	11	9	SCOOP 2000
Italy (1996)	26	32	7	35	-	SCOOP 2000
Japan (2000)	2	*12	-	71	15	Tsutsumi et al 2001
Korea (1999)	1	4	5	39	51	Kim et al 2000
Norway (1997)	22	14	12	46	6	SCOOP 2000
Spain (2000)	27	13	2	30	28	Llobet et al 2003
Sweden (1999)	19	15	1	36	51	Lind et al 2002
Sweden (1999)	19	31	2	34	14	SCOOP 2000
The Netherlands (1999)	24	21	5	10	40	Freijer et al 2001
The Netherlands (1999)	39	20	4	2	35	SCOOP 2000
The United Kingdom (1992)	25	20	4	6	45	SCOOP 2000
The United Kingdom (2001)	44	18	1	6	31	FSA UK, report 38/03 2003

Country and study period	Dairy (%)	Meat, poultry (%) (where shown with an * means the data includes meat, poultry & eggs)	Eggs (%)	Fish (%)	Cereals, cereal products, vegetables, fruit, vegetable fats and oils (%)	References
USA (1995)	29	30	7	6	28	Schecter et al 2001
Average	22	18	6	**37**	24	

Table 5.11. Total PCDD, PCDF and PCBs, expressed as pg TEQ/g, in fish from freshwater, estuarine, and marine locations in Australia

[Gatehouse 2004; *http://www.environment.gov.au/system/files/pages/8fe9c060-057b-4271-aab5-42fee08f78dd/files/report-11.pdf*; ***TEQ*** *= toxic equivalent (calculated by multiplying the concentration of a congener by the toxic equivalency factor (TEF));* ***pg*** *= picogram (1 pg = 0.001 nanogram);* ***dioxins food safety guidelines***[16]*; file:///C:/Users/User1/Downloads/CXP_062e.pdf*]].

Species	Location	pg TEQ/g	
		Wet weight	Lipid weight
Australian Salmon, *Arripis truttaceus*	Hobart - Storm Bay	0.08	1.3
Barramundi, *Lates calcarifer*	Cairns Region	0.054	5.4
Barramundi, *Lates calcarifer*	Darwin Region	0.078	6.5
Barramundi, *Lates calcarifer*	Gulf of Carpentaria	0.24	60
Barramundi, *Lates calcarifer*	Roebuck Bay	0.025	1.9
Bream, *Dentex* spp.	Moreton Bay	0.061	6.1
Bream, *Dentex* spp.	Port Jackson	0.009	0.69

[16] ***dioxins food safety guidelines*** *(maximum content) in fish, other aquatic animals, their products and by-products (with the exception of fish oil and fish protein hydrolysates containing more than 20 % fat): Sum of dioxins and dioxin-like PCBs (sum of polychlorinated dibenzo-para-dioxins (PCDDs), polychlorinated dibenzofurans (PCDFs) and polychlorinated biphenyls (PCBs) expressed in World Health Organisation (WHO) toxic equivalents, using the WHO-TEFs (toxic equivalency factors): 4.5 ng WHO-PCDD/F-PCB-TEQ/kg https://eur-lex.europa.eu/LexUriServ/LexUriServ.do?uri=OJ:L:2006:032:0044:0053:EN:PDF*

Species	**Location**	**pg TEQ/g**	
		Wet weight	**Lipid weight**
Flathead, *Platycephalus* spp.	Adelaide Region	0.01	0.7
Flathead, *Platycephalus* spp.	Albany Region	0.04	4
Flathead, *Platycephalus* spp.	Gippsland Lakes	0.089	1.4
Flathead, *Platycephalus* spp.	Melbourne Region	0.043	4.7
Flathead, *Platycephalus* spp.	Moreton Bay	0.04	4.9
Flathead, *Platycephalus* spp.	Port Jackson	0.03	1.5
Flathead, *Platycephalus* spp.	Triabunna Region	0.04	1.2
Garfish, *Hyporhamphus australis*	Adelaide Region	0.017	0.68
Golden Perch, *Macquaria ambigue*	Coopers Creek	0.36	18
Golden Perch, *Macquaria ambigue*	Murray River	0.036	2.8
Green Backed Flounder, *Rhombosolea tapirina*	Melbourne Region	0.04	2.1
King George Whiting, *Sillaginodes punctate*	Melbourne Region	0.096	5.6
King George Whiting, *Sillaginodes punctate*	Port Jackson	0.49	10.7
River Cobbler, *Cnidoglanis macrocephalus*	Albany Region	0.10	10
Sand Whiting, *Sillago cilitia*	Moreton Bay	0.058	9.5
Short-finned Eel, *Anguilla australis*	Latrobe River	0.033	2.8
Average		**0.09**	**7.1**

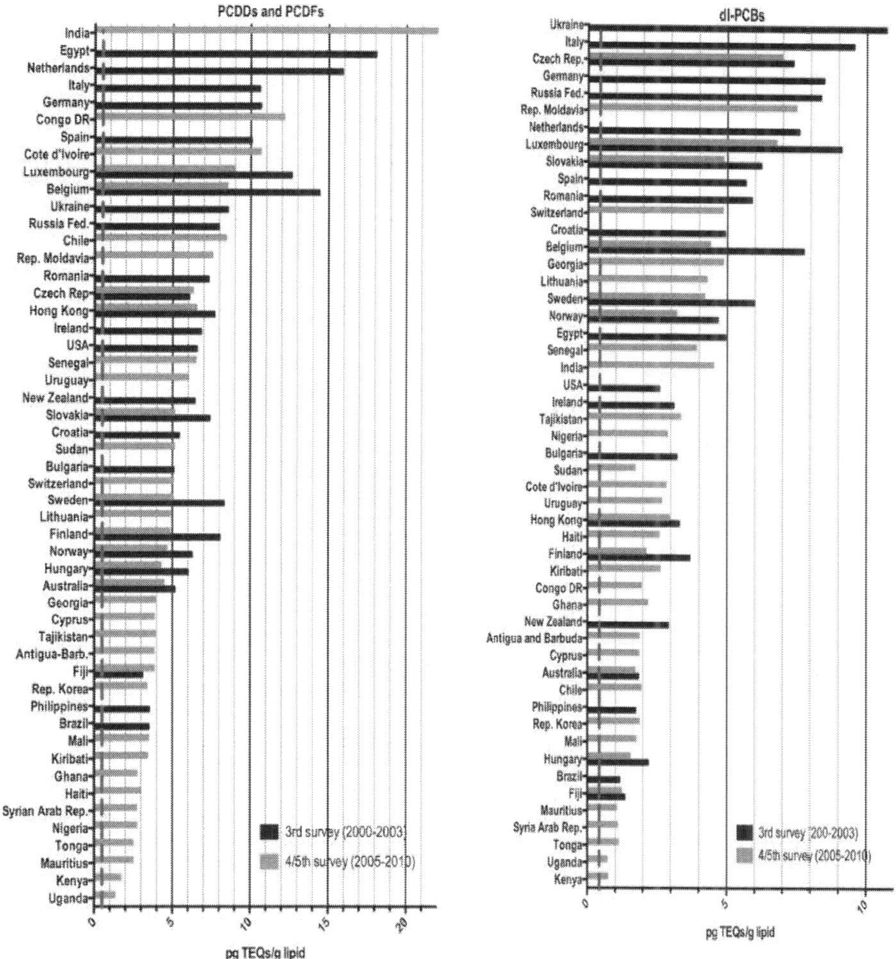

Fig. 5.4. Results of the WHO/ UNEP surveys for PCDDs and PCDFs (left Figure) and dl-PCBs (dl = dioxin-like) (right Figure) in TEQs (pg/g lipid) in pooled human breast milk samples from different countries

[dotted horizontal red/pink line represents the calculated safe level of these compounds for the breastfed infant (van den Berg et al 2017, pages 87 and 88)].

Table 5.12. Health hazards associated with POPs in food (after Guo et al 2019).

POP types	Associated health hazards	References
Dioxins/furans) (PCDDs, PCDFs	Language delay, disturbances in mental and motor development, cancer, diabetes, endocrine disruption, high blood pressure, glucose intolerance and Cardiovascular problems.	Alharbi et al 2018

POP types	Associated health hazards	References
Polychlorinated biphenyls (PCBs)	Endocrine disruption, neurological disorders, liver injury, diabetes, cancer (breast, prostate, testicular, kidney, ovarian and uterine cancers), cardiovascular problems and obesity.	Wang et al 2008, Domingo 2017,
Polybrominated diphenyl ethers (PBDEs)	Reproductive problems, cancer (testicular), diabetes, obesity and cardiovascular problems.	Wang et al 2008, Domingo 2017, Alharbi et al 2018
PFOS and PFOA	Breast cancer.	Alharbi et al 2018
Organochlorine pesticides (OCP)	Neurological symptoms, endocrine disruption, infertility and foetal malformation, diabetes, cancer (breast cancer, testicular, prostate and kidney cancer), reproductive problems, cardiovascular problems, high blood pressure, glucose intolerance and obesity.	Alharbi et al 2018

5.3.2. Impacts of polybrominated diphenyl ethers (PBDEs)

PBDEs have been detected in air, sediments, surface water, fish and other marine animals (ATSDR 2015a). PBDEs can enter the soil from discarded products (e.g., in landfills). Bio-solids may also contain PBDEs. People can be exposed to PBDEs by **i.** inhalation; **ii.** ingestion of contaminated dust in indoor environments; **iii**. eating contaminated foods (**fatty fish).** In breastfeeding infants, breast milk may be a major source of PBDE exposure because PBDEs can accumulate in **breast milk** (ATSDR 2017). Exposure of PBDEs could cause reproductive problems, cancer (testicular), diabetes, obesity and cardiovascular problems (Alharbi et al 2018) (see Table 5.13 for PBDE concentrations in different environmental matrices (air, dust, soil, sediments, sewage sludge and aquatic biota) in different countries).

5.3.2.1. Some examples of the impact of PBDEs

- Higher levels of PBDE have been detected in the air in offices/working places that use electronic devices such as computers (electronic devices are the source of PBDE particles in indoor air) (Bergman et al 1997).
- PBDEs concentration in Canadian homes (indoor air) was approximately 50 times higher than the concentrations in outdoor air (Wilford et al 2004).
- Presence of PBDEs in marine fish, marine animals and shell-fish has been confirmed (e.g. in marine fish and shell-fish in Osaka Bay, Japan (Watanabe

et al 1987); whales blubber in the Faroe Islands, in the North Atlantic (Lindstrom et al 1999); and in the Canadian Arctic, in ringed seal blubber (Ikonomou et al 2002b).

- In Australia, research carried out found a trend/correlation between land-use and ΣPBDEs concentrations, for example, ΣPBDEs was higher in the **industrial/ urban areas** (nd = not detected; dw = dry weight; to 61,000 pg/g dw) and followed in descending order by industrial (nd to 25,000 pg/g dw), STPs (STP=sewage treatment plants) (380-7,700 pg/g dw), urban (nd to 2,800 pg/g dw), remote areas (nd to 480 pg/g dw), agricultural (nd to250 pg/g dw), agricultural/remote (37-57 pg/g dw) and agricultural/urban/ industrial (33 pg/g dw)) (Toms et al 2006).
- In Australia, a total of 35 food items were analysed including meat, dairy, oils and spreads, bread and bakery products, and vegetables. The majority (69%) of the congener measurements analysed reported 'non-detects'. The following foods including tap water, full-fat milk, low-fat milk, canola oil and iodised table salt had no detectable PBDE congeners. The highest levels of PBDEs were detected in boiled eggs (932 pg/g or 0.93 ng/g), grilled pork chops (699 pg/g or 0.7 ng/g), bacon (541 pg/g or 0.54 ng/g) and cream (490 pg/g or 0.49 ng/g), and fish fillets (190 pg/g or 0.19 ng/g).

  https://www.foodstandards.gov.au/science/surveillance/documents/PBDE_Report_Dec_07.pdf
- PBDEs caused alteration of thyroid hormones in fish, inhibiting sperm production, reduction in egg production, gonad development, and ovarian growth, swimming activity, fertilization and increase in larval mortality (Nugegoda and Kibria 2017).
- See Table 5.13 for PBDE concentrations in air, dust, soil, sediments, sewage sludge and aquatic biota in different countries.

Table 5.13. PBDE concentrations in different environmental matrices (air, dust, soil, sediments, sewage sludge and aquatic biota) in different countries
(after Taheran et al 2017).

[ww = wet weight; nd = not detected; pg = picogram, 1 picogram = 10^{-12}g].

Matrix	Countries	Concentrations range (mean)	References
Indoor/outdoor air (pg/m^3)	Arctic (Alert, Tagish)	(240-424)	de Wit et al 2004
	Canada (Toronto)	31.2-59.2 (42.1)	Butt et al 2004
	Canada (Toronto)	2-3600 (120)	Wilford et al 2004
	China (different	0.13-340	Jaward et al 2005

Matrix	Countries	Concentrations range (mean)	References
	locations)		
	Germany (Munich)	8.2-477 (73.1)	Fromme et al 2009
	Japan (different locations)	5.0-71	Jaward et al 2005
	South Korea (different locations)	2-.027	Jaward et al 2005
Indoor (dust) (ng/g)	Australia (Brisbane)	219-3062 (527)	Toms et al 2009
	Canada (Ottawa)	170-170000	Wilford et al 2005
	Germany (Munich)	36.5-1580 (438)	Fromme et al 2009
	Kuwait	1.0-393 (148.8)	Gevao et al 2006
	UK (Birmingham)	16.2-625.4 (215.2)	Harrad et al 2006
	USA (Massachusetts)	3020-192100 (137323)	Allen et al 2008
	USA (Washington)	780-30,100 (590)	Stapleton et al 2005
Soil and sediments (ng/g wet weight)	Canada (British Columbia)	3.8-90.9	Rayne et al 2003
	China (Guiyu)	4434-16088 (9357)	Luo et al 2007
	South Korea (different locations)	0.45-494 (27.8)	Moon et al 2007
	USA (San Francisco)	nd-212	Oros et al 2005
Sewage sludge (ng/g dry weight)	Australia (different urban locations)	230-4230 (1308)	Clarke et al 2008
	Canada (different locations)	230-82000 (990)	Kim et al 2013
	China (Guangzhou)	158-23750	Peng et al 2009
	China (Shanghai)	30.9-35330 (2430)	Yang et al 2011
	Kuwait (Umm Haylaman)	23.4-1599.8 (376.7)	Gevao et al 2006

Matrix	Countries	Concentrations range (mean)	References
	Spain (different locations)	57.5-2606 (488)	Dela Torre et al 2011
	USA (North Carolina)	37500-97400	La Guardia et al 2007
Aquatic biota (ng/g)	Canada (Fraser River) different biota	4-2300	Ikonomou et al 2002b
	Japan (Osaka Bay) fish and shell-fish	(14.6)	Watanabe et al 1987
	Sweden (Viscan River)-fish	150-22000	Andersson and Blomkvist 1981
	USA (Atlantic coasts) – bottlenose dolphins	(200)	Kuehl et al 1991
	USA (Washington)-Mountain whitefish	(1250)	Johnson and Olson 2001
	USA (Washington)-rainbow trout	(1.4)	Johnson and Olson 2001

5.3.3. Impacts of Perfluorooctane Sulfonate (PFOS) and Perfluorooctanoic acid (PFOA)

5.3.3.1. Impacts on the environmental matrices

Perfluorinated substances are very stable, do not biodegrade, and hence can be found throughout the environment globally (Debb et al 2018). PFOS and PFOA can be transported to surface waters and sub-surface waters (groundwater) as a result of run-off and leaching. Atmospheric transport of precursor compounds and long-range transport via ocean currents may also occur (US EPA 2014). PFOS has been detected in surface water, sediment, wastewater treatment plant, effluent, sewage sludge and landfill leachate (OECD 2002, Oliaei et al 2013), biota (in many parts of the world), and oceans (ATSDR 2015b). Trace amounts of perfluorinated substances (PFOS, PFOA) have been detected at remote locations like the Arctic caps, far from potential point sources (Lau et al 2007, Martin et al 2004a, Young et al 2007).

Both PFOS and PFOA have been detected widely in the Australian environment, in particular, **fire training** grounds, **landfills** and bio-solids are potential sources of PFOS and PFOA (Table 5.14). Newer (recent) landfills and landfills accepting primarily construction and demolition wastes had higher PFAS concentrations (Gallen et al 2017; Table 5.14).

Table 5.14. Examples of contamination concentration range of PFOS and PFOA in Australia in a range of matrices (CRC Care 2017, Gallen et al 2017)

*[**LOR** = limit of reporting, **Leachate** = liquid that drains or 'leaches' from a landfill; **na** = data is not available; **PFAS** = Perfluoroalkyl substances; **PFOA** = Perfluorooctanoate/ Perflurooctanoic acid; **PFOS** = Perfluorooctane sulfonate/ Perflurooctane sulfonic acid; **International screening levels** for PFOS and PFOA: **Soil** = 6 mg/kg (PFOS), 16 mg/kg (PFOA; **Drinking water** = 0.2 µg/L (PFOS), 0.4 µg/L (PFOA) (USEPA 2009); **Aquatic** = <1 µg/L (freshwater), 2.5 µg/L (marine) (UK EA 2007)].*

Sites/location	Soil/sediment/bio-solids (mg/kg)		Waters (µg/L)		References
	PFOS	PFOA	PFOS	PFOA	
Fire training grounds (soil, ground-water)	Soil (<0.01-460)	Soil (<LOR-3.2)	Groundwater (0.07-2100)	Groundwater (LOR-160)	CRC Care 2017
Landfills (operating with municipals waste (MW) and C&D (construction and demolition) waste	na	na	Landfills (MW waste): 300 ng/L Landfills (C&D waste): 1100 ng/L	Landfills (MW waste): 510 ng/L Landfills (C&D waste): 1400 ng/L	Gallen et al 2017
Landfills (Leachate from 27 landfills-Australia wide)	na	na	Leachate Range: 13–2700 ng/L (mean 310 ng/L)	Leachate Range: 17–7500 ng/L (mean 690 ng/L)	Gallen et al 2017
Landfills (leachates/ evaporation/ aeration pond)	na	na	>LOR-1.9	>LOR-0.88	ALS 2015
Landfills (untreated landfill leachate)	na	na	Median (0.45)	Median (0.306)	Gallen et al 2014

	Soil/sediment/bio-solids (mg/kg)		**Waters (µg/L)**		
Sewage treatment plants (leachate and bio-solids)	Bio-solids (<LOR – 0.19)	Bio-solids (<LOR – 0.016)	Leachate (<LOR – 1.1)	Leachate (0.016 – 2.1)	Gallen et al 2014
Surface water (Brisbane-Moreton Bay and four influent rivers)	na	na	Surface water (0.00018 – 0.015)	Surface water (0.00013 – 0.0062)	Gallen et al 2014
Surface water (Sydney Harbour-Homebush Bay)	Sediment (0.0008 – 0.0062)	Sediment (<LOR – 0.00016)	Surface water (0.0075 – 21)	Surface water (0.0042 – 0.0064)	Thompson et al 2011

5.3.3.2. Impacts on aquatic biota

PFOS and PFOA are known to bioaccumulate and biomagnify in food webs, as evidenced by PFOS detection in humans, wildlife and fish (Konwick et al 2008, Sharpe et al 2010, Hazelton et al 2012). Perfluorinated compounds/ Perfluoroalkylated substances (PFAS), unlike most other POPs, are lipophobie[17] (i.e. they do not absorb/ accumulate in fatty tissues) instead, they bind to proteins and accumulate in the liver, kidney, and blood (Wikoff et al 2012, Concawe 2016). PFOS and PFOA have been detected in **algae, birds, clams, crabs, fish, frog, mammals, oysters, shrimps, starfish, turtles and zooplankton** (Table 5.15). All most all aquatic higher trophic animals such as fish, dolphins, seals, polar bears and birds have had elevated PFOS concentrations (Butt et al 2008, Conder et al 2008, Thompson et al 2011, Fair et al 2013, Greaves et al 2013, Lucia et al 2015). For example, PFOS concentrations detected in the liver of the Canadian Arctic polar bear ranged from 1,700 to more than 4,000 ng/g (Martin et al 2004a, Lau et al 2007, Young et al 2007). In fish, elevated concentrations of PFOS has been detected in liver, muscle and blood samples from various species and locations worldwide (Table 5.15). PFOS was the major PFAS found in the cod livers. Out of the 200 cod (*Gadus morhua*) livers analysed, 144 livers (72%) had a PFOS concentration above the LOQ (limit of quantification) of the method (1.5 µg/ kg wet weight) (Valdersnes et al 2017). PFOS was found more toxic than PFOA to Zebra fish (*Danio rerio*) and also caused sub-lethal effects

[17]***Lipophobicity*** *or lipophobia is a chemical property of chemical compounds which means "fat rejection"/ "fear of fat". Lipophobic compounds are those not soluble in lipids or other non-polar solvents.*

(pericardial edema (buildup of fluid), malformation like spinal curvature)) (Ulhaq et al 2013).

Table 5.15. Levels of PFOS and PFOA detected in biota (algae, birds, clams, crabs, fish, frogs, mammals, oysters, shrimps, starfish, turtles and zooplankton) from selected countries (after OECD 2002, Houde et al 2006, D'Hollander et al 2010) *[LOD=limit of detection; **na** = data is not available; **ppb** = parts per billion (1 ppb = 1 nanogram)].*

Species and sampling location	PFOS	PFOA	References
Algae (Benthic) (Great Lakes, USA)	2.4-3.1 ng/g ww (mean-whole)	<0.2 ng/g ww	Kannan et al 2005
Birds (eagle and others),USA	1,047 ppb or 1,047 ng/ml or 1.0 mg/L (plasma)	na	3M 2000
Birds (eagle, cormorant), (USA)	2,200 ng/mL (plasma)	na	Giesy and Kannan 2001b
Birds (Brandts cormorant), (USA)	1,780 ng/g ww (liver)	na	Giesy and Kannan 2001b
Birds (six bird species), (USA)	2,055 ppb (liver)	na	3M 2000
Clams (Ariake Sea, Japan, 2003)	<0.3 ng/g ww (mean)	7.5 ng/g ww (mean)	Nakata et al 2006
Clam & Fish (upstream and downstream of 3M facility in Decatur, Alabama, USA)	Clam: 15.6 µg/kg ww. (upstream); Fish: 14.1 µg/kg ww (downstream)	na	Giesy and Kannan 2001e
Crab (*Carcinus maenas*) (W. Scheldt-North Sea)	24-877 ng/g ww (soft tissue)	na	Van de Vijver et al 2003
Fish (Belgian estuary)	923 ng/g ww (muscle)	na	Giesy and Kannan 2001a
Fish (Lake Biwa, Japan)	Fish: 345 ng/mL (average blood levels)	na	Taniyasu et al 2002
Fish (upstream and downstream of 3M facility in Decatur, Alabama, USA)	59.1 µg/kg wet wt. (whole body-upstream) 1.3 mg/kg ww (whole body-downstream)	na	Giesy and Kannan 2001e

Species and sampling location	PFOS	PFOA	References
Fish- carp *(Cyprinus carpio)* (Michigan, USA)	124 ng/g ww (muscle)	< 36 ng/g ww (muscle)	Kannan et al2005
Fish, *Salvelinus namaycush* (lake trout) (Lake Ontario, North America, Canada, 1980–2001)	170 ng/g ww (mean) (whole fish)	ng/g ww (mean) (whole fish)	Martin et al 2004b
Fish, *Boreogadus saida* (Arctic cod) (Davis strait, Artic Ocean, 2000–2001)	1.3 ng/g ww (mean) (whole fish)	0.16 ng/g ww (mean)(whole fish)	Tomy et al 2004
Fish, Carp (Rivers of Michigan, USA, 1998–1999)	124 ng/g ww (mean) (muscle)	<36 ng/g ww (mean) (muscle)	Kannan et al 2005
Fish, *Pampus argenteus*(silvery pomfret Guangzhou, China, 2004)	0.67ng/g ww (mean) (muscle)	<0.25ng/g ww (mean) (muscle)	Gulkowska et al 2006
Fish, *Pneumatophorus japonicas* (Japanese mackerel) (Guangzhou, China, 2004)	2.2 ng/g ww (mean) (muscle)	<0.25 ng/g ww (mean) (muscle)	Gulkowska et al 2006
Fish, *Pseudosciaena plyacti* (small yellow croaker) Guangzhou, China, 2004)	2.9 ng/g ww (mean) (muscle)	<0.2 ng/g ww (mean) (muscle)	Gulkowska et al 2006
Fish-Arctic cod (*Gadus morhua*), Sweden	6.4-62 ng/g ww (liver)	<0.4 ng/g ww (liver)	Kallenborn et al 2004
Fish-Atlantic salmon (*Salmo salar*) (Baltic Sea)	<8 ng/g ww (liver)	<19 ng/g ww (liver)	Kannan et al 2002b
Fish-bluefin tuna (*Thunnus thynnus*) (Mediterranean Sea, Italy)	40 ng/g ww (blood)	<2.5 ng/g ww (blood)	Kannan et al 2002b
Fish-brown trout *(Salmo trutta)* (Michigan, USA)	64 ng/g ww (eggs)	<18 ng/g ww (eggs)	Kannan et al 2005

Species and sampling location	PFOS	PFOA	References
Fish-carp (Belgium)	934 ng/g ww (liver)	na	Hoff et al2005
Fish-carp (US Great Lakes)	96 ng/g ww (muscle)	na	Giesy and Kannan 2001a
Fish-common shiner (*Notropuscornutus*) (Etobicoke creek, ON, Canada)	2,000-72,900 ng/g ww (liver) (**accidental release of firefighting foam**)	6-91ng/g ww (liver)	Moody et al 2002
Fish-Perch (in Lake Malaren and near Stockholm) (Sweden)	20-44 ng/g	na	Jarnberg and Holmstrom 2003
Fish-Perch (Swedish urban sites in the vicinity of municipal STPs)	3 - 8 ng/g	na	Jarnberg and Holmstrom 2003
Frogs (green) (*Rana clamitans*) (Michigan, USA)	<35-285 ng/g ww (liver)	<72 ng/g ww (liver)	Kannan et al 2005
Mammals (Bottlenose dolphin) (Florida, USA)	1,520 ng/g ww (liver)	na	OECD 2002
Mammals (Harbor seal) (Dutch Wadden Sea)	2,725 ng/g ww (2.7 μg/g) (muscle)	<LOD	van de Vijver et al 2005
Mammals (Mink) (USA)	4,800 ng/g ww (liver)	na	Giesy and Kannan 2001c
Mammals (Polar bear)	3,100 ng/g (liver)		Martin et al 2004a
Mammals (Ringed seal) (Northern Baltic Sea)	475 ng/mL (blood)	na	OECD 2002
Mammals (River otter) (USA)	994 ng/g ww	na	Giesy and Kannan 2001c
Mussel, *Mytilus galloprovincialis*	37-126 (72.5 mean) ng/g ww	na	Cunha et al 2005

Species and sampling location	PFOS	PFOA	References
(North-central estuaries, Portugal)			
Oysters (Chesapeake Bay and Gulf of Mexico, USA)	1,225 ng/g dry weight	na	Giesy and Kannan 2001d
Oysters (*Crassostrea virginica*) (Mississippi, USA)	417 ng/g ww (mean-whole)	na	Kannan et al 2002a
Shrimp, *Crangon* (Western Scheldt, Belgium, 2001)	19-520 (215 mean) ng/g ww	na	Van de Vijver et al 2003
Squid, *Loligo vulgaris (*Mediterranean Sea)	2 ng/g ww (mean)	2.5 ng/g ww (mean)	Nania et al 2009
Starfish (*Asterias rubens*) (W. Scheldt, North Sea)	9-176 ng/g ww (soft tissue)	na	Van de Vijver et al 2003
Turtle (Yellow-blotched map) (Mississippi, USA)	190 ng/g ww (mean-liver)	na	Kannan et al 2005
Zooplankton (East Arctic)	1.8 ng/g ww (mean-whole)	2.6 ng/g ww (mean)	Tomy et al 2004

5.3.3.3. Impacts on food

The human diet is recognised as one possible major exposure route to the overall perfluorinated alkylated substances (PFAS) burden to the human population. Possible exposure of PFAS may result directly from the contamination of unprocessed and processed food and migration from food packaging or cookware (Herzke et al 2013). Noorlander et al 2011 analysed 15 common food items from the Netherlands (bakery products, beef, butter, chicken, cheese, crustaceans, eggs, fatty fish, flour, industrial oil, lean fish, milk, pork, vegetables/fruit, vegetable oil) for PFOS and PFOA. The highest concentration was found in **crustaceans and lean fish** (Fig. 5.5), therefore, fish may be an important dietary source of PFOS and PFOA for fish consumers. In a Canadian study, PFOS was detected in beef steak, ground beef, popcorn, and marine and freshwater fish (Tittlemier et al 2007). The British Food Safety Agency (FSA UK 2009) analysed a variety of food items (bread, cereals, dairy products, eggs, fish, jams, offal, meat, milk, popcorn, vegetables) on sale in the UK for PFOS and PFOA. The two highest concentrations of PFOS were in smoked eel and whitebait. PFOS was not detected in any samples of meat, potatoes, potato products, popcorn or other cereals, vegetable oils or fish oil dietary supplements. PFOA was detected only in one sample of whitebait, six of crab and three of the liver, but all at low concentration (FSA UK 2009; http://fluoridealert.org/news/uk-survey-of-pfos-and-pfoa-in-food/). The above results may confirm that diet could be an important source of exposure of humans to PFOS and PFOA compounds.

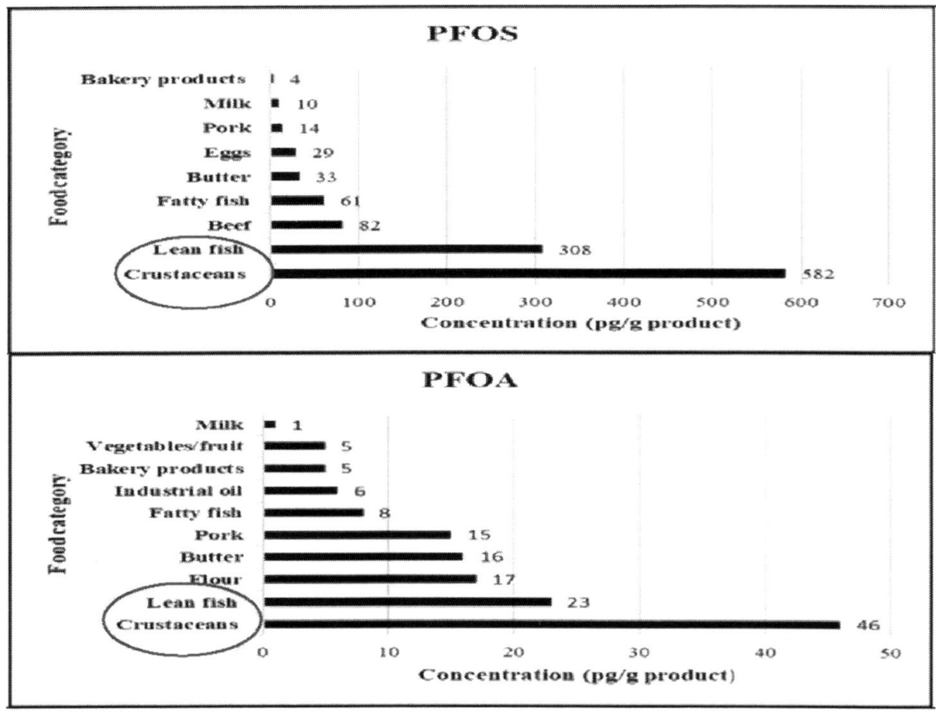

Fig. 5.5. Concentrations of PFOS and PFOA in food categories in the Netherlands (data plotted based on data of Noorlander et al 2011).

5.3.3.4. *Impacts on human health*

Humans can be exposed to Perfluoroalkyl substances through contaminated food (Fig. 5.5) and drinking water (Table 5.16), or through ingestion of soil, inhalation of contaminated air or contaminated dust, exposure to carpets treated with perfluoroalkyls (children), or use of commercial products. Consumption of fish and fish products can be one of the primary sources of human exposure to PFOS (ATSDR 2015b). High levels of PFOS and PFOA are toxic for reproduction and development of the foetus (such as reducing birth weight) (Fei et al 2009) and lowering semen quality (Joensen et al 2009). Researchers found probable links between elevated PFOA blood levels and several diseases including high cholesterol (hyper-cholesteremia[18]), thyroid diseases, testicular cancer, kidney cancer and elevated blood pressure during pregnancy (reviewed by OECD 2013). The IARC has classified PFOA as a Group 2B carcinogen, i.e., possibly carcinogenic to humans (IARC 2019).

Table 5.16.The observed range of PFOS and PFOA in drinking water (after D'Hollander et al 2010).

Country and sampling sites	PFOS (ng/L)	PFOA (ng/L)	References
Germany (Rhine, Ruhr Moehne area, 2006)	<1-22	<1-519	Skutlarek et al 2006
Italy (area of Lake Maggiore, 2007)	6.2-9.7	1.0-2.9	Loos et al 2007
Japan (Osaka and Tohoku, Area, 2003)	<0.1-12	0.7-50	Saito et al 2004
Japan (Osaka, 2006-2007)	<0.1-22	2.3-84	Takagi et al 2008
Japan (Tokyo)	0.1-51	-	Harada et al 2003
Spain (Catalonia, 2007)	0.4-0.9	0.3-6.3	Ericson et al 2008
Spain (Catalonia, 2007) (bottled water)	<0.24	0.2-0.7	Ericson et al 2008

5.4. Linkages between POPs and climate change

Projected climate change has the potential to directly and indirectly affect POP transport and its fate in the environment. The projected rise in temperatures, melting of ice (sea–ice reduction), and extreme weather events etc. can increase primary emissions of POPs (Gusev et al 2012). *Firstly*, the increase in surface temperatures could result in the volatilization/evaporation of POPs (from soil into the air) from current temperate and tropical source regions and their deposition in colder regions (high-elevation and high-latitude ecosystems). An increase in 1°C could result in about 10% to 15% increase in the volatility of typical POP such as PCBs; *https://www.amap.no/documents/download/3237/inline*. *Secondly*, the melting of glaciers may result in the release of POPs stored decades ago in the soil and vegetation, which may also be re-released during fire events. *Thirdly,* the decreases in sea-ice cover and increases in ocean temperature have the potential

[18] ***Hypercholesterolemia,*** *also called high cholesterol, is the presence of high levels of cholesterol in the blood.*

to result in the redistribution of the more volatile POPs stored in ocean water (NRC 2010; *https://www.nap.edu/read/12743/chapter/7*). *Fourthly*, the anticipated rise in temperatures could cause proliferation of insects and vector-borne diseases which could lead to greater use of pesticides. All these situations could lead to increasing emissions of pesticides. *Fifthly*, the projected increase in the frequency and intensity of droughts and less rainfall could lead to an increase of wildfires/ bushfires and thus higher emissions of POPs like dioxins and furans (forest/ wildfires accounts for a major and a natural source of dioxins). *Finally*, where heavier rainfall is expected (due to climate change) there it can lead to greater surface deposition of airborne POPs and increased run-off of POPs (Kibria et al 2010, Kibria 2014).

5.5. Management of POPs

To reduce risks and impacts of POPs (dioxins, furans, PCBs, PBDEs, PFAS) on the environment, and human health, the following strategies can be applied.

- The Regular monitoring programs should be in place to monitor priority pollutants like dioxins, furans, PCBs, PBDEs, PFAS levels in all the environmental matrices possible (water, soil, sediments, biota, food (crops, vegetables, fish, meat, milk), human breast milk and compare with guidelines threshold values (Table 5.17) (in particular, food and feed contamination monitoring systems must be in place to ensure that tolerance levels are not exceeded).
- Reduce the production, use and sale of POPs listed under the Stockholm Convention (eliminate and restrict; see Table 5.1).
- Reduce release of dioxins and furans, PCBs, PBDEs, PFOS and PFOA to land, air and water from point- and non-point sources.
- Implement the Basel Convention on the control of transboundary movements of POPs related hazardous wastes and their disposal (*The Basel Convention on the Control of Transboundary Movements of Hazardous Wastes and Their Disposal is an international treaty that was designed to reduce the movements of hazardous waste between nations; it is to protect human health and the environment against the adverse effects of hazardous wastes*).
http://www.basel.int/TheConvention/Overview/tabid/1271/Default.aspx
- Develop remedial measures/ technologies of POPs comprising bioremediation methods, adsorption, electrochemical, and membrane technologies etc. (the currently available techniques for fighting with POPs are solvent extraction, alkali metal reduction (sodium reduction), bioremediation, verification, incineration, solidification, gas phase chemical reduction, stabilization, pyrolysis, established ball milling, mechano-chemical dehalogenation, super critical extraction, soil washing chemical dehalogenation and thermal desorption) (Alharbi et al 2018).

Table 5.17. Water, food and health quality guideline threshold values for dioxins, PCBs, PBDEs, PFOS and PFOA

*[**MCL** = maximum contaminant level; **MPL** = maximum permissible concentration (the MPC is the level at which no harmful effects are expected, the consumption of fish by humans is the most critical route of exposure to PFOS (RIMV 2010); **na** = data is not available].*

POPs		Human Health					Ecological	
		Drinking water	Recreational water quality	Food	Soil	Tolerable daily intakes	Aquatic species protection	Sediments
Dioxins, furans and PCBs	Dioxins, furans	0.001 ng/L (Japan) (EPHC 2005)	na	(ΣPCDDs+PCDFs) 3 pg/g fat (meat -cattle and sheep) 4 pg/g fresh weight (fish and fishery products) 3 pg/g fat (eggs, raw milk and dairy products) (FSA Ireland 2009)	1 ng/kg (Japan) (EPHC 2005)	1 pg TCDD/ kg bw/day (Germany) (EPHC 2005)	na	55 ng/kg (UK) (EPHC 2005)
	PCBs	0.016 ng/L (PCBs 105, 118, 156) (NWQMS 2008) 0.0005 mg/L (MCL)	na	0.2-3.0 mg/L (all foods) 2.0 mg/L (fish) (ATSDR 2018)	na	1-4 pg kg/bw/day dioxins/furans and dioxin-like PCBs (WHO 1998)	na	na

POPs		**Human Health**					**Ecological**	
		Drinking water	**Recreational water quality**	**Food**	**Soil**	**Tolerable daily intakes**	**Aquatic species protection**	**Sediments**
		(ATSDR 2018)						
PBDEs	DecaBDE 209	110 µg/L (USEPA 2017a)	na	na	440 mg/kg (residential) 3,300 mg/kg (industrial) (USEPA 2017a)	na	na	na
	PentaBDE	40 µg/L (USEPA 2017a)	na	na	160 mg/kg (residential) 2,300 mg/kg (industrial) (USEPA 2017a)	na	na	na
	OctaBDE	61 µg/L (USEPA 2017a)	na	na	190 mg/kg (residential) 2,500 mg/kg (industrial)	na	na	na

POPs		Human Health					Ecological	
		Drinking water	Recreational water quality	Food	Soil	Tolerable daily intakes	Aquatic species protection	Sediments
					(USEPA 2017a)			
PFOS, PFOA	PFOS	0.2 µg/L (USEPA 2009) 0.07 µg/L (NHMRC 2011)	0.7 µg/L (NSW 2017)	0.65 ng/L (MPC-fresh water fish consumption) (RIMV 2010)	6 mg/kg (USEPA 2009)	0.02 µg/g/ day (NSW 2017)	<1 µg/L (fresh), 2.5 µg/L (marine) (UK EA 2007)	na
	PFOA	0.4 µg/L (USEPA 2009) 0.56 µg/L (NHMRC 2011)	5.6 µg/L (NHMRC 2011)	na	16 kg/kg (USEPA 2009)	0.16 µg/g/ day (NSW 2017)	na	na

- Use of natural coagulants such as guar gum[19] and xanthan (thickening agent) could be an option for treating POPs, because it is a biodegradable biopolymer, is non-toxic, and has low treatment costs, is easily available and is produced in abundance. Such technology can be used for removing POPs from farm effluent and landfill leachate (Pariatamby and Kee 2016).

5.6. Conclusion

POPs are a group of chemical pollutants and listed under the Stockholm Convention for global elimination and restriction. They are potentially hazardous to living organism, persist in the environment for years or decades, bioaccumulate in the fatty tissues of living organisms (including humans) and biomagnify through the food chain. POPs are highly toxic to humans and wildlife and can transport over large distances through soil, water, air and can be found worldwide including the Arctic and the Antarctic environment. This chapter emphasised the risks and impacts of common POPs, principally, polychlorinated dibenzo-p-dioxins (PCDDs/Dioxins), polychlorinated dibenzofurans (PCDFs/Furans), polychlorinated biphenyls (PCBs), polybrominated diphenyl ethers (PBDEs), as well as fluorinated compounds (perfluorooctane sulfonic acid (PFOA) and perfluorooctanoic acid (PFOA). PCDDs/PCDFs, PCBs, PBDEs and PFOS, PFOA have been detected in all the environmental matrices (air, soil, sediments, water, biota/wildlife, terrestrial food (crops, vegetables, meat, eggs) and seafood (fish, prawns, mussels) and humans (humans blood and human breast milk)). Food is the main source of PCDDs, PCDFs, PCBs exposure for humans, accounting for 90% to 98% of the total intake. The presence of PBDEs in marine fish, marine animals and shell-fish has been confirmed. PFOS and PFOA have been detected in algae, birds, clams, crabs, fish, frogs, mammals, oysters, shrimps, starfish, turtles and zooplankton. The International Agency for Research on Cancer (IARC) has classified PCDDs/PCDFs and PCBs as Group 1 (carcinogenic to humans); the PBDEs as group 2A (probably carcinogenic to humans) and the PFOA as Group 2B (possibly carcinogenic). It is, therefore, imperative that regular monitoring programs should be in place to monitor priority pollutants like dioxins, furans, PCBs, PBDEs, PFOS levels in all the environmental matrices to reduce risks and impacts on the environment and human health. Reducing the production, use and sale of POPs listed under the Stockholm Convention should also be prioritised.

5.7. References

3M Company 1999 Perfluorooctane sulfonate: Current summary of human sera, health and toxicology data. 3M Company, January 21, 1999.

http://www.chemicalindustryarchives.org/dirtysecrets/scotchgard/pdfs/226-0548.pdf

[19] ***The guar*** *gum or Lond bean, with the botanical name Cyamopsis tetragonoloba, is an annual legume and the source of guar gum, used for both human consumption and cattle feed;*

3M Company 2000 Sulfonated perfluorochemicals in the environment: sources, dispersion, fate and effects (AR226-0545). 3M Company, St Paul, MN. (*In*: OECD 2002).

3M Company 2001 Solubility of PFOS in water. Laboratory Project Number E00-1716. EPA Docket AR226-1030a025. 3M Company, 3M Environmental Laboratory, St Paul, MN.

AHPPC (Australian Health Protection Principal Committee) 2016 Per- and poly-fluoroalkyl substances (PFAS) Fact Sheet.

https://www1.health.gov.au/internet/main/publishing.nsf/Content/A12B57E41EC9F326CA257BF0001F9E7D/$File/PFAS-factsheet-15June2016.pdf

Akortia E, Okonkwo JJO, Lupankwa M, et al 2016 A review of sources, levels and toxicity of polybrominated diphenyl ethers (PBDEs) and their transformation and transport in various environmental compartments. *Environmental Reviews* **24(3)** 253-273 *doi 10.1139/er-2015-0081*

Alharbi OML, Basheer AA, Khattab RA and Ali I 2018 Health and environmental effects of persistent organic pollutants. *Journal of Molecular Liquids* **263** 442–453

ALS 2015 Perfluorinated compounds (PFAS) in landfill leachate – trial study, ALS Environmental, Enviro Mail **#86,** *https://www.alsglobal.com/en-au/myals/news/2015/2/enviromail-86-perfluorinated-compounds-in-landfill-leachate*

Andersson O and Blomkvist G 1981 Polybrominated aromatic pollutants found in fish in Sweden. *Chemosphere* **10** 1051–1060

ATSDR (Agency for Toxic Substances and Disease Registry) 2015a "Draft Toxicological Profile for Polybrominated Diphenyl Ethers." *ww.atsdr.cdc.gov/toxprofiles/tp207.pdf*

ATSDR (Agency for Toxic Substances and Disease Registry) 2015b "Draft Toxicological Profile for Perfluoroalkyls." *ww.atsdr.cdc.gov/toxprofiles/tp200.pdf*

ATSDR (Agency for Toxic Substances and Disease Registry) 2017 Public health statement. Polybrominated Diphenyl Ethers (PBDEs). h*ttps://www.atsdr.cdc.gov/phs/phs.asp?id=1449&tid=183*

ATSDR (Agency for Toxic Substances and Disease Registry) 2018 Polychlorinated Biphenyls (PCBs) Toxicity. What Standards and Regulations Exist for PCB Exposure? *https://www.atsdr.cdc.gov/csem/csem.asp?csem=30&po=8 (accessed 14 January 2020)*

Allen JG, McClean MD, Stapleton HM and Webster TF 2008 Critical factors in assessing exposure to PBDEs via house dust. *Environment International* **34** 1085–1091

BacciE, Cerejeira MJ, Gaggi G et al 1990 Bioconcentration or organic chemical vapours in plant leaves: the azalea model. *Chemosphere* **21** 525-525

Beach SA, Newsted JL, Coady KK and Giesy JP 2006 Ecotoxicological Evaluation of Perfluorooctanesulfonate (PFOS) *Reviews of Environmental Contamination & Toxicology* **186** 133-74 *DOI: 10.1007/0-387-32883-1_5*

Bergman A, Ostman C, Nyborn R et al 1997 Flame retardants and plasticisers on particulate in the modern computerized indoor environment. *Organohalogen Compounds* **33** 414–419

Birch GF, Harrington C, Symons R K and Hunt JW 2007 The source and distribution of polychlorinated dibenzo-*p*-dioxin and polychlorinated dibenzoofurans in sediments of Port Jackson, Australia. *Marine Pollution Bulletin* **54** 295-308

Boas M, Feldt-Rasmussen U and Main KM 2012 Thyroid effects of endocrine disrupting chemicals. *Reviews Molecular and Cellular Endocrinology* **355** 240–248

Booth S, Hui J, Alojado Z et al 2013 Global deposition of airborne dioxin. *Marine Pollution Bulletin* **75(1-2)** 182-186

Boudreau TM, Sibley PK, Mabury SA, Muir DGC and Solomon KR 2003 'Laboratory evaluation of the toxicity of Perfluorooctane sulfonate (PFOS) on *Selenastrum capricornutum, Chlorella vulgaris, Lemna gibba, Daphnia magna* and *Daphnia pulicaria. Archives of Environmental Contamination and Toxicology* **44** 307– 313

Brooke D, Footitt A and Nwaogu TA 2004 Building Research Establishment Ltd and Risk and Policy Analysts Ltd for Environment Agency, Bristol. ISBN: 978-1-84911-124-9

https://assets.publishing.service.gov.uk/government/uploads/system/uploads/attachment_data/file/290857/scho1009brbl-e-e.pdf

Butt CM, Diamond ML, Truong J et al 2004 Spatial distribution of polybrominated diphenyl ethers in Southern Ontario as measured in indoor and outdoor window organic films. *Environmental Science and Technology* **38** 724–731

Butt CM, Mabury SA, Kwan M, Wang, X and Muir D 2008 'Spatial trends of perfluoroalkyl compounds in ringed seals (*Phoca hispida*) from the Canadian arctic'. *Environmental Toxicology and Chemistry* **27(3)** 542–553

Clarke B, Porter N, Symons RK et al 2008 Polybrominated diphenyl ethers and polybrominated biphenyls in Australian sewage sludge. *Chemosphere* **73** 980–989

Concawe (Conservation of Clean Air and Water in Europe) 2016 Environmental fate and effects of poly- and perfluoroalkyl substances (PFAS). Report no. 8/16.

h*ttps://www.enviro.wiki/images/7/75/Concawe-2016-Environmental_fate_and_effects_of_PFAS.pdf*

Conder JM, Hoke RA, Wolf WD et al 2008 Are PFCAs bioaccumulative? A critical review and comparison with regulatory criteria and persistent lipophilic compounds. *Environmental Science & Technology* **42(4)** *995-1003 doi: 10.1021/es070895g*

Commonwealth of Australia 2004 National dioxins program. Dioxins in Australia: a summary of studies conducted from 2001 to 2004. Australian Government. Department of the Environment and Heritage. Canberra. 16 p. *http://www.deh.gov.au/indutry/chemicals/dioxins/index.html.*

CRC CARE (Cooperative Research Centre for Contamination Assessment and Remediation of the Environment) 2017 Assessment, management and remediation guidance for perfluorooctanesulfonate (PFOS) and perfluorooctanoic acid (PFOA) – Part 1: background, CRC CARE Technical Report no. CRC CARE 2017.

https://www.crccare.com/files/dmfile/CRCCARETechReport38Part1_AssessmentmanagementandremediationforPFOSandPFOA_background2.pdf

Crawford CB and Quinn B 2017 *Microplastic pollutants*. Chapter 6. The interactions of microplastics and chemical pollutants. Persistent organic pollutants (POPs). p. 131-156 Elsevier Inc

Cunha I, Hoff P, Van de Vijver K et al 2005 Baseline study of perfluorooctane sulfonate occurrence in mussels, *Mytilus galloprovincialis* from north-central Portuguese estuaries. *Marine Pollution Bulletin* **50** 1121–1145

DEPA (Danish Environmental Protection Agency) 2015 Perfluoroalkylated substances: PFOA, PFOS and PFOSA, Evaluation of health hazards and proposal of a health-based quality criterion for drinking water, soil and groundwater, Environmental project No. 1665, The Danish Environmental Protection Agency, Copenhagen, Denmark.

Deeb R, Field J, Hawley E and Higgins C 2018 Perfluoroalkyl and Polyfluoroalkyl Substances (PFASs). *https://www.enviro.wiki/index.php?title=Perfluoroalkyl_and_Polyfluoroalkyl_Substances_(PFASs)*

de Wit C, Alaee M and Muir D 2004 Brominated flame retardants in the Arctic–an overview of spatial and temporal trends. *Organohalogen Compounds* **66** 3764–3769

De la Torre A, Alonso E, Concejero MA et al 2011 Sources and behaviour of polybrominated diphenyl ethers (PBDEs), polychlorinated dibenzo-p-dioxins and dibenzofurans (PCDD/Fs) in Spanish sewage sludge. *Waste Management* **31** 1277-1284

D'Hollander W, de Voogt P, De Coen W and Bervoets L 2010 Perfluorinated Substances in human food and other sources of human exposure. *In*: P. de Voogt (ed.), *Reviews of Environmental Contamination and Toxicology* **208** Perfluorinated alkylated substances: 179-215

Ding GH, Wouterse M, Baerselman R and Peijnenburg WJ 2012a Toxicity of poly- and perfluorinated compounds to lettuce (*Lactuca sativa*) and green algae (*Pseudokirchneriella subcapitata*). *Archives of Environmental Contamination & Toxicology* **62** 49–55

Ding GH, van den Brandhof EJ, Baerselman R, Peijnenburg WJ 2012b Acute toxicity of poly- and perfluorinated compounds to two cladocerans, *Daphnia magna* and *Chydorus sphaericus*. *Environmental Toxicology and Chemistry* **31** 605–610

Ding G, Zhang J, Chen Y et al 2013 Combined effects of PFOS and PFOA on Zebrafish (*Danio rerio*) Embryos. *Archives of Environmental Contamination and Toxicology* **64** 668–675 *DOI 10.1007/s00244-012-9864-2*

Domingo JL2017 Concentrations of environmental organic contaminants in meat and meat products and human dietary exposure: A review. *Food and Chemical Toxicology* **107** 20–26

Drottar KR and Krueger HO 2000 PFOS: A 48-hr static acute toxicity test with the cladoceran (*Daphnia magna*). Project No. 454A-104. EPA Docket AR226-0087. Wildlife International Ltd., Easton, MD

EPHC (Environment Protection and Heritage Council) 2005 National Dioxins Program: National Action Plan for Addressing Dioxins in Australia. ISBN Web: 0 642 323 89 5.

http://www.nepc.gov.au/system/files/resources/74b7657d-04ce-b214-d5d7-51dcbce2a231/files/cmgt-rpt-national-dioxins-program-national-action-plan-addressing-dioxins-australia-final-200507.pdf

ES (Endocrine Society) 2014 Introduction to endocrine disrupting chemicals (EDCs). A guide for public interest organisations and policy-makers *https://www.endocrine.org/-/media/endosociety/files/advocacy-and-outreach/important-documents/introduction-to-endocrine-disrupting-chemicals.pdf*

Ericson I, Nadal M, van Bavel B, Lindström G and Domingo JL 2008 Levels of perfluoro chemicals in water samples from Catalonia, Spain: is drinking water a significant contribution to human exposure? *Environmental Science and Pollution Res*earch **15** 614–619

Fair PA, Romano T, Schaefer AM, et al 2013 'Associations between perfluoroalkyl compounds and immune and clinical chemistry parameters in highly exposed bottlenose dolphins (*Tursiops truncatus*)'. *Environmental Toxicology and Chemistry* **32(4)** 736–746

Fei C, McLaughlin JK, Lipworth L and Olsen J 2009 Maternal levels of perfluorinated chemicals and sub-fecundity. *Human Reprod*uction **24** 1200–1205

Focant JF, Eppe G, Pirard C et al 2002 Levels and congener distributions of PCDDs, PCDFs and non-ortho PCBs in Belgian foodstuffs Assessment of dietary intake. *Chemosphere* **48** 167–179

FSA UK (Food Standards Agency) 2003 Dioxins and dioxin-like PCBs in the UK diet: 2001 total diet study samples. Report **38/03**

FSA Ireland (Food Safety Authority Ireland) 2009 Dioxins and PCBs in Food. *https://www.fsai.ie*

FSA UK (Food Standards Agency UK) 2009 Fluorinated Chemicals in Food; Food Survey Information Sheet 05/09; Food Standards Agency, UK, 2009. *http://www.food.gov. uk/science/surveillance/fsisbranch2009/fsis0509; http://fluoridealert.org/news/uk-survey-of-pfos-and-pfoa-in-food/*

Fries GF 1995 A review of the significance of animal food products as potential pathways of human exposures to dioxins. *Journal of Animal Science* **73** 1639–1650

Freijer JI, Hoogerbrugge R, van Klaveren JD, Traag WA, Hoogenboom LAP, Liem AKD 2001 Dioxins and dioxin-like PCBs in foodstuffs: Occurrence and dietary intake in The Netherlands at the end of the 20th century. RIVM report 639102022, Bilthoven, The Netherlands

Fromme H, Korner W, Shahin N et al 2009 Human exposure to polybrominated diphenyl ethers (PBDE), as evidenced by data from a duplicate diet study, indoor air, house dust, and biomonitoring in Germany. *Environment International* **35** 1125–1135

FSANZ (Food Standards Australia and New Zealand) 2007 Polybrominated diphenyl ethers (PBDE) in food in Australia. Study of concentrations in foods including dietary exposure assessment and risk characterisation. *https://www.foodstandards.gov.au/science/surveillance/documents/PBDE_Report_Dec_07.pdf*

FSANZ (Food Standards Australia and New Zealand) 2012 Dioxins. Food Standards Australia and New Zealand FSANZ).

https://www.foodstandards.gov.au/consumer/chemicals/dioxins/Pages/default.aspx

Gallen C, Drage D, Broomhall S et al 2014, National snap-shot of POPs in Australian-sourced landfill leachate and biosolids, ENTOX, The University of Queensland, Australia

Gallen C, Dragea D, Eagleshama G et al 2017 Australia-wide assessment of perfluoroalkyl substances (PFASs) in landfill leachates. *Journal of Hazardous Materials* **331** 132–141

Gatehouse R 2004 *Ecological risk assessment of dioxins in Australia-National Dioxins Program.* Technical report no. 11. Australian Government. Department of the Environment and Heritage, Canberra. Commonwealth of Australia. 146 p.

http://www.environment.gov.au/settlements/publications/chemicals/dioxins/report-11/pubs/report-11.pdf

Gevao B, Al-Bahloul M, Al-Ghadban A et al 2006 House dust as a source of human exposure to polybrominated diphenyl ethers in Kuwait. *Chemosphere* **64** 603–608

Geyer HJ, Schramm K-W, Feicht EA et al 2002 Half-lives of tetra-, penta-, hexa-, hepta-, and octachlorodibenzo-*p*-dioxin in rats, monkeys, and humans-a critical review. *Chemosphere* **48** 631-644

Giesy JP and Kannan K 2001 Global distribution of perfluorooctane sulfonate in wildlife. *Environ. Sci. Technol.* **35** 1339-1342

Giesy JP and Kannan K 2001a Accumulation of perfluorooctanesulfonate and related fluorochemicals in marine mammals. Prepared for 3M, St Paul, MN. In US EPA Administrative Record AR226-1030A. (*In*: OECD 2002)

Giesy JP and Kannan K 2001b Perfluorooctanesulfonate and related fluorochemicals in fish-eating water birds. Prepared for 3M, St Paul, MN. *In:* US EPA Administrative Record AR226-1030A. (*In*: OECD 2002).

Giesy JP and Kannan K 2001c Accumulation of perfluorooctanesulfonate and related fluorochemicals in fish tissues. Prepared for 3M, St Paul, MN. *In*: US EPA Administrative Record AR226-1030A. (*In*: OECD 2002).

Giesy JP and Kannan K 2001d Accumulation of perfluorooctanesulfonate and related fluorochemicals in mink and river otters. Prepared for 3M, St Paul, MN. *In*: US EPA Administrative Record AR226-1030A. (*In*: OECD 2002).

Giesy JP and Kannan K 2001e Perfluorooctanesulfonate and related fluorochemicals in oyster, Crassostrea virginica, from the Gulf of Mexico and Chesapeake Bay. Prepared for 3M, St Paul, MN. In US EPA Administrative Record AR226-1030A. (*In*: OECD 2002).

Giesy JP, Mabury SA, Martin JW et al 2006 Perfluorinated compounds in the Great Lakes. Handbook of Environmental *Chemistry* **5(N)** 391–438 *DOI 10.1007/698_5_046*

Giesy JP, Naile JE, Khim JS, Jones PD, and John L. Newsted 2010 Aquatic toxicology of perfluorinated chemicals. D.M. Whitacre (ed.), Reviews of Environmental Contamination and Toxicology. 52p. in: *Reviews of Environmental Contamination and Toxicology* **202** *DOI 10.1007/978-1-4419-1157-5_1,* _ Springer Science Business Media, LLC 2010

Gill U, Chu I, Ryan JJ and Feeley M 2004. Polybrominated diphenyl ether: human tissue levels and toxicology. *Reviews in Environmental Contamination and Toxicology* **182** 55-96

Gulkowska A, Jiang Q, So MK, Taniyasu S, Lam PKS and Yamashita N 2006 Persistent perfluorinated acids in seafood collected from two cities of China. *Environmental Science and Technol*ogy **40** 3736–3741

Gouin T, Mackaya D, Jones KC et al 2004 Evidence for the “grasshopper” effect and fractionation during long-range atmospheric transport of organic contaminants *Environmental Pollution* **128(1-2)**139-148

Guo W, Pan B, Sakkiah S et al 2019 Review Persistent organic pollutants in food: contamination sources, health effects and detection methods. *International Journal of Environmental Research & Public Health* **16** 4361 *doi:10.3390/ijerph16224361*

Gusev A, MacLeod M and Bartlett P 2012 Intercontinental transport of persistent organic pollutants: a review of key findings and recommendations of the task force on hemispheric transport of air pollutants and directions for future. *Atmospheric Pollution Research* **3** 463–465

Greaves AK, Letcher RJ, Sonne C and Dietz R 2013 'Brain region distribution and patterns of bioaccumulative perfluoroalkyl carboxylates and sulphonates in East Greenland polar bears (*Ursus maritumis*)'.*Environmental Toxicology and Chemistry* **32(3)** 713–722

Hahn ME and Stegeman JJ 1992 Phylogenetic distribution of the Ah receptor in non-mammalian species: implications for dioxin toxicity and Ah receptor evolution. *Chemosphere* **25** 931-937

Harada K, Saito N, Sasaki K, Inoue K and Koizumi A 2003 Perfluorooctane sulfonate contamination of drinking water in the Tama River, Japan: estimated effects on resident serum levels. *Bulletin of Environmental Contamination & Toxicology* **71** 1–36

Harrad S, Hazrati S and Ibarra C 2006 Concentrations of polychlorinated biphenyls in indoor air and polybrominated diphenyl ethers in indoor air and dust in Birmingham, United Kingdom: implications for human exposure. *Environmental Science and Technology* **40** 4633–4638

Hazelton PD, Cope WG, Pandolfo TJ et al 2012 'Partial life-cycle and acute toxicity of perfluoroalkyl acids to freshwater mussels. *Environmental Toxicology and Chemistry* **31(7)** 1611–1620

Hawker DW and Connell DW 1988 Octanol-water partition coefficients of polychlorinated biphenyl congeners. *Environmental Science and Technology* **22** 382-387

Herzke D, Huber S, Bervoets L et al 2013 Perfluorinated alkylated substances in vegetables collected in four European countries; occurrence and human exposure estimations. *Environmental Science and Pollution Research* **20** 7930–7939

Hoff PT, Van Campenhout K, Van de Vijver K et al 2005 Perfluorooctane sulfonic acid and organohalogen pollutants in liver of three freshwater fish species in Flanders (Belgium): relationships with biochemical and organismal effects. *Environment Pollution* **137** 324-333

Houde M, Martin JW, Letcher RJ, Solomon KR and Muir DC 2006 "Biological monitoring of polyfluoroalkyl substances: A review". *Environmental Science and Technology* **40(11)** 3463–73

IARC (International Agency for Research on Cancer) 2019 IARC monographs on the identification of carcinogenic hazards to humans. Lists of Classifications, Volumes 1 to 116. *https://monographs.iarc.fr/agents-classified-by-the-iarc/*

Ikonomou MG Rayne S and Addison RF 2002a Exponential increases of the brominated flame retardants, polybrominated diphenyl ethers, in the Canadian Arctic from 1981 to 2000 *Environmental Science and Technology* **36 (9)** 1886 – 1892

Ikonomou MG, Rayne S, Fischer M et al 2002b Occurrence and congener profiles of polybrominated diphenyl ethers (PBDEs) in environmental samples from coastal British Columbia, Canada. *Chemosphere* **46** 649–663

IUPAC (International Union of Pure and Applied Chemistry) 1997 Compendium of Chemical Terminology, 2nd ed. (the "Gold Book") (1997). Online corrected version: (2006–) "congener". *doi:10.1351/goldbook.CT06819*

Jarnberg U and Holmstrom K 2003 Perfluorooctane sulphonate concentrations in Swedish urban and background fish samples (WEP/18). Paper presented at Society for Environmental Toxicology and Chemistry Europe 13th Annual Meeting, 27 April – 1 May 2003, Hamburg.

JawardFM, Zhang G, Nam JJ et al 2005 Passive air sampling of polychlorinated biphenyls, organochlorine compounds, and polybrominated diphenyl ethers across Asia. *Environmental Science and Technology* **39** 8638–8645

JECFA (Joint Expert Committee on Food Additives) 2006 Evaluation of certain food contaminants (Sixty-fourth report of the Joint FAO/WHO Expert Committee on Food Additives). WHO Technical Report Series, No. 930.

Jensen J, Ingvertsen ST and Magid, J 2012 Risk evaluation of five groups of persistent organic contaminants in sewage sludge. Environmental Project No. 1406. Danish EPA, Copenhagen.

Joensen UN, Bossi R, Leffers H et al 2009 Do perfluoroalkyl compounds impair human semen quality? *Environmental Health and Perspectives* **117** 923–927

Johnson A and Olson N 2001 Analysis and occurrence of polybrominated diphenyl ethers in Washington State freshwater fish. *Archives of Environmental Contamination and Toxicology* **41** 339–344

Kallenborn R, Berger U and Jarnberg U 2004 Perfluorinated Alkylated Substances (PFAS) in the Nordic Environment; Nordic Council of Ministers: Copenhagen, Norway.

Kannan K, Hansen KJ, Wade TL and Giesy JP 2002a Perfluorooctane sulfonate in oysters, *Crassostrea virginica*, from the Gulf of Mexico and the Chesapeake Bay. *Archives of Environmental Contamination and Toxicology* **42** 313-318

Kannan K, Corsolini S, Falandysz J et al 2002b Perfluorooctanesulfonate and related fluorinated hydrocarbons in marine mammals, fishes, and birds from coasts of the Baltic and the Mediterranean Seas. *Environmental Science and Technology* **36** 3210-3216

Kannan K, Tao L, Sinclair E, Pastva SD, Jude DJ, Giesy JP 2005 Perfluorinated compounds in aquatic organisms at various trophic levels in a Great Lakes food chain. *Archives of Environmental Contamination and Toxicology* **48** 559–566

Kennedy GL, Butenhoff JL, Olsen GW et al 2004 The toxicology of perfluorooctanoate. *Critical Reviews in Toxicology* **34** 351-384

Kibria G 2014 Climate Change and Chemical Contaminants. *DOI: 10.13140/RG.2.1.2733.6169*

Kibria G, Haroon AKY, Nugegoda D and Rose G 2010 Climate change and chemicals: Environmental Biological aspects. New India Publishing Agency. New Delhi. 470 p. ISBN: 9789380235301;

https://www.researchgate.net/publication/261216635_Climate_Change_and_Chemicals_Environmental_and_Biological_Aspects

Kim JG, Kim KS, Joo CH and You JC 2000 Exposure of PCDD/ DFs via air and food in Koreans. *Organohalogen Compounds* **47** 314–317

Kim M, Guerra P, Theocharides M et al 2013 Polybrominated diphenyl ethers in sewage sludge and treated biosolids: Effect factors and mass balance. *Water Research* **47** 6496–6505

Kiviranta H, Hallikainen A, Ovaskainen ML, Kumpulainen J and Vartiainen T 2001 Dietary intakes of polychlorinated dibenzo-p-dioxins, dibenzofurans and polychlorinated biphenyls in Finland. *Food Additives and Contaminants* **18** 945–953

Kiviranta H, Ovaskainen ML and Vartiainen T 2004 Market basket study on dietary intake of PCDD/Fs, PCBs, and PBDEs in Finland. *Environment International* **30** 923–932

Konwick BJ, Tomy GT, Ismail N et al 2008 'Concentrations and patterns of perfluoroalkyl acids in Georgia, USA, surface waters near and distant to a major use source', *Environmental Toxicology and Chemistry* **27(10)** 2011–2018

Kuehl DW, Haebler R and Potter C 1991 Chemical residues in dolphins from the U.S. Atlantic coast including Atlantic bottlenose obtained during the 1987/88 mass mortality. *Chemosphere* **22** 1071–1084

Kulkarni PS, Crespo JG and Afonso AMC 2008 Dioxins sources and current remediation technology- A review. *Environment International* **34** 139-153

La Guardia MJ, Hale RC and Harvey E 2007 Evidence of debromination of decabromodiphenyl ether (BDE-209) in biota from a wastewater receiving stream. *Environmental Science and Technology* **41** 6663–6670

Lakind JS 2007 Recent global trends and physiological origins of dioxins and furans in human milk. *Journal of Exposure & Environmental Epidemiology* **17** 510-524

Lau, C, Anitole, K, Hodes, C et al 2007 "Perfluoroalkyl Acids: A Review of Monitoring and Toxicological Findings." *Toxicological Sciences* **99(2)** 366 to 394

Li Q,Loganath A, Seng Chong Y, Tan J and Philip Obbard J 2006 Persistent organic pollutants and adverse health effects in humans. *Journal of Toxicology and Environmental Health A* **69** 1987–2005

Li J, Zhang L, Wu Y et al 2009 A national survey of polychlorinated dioxins, furans (PCDD/Fs) and dioxin-like polychlorinated biphenyls (di-PCBs) in human milk in China. *Chemosphere* **75** 1236-1242

Lindstrom G, Wingfors H, Dam M and Bavel BV 1999 Identification of 19 polybrominated diphenyl ethers (PBDEs) in long-finned pilot whale (*Globicephala melas*) from the Atlantic. *Archives of Environmental Contamination and Toxicology* **36** 355–363

Lind Y, Darnerud PO, Aune M and Becker W 2002 Exponering fo˜r organiska miljo¨kontaminanter via livsmedel. Report from the Swedish NFA (in Swedish), report 26

Llobet JM, Domingo JL, Bocio A, Casas C, Teixido´ A and Müller L 2003 Human exposure to dioxins through the diet in Catalonia, Spain: carcinogenic and non-carcinogenic risk. *Chemosphere* **50** 1193–1200

Loos R, Wollgast J, Huber T and Hanke G 2007 Polar herbicides, pharmaceutical products, perflurooctanesulfonate (PFOS), perfluorooctanoate (PFOA), and nonylphenol and its carboxylates and ethoxylates in surface and tap waters around lake Maggiore in Northern Italy. *Analytical & Bioanalytical Chemistry* **387** 1469–1478

Lucia M, Verboven N, Strom H et al 2015 'Circumpolar contamination in eggs of the high-arctic ivory gull *Pagophila eburnean'.Environmental Toxicology and Chemistry* **34(7)** 1552–1561

Luo Q, Cai ZW and Wong MH 2007 Polybrominated diphenyl ethers in fish and sediment from river polluted by electronic waste. *Science of the Total Environment* **383** 115–127

Mhadhbi L, Fumega J, Boumaiza M and Ricardo Beiras 2012 Acute toxicity of polybrominated diphenyl ethers (PBDEs) for turbot (*Psetta maxima*) early life stages (ELS). *Environmental Science & Pollution Research* **19** 708–717

Mai TA, Doan TV, Terradellas J, de Alencastro LF and Grandjean D 2007 Dioxin contamination in soils of Southern Vietnam. *Chemosphere* **67** 1802-1807

Malarvannan D, Kuisue T, Isobe T et al 2009 Organohalogen compounds in human breast milk from mothers living in Payatas and Malate, the Philippines: Levels, accumulation kinetics and infant health risk. *Environmental Pollution* **157** 1924-1932

Martin JW, Smithwick MM, Braune BM et al 2004a "Identification of Long Chain Perfluorinated Acids in Biota from the Canadian Arctic" *Environmental Science and Technology* **38** 373-380

Martin JW, Whittle DM, Muir DCG, Mabury SA 2004b Perfluoroalkyl contaminants in a food web from Lake Ontario. *Environmental Science and Technology* **38** 5379–5385

Moody CA, Martin JW, Kwan WC, Muir DCG, Mabury SA 2002 Monitoring perfluorinated surfactants in biota and surface water samples following an accidental release of fire-fighting foam into Etobicoke creek. *Environmental Science and Technology* **36** 545- 551

Moon H, Kannan K, Lee S and Choi M 2007 Polybrominated diphenyl ethers (PBDEs) in sediment and bivalves from Korean coastal waters. *Chemosphere* **66** 243–251

Muirhead EK, Skillman AD, Hook SE and Schultz IR 2006 Oral exposure of PBDE- 47 in fish: toxicokinetics and reproductive effects in Japanese Medaka (*Oryzias latipes*) and fathead minnows (*Pimephales promelas*). *Environmental Science and Technology* **40** 523–528

Nakata H, Kannan K, Nasu T et al A 2006 Perfluorinated contaminants in sediments and aquatic organisms collected from shallow water and tidal flat areas of the Ariake Sea, Japan: environmental fate of perfluorooctane sulfonate in aquatic ecosystems. *Environmental Science and Technology* **40** 4916–4921

Nania V, Pellegrini GE, Fabrizi L et al 2009 Monitoring of perfluorinated compounds in edible fish from the Mediterranean Sea. *Food Chemistry* **115** 951–957

Noorlander CW, van Leeuwen SPJ, Biesebeek JD et al 2011 Levels of Perfluorinated compounds in food and dietary intake of PFOS and PFOA in The Netherlands. *Journal of Agricultural and Food Chemistry* **59** 7496–7505 *DOI: 10.1021/jf104943p*

Norström K, Viktor T, Cousins AP and Rahmberg, IVL 2015 Risks and effects of the dispersion of PFAS on aquatic, terrestrial and human populations in the vicinity of international airports, IVL Swedish Environmental Research Institute Ltd, Stockholm, Sweden.

NHMRC (National Health and Medical Research Council) 2011 *Australian Drinking Water Guidelines. Paper 6 National Water Quality Management Strategy*. National Health and Medical Research Council, National Resource Management Ministerial

Council, Commonwealth of Australia, Canberra. National Resource Management Ministerial Council, Commonwealth of Australia, Canberra.

https://www.nhmrc.gov.au/sites/default/files/documents/NHMRC%20ADWG%206%20-%20Version%203.5%20-%20Proof%203.pdf

NRC (National Research Council) 2010 National Research Council. 2010. Global sources of local pollution: An assessment of long-range transport of key air pollutants to and from the United States. Washington, DC: The National Academies Press. doi: 10.17226/12743.

https://www.nap.edu/read/12743/chapter/7#122

NSW (New South Wales) 2017 PFOS and PFOA.

https://www.health.nsw.gov.au/environment/factsheets/Pages/pfos.aspx; (accessed 14 January 2020)

Nugegoda D and Kibria G 2017 Effects of environmental chemicals on fish thyroid function: Implications for fisheries and aquaculture in Australia. *General and Comparative Endocrinology* **244** 40-53 *doi: 10.1016/j.ygcen.2016.02.021. Epub 2016 Feb 2*

NWQMS 2008 National Water Quality Management Strategy (NWQMS) Australian Guidelines for Water Recycling: Managing Health and Environmental Risks (Phase 2) — Augmentation of Drinking Water Supplies. Natural Resource Management Ministerial Council. Environment Protection and Heritage Council, National Health and Medical Research Council. Canberra (159 p.).

file:///C:/Users/User1/Downloads/Aus-guidelines-for-water-recycling-phase-2%20(1).pdf

OECD (Organisation for Economic Co-operation and Development) 2002 Hazard assessment for perfluoro octane sulfonate (PFOS) and its salts. ENV/JM/RD (2002) 17/FINAL. *https://www.oecd.org/chemicalsafety/risk-assessment/2382880.pdf*

OECD (Organisation for Economic Co-operation and Development) 2013 OECD/UNEP Global PFC Group, Synthesis paper on per- and polyfluorinated chemicals (PFCs), Environment, Health and Safety, Environment Directorate, OECD. *https://www.oecd.org/env/ehs/risk-management/PFC_FINAL-Web.pdf*

Oliaei, F, Kriens, D, Weber, R and A. Watson A 2013 "PFOS and PFC Releases and associated pollution from a PFC production plant in Minnesota (USA).*Environmental Science & Pollution Research* **20(4)** 1977-1992 Link: *springer.com/article/10.1007/s11356- 012-1275-4*

Oros DR, Hoover D, Rodigari F et al 2005 Levels and distribution of polybrominated diphenyl ethers in water, surface sediments, and bivalves from the San Francisco estuary. *Environmental Science & Technology* **39** 33–41

Pariatamby A and Kee YL 2016 The Tenth International Conference on Waste Management and Technology (ICWMT) Persistent organic pollutants management and remediation. *Procedia Environmental Sciences* **31** 842 – 848

Peng X, Tang C, Yu Y et al 2009 Concentrations, transport, fate, and releases of polybrominated diphenyl ethers in sewage treatment plants in the Pearl River Delta, South China. *Environment International* **35** 303–309

PHE (Public Health, England) (2009). PFOS and PFOA: Toxicological Overview (Version 1). Toxicology Department, PHE

PubChem MSDS 2005 (updated 2020) MSDS Perfluorooctanesulfonic acid 2005 Updated 2020

https://pubchem.ncbi.nlm.nih.gov/compound/Perfluorooctanesulfonic-acid

Nationhttps://pubchem.ncbi.nlm.nih.gov/compound/Perfluorooctanesulfonic-acidal Library of Medicine

Rappe C 1996 Sources and environmental concentrations of dioxins and related compounds. *Pure & Applied Chemistry* **68(9)** 1781-1789

Rayne S, Ikonomou MG and Antcliffe B 2003 Rapidly increasing polybrominated diphenyl ether concentrations in the Columbia River System from 1992 to 2000. *Environmental Science and Technology* **37** 2847–2854

Reeuwijk NM, Talidda A, Malisch R et al 2013 Dioxins (polychlorinated dibenzo-p-dioxins and polychlorinated dibenzo-furans) in traditional clay products used during pregnancy. *Chemosphere* **90(5)** 1678–1685

Reiner EJ, Clement RE and Okey AB 2006 Advances in analytical techniques for polychlorinated dibenzo-p-dioxins, polychlorinated dibenzofurans and dioxin-like PCBs. *Analytical and Bio-analytical Chemistry* **386** 791-806

RIMV (The National Institute for Public Health and the Environment (RIVM) 2010 Environmental risk limits for PFOS. A proposal for water quality standards in accordance with the Water Framework Directive. RIVM Report 601714013/2010; *https://www.rivm.nl/bibliotheek/rapporten/601714013.pdf*

Robertson JC 1986 Potential for environmental impact of AFA-6 surfactant. EPA Docket AR226-1030a043.Beak

Saito N, Harada K, Inoue K et al 2004 Perfluorooctanoate and perfluorooctane sulfonate concentrations in surface water in Japan. *Journal of Occupational Health* **46** 49–59

Scanlan LD, Loguinov AV, Teng Q et al 2015 Gene transcription, metabolite and lipid profiling in eco indicator *Daphnia magna* indicate diverse mechanisms of toxicity by legacy and emerging flame-retardants. *Environ. Sci. Technol.* **49** 7400–7410

Schuhmacher M, Kiviranta H, Ruokojärvi P, Nadal M and Domingo JL 2009 Concentrations of PCDD/Fs, PCBs and PBDEs in breast milk of women from Catalonia Spain: A follow-up study. *Environmental International* **35** 607-613

Schecter A, Cramer P, Boggess K, Stanley J, Pa¨pke O, Olson J et al 2001 Intake of dioxins and related compounds from food in the U.S. population. *Journal of Toxicology and Environmental Health Sciences* **63** 1–18

SCOOP (Scientific Co-operation) 2000 Scientific co-operation on questions relating to food ''assessment of dietary intake of dioxins and related PCBs by the population of EU member states'' Task 3.2.5-Final report- 7 June

Sharpe RL, Benskin JP, Laarman et al 2010 Perfluorooctane sulfonate toxicity, isomer-specific accumulation and maternal transfer in zebrafish (Danio rerio) and rainbow trout (*Oncorhynchus mykiss*). *Environmental Toxicology & Chemistry* **29(9)** 1957–1966

Skutlarek D, Exner M and Färber H 2006 Perfluorinated surfactants in surface and drinking waters. *Environmental Science & Pollution Research* **13** 299–307

Srogi K 2008 Levels and congener distributions of PCDDs, PCDFs and dioxin-like PCBs in environmental and human samples: A review. *Environmental Chemistry Letter* **6** 1-28

Stapleton HM, Dodder NG, Offenberg JH et al 2005 Polybrominated diphenyl ethers in house dust and clothes dryer lint. *Environmental Science and Technology* **39** 925–931

Sutherland CA and Krueger HO 2001 PFOS: A 96-hour toxicity test with the freshwater diatom (*Navicula pelliculosa*). Wildlife International Ltd. Project number 454A-112. EPA Docket AR226-1030a055. Wildlife International, Ltd. Easton, MD Consultants Ltd., Missassauga, Ontario, Canada

Takagi S, Adachi F, Miyana K et al 2008 Perfluorooctanesulfonate and perfluoro-octanoate in raw and treated tap water from Osaka, Japan. *Chemosphere* **72** 1409–1412

Tanabe S and Kunisue T 2007 Persistent organic pollutants in human breast milk from Asian countries (review). *Environmental Pollution* **146** 400-413

Taniyasu S, Kannan K, Horii Y and Yamashita N 2002 The first environmental survey of perfluorooctane sulphonate (PFOS) and related compounds in Japan. *Organohalogen Compounds* **59** 311-314

Taheran M, Komtchou S, Lonappan L et al 2017 Environmental issues of Polybrominated Diphenyl Ethers, *Critical Reviews in Environmental Science and Technology* **47(13)** 1107-1142

Thompson J, Roach A, Eaglesham G et al 2011 'Perfluorinated alkyl acids in water, sediment and wildlife from Sydney Harbour and surroundings'. *Marine Pollution Bulletin* **62** 2869–2875

Thrasher JM 2014 Sources and pathways of dioxins to humans – Diminished contribution of modern WTE facilities.

http://www.seas.columbia.edu/earth/wtert/sofos/Jay_Thrasher_2014.pdf

Tittlemier SA, Pepper K, Seymour C et al 2007 Dietary exposure of Canadians to perfluorinated carboxylates and perfluorooctane sulfonate via consumption of meat, fish, fast foods, and food items prepared in their packaging. *Journal of Agricultural and Food Chem*istry **55** 3203–3210.

Toms L, Mueller J, Mortimer M, Symons R, Stevenson G and Gaus C 2006 Assessment of concentrations of polybrominated diphenyl ether flame retardants in aquatic environments in Australia, Australian Government Department of the Environment and Heritage, Canberra.

https://www.environment.gov.au/system/files/resources/6aab3116-5a64-498b-a408-a338c944379e/files/bfr-aquatic.pdf

Toms L, Hearn L, Kennedy K et al 2009 Concentrations of polybrominated diphenyl ethers (PBDEs) in matched samples of human milk, dust and indoor air. *Environment International* **35** 864–869

Tomy GT, Budakowski W, Halldorson T et al 2004 Fluorinated organic compounds in an Eastern Arctic marine food web. *Environmental Science and Technology* **38** 6475–6481

Tsang HL, Wu SC, Wong CKC et al 2009 Risk assessment of PCDD/Fs levels in human tissues related to major food items based on chemical analysis and micro-EROD assay. *Environment International doi:10.1016/j.envint.2009.05.005*

Tsutsumi T, Yanagi T, Nakamura M, Kono Y, Uchibe H and Iida T et al 2001 Update of daily intake of PCDDs, PCDFs, and dioxin-like PCBs from food in Japan. *Chemosphere* **45** 1129–1137

UK EA (UK Environment Agency) 2007 Draft derivation of soil screening values for priority contaminants total petroleum hydrocarbons. Unpublished report. UK Environment Agency, Bristol.

Ulhaq M, Carlsson G, Örn S and Norrgren L 2013 Comparison of developmental toxicity of seven perfluoroalkyl acids to zebra fish embryos. *Environmental Toxicology and Pharmacology* **36** 423-426

UNEP 2011 Climate change and POPS: predicting the impacts. Report of the UNEP/AMAP Expert Group Secretariat of the Stockholm Convention, Geneva, Switzerland.

UNEP 2017 Stockholm Convention on Persistent Organic Pollutants (POPs). The 16 New POPs. An introduction to the chemicals added to the Stockholm Convention as Persistent Organic Pollutants by the Conference of the Parties. *file:///C:/Users/User1/Downloads/UNEP-POPS-PUB-Brochure-16NewPOPs-201706.English%20(2).pdf*

USEPA (US Environment Protection Agency) 1993 Interim report on data and methods for assessment of 2,3,7,8- tetrachlorodibenzo-*p*-dioxin risk to aquatic life and associated wildlife. EPA/600/R-93/055.

USEPA (US Environment Protection Agency) 1994 Health assessment document for 2,3,7,8-tertachlorodibenzo-p-dioxin (TCDD) and related compounds. EPA/600/Bp-92/001c estimating exposure to dioxin-like compounds, epa/600/6-88/005cb. Washington, DC: Office of research and development.

USEPA (US Environment Protection Agency) 2009 Provisional Health Advisories for Perfluorooctanoic Acid (PFOA) and Perfluorooctane Sulfonate (PFOS), US EPA Region 4, US EPA 2009

water.epa.gov/action/advisories/drinking/upload/2009_01_15_criteria_drinking_pha-PFOA_PFOS.pdf

USEPA (US Environment Protection Agency) 2014 United States Environmental Protection Agency (US EPA) *Emerging contaminants – perfluorooctane sulfonate (PFOS) and perfluorooctanoic acid (PFOA)*, Emerging contaminants fact sheet – PFOS and PFOA,

www2.epa.gov/sites/production/files/2014-04/documents/factsheet_contaminant_pfos_pfoa_march2014.pdf

USEPA (US Environment Protection Agency) 2016 A summary of literature on the chemical toxicity of plastics pollution to aquatic life and aquatic-dependent wildlife. EPA-822-R-16-009. Environmental Protection December 2016;

https://www.epa.gov/sites/production/files/2016-12/documents/plastics-aquatic-life-report.pdf

USEPA (US Environment Protection Agency) 2017a Technical fact sheet – Polybrominated Diphenyl Ethers (PBDEs). EPA 505-F-17-015. US United States Environmental Protection Agency.

https://www.epa.gov/sites/production/files/2014-03/documents/ffrrofactsheet_contaminant_perchlorate_january2014_final_0.pdf

USEPA (US Environment Protection Agency) 2017b Technical Fact Sheet – Perfluorooctane Sulfonate (PFOS) and Perfluorooctanoic Acid (PFOA) November2017. EPA 505-F-17-001.

https://www.epa.gov/sites/production/files/2017-12/documents/ffrrofactsheet_contaminants_pfos_pfoa_11-20-17_508_0.pdf

ValdersnesS, Nilsen BM, Breivik JF, Borge A and Maage A 2017 Geographical trends of PFAS in cod livers along the Norwegian coast. *PLoS ONE* **12(5)** e0177947. *https://doi.org/ 10.1371/journal.pone.0177947*

Van de Vijver KI, Hoff PT, Van Dongen W et al 2003 Exposure patterns of perfluorooctane sulfonate in aquatic invertebrates from the Western Scheldt estuary and the Southern North Sea. *Environmental Toxicology & Chemistry* **22** 2037-2041

Van den Berg M, Birnbaum L, Bosveld AT et al 1998 Toxicity equivalency factors (TCFs) for PCBs, PCDDs, PCDFs for humans and wildlife. *Environmental Health Perspectives* **106(12)** 775-792

van den Berg M, Kypke K, Kotz A et al 2017 WHO/UNEP global surveys of PCDDs, PCDFs, PCBs and DDTs in human milk and benefit–risk evaluation of breastfeeding. *Archives of Toxicology* **91** 83–96

Van Hoven RL and Nixon WB 2000 Determination of the water solubility of PFOS by the shake flask method. Project 454C-107, EPA Docket AR226-0052. Wildlife International Ltd., Easton, MD

Van de Vijver IK, Hoff PT, Van Dongen W et al 2003 Exposure patterns of perfluorooctane sulfonate in aquatic invertebrates from the Western Scheldt estuary and the southern North Sea. *Environmental Toxicology and Chemistry* **22** 2037–2041

Wada O 2002 Health impact of dioxins *JMAJ* **45(7)** 312–315.

http://www.med.or.jp/english/pdf/2002_07/312_315.pdf

Walker MK and Peterson RE 1994 Toxicity of 2,3,7,8-tetrachlorobenzo-*P*-dioxin to brook trout (*Salvelinus fontinalis*) during early development. *Environmental Toxicology and Chemistry* **13(5)** 817-820

Walsh GE, Yoder MJ, McLaughlin LL and Lores EM 1987 Response of marine unicellular algae to brominated organic compounds in six growth media. *Ecotoxicology and Environmental Safety* **14(3)** 215-222

Wang S-L, Tsai P-C, Yang C.-Y and Leon Guo Y 2008 Increased risk of diabetes and polychlorinated biphenyls and dioxins: A 24-year follow-up study of the Yucheng cohort. *Diabetes Care* **31** 1574–1579

Wang I-C, Wu Y-L, Lin L-F and Chang-Chien, G-P 2009 Human dietary exposure to polychlorinated dibenzo-*p*-dioxins and polychlorinated dibenzofurans in Taiwan. *Journal of Hazardous Materials* **164** 621-626

Watanabe I, Kashimoto T and Tatsukawa R 1987 Polybrominated biphenyl ethers in marine fish, shellfish and river and marine sediments in Japan. *Chemosphere* **16** 2389–2396

Wikoff D, Fitzgerald L and Birnbaum L2012 Persistent organic pollutants: An overview (pages: 1-35). *In*: *Dioxins and Health*: *Including Other Persistent Organic Pollutants & Endocrine Disruptors*. Third Edition Arnold Schecter (ed). John Wiley & Sons, Inc.

Wilford BH, Harner T, Zhu J et al 2004 Passive sampling survey of polybrominated diphenyl ether flame retardants in indoor and outdoor air in Ottawa, Canada: implications for sources and exposure. *Environmental Science and Technology* **38** 5312–5318

Wilford BH, Shoeib M, Harner T et al 2005 Polybrominated diphenyl ethers in indoor dust in Ottawa, Canada: implications for sources and exposure. *Environmental Science and Technology* **39** 7027–7035

WHO (World Health Organisation) 1998 Assessment of the health risk of dioxins: re-evaluation of the Tolerable Daily Intake (TDI) WHO Consultation May 25-29 1998, Geneva, Switzerland. WHO European Centre for Environment and Health International Programme on Chemical Safety.

https://www.who.int/ipcs/publications/en/exe-sum-final.pdf

WHO (World Health Organisation) 2010 Persistent organic pollutants: Impact on child health. ISBN 978 92 4 150110 1.

https://www.who.int/ceh/publications/persistent_organic_pollutant/en/

WHO (World Health Organisation) 2012 State of the Science of endocrine disrupting chemicals 2012. An assessment of the state of the science of endocrine disruptors prepared by a group of experts for the United Nations Environment Programme and World Health Organization.

https://apps.who.int/iris/bitstream/handle/10665/78102/WHO_HSE_PHE_IHE_2013.1_eng.pdf;jsessionid=571575BB005691E201D6061C0FC637AB?sequence=1

WHO (World Health Organisation) 2016 Dioxins and their effects on human health. *https://www.who.int/news-room/fact-sheets/detail/dioxins-and-their-effects-on-human-health*

Wu Y, Li J, Zhao Y, Chen Z, Li W and Chen J 2002 Dietary intake of polychlorinated dibenzo-p-dioxins (PCDDs) and dibenzofurans (PCDFs) in populations from China. *Organohalogen Compounds* **57** 221–223

Yang C, Meng X, Chen, L and Xia S 2011 Polybrominated diphenyl ethers in sewage sludge from Shanghai, China: Possible ecological risk applied to agricultural land. *Chemosphere* **85** 418–423

Young CJ, Furdui VI, Franklin, J et al 2007 "Perfluorinated Acids in arctic snow: New Evidence for atmospheric formation." *Environmental Science & Technology* **41(10)** 3455-3461 Doi: 10.1021/es0626234

Yu L, Wang S and Sun B-G 2014 Food safety chemistry: Toxicant occurrence, analysis and mitigation. CRC Press. ISBN 9781138033818

Zhao G, Wang Z, Zhou H and Zhao Q 2009 Burdens of PBBs, PBDEs, and PCBs in tissues of the cancer patients in the e-waste disassembly sites in Zhejiang, China. *Science of the Total Environment* **407** 4831-4837

■■■■

6 Oil Spill and Oil Pollution Risks, Impacts and Management

Summary

An oil spill is an environmental disaster and results in pollution of a wide area around the spill. Oil floats on water and spreads very quickly. Oil compounds (PAHs, BTEX) are persistent, bio-accumulative and toxic and some are also carcinogenic and endocrine-disrupting. Oil spills and the resulting pollution can have a significant impact on social, economic and environmental sectors. Oil exposure to fish causes reduced larval survival, decreased growth and reproduction. Oil contaminated seafood can have a taste and smell of petroleum. Oil can foul boats, fishing gears and mariculture facilities. The tourism industry can be impacted by the direct impact of oil spills to beaches and waterfront properties. Wildlife can be harmed via exposure to oil through ingestion/inhalation or through skin and eye irritation. Different ecosystems (corals, sea grasses and mangroves which are habitats for fishes, sponges, molluscs, crustaceans) are sensitive to oil. Humans can be exposed to oil from eating contaminated seafood and from breathing oil vapours. This chapter reviews **i.** the chemistry, toxicity and ecotoxicity of oil compounds (PAHs, BTEX); **ii.** effects of the oil spill and oil pollution on the social, economic and environmental sectors including fish, seafood, mariculture, fishing, tourism, wildlife, ecosystems, human's health, industrial water intakes, salt production, and agriculture; and **iii**. management/measures to minimise and reduce the impact of oil spill and oil pollution.

Contents

List of Tables

List of Figures

6.1. Introduction

An oil spill (spill of crude oil or refined products like gasoline/petrol, diesel fuels, jet fuels, and kerosene) is an environmental disaster and results in pollution around the spill. An oil spill can happen accidentally (blow out, natural disaster), collisions of ships/tankers, accidents related to drilling rigs & wells and during transportation of oil (road and rail) or deliberate/intentional release of oil (Fig. 6.1).

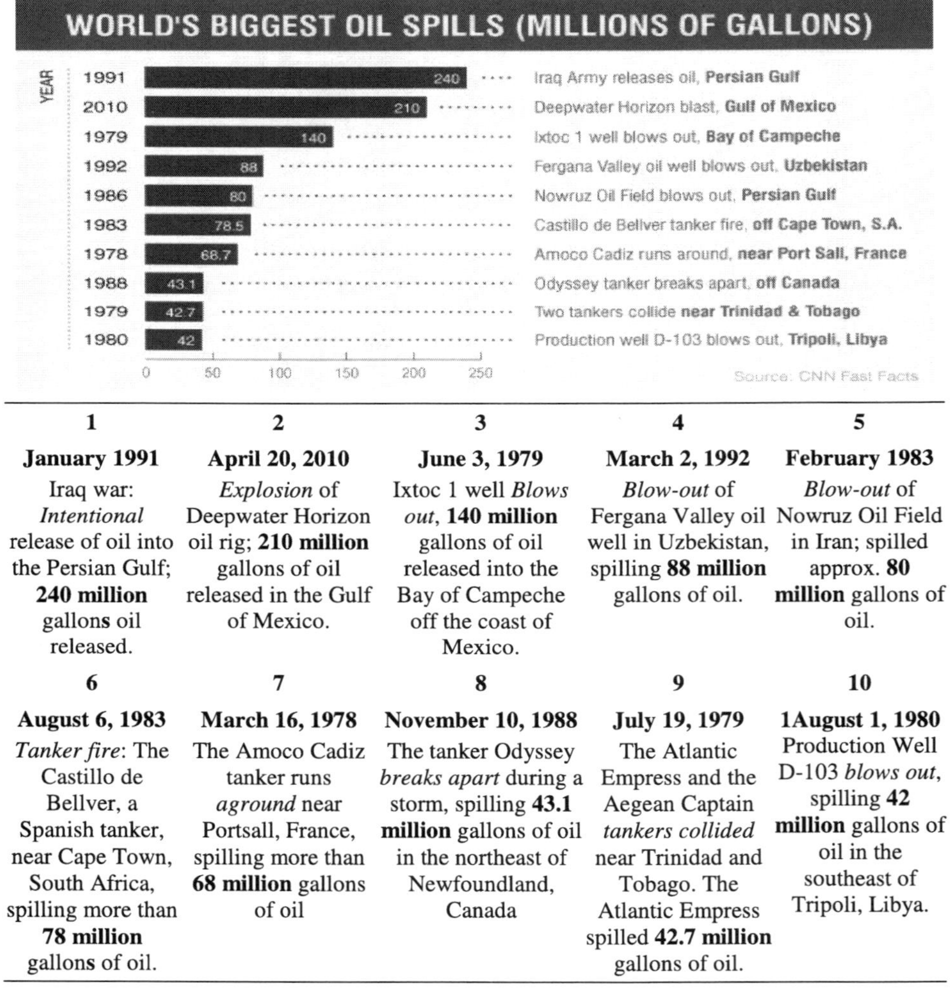

1	2	3	4	5
January 1991	**April 20, 2010**	**June 3, 1979**	**March 2, 1992**	**February 1983**
Iraq war: *Intentional* release of oil into the Persian Gulf; **240 million** gallons oil released.	*Explosion* of Deepwater Horizon oil rig; **210 million** gallons of oil released in the Gulf of Mexico.	Ixtoc 1 well *Blows out*, **140 million** gallons of oil released into the Bay of Campeche off the coast of Mexico.	*Blow-out* of Fergana Valley oil well in Uzbekistan, spilling **88 million** gallons of oil.	*Blow-out* of Nowruz Oil Field in Iran; spilled approx. **80 million** gallons of oil.
6	**7**	**8**	**9**	**10**
August 6, 1983	**March 16, 1978**	**November 10, 1988**	**July 19, 1979**	**1August 1, 1980**
Tanker fire: The Castillo de Bellver, a Spanish tanker, near Cape Town, South Africa, spilling more than **78 million** gallons of oil.	The Amoco Cadiz tanker runs *aground* near Portsall, France, spilling more than **68 million** gallons of oil	The tanker Odyssey *breaks apart* during a storm, spilling **43.1 million** gallons of oil in the northeast of Newfoundland, Canada	The Atlantic Empress and the Aegean Captain *tankers collided* near Trinidad and Tobago. The Atlantic Empress spilled **42.7 million** gallons of oil.	Production Well D-103 *blows out*, spilling **42 million** gallons of oil in the southeast of Tripoli, Libya.

Fig. 6.1. Records of world's biggest ten oil spills (CNN 2019).

Oil spills can have a devastating effect on several sectors including the following **i.** principally on fish, seafood, mariculture, fisheries/fishing, tourism, wildlife, and ecosystems; **ii**. but also, on human's health, industrial seawater intakes, salt production. Persistent oils are heavier oils (such as crude oils, bunker C[1]) and do not dissipate quickly through evaporation, whereas non-persistent oils (gasoline/petrol, kerosene and diesel) are lighter oils which dissipate quickly through evaporation (ITOPF 2001; Fig. 6.2).

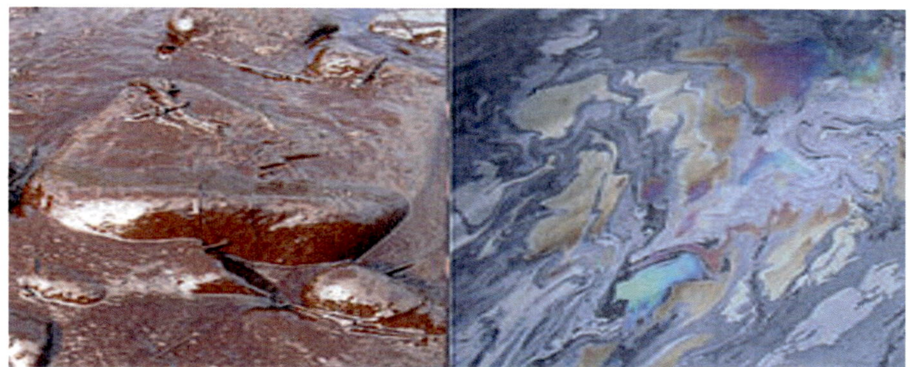

Fig. 6.2. Persistent oil/ heavy oil (left) and non-persistent oil/ light oil (right) *[https://www.itopf.org/knowledge-resources/documents-guides/document/persistent-vs-non-persistent-oils-what-you-need-to-know-2001/].*

Oil spills may impact the environment through one or more of the following mechanisms:

- Physical smothering/coating of marine biota (birds, sea turtles, mammals) by oils;
- Chemical toxicity[2] of oils causing lethal (fish mortality) or sub-lethal (reduced fish growth, reproduction) effects or impairment of cellular functions (tissue damage or tissue death);
- Loss of keystone (flagship) species (plant or animal species that plays a unique and crucial role in ecosystem functions);
- The loss of ecological habitats or shelters (mortality or decreased growth of mangroves, seagrasses, corals);
- Loss of seafood security (tainting or bioaccumulation of oil compounds in fish, oysters, mussels, crabs, prawns, shrimps, lobsters); and
- Loss of tourism (disruption of recreational activities due to contamination of beaches and waterfront).

---

[1] ***Bunker C fuel*** *is any fuel used on board a ship. The most commonly used type is residual fuel oil bunker or Bunker C (heavy oil with high viscosity). They are low acutely toxic to biota but can cause long-term (chronic) effects on biota and ecosystems via smothering or coating.*

[2]***Toxicity*** *is the degree to which a chemical substance or a particular mixture of substances can harm/damage an organism.*

This chapter emphasises the consequences of marine oil spill/ oil pollution on social, economic and environmental sectors.

6.2. Chemistry, toxicity and ecotoxicity[3] of oil

Oils are fossil fuels (i.e. they are formed from the remains of plants and animals which died millions of years ago). Most of the compounds in crude oil consist of molecules made up of hydrogen and carbon (HC) atoms (about 82- 87% carbons by weight and 12-15% hydrogen by weight) *https://www.britannica.com/ science/crude-oil*. Crude oils are mixtures of: **a.** saturated HC/alkanes (hydrocarbons containing single bonds between carbons such as methane, ethane), and **b.** unsaturated HC/alkenes (hydrocarbons that have double or triple covalent bonds between adjacent carbon atoms and known as aromatics). The aromatic compounds include: **i. BTEX** (benzene, toluene, ethylbenzene and xylenes - single ring); **ii.** polycyclic aromatic hydrocarbons (**PAHs**) (two or more rings, double bonds, such as naphthalene - double rings; phenanthrene - three rings; pyrene - four rings) and **iii.** polar poly-aromatics/ heterocyclic aromatics (containing sulfur, nitrogen, or oxygen) (Killops and Killops 1993). PAHs that can be released in to the environment following oil spillage include acenaphthene, anthracene, naphthalene, phenanthrene, benzo[a]pyrene and are often a result of environmental pollution by oil and oil spills (Nugegoda and Kibria 2017).

Crude oils are refined to produce gasoline/ petrol, diesel and various forms of petrochemicals. Oils are less dense than water, therefore, float. The densities of freshwater and seawater are 1.0 g/mL and 1.025 g/mL, respectively. The density of crude oils and refined petroleum is <1 g/mL (Table 6.1). Oils are classified into very **light crude oils** [gasoline/ petrol-like products; light crude oils (diesel-like products; medium crude oils (intermediate products)] and **heavy crude oils** based on various oil properties (specific gravity, API (The American Petroleum Institute, API density), volatility, evaporability, viscosity, and toxicity to animals and plants Table 6.1).

Light oils (gasoline/ petrol-like products) are very volatile, flammable, low viscous, and can spread and evaporate rapidly but are highly and acutely toxic to animals and plants. The USEPA has classified these pollutants as priority pollutant (Fayemiwo et al 2017). **Heavier oils** (bunker oils used by larger ships) are more viscous than light oils and have low acute toxicity[4] to biota but can cause long-term (chronic) effects on biota and ecosystems via smothering or coating (birds). Heavier oils tend to ultimately sink when not remediated and can also coat or combine with sediments. Lighter oil is unlikely to coat surfaces of

---

[3]***Ecotoxicity*** *refers to the potential for biological, chemical or physical stressors to affect ecosystems.*

[4]***Acute toxicity*** *describes the adverse effects of a substance that result either from a single exposure or from multiple exposures in a short period of time (usually less than 24 hours).*

animals but can be taken up by diffusion across the gills by larger organisms (fish) or across the cutaneous/ skin layer in smaller organisms (plankton) (Escher and Hermens 2002, Hook et al 2016).

Table 6.1. Oil groups and their characteristics
(NOAA 2014, Hook et al 2016, Kibria 2018).

*[**API**= the American Petroleum Institute gravity scale (oils grouped); **API** = (141.5/density) – 131.5; API gravity, is a measure of how heavy or light a petroleum liquid is compared to water: if its API gravity is greater than 10, it is lighter and floats on water; if it is less than 10, it is heavier and sinks. API gravity is thus an inverse measure of a petroleum liquid's density relative to that of water (also known as specific gravity https://en.wikipedia.org/wiki/API_gravity); the density of freshwater water is 1.0 g/mL, seawater 1.025 g/mL].*

Class	Density	API[5] density (g/mL)	Volatility	Evaporability	Viscosity[6]	Toxicity to Animals and Plants
Very light oils (Gasoline/ petrol products, jet fuels)	< 0.80	>45	Very volatile and highly flammable	Evaporate and dissolve rapidly	Low viscosity oil, spreads rapidly into thin sheens[7]	Will penetrate substrates but are not sticky High acute toxicity to animals and plants
Light crude oils (Diesel, number 2 fuel oil)	0.80-0.85	35-45	Moderately volatile and soluble	Refined products can evaporate leaving no residue	Low to moderate viscosity Spreads rapidly into thin slicks[8]	More bio-available than lighter oils (because they persist longer) More likely to affect animals in water and sediments
Medium Crude Oils (Intermediate Products, most crude oils)	0.85-0.95	17.5-35	Moderately volatile	Up to one-third will evaporate in the first 24 hours	Moderate to high viscosity; will spread into thick slicks	More bioavailable than lighter oils More likely to affect animals and plants in water and sediments Can form stable emulsions and Can cause long-term effects via smothering or coating

[5] ***API*** *= American Petroleum Institute gravity scale*

[6] ***Viscosity*** *= thickness of the oil, is a measure of a fluid's resistance to flow*

[7] ***Sheen*** *= a very thin layer of oil which is < 1 μm, according to their thickness, sheens vary in colour from rainbows (for the thicker layers), to silver/grey (for thinner layers) to almost transparent (for the thinnest layers) (Fingas 2018).*

[8] ***Slick*** *= A mass of floating oil covering an area of water*

Class	Density	API[5] density (g/mL)	Volatility	Evaporability	Viscosity[6]	Toxicity to Animals and Plants
Heavy crude oils (number 6 crude oil, Bunker C oil)	0.95-1.0	10-17.5	Very little product loss by evapo-ration or dissolution	Very little product loss by evapo-ration or dissolution	Very viscous to semi-solid	Low acute toxicity to biota. Can cause long-term effects via smothering or coating, or as residues on or in sediments.

6.2.1. The fate of oils in the environment (weathering processes)

As soon as the oil is spilled, its chemical and physical properties begin to change through a step-wise process known as weathering. It includes dissipation, evaporation, dispersion, emulsification, dissolution, photo-oxidation, sedimentation and biodegradation (SINTEF 2011, Hook et al 2016). The **early stage** of an oil spill is dissipation, evaporation, dispersion, emulsification and dissolution and the **later stage** is photo-oxidation, sedimentation and biodegradation. The fate and persistence of spillage are influenced by a range of factors, such as the type of oil (light or heavy), the environment where it spilled (open sea or enclosed inlet), and the prevailing conditions (waves, currents, winds, sunlight) at the time of the spill. The biodegradation of oil will be faster where there is the presence of high microbial population (bacteria). Fate of oils in the environment, i.e. the weathering processes of oil is summarised below (NRC 2003, SINTEF 2011, Hook et al 2016, Kibria 2019; *https://www.itopf.org/ knowledge-resources/documents-guides/fate-of-oil-spills/weathering/*).

i. Dissipation (*spreading*)

- Begins as soon as the oil is spilled.
- Low-viscosity oils (petrol) spread much quicker than high viscosity oil (heavy oil) (Table 6.2).
- Higher ambient temperatures, stronger winds and sea surface currents would result in a more rapid rate of spreading.

Table 6.2. Expected fate matrix of spreading for different spilled oils (Hook et al 2016).

Fate matrix	Light oil	Medium oil	Heavy oil	Extra-heavy oil
Spreading	Rapid	Rapid	Moderate-rapid	Slow-moderate

ii. Evaporation *(changing from a liquid to a gas or vapour)*

- Depends upon the volatility of the oil.

- Rougher seas, high wind speeds and temperatures would increase the rate of evaporation.
- Evaporation is fast with light oil products (up to 100%, gasoline/ petrol can be lost via evaporation) and evaporation is very slow/ low with heavy fuel (up to 20% Bunker C oil can be lost via evaporation) (Figure 6.3).

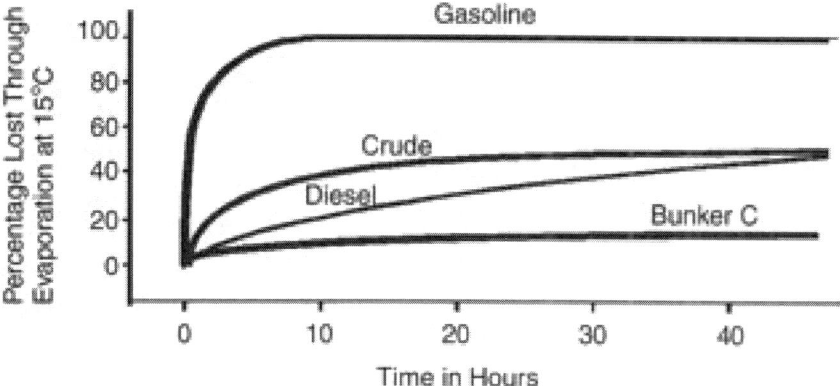

Fig. 6.3. Evaporation rates for different types of oil at 15°C (NRC 2003).

iii. Dispersion *(spreading out into the water column via currents and waves)*

Light and low viscous oils (gasoline, kerosene) disperse readily with water turbulence (wave, water currents and wind energy) (Table 6.3)

Table 6.3. Expected fate matrix of dispersion for different spilled oils (Hook et al 2016)

Fate matrix	Light oil	Medium oil	Heavy oil	Extra heavy oil
Dispersion	High	Moderate-high	Moderate-rapid	Slowly-moderate

iv. Emulsification *(water mixing into oil, creating an emulsion or mousse)*

- Water mixing into oil, creating an emulsion or mousse (appearance of "chocolate mousse' or looks like chocolate pudding) (Table 6.4; Fig. 6.4).
- Mousse formation occurs mostly in oils with lower viscosities with higher (at least 5%) asphaltene content (S, N, O).
- Stable emulsions contain between 60 and 85 percent water.

Table 6.4. Expected fate matrix of emulsification for different spilled oils (Hook et al 2016).

Fate matrix	Light oil	Medium oil	Heavy oil	Extra heavy oil
Emulsification	Slight	Moderate	Moderate-high	High

Fig. 6.4. 'Chocolate mousse' (SINTEF 2011, Hook et al 2016).

v. Dissolution *(dissolving via mixing into the water)*

- The degree to which the oil will dissolve in water is highly dependent on its chemical constituents and a variety of environmental factors, such as water temperature, turbulence (e.g. wind and wave energy) and physical dispersion within the water column (NRC 2003, ITOPF 2012).
- The majority of oil components have very low solubility (Table 6.5), except for some light aromatic hydrocarbons such as, benzene and toluene which are soluble in water.

Table 6. 5. The solubility of some oil components/ oils (NRC 2003) [0 salinity = distilled water].

Oil	Solubility (mg/L)	Oil	Solubility (mg/L)
Benzene	1,700.0	Bunker C oil (22°C, 0 salinity)	6.0
Toluene	530.0	Diesel fuel (20°C, 0 salinity)	3.0
Ethylbenzene	170.0	Fluorene	2.0
p-Xylene	150.0	Phenanthrene	1.0
Gasoline/petrol (22°C, 0 salinity)	98.0	Chrysene	0.002
Naphthalene	30.0		

vi. Photo-oxidation *(the degradation of a compound in the presence of oxygen that is facilitated by radiant energy such as UV light)*

- Photo-oxidation plays only a minor role in oil weathering.
- However, oils react chemically with oxygen forming persistent compounds called tars/ tar balls (dark coloured, hard and crusty on the outside) (Table 6.6; Fig.6.5).
- Formation of tars is caused by the oxidation of thick layers of high viscosity oils or emulsions.
- Contact with tar ball or oil may cause an allergic reaction or develop rashes to people (cause an allergic skin reaction)

Table 6.6. Expected fate matrix of photo-oxidation for different spilled oils (Hook et al 2016).

Fate matrix	Light oil	Medium oil	Heavy oil	Extra heavy oil
Photo-oxidation	Little	Little	Little	Little

Fig. 6.5. A tar ball
[https://response.restoration.noaa.gov/oil-and-chemical-spills/oil-spills/resources/tarballs.html].

vii. Sedimentation/ sinking *(Oil binding to particles and sinking to the bottom sediments)*

- When oil comes in contact with suspended sediments, it can lead to sedimentation (Table 6.7).
- Burning of oil at the sea can also led to an increase in the density of the residue, increasing the likelihood of sedimentation (ITOPF 2012).
- Very few oils sink in the marine environment.

Table 6.7. Expected fate matrix of sedimentation for different spilled oils (Hook et al 2016).

Fate matrix	Light oil	Medium oil	Heavy oil	Extra heavy oil
Sedimentation	Low	Little	Moderate-little	Moderate

viii. Biodegradation (*broken down by bacteria*)

- Micro-organisms (primarily bacteria, fungi and micro-algae oxidize hydrocarbons (micro-organisms use hydrocarbons as a source of energy) (Table 6.8).
- Biodegradation of hydrocarbons has been considered as one of the principal removal mechanisms in the aquatic environment.
- Saturated alkanes (*note:* saturate HC is alkanes and unsaturated HC is alkenes) are more quickly degraded by micro-organisms than aromatic compounds (BTEX, PAHs) (NRC 2003).
- The microbial population, temperature and the availability of nutrients and oxygen concentrations will determine the biodegradation processes (in low-temperature environments, oil viscosity increases, and water solubility decreases) (NRC 2003).
- If oil spill occurs in an area where there are the presences of significant microbial organisms (bacteria), the oil degradation will be much faster.

Table 6.8. Expected fate matrix of biodegradation for different spilled oils (Hook et al 2016).

Fate matrix	**Light oil**	**Medium oil**	**Heavy oil**	**Extra heavy oil**
Biodegradation	Not important	Potentially important	Potentially important	Potentially important

6.2.2. Toxicity and ecotoxicity of oils

Oil compounds are persistent, bio-accumulative and toxic and some are also carcinogenic and endocrine-disrupting. The **log K_{ow}** (logarithm to the base 10 of the octanol/ water partition coefficient values) of several PAHs indicates that they are highly hydrophobic (less soluble in water) and bio-accumulative [log K_{ow} in between 3.3 and 7.0 (Table 6.9); *note*: chemicals with a log K_{ow} value of more than 3.0 is regarded as highly bio-accumulative (Kibria et al 2010)]. Hydrophobic oil compounds (in particular PAHs) can accumulate in different parts/organs of biota (fish) causing lethal or sub-lethal effects to aquatic organisms (see section 6.3 for impacts of oil compounds on fish and tainting in seafood). It is known that compounds with a log K_{ow} <4 will be preferentially accumulated from water, whereas those with a log K_{ow} >6 will be taken up primarily from food and compounds with intermediate values can be accumulated from both food and water (McKim 1994, Law and Hellou 1999). Low molecular weight PAH compounds with 2 and 3 rings (acenaphthene, acenaphthylene, anthracene, fluorene) are more water-soluble, and less bio-accumulative than high molecular PAHs (4 or 5 ring PAHs such as Benzo[a]pyrene, Benzo[*b*]fluoranthene, Chrysene) (Law and Hellou 1999) (see Table 6.9 for water solubility for PAHs and BTEX).

Table 6.9. Chemistry, toxicity and ecotoxicity of oil compounds (PAHs and BTEX)

*[**h**= hours; **IARC**= International Agency for Research on Cancer; **EC_{50}**= (median effective concentration, the concentration of test substance which results in a 50 percent reduction in either algae growth/ algae growth rate or Daphina immobilization ; **LD_{50}**= Lethal Dose at which 50% of the test animals are expected to die; **LC_{50}**= Lethal Concentration at which 50% of the test animals are expected to die; **na**= data is not available; https://www.chemsafetypro.com/Topics/CRA/Toxicology_Dose_Descriptors.html; **Toxicity classification: Rat** LD_{50} < 50 mg/kg=extremely toxic; 50-500 mg/kg= very toxic; 500-5,000 mg/kg=moderately toxic; **Fish** LC_{50}< 1 mg/L= extremely toxic; 1-10 mg/L= very toxic; 10-100 mg/L= moderately toxic].*

Oil compounds and chemical formula	**Log K_{ow} and Bioaccumulation potential** (Neff et al 2005)	**Water solubility (mg/L)** (Neff et al 2005)	**IARC carcinogenic group** (IARC 2019)	**Toxicity** **Toxicity to mammal (rat)** LD_{50} (rat) oral	**Ecotoxicity** **Acute aquatic toxicity** EC_{50} (algae/ daphnia); LC_{50} (fish/ crustacean, frog, worm)
PAH (polycyclic aromatic hydrocarbons) compounds					
Acenaphthene ($C_{12}H_{10}$)	Log K_{ow}: 3.92 Bioaccumulation: High	16.9	3	Rat, LD_{50} = 10000 mg/kg (MSDS-Thermo Fisher Scientific 2010- Acenaphthene)	Algae, *Pseudokirchneriella subcapitata,* EC_{50}: 0.23 - 1.15 mg/L for 96 h Fish, *Lepomis macrochirus* LC_{50}: 1.3 - 2.1 mg/L for 96 h static Fish, *Oncorhynchus mykiss*, LC_{50}: 0.6 - 0.75 mg/L for 96 h flow-through; Fish, *Pimephales promelas*, LC_{50}: 0.509 mg/L for 96 h flow-through (MSDS-Thermo Fisher Scientific 2010- Acenaphthene)
Acenaphthylene ($C_{12}H_8$)	4.07 High	16.68	na	Rat, LD_{50}: 1760 mg/kg (MSDS-PubChem 2004 - Acenaphthylene)	na

Oil compounds and chemical formula	**Log K_{ow} and Bioaccumulation potential** (Neff et al 2005)	**Water solubility (mg/L)** (Neff et al 2005)	**IARC carcinogenic group** (IARC 2019)	**Toxicity** **Toxicity to mammal (rat)** LD_{50} (rat) oral	**Ecotoxicity** **Acute aquatic toxicity** EC_{50} (algae/ daphnia); LC_{50} (fish/ crustacean, frog, worm)
Anthracene ($C_{14}H_{10}$)	4.54 High	0.790	3	Rat, LD50>16000 mg/kg (MSDS-Thermo Fisher Scientific 2018-Anthracene)	Algae, *Pseudokirchneriella subcapitata*: EC_{50}: 0.0039 - 0.0374 mg/L, for 22 h Daphnia, *Daphnia magna*: EC_{50}: 0.081 - 0.112 mg/L for 48 h Fish-bluegill, *Lepomis macrochirus*, LC_{50}: 0.00278 mg/L, for 96 h (MSDS-Thermo Fisher Scientific, 2018- Anthracene
Benz[*a*]anthracene ($C_{18}H_{12}$)	5.91 High	0.0147	2B	na	Daphnia, *Daphnia pulex* LC_{50}: 0.01 mg/L for 96 h Daphnia, *Daphnia magna* EC_{50}: 0.0042 mg/L for 48 h (MSDS-Thermo Fisher Scientific 2018-Benzo[*a*]anthracene)
Benzo[a]pyrene ($C_{20}H_{12}$)	6.04 High	0.0014	1	Rat, LD_{50}: >50 mg/kg (MSDS-PubChem 2004-Benzo[a]pyrene)	Algae, *Pseudokirchneriella subcapitata* (freshwater, static): EC_{50}: 15 µg/L for 72 h Daphnia, *Daphnia pulix*, LC_{50}: 0.005 mg/L for 96 h Fish, *Poeciliopsis lucida,* LC_{50}: 1.2-3.7 mg/L for 24 h Frog, *Rana pipiens*, LC_{50}: >6.7 mg/L for 24 h Worm, *Neanthes arenaceodentata* (annelid), LC_{50}: >1.0 mg/L for 96 h (MSDS-PubChem 2004-Benzo[a]pyrene)

Oil compounds and chemical formula	**Log K_{ow} and Bioaccumulation potential** (Neff et al 2005)	**Water solubility (mg/L)** (Neff et al 2005)	**IARC carcinogenic group** (IARC 2019)	**Toxicity** **Toxicity to mammal (rat)** LD_{50} (rat) oral	**Ecotoxicity** **Acute aquatic toxicity** EC_{50} (algae/ daphnia); LC_{50} (fish/ crustacean, frog, worm)
Benzo[*b*]fluoranthene ($C_{20}H_{12}$)	5.8 High	0.0041	2B	na	na
Benzo[*e*]pyrene ($C_{20}H_{12}$)	6.04 High	0.004	3	na	na
Benzo[*g, h, i*]perylene ($C_{22}H_{12}$)	6.5 High	0.0003	3	na	na

Oil compounds and chemical formula	**Log K_{ow} and Bioaccumulation potential** (Neff et al 2005)	**Water solubility (mg/L)** (Neff et al 2005)	**IARC carcinogenic group** (IARC 2019)	**Toxicity Toxicity to mammal (rat)** LD_{50} (rat) oral	**Ecotoxicity Acute aquatic toxicity** EC_{50} (algae/ daphnia); LC_{50} (fish/ crustacean, frog, worm)
Chrysene ($C_{18}H_{12}$)	5.86 high	0.006	2B	na	Daphnia, *Daphnia magna,* EC_{50}: 0288 mg/L for 48 h (MSDS-Merck 2006-Chrysene)
Dibenz[*a,h*]anthracene ($C_{22}H_{14}$)	6.75 High	0.0005	2A	na	na
Fluoranthene ($C_{16}H_{10}$)	5.22 High	0.261	3	Rat, LD_{50}: 2000 mg/kg	Daphnia, EC_{50}: 0.78 mg/L for 20 h Fish, Rainbow trout: LC_{50}: 0.0077 mg/L for 96 h (MSDS-Thermo- Fisher Scientific 2019- Fluoranthene)
Fluorene ($C_{13}H_{10}$)	4.18 High	2.04	3	na	Algae, EC_{50}: 3.4 mg/L for 96 h (MSDS-Thermo Fisher Scientific 2016- Fluorene) Daphnia, *Daphnia magna*, EC_{50}: 0.43 mg/L for 48 h (MSDS. Fisher-Scientific 1999- Fluorene) Fish, LC_{50}: 0.82 mg/L for 96 h (MSDS-Thermo Fisher Scientific 2016- Fluorene)

Oil compounds and chemical formula	**Log K_{ow} and Bioaccumulation potential** (Neff et al 2005)	**Water solubility (mg/L)** (Neff et al 2005)	**IARC carcinogenic group** (IARC 2019)	**Toxicity** **Toxicity to mammal (rat)** LD_{50} (rat) oral	**Ecotoxicity** **Acute aquatic toxicity** EC_{50} (algae/ daphnia); LC_{50} (fish/ crustacean, frog, worm)
					Fish, fathead minnow toxicity, LC_{50}: > 100 mg/L for 48 h; Fish, bluegill sunfish, goldfish. LC_{50}: +/-5 mg/L for 24 h; Fish, himedaka killifish, LC_{50}: 3.3 mg/L Worm, *Neanthes arenacoedentata,* LC_{50}: for 1 mg/L for 96 h (MSDS. Fisher-Scientific 1999- Fluorene)
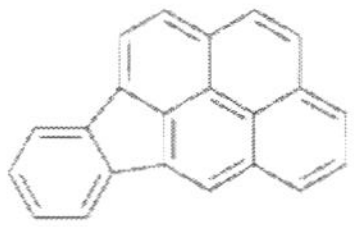 Indeno[*1,2,3,-c,d*]pyrene ($C_{22}H_{12}$)	7.0 High	0.006	2B	na	na
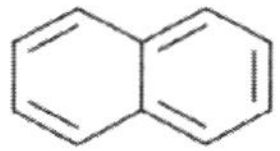 Naphthalene ($C_{10}H_8$)	3.37 High	33.72	2B	Rat, LD_{50}: 1800 mg/kg (MSDS-Chem Supply 2017- Naphthalene)	Algae, EC_{50}: 33 mg/L for 24 h Fish, pink salmon, *Oncorhynchus gorbuscha*, LC_{50}: 1.4 mg/L for 96 h Fish, Fathead minnow, *Pimephales promelas:* 7.76 mg/L for 24 h (MSDS-Chem Supply, 2017- Naphthalene)

Oil compounds and chemical formula	Log K_{ow} and Bioaccumulation potential (Neff et al 2005)	Water solubility (mg/L) (Neff et al 2005)	IARC carcinogenic group (IARC 2019)	Toxicity Toxicity to mammal (rat) LD_{50} (rat) oral	Ecotoxicity Acute aquatic toxicity EC_{50} (algae/ daphnia); LC_{50} (fish/ crustacean, frog, worm)
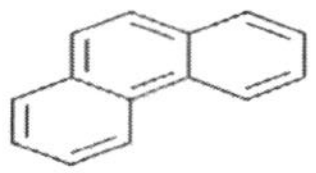 Phenanthrene ($C_{14}H_{10}$)	4.46 High	1.10	3	Rat, LD_{50}: 1800 mg/kg (MSDS-Merck 2006-. Phenanthrene)	Daphnia, *Daphnia pulex* (Water flea), EC_{50}: 0.35 mg/L for 48 h Fish, Rainbow trout, *Oncorhynchus mykiss*, LC_{50}: 3.2 mg/L, for 96 h (MSDS-Merck 2006-Phenanthrene)
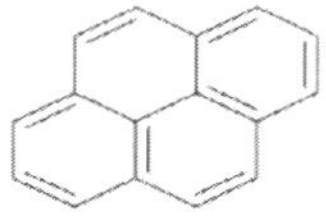 Pyrene ($C_{16}H_{10}$)	5.18 High	0.134	3	Rat, LD_{50}: 2,700 mg/kg (MSDS-Sigma-Aldrich 2006-Pyrene)	Daphnia, *Daphnia magna* (Water flea), EC_{50}: 0.002 - 0.003 mg/L for 48 h Fish, Rainbow trout, *Oncorhynchus mykiss*, LC_{50}: > 2 mg/L for 96 h (MSDS-Sigma-Aldrich 2006-Pyrene)
BTEX compounds					
 Benzene (C_6H_6)	2.13 Medium	1700*	1	Rat, LD_{50}: 930 mg/kg (MSDS-NIS 2015-Benzene)	Fish, Rainbow trout, LC_{50}: 5.3 mg/L for 96 h, flow-through at 25°C (pH 7.9-8.0) Fish: Bluegill/ Sunfish: LD_{50}: 20 mg/L for 24-48 h (MSDS Thermo Fisher Scientific 1999-Benzene)

Oil compounds and chemical formula	Log K_{ow} and Bioaccumulation potential (Neff et al 2005)	Water solubility (mg/L) (Neff et al 2005)	IARC carcinogenic group (IARC 2019)	Toxicity Toxicity to mammal (rat) LD_{50} (rat) oral	Ecotoxicity Acute aquatic toxicity EC_{50} (algae/ daphnia); LC_{50} (fish/ crustacean, frog, worm)
CH_3 Ethylbenzene (C_7H_6)	3.27 Medium	15**	2B	Rat, LD_{50}: 3500 mg/kg (MSDS-Thermo Fisher Scientific 2016-Ethylbenzene)	Algae, *Pseudokirchneriella subcapitata*, EC_{50}: 2.6 - 11.3 mg/L for 72 h, static Fish, Rainbow trout, *Oncorhynchus mykiss*, LC_{50}: 11.0 - 18.0 mg/L for 96 h, static Fish, *Oncorhynchus mykiss*, LC_{50}: 4.2 mg/L for 96 h, semi-static Fish, bluegill, *Lepomis macrochirus*, LC_{50}: 32 mg/L for 96 h, static (MSDS-Thermo Fisher Scientific, 2016-Ethylbenzene)
CH_3 Toluene	2.73 Medium	57-59 (MSDS- Lab Chem, 2014). Toluene	3	Rat, LD_{50}: 5580 mg/kg (MSDS- Lab Chem 2014- Toluene)	Fish, Coho salmon, *Oncorhynchus kisutch,* LC_{50}: 5.5 mg/L for 96 h, flow-through system, fresh water (MSDS- Lab Chem 2014- Toluene)
CH_3 CH_3 Xylene	3.1 Medium	na	3	Rat, LD_{50}: 50>2000 mg/kg (MSDS- Lab Tech 2013). Xylene)	Daphnia, *Daphnia magna*, EC_{50}: 318 mg/L for 24 h Fish, Fathead minnow, LC_{50}: 56-64 mg/L for 96h (MSDS- Lab Tech 2013- Xylene)

**https://en.wikipedia.org/wiki/Benzene; **https://en.wikipedia.org/wiki/Ethylbenzene*

According to the US Agency for Toxic Substances and Disease Registry (ATSRD) and USEPA, several oil compounds including four BTEX, 16 PAHs compounds (as listed in Table 6.9) are harmful to human health and the environment (USEPA 2014, ATSDR 2017). Humans can be exposed to PAHs via eating oil-contaminated seafood (see section 6.3). Several oil compounds are hazardous to humans (identified by the IARC) as carcinogenic (cancer-causing) such as, benzo[*a*]pyrene, benz[*a*]anthracene, benzo[*b*]fluoranthene, chrysene, dibenz[*a,h*]anthracene, naphthalene, benzene, ethylbenzene (Table 6.9). The above oil compounds fall within the following IARC carcinogenic groups:

- IARC group 1: carcinogenic to humans (e.g. benzo[*a*]pyrene, benzene; see Table 6.9).
- IARC group 2A: probably carcinogenic (e.g. dibenz[*a,h*]anthracene; see Table 6.9).
- IARC group 2B: possibly carcinogenic (e.g. benz[*a*]anthracene, benzo[*b*]fluoranthene, chrysene, indeno[*1,2,3,-c,d*]pyrene, naphthalene, ethylbenzene; see Table 6.9).
- IARC group 3: not classifiable as carcinogenic to humans (e.g. acenap-hthene, anthracene, benzo[*e*]pyrene, benzo[*g,h,i*]perylene, fluoranthene, fluorene, phenanthrene, pyrene, toluene, xylene; see Table 6.9).

Most of PAHs and BTEX compounds are very highly to moderately acute toxic to aquatic organisms (fish). Anthracene, acenaphthene, fluoranthene and fluorene are highly toxic having LC_{50} of fish <1 mg/L (range: 0.000278-0.82 mg/L). Benzo[a]pyrene, naphthalene, phenanthrene, pyrene, benzene, ethylbenzene, toluene has an LC_{50} of fish <5 mg/L (range: 1.2-5.5 mg/L; Table 6.9) (see section 6.3 for impacts of oil on social, economic and environmental sectors). Some oil compounds act as endocrine-disrupting in fish (e.g. benzo[a]pyrene and water-soluble BTEX) (Nugegoda and Kibria 2017) which has been discussed in section 6.3. The **lighter crude oils** (petrol and kerosene) contain relatively high proportions of the low molecular weight aromatic hydrocarbons that can cause acute toxic effects. The exposure of the toxic components of oil can impair various physiological functions in marine organisms (growth, reproduction, tissue damage, etc.) as discussed in section 6.3.

6.3. Impacts of oil spill and oil pollution on social, economic and environmental sectors

The impacts of spilled oil are determined by its physical and chemical characteristics (density, viscosity, evaporability), the type of oil, volume of oil spilled, the chemical composition of the oil, the ambient conditions and the sensitivity of the affected organisms. Historic oil spills indicate that some industries regularly experience losses caused by direct damage or market impairment including **a.** impacts on major social, economic and environmental sectors including fish, seafood, mariculture, fishing, tourism, wildlife, ecosystems and **b.** impacts on minor social, economic and environment sectors including human's health, industrial water intakes, salt production, and agriculture.

6.3.1 Fish toxicity, seafood security, mariculture, fisheries and fishing

6.3.1.1. Fish toxicity

The **uptake of oil** compounds (PAHs) by biota (fish) proceeds via two major pathways: directly from water and through the diet with uptake by the former route being more rapid (Law and Hellou 1999). Light oils (diesel, petrol, jet fuel, kerosene) are highly acutely toxic to animals and can cause instant fish kills (NOAA 2014). For example, fish embryos (zebra fish) exposed to crude oil (PAHs only) cause reduced larval survival, changes in heart shape, decreased swimming speeds and decreased ability to capture prey (Hicken et al 2011). In fact, oil exposure causes build-up of fluid in fish embryos (edema) (Fig. 6.6), necrosis (tissue death) (Incardona et al 2012a and 2012b, Hook et al 2016), neoplasia (tissue damage), decreased reproductive success and immune capacity (Collier et al 2013) and reduce normal development and growth in fish (Hook et al 2016). In addition, oil can kill fishes food (zooplankton) leading to an increased risk of starvation for larval fishes (Fleeger et al 2003).

Incardona et al 2012b assessed the health and viability of herring fish embryos from oiled (bunker oil) and un-oiled locations soon after the 2007 Cosco Busan oil spill in San Francisco Bay, USA. The Pacific herring (*Clupea pallasii*) embryos at oil spilled sites showed sub-lethal cardiac toxicity, high rates of tissue necrosis and other effects (body axis defect, tissue opacity, edema, arrhythmia), however, no toxicity was observed in embryos from un-oiled sites) (see Table 6.10).

Table 6.10. Incidence of abnormalities in the Pacific herring embryos from oiled and non-oiled sites in 2008 (exposed to the 2007 Cosco Busan oil spill in San Francisco Bay) (Incardona et al 2012b).

Site types	**Body axis defect**	**Tissue opacity**	**Edema** (fluid in embryo)	**Arrhythmia** (abnormal heart beat)
Oiled sites (mean of three sites)	83%	74%	18%	70%
Non-oiled site	0%	0%	1%	0%

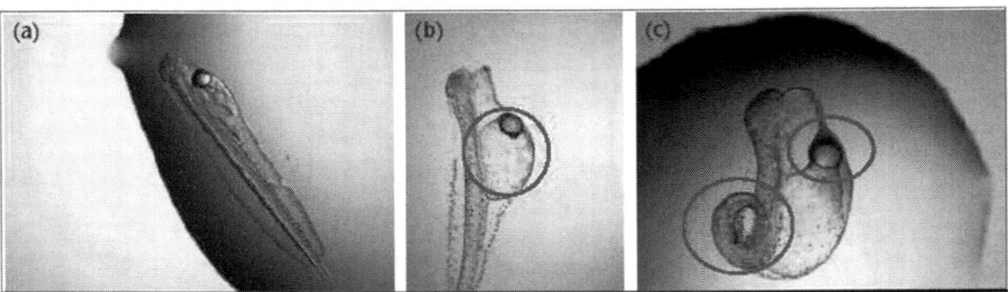

Fig. 6.6. Examples of developmental abnormalities caused by exposure to crude oil in yellowtail kingfish embryos. (**a**) Control embryo showing normal development; (**b**) oil exposed embryos showing pericardial edema; (**c**) oil exposed embryo showing both pericardial edema and spinal curvature (Hook et al 2016, page 25).

Incardona et al 2014 studied the effects of field-collected Deepwater Horizon disaster (Gulf of Mexico) oil samples on embryos of three pelagic fish: blue-fin tuna, yellow-fin tuna, and amberjack. Environmentally realistic exposures (1–15 μg/L total PAH) caused specific dose-dependent defects in cardiac function in all the three species, with circulatory disruption culminating in pericardial edema and other secondary malformations (a lack of actinotrichia - fin ray precursors), reduction in the outgrowth of the fin folds or fin fold blisters, a dorsal or upward curvature of the body axis, and a marked reduction in the growth of the eye and an irregular heart beat/abnormal heart beat. Endocrine disruption has also been demonstrated in some fish. Benzo[a]pyrene elevated testosterone caused a significant increase in plasma free T3 (hormone 3,3',5-triiodo-L-thyronine) concentration in Areolated grouper fish (*Epinephelus areolatus*) (Wu et al 2003) and impaired ovarian growth and decreased circulating sex steroids in Atlantic croaker (*Micropogonias undulatus*) (Thomas 1990; reviewed by Nugegoda and Kibria 2017). Water-soluble crude oil (BTEX) increased whole-body concentrations of T4 (hormone- 3,3',5,5'-tetraiodo-L-thyronine), increased fish larval mortalities and affected swimming activity of Turbot larvae (*Scophthalmus maximus*) (Stephens et al 1997; reviewed by Nugegoda and Kibria 2017).

6.3.1.2. Seafood security

The rapid uptake of oil and polycyclic aromatic hydrocarbons (PAHs) by exposed fish and shell-fish (Table 6.11), resulting from large oil spillages in coastal waters (as discussed in the fish toxicity section 6.3.1.1 above), poses a potential threat to consumers of fish and shell-fish (molluscs and crustaceans) and also affect the marketability of catches (Law and Hellou 1999). Seafood animals and plants (fish, prawn, lobsters, bivalves, seaweeds) may be tainted through exposure to oil droplets suspended in the water column. Oil contaminated seafood can have the taste and smell of petroleum (tainting/ off flavour). Bivalve, molluscs and filter-feeding, sedentary animals are vulnerable to tainting. Filter-feeding animals (bivalves, fish) are at a greater risk of ingesting dispersed oil droplets and oiled particles suspended in the water column. Fish and aquatic organisms with a high fat content have a greater tendency to accumulate and retain petroleum hydrocarbons in their tissue (Kibria et al 2010, Kibria et al 2013, ITOPF 2014a, Kibria 2019). Some of the fatty fishes and zooplankton include salmon *(Oncorhynchus* spp), mackerel (*Scomber scombrus*), sardines (*Sardina pilchardus*), herring (*Clupea harengus*), hilsa (*Tenualosa ilisha*), trout (*Oncorhynchus mykiss*), anchovies (*Engraulis encrasicolus*), bluefin tuna (*Thunnus thynnus*), swordfish (*Xiphias gladius*) and Antarctic zooplankton (krill) (*Euphausia superba*). Oil contains the most toxic and carcinogenic PAH compounds such as benzo[e]pyrene, which can be accumulated in seafood (Moller et al 1999, Topping et al 1997). Mussels, cockles, oysters tend to concentrate PAH in hepatopancreas, gonads, fatty tissues because of slow depuration (Moller et al 1999, Topping et al 1997, Table 6.11). Hence, tainted seafood cannot be sent to markets or sold as people may suspect that seafood may have been contaminated with toxic and carcinogenic oil compounds.

Table 6.11. PAH concentrations in shell-fish after oil spill (Topping et al 1997)

Shell-fish/organ	Contaminated shell-fish ng*/g	Uncontaminated shell-fish ng/g
Scallop muscle	223-3,580	12.8-289
Scallop gonads	90.2-20,800	25.7-2,030
Lobster meat	11.2-1,060	2.7-24.9
Crab white meat	90.2-20,800	25.7-2,030

* *ng = nanogram*

6.3.1.3. Mariculture

Caged marine animals and plants (fish, oysters, seaweeds) that are cultivated in fixed coastal locations are at greater risk since they are unable to avoid oil contamination in the surrounding water (Kibria 2019). Edible seaweeds and shell-fish are sensitive to both smothering and oil toxicity and can be coated by oil. The cultivation of seaweeds, fin-fish and marine animals (crustaceans, molluscs, echinoderms) require an intake of clean seawater. Such intake could be threatened by sunken oil or dispersed oil droplets which may cause contamination of pipes and rearing tanks resulting in the loss of cultivated stock. The presence of oil in rearing tanks also increases stress on aquacultured animals (ITOPF 2014a). Consumers can be reluctant to purchase seafood products from an oil-affected region and there could be a loss of the market confidence, the consequence of which may cause an economic loss of seafood industries (even if there is no actual contamination of the produce).

6.3.1.4. Fisheries and fishing

Fishing gear and fish cultivation (mariculture) equipment may be oiled, thus, halting fishing activities. Oil can foul boats, fishing gears and mariculture facilities (e.g. marine fish hatcheries requiring intake of large volumes of clean seawater). Floating equipment (buoys, floats, lift nets, cast nets, fixed traps/ pots, oyster racks) is particularly at risk of contamination by floating oil (ITOPF 2014a). Fishing fleets can be affected by spilled oil, either as a result of contamination of vessels and by fishing bans. A major oil spill can lead to the imposition of fishing restrictions such as harvesting bans and closures, thus impacting the local economy and livelihoods of people associated with fishing (Chang et al 2014, ITOPF 2014a). Commercial fisheries and aquaculture businesses can be affected by loss of seafood product, caused by direct mortality of fish, molluscs and prawns due to toxic effects of oil (light and heavy oils, or oil compounds - PAHs, BTEX) or habitat loss (mangroves, coral reefs, seagrass/ seaweed beds), or by loss of access to fishing grounds caused due to harvesting bans and closures.

6.3.2. Tourism

The tourism industry can be impacted by the direct impact of oil spills (U.S. Department of Commerce 1983, Oxford Economics 2010) since it can disrupt recreational activities such as swimming, boating, angling and diving (ITOPF 2014b) due to contamination of beaches and water front and loss of aesthetic value. The consequences would be losses in tourism sub-sectors such as, accommodation, transportation, guides, recreational fishing (McDowell Group 1990, Moore et al 1998), the retail and restaurant establishments (U.S. Department of Commerce 1983, Martin 1999, Loureiro et al 2005, Chang et al 2014). Oil spills may further impact sailing schools, campsites, caravan parks, aquariums, seawater swimming pools resulting in losses in sales and wages (ITOPF 2014b).

6.3.3. Wildlife

Wildlife can be harmed via exposure to oil through ingestion/ inhalation or through skin and eye irritation. Oil (heavy oil-such as bunker oils, which are black and sticky) can coat feathers and fur, preventing gas exchange, suffocating the organisms and reducing the birds' and mammals' ability to maintain their body temperature. Oil can smother/ cover small fish or invertebrates (*https://oceanservice.noaa.gov/facts/oilimpacts.html*). Benthic species such as, amphipods, filter-feeding bivalves, sea- urchins are sensitive to oil, they can be killed, and their abundance and richness can be impacted from exposure of oil (Gomez Gesteria and Dauvin 2005). Fertilisation success in sea urchins can be inhibited from exposure to oil (Rial et al 2014). Exposure to PAHs has been shown to inhibit the settlement of sponge larvae (Cebrian and Uriz 2007). The overall health, growth and survival of sea turtles can be hindered when they inhale or ingest spilled oil or are covered by oil (*https://response.restoration.noaa.gov/about/media/how-do-oil-spills-affect-sea-turtles.html*).

6.3.4. Ecosystems (corals, seagrasses, mangroves)

An oil spill can impact the biodiversity of numerous biomes (lakes, lagoons, coasts). ***Coral reefs,*** which are habitats for fish, sponges, molluscs, crustaceans can be affected by oil exposure resulting in reduced growth rates, mortality, fertilization, lipid content, survival of larvae, gonadal development and increased bleaching (Peters et al 1997, Negri and Heyward 2000, NOAA 2013). Oil can cause severe impacts on ***seagrasses*** (nursery and feeding areas of marine invertebrates, fish, sea turtles and dugongs) via decreased growth rates, coating of leaves, changes in the chlorophyll content of leaves, and reduced photosynthetic ability (Peters et al 1997). ***Mangroves*** are most sensitive to oil. Mangroves act as a nursery for fish (Peters et al 1997) and habitat for both aquatic and terrestrial organisms. The possible impact of oil spill on mangroves are **i)** covering mangrove plants and their roots (pneumatophores) with oil (smothering and coating impacts by heavier oil; Hook et al 2016), **ii)** killing mangrove plants within a few weeks to several months (Naidoo et al 2010, Lewis et al 2011); **iii)** suffocating/ coating mangrove plants (reducing O_2 flow) (Lewis et al 2011); **iv)** causing starvation, defoliation, reducing and coating pneumatophores and efficiency of photosynthesis (Proffitt et al 1997); **v)** yellowing of leaves, germination failure, and decreasing the canopy cover (Proffitt et al 1997, NOAA 2014).

6.3.5. Humans health

Trophic transfer of oil to humans: Oil can be accumulated in algae (Wolfe et al 1998), zooplankton (Almeda et al 2013), mussels (Rust et al 2004), fish (Escher and Hermens 2002) and birds (Velando et al 2010) and can be transferred from lower trophic level (algae) to higher level up the food chain (fish) to humans (Kibria 2019). *Human health and the well-being* of communities can be affected by oil spills via **i)** eating seafood contaminated with bio-accumulated oil toxins; **ii)** loss of fish stocks to fishers from oil spill; **iii)** breathing oil vapours. The direct contact with crude oil, or indirect contact through, for example, inhalation of vapours or consumption of tainted seafood can cause deleterious health effects ranging from dizziness and nausea to certain types of cancers (as a result of chronic toxicity) and central nervous system (reviewed by Change et al 2014)

6.3.6. Industrial water intakes

Seawater is used as a coolant for thermal and nuclear power stations, refineries, desalination plants, mariculture facilities, seafood processing plants. Submerged water intakes are at risk of contamination from light crude oils (which disperse easily into the water column than heavy viscous oils). Water intakes may have to be shut down as a precaution against damage to machinery following an oil spill. Oil contamination could foul the surface of the very expensive reverse osmosis (RO) systems in desalination plants (ITOPF 2014b).

6.3.7. Salt production

Communities reliant on salt production through the evaporation of seawater could be vulnerable to oil contamination. For, example, if an oil spill occurs when water is let into the salt pans (ITOPF 2014b).

6.3.8. Agriculture

Contamination of crops and farmed animals (although rare) can occur due to marine oil spills, if the spills coincide with high tides and onshore winds and water levels can rise sufficiently allowing to contaminate coastal banks where animals are grazed. Strong wind and waves can also result in oil being blown ashore in sea spray to contaminate coastal crop lands (such as rice lands in Bangladesh, Myanmar, Sri Lanka, Vietnam). If the oil spills occur in rivers (where river water is used for agriculture), rice crops and vegetables can be contaminated via irrigation of oil contaminated water (ITOPF 2014b).

6.4. Management of oil spill and oil pollution

To reduce impacts from oil spills and oil pollution, the following techniques/ methods can be applied:

- Use of **floating devices** (boomer to prevent spreading) and skimmers (to suck and remove spilled oil);
- Use of **dispersants** (to facilitate to form smaller droplets and disperse by wind, wave and currents) (NOAA 2014) (see also Table 6.12 on policies for oil spill cleanup in different countries);
- Offshore **burning** (to remove large quantity of light oil quickly, to minimize impacts on wildlife and to prevent oil from reaching sensitive near-shore habitats) (NOAA 2014);
- Use of **sorbents** (use of peat, wood, cotton, straw, corncob, shredded sugarcane, clay, wool, sand); and remove by waves and currents such as light oils) (NOAA 2014);
- Manual **oil removals** (using hand tools and manual labour to remove stranded heavier oils and limiting sediment contamination);
- **Vacuuming** (remove pooled oil or thick oil from sediment surface, depressions and channels);

- Applying **bioremediation** techniques (seeding contaminated sites with micro-organisms to degrade and metabolize oil) (NOAA 2014);
- **Monitoring** ecosystems affected by oil spill and then comparing with water quality guidelines for the protection of aquatic ecosystems and drinking water (Table 6.13);
- **Biomonitoring** human populations exposed to spilled oils, especially those individuals involved in the clean-up, in order to evaluate not only the possible immediate consequences for their health but also the medium- and long-term effects, and the effectiveness of the protective devices used (Aguilera et al 2010); and
- Where feasible, **contingency plan** should consider measures that could be used to mitigate impacts.

Table 6.12. Policies for oil spill cleanup in different countries
(Tewari and Sirvaiya 2015).

Country	Policy for clean up	Responsibility at sea	Responsibility at shore
France	Containment and recovery preferred, use of dispersants in designated areas.	Maritime perfect (Navy)	Coastal Communes
Germany	Containment and recovery preferred, use of dispersants	Federal Board of Waterways and navigation; coastal states	Coastal states
India	Skimmers, Boomers, and dispersants.	Coastal guard	Environmental Government Agencies, NGOs
Netherlands	Containment and recovery	North Sea Directorate of State waterways Board	Coastal provincial and municipal states.
Sweden	Containment and recovery preferred, dispersants used in certain permissible conditions	Coast Guard services	Municipal fire brigades; provincial authorities
United Kingdom	Aerial application of dispersants; containment and recovery where applicable	Marine Pollution Control Unit of Martine Directorate	Marine Pollution Control Unit of Martine Directorate; coastal unit

Table 6.13. Guidelines for oil compounds for the protection of aquatic ecosystems, drinking water and seafood

[na= not available; TV= trigger value; ww= weight wet].

Oil compounds	**Aquatic ecosystems protection** TV (trigger values)[†] for 99%, 95%, 90%,and 80% aquatic species protection; µg/L [(ANZECC 2000a, 2000b, Leusch and Bartkow 2010; https://environment.des.qld.gov.au/management/coal-seam-gas/pdf/btex-report.pdf].	**Drinking water** (NHMRC 2004)	**Maximum permitted levels in marine products**
i. Polycyclic aromatic hydrocarbons (PAH)			
Anthracene	TV for **fresh and marine water:** 0.01 µg/L (99% protection); 0.4 µg/L (95% protection); 1.5 µg/L (90% protection); and 7.0 µg/L (80% protection) (ANZECC 2000b).	na	na
Benzo[*a*]-pyrene	TV for **fresh and marine water**: 0.1 µg/L (99% protection); 0.2 µg/L (95% protection); 0.4 µg/L (90% protection); 0.7 µg/L (80% protection) (ANZECC 2000b).	0.00001 mg/L	<2 µg/kg ww (fish); <5 µg/kg ww (crustaceans and cephalopods); <10 µg//kg ww (shell-fish/molluscs); (IOTPF 2014a)

[†] ***TVs** are the concentration of key performance indicators measured for the ecosystem, below which there exists a low risk and adverse biological (ecological) effects. The TVs protection levels signify the percentage of species to be protected; **i.** 99% protection figure should be applied where the toxicant/ chemical has the potential to bioaccumulate, or for ecosystems which have high conservation/ ecological values (viz. Ramsar site); **ii.** 95% protection figure can be applied to slightly-moderately disturbed ecosystems and **iii.** 90% and 80% protection can be applied to highly disturbed ecosystem (ANZECC 2000a); **iv.** for the protection of aquaculture species, the recommended guidelines for oils and greases (including petrochemicals) is <300.0 µg/L (ANZECC 2000a).*

Fluoranthene	TV for **fresh and marine water**: 1.0 µg/L (99% protection); 1.4 µg/L (95% protection); 1.7 µg/L (90% protection); and 2.0 µg/L (80% protection) (ANZECC 2000b).	na	na
Naphthalene	TV for **freshwater**: 2.5 µg/L (99% protection); 16.0 µg/L (95% protection); 37.0 µg/L (90% protection); and 85.0 µg/L (80% protection) (ANZECC 2000a) TV for **marine water**: 50.0 µg/L (99% protection); 70.0 µg/L (95% protection); 90.0 µg/L (90% protection); and 120.0 µg/L (80% protection) (ANZECC 2000a).	na	na
Phenanthrene	TV for **fresh and marine water**: 0.6 µg/L (99% protection); 2.0 µg/L (95% protection); 4.0 µg/L (90% protection); and 8.0 µg/L (80% protection) (ANZECC 2000b).	na	na
ii. BTEX (benzene, toluene, ethylbenzene, xylene) compounds			
Benzene	TV for **freshwater**: 600.0 µg/L (99% protection); 950.0 µg/L (95% protection); 1300.0 µg/L (90% protection); and 2000.0 µg/L (80% protection) (ANZECC 2000a). TV for **marine water**: 500.0 µg/L (99% protection); 700 .0µg/L (95% protection); 900.0 µg/L (90% protection); and 1300.0 µg/L (80% protection) (ANZECC 2000a).	1.0 µg/L	na
Ethylbenzene	50.0 µg/L (99% protection) (Leusch and Bartkow 2010)	300.0 µg/L	na
Toluene	180.0 µg/L (99% protection) (Leusch and Bartkow 2010)	800.0 µg/L	na

Xylene o-xylene)	TV for **freshwater**: 200.0 µg/L (99% protection); 350.0 µg/L (95% protection); 470.0 µg/L (90% protection); and 640.0 µg/L (80% protection) (ANZECC 2000a)	600.0 µg/L	na
iii. Oil spill dispersants			
Corexit 9527	TV for **marine water**: 230 .0µg/L (99% protection); 1100.0 µg/L (95% protection); 2200.0 µg/L (90% protection); and 4400.0 µg/L (80% protection) (ANZECC 2000a)	na	na

6.4.1. Recovery and restoration of ecosystems after oil spills

After an oil spill at sea or river, it is essential to contain the floating oil with booms and remove the spilled oil with floating skimmers before it reaches the sensitive ecosystems. The recovered oil can be burned. The disposal of the oil would be a critical step to mitigate environmental damage if spilled oil is not removed. As part of restoration and recovery of the ecosystem, the following three steps can be undertaken (based on NOAA 2010, NOAA 2014, *https://www.gulfspillrestoration.noaa.gov/sites/default/files/wp-content/uploads/2010/10/Oil-Spill-Assessment-and-Restoration.pdf):*

i. **Preassessment** of the affected ecosystems: determine whether injury to natural resources has occurred; and determine the extent and severity of injury to the ecosystem.

ii. **Injury assessment** and restoration planning of the ecosystem: carryout some scientific and economic studies (collect data to assess and quantify the injuries and the loss of ecosystem services due to oil spill), and plan and develop a restoration plan to identify restoration projects aimed at returning the affected ecosystem to the condition it was in before the spill .

iii. **Restoration:** Examples of restoration efforts could be replanting mangroves or creating new wetland habitats, enhancing beaches and shorelines and restoring coral reefs, seagrass beds and fisheries and aquaculture habitats that have been impacted by oil spills. In some cases, natural regeneration of mangroves or wetlands can be encouraged (when clean-up would cause more harm than benefit to mangroves or other associated habitats, or when shorelines are inaccessible) (NOAA 2014). The restoration process can be implemented in cooperation and coordination with Government and private institutions and local communities for a good outcome.

6.5. Conclusion

An oil spill is an environmental disaster, resulting in pollution of the surrounding environment. It can have severe environmental impacts on marine and aquatic ecosystems. Oil compounds (PAHs, BTEX) are persistent, bio-accumulative and toxic and some are also carcinogenic and endocrine-disrupting. Light oils (gasoline/ petrol-like products) are very volatile, flammable, low viscous, spread and evaporate rapidly and cause acute toxicity to animals and plants. Heavier oils (bunker oils used by larger ships) are more viscous than light oils and are of low acute toxicity to biota but can cause long-term (chronic) effects on biota and ecosystems via smothering or coating of birds, turtles, and other animals and plants. The fate and persistence of oil spills are influenced by a range of factors, such as the type of oil (light or heavy), the environment where it spilled (open sea or enclosed inlet), and the prevailing conditions (wave, currents, winds, sunlight) at the time of spill. Oil spill can have social, economic and environmental impacts. Social and economic impacts include the recreational and professional fishing, tourism, mariculture, wildlife, ecosystems, human's health, industrial water intakes, salt production, and coastal agriculture. Oil exposure to fish cause reduced larval survival, decreased swimming speeds, edema (buildup of fluid), necrosis (tissue death), neoplasia (tissue damage), arrhythmia (abnormal heartbeat), decreased reproductive success and immune capacity, reduced growth in fish. Seafood animals (fish, prawn, lobsters, and bivalves) may be tainted through exposure to oil droplets suspended in the water column. Oil contaminated seafood can have the taste and smell of petroleum. Caged mariculture animals and plants (fish, oysters, seaweed) that are cultivated in fixed coastal locations are at greater risks since they are unable to avoid oil contamination in the surrounding water. Spilled oil and lubricants can foul boats, fishing gears and mariculture facilities. The tourism industry can be impacted by the direct impact of spills to beaches and waterfront properties and by the disruption of recreational activities (swimming, boating, angling and diving). Wildlife can be harmed via exposure to oil through ingestion/ inhalation or through the skin and eye irritation. Different ecosystems (corals, seagrasses and mangroves which are habitats for fish, sponges, molluscs, and crustaceans) are sensitive to oils. Humans can be exposed to oil from eating contaminated seafood and from breathing oil vapours. Clean seawater used by various industries (thermal and nuclear power stations, refineries, desalination plants, mariculture facilities, seafood processing plants) could be at risk if water is contaminated with oil and lubricants. Communities reliant on salt production through sun-evaporation of seawater could be vulnerable to oil contamination. Marine oil spills may contaminate coastal crops and livestock farms. To reduce impacts from oil spills and pollution, use of the following techniques/ methods can be applied (as appropriate and feasible): floating devices (boomers and skimmers), dispersants, offshore burning, sorbents, bioremediation using microbes, natural recovery, and monitoring.

6.6. References

Aguilera F, Méndez J, Pásaroa E and Laffona B 2010 Review on the effects of exposure to spilled oils on human health. *Journal of Applied Toxicology* **30** 291–301

Almeda R, Wambaugh Z, Wang Z et al 2013 Interactions between zooplankton and crude oil: toxic effects and bioaccumulation of polycyclic aromatic hydrocarbons. *PLoS One* **8** e67212. *doi:10.1371/journal.pone.0067212*

ANZECC and ARMCANZ 2000a Australian and New Zealand Environment and Conservation Council (ANZECC) and Agriculture and Resource Management Council of Australia and New Zealand (ARMCANZ). Australian and New Zealand Guidelines for Fresh and Marine Water Quality, Volume 1. The Guidelines Chapters 1-7. *http://www.waterquality.gov.au/anz-guidelines/resources/previous-guidelines/anzecc-armcanz-2000*

ANZECC and ARMCANZ 2000b Australian and New Zealand Environment and Conservation Council (ANZECC) and Agriculture and Resource Management

Council of Australia and New Zealand (ARMCANZ). Australian and New Zealand Guidelines for Fresh and Marine Water Quality, Volume 2. The Guidelines Chapter 8. Aquatic Ecosystems — Rationale and Background Information (Chapter 8). *http://www.waterquality.gov.au/anz-guidelines/Documents/ANZECC-ARMCANZ-2000-guidelines-vol2.pdf*

ATSDR (Agency for Toxic Substances and Disease Registry) 2017 ATSDR's Substance Priority List. *https://www.atsdr.cdc.gov/spl/index.html#2017spl*

Cebrian E and Uriz MJ 2007 Contrasting effects of heavy metals and hydrocarbons on larval settlement and juvenile survival in sponges. *Aquatic Toxicology* **81** 137–145 *doi:10.1016/j.aquatox.2006.11.010*

Chang SE, Stone J. Demes K and Piscitelli M 2014 Consequences of oil spills: a review and framework for informing planning. *Ecology and Society* **19(2)** 26 *http://dx.doi.org/10.5751/ES-06406-190226*

CNN 2019 Oil spills fast facts. CNN Library. Updated 1705 GMT (0105 HKT) April 22, 2019. *https://edition.cnn.com/2013/07/13/world/oil-spills-fast-facts/index.html*

Collier TK, Anulacion BF, Arkoosh MR et al 2013 Effects on fish of polycyclic aromatic hydrocarbons (PAHs) and naphthenic acid exposures. In: Organic Chemical Toxicology of Fishes. (Tierney, KB ed.) pp. 195–255. Elsevier, London, UK.

Escher BI and Hermens JLM 2002 Modes of action in ecotoxicology: their roles in body burdens, species sensitivity, QSARs, and mixture effects. *Environmental Science and Technology* **36** 4201–4217 *doi:10.1021/es015848h*

Fayemiwo OM, Daramola MO and Moothi K 2017 BTEX compounds in water – future trends and directions for water Treatment. *http://dx.doi.org/10.4314/wsa.v43i4.08;* Available on website *http://www.wrc.org.za*; ISSN 1816-7950 (Online) = Water SA Vol. 43 No. 4 October 2017.

Fingas M 2018 The challenges of remotely measuring oil slick thickness. *Remote Sensing* **10** 319 *doi:10.3390/rs10020319*

Fleeger JW, Carman KR and Nisbet RM 2003 Indirect effects of contaminants in aquatic ecosystems. *Science of the Total Environment* **317(1-3)** 207-233

Gomez Gesteria JL and Dauvin J-C 2005 Impact of the Aegean sea oil spill on the subtidal fine sand macrobenthic community of the Ares-Betanzos Ria (Northwest Spain). *Marine Environmental Research* **60** 289–316 *doi:10.1016/j.marenvres.2004.11.001*

Hicken CE, Linbo TL, Baldwin DH et al 2011 Sub-lethal exposure to crude oil during embryonic development alters cardiac morphology and reduces aerobic capacity in adult fish. *Proceedings of the National Academy of Sciences of the United States of America* **108** 7086–7090 *doi:10.1073/pnas.1019031108*

Hook S, Batley G, Holloway M, Irving P and Ross A 2016 Oil spill monitoring handbook. In: Sharon Hook, Graeme Batley, Michael Holloway, Paul Irving and Andrew Ross (eds.). CSIRO Publishing Locked Bag 10, Clayton South VIC 3169, Australia. 9781486306367 (epub)

IARC (International Agency for Research on Cancer) 2019 Agents Classified by the IARC Monographs, Volumes 1–123. *https://monographs.iarc.fr/wp-content/uploads/2018/09/ClassificationsAlphaOrder.pdf*

Incardona JP, Vines CA, Linbo TL et al 2012a Potent phototoxicity of marine bunker oil to translucent herring embryos after prolonged weathering. *PLoS One* **7** e30116 *doi:10.1371/journal.pone.0030116*

Incardona JP, Vines CA, Anulaciona BF, Baldwina DH et al 2012b Unexpectedly high mortality in Pacific herring embryos exposed to the 2007 Cosco Busan oil spill in San Francisco Bay. *Proceedings of the National Academy of Science of the USA* **109(2)** E51–E58 *doi: 10.1073/pnas.1108884109. Epub 2011 Dec 27*

Incardona JP, Gardner LD, Linbo TL et al 2014 Deepwater Horizon crude oil impacts the developing hearts of large predatory pelagic fish. *PNAS* **111(15)** E1510-E1518 *https://doi.org/10.1073/pnas.1320950111*

ITOPF (International Tanker Owners Pollution Federation Ltd) 2012 International Tanker Owners Pollution Federation Technical Information Papers. ITOPF, London, UK, *www.itopf.com*

ITOPF (International Tanker Owners Pollution Federation Ltd) 2001 Persistent vs non-persistent oils: What you need to know. The International Tanker Owners Pollution Federation Ltd. 2 p.

https://www.itopf.org/fileadmin/data/Documents/Papers/persistent.pdf

ITOPF (International Tanker Owners Pollution Federation Ltd) 2012 International Tanker Owners Pollution Federation Technical Information Papers. ITOPF, London, UK, *www.itopf.com*

ITOPF (International Tanker Owners Pollution Federation Ltd) 2014a Effects of oil pollution on fisheries and mariculture. Technical Information paper 11. *https://www.itopf.org/knowledge-resources/documents-guides/document/tip-11-effects-of-oil-pollution-on-fisheries-and-mariculture/*

ITOPF (International Tanker Owners Pollution Federation Ltd) 2014b Technical Information Paper (TIP) 12: Effects of oil pollution on social and economic activities. *https://www.itopf.org/knowledge-resources/documents-guides/document/tip-12-effects-of-oil-pollution-on-social-and-economic-activities/*

Kibria G 2018 Presentation: Oil spill and oil pollution- chemistry, ecotoxicology, impact on biodiversity and ecosystems and management. *DOI: 10.13140/RG.2.2.14293.06885/1*

https://www.researchgate.net/publication/329074954_Presentation_Oil_Spill_Oil_Pollution-

Kibria, G. 2019. Oil spill and oil Pollution- A threat to wildlife, fisheries, seafood security and ecosystems. 5 p.

https://www.researchgate.net/publication/333091943_Oil_Spill_and_Oil_Pollution-_A_Threat_to_Wildlife_Fisheries_Seafood_Security_and_Ecosystems

Kibria G, Haroon AKY, Nugegoda D and Rose G 2010 Chemicals and Climate Change: Environmental and Biological Aspects. NIPA, India, 460 p. ISBN: 93-80235-30-1. *https://www.researchgate.net/publication/261216635_Climate_Change_and_Chemicals_Environmental_and_Biological_Aspects*

Kibria G, Haroon AKY and Nugegoda D 2013 Agriculture and Climate Change: Impacts, vulnerabilities and Remedies. NIPA, India, 300 p. ISBN 97-89381-45051-2. *https://www.researchgate.net/publication/261178054_Climate_Change_and_Agricultural_Food_Production_Impacts_Vulnerabilities_and_Remedies*

Killops SD and Killops VJ. 1993 An Introduction to Organic Geochemistry. Longman Scientific and Technical, New York, USA.

Law RJ and Hellou J 1999 Contamination of fish and shellfish following oil spill incidents. Environmental Geosciences **6(2)** 90-98

https://www.researchgate.net/profile/Jocelyne_Hellou/publication/227772914_Contamination_of_Fish_and_Shellfish_Following_Oil_Spill_Incidents/links/577e539208aeaee3b2834c5a.pdf

Leusch F and Bartkow M 2010 A short primer on benzene, toluene, ethylbenzene and xylenes (BTEX) in the environment and in hydraulic fracturing fluids, Griffith University – Smart Water Research Centre 17 Nov 2010. *https://environment.des.qld.gov.au/management/coal-seam-gas/pdf/btex-report.pdf*

Lewis M, Pryor R and Wilking L 2011 Fate and effects of anthropogenic chemicals in mangrove ecosystems: a review. *Environmental Pollution* **159** 2328–*2346 doi:10.1016/j.envpol.2011.04.027*

Loureiro ML, Ribas A, López E and Ojea E 2005 Estimated costs and admissible claims linked to the Prestige oil spill. *Ecological Economics* **59** 48-63 *http://dx.doi.org/10.1016/j.ecolecon.2005.10.001*

Martin G 1999 Valdez spill leaves bitter residue: oil is gone after 10 years, but ecological, economic fallout continues. San Francisco Chronicle, San Francisco, California, USA.

http://www.sfgate.com/news/article/Valdez-Spill-Leaves-Bitter-Residue-Oil-is-gone-2940197.php

McDowell Group 1990 An assessment of the impact of the Exxon Valdez oil spill on the Alaska tourism industry. McDowell Group, Juneau, Alaska, USA. *http://www.evostc.state.ak. us/Universal/Documents/Publications/Economic/Econ_Tourism.*

McKim JM 1994 Physiological and biochemical mechanisms that regulate the accumulation and toxicity of environmental chemicals in fish. In: J. L. Hamelink, P. F. Landrum, H. L. Bergman, and W. H. Benson (Eds.), Bioavailability: Physical, biological and chemical interactions (pp. 179–201). Boca Raton, FL: CRC Press

Moller TH, Dicks B, Whittle K and Girin M. 1999 Fishing and harvesting bans in oil spill response. In: Proceedings of the 1999 International Oil Spill Conference, pp. 693-699

Moore LY, Footitt AJ, Reynolds LM et al 1998 Sea Empress cost-benefit project - final report research and development. Environment Agency, Bristol, UK. *http://www.rpaltd.co.uk/documents/J200-SeaEmpress.pdf*

MSDS-Chem Supply 2017. Naphthalene. 7 p.

https://www.chemsupply.com.au/documents/NL0171CH91.pdf

MSDS-Fisher-Scientific 1999 Fluorene (revised 2008).

https://fscimage.fishersci.com/msds/96869.htm

MSDS-Lab Tech 2013 Xylene. 8 p. *https://www.labtech.com.au/msds/17-xylene/file*

MSDS-Lab Chem 2014 Toluene (revised February 2018).

http://www.labchem.com/tools/msds/msds/LC26170.pdf

MSDS-Merck 2006 Phenanthrene. 11 p, (revised June 2019)

http://www.merckmillipore.com/AU/en/product/msds/MDA_CHEM-822122

MSDS-Merck 2006 Chrysene 11 p, (revised September 2018) *http://www.merckmillipore.com/AU/en/product/msds/MDA_CHEM-841690?ReferrerURL=https%3A%2F%2Fwww.google.com%2F&bd=1*

MSDS-NIS 2015 Benzene. 16 p (revised March 2016) *https://www.nis.eu/en/wp-content/uploads/sites/3/2016/07/EN_SDS_Benzene___SDS-GHS-ANNEXII-CLP_2016311_10_NIS-a.d.-Novi-Sad____.pdf*

MSDS-PubChem 2004 Acenaphthylene. (revised, June 2019) *https://pubchem.ncbi.nlm.nih.gov/compound/acenaphthylene*

MSDS-PubChem 2004 Benzo[a]pyrene. *https://pubchem.ncbi.nlm.nih.gov/compound/benzo_a_pyrene*

https://pubchem.ncbi.nlm.nih.gov/compound/benzo_a_pyrene#section=Exposure-Routes;

http://datasheets.scbt.com/sc-257130.pdf

MSDS-Sigma-Aldrich 2006 Pyrene. 11p. (revised February 2019) *file:///C:/Users/User1/Downloads/821051_SDS_AU_EN%20(1).PDF*

MSDS-Thermo Fisher Scientific 1999 Benzene (revised 2008) *https://fscimage.fishersci.com/msds/02610.htm*

MSDS-Thermo Fisher Scientific 2010 Acenaphthene. 7p. (revised January 2018) *https://www.fishersci.com/store/msds?partNumber=AC201340050&productDescription=ACENAPHTHENE%2C+99%25+5GR&vendorId=VN00032119&countryCode=US&language=en*

MSDS-Thermo Fisher Scientific 2016 Ethylbenzene. 8 p. (revised January 2018) *https://www.fishersci.com/store/msds?partNumber=O27511&productDescription=ETHYLBENZENE+CERT+1L&vendorId=VN00033897&countryCode=US&language=en*

MSDS-Thermo Fisher Scientific 2016 Fluorene. 7 p. (revised January 2018) *https://www.fishersci.com/store/msds?partNumber=AC156130250&productDescription=FLUORENE+98%25+25GR&vendorId=VN00032119&countryCode=US&language=en*

MSDS-Thermo Fisher Scientific 2018 Anthracene. 7 p. (revised January 2018) *https://www.fishersci.com/store/msds?partNumber=AC104860050&productDescription=ANTHRACENE%2C+99%25+5GR&vendorId=VN00032119&countryCode=US&language=en (accessed 6 June 2019)*

MSDS-Thermo Fisher Scientific 2018 Benzo[a]anthracene. 7 p. *https://www.fishersci.com/store/msds?partNumber=AC105250010&productDescription=1%2C2-BENZANTHRACENE%2C+99%25+1GR&vendorId=VN00032119&countryCode=US&language=en*

MSDS-Thermo Fisher Scientific 2019 Fluoranthene. 7 p *https://www.fishersci.com/shop/msdsproxy?productName=AC119175000&productDescription=*

Naidoo G and Naidoo Y and Achar P 2010 Responses of the mangroves *Avicennia marina* and *Bruguiera gymnorrhiza* to oil contamination. *Flora* **205** 357–362 *doi:10.1016/j.flora.2009.12.033*

Neff JM, Stout SA, Donald G et al 2005 Ecological Risk Assessment of Polycyclic Aromatic Hydrocarbons in Sediments: Identifying Sources and Ecological Hazard. *Integrated Environmental Assessment and Management* **1(1)** 22–33 22 *https://doi.org/10.1897/IEAM_2004a-016.1*

Negri AP and Heyward AJ 2000 Inhibition of fertilization and larval metamorphosis of the coral Acropora millepora (Ehrenberg, 1834) by petroleum products. *Marine Pollution Bulletin* **41** 420–427 *doi:10.1016/S0025-326X(00)00139-9*

NHMRC 2004 Australian drinking water guidelines. National Health and Medical Research Council and Natural Resource Management Ministerial Council, Canberra, Australia. *https://www.nhmrc.gov.au/about-us/publications/australian-drinking-water-guidelines*

NOAA 2010 Oil spills: assessment and restoration.

https://www.gulfspillrestoration.noaa.gov/sites/default/files/wp-content/uploads/2010/10/Oil-Spill-Assessment-and-Restoration.pdf

NOAA 2013 How do oil spills affect coral reefs?

https://response.restoration.noaa.gov/about/media/how-do-oil-spills-affect-coral-reefs.html

NOAA 2014 Oil Spills in Mangroves. planning and response considerations. U.S. Department of Commerce. D. Rebecca Hoff and Jacqueline Michel (Editors). *https://response.restoration.noaa.gov/sites/default/files/Oil_Spill_Mangrove.pdf*

NRC 2003 Oil in the Sea III: Inputs, Fates, and Effects. Committee on Oil in the Sea; Inputs, Fates and Effects and Ocean Studies Board and Marine Board, Divisions of Earth and Life Studies and Transportation Research Board. National Research Council. The National Academies Press, Washington, DC, USA. *https://doi.org/10.17226/10388*

Nugegoda D and Kibria G 2017 Effects of environmental chemicals on fish thyroid function: Implications for fisheries and aquaculture in Australia. *General and Comparative Endocrinology* **244** 40-53 *doi: 10.1016/j.ygcen.2016.02.021. Epub 2016 Feb 2*

Oxford Economics 2010 Potential impact of the Gulf oil spill on tourism. Oxford Economics, London, UK.

http://www.mississippiriverdelta.org/blog/files/2010/10/Gulf_Oil_Spill_Analysis_Oxford_Economics_710.pdf

Peters EC, Gassman NJ, Firman et al 1997 Ecotoxicology of tropical marine ecosystems. Environmental Toxicology and Chemistry 16 12–40 *doi:10.1002/etc.5620160103*

Proffitt CE (ed.) 1997 Managing Oil spills in mangrove ecosystems: effects, remediation, restoration, and modeling. OCS Study MMS 97-0003. New Orleans: U.S. Department of the Interior, Minerals Management Service, Gulf of Mexico OCS Region. 76 p.

Rial D, Vazquez JA and Murado MA 2014 Toxicity of spill-treating agents and oil to sea urchin embryos. *The Science of the Total Environment* **472** 302–308 *doi:10.1016/j.scitotenv.2013. 11.030*

Rust AJ, Burgess RM, McElroy AE, Cantwell MG and Brownawell BJ 2004 Role of source matrix in the bioavailability of polycyclic aromatic hydrocarbons to deposit-feeding benthic invertebrates. *Environmental Toxicology and Chemistry* **23** 2604–2610 *doi:10.1897/03-353*

SINTEF 2011 Weathering Properties of Bream Crude Oil Related to Oil Spill Response. SINTEF, Trondheim, Norway, *http://www.sintef.com/home/topics/climate-and-environment/environment/#/>.*

Stephens SM, Alkind AYA, Waring CP and Brown J 1997 Corticosteroid and thyroid responses of larval and juvenile turbot exposed to the water-soluble fraction of crude oil. *Journal of Fish Biology* **50** 953–964

Tewari S and Sirvaiya A 2015 Oil spill remediation and its regulation. International Journal of Research In Science & Engineering **1(6)** e-ISSN: 2394-8299 IJRISE| *www.ijrise.org|editor@ijrise.org [01-07]*

Thomas P 1990 Teleost model for studying the effects of chemicals on the female reproductive endocrine function. *Journal of Experimental Zoology* **Suppl 4** 126–128

Topping G , Davies JM, Mackie PR and Moffat CF 1997 The impact of the Braer spill on commercial fish and shell-fish. In: J.M. Davies and G. Topping (Eds.). The impact of an oil spill in turbulent waters: The Braer (pp. 121–143). Edinburgh: The Stationery Office

USEPA (United States Environmental Protection Agency) 2014 Priority Pollutant List *https://www.epa.gov/sites/production/files/2015-09/documents/priority-pollutant-list-epa.pdf*

U.S. Department of Commerce 1983 Assessing the social costs of oil spills: the Amoco Cadiz case study. U.S. Department of Commerce, Washington, D.C., USA *https://ia600600.us.archive.org/15/items/assessingsocialc00unit/assessingsocialc00unit.pdf*

Velando A, Munilla I, Lopez-Alonso M, Freire J and Perez C 2010 EROD activity and stable isotopes in seabirds to disentangle marine food web contamination after the Prestige oil spill. *Environmental Pollution* **158** 1275–1280 *doi:10.1016/j.envpol.2010.01.029*

Wolfe MF, Schlosser JA, Schwartz GFB et al 1998 Influence of dispersants on the bioavailability and trophic transfer of petroleum hydrocarbons to primary levels of a marine food chain. *Aquatic Toxicology* **42** 211–227 *fdoi:10.1016/S0166-445X(97)00096-9*

Wu RSS, Pollino CA, Au DWT et al 2003 Evaluation of biomarkers of exposure and effect in juvenile areolated grouper (*Epinephelus areolatus*) on food borne exposure to benzo[a]pyrene. *Environmental Toxicology and Chemistry* **22** 1568–1573 *http://dx.doi.org/10.1002/etc.5620220720*

■■■■

7

Pesticides Risks, Impacts and Management

Summary

Pesticides are chemicals that are broadly used to protect crops, livestock, animals and plants from pests and diseases. They are classified according to the type of pest to be controlled (algicides, herbicides, fungicides and insecticides). Pesticides are persistent, toxic and bioaccumulative but some of them are also endocrine-disrupting as well as carcinogenic. About 109 pesticides have been identified as endocrine disruptors. Pesticide residues may enter the environment via spray drift, vaporisation, surface run-off, unlawful acts, spills and drainage discharges, and through leaching or soil dust. China, USA, Argentina, Thailand, Brazil are the top pesticide consumers in the world. This chapter reviews: **i.** chemistry, toxicity and ecotoxicity of pesticides including persistence, bioaccumulation potential, the toxicity of mammals (LD_{50}) and ecotoxicity (EC_{50}, LC_{50}) of birds, bees, algae, daphnia and fish; **ii.** impacts of pesticides on social, economic and environmental sectors including residues in fruit, vegetables, seafood/ fish; pesticide residues in the surface, groundwater and sediments; and pesticide residues in humans (women) breast milk; **iii.** impacts of pesticides on non-target organisms including birds, earthworms, fish, frogs, bees; **iv.** impacts of pesticides on human's health; **v.** three case studies related to glyphosate threat to human health, insecticides in global waterways and the threat of neonicotinoid pesticide to bees/pollinators; **vi**. linkages between pesticides and climate change and **vii.** management of pesticides risk.

Contents

List of Tables

List of Figures

List of Boxes

7.1. Introduction

Pesticides are chemical substances or mixtures of substances that are mainly used in agriculture to protect crops, livestock (as ectoparasiticides)[1], other animals and plants from pests/insects, weeds and diseases and to protect humans from vector[2]-borne diseases, such as malaria, dengue fever, and schistosomiasis[3]. Green revolution era experienced maximum use of various kinds of pesticides to

[1]***Ectoparasiticide*** = *an antiparasitic drug used in the treatment of ectoparasitic infestations. These drugs are used to kill the parasites that live on the body surface.*

[2]***Vector*** = *an organism, typically a biting insect or tick, that transmits a disease or parasite from one animal or plant to another.*

[3]***Schistosomiasis****, also known as snail fever and bilharzia, is a disease caused by parasitic flatworms called schistosomes.*

increase the production of food crops. They are also used for improvement and maintenance of public urban green areas and sports fields, lakes (to control algae). They are often used for less known applications such as in producing pet shampoos, building materials, and coating boat bottoms (for boat antifouling) in order to eliminate or prevent the growth of unwanted organisms (Kibria et al 2010, Nicolopoulou-Stamatiet al 2016). Pesticides include insecticides, herbicides, fungicides (Fig. 7.1), rodenticides, molluscicides, acaricides, nematicides, plant growth regulator. Herbicides are the most common pesticides that are used globally (glyphosate has the highest global production in volume of all the herbicides) (*https://foodprint.org/issues/pesticides/?cid=263*).

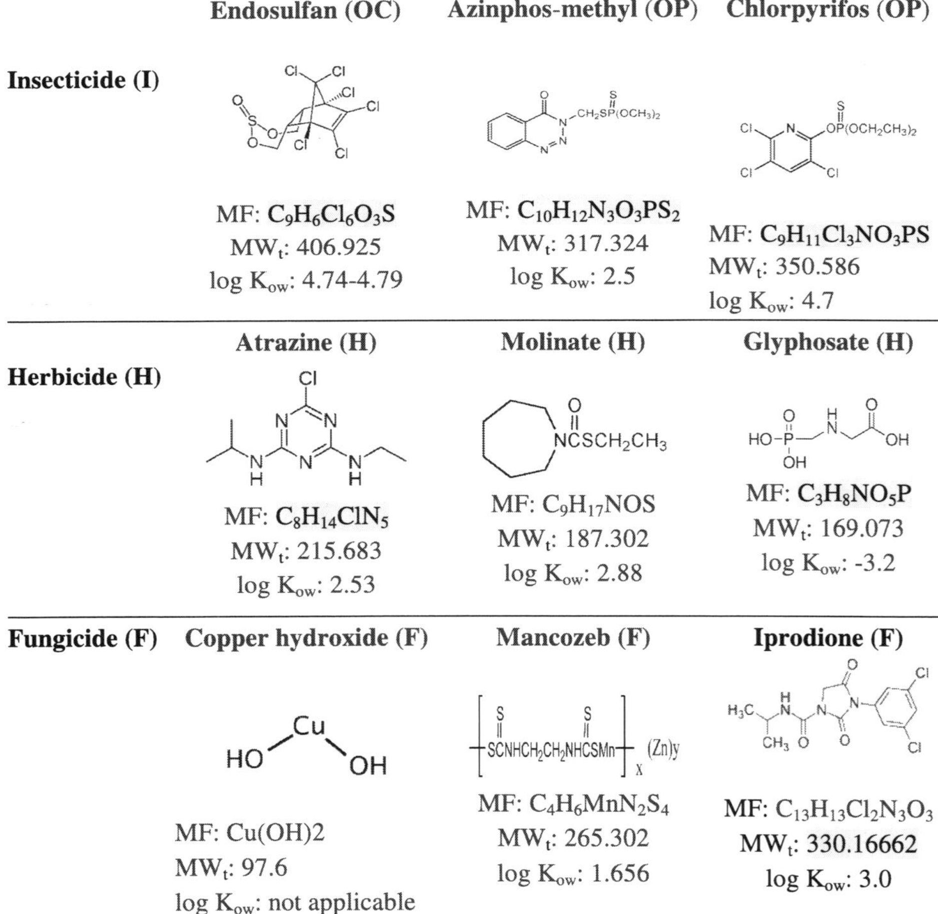

Fig. 7.1. Chemical structure of some common pesticides
[Structures are based on https://www.google.com/search; https://en.wikipedia.org/wiki/; log K_{ow} = Octanol/ water partition coefficient; MF = molecular formula; MW_t = molecular weight; OC = Organochlorine; OP = organophosphate].

Pesticides are made up of active ingredients (which damage the pest) and sometimes with inert ingredients (solvent in the solution). All pesticides must be registered with the appropriate authority (e.g. USEPA-USA or APVMA[4]-Australia) before selling/distribution. Many tests are required before the pesticides can be approved (e.g. assessing effects via lethal dose or lethal concentration- LD_{50}, LC_{50}, EC_{50}; see Table 7.4) to establish "tolerances" or maximum residue limits (**MRL**)[5] (the amount of a given pesticide that can safely remain in or on a food). Pesticide residues may enter the environment via spray drift, vaporisation, surface run-off, unlawful acts, spills and drainage discharges, and through leaching or soil dust. The appropriate Government authority is responsible for monitoring pesticide residue levels on fruits, vegetables, eggs, meat, dairy products, seafood, water, soil, and human breast milk.

7.1.1. Classification of pesticides

Pesticides are classified according to the type of pest to be controlled (herbicides, fungicides and insecticides) (Table 7.1). They are also classified based on their chemical classes (Table 7.2) and mode of action. Biological or biopesticides include micro-organisms (bacteria, fungi, viruses and protozoa) that are effective in controlling target pests. These agents usually do not have toxic effects on animals and humans and do not leave toxic or persistent chemical residues in the environment. Broad-spectrum pesticides are those that kill an array of species, while narrow-spectrum or selective pesticides only kill a small group of species. Pesticides are also classified into systemic and non-systemic pesticides. ***Systemic pesticides*** (water-soluble) penetrate plant tissues and are translocated[6] to other parts of plant via diffusion through the cuticle[7] of leaves or through the outer layer of roots (soil applications) into the epidermal cells, into either the xylem or phloem. ***Non-systemic*** *or contact pesticides* have little or no ability to penetrate plant tissues (leaves or roots) and translocate from the site of contact to distal parts of the plant (Kibria et al 2010).

---

[4] ***APVMA*** = *Australian Pesticides and Veterinary Medicines Authority.*

[5] ***Maximum residue limit (MRL)*** = *A maximum residue limit (MRL) is the highest amount of an agricultural or veterinary (agro-vet) chemical residue that is legally allowed in a food product sold in Australia whether it is produced domestically or imported. MRLs help enforcement agencies to monitor whether an agro-vet chemical has been used as directed to control pests and diseases in food production (https://www.foodstandards.gov.au/consumer/chemicals/maxresidue/pages/default.aspx)* or *MRL is the maximum concentration of a pesticide/ chemical residue that is legally permitted or recognised as acceptable in or on a food, agriculture commodity or animal feed and are set at levels that pose no risks to human health.*

[6] ***Translocation*** = *Transport (of a dissolved substance) within an organism, especially in the phloem of a plant, or actively across a cell membrane.*

[7] ***Cuticle*** = *In plants, the cuticle is the waxy covering on the surface of many of plant organs, e.g. leaves and young shoots, This waxy, hydrophobic covering protects the plant by minimizing water loss and curbing pathogen entry; https://www.biologyonline.com/dictionary/cuticle*

Table 7.1. Major classification of pesticides.

Pesticides	Purpose of use
Algicides or Algaecides	Control of algae
Avicides	Control of birds
Bactericides	Control of bacteria
Fungicides	Control of fungi and oomycetes
Herbicides	Control of weeds
Insecticides	Control of insects including the killing of ovicides (eggs), larvicides (larvae) or adulticides (adult insects)
Miticides or Acaricides	Control of mites
Molluscicides	Control of slugs and snails
Nematicides	Control of nematodes
Rodenticides	Control of rodents
Virucides	Control of viruses

Table 7.2. Examples of chemical classes of pesticides.

[Tomlin 2000 and 2006, note: Carbamates, Pyrethroids and Neonicotinoids are new generation pesticides; *https://www.atsdr.cdc.gov/substances/toxchemicallisting.asp?sysid=39; https://www.ncbi.nlm.nih.gov/pmc/articles/PMC5464684/pdf/ITX-9-90.pdf; https://www.cdc.gov/nceh/clusters/fallon/orgranophosfaq.htm)*].

Pesticides group	Examples ***Symbols used in column 2*** [a] ***= Australia*** *(pesticides that were detected in different environmental compartments in Australia) (based on Table 7.8);* [b] ***= Banned*** *(pesticides those are globally banned*[8]*) (UTZ 2015);* [p] ***=POPs*** *(pesticides which are POPs)*[9] *(Wikoff et al 2012);* [w]***= Waters*** *(pesticides which are a threat to global surface waters) (Stehle and Schulz 2016).*

[8]***Reason for banning*** *of individual pesticides are either international convention or due to effects related to acute toxicity or carcinogenicity or mutagenicity or reproductive toxicity or endocrine disruption (UTZ 2015).*

[9]***POPs*** *= Persistent organic pollutants (see Chapter 5).*

Pesticides group	Examples ***Symbols used in column 2*** [a] = ***Australia*** *(pesticides that were detected in different environmental compartments in Australia) (based on Table 7.8);* [b] = ***Banned*** *(pesticides those are globally banned*[8]*) (UTZ 2015);* [p] =***POPs*** *(pesticides which are POPs)*[9] *(Wikoff et al 2012);* [w]= ***Waters*** *(pesticides which are a threat to global surface waters) (Stehle and Schulz 2016).*
i. Insecticides	
Amidine	[b]Amitraz
Bacterium *(microbial pesticides)*[10]	*Bacillus sphaericus, B. subtilis, B. thuringiensis (Bt)*[11], *Pseudomonas* spp., *Streptomyces griseoviridis*
Carbamates[12]	[b]Aldicarb, Asulam, Bendiocarb, Butocarboxim, [b]Butoxycarboxim, [a,b]Carbaryl, Carbetamide, [b,w]Carbofuran, [b]Carbosulfan, [a]Chlorpropham, [b]Ethiofencarb, Fenoxycarb, [b]Formetanate, [b]Furathiocarb, Isoprocarb, [a,b]Methiocarb, [b]Methomyl, [b]Oxamyl, [a,b]Pirimicarb, Propamocarb, Propham, Propoxur, [b]Thiofanox, [b]Thiodicarb
Carbamoyltriazole	[b]Triazamate
Diacylhydrazine	Chromafenozide, Halofenozide, Methoxyfenozide, [a]Tebufenozide
Neonicotinoids[13]	[w]Acetamiprid, Clothianidin, [a,w]Imidacloprid, Nitenpyram, [w]Thiacloprid, [w]Thiamethoxam
Organochl-orines[14]	[a,p]Aldrin, [a,b]BHC (Benzene hexachloride), [b,p]HCH (Hexachlorocyclohexane), [a,b,p]Chlordane, [b]Dicofol, [a,b,p]DDT

[10]***Microbial pesticides*** *consist of a microorganism (e.g. a bacterium, fungus, virus or protozoan) as the active ingredient.*

[11]***Bacillus thuringiensis (Bt)*** *is a gram-positive, soil-dwelling bacterium, commonly used as a biological pesticide. Bt is toxic to caterpillars, some fly larvae, and some beetle larvae. Bt is a microbial biopesticide that is very specific to certain insects. It causes insects to stop feeding in a few hours and usually kills insects in a few days (https://www.lsuagcenter.com/articles/page1508333080784).*

[12]***Carbamates*** *are organic compounds derived from carbamic acid (NH_2COOH). These insecticides kill insects by reversibly inactivating the enzyme acetylcholinesterase like organophosphates.*

[13]***Neonicotinoids (NEs)*** *= An agricultural insecticide resembling nicotine. NEs (e.g., imidacloprid, acetamiprid, thiacloprid, and clothianidin) are the most recently introduced class of insecticides and have become the most widely used insecticides worldwide. They are broad spectrum insecticides used for urban pest control, in veterinary products, and for pest control in fish farming. NEs are systemic in that they translocate to all parts of plants onto which they are applied; https://www.sciencedirect.com/topics/pharmacology-toxicology-and-pharmaceutical-science/thiacloprid*

[14] ***Organochlorines*** *are chlorinated organic compounds belonging to the class of persistent organic pollutants (**POPs**) i.e. they are persistent, bioaccumulative and toxic (see **Chapter 5**).*

Pesticides group	Examples ***Symbols used in column 2*** [a] ***= Australia*** *(pesticides that were detected in different environmental compartments in Australia) (based on Table 7.8);* [b] ***= Banned*** *(pesticides those are globally banned[8]) (UTZ 2015);* [p] ***=POPs*** *(pesticides which are POPs)[9] (Wikoff et al 2012);* [w] ***= Waters*** ***(****pesticides which are a threat to global surface waters) (Stehle and Schulz 2016).*
	(Dichlorodiphenyltrichloroethane), [a,b,p]Dieldrin, [a,b,p] Endrin,[a,p,w]Endosulfan, [b]Epichlorohydrin, [a,b,p]Heptachlor, [b,p]Lindane, Methoxychlor, [b,p]Mirex, [a]Oxychlordane, [b,p]Pentachlorophenol, [b,p]Toxaphene
Organophos-phates[15]	[b]Acephate, Azamethiphos, [a,b]Azinphos-ethyl, [a,b,w]Azinphos-methyl, [b]Cadusafos, [b] Chlorethoxyfos, [b]Chlorfenvinphos, [b]Chlormephos, [a,w]Chlorpyrifos, [b]Coumaphos, Cyanophos, [a,w]Diazinon, [b]Demeton-S-methyl, Diazinon,[b]Dichlorvos, [b]Dicrotophos, [a]Dimethoate, Dimethylvinphos, [b]Disulfoton, [b]EPN[16], Ethion, [b]Ethoprophos, [b]Famphur, [a,b]Fenamiphos, Fenitrothion, [b]Fenthion, [b]Heptenophos, Isofenfos, [b]Isoxathion, Mecarbam, [a,w]Malathion, [b]Methamidophos, [b]Methidathion, [b]Mevinphos, [b]Monocrotophos, [b]Omethoate, [b]Oxydemeton-methyl, [w]Parathion-ethyl, [a,b,w]Parathion-methyl, Phenothoate, [b]Phorate, [b]Phosalone, Phosmet, [b]Phosphamidon, Phoxim, Primiphos-methyl, Profenopos, Propaphos, [b]Propetamphos, [b]Propham, Prothiofos, Quinalphos, Temephos, Tetrachlorvinphos, [b]Terbufos, [b]Thiometon, [b]Triazophos, [b]Trichlorfon, [b]Vamidothion
Oxadiazine	[a]Indoxacarb
Pheromones[17] *(Biochemical*	Codlemone, Cuelure, Disparule, Gossyplure, Grandilure, Japonilure, Muscalure

[15] ***Organophosphates*** *are organic compound whose molecule contains one or more phosphate ester groups; organophosphate works by damaging an enzyme in the body called acetylcholinesterase. The damage to this enzyme kills pests causing the nervous system to stop working properly.*

[16] ***EPN=*** *O-Ethyl O-(4-nitrophenyl) phenylphosphonothioate*

[17] ***Pheromone*** *= A pheromone is a chemical that an animal produces which changes the behaviour of another animal of the same species. Pheromones, unlike most other hormones, are ectohormones. Most insects use pheromones to communicate or mating (https://www.merriam-webster.com/dictionary/pheromone);* ***mating disruption*** *(MD) is a pest management technique designed to control certain insect pests by introducing artificial stimuli that confuse the individuals and disrupt mate localisation and/or courtship, thus preventing mating and blocking the reproductive cycle. It usually involves the use of synthetic sex pheromones. The use of mating disruption to control key pests could help reduce the already insecticide residues on fruit; https://en.wikipedia.org/wiki/Mating_disruption; http://treefruit.wsu.edu/crop-protection/opm/ mating-disruption/*

Pesticides group	Examples
	Symbols used in column 2
	[a] = ***Australia*** *(pesticides that were detected in different environmental compartments in Australia) (based on Table 7.8);*
	[b] = ***Banned*** *(pesticides those are globally banned[8]) (UTZ 2015);*
	[p] =***POPs*** *(pesticides which are POPs)[9] (Wikoff et al 2012);*
	[w]= ***Waters*** *(pesticides which are a threat to global surface waters) (Stehle and Schulz 2016).*
pesticides)[18]	
Phenylpyrazole insecticide	[a]Fipronil, Pyriprole
Pyrazole	[b]Chlorfenpyr
Pyrethroids[19]	Allethrins, [w]Acrinatrin, Bonthrin, [a,w]Bifenthrin, Bioallerthin, Bioresmethrin, Cycloprothrin, [b,w]Cyfluthrin, [b,w]β-cyfluthrin, [b,w]Cyhalothrin, [a,w]Cypermethrin (alpha, beta), [w]Deltamethrin, Dimethrin, [a,w]Esfenvalerate, [w]Fenpropathrin, [b,w]Fenvalerate, [b]Flucythrinate, Flumethrin, Imiprothrin, [a,b,w]Permethrin, Phenothrin, Prallethrin, Pyrethrum/ Pyrethrins, Resmethrin, [b]Silafluofen, [b]Sulfotep, [a,w]Tau-fluvalinate, [b]Tebupirimifos, [b,w] Tefluthrin, Tetramethrin, [w]Tralomethrin, Transfluthrin
Ungrouped	[b]Borax, [b]Nicotine, Rotenone
ii. Herbicides	
Amides[20]	Carboxin, Flutolanil, Oxycarboxin, Propanil
Amidines	[b]Amitraz
Aryloxyphen-oxypropionate	Clodinafop-propargyl, Cyhalofop-butyl, Diclof-methyl, Fenoxaprop-P-ethyl, [b]Fluazifop-butyl, Fluazifop-P-butyl, [b]Haloxyfop, Propaquizafop, Quizalofop, Quizalofop-P
Benzoic acids[21]	Chloramben, Chloroambin, [a]Dicamba, Tricamba

[18]***Biochemical pesticides*** *include substances that interfere with mating, such as insect sex pheromones, as well as various scented plant extracts that attract insect pests to trap (https://www.epa.gov/ingredients-used-pesticide-products/what-are-biopesticides#classes).*

[19]***Pyrethroid*** *(pyrethrum-like) is an organic compound similar to the natural pyrethrins, which are produced from the flowers of pyrethrums (Chrysanthemum cinerariaefolium* and *C. coccineum); pyrethroids affect the sodium channels and lead to paralysis of the organism.*

[20]***Amide herbicides*** *have diverse biological properties. Amides are used to control the major grass weeds; broad leaves and yellow nutsedge in corn, soybeans, peanuts, dry beans and sorghum as a selective pre-emergence herbicide (https://ipmworld.umn.edu/ware-intro-herbicides).*

[21]***Benzoic acid*** *= They are classed as growth hormone herbicides, and move both from leaves to the terminal meristems of the leaf, shoot and root, and also move in the transpiration stream, permitting them to also be soil-applied (https://ipmworld.umn.edu/ware-intro-herbicides).*

Pesticides group	Examples ***Symbols used in column 2*** [a] = ***Australia*** *(pesticides that were detected in different environmental compartments in Australia) (based on Table 7.8);* [b] = ***Banned*** *(pesticides those are globally banned[8]) (UTZ 2015);* [p] =***POPs*** *(pesticides which are POPs)[9] (Wikoff et al 2012);* [w]= ***Waters (****pesticides which are a threat to global surface waters) (Stehle and Schulz 2016).*
Bipyridiliums[22]	Diquat, [b]Paraquat
Chloroacetamide	Acetochlor, [b]Alachlor, Butachlor, Dimethachlor, [b]Dimethenamid, Metazachlor, [a]Metolachlor, Pretilachlor, Propachlor, Propisochlor, S-metolachlor
Chlorphenoxy[23] compounds	[a]2,4-D (2,4-Dichlorophenoxyacetic acid), 2,4 DPA (Dalapon or 2,2 dichloropropionic acid), 2,4-DB (2,4-dichlorophenoxy) butyric acid), 2,4,5-T (2,4,5-trichlorophenoxyacetic acid or agent orange), Dichloprop (2,4-dichlorophenoxypropionic acid; 2,4-DP), Fenoprop (2,4,5-trichlorophenoxypropionic acid; 2,4,5-TP; silvex), [a]MCPA (2-methyl-4-chlorophenoxyacetic acid), Mecoprop (2(2-methyl-4-chlorophenoxy) propionic acid)
Contact[24]	[b]Acrolein (Magnacide H)
Cyclohexane-dione oxime	Alloxydim, BAS 625 H, Butroxydim, Clethodim, Cycloxydim, [b]Profoxydim, Sethoxydim, [b]Tepraloxydim, Tralkoxydim
Dinitroaniline	Benfluralin, Butralin, Dinitramine, Ethafluralin, Oryzalin, [a]Pendimethalin, Trifluralin
Glycine derivatives *(amino acid)*	[a]Glyphosate[25]

[22]***Bipyridiliums*** *consists of two pyridyl rings. They are contact herbicides that damage plant tissues quickly, causing the plants to be injured because of cell membrane destruction.*

[23]***Chlorophenoxy*** *compounds are derived from chlorophenols. These herbicides are weed killers that act as synthetic auxins (plant "hormones") and cause plant death by disrupting nutrient transport and growth (https://www.who.int/water_sanitation_health/dwq/chemicals/ chlorophenoxyherb.pdf; https://link.springer.com/referenceworkentry/10.1007%2F978-3-319-17900-1_6).*

[24]***Contact herbicides*** *(acrolein) are used to control submerged and floating weeds, as well as algae, in irrigation channels. Acrolein is a restricted chemical product, used extensively worldwide to control weed growth in irrigation channel (file:///C:/Users/User1/Downloads/ Fact%20Sheet%20Weeds%20Management%20and%20Acrolein%20(3).pdf).*

[25]***Glyphosate*** *is one of the most widely used herbicides with applications in agriculture, forestry, industrial weed control, lawn, garden, and aquatic environments.*

Pesticides group	Examples ***Symbols used in column 2*** [a] ***= Australia*** *(pesticides that were detected in different environmental compartments in Australia) (based on Table 7.8);* [b] ***= Banned*** *(pesticides those are globally banned[8]) (UTZ 2015);* [p] ***=POPs*** *(pesticides which are POPs)[9] (Wikoff et al 2012);* [w]***= Waters*** *(pesticides which are a threat to global surface waters) (Stehle and Schulz 2016).*
Nitrile/ Hydro benzo nitrile[26]	[b]Bromoxynil, Dichlobenil, Ixoxynil
N-phenylphthal-imide	Cinidon-ethyl, Flumiclorac-pentyl, [b]Flumioxazin
Organoarsenic	Dimethylarsinic acid, Methylarsonic acid
Phosphinic acids	Bilanafos, Glufosinate-ammonium
Plant growth regulator	[b]Daminozide, [b]Maleic hydrazide
Pyrazole	Benzofenap, Pyrazolynate, Pyrazoxyfen
Pyridine carboxylic acid/ Picolinic acid[27]	Aminopyralid, [a]Clopyralid, Diflufenican, [a]Fluroxyper, [a]Picloram, Picolinafen, [a]Triclopyr
Sulfonylureas[28]	Amidosulfuron, Azimsulfuron, Bensulfuron, [a]Chlorsulfuron, Halosulfuron -Methyl, Iodosulfuron-methyl-sodium, [a]Metsulfuron-methyl, Oxasulfuron, Prosulfuron, Primisulfuron- ethyl, Rimsulfuron, [a]Sulfometuron-methyl, Sulfosulfuron, Thifensulfuron-Methyl, [a]Triasulfuron, Tribenuron-Methyl
Thiocarbamate	Cycloate, EPTC, Esprocarb, [a,b]Molinate, Thiobencarb, Vernolate
Triazines[29]	[a]Ametryn, [a,b]Atrazine, Cyanazine, Dimethametryn, [a]Prometryn, [a]Propazine, [a,b]Simazine, Terbumeton, [a]Terbutryn

---

[26]***Nitrile*** *are contact herbicides which control broad leaf weeds in grass-type crops (https://ipmworld.umn.edu/ware-intro-herbicides).*

[27]***Pyridine carboxylic acid/ Picolinic acid*** *is used for control of broad leaf weeds, especially thistles and clovers.*

[28]***Sulfonylureas*** *are widely used for control of annual and perennial broad-leaved weeds in many crops (i.e. wheat, barley, soybean, maize, oats and rice). They are the most important acetolactate synthase (ALS) inhibition enzyme inhibitors. ALS is a key enzyme in the biosynthesis of essential amino acids valine, leucine and isoleucine (file:///C:/Users/User1/Downloads/Azcarateetal2015.pdf).*

[29]***Triazine*** *= any of a group of compounds whose molecules contain an unsaturated ring of three carbon and three nitrogen atoms ($C_3H_3N_3$). Triazines are pre- and post-emergence herbicides used to control broad-leafed weeds and some annual grasses. These herbicides inhibit the photosynthetic electron transport in certain plants.*

Pesticides group	Examples *Symbols used in column 2* [a] = ***Australia*** *(pesticides that were detected in different environmental compartments in Australia) (based on Table 7.8);* [b] = ***Banned*** *(pesticides those are globally banned[8]) (UTZ 2015);* [p] =***POPs*** *(pesticides which are POPs)[9] (Wikoff et al 2012);* [w]= ***Waters*** *(pesticides which are a threat to global surface waters) (Stehle and Schulz 2016).*
Triazolinone	Amicarbazone, [b]Azafendin, Carfentrazone-ethyl, Sulfentrazone
Triazinones[30]	[a]Hexazinone, Metamitron, [a]Metribuzin
Triazoles	[a]Amitrole
Uracils	[a]Bromacil, Lenacil, Terbacil
Ureas[31]	[b]Chlorotoluron, Dimefuron, [a]Diuron, Fenuron, [a]Fluometuron, Isoproturon, Isouron, [a,b]Linuron, Methabenzthiazuron, Metobenzuron, Metobromuron, Metoxuron, [b]Monolinuron, Neburon, Siduron, [a]Tebuthiuron
Ungrouped	[b]Borax, [b]Fenchlorazole-ethyl
iii. Fungicides	
Antibiotic	[b]Blasticidin-S
Anilinopyr-imidine	Cyprodinil, Mepanipyrim, [a]Pyrimethanil
Aromatic hydrocarbon (chlorophenyl)	Dicloran, [b]Quintozene, [b]Tecnazene, Tolclofos-methyl
Benzimidizole[32]	[b]Benomyl, Carbendazim, Thiabendazole, Thiophanate methyl
Carboxamide[33]	[a]Boscalid, Carboxin, Flutolanil, Oxycarboxin

[30]***Triazinones*** *(metribuzin) are widely used for the control of grasses and broad-leaved weeds in soybeans, sugarcane, and numerous other crops.*

[31]***Urea herbicides****, which act by inhibiting photosynthesis was introduced in 1952 and are now used as pre- and post-emergence herbicides for general weed control in agricultural and non-agricultural practices (https://link.springer.com/article/10.2165/00139709-200524030-00008).*

[32]***Benzimidazole*** *is a heterocyclic aromatic organic compound ($C_7H_6N_2$). It is composed of a benzene ring fused with an imidazole ring.*

Pesticides group	**Examples** ***Symbols used in column 2*** [a] ***= Australia*** *(pesticides that were detected in different environmental compartments in Australia) (based on Table 7.8);* [b] ***= Banned*** *(pesticides those are globally banned*[8]*) (UTZ 2015);* [p] ***=POPs*** *(pesticides which are POPs)*[9] *(Wikoff et al 2012);* [w]***= Waters*** *(pesticides which are a threat to global surface waters) (Stehle and Schulz 2016).*
Chloronitrile	[a]Chlorothalonil
Cinnamic acid	[a]Dimethomorph
Copper based *(inorganic)*	Copper hydroxide, Copper octanoate, Copper oxychloride, Copper oxide, Copper sulfate
Dicarboximide	[b]Chlozolinate, [a]Iprodione, Procymidone, [a]Vinclozolin
Dithiocarba-mates[34]	[b]Ethylene thiourea, [b]Mancozeb, [b]Maneb, [b]Zineb, [b]Ziram
Dimethyldi-thiocarbamate	[b]Ferbam, [b]Thiram, Ziram
DMI-imidazole	Imazalil, O xapoconazole fumarate, Perfurazoate, [a]Prochloraz, Traflumizole,
DMI[35]-Pyrimidine	[a]Fenarimol, Nuarimol
Dinitrophenol	[b]Dinoterb, [b]DNOC (4,6-Dinitro-o-cresol)
FRAC[36]	[b]Dinocap
Imidazole	Imazalil, Prochloraz, Triflumizole
Morpholine	Fenpropimorph, [b]Tridemorph
Organotin[37] compounds	Azocyclotin, Cyhexatin, [b]Fentin acetate; [b]Tributyltin, [b]Triphenyltin acetate, [b]Triphenyltin hydroxide, Tricyclohexyltin hydroxide

[33]***Carboxamide*** *= any amide of a carboxylic acid - RC(=O)NR (A carboxylic acid contains the –COOH functional group, and in an amide the –OH part of that group is replaced by an* $-NH_2$ *group. So, amides contain the* $-CONH_2$ *group (https://en.wikipedia.org/wiki/Amide).*

[34]***Dithiocarbamate*** **(CH_2NS_2)**. *It is the analogue (related) of a carbamate in which both oxygen atoms are replaced by sulfur atoms.*

[35]***DMI*** *= DeMethylation Inhibitors.*

[36]***FRAC*** *= Fungicide Resistance Classification.*

Pesticides group	Examples *Symbols used in column 2* [a] = ***Australia*** *(pesticides that were detected in different environmental compartments in Australia) (based on Table 7.8);* [b] = ***Banned*** *(pesticides those are globally banned*[8]*) (UTZ 2015*[);] [p] =***POPs*** *(pesticides which are POPs)*[9] *(Wikoff et al 2012);* [w]= ***Waters*** *(pesticides which are a threat to global surface waters) (Stehle and Schulz 2016).*
	Fentin hydroxide; Triphenyltin hydroxide
Oxathiin[38]	Carboxin, Flutolanil
Phenylamide-acylalanine	Benalaxyl, Furalaxyl, [a]Metalaxyl, Metalaxyl-M
Phenylamide-oxazolidinone	[a]Oxadixyl
Phenylpyrazole	Fipronil
Phosphorothiolate	[b]Edifenphos, Iprobenfos, Isoprothiolane
Phosphorothiolate	Edifenphos, Iprobenfos, Isoprothiolane, [b]Pyrazophos
Phthalimide	Captan, [b]Captafol, Folpet
Strobilurin	[a]Azoxystrobin, [b]Dimoxystrobin,-Kresoxim-methyl, [a]Pyraclostrobin, [a]Trifloxystrobin
Triazoles[39]	Azaconazole, Bitertanol, Bromuconazole, [a]Cyproconazole,

[37]***Organotin*** *compounds are chemical compounds based on the tin with hydrocarbon substituents. For example, "**Tributyltin**s" are used as industrial biocide as **antifungal** agents in textiles and paper, wood pulp and paper mill systems, breweries, and industrial cooling systems. **Triphenyltin** derivatives are used as active components of **antifungal paints** and agricultural fungicides. Tributyltin oxide has been extensively used as a wood preservative. Tributyltin compounds were once widely used as **marine anti-biofouling** agents to improve the efficiency of ocean-going ships.*

[38]***Oxathiin*** *= The systemic fungicide. The systemic pesticides (water-soluble) penetrate plant tissues.*

[39]***Triazole*** *= compound whose molecule contains a ring of three nitrogen and two carbon atoms* ($C_2H_3N_3$).

Pesticides group	Examples ***Symbols used in column 2*** [a] ***= Australia*** *(pesticides that were detected in different environmental compartments in Australia) (based on Table 7.8);* [b] ***= Banned*** *(pesticides those are globally banned[8]) (UTZ 2015);* [p] ***=POPs*** *(pesticides which are POPs)[9] (Wikoff et al 2012);* [w]***= Waters*** *(pesticides which are a threat to global surface waters) (Stehle and Schulz 2016).*
	Diclobutrazol, [a]Difenoconazole, Diniconazole, Epoxyconazole, Fenbuconazole, Fluquinconazole, [b]Flusilazole, Flutriafol, Hexaconazole, Metconazole, [a]Myclobutanil, Paclobutrazol. [a]Penconazole, Propiconazole, [a]Tebuconazole, [a]Tetraconazole, Triadimefon, [a]Triadimenol, Triticonazole
Ungrouped	[b]Binapacryl, [b]Borax, [b]Formaldehyde, [a,p]HCB (Hexachlorobenzene)
iv. Rodenticides	
Indandione anticoagulant	[b]Chlorophacinone, Diaphacinone, Pindone
Coumarin anticoagulant	[b]Brodifacoum, [b]Bromadiolone, [b]Coumatetralyl, [b]Difenacoum, [b]Difethialone, [b]Flocoumafen, [b]Warafin
Indandione anticoagulant	Chlorophacinone, [b]Diphacinone, Pindone
Others	[b]Alpha-chlorohydrin, [b]Bromethalin, [b]Fluoroacetamide, Phosphine, [b]Pyriminil, [b]Sodium fluoroacetate, [b]Strychnine

7.1.2. History of pesticides use

People have been using pesticides for thousands of years to control various types of pests such as insects, weeds, bacteria, rodents and other biological organisms. The earliest pesticides were following inorganic substances:

- ***Arsenic*** (wood preservatives).
- ***Ash*** (various purposes).
- ***Chromium*** (wood preservatives).
- ***Copper*** (algicide, wood preservatives).
- ***Lead*** (insecticide).

The Sumerians used sulfur to control insects and mites 5,000 years ago. While the Chinese used mercury and arsenic compounds to control body lice and other pests. The Greeks and Romans used oil, ash, sulfur, and other materials to protect themselves, their livestock and crops from various pests. People in various cultures have used smoke, salt, spices and insect-repelling plants to preserve food and keep pests away. By the 15th century, toxic chemicals such as arsenic (As), mercury (Hg) and lead (Pb) were being applied to crops to kill pests. In the 17th century, nicotinesulfate was extracted from tobacco leaves for use as an insecticide. The 19th century saw the introduction of two more natural pesticides, pyrethrum which is derived from chrysanthemums (a kind of flowering plant) and rotenone which is derived from the roots of the tropical climber (*Derris* spp.). In 1939, it was discovered that dichlorodiphenyltri-chloroethane (DDT) is a very effective insecticide. It quickly became the most widely-used pesticide in the world. In the 1940s manufacturers began to produce large amounts of synthetic pesticides and their use became widespread. About 20 years later (1962), **Ms Rachel Louise Carson** (a USA Marine Biologist) wrote the book "**Silent Spring**" where she shed light on the devastating effects of DDT on the environment. Later serious concerns about its impact on human and animal health led the USA and 80 more countries to ban its use. Rachel Carson reported that DDT was causing egg-shell thinning in bird eggs and thus was leading to the near extinction of bird species such as peregrine falcons and bald eagles (see Table 7.3 for a chronology of the history of pesticide use).

Table 7.3. A brief history of pesticide use
(Stehle and Schulz 2015, Arnoult and Lukac 2018)

*[**BCE** = Before the Common Era or Before the Current Era/ Before Christ (BC);*
***CE** = Common Era or Anno Domini (AD)].*

First Steps and Early Interactions	
2500 BCE	Sumerians used **sulphur** to control mites and insects.
1000 BCE	Ancient Egyptians used **plant extracts** as insecticide.
c700 BCE	Homer referred to sulphur **fumigation**[40] as pest control.
400 BCE	**Pyrethrum**[41] used in Persia as an insecticide.
307 CE	Chinese introduced **ants** in citrus trees to manage pests.

[40]***Fumigation*** *is a method of pest control that completely fills an area with gaseous pesticides or fumigants to suffocate or poison the pests within.*

[41]***Pyrethrum*** *is a natural insecticide made from the dried flower heads of Chrysanthemum cinerariifolium and C. coccineum. Its active ingredient is pyrethrin.*

77 CE	Pliny Secundus (a Roman, 23-79 AD) listed **arsenic** and olive residue as pesticides.
1100 CE	Chinese used **soap** as an insecticide.
Renaissance[42] and Enlightenment	
1500s on wards	Insecticidal plants brought back from the New World: Sabadilla[43], Nicotine, Quassin, Heliopsin, Ryanodine.
1650-1780	Entomology flourished, both systematic and developmental subjects began to flourish.
Industrial Revolution control of caterpillars and Early Modern Age	
1763	Carl Linnaeus (a Swedish, 1707-1778)[44] on mechanical and biological control of caterpillars.
1865	Paris green (copper acetoarsenite)[45], to control the Colorado potato beetle.
1880	First commercial pesticide spraying machine evolved.
1880s	Bordeaux mixture discovered to control downy mildew in vineyards.
1896	Iron sulphate, first selective herbicide became known.
1921	First aircraft spray (Ohio, USA) of insecticide/ pesticide began.

[42]***Renaissance*** *is a period from the 14th to the 17th century, considered the bridge between the middle Ages and Modern history. It started as a cultural movement in Italy in the late medieval period and later spread to the rest of Europe.*

[43]***Sabadilla***= *a Mexican plant of the lily family*

[44]***Carl Linnaeus*** *was a Swedish biologist, and physician who formalised binomial nomenclature, the modern system of naming organisms. He is known as the "father of modern taxonomy.*

[45]***Paris green*** *is known as copper(II) acetoarsenite* $[Cu(CH_3COO)_2 \cdot 3Cu(AsO_2)_2]$. *It has copper, acetate, and arsenite ions in it. Paris green was first used to kill rats in sewers in Paris. It was later used as an insecticide and sprayed by aeroplanes to stop malaria in Italy. It was further used as paints in ships to prevent growing of barnacles. In many places, Paris green is now banned because of its toxic effects on non-target organisms (https://simple.wikipedia.org/wiki/Paris_green; https://www.sciencedirect.com/topics/agricultural-and-biological-sciences/paris-green).*

1939/1940	Development of **DDT**, the start of the modern insecticide era.
1950/1962	Development of **organophosphates** and **carbamates** (see Table 7.2 for examples of organophosphates and carbamates).
1962	**Ms Rachel Carson's** (an American, 1907-1964), published the book *Silent Spring*, where she reported that DDT was causing egg-shell thinning in bird eggs and leading to the near-extinction of bird species (peregrine falcons and bald eagles).
1972	**DDT banned** in the USA.
1973	**Pyrethroids** emerged (see Table 7.2 for examples of pyrethroids).
1991	**Neonicotinoids** emerged (see Table 7.2 for examples of neonicotinoids).
1996	First transgenic "GMOs" (genetically modified organisms); and **Bt** (*Bacillus thuringiensis*) cotton crop in the USA.

7.1.3. Current pesticide use

China, the USA, Argentina, Thailand, Brazil are the **top pesticide users/** consumers in the world. China, the USA and Argentina alone accounts for 70% of the world, pesticides use in agriculture (with China alone using half of the pesticides worldwide) (Pretty and Bharucha 2015; Fig. 7.2). Six countries each consumes between 50–100 millions of kg (Thailand, Brazil, Italy, France, Canada and Japan) and 13 countries between 10–50 millions kg (India, Spain, Germany, Bangladesh, Turkey, South Africa, Russia, Chile, Vietnam, UK, Ghana, Cameroon, and Pakistan). In the past 20 years, there have been substantial increases in pesticide use, with consumption in China growing four-fold, Argentina eight-fold, Brazil three-fold, Bangladesh five-fold, and in Thailand four-fold. Countries starting from a low base level have seen greater increases. In Burkina Faso, pesticide use increased by 50-fold, Ethiopia 13-fold, Ghana 17-fold, and Cameroon eight-fold. Aggregate use has been stable in Germany and the USA; there have been notable reductions in pesticides used in the UK (down 44%), France (down 38%), Japan (down 32%), Italy (down 26%), Vietnam (down 24%) and Denmark (down 21%). Some of these declines were provoked by policy changes, particularly in the EU (Pretty and Bharucha 2015).

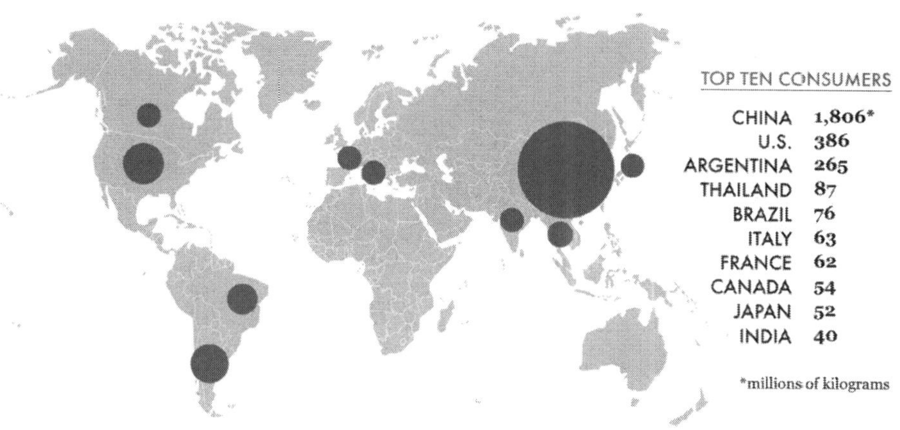

Fig. 7.2. Annual pesticide consumption
(Pariona 2017; *https://www.worldatlas.com/articles/top-pesticide-consuming-countries-of-the-world.html)*.

7.1.4. Merits and demerits of pesticide usage

Pesticides use has greatly helped to increase and improve world food production. These are mainly used to increase farmer's profit by preventing crop losses due to pest attacks and weed infestation. In the USA alone, farmers get an estimated four-fold return on the money they spend on pesticides. Application of pesticides further helps to protect human health by killing mosquitoes (which cause diseases such as Nile virus, dengue and malaria) and preventing moulds those grow in human food. Applications of pesticides reduce the risk of environmental damages through eradicating invasive[46] weeds in parks and wilderness and controlling algae in ponds and lakes (which interfere in swimming, fishing and produces unpleasant odour). Despite the above benefits, there are some concerns about the adverse effects of pesticides on non-target organisms including

[46]***An invasive weed** is a plant species that is not native to a specific location. A weed is often plants that have been introduced from outside. Invasive plants/ weeds are among the most serious threats to Australia's natural environment and primary production industries. They displace native species, contribute significantly to land degradation, and reduce farm and forest productivity. Some of the **invasive plant species found in Victoria, Australia** are African boxthorn (Lycium ferocissimum); Alligator weed (Alternanthera philoxeroides); Arrow head (Sagittaria spp.); Bathurst burr (Xanthium Spinosum); Black berry (Rubus* spp.*); Paterson's curse (Echium plantagineum); Parrots feather (Myriophyllum aquaticum); Water hyacinth (Eichhornia crassipes) and Water lettuce (Pistia stratiotes) (http://vro.agriculture vic.gov.au/ dpi/vro/vrosite.nsf/pages/lwm_invasive-plants_common-name).*

humans, fish and wildlife. For example, non-target pesticide poisoning could cause fish kills; reproductive failure in birds and illness in humans and possible endocrine disruption (see section 7.3.2). Systemic insecticides which poison pollen and nectar in the flowers may kill needed pollinators such as honey bees. More than 98% of sprayed pesticides reach destinations other than their target species. This causes pollution of air, water, and soil (*https://www.worldatlas.com/articles/top-pesticide-consuming-countries-of-the-world.html*). Several pesticides can persist in the environment for a long period (persistent), bioaccumulative and toxic (hence termed as PBT). In addition, some are also carcinogenic and endocrine-disrupting (see section 7.2).

7.2. Chemistry, toxicity and ecotoxicity of pesticides

A number of pesticides are persistent, bioaccumulative and toxic (PBT) as well as endocrine-disrupting and carcinogenic. PBT criteria can be used to assess risks and harmful effects posed by organic contaminants on aquatic and terrestrial biota (EPA 2012, Strempel et al 2012). Such PBT based risk assessment strategy (half-life, log K_{ow}, toxicity (LD_{50} for rats, birds) and ecotoxicity (EC_{50}/ LC_{50} of bacteria, algae, daphnia and fish) has been applied throughout this book.

7.2.1. Persistence

Several pesticides are "Persistent" (persistence is measured as the time it takes for half of the initial amount of a pesticide to breakdown). Persistent pesticides are those that do not breakdown easily (having longer half-life, $T_{1/2}$) and stay long in the environment, they are hydrophobic (insoluble in water). Thus, if a pesticide's half-life is 30 days, half of it will be left after 30 days, one-fourth after 60 days, and one-eighth after 90 days and so on. Pesticides having a short half-life and very soluble in water can move rapidly through certain soils. In general, a pesticide which is highly soluble in water is more easily carried-off with rainwater as run-off. Such chemicals can have adverse effects on fish and other aquatic biota. The greater the half-life of a pesticide, the greater is its persistence in the environment. Persistent chemicals tend to have a half-life in soil of > 2-3 weeks. The half-life of some pesticides are listed below (based on Table 7.4): *Aldicarb* (up to few months), ***Aldrin and Dieldrin*** (up to five years), *atrazine* (up to 105 days), *Azinphos-methyl* (up to several weeks), *Boscalid* (up to 578 days, 1.58 years), *Bromacil* (up to five months), *Carbendazim* (up to 12 months), *Chlordane* (up to one year), *Chlorfluazuron* (up to few months), *Cyhalothrin* (up to 12 weeks), ***DDT*** (up to 30 years), *Dieldrin* (up to five years), *Diuron* (up to eight months), *Endosulfan* (up to eight months), ***Endrin*** (up to 12 years), *Fenarimol* (> 365 days), *Heptachlor* (up to 10 months), *Imazalil* (up to 6.5 months), *Imazapyr* (up to two years), Indosulfuron-methyl-sodium (up to 365 days), *Lindane* (up to 40 weeks or 9.3 months), ***Mirex*** (up to 12 years), Pendimethalin (up to four months), Quintozene (up to 10 months), *Tetradifon* (up to 36 weeks or 4.8 months), *Thiobencarb* (up to eight months), ***Toxaphene*** (up to

12 years). The above list shows that most of organochlorine pesticides (aldrin, DDT, dieldrin, endirn, endsulfan, lindane, mirex and toxaphene) are very persistent. These organochlorine pesticides fall under the category of persistent organic pollutants (POPs) and their uses have now been restricted or banned globally (see Table 7.2, see also chapter 5 on POPs).

7.2.2. Bioaccumulation

Several pesticides are bioaccumulative[47] having a log K_{ow}[48] > 3 (Table 7.4) The log K_{ow} is the ratio of the concentration of a chemical in *n*-octanol (a surrogate/ a substitute for animal lipid such as fish) to the concentration in water both at equilibrium and at a constant temperature (also called (*Octanol/ Water Partition Coefficient*). Some of the highly bioaccumulative pesticides (log K_{ow} > 4) under different pesticide classes are listed below:

- **Acaricides (insecticides):** Amitraz (log K_{ow}: 5.5), Chlorfenapyr (log K_{ow}: 4.83), Etoxazole (log K_{ow}: 5.59), Fenpyroximate (log K_{ow}: 5.01), Indoxacarb (log K_{ow}: 4.65), Tebufenozide (log K_{ow}: 4.25), Tetradifon (log K_{ow}: 4.61).
- **Carbamates:** Fenoxycarb (log K_{ow}: 4.07).
- **Herbicides:** Clethodim (log K_{ow}: 4.14), Diclofop-methyl (log K_{ow}: 4.58), Fenoxaprop-p-ethyl (log K_{ow}: 4.58), Pendimethalin (log K_{ow}: 5.2), Thiobencarb (log K_{ow}: 4.23), Trifluralin (log K_{ow}: 4.83).
- **Fungicides**: Difenoconazole (log K_{ow}: 4.4), Fluazuron (log K_{ow}: 5.1), Hexachlorobenzene (log K_{ow}: 5.86), Quintozene (log K_{ow}: 5.1), Trifloxystrobin (log K_{ow}: 4.5).
- **Insecticides:** Chlorfluazuron (log K_{ow}: 5.9), Fipronil (log K_{ow}: 4.0), Methoprene (log K_{ow}: >6.0), Methoxychlor (log K_{ow}: 5.08), Triflumuron (log K_{ow}: 6.1).
- **Organochlorines:** Aldrin and Dieldrin (log K_{ow}: 4.32), Chlordane (log K_{ow}: 6.0), DDT (log K_{ow}: 6.9), Dicofol (log K_{ow}: 4.58), Dieldrin (log K_{ow}: 4.32), Endosulfan (log K_{ow}: 4.79), Endrin (log K_{ow}: 5.6), Heptachlor (log K_{ow}: 5.27), Mirex (log K_{ow}: 5.28), Toxaphene (log K_{ow}: 6.44).

---

[47]***Bioaccumulation** is the accumulation and enrichment of contaminants (pesticides or other chemicals) in aquatic organisms (fish, prawns, and mussels), relative to that in the environment. It is the net result of all uptake and loss processes, such as respiratory and dietary uptake, and loss by egestion, passive diffusion, metabolism, transfer to offspring and growth (https://www.sciencedirect.com/topics/pharmacology-toxicology-and-pharmaceuticalscience/ bioaccumulation).* Bioaccumulation is also called **'secondary poisioing'.**

[48]***log K_{ow}** = Octanol–water partition coefficient. It is a measure of hydrophilicity/ lipophilicity of a compound. log K_{ow} is also called log P. The biological properties such as **bioaccumulation and toxicity** are largely determined by log K_{ow} (Cumming and Rücker 2017).*

Table 7.4. Chemistry, toxicity and ecotoxicity of selected harmful pesticides (Tomlin 2000 and 2006, Kibria et al 2010)

*[**A** = Acaricide; **C** = carbamate; **ED** = Endocrine disruptor; **F** = fungicide; **H** = herbicide; **I** = insecticide; **IARC** = International Agency for Research on Cancer; **Log K_{ow}** = Octanol/ Water Partition Coefficient; **M** = molluscicide; **N** = nematicide; **na** = data is not available; **Neo** = neonicotinoid; **OC** = Organochlorine; **OP** = organophosphate; **R** = Rodenticide; **SP** = synthetic pyrethroid; **MRL** = maximum residue limits; **Guidelines for risk assessment**: **persistence**: half-life in soil six months; water > two months (Australian Government 2018), **Toxicity** (see Table 7.5; https://www.nicnas.gov.au/notify-your-chemical/types-of-assessments/permit-categories/controlled-use-permit/controlled-use-eligibility-criteria/Persistent-organic-pollutants-criteria].*

Pesticides	Chemistry (half-life, log K_{ow}, solubility) (Tomlin 2000 and 2006)			Toxicity (Cancer, Endocrine disruption, LD_{50} of mammals)			Ecotoxicity (LD_{50} of birds, bees and EC_{50}/ LC_{50} of algae, *Daphnia*, fish)		Remarks
	Half-life *(d = days, h = hours, mo = months, w = weeks, Y = year)*	**log K_{ow}** *(bioaccumulation potential)*	**Water solubility**	**Cancer group** (IARC 2019)	**Endocrine Disruptor (ED)** (McKinlay et al 2008, Mnif et al 2011)	**Toxicity Mammal** LD_{50} (Rats) oral (Tomlin 2000 and 2006)	**Toxicity birds and bees** LD_{50} birds-mallard ducks/ quail; LD_{50} bees oral/ contact (Tomlin 2000 and 2006)	**Ecotoxicity to aquatic organisms** Algae, *Selenastrum capricornutum;* Daphnia, *Daphnia magna*; Fish, rainbow trout (Tomlin 2000 and 2006)	
2,4-D (H)	< 7 d (soil)	2.58-2.83 Moderate	45,000 mg/L	2B	ED	Rat: 639-764 mg/kg	Birds: >5,620 mg/kg Bees: 104 µg/bee	Algae, EC_{50}: 33.2 mg/L for 5 d *Daphnia*, LC_{50}: 235 mg/L for 21 d Fish, LC_{50}: >100 mg/L for 96 h	Possibly carcinogenic, ED, moderately toxic to mammals, slightly toxic to birds, and fish.
Acephate (OP)	2-7 d (soil)	-0.89 Low	790 g/L	na	ED	Rat: 1,030-1,447 mg/kg	Birds: 350 mg/kg Bees: 1.2 µg/bee	Algae, EC_{50}: >980 mg/L for 72 h *Daphnia*, EC_{50}: 67.2 mg/L for 48 h Fish, LC_{50}: >1,000 mg/L for 96 h	ED, Moderately toxic to birds and bees.

Pesticides	Chemistry (half-life, log K_{ow}, solubility) (Tomlin 2000 and 2006)			Toxicity (Cancer, Endocrine disruption, LD_{50} of mammals)			Ecotoxicity (LD_{50} of birds, bees and EC_{50}/ LC_{50} of algae, *Daphnia*, fish)		Remarks
	Half-life *(d = days, h = hours, mo = months, w = weeks, Y = year)*	**log K_{ow}** *(bioaccumulation potential)*	**Water solubility**	**Cancer group** (IARC 2019)	**Endocrine Disruptor (ED)** (McKinlay et al 2008, Mnif et al 2011)	**Toxicity Mammal** LD_{50} (Rats) oral (Tomlin 2000 and 2006)	**Toxicity birds and bees** LD_{50} birds- mallard ducks/ quail; LD_{50} bees oral/ contact (Tomlin 2000 and 2006)	**Ecotoxicity to aquatic organisms** Algae, *Selenastrum capricornutum;* Daphnia, *Daphnia magna*; Fish, rainbow trout (Tomlin 2000 and 2006)	
Acrolein (H)	7.5-10 h	1.08 Low	208 g/kg	3	na	Rat: 29 mg/kg	Birds: 19-20.2 mg/kg	Algae, EC_{50}: 0.050 mg/L for 5 d *Daphnia*, LC_{50}: 22 mg/L for 48 h Fish, LC_{50}: >0.15 mg/L for 96 h	Highly toxic to mammal, fish and algae.
Aldicarb (I, C, A, N) [POPs)	1 d to few months (water) (Extoxnet 1993a)	1.15 Low	4.93 g/L	3	ED	Rat: 0.93 mg/kg	Birds: 71 mg/kg Bees: 0.285 µg/bee	Algae, EC_{50}: 1.4 mg/L for 5 d *Daphnia*, LC_{50}: 0.18 mg/L for 48 h Fish, LC_{50}: 0.88 mg/L for 96 h	ED, extremely toxic to mammals, bees, *Daphnia* and fish.
Aldrin and Dieldrin (OC) [POPs]	5 yrs (soil) (ATSDR 2002)	4.32 High	0.17 mg/L	2A	ED	Rat: 38-46 mg/kg (ATSDR 2002)	Birds: 50 mg/kg	Fish, LC_{50}: 0.0037-5.2 mg/L for 96 h	Very persistent, highly bioaccumulative, ED, probably carcinogenic, extremely toxic to mammals, birds and fish.
Amitraz (I, A)	<1 d (soil)	5.5 High	<0.1 mg/L	na	na	Rat: 650 mg/kg	Birds: 788 mg/kg Bees: Low toxic	Algae, EC_{50}: >12 mg/L *Daphnia*, LC_{50}: 0.035 mg/L for 48 h Fish, LC_{50}: 0.74 mg/L for 96 h	Highly bioaccumulative, extremely toxic to fish and *Daphnia*.

Pesticides	Chemistry (half-life, log K_{ow}, solubility) (Tomlin 2000 and 2006)			Toxicity (Cancer, Endocrine disruption, LD_{50} of mammals)			Ecotoxicity (LD_{50} of birds, bees and EC_{50}/ LC_{50} of algae, *Daphnia,* fish)		Remarks
	Half-life *(d = days, h = hours, mo = months, w = weeks, Y = year)*	**log K_{ow}** *(bioac-cumul-ation potential)*	**Water solubility**	**Cancer group** (IARC 2019)	**Endocrine Disruptor (ED)** (McKinlay et al 2008, Mnif et al 2011)	**Toxicity Mammal** LD_{50} (Rats) oral (Tomlin 2000 and 2006)	**Toxicity birds and bees** LD_{50} birds- mallard ducks/ quail; LD_{50} bees oral/ contact (Tomlin 2000 and 2006)	**Ecotoxicity to aquatic organisms** Algae, *Selenastrum capricornutum;* Daphnia, *Daphnia magna*; Fish, rainbow trout (Tomlin 2000 and 2006)	
Amitrole (H)	2-4 w (soil)	-0.97 Very Low	<1,385 g/L	3	ED	Rat: >10,000 mg/kg	Birds: > 5,000 mg/kg Bees: Nontoxic; >10 µg/bee	Algae, EC_{50}: 2.3-5.1 mg/L for 72 h *Daphnia*, LC_{50}: >10 mg/L for 48 h Fish, LC_{50}: >1,000 mg/L for 96 h	ED, non-toxic to mammals, birds, fish, *Daphnia* and algae.
Atrazine (H)	16-77 (soil), 10-105 d (natural waters)	2.5 Medium	33 mg/L	3	ED	Rat: 1,869-3,090 mg/kg	Birds: > 2,000 mg/kg Bees: 97 µg/bee	Algae, EC_{50}: 0.13 mg/L for 72 h *Daphnia*, LC_{50}: 87 mg/L for 24 h Fish, LC_{50}: 4.5-11 mg/L for 96 h	ED, extremely toxic to algae.
Azamethiphos (OP, I)	6 h (soil)	1.05 Low	1.1 g/L	na	na	Rat: 1,180 mg/kg	Birds: 48.4 mg/kg Bees: Toxic, <0.1 µg/bee	*Daphnia*, LC_{50}: 0.00067 mg/L for 24 h Fish, LC_{50}: 0.115-0.2 mg/L for 96 h	Extremely toxic to birds, *Daphnia* and fish.
Azinphos-methyl (OP)	Several weeks (soil)	2.96 Medium	28 mg/L	na	na	Rat: 9 mg/kg	Birds: 32 mg/kg Bees: Toxic	Algae, EC_{50}: 7.15 mg/L for 72 h *Daphnia*, LC_{50}: 0. 0011 mg/L for 48 h Fish, LC_{50}: 0.02 mg/L for 96 h	Extremely toxic to mammals, birds, *Daphnia* and fish.
Azoxystrobin (F)	1-8 w (soil)	2.5 Medium	6 mg/L	na	na	Rat: >5,000 mg/kg	Birds: > 2,000 mg/kg Bees: >200 µg/bee	Algae, EC_{50}: 0.12 mg/L for 120 h *Daphnia*, EC_{50}: 0.08 mg/L for 48 h Fish, LC_{50}: 0.47 mg/L for 96 h	Persistent, extremely toxic to algae, *Daphnia* and fish.

Pesticides	Chemistry (half-life, log K_{ow}, solubility) (Tomlin 2000 and 2006)			Toxicity (Cancer, Endocrine disruption, LD_{50} of mammals)			Ecotoxicity (LD_{50} of birds, bees and EC_{50}/ LC_{50} of algae, *Daphnia*, fish)		Remarks
	Half-life *(d = days, h = hours, mo = months, w = weeks, Y = year)*	**log K_{ow}** *(bioac-cumul-ation potential)*	**Water solubility**	**Cancer group** (IARC 2019)	**Endocrine Disruptor (ED)** (McKinlay et al 2008, Mnif et al 2011)	**Toxicity Mammal** LD_{50} (Rats) oral (Tomlin 2000 and 2006)	**Toxicity birds and bees** LD_{50} birds-mallard ducks/ quail; LD_{50} bees oral/ contact (Tomlin 2000 and 2006)	**Ecotoxicity to aquatic organisms** Algae, *Selenastrum capricornutum;* Daphnia, *Daphnia magna*; Fish, rainbow trout (Tomlin 2000 and 2006)	
Benomyl (F)	19 h (soil)	1.37 Low	1.9-3.6 µg/L (pH related)	na	ED	Rat: >5,000 mg/kg	Birds: > 10,000 mg/kg Bees: non-toxic; >50 µg/bee	Algae, EC_{50}: 2.1-3.1 mg/L for 72 h *Daphnia*, LC_{50}: 0.64 mg/L for 48 h Fish, LC_{50}: 0.27 mg/L for 96 h	ED, extremely toxic to *Daphnia* and fish.
Bifenthrin (SP)	65-125 d (soil)	>6 Very high	<1 µg/L	na	ED	Rat: 54.5 mg/kg	Birds: 2,150 mg/kg Bees: 0. 1 µg/bee	*Daphnia*, LC_{50}: 0.00016 mg/L for 48 h Fish, LC_{50}: 0.00015 mg/L for 96 h	Persistent, highly bioaccmulative, extremely toxic to mammals, bees, *Daphnia* and fish.
Boscalid (F)	96-578 d (soil) (MSDS Pubchem 2009)	2.96 Medium	4.6 mg/L	na	na	Rat: >5,000 mg/kg	Birds: >2,000 mg/kg Bees: 166 µg/bee	Algae, EC_{50}: 3.75 mg/L for 72 h *Daphnia*, LC_{50}: 5.33 mg/L for 48 h Fish, LC_{50}: 2.7 mg/L for 96 h	Very persistent, highly toxic to fish, *Daphnia* and algae.
Bromacil (H)	5 mo (soil)	1.88 Low	700-1,287 mg/L (pH related)	na	na	Rat: 1,300-2,000 mg/kg	Birds: 2,250 mg/kg Bees: non- toxic	Algae, EC_{50}: 6.8 mg/L for 72 h *Daphnia*, LC_{50}: 121 mg/L for 48 h Fish, LC_{50}: 36 mg/L for 48 h	Very persistent, slightly toxic to fish and algae.
Bromoxynil (H)	10 d (soil)	1.04 Low	89 mg/L	na	na	Rat: 200 mg/kg	Birds: 217 mg/kg Bees: non-toxic; 150 µg/bee	Algae, EC_{50}: 0.65-44 mg/L for 72 h *Daphnia*, LC_{50}: 12.5 mg/L for 48 h Fish, LC_{50}: 29.2 mg/L for 48 h	Moderately toxic to mammal, highly toxic to *Daphnia*.

Pesticides	Chemistry (half-life, log K_{ow}, solubility) (Tomlin 2000 and 2006)			Toxicity (Cancer, Endocrine disruption, LD_{50} of mammals)			Ecotoxicity (LD_{50} of birds, bees and EC_{50}/ LC_{50} of algae, *Daphnia*, fish)		Remarks
	Half-life *(d = days, h = hours, mo = months, w = weeks, Y = year)*	**log K_{ow}** *(bioac-cumul-ation potential)*	**Water solubility**	**Cancer group** (IARC 2019)	**Endocrine Disruptor (ED)** (McKinlay et al 2008, Mnif et al 2011)	**Toxicity Mammal** LD_{50} (Rats) oral (Tomlin 2000 and 2006)	**Toxicity birds and bees** LD_{50} birds- mallard ducks/ quail; LD_{50} bees oral/ contact (Tomlin 2000 and 2006)	**Ecotoxicity to aquatic organisms** Algae, *Selenastrum capricornutum;* Daphnia, *Daphnia magna*; Fish, rainbow trout (Tomlin 2000 and 2006)	
Bupirimate (F)	35-90 d (flooded soil)	3.9 High	22 mg/L	na	ED	Rat: >4,000 mg/kg	Birds: >5,200 mg/kg Bees: 0.2 µg/bee	*Daphnia*, LC_{50}: 7.3 mg/L for 48 h Fish, LC_{50}: 1.4 mg/L for 96 h	ED, extremely toxic to bees and highly toxic to fish.
Captan (F)	1 d (soil)	2.8 Medium	3.3 mg/L	3	ED	Rat: 9,000 mg/kg	Birds: >5,000 mg/kg Bees: 91 µg/bee	*Daphnia*, LC_{50}: 7-10 mg/L for 48 h Fish, LC_{50}: 0.034 mg/L for 96 h	ED, extremely toxic to fish.
Carbaryl (I-C)	7-28 d (soil)	1.85 Low	120 mg/L	3	ED	Rat: 246-500 mg/kg	Birds: >2,179 mg/kg Bees: Toxic; 1 µg/bee	*Daphnia*, LC_{50}: 0.006 mg/L for 48 h Fish, LC_{50}: 1.3 mg/L for 96 h	ED, Moderately toxic to mammals, extremely toxic to *Daphnia*.
Carbendazim (F)	6-12 mo (bare soil), 2-25 mo (water- aerobic - anaerobic conditions)	1.38 Low	7-29 mg/L	na	na	Rat: >15,000 mg/kg	Birds: 5,826- 15,595 mg/kg Bees: >50 µg/bee	Algae, EC_{50}: 1.3 mg/L for 72 h *Daphnia*, LC_{50}: 0.13-0.22 mg/L for 48 h Fish, LC_{50}: 0.83 mg/L for 96 h	Very persistent, extremely toxic to *Daphnia* and fish
Carbofuran (I, C)	30-60 d	1.52 Low	320-351 mg/L	na	ED	Rat: 8 mg/kg	Birds: 2.5-5 mg/kg Bees: Toxic	*Daphnia*, LC_{50}: 0.038 mg/L for 48 h Fish, LC_{50}: 22-29 mg/L for 96 h	ED, extremely toxic to mammals, birds, and *Daphnia*.

Pesticides	Chemistry (half-life, log K_{ow}, solubility) (Tomlin 2000 and 2006)			Toxicity (Cancer, Endocrine disruption, LD_{50} of mammals)			Ecotoxicity (LD_{50} of birds, bees and EC_{50}/ LC_{50} of algae, *Daphnia,* fish)		Remarks
	Half-life *(d = days, h = hours, mo = months, w = weeks, Y = year)*	**log K_{ow}** *(bioac-cumul-ation potential)*	**Water solubility**	**Cancer group** (IARC 2019)	**Endocrine Disruptor (ED)** (McKinlay et al 2008, Mnif et al 2011)	**Toxicity Mammal** LD_{50} (Rats) oral (Tomlin 2000 and 2006)	**Toxicity birds and bees** LD_{50} birds-mallard ducks/ quail; LD_{50} bees oral/ contact (Tomlin 2000 and 2006)	**Ecotoxicity to aquatic organisms** Algae, *Selenastrum capricornutum;* Daphnia, *Daphnia magna*; Fish, rainbow trout (Tomlin 2000 and 2006)	
Carfentrazone-ethyl (H)	< 7 d (soil)	3.36 Medium	12 µg/L	na	na	Rat: 5,143 mg/kg	Birds: >1,000 mg/kg Bees: >35 µg/bee	Algae, EC_{50}: 0.012 mg/L *Daphnia*, EC_{50}: 9.8 mg/L for 48 h Fish, LC_{50}: 1.6-43 mg/L for 96 h	Extremely toxic to algae.
Chlordane (I, OC)	1 y	6 Very high	0.1 mg/L	2B	ED	Rat: 133-649 mg/kg	Birds: 795 mg/kg Bees: Toxic	*Daphnia*, LC_{50}: 0.59 mg/L for 48 h Fish, LC_{50}: 0.09 mg/L for 96 h	Very persistent, highly bioaccumu-altive, possibly carcinogenic, ED, extremely toxic to fish and *Daphnia.*
Chlorfenapyr (A)	230-250 d (soil), 250 d (sediment) (MSDS Pubchem 2005b)	4.83 High	Insoluble	na	na	Rat: 441-1,152 mg/kg	Birds: 10 mg/kg Bees: 0.2 µg/bee	Algae, EC_{50}: 0.0013 mg/L *Daphnia*, LC_{50}: 0.0061 mg/L for 48 h Fish, LC_{50}: 0.0074 mg/L for 96 h	Persistent, bioaccumulative, extremely toxic to birds, bees, algae, *Daphnia* and fish.
Chlorfenv-inphos (I, OP)	> 150 d (peat soil) (MSDS Pubchem 2005c)	3.85-4.22 High	145 mg/L	na	ED	Rat: 10 mg/kg	Birds: 16-107 mg/kg Bees: 0.55 µg/bee	Algae, EC_{50}: 1.6 mg/L *Daphnia*, LC_{50}: 0.0003 mg/L for 48 h Fish, LC_{50}: 0.3-1.6 mg/L for 96 h	ED, Bioaccumu-lative, extremely toxic to mammal, bees, birds, *Daphnia* and fish.

Pesticides	Chemistry (half-life, log K_{ow}, solubility) (Tomlin 2000 and 2006)			Toxicity (Cancer, Endocrine disruption, LD_{50} of mammals)			Ecotoxicity (LD_{50} of birds, bees and EC_{50}/ LC_{50} of algae, *Daphnia,* fish)		Remarks
	Half-life *(d = days, h = hours, mo = months, w = weeks, Y = year)*	**log K_{ow}** *(bioac-cumulation potential)*	**Water solubility**	**Cancer group** (IARC 2019)	**Endocrine Disruptor (ED)** (McKinlay et al 2008, Mnif et al 2011)	**Toxicity Mammal** LD_{50} (Rats) oral (Tomlin 2000 and 2006)	**Toxicity birds and bees** LD_{50} birds- mallard ducks/ quail; LD_{50} bees oral/ contact (Tomlin 2000 and 2006)	**Ecotoxicity to aquatic organisms** Algae, *Selenastrum capricornutum;* Daphnia, *Daphnia magna*; Fish, rainbow trout (Tomlin 2000 and 2006)	
Chlorfluazuron (I)	Few weeks to a few months (soil)	5.9 Very high	0.012 mg/L	na	na	Rat:> 8,500 mg/kg	Birds: >2,510 mg/kg Bees: >100 μg/bee	Algae, EC_{50}: >1.8 mg/L *Daphnia*, LC_{50}: 0.908 μg/L for 48 h Fish, LC_{50}: >300 mg/L for 48 h	Very persistent and extremely toxic to *Daphnia.*
Chlorothalonil (F)	5-36 d (soil)	2.92 Medium	0.81 mg/L	2B	ED	Rat: > 5,000 mg/kg	Birds: >4,640 mg/kg Bees: Slightly toxic; >63 μg/bee	Algae, EC_{50}: 0.21 mg/L for 120 h *Daphnia*, LC_{50}: 0.07 mg/L for 48 h Fish, LC_{50}: 0.047 mg/L for 96 h	Possibly carcinogenic, ED, extremely toxic to algae, *Daphnia* and fish.
Chlorpyrifos (I, OP)	33-56 d (soil)	4.7 High	1.4 mg/L	na	na	Rat:> 135-165 mg/kg	Birds: 490 mg/kg Bees: Toxic; 360 ng/bees	Algae, EC_{50}: >0.4 mg/L for 120 h *Daphnia*, LC_{50}: 0.0017 mg/L for 48h Fish, LC_{50}: 0.007-0.051 mg/L for 96 h	Highly bioaccumulative, extremely toxic to algae, *Daphnia* and fish.
Chlorpyrifos-methyl (I, OP)	1.5-33 d (soil)	4.24 High	2.6 mg/L	na	ED	Rat: 12,800 mg/kg	Birds: >10,000 mg/kg Bees: Highly toxic; 0.38 μg/bee	*Daphnia*, LC_{50}: 5.1 mg/L for 48 h Fish, LC_{50}: 3.2 mg/L for 96 h	Highly bioaccumulative; ED, highly toxic to bees, *Daphnia* and fish.

Pesticides	Chemistry (half-life, log K_{ow}, solubility) (Tomlin 2000 and 2006)			Toxicity (Cancer, Endocrine disruption, LD_{50} of mammals)			Ecotoxicity (LD_{50} of birds, bees and EC_{50}/ LC_{50} of algae, *Daphnia,* fish)		Remarks
	Half-life *(d = days, h = hours, mo = months, w = weeks, Y = year)*	**log K_{ow}** *(bioaccumulation potential)*	**Water solubility**	**Cancer group** (IARC 2019)	**Endocrine Disruptor (ED)** (McKinlay et al 2008, Mnif et al 2011)	**Toxicity Mammal** LD_{50} (Rats) oral (Tomlin 2000 and 2006)	**Toxicity birds and bees** LD_{50} birds-mallard ducks/ quail; LD_{50} bees oral/ contact (Tomlin 2000 and 2006)	**Ecotoxicity to aquatic organisms** Algae, *Selenastrum capricornutum;* Daphnia, *Daphnia magna*; Fish, rainbow trout (Tomlin 2000 and 2006)	
Chlorsulfuron (H)	4-6 weeks (soil)	-0.99 Very low	590-31,800 mg/L (pH related)	na	na	Rat: 5,545 mg/kg	Birds: >5,000 mg/kg Bees: >25 µg/bee	Algae, EC_{50}: 0.05 mg/L for 120 h *Daphnia*, LC_{50}: 370 mg/L for 48 h Fish, LC_{50}: >250 mg/L for 96 h	Persistent, extremely toxic to algae.
Clethodim (H)	1-3 d (soil)	4.14 High	na	na	na	Rat: 1,360-1,630 mg/kg	Birds: >2,000 mg/kg Bees: non-toxic; >100 µg/bee	Algae, EC_{50}: 57.8 mg/L for 5 d *Daphnia*, LC_{50}: >120 mg/L for 48 h Fish, LC_{50}: 67 mg/L for 96 h	Slightly toxic to mammals, birds, algae, *Daphnia* and fish.
Clodinafop-propargyl (H)	5-20 d (soil)	3.9 Medium	4 mg/L	na	na	Rat: 1,829 mg/kg	Birds: >2,000 mg/kg Bees: non-toxic; >100 µg/bee	Algae, EC_{50}: 25 mg/L for 96-120 h *Daphnia*, LC_{50}: >74 mg/L for 48 h Fish, LC_{50}: 0.39 mg/L for 96 h	Bioaccumulative, extremely toxic to fish.
Clopyralid (H)	2-94 d (soil)	-1.81 Very low	118-143 g/L (pH related)	na	na	Rat: 2,675-3,738 mg/kg	Birds: 1465 mg/kg Bees: non-toxic; >100 µg/bee	Algae, EC_{50}: 6.9-7.3 mg/L for 96 h *Daphnia*, EC_{50}: 225 mg/L for 48 h Fish, LC_{50}: 103.5 mg/L for 96 h	Slightly toxic to mammals, birds, algae and fish.
Coumaphos (I, OP)	23.8 d (soil)	4.13 High	1.5 mg/L	na	na	Rat: 16-41 mg/kg	Birds: >29.8 mg/kg Bees: -	*Daphnia*, LC_{50}: 0.001 mg/L for 48 h Fish, LC_{50}: 0.34-0.84 mg/L for 96 h	Highly bioaccumulative, extremely toxic to mammals, birds, *Daphnia* and fish.

Pesticides	Chemistry (half-life, log K_{ow}, solubility) (Tomlin 2000 and 2006)			Toxicity (Cancer, Endocrine disruption, LD_{50} of mammals)			Ecotoxicity (LD_{50} of birds, bees and EC_{50}/ LC_{50} of algae, *Daphnia*, fish)		Remarks
	Half-life *(d = days, h = hours, mo = months, w = weeks, Y = year)*	**log K_{ow}** *(bioaccumulation potential)*	**Water solubility**	**Cancer group** (IARC 2019)	**Endocrine Disruptor (ED)** (McKinlay et al 2008, Mnif et al 2011)	**Toxicity Mammal** LD_{50} (Rats) oral (Tomlin 2000 and 2006)	**Toxicity birds and bees** LD_{50} birds- mallard ducks/ quail; LD_{50} bees oral/ contact (Tomlin 2000 and 2006)	**Ecotoxicity to aquatic organisms** Algae, *Selenastrum capricornutum;* Daphnia, *Daphnia magna*; Fish, rainbow trout (Tomlin 2000 and 2006)	
Cyfluthrin (I, SP)	na	6 Very high	2.2-2.5 µg/L	na	na	Rat: 500 mg/kg	Birds: >2,000 mg/kg Bees: Toxic	Algae, EC_{50}: >10 mg/L *Daphnia*, LC_{50}: 0.00016-0.0027 mg/L for 48 h Fish, LC_{50}: 0.0006-0.003 mg/L for 96 h	Highly bioaccumulative, extremely toxic to *Daphnia* and fish.
Cyhalothrin (I, SP)	4-12 w	6.9 Very high	0.005 mg/L	na	ED	Rat: 114-166 mg/kg	Birds: >5,000 mg/kg Bees: -	*Daphnia*, LC_{50}: 0.00038 mg/L for 48h Fish, LC_{50}: 0.00054 mg/L for 96 h	Persistent, highly bioaccumulative; ED, extremely toxic to *Daphnia* and fish.
Cypermethrin (I, SP)	60 d (soil)	6.6 Very high	0.004 mg/L	na	ED	Rat: 250-4,150 mg/kg	Birds: >10,000 mg/kg Bees: Highly toxic 0.035 µg/bee	*Daphnia*, LC_{50}: 0.00015 mg/L for 48 h Fish, LC_{50}: 0.69 mg/L for 96 h	Highly bioaccumulative; ED, extremely toxic to *Daphnia*, fish and bees.
Cyproconazole (F)	3 m (soil)	3.1 Medium	93 mg/L	na	ED	Rat: 1,020-1,333 mg/kg	Birds: 1,197 mg/kg Bees: >0.1 mg/bee	*Daphnia*, LC_{50}: 26 mg/L for 48 h Fish, LC_{50}: 19 mg/L for 96 h	Persistent, slightly toxic to fish and *Daphnia*.
Cyprodinil (F)	20-60 d (soil)	3.9 Medium	13-20 mg/L (pH related)	na	na	Rat: >2,000 mg/kg	Birds: >2,000 mg/kg Bees: >316 µg/bee	Algae, EC_{50}: 0.75 mg/L for 72 h *Daphnia*, LC_{50}: 0.033-0.10 mg/L for 48 h Fish, LC_{50}: 0.98-2.41 mg/L for 96 h	Bioaccumulative, extremely toxic to algae, *Daphnia* and fish.

Pesticides	Chemistry (half-life, log K_{ow}, solubility) (Tomlin 2000 and 2006)			Toxicity (Cancer, Endocrine disruption, LD_{50} of mammals)			Ecotoxicity (LD_{50} of birds, bees and EC_{50}/ LC_{50} of algae, *Daphnia*, fish)		Remarks
	Half-life *(d = days, h = hours, mo = months, w = weeks, Y = year)*	**log K_{ow}** *(bioac-cumul-ation potential)*	**Water solubility**	**Cancer group** (IARC 2019)	**Endocrine Disruptor (ED)** (McKinlay et al 2008, Mnif et al 2011)	**Toxicity Mammal** LD_{50} (Rats) oral (Tomlin 2000 and 2006)	**Toxicity birds and bees** LD_{50} birds- mallard ducks/ quail; LD_{50} bees oral/ contact (Tomlin 2000 and 2006)	**Ecotoxicity to aquatic organisms** Algae, *Selenastrum capricornutum;* Daphnia, *Daphnia magna*; Fish, rainbow trout (Tomlin 2000 and 2006)	
DDT (I, OC) [POPs]	4-30 y (soil temperate)	4.9-6.9 High-very high	Insoluble	2A	ED	Rat: 113-118 mg/kg	Birds: (595 mg/kg quail) moderately toxic Bees: 5 µg/bee	*Daphnia*, LC_{50}: 0.0011 mg/L for 48 h Fish, LC_{50}: 0.007 mg/L for 96 h	Very persistent, highly bioaccumulaitve, probably carcinogenic, ED, moderately toxic to mammals, birds, extremely toxic to *Daphnia* and fish
Deltamethrin (I, SP)	1-2 w (soil)	4.6 High	<0.2 µg/L	3	ED	Rat: 135-5,000 mg/kg	Birds: >4,640 mg/kg Bees: Toxic; 0.079 µg/bee	Algae, EC_{50}: >9.1 mg/L *Daphnia*, LC_{50}: 0.0035 mg/L for 48 h Fish, LC_{50}: 0.91 mg/L for 96 h	Highly bioaccum-ulative, ED, extremely toxic to *Daphnia* and fish, highly toxic to bees.
Demeton (I, A)	na	1.3 Low	22 g/L	na	na	Rat: 30 mg/kg	Birds: 54 mg/kg Bees: Toxic	Algae, EC_{50}: 22.1 mg/L *Daphnia*, LC_{50}: 0.023 mg/L for 48 h Fish, LC_{50}: 6.4 mg/L for 96 h	Extremely toxic to mammal, birds, *Daphnia*, and toxic to bees.
Diazinon (I, OP)	na	3.3 Medium	60 mg/L	2A	ED	Rat: 1,250 mg/kg	Birds: 2.7 mg/kg Bees: Highly toxic	Algae, EC_{50}: >1 mg/L *Daphnia*, LC_{50}: 0.00096 mg/L for 48h Fish, LC_{50}: 2.6-3.2 mg/L for 96 h	Probably carcinogenic, ED, extremely toxic to birds, bees and *Daphnia*.

Pesticides	Chemistry (half-life, log K_{ow}, solubility) (Tomlin 2000 and 2006)			Toxicity (Cancer, Endocrine disruption, LD_{50} of mammals)			Ecotoxicity (LD_{50} of birds, bees and EC_{50}/ LC_{50} of algae, *Daphnia,* fish)		Remarks
	Half-life (*d = days, h = hours, mo = months, w = weeks, Y = year*)	**log K_{ow}** (*bioaccumulation potential*)	**Water solubility**	**Cancer group** (IARC 2019)	**Endocrine Disruptor (ED)** (McKinlay et al 2008, Mnif et al 2011)	**Toxicity Mammal** LD_{50} (Rats) oral (Tomlin 2000 and 2006)	**Toxicity birds and bees** LD_{50} birds- mallard ducks/ quail; LD_{50} bees oral/ contact (Tomlin 2000 and 2006)	**Ecotoxicity to aquatic organisms** Algae, *Selenastrum capricornutum;* Daphnia, *Daphnia magna*; Fish, rainbow trout (Tomlin 2000 and 2006)	
Dicamba (H)	< 14 d (soil)	0.55 Very low	6 g/L	na	na	Rat: 1,707 mg/kg	Birds: 2,000 mg/kg Bees: Non-toxic; >100 µg/bee	Algae, EC_{50}: >250 mg/L *Daphnia*, LC_{50}: 111 mg/L for 48 h Fish, LC_{50}: 135 mg/L for 96 h	Slightly toxic to mammal, birds, fish and *Daphnia*;
Dichlorvos (I, OP)	<1 d (soil and water)	1.9 Low	18 g/L	na	ED	Rat: 50 mg/kg	Birds: 24 mg/kg Bees: 0.29 µg/bee	Algae, EC_{50}: 52.8 mg/L *Daphnia*, LC_{50}: 0.0019 mg/L for 48 h Fish, LC_{50}: 0.2 mg/L for 96 h	Extremely toxic to mammal, birds, bees, *Daphnia* and fish; ED.
Diclofop-methyl (H)	1-57 d (soil)	4.58 High	0.8 g/L	na	na	Rat: 481-693 mg/kg	Birds: 1,100 mg/kg Bees: Non-toxic	Algae, EC_{50}: 1.5 mg/L for 120 h *Daphnia*, LC_{50}: 0.23 mg/L for 48 h Fish, LC_{50}: 0.23 mg/L for 96 h	Highly bioaccumulative, extremely toxic to *Daphnia* and fish.
Dicofol (I, OC, A)	30 d (soil)	4.3 High	0.8 mg/L	3	ED	Rat: 578-595 mg/kg	Birds: 1,651 mg/kg Bees: Non-toxic; > 10 µg/bee	Algae, EC_{50}: 0.075 mg/L for 120 h *Daphnia*, LC_{50}: 0.14 mg/L for 48 h Fish, LC_{50}: 0.12 mg/L for 96 h	Bio-accumulative, ED, extremely toxic to algae, *Daphnia* and fish.
Dieldrin (I, OC) [POPs)	5 y (ATSDR 2002)	4.32 High	0.17 mg/L	na	ED	Rat: 38-46 mg/kg	Birds: 50 mg/kg Bees: na	*Daphnia*, LC_{50}: 11.9 mg/L for 48 h Fish, LC_{50}: 0.0037-5.3 mg/L for 96 h	Very persistent, highly bioaccumulative, extremely toxic to mammals, birds and fish.

Pesticides	Chemistry (half-life, log K_{ow}, solubility) (Tomlin 2000 and 2006)			Toxicity (Cancer, Endocrine disruption, LD_{50} of mammals)			Ecotoxicity (LD_{50} of birds, bees and EC_{50}/ LC_{50} of algae, *Daphnia,* fish)		Remarks
	Half-life *(d = days, h = hours, mo = months, w = weeks, Y = year)*	**log K_{ow}** *(bioac-cumul-ation potential)*	**Water solubility**	**Cancer group** (IARC 2019)	**Endocrine Disruptor (ED)** (McKinlay et al 2008, Mnif et al 2011)	**Toxicity Mammal** LD_{50} (Rats) oral (Tomlin 2000 and 2006)	**Toxicity birds and bees** LD_{50} birds- mallard ducks/ quail; LD_{50} bees oral/ contact (Tomlin 2000 and 2006)	**Ecotoxicity to aquatic organisms** Algae, *Selenastrum capricornutum;* Daphnia, *Daphnia magna*; Fish, rainbow trout (Tomlin 2000 and 2006)	
Difenocon-azole (F)	145 d (sunlight)	4.4 High	15 mg/L	na	na	Rat: 1,453 mg/kg	Birds: >2,150 mg/kg Bees: Non-toxic; >187 µg/bee	Algae, EC_{50}: 1.2 mg/L for 72 h *Daphnia*, LC_{50}: 0.77 mg/L for 48 h Fish, LC_{50}: 0.8 mg/L for 96 h	Bioaccumulative, ED; extremely toxic to *Daphnia* and fish.
Diflubenzuron (I)	7 d (soil)	3.89 Medium	0.08 mg/L	na	ED	Rat: >4,640 mg/kg	Birds: >4,640 mg/kg Bees: Non-toxic; >100 µg/bee	Algae, EC_{50}: >0.3 mg/L *Daphnia*, LC_{50}: 0.0071 mg/L for 48 h Fish, LC_{50}: >0.2 mg/L for 96 h	ED, extremely toxic to fish, *Daphnia* and algae.
Dimethoate (I, OP, A)	2-4.1 d (soil)	0.704 Very low	23.3-25 g/L (pH related)	na	ED	Rat: 387 mg/kg	Birds: 10.5-42 mg/kg Bees: Toxic; 0.1-0.2 µg/bee	Algae, EC_{50}: 282.3 mg/L for72 h *Daphnia*, LC_{50}: 4.7 mg/L for 48 h Fish, LC_{50}: 6.2 mg/L for 96 h	ED, extremely toxic to birds and bees.
Diquat dibr-omide (H)	<1 w	-4.6 Very low	700 g/L	na	na	Rat: 408 mg/kg	Birds: 155 mg/kg Bees: 22 µg/bee	Algae, EC_{50}: 0.021 mg/L for72 h *Daphnia*, LC_{50}: 0.0022 mg/L for 48 h Fish, LC_{50}: 39 mg/L for 96 h	Extremely toxic to *Daphnia* and algae.
Diuron (H)	4-8 mo (soil)	2.85 Medium	37.4 mg/L	na	ED	Rat: 3,400 mg/kg	Birds: >5,000 mg/kg Bees: Non-toxic	*Daphnia*, LC_{50}: 12 mg/L for 48 h Fish, LC_{50}: 5.6 mg/L for 96 h	Very persistent, ED; moderately toxic to fish and *Daphnia*.

Pesticides	Chemistry (half-life, log K_{ow}, solubility) (Tomlin 2000 and 2006)			Toxicity (Cancer, Endocrine disruption, LD_{50} of mammals)			Ecotoxicity (LD_{50} of birds, bees and EC_{50}/ LC_{50} of algae, *Daphnia*, fish)		Remarks
	Half-life *(d = days, h = hours, mo = months, w = weeks, Y = year)*	**log K_{ow}** *(bioac-cumul-ation potential)*	**Water solubility**	**Cancer group** (IARC 2019)	**Endocrine Disruptor (ED)** (McKinlay et al 2008, Mnif et al 2011)	**Toxicity Mammal** LD_{50} (Rats) oral (Tomlin 2000 and 2006)	**Toxicity birds and bees** LD_{50} birds- mallard ducks/ quail; LD_{50} bees oral/ contact (Tomlin 2000 and 2006)	**Ecotoxicity to aquatic organisms** Algae, *Selenastrum capricornutum;* Daphnia, *Daphnia magna*; Fish, rainbow trout (Tomlin 2000 and 2006)	
Dodine (F)	na	1.65 Low	630 mg/L	na	na	Rat: 1,000 mg/kg	Birds: 1,142 mg/kg Bees: >0.011 mg/bee	Algae, EC_{50}: 0.0051 mg/L *Daphnia*, LC_{50}: 0.13 mg/L for 48 h Fish, LC_{50}: 0.53-0.6 mg/L for 96 h	Extremely toxic to algae, *Daphnia* and fish.
Endosulfan (I, OC, A) [POPs]	5-8 mon (soil)	4.74-4.79	0.32-0.33 mg/L	na	ED	Rat: 44.9-70 mg/kg	Birds: 205-245 mg/kg Bees: Non-toxic	Algae, EC_{50}: 0.51 mg/L for 72 h *Daphnia*, LC_{50}: 0.0075-0.75 mg/L for 48 h Fish, LC_{50}: 0.002 mg/L for 96 h	Very persistent, highly bioaccumulative, ED; extremely toxic to mammals, algae, *Daphnia* and fish.
Endrin (I, OC) [POPs]	12 y (soil) (Hatfield 1995b)	5.6 High	Insoluble	3	ED	Rat: 17.8 mg/kg	Birds: na Bees: na	*Daphnia*, LC_{50}: 0.00004-0.333 mg/L for 48 h Fish, LC_{50}: 0.0003-3.4 mg/L for 96 h (USEPA 1980)	Very persistent, highly bioaccum-ulative, ED; extremely toxic to mammals, *Daphnia* and fish.
Epoxiconazole (F)	2-3 mo (soil)	3.33 Medium	na	na	ED	Rat: >5,000 mg/kg	Birds: >2,000 mg/kg Bees: >100 µg/bee	Algae, EC_{50}: 2.3 mg/L for 72 h *Daphnia*, LC_{50}: 8.7 mg/L for 48 h Fish, LC_{50}: 2.2-4.6 mg/L for 96 h	ED, moderately toxic to algae, *Daphnia* and fish.

Pesticides	Chemistry (half-life, log K_{ow}, solubility) (Tomlin 2000 and 2006)			Toxicity (Cancer, Endocrine disruption, LD_{50} of mammals)			Ecotoxicity (LD_{50} of birds, bees and EC_{50}/ LC_{50} of algae, *Daphnia,* fish)		Remarks
	Half-life (*d = days, h = hours, mo = months, w = weeks, Y = year*)	**log K_{ow}** (*bioac-cumul-ation potential*)	**Water solubility**	**Cancer group** (IARC 2019)	**Endocrine Disruptor (ED)** (McKinlay et al 2008, Mnif et al 2011)	**Toxicity Mammal** LD_{50} (Rats) oral (Tomlin 2000 and 2006)	**Toxicity birds and bees** LD_{50} birds-mallard ducks/ quail; LD_{50} bees oral/ contact (Tomlin 2000 and 2006)	**Ecotoxicity to aquatic organisms** Algae, *Selenastrum capricornutum;* Daphnia, *Daphnia magna*; Fish, rainbow trout (Tomlin 2000 and 2006)	
Esfenvalerate (I, SP)	287 d	6.22 Very high	0.002 mg/L	3	na	Rat: > 75-88 mg/kg	Birds: 5,247 mg/kg Bees: 0.017 µg/bee	*Daphnia*, LC_{50}: 0.00024 mg/L for 48 h Fish, LC_{50}: 0.00026 mg/L for 96 h	Highly bioaccumulative, extremely toxic to bees, *Daphnia* and fish.
Ethion (I, OP, A)	90 d (soil)	4.28 Very high	2 mg/L	na	na	Rat: 208 mg/kg	Birds: >2,000 mg/kg Bees: Toxic	*Daphnia*, LC_{50}: 0.056 mg/L for 48 h Fish, LC_{50}: 0.72 mg/L for 96 h	Bio-accumulative, extremely toxic to *Daphnia* and fish.
Ethofumesate (H)	84-407 d (soil)	2.7	50 mg/L	na	na	Rat: >5,000 mg/kg	Birds: >3,552 mg/kg Bees: >50 µg/bee	Algae, EC_{50}: 3.9 mg/L *Daphnia*, LC_{50}: 13.5-22 mg/L for 48 h Fish, LC_{50}: 11.91-20.2 mg/L for 96 h	Persistent, moderately toxic to algae, *Daphnia* and fish.
Ethoprofos (I, N)	14-87 d (soil, varies with pH)	3.59 Medium	700 mg/L	na	na	Rat: 62 mg/kg	Birds: 61 mg/kg Bees: Non-toxic	Fish, LC_{50}: 13.8 mg/L for 96 h	Moderately toxic to mammals, birds and fish.
Etoxazole (A)	19 d (soil)	5.59 Very high	75.4 µg/L	na	na	Rat: >5,000 mg/kg	Birds: >2,000 mg/kg Bees: >200 µg/bee	Algae, EC_{50}: >1 mg/L *Daphnia*, LC_{50}: > 40 mg/L for 48 h Fish, LC_{50}: >40 mg/L for 96 h	Highly bioaccumu-lative, and highly toxic to algae.

Pesticides	Chemistry (half-life, log K_{ow}, solubility) (Tomlin 2000 and 2006)			Toxicity (Cancer, Endocrine disruption, LD_{50} of mammals)			Ecotoxicity (LD_{50} of birds, bees and EC_{50}/ LC_{50} of algae, *Daphnia,* fish)		Remarks
	Half-life *(d = days, h = hours, mo = months, w = weeks, Y = year)*	**log K_{ow}** *(bioac-cumul-ation potential)*	**Water solubility**	**Cancer group** (IARC 2019)	**Endocrine Disruptor (ED)** (McKinlay et al 2008, Mnif et al 2011)	**Toxicity Mammal** LD_{50} (Rats) oral (Tomlin 2000 and 2006)	**Toxicity birds and bees** LD_{50} birds-mallard ducks/ quail; LD_{50} bees oral/ contact (Tomlin 2000 and 2006)	**Ecotoxicity to aquatic organisms** Algae, *Selenastrum capricornutum;* Daphnia, *Daphnia magna;* Fish, rainbow trout (Tomlin 2000 and 2006)	
Fenarimol (F)	> 365 d (soil)	3.69 Medium	13.7 mg/L	na	ED	Rat: 2,500 mg/kg	Birds: >2,000 mg/kg Bees: >10 µg/bee	Fish, LC_{50}: 4.1 mg/L for 96 h *Daphnia*, EC_{50}: 5.1 mg/L for 48 h Algae, EC_{50}: 5.1 mg/L	Persistent, ED; moderately toxic to algae, *Daphnia* and fish.
Fenitrothion (I-OP)	18-28 d (soil)	3.43 Medium	14 mg/L	na	ED	Rat: 1,700-1,720 mg/kg	Birds: 1,190 mg/kg Bees: Toxic	Algae, EC_{50}: 1.3 mg/L *Daphnia*, EC_{50}: 0.0086 mg/L for 48 h Fish, LC_{50}: 1.3 mg/L for 96 h	ED, extremely toxic to *Daphnia* and highly toxic to bees, algae and fish.
Fenoxaprop-p-ethyl (H)	1-10 d (soil)	4.58 High	0.7 mg/L	na	na	Rat: 3,150-4,000 mg/kg	Birds: >2,000 mg/kg Bees: Toxic	Algae, EC_{50}: 0.51 mg/L for 96 h *Daphnia*, LC_{50}: 0.56 mg/L for 48 h Fish, LC_{50}: 0.46 mg/L for 96 h	Highly bioaccum-ulative, extremely toxic to algae, *Daphnia* and fish.
Fenoxycarb (I, C)	31 d (soil)	4.07 High	7.9 mg/L	na	ED	Rat: >10,000 mg/kg	Birds: >7,000 mg/kg Bees: Non-toxic	Algae, EC_{50}: 1.10 mg/L for 96 h *Daphnia*,LC_{50}: 0.4 mg/L for 48 h Fish, LC_{50}: 1.6 mg/L for 96 h	Highly bioaccumulative, ED; extremely toxic to *Daphnia*.
Fenpyroximate (A)	26.3-49.7 d (soil)	5.01 High	$1.45x10^{-2}$ mg/L	na	na	Rat: 480 mg/kg	Birds: >2,000 mg/kg Bees: Non-toxic	Algae, EC_{50}: 9.98 mg/L *Daphnia*, LC_{50}: 0.0032 mg/L for 48 h Fish, LC_{50}: 0.0055 mg/L for 96 h	Highly bioaccumulative, and extremely toxic to *Daphnia* and fish.

Pesticides	**Chemistry** (half-life, log K_{ow}, solubility) (Tomlin 2000 and 2006)			**Toxicity** (Cancer, Endocrine disruption, LD_{50} of mammals)			**Ecotoxicity** (LD_{50} of birds, bees and EC_{50}/ LC_{50} of algae, *Daphnia*, fish)		**Remarks**
	Half-life (*d = days, h = hours, mo = months, w = weeks, Y = year*)	**log K_{ow}** (*bioaccumulation potential*)	**Water solubility**	**Cancer group** (IARC 2019)	**Endocrine Disruptor (ED)** (McKinlay et al 2008, Mnif et al 2011)	**Toxicity Mammal** LD_{50} (Rats) oral (Tomlin 2000 and 2006)	**Toxicity birds and bees** LD_{50} birds- mallard ducks/ quail; LD_{50} bees oral/ contact (Tomlin 2000 and 2006)	**Ecotoxicity to aquatic organisms** Algae, *Selenastrum capricornutum;* Daphnia, *Daphnia magna*; Fish, rainbow trout (Tomlin 2000 and 2006)	
Fenthion (I, OP)	1.5 d (soil)	4.84 High	4.2 mg/L	na	na	Rat: 250 mg/kg	Birds: 7.2 mg/kg Bees: 0.36 µg/bee	Algae, EC_{50}: 1.79 mg/L *Daphnia*, LC_{50}: 0.0057 mg/L for 48 h Fish, LC_{50}: 0.8-1.0 mg/L for 96 h	Highly bioaccumulative, extremely toxic to birds, *Daphnia* and fish
Fenvalerate (I, SP, A)	75-80 d (soil)	5.01 High	<10 µg/L	na	ED	Rat: 451 mg/kg	Birds: 9,932 mg/kg Bees: Toxic, 0.23 µg/bee	Fish, LC_{50}: 0.0036 mg/L for 96 h	Highly bioaccumulative, ED, extremely toxic to bees and fish.
Fipronil (I)	na	4 High	1.9-2.4 mg/L	na	ED	Rat: 97 mg/kg	Birds: >2,000 mg/kg Bees: Highly toxic	Algae, EC_{50}: 0.0048 mg/L for 96 h *Daphnia*, LC_{50}: 0.19 mg/L for 48 h Fish, LC_{50}: 0.248 mg/L for 96 h	Highly bioaccumulative, ED, extremely toxic to algae, *Daphnia* and fish.
Flamprop-m-methyl (H)	na	3.09 Medium	12 mg/L	na	na	Rat: >4,000 mg/kg	Birds: >2,000 mg/kg Bees: Non-toxic	Algae, EC_{50}: 6.8 mg/L for 96 h *Daphnia*, LC_{50}: Slightly to moderately toxic Fish, LC_{50}: 2.4 mg/L for 96 h	Bioaccumulative, moderately toxic to algae *Daphnia* and fish
Fluazinam (F)	33-62 d (soil)	3.56 High	1.7 mg/L	na	na	Rat: >5,000 mg/kg	Birds: 4,190 mg/kg Bees: >100 µg/bee	*Daphnia*, LC_{50}: 0.22 mg/L for 48 h Fish, LC_{50}: 0.11 mg/L for 96 h	Bioaccumulative, extremely toxic to *Daphnia* and fish.

Pesticides	Chemistry (half-life, log K_{ow}, solubility) (Tomlin 2000 and 2006)			Toxicity (Cancer, Endocrine disruption, LD_{50} of mammals)			Ecotoxicity (LD_{50} of birds, bees and EC_{50}/ LC_{50} of algae, *Daphnia*, fish)		Remarks
	Half-life *(d = days, h = hours, mo = months, w = weeks, Y = year)*	**log K_{ow}** *(bioaccumulation potential)*	**Water solubility**	**Cancer group** (IARC 2019)	**Endocrine Disruptor (ED)** (McKinlay et al 2008, Mnif et al 2011)	**Toxicity Mammal** LD_{50} (Rats) oral (Tomlin 2000 and 2006)	**Toxicity birds and bees** LD_{50} birds-mallard ducks/ quail; LD_{50} bees oral/ contact (Tomlin 2000 and 2006)	**Ecotoxicity to aquatic organisms** Algae, *Selenastrum capricornutum;* Daphnia, *Daphnia magna*; Fish, rainbow trout (Tomlin 2000 and 2006)	
Fluazuron (F,H) (Ectoparasiticide)	na	5.1 High	<0.002 mg/L	na	na	Rat: >5,000 mg/kg	Birds: >2,000 mg/kg Bees: Not-toxic	Algae, EC_{50}: 27.9 mg/L (non-toxic) *Daphnia*, LC_{50}: 0.0006 mg/L for 48 h Fish, LC_{50}: 15 mg/L for 96 h	Highly bioaccumulative, extremely toxic to *Daphnia*.
Fluquinconazole (F)	50-300 d (soil)	3.24 Medium	1 mg/L	na	na	Rat: 112 mg/kg	Birds: >2,000 mg/kg Bees: na	Algae, EC_{50}: 0.014 mg/L *Daphnia*, LC_{50}: >5 mg/L for 48 h Fish, LC_{50}: 1.9 mg/L for 96 h	Persistent, bioaccumulative, extremely toxic to algae.
Flusilazole (F)	95 d (soil)	3.74 Medium	45-900 mg/L	na	ED	Rat: 674-1,100 mg/kg	Birds: >1,590 mg/kg Bees: Non-toxic	*Daphnia*, LC_{50}: 3.4 mg/L for 48 h Fish, LC_{50}: 1.2 mg/L for 96 h	Bioaccumulative, ED, moderately toxic to *Daphnia* and fish.
Flutriafol (F)	na	2.3 Low	130 mg/L	na	ED	Rat: 1,140-1,480 mg/kg	Birds: >5,000 mg/kg Bees: Non-toxic	*Daphnia*, LC_{50}: 78 mg/L for 48 h Fish, LC_{50}: 61 mg/L for 96 h	ED, slightly toxic to fish and *Daphnia*.
Glufosinate-ammonium (H)	7-20 d (soil)	<0.1 Very low	1370 g/L	na	na	Rat: 1,620-2,000 mg/kg	Birds: >5,000 mg/kg Bees: Non-toxic; >100 µg/bee	Algae, EC_{50}: >1,000 mg/L *Daphnia*, LC_{50}: 560-1,000 mg/L for 48 h Fish, LC_{50}: 710 mg/L for 96 h	Non-toxic to most organisms (mammal, bird, bees, fish, *Daphnia*, algae).

Pesticides	Chemistry (half-life, log K_{ow}, solubility) (Tomlin 2000 and 2006)			Toxicity (Cancer, Endocrine disruption, LD_{50} of mammals)			Ecotoxicity (LD_{50} of birds, bees and EC_{50}/ LC_{50} of algae, *Daphnia*, fish)		Remarks
	Half-life (*d = days, h = hours, mo = months, w = weeks, Y = year*)	**log K_{ow}** (*bioaccumulation potential*)	**Water solubility**	**Cancer group** (IARC 2019)	**Endocrine Disruptor (ED)** (McKinlay et al 2008, Mnif et al 2011)	**Toxicity Mammal** LD_{50} (Rats) oral (Tomlin 2000 and 2006)	**Toxicity birds and bees** LD_{50} birds- mallard ducks/ quail; LD_{50} bees oral/ contact (Tomlin 2000 and 2006)	**Ecotoxicity to aquatic organisms** Algae, *Selenastrum capricornutum;* Daphnia, *Daphnia magna*; Fish, rainbow trout (Tomlin 2000 and 2006)	
Glyphosate (H)	3-174 d (soil)	<-3.2 Very low	144 g/L	2A	ED	Rat: >5,600 mg/kg	Birds: 4,641 mg/kg Bees: >100 µg/bee	Algae, EC_{50}: 485 mg/L for 72 h *Daphnia*, LC_{50}: 780 mg/L for 48 h Fish, LC_{50}: 86 mg/L for 96 h	Probably carcinogenic, ED; non-toxic to bees, fish, *Daphnia* and algae (see section 7.4 for a case study).
HCB Hexachlor-obenzene (F)	na	5.86 High	Insoluble	2B	ED	Rat: 10,000 mg/kg	Birds: na Bees: Non-toxic	Fish, LC_{50}: 0.05-0.2 mg/L for 96 h	Highly bioaccum-ulative, Probably carcinogenic, ED; non-toxic to mam-mals, bees, but extr-emely toxic to fish.
Heptachlor (I, OC) [POPs]	9-10 mo (soil)	5.27 High	0.056 mg/Lg	2B	Ed	Rat: 147-220 mg/kg	Birds: >2,000 mg/kg Bees: na	Fish, LC_{50}: 0.007 mg/L for 96 h	Very persistent, possibly carcinog-enic, ED; extremely toxic to fish.
Hexythiazox (I, A)	8 d (soil)	2.53 Medium	0.5 mg/L	na	na	Rat: >5,000 mg/kg	Birds: >2,510 mg/kg Bees: non-toxic; >200 µg/bee	*Daphnia*, LC_{50}: 1.2 mg/L for 48 h Fish, LC_{50}:>300 mg/L for 96 h	Highly toxic to *Daphnia*.

Pesticides	Chemistry (half-life, log K_{ow}, solubility) (Tomlin 2000 and 2006)			Toxicity (Cancer, Endocrine disruption, LD_{50} of mammals)			Ecotoxicity (LD_{50} of birds, bees and EC_{50}/ LC_{50} of algae, *Daphnia*, fish)		Remarks
	Half-life *(d = days, h = hours, mo = months, w = weeks, Y = year)*	**log K_{ow}** *(bioac-cumul-ation potential)*	**Water solubility**	**Cancer group** (IARC 2019)	**Endocrine Disruptor (ED)** (McKinlay et al 2008, Mnif et al 2011)	**Toxicity Mammal** LD_{50} (Rats) oral (Tomlin 2000 and 2006)	**Toxicity birds and bees** LD_{50} birds-mallard ducks/ quail; LD_{50} bees oral/ contact (Tomlin 2000 and 2006)	**Ecotoxicity to aquatic organisms** Algae, *Selenastrum capricornutum;* Daphnia, *Daphnia magna*; Fish, rainbow trout (Tomlin 2000 and 2006)	
Imazalil (F)	6.5 mo (soil) (Extoxnet 1993c)	3.82	0.18 g/L	na	na	Rat: 227-343 mg/kg	Birds: >2,510 mg/kg Bees: non-toxic; 40 µg/bee	*Daphnia*, LC_{50}: 3.5 mg/L for 48 h Fish, LC_{50}:1.5 mg/L for 96 h	Very persistent, moderately toxic to mammals, but highly toxic to fish and *Daphnia*.
Imazapyr (H)	6 mo to 2 y	0.11 Very low	9.74-11.3 g/L (pH related)	na	na	Rat: >5,000 mg/kg	Birds: >2,150 mg/kg Bees: >100 µg/bee	*Daphnia*, LC_{50}: >100 mg/L for 48 h Fish, LC_{50}:>100 mg/L for 96 h	Very persistent, non-toxic to mammals, slightly toxic to bees, *Daphnia* and fish.
Imidacloprid (I, Neo)	40-124 d (soil)	0.56 Very low	0.61 g/L	na	na	Rat: 450 mg/kg	Birds: 5,000 mg/kg Bees: Toxic	Algae, EC_{50}: >100 mg/L *Daphnia*, LC_{50}: 85 mg/L for 48 h Fish, LC_{50}: 211 mg/L for 96 h	Moderately toxic to mammals, but toxic to bees.
Indosulfuron-methyl-sodium (H)	50-365 (soil varies with pH)	1.07	0.16-65 g/L	na	na	Rat: 2,678 mg/kg	Birds: na Bees: na	Algae, EC_{50}: 0.02 mg/L Fish, LC_{50}: 100 mg/L for 96 h	Extremely toxic to algae.
Indoxacarb (A)	3.8-17 d (soil)	4.65 High	0.2 mg/L	na	na	Rat: 1,732 mg/kg	Birds: 98 mg/kg Bees: 23.33 µg/bee	*Daphnia*, LC_{50}: 0.60 mg/L for 48 h Fish, LC_{50}: 0.65 mg/L for 96 h	Highly bioaccumulative, extremely toxic to fish and *Daphnia*.

Pesticides	Chemistry (half-life, log K_{ow}, solubility) (Tomlin 2000 and 2006)			Toxicity (Cancer, Endocrine disruption, LD_{50} of mammals)			Ecotoxicity (LD_{50} of birds, bees and EC_{50}/ LC_{50} of algae, *Daphnia*, fish)		Remarks
	Half-life (*d = days, h = hours, mo = months, w = weeks, Y = year*)	**log K_{ow}** (*bioac-cumul-ation potential*)	**Water solubility**	**Cancer group** (IARC 2019)	**Endocrine Disruptor (ED)** (McKinlay et al 2008, Mnif et al 2011)	**Toxicity Mammal** LD_{50} (Rats) oral (Tomlin 2000 and 2006)	**Toxicity birds and bees** LD_{50} birds- mallard ducks/ quail; LD_{50} bees oral/ contact (Tomlin 2000 and 2006)	**Ecotoxicity to aquatic organisms** Algae, *Selenastrum capricornutum;* Daphnia, *Daphnia magna*; Fish, rainbow trout (Tomlin 2000 and 2006)	
Iprodione (F)	20-160 d (soil)	3.0 High	13 mg/L	na	ED	Rat: >2,000 mg/kg	Birds: 10,400 mg/kg Bees: >0.4 µg/bee	Algae, EC_{50}: 1.9 for 120 h *Daphnia*, LC_{50}: 0.25 mg/L for 48 h Fish, LC_{50}: 4.1 mg/L for 96 h	Persistent, ED, extremely toxic to *Daphnia*.
Kreoxsim-methyl- sodium (F)	na	3.4 Medium	2 mg/L	na	na	Rat: >5,000 mg/kg	Birds: 2,150 mg/kg Bees: >20 µg/bee	Algae, EC_{50}: 0.63 mg/L for 120 h *Daphnia*, LC_{50}: 0.186 mg/L for 48 h Fish, LC_{50}: 0.19 mg/L for 96 h	Extremely toxic to algae, *Daphnia* and fish.
Lindane (gamma-HCH)/ Hexachlorocyclo hexanes (I, OC) [POPs]	40 w (Extoxnet 1993d)	3.99 Medium	8.35-8.52 mg/L	1	ED	Rat: 88-270 mg/kg	Birds: 120-130 mg/kg Bees: 0.011 µg/bee	Fish, LC_{50}: 0.022-0.028 mg/L for 96 h Algae, EC_{50}: 0.78 mg/L for 120 h	Very persistent, Bioaccumulative, carcinogenic, ED; toxic to mammals, birds, bees, and extremely toxic to algae and fish.
Linuron (H)	38-67d (soil)	3.00 Medium	63.8 mg/L	na	ED	Rat: 1,500-4,000 mg/kg	Birds: 940 mg/kg Bees: >1,600 µg/bee	*Daphnia*, LC_{50}: 0.75 mg/L for 48 h Fish, LC_{50}: 3.15 mg/L for 96 h	ED, extremely toxic to *Daphnia*.

Pesticides	**Chemistry** (half-life, log K_{ow}, solubility) (Tomlin 2000 and 2006)			**Toxicity** (Cancer, Endocrine disruption, LD_{50} of mammals)			**Ecotoxicity** (LD_{50} of birds, bees and EC_{50}/ LC_{50} of algae, *Daphnia*, fish)		**Remarks**
	Half-life *(d = days, h = hours, mo = months, w = weeks, Y = year)*	**log K_{ow}** *(bioac-cumul-ation potential)*	**Water solubility**	**Cancer group** (IARC 2019)	**Endocrine Disruptor (ED)** (McKinlay et al 2008, Mnif et al 2011)	**Toxicity Mammal** LD_{50} (Rats) oral (Tomlin 2000 and 2006)	**Toxicity birds and bees** LD_{50} birds- mallard ducks/ quail; LD_{50} bees oral/ contact (Tomlin 2000 and 2006)	**Ecotoxicity to aquatic organisms** Algae, *Selenastrum capricornutum;* Daphnia, *Daphnia magna*; Fish, rainbow trout (Tomlin 2000 and 2006)	
Malathion (OP)	na	2.75 Medium	145 mg/L	2A	ED	Rat: 1,375-2,800 mg/kg	Birds: 3,500 mg/kg Bees: Toxic 0.71 µg/bee	Algae, EC_{50}: 13 mg/L *Daphnia*, LC_{50}: 0.001 mg/L for 48 h Fish, LC_{50}: 0.1 mg/L for 96 h	Probably carcinogenic, ED, extremely toxic to *Daphnia* and fish.
Mancozeb (F)	6-15 d (soil)	0.26 Very low	6.2 mg/L	na	na	Rat: >5,000 mg/kg	Birds: 6,400 mg/kg Bees: na	Fish, LC_{50}: 2.2 mg/L for 96 h	Moderately toxic to fish.
MCPA (H)	< 7 d (soil)	2.75 Medium	26.2-395 mg/L (pH related)	na	na	Rat: 700-1,160 mg/kg	Birds: 377 mg/kg Bees: 0.104 µg/bee	Algae, EC_{50}: >392 mg/L *Daphnia*, LC_{50}: >100 mg/L Fish, LC_{50}: 232 mg/L for 96 h	Moderately toxic to birds, bees.
Methidathion (I, OP)	3-18 d (soil)	2.2 Medium	200 mg/L	na	na	Rat: 25-54 mg/kg	Birds: 23.6-28 mg/kg Bees: Slightly toxic	Algae, EC_{50}: 22 mg/L for 72 h Fish, LC_{50}: 0.01 mg/L for 96 h	Extremely toxic to mammals and fish.
Methomyl (I, C)	na	0.093 Very low	57.9 mg/L	na	ED	Rat: 30-34 mg/kg	Birds: 15.9 mg/kg Bees: Toxic; 0.1 µg/bee	Algae, EC_{50}: 0.0317 mg/L for 72 h *Daphnia*, LC_{50}: 0.0317 mg/L for 48 h Fish, LC_{50}: 3.4 mg/L for 96 h	ED, extremely toxic to mammals, birds, bees, algae and *Daphnia.*

Pesticides	Chemistry (half-life, log K_{ow}, solubility) (Tomlin 2000 and 2006)			Toxicity (Cancer, Endocrine disruption, LD_{50} of mammals)			Ecotoxicity (LD_{50} of birds, bees and EC_{50}/ LC_{50} of algae, *Daphnia*, fish)		Remarks
	Half-life *(d = days, h = hours, mo = months, w = weeks, Y = year)*	**log K_{ow}** *(bioac-cumul-ation potential)*	**Water solubility**	**Cancer group** (IARC 2019)	**Endocrine Disruptor (ED)** (McKinlay et al 2008, Mnif et al 2011)	**Toxicity Mammal** LD_{50} (Rats) oral (Tomlin 2000 and 2006)	**Toxicity birds and bees** LD_{50} birds- mallard ducks/ quail; LD_{50} bees oral/ contact (Tomlin 2000 and 2006)	**Ecotoxicity to aquatic organisms** Algae, *Selenastrum capricornutum;* Daphnia, *Daphnia magna*; Fish, rainbow trout (Tomlin 2000 and 2006)	
Methoprene (I)	10 d (soil)	>6 Very high	0.76 mg/L	na	na	Rat: >5,000 mg/kg	Birds: na Bees: Non-toxic; >1,000 µg/bee	Algae, EC_{50}: 1.33 mg/L for 72 h *Daphnia*, LC_{50}: 0.036 mg/L for 48 h Fish, LC_{50}: 4.4 mg/L for 96 h	Highly bioaccumulative and extremely toxic to *Daphnia*.
Methoxychlor (I)	46 d	4.83-5.08 High	0.1-0.2 mg/L	3	ED	Rat: 6,000 mg/kg	Birds: >2,000 mg/kg Bees: na	*Daphnia*, LC_{50}: 0.00078 mg/L for 48 h Fish, LC_{50}: 0.052 mg/L for 96 h	Bioaccumulative, ED, extremely toxic to *Daphnia* and fish.
Metiram (F)	na	0.3 Very low	Insoluble	na	ED	Rat: >10,000 mg/kg	Birds: Slightly toxic Bees: >50 µg/bee	Algae, EC_{50}: 0.3 mg/L for 96 h *Daphnia*, LC_{50}: 2.55 mg/L for 48 h Fish, LC_{50}: 1.1 mg/L for 96 h	ED, Extremely toxic to algae; highly toxic to *Daphnia* and fish.
Metolachlor (H)	20 d	2.9 Medium	488 mg/L	na	ED	Rat: 2,780 mg/kg	Birds: >2,150 mg/kg Bees: >100 µg/bee	Algae, EC_{50}: 0.1 mg/L *Daphnia*, LC_{50}: 25 mg/L for 48 h Fish, LC_{50}: 3.9 mg/L for 96 h	ED, extremely toxic to algae.
Metosulam (H)	25 d (soil)	0.978 Very low	100-5,000 mg/L (pH related)	na	na	Rat: >5,000 mg/kg	Birds: >2,000 mg/kg Bees: Not-toxic; >50 µg/bee	Algae, EC_{50}: 0.075 mg/L	Extremely to toxic to algae.

Pesticides	Chemistry (half-life, log K_{ow}, solubility) (Tomlin 2000 and 2006)			Toxicity (Cancer, Endocrine disruption, LD_{50} of mammals)			Ecotoxicity (LD_{50} of birds, bees and EC_{50}/ LC_{50} of algae, *Daphnia*, fish)		Remarks
	Half-life *(d = days, h = hours, mo = months, w = weeks, Y = year)*	**log K_{ow}** *(bioac-cumul-ation potential)*	**Water solubility**	**Cancer group** (IARC 2019)	**Endocrine Disruptor (ED)** (McKinlay et al 2008, Mnif et al 2011)	**Toxicity Mammal** LD_{50} (Rats) oral (Tomlin 2000 and 2006)	**Toxicity birds and bees** LD_{50} birds-mallard ducks/ quail; LD_{50} bees oral/ contact (Tomlin 2000 and 2006)	**Ecotoxicity to aquatic organisms** Algae, *Selenastrum capricornutum;* Daphnia, *Daphnia magna*; Fish, rainbow trout (Tomlin 2000 and 2006)	
Metsulfuron-methyl (H)	1-5 w (soil)	-1.74 Very low	0.55-213 g/L (pH related)	na	na	Rat: >5,000 mg/kg	Birds: >5,000 mg/kg Bees: Non-toxic; >25 µg/bee	Fish, LC_{50}: >150 mg/L for 96 h *Daphnia*, LC_{50}: >150 mg/L for 48 h	Non-toxic to mammals, bees, fish, *Daphnia* and algae.
Mirex (I, OC) [POPs]	12 y (soil) (CDC 2017)	5.28 High	Insoluble	2B	ED	Rat: 235 mg/kg (MSDS-Sigma-Mirex 2012)	Birds: na Bees: na	*Daphnia*, LC_{50}: 2.6 mg/L for 48 h Fish, LC_{50}: 0.023 mg/L for 96 h	Very persistent, possibly carcinogenic, highly bioaccumulative, ED, extremely toxic to fish.
Molinate (H)	40-160 d (flooded soil)	2.88 Medium	990 mg/L	na	ED	Rat: 369-450 mg/kg	Birds: na Bees: na	Fish, LC_{50}: 13 mg/L for 96 h	ED, moderately toxic to mammals.
Monocrotophos (I, OP, A)	1-5 d (soil)	0.22 Very low	Very soluble	na	na	Rat: 18-20 mg/kg	Birds: 4.8 mg/kg Bees: Highly toxic	*Daphnia*, LC_{50}: 0.00024 mg/L for 48 h Fish, LC_{50}: 7 mg/L for 96 h	Extremely toxic to mammals, birds, bees, and *Daphnia*.
Myclobutanil (F)	66 d (soil)	2.94 Medium	142 mg/L	na	ED	Rat: 1,600-2,290 mg/kg	Birds: 510 mg/kg Bees: Non-toxic	*Daphnia*, LC_{50}: 11 mg/L for 48 h Fish, LC_{50}: 4.2 mg/L for 96 h	ED, moderately toxic to fish.

Pesticides	Chemistry (half-life, log K_{ow}, solubility) (Tomlin 2000 and 2006)			Toxicity (Cancer, Endocrine disruption, LD_{50} of mammals)			Ecotoxicity (LD_{50} of birds, bees and EC_{50}/ LC_{50} of algae, *Daphnia,* fish)		Remarks
	Half-life *(d = days, h = hours, mo = months, w = weeks, Y = year)*	**log K_{ow}** *(bioaccumulation potential)*	**Water solubility**	**Cancer group** (IARC 2019)	**Endocrine Disruptor (ED)** (McKinlay et al 2008, Mnif et al 2011)	**Toxicity Mammal** LD_{50} (Rats) oral (Tomlin 2000 and 2006)	**Toxicity birds and bees** LD_{50} birds- mallard ducks/ quail; LD_{50} bees oral/ contact (Tomlin 2000 and 2006)	**Ecotoxicity to aquatic organisms** Algae, *Selenastrum capricornutum;* Daphnia, *Daphnia magna*; Fish, rainbow trout (Tomlin 2000 and 2006)	
Omethoate (I, OP, A)	Few days (soil)	-0.74 Very low	Soluble	na	na	Rat: 25 mg/kg	Birds: 79.7 mg/kg Bees: Toxic	Algae, EC_{50}: 167.5 mg/L *Daphnia*, LC_{50}: 0.022 mg/L for 48 h Fish, LC_{50}: 9.1 mg/L for 96 h	Extremely toxic to mammals, *Daphnia.*
Paraquat dichloride (H)	< 1 w (soil)	-4.5 Very low	620 g/L	na	na	Rat: 129-157 mg/kg	Birds: 199 mg/kg Bees: 36 µg/bee	Algae, EC_{50}: 0.28 mg/L for 96 h *Daphnia*, LC_{50}: 6.1 mg/L for 48 h Fish, LC_{50}: 26 mg/L for 96 h	Moderately toxic to mammals and birds and extremely toxic to algae.
Parathion (I, OP, A)	1-10 d (water) (Extoxnet 1993e)	3.83 Medium	11 mg/L	2B	ED	Rat: 2 mg/kg	Birds: na Bees: Toxic	Fish, LC_{50}: 1.5 mg/L for 96 h *Daphnia*, LC_{50}: 0.0025 mg/L for 10 h Algae, EC_{50}: 0.5 mg/L	Possibly carcinogenic, ED, extremely toxic to mammals, *Daphnia* and algae.
Parathion-methyl (I, OP)	na	3	55 mg/L	na	na	Rat: 3 mg/kg	Birds: na Bees: Toxic	Algae, EC_{50}: 3 mg/L *Daphnia*, LC_{50}: 0.0073 mg/L for 48 h Fish, LC_{50}: 2.7 mg/L for 96 h	Extremely toxic to mammals and *Daphnia.*
Penconazole (F)	133-343 d (soil)	3.72 Medium	73 g/m^3	na	ED	Rat: 2,125 mg/kg	Birds: >1,590 mg/kg Bees: Non-toxic	Algae, EC_{50}: 3.0 mg/L for 5d *Daphnia*, LC_{50}: 7-11 mg/L for 48 h Fish, LC_{50}: 1.7-4.3 mg/L for 96 h	Bio-accumulative, ED, moderately toxic to fish, *Daphnia* and algae.

Pesticides	Chemistry (half-life, log K_{ow}, solubility) (Tomlin 2000 and 2006)			Toxicity (Cancer, Endocrine disruption, LD_{50} of mammals)			Ecotoxicity (LD_{50} of birds, bees and EC_{50}/ LC_{50} of algae, *Daphnia,* fish)		Remarks
	Half-life (*d = days, h = hours, mo = months, w = weeks, Y = year*)	**log K_{ow}** (*bioac-cumul-ation potential*)	**Water solubility**	**Cancer group** (IARC 2019)	**Endocrine Disruptor (ED)** (McKinlay et al 2008, Mnif et al 2011)	**Toxicity Mammal** LD_{50} (Rats) oral (Tomlin 2000 and 2006)	**Toxicity birds and bees** LD_{50} birds-mallard ducks/ quail; LD_{50} bees oral/ contact (Tomlin 2000 and 2006)	**Ecotoxicity to aquatic organisms** Algae, *Selenastrum capricornutum;* Daphnia, *Daphnia magna*; Fish, rainbow trout (Tomlin 2000 and 2006)	
Pendimethalin (H)	3-4 mo (soil)	5.2 High	0.33	na	na	Rat: 1,050-1,250 mg/kg	Birds: na Bees: >50 µg/bee	*Daphnia*, LC_{50}: 0.28 mg/L for 48 h Fish, LC_{50}: 0.14 mg/L for 96 h	Highly bioaccumulative, and extremely toxic to fish and *Daphnia*.
Permethrin (I, SP)	< 38 d (soil)	6.1 Very high	$6x10^{-3}$ mg/L	na	ED	Rat: 430-4,000 mg/kg	Birds: >9,800 mg/kg Bees: Toxic; 0.098 µg/bee	*Daphnia*, LC_{50}: 0.0006 mg/L for 48 h Fish, LC_{50}: 0.0025 mg/L for 96 h	Highly bioaccum-ulative; extremely toxic to bees, fish and *Daphnia*.
Phenothrin (I, SP)	na	6.01 Very high	<9.7 µg/L	na	na	Rat: >5,000 mg/kg (MSDS-Sumitomo-Chem 2015)	Birds: 2,500 mg/kg Bees: na	*Daphnia*, LC_{50}: 0.0043 mg/L for 48 h Fish, LC_{50}: 0.0027 mg/L for 96 h	Highly bioaccumulative, extremely toxic to *Daphnia* and fish.
Phosmet (I, OP)	na	2.95 Medium	25 mg/L	na	na	Rat: 113-160 mg/kg	Birds: > 5,000 mg/kg Bees: Highly toxic; 0.001 mg/bee	*Daphnia*, LC_{50}: 0.0085 mg/L for 48 h Fish, LC_{50}: 0.23 mg/L for 96 h	Extremely toxic to bees, *Daphnia* and fish.
Phosphine (I, R)	na	na	26 cm^3/100 ml	na	na	Rat: 8.7 mg/kg	Birds: 37.5 mg/kg Bees: >50 µg/bee	*Daphnia*, LC_{50}: 0.2 mg/L for 24 h Fish, LC_{50}: 0.5 mg/L for 96 h	Extremely toxic to mammals, birds, *Daphnia* and fish.

Pesticides	Chemistry (half-life, log K_{ow}, solubility) (Tomlin 2000 and 2006)			Toxicity (Cancer, Endocrine disruption, LD_{50} of mammals)			Ecotoxicity (LD_{50} of birds, bees and EC_{50}/ LC_{50} of algae, *Daphnia*, fish)		Remarks
	Half-life *(d = days, h = hours, mo = months, w = weeks, Y = year)*	**log K_{ow}** *(bioac-cumul-ation potential)*	**Water solubility**	**Cancer group** (IARC 2019)	**Endocrine Disruptor (ED)** (McKinlay et al 2008, Mnif et al 2011)	**Toxicity Mammal** LD_{50} (Rats) oral (Tomlin 2000 and 2006)	**Toxicity birds and bees** LD_{50} birds-mallard ducks/ quail; LD_{50} bees oral/ contact (Tomlin 2000 and 2006)	**Ecotoxicity to aquatic organisms** Algae, *Selenastrum capricornutum;* Daphnia, *Daphnia magna*; Fish, rainbow trout (Tomlin 2000 and 2006)	
Picloram (H)	30-90 d (soil)	1.9 Low	430 mg/L	na	na	Rat: >5,000 mg/kg	Birds: >5,000 mg/kg Bees: >100 µg/bee	Algae, EC_{50}: 36.9 mg/L *Daphnia*, LC_{50}: 34.4 mg/L Fish, LC_{50}: 5.5 mg/L for 96 h	Persistent, moderately toxic to bees, algae, *Daphnia* and fish.
Primicarb (I, C)	69 d (soil) (MSDS Pubchem 2005d)	1.7 Low	3 g/L	na	ED	Rat: 142 mg/kg	Birds: 28.5 mg/kg Bees: not-toxic	Algae, EC_{50}: 140 mg/L for 96 h *Daphnia*, LC_{50}: 0.017 mg/L for 48 h Fish, LC_{50}: 79 mg/L for 96 h	ED, extremely toxic to birds and *Daphnia*.
Primiphosmethyl (I, OP)	na	4.2 High	9.7-11 mg/L (pH related)	na	na	Rat: 1,414 mg/kg	Birds: 1,695 mg/kg Bees: Toxic	Algae, EC_{50}: 1.0 mg/L *Daphnia*, LC_{50}: 0.00021 mg/L for 48 h Fish, LC_{50}: 0.64 mg/L for 96 h	Highly bioaccumulative, extremely toxic to, *Daphnia* and fish.
Procymidone (F)	4-12 w (soil)	3.14 Medium	4.5 mg/L	na	ED	Rat: 6,800 mg/kg	Birds: na Bees: not -toxic	Fish, LC_{50}: 7.2 mg/L for 96 h	Bioaccumulative, and ED.
Profenopos (I, OP)	1w (soil)	4.44 High	28 mg/L	na	na	Rat: 358 mg/kg	Birds: 150-612 mg/kg Bees: Toxic; 0.102 µg/bee	Algae, EC_{50}: Toxic; 1.16 mg/L *Daphnia*, LC_{50}: 0.001 mg/L Fish, LC_{50}: 0.08 mg/L for 96 h	Highly bioaccumu-lative, extremely toxic to bees, algae, *Daphnia* and fish.

Pesticides	Chemistry (half-life, log K_{ow}, solubility) (Tomlin 2000 and 2006)			Toxicity (Cancer, Endocrine disruption, LD_{50} of mammals)			Ecotoxicity (LD_{50} of birds, bees and EC_{50}/ LC_{50} of algae, *Daphnia*, fish)		Remarks
	Half-life *(d = days, h = hours, mo = months, w = weeks, Y = year)*	**log K_{ow}** *(bioac-cumul-ation potential)*	**Water solubility**	**Cancer group** (IARC 2019)	**Endocrine Disruptor (ED)** (McKinlay et al 2008, Mnif et al 2011)	**Toxicity Mammal** LD_{50} (Rats) oral (Tomlin 2000 and 2006)	**Toxicity birds and bees** LD_{50} birds-mallard ducks/ quail; LD_{50} bees oral/ contact (Tomlin 2000 and 2006)	**Ecotoxicity to aquatic organisms** Algae, *Selenastrum capricornutum;* Daphnia, *Daphnia magna*; Fish, rainbow trout (Tomlin 2000 and 2006)	
Propachlor (H)	4-6 w (soil)	1.4-2.3 Low	580 mg/L	na	na	Rat: 500-1,800 mg/kg	Birds: 91 mg/kg Bees: >197 µg/bee	Algae, EC_{50}: 0.015 mg/L for 72 h *Daphnia*, LC_{50}: 7.8 mg/L for 48 h Fish, LC_{50}: 0.17 mg/L for 96 h	Persistent, extremely toxic to algae and fish.
Propanil (H)	Few days	3.3 Medium	130 mg/L	na	ED	Rat: >2,500 mg/kg	Birds: 375 mg/kg Bees: na	Algae, EC_{50}: na *Daphnia*, LC_{50}: 4.8 mg/L for 48 h Fish, LC_{50}: 8-11 mg/L for 96 h	Bioaccumulative, ED, moderately toxic to birds, fish and *Daphnia*.
Propargite (A)	7-14 w (soil)	3.93 Medium	632 mg/L	na	na	Rat: 2,800 mg/kg	Birds: 4,640 mg/kg Bees: 15 µg/bee	Algae, EC_{50}: >1.08 mg/L for 96 h *Daphnia*, LC_{50}: 0.092 mg/L for 48 h Fish, LC_{50}: 0.118 mg/L for 96 h	Bioaccumulative, extremely toxic to algae, *Daphnia* and fish.
Propiconazole (F)	40-70 d (soil)	3.72 Medium	100 mg/L	na	ED	Rat: 1,517 mg/kg	Birds: >2,510 mg/kg Bees: Non-toxic; >100 µg/bee	Algae, EC_{50}: 0.02-13.6 mg/L *Daphnia*, LC_{50}: 4.8 mg/L for 48 h Fish, LC_{50}: 5.3 mg/L for 96 h	Bioaccumulative, ED extremely toxic to algae.
Prothiofos (OP)	1-2 mo (soil)	5.67 High	0.07 mg/L	na	na	Rat: 1,569 mg/kg	Birds: 100-200 mg/kg Bees: Non-toxic	Algae, EC_{50}: 2.3 mg/L *Daphnia*, LC_{50}: 0.014 mg/L for 48 h Fish, LC_{50}: 0.5-1 mg/L for 96 h	Bioaccumulative, extremely toxic to *Daphnia* and fish.

Pesticides	**Chemistry** (half-life, log K_{ow}, solubility) (Tomlin 2000 and 2006)			**Toxicity** (Cancer, Endocrine disruption, LD_{50} of mammals)			**Ecotoxicity** (LD_{50} of birds, bees and EC_{50}/ LC_{50} of algae, *Daphnia*, fish)		**Remarks**
	Half-life (*d = days, h = hours, mo = months, w = weeks, Y = year*)	**log K_{ow}** (*bioac-cumul-ation potential*)	**Water solubility**	**Cancer group** (IARC 2019)	**Endocrine Disruptor (ED)** (McKinlay et al 2008, Mnif et al 2011)	**Toxicity Mammal** LD_{50} (Rats) oral (Tomlin 2000 and 2006)	**Toxicity birds and bees** LD_{50} birds- mallard ducks/ quail; LD_{50} bees oral/ contact (Tomlin 2000 and 2006)	**Ecotoxicity to aquatic organisms** Algae, *Selenastrum capricornutum;* Daphnia, *Daphnia magna*; Fish, rainbow trout (Tomlin 2000 and 2006)	
Pyraclostrobin (F)	na	3.49 Medium	0.082 mg/L	na	na	Rat: >5,000 mg/kg	Birds: >2,000 mg/kg Bees: >73.1 µg/bee	Algae, EC_{50}: 0.843 mg/L *Daphnia*, LC_{50}: 0.016 mg/L for 48 h Fish, LC_{50}: 0.006 mg/L for 96 h	Bioaccumulative, extremely toxic to algae, *Daphnia* and fish.
Quintozene (F)	4-10 m (soil)	5.1 High	0.1 mg/L	na	na	Rat: >5,000 mg/kg	Birds: 2,000 mg/kg Bees: >100 µg/bee	Algae, EC_{50}: 0.058 mg/L *Daphnia*, LC_{50}: 0.77 mg/L for 48 h Fish, LC_{50}: 0.55 mg/L for 96 h	Very persistent, bioaccumulative, extremely toxic to algae, *Daphnia* and fish.
Simazine (H)	27-102 d	2.1 Low	6.2 mg/L	na	ED	Rat: 500-10,000 mg/kg	Birds: >2,000 mg/kg Bees: >99 µg/bee	Algae, EC_{50}: 0.042 mg/L for 72 h *Daphnia*, LC_{50}: >100 mg/L for 48 h Fish, LC_{50}: >100 mg/L for 96 h	Persistent, ED, extremely toxic to algae.
Spinosad (I)	8.7-250 d (soil)	2.8-5.2 (pH dep-endant) Low to High	89-235 mg/L	3	na	Rat: 3,783-5,000 mg/kg	Birds: >2,000 mg/kg Bees: Highly toxic; >0.0029 µg/bee	Algae, EC_{50}: 105.5 mg/L *Daphnia*, LC_{50}: 14 mg/L for 48 h Fish, LC_{50}: >30 mg/L for 96 h	Bioaccumulative and highly toxic to bees.

Pesticides	Chemistry (half-life, log K_{ow}, solubility) (Tomlin 2000 and 2006)			Toxicity (Cancer, Endocrine disruption, LD_{50} of mammals)			Ecotoxicity (LD_{50} of birds, bees and EC_{50}/ LC_{50} of algae, *Daphnia*, fish)		Remarks
	Half-life *(d = days, h = hours, mo = months, w = weeks, Y = year)*	**log K_{ow}** *(bioac-cumul-ation potential)*	**Water solubility**	**Cancer group** (IARC 2019)	**Endocrine Disruptor (ED)** (McKinlay et al 2008, Mnif et al 2011)	**Toxicity Mammal** LD_{50} (Rats) oral (Tomlin 2000 and 2006)	**Toxicity birds and bees** LD_{50} birds- mallard ducks/ quail; LD_{50} bees oral/ contact (Tomlin 2000 and 2006)	**Ecotoxicity to aquatic organisms** Algae, *Selenastrum capricornutum;* Daphnia, *Daphnia magna*; Fish, rainbow trout (Tomlin 2000 and 2006)	
Tebuconazole (F)	1-4 w	3.7 Medium	36 mg/L	na	na	Rat: 1,700-4,000 mg/kg	Birds: 2,912- 4,438 mg/kg Bees: 0.6 µg/bee	Algae, EC_{50}: 4.01 mg/L for 96 h *Daphnia*, LC_{50}: 4.2 mg/L for 48 h Fish, LC_{50}: 4.4 mg/L for 96 h	Bioaccumulative; moderately toxic to bees, highly toxic to algae, *Daphnia* and fish.
Tebufenozide (A)	7-179 d (soil)	4.25 High	0.83 mg/L	na	na	Rat: >5,000 mg/kg	Birds: >2,150 mg/kg Bees: >234 µg/bee	Fish, LC_{50}: 5.7 mg/L for 96 h Algae, EC_{50}: 0.64 mg/L for 96 h *Daphnia*, LC_{50}: 3.8 mg/L for 48 h	Persistent, bio-accumulative, extremely toxic to *Daphnia*.
Tebufenpyrad (A)	20-30 d	4.93 High	2.61 mg/L	na	na	Rat: 595-997 mg/kg	Birds: >2,000 mg/kg Bees: Low toxicity	Algae, EC_{50}: 0.54 mg/L *Daphnia*, LC_{50}: 0.046 mg/L for 48 h Fish, LC_{50}: 0.03 mg/L for 96 h	Bioaccumulative; extremely toxic to algae *Daphnia* and fish.
Tebuthiuron (H)	na	1.82 Low	2.5 mg/L	na	na	Rat: 644 mg/kg	Birds: >500 mg/kg Bees: >100 µg/bee	Algae, EC_{50}: 0.05 mg/L *Daphnia*, LC_{50}: 297 mg/L for 48 h Fish, LC_{50}: 144 mg/L for 96 h	Extremely toxic to algae.
Temephos (OP, I)	50 d	4.91 High	0.03 mg/L	na	na	Rat: 4,204-10,000 mg/kg	Birds: na Bees: Highly toxic; 1.55 µg/bee	Fish, LC_{50}: 31.8 mg/L for 96 h	Bioaccumulative, highly toxic to bees.

Pesticides	Chemistry (half-life, log K_{ow}, solubility) (Tomlin 2000 and 2006)			Toxicity (Cancer, Endocrine disruption, LD_{50} of mammals)			Ecotoxicity (LD_{50} of birds, bees and EC_{50}/ LC_{50} of algae, *Daphnia*, fish)		Remarks
	Half-life *(d = days, h = hours, mo = months, w = weeks, Y = year)*	**log K_{ow}** *(bioac-cumul-ation potential)*	**Water solubility**	**Cancer group** (IARC 2019)	**Endocrine Disruptor (ED)** (McKinlay et al 2008, Mnif et al 2011)	**Toxicity Mammal** LD_{50} (Rats) oral (Tomlin 2000 and 2006)	**Toxicity birds and bees** LD_{50} birds-mallard ducks/ quail; LD_{50} bees oral/ contact (Tomlin 2000 and 2006)	**Ecotoxicity to aquatic organisms** Algae, *Selenastrum capricornutum;* Daphnia, *Daphnia magna*; Fish, rainbow trout (Tomlin 2000 and 2006)	
Terbufos (OP, I, N)	9-27 d (soil)	2.77 Low	4.5 mg/L	na	na	Rat: 1.6-5.4 mg/kg	Birds: - Bees: Non-toxic; 4.1 μg/bee	Fish, LC_{50}: 0.01 mg/L for 96 h	Extremely toxic to mammals and fish.
Tetradifon (A)	36 w (IPCS 1986)	4.61 High	0.078 mg/L	na	na	Rat: >14,700 mg/kg	Birds: >5,000 mg/kg Bees: Non-toxic	Algae, EC_{50}: >100 mg/L for 96 h *Daphnia*, LC_{50}: >2 mg/L for 48 h Fish, LC_{50}: 1,200 mg/L for 96 h	Highly persistent, bioaccumulative; highly toxic to *Daphnia.*
Thiabendazole (F)	33-120 d (soil)	2.39 Low	0.03-0.16 mg/L (pH related)	na	na	Rat: 3,600 mg/kg	Birds: >2,250 mg/kg Bees: Non-toxic	Algae, EC_{50}: 3.2 mg/L for 96 h *Daphnia*, LC_{50}: 0.81 mg/L for 48 h Fish, LC_{50}: 0.055 mg/L for 96 h	Extremely toxic to *Daphnia*; and fish.
Thiacloprid (I, Ne)	12-142 d (MSDS Pubchem 2005a)	1.26 Low	185 mg/L	na	na	Rat: 396-836 mg/kg	Birds: 2,716 mg/kg Bees: 17.32 μg/bee	Algae, EC_{50}: 97 mg/L for 96 h *Daphnia*, LC_{50}: na Fish, LC_{50}: 30.5 mg/L for 96 h	Persistent, moderately toxic to mammal.
Thiobencarb (H)	3 w to 8 mo	4.23 High	16.7 mg/L	na	na	Rat: 1,033-1,130 mg/kg	Birds: >10,000 mg/kg Bees: >100 μg/bee	Algae, EC_{50}: 0.017 mg/L for 120 h *Daphnia*, LC_{50}: 0.1 mg/L for 48 h Fish, LC_{50}: 2.4 mg/L for 96 h	Highly persistent, bioaccumulative, extremely toxic to algae and *Daphnia.*

Pesticides	Chemistry (half-life, log K_{ow}, solubility) (Tomlin 2000 and 2006)			Toxicity (Cancer, Endocrine disruption, LD_{50} of mammals)			Ecotoxicity (LD_{50} of birds, bees and EC_{50}/ LC_{50} of algae, *Daphnia*, fish)		Remarks
	Half-life *(d = days, h = hours, mo = months, w = weeks, Y = year)*	**log K_{ow}** *(bioac- cumul- ation potential)*	**Water solubility**	**Cancer group** (IARC 2019)	**Endocrine Disruptor (ED)** (McKinlay et al 2008, Mnif et al 2011)	**Toxicity Mammal** LD_{50} (Rats) oral (Tomlin 2000 and 2006)	**Toxicity birds and bees** LD_{50} birds- mallard ducks/ quail; LD_{50} bees oral/ contact (Tomlin 2000 and 2006)	**Ecotoxicity to aquatic organisms** Algae, *Selenastrum capricornutum;* Daphnia, *Daphnia magna*; Fish, rainbow trout (Tomlin 2000 and 2006)	
Thiodicarb (C, I, M)	3-8 d (soil)	1.7 Low	19.1 mg/L	na	na	Rat: 66 mg/kg	Birds: 2,023 mg/kg Bees: Moderate toxic	*Daphnia*, LC_{50}: 0.053 mg/L for 48 h Fish, LC_{50}: 2.55 mg/L for 96 h	Extremely toxic to mammals and *Daphnia*.
Thiram (F)	0.5 (soil)	1.73 Low	18 mg/L	3	ED	Rat: 2,600 mg/kg	Birds: >2,800 mg/kg Bees: 73.5 µg/bee	Algae, EC_{50}: 0.065 mg/L for 120 h *Daphnia*, LC_{50}: 0.21 mg/L for 48 h Fish, LC_{50}: 0.128 mg/L for 96 h	ED, Extremely toxic to fish, *Daphnia* and algae.
Toxaphene (I, OC) [POPs]	100 d- 12 yr (Hatfield 1995a)	3.33-6.44 High	Negligible	2B	ED	Rat: 12.27 mg/kg (MSDS- Dunrite Toxaphene 2000)	Birds: na Bees: na	Fish, LC_{50}: 0.56 mg/L for 96 h	Highly persistent and bioaccumulative, carcinogenic and extremely toxic to mammals and fish.
Tralkoxydim (H)	2-5 d (water)	2.1 Low	6.7 mg/L	na	na	Rat: 934-1,231 mg/kg	Birds: >3,020 mg/kg Bees: Non-toxic	Algae, EC_{50}: >7.6 mg/L *Daphnia*, LC_{50}: 175 mg/L for 48 h Fish, LC_{50}: 7.2 mg/L for 96 h	Moderately toxic to fish; non- toxic to bees.
Triadimenol (F)	110-275 d (soil)	3.08-3.28	33-62 mg/L	na	ED	Rat: 700 mg/kg	Birds: > 2,000 mg/kg Bees: Non-toxic	Algae, EC_{50}: >3.7 mg/L *Daphnia*, LC_{50}: 51 mg/L for 48 h Fish, LC_{50}: 1.4-23.5 mg/L for 96 h	Persistent, bioaccumulative, toxic to algae and fish.

Pesticides	Chemistry (half-life, log K_{ow}, solubility) (Tomlin 2000 and 2006)			Toxicity (Cancer, Endocrine disruption, LD_{50} of mammals)			Ecotoxicity (LD_{50} of birds, bees and EC_{50}/ LC_{50} of algae, *Daphnia*, fish)		Remarks
	Half-life (*d = days, h = hours, mo = months, w = weeks, Y = year*)	**log K_{ow}** (*bioaccumulation potential*)	**Water solubility**	**Cancer group** (IARC 2019)	**Endocrine Disruptor (ED)** (McKinlay et al 2008, Mnif et al 2011)	**Toxicity Mammal** LD_{50} (Rats) oral (Tomlin 2000 and 2006)	**Toxicity birds and bees** LD_{50} birds-mallard ducks/ quail; LD_{50} bees oral/ contact (Tomlin 2000 and 2006)	**Ecotoxicity to aquatic organisms** Algae, *Selenastrum capricornutum;* Daphnia, *Daphnia magna*; Fish, rainbow trout (Tomlin 2000 and 2006)	
Triasulfuron (H)	3 d (soil)	-0.59-1.1 V. low-Low	32-13,500 mg/L (pH related)	na	na	Rat: >5,000 mg/kg	Birds: > 2,150 mg/kg Bees: Non-toxic; >100 µg/bee	Algae, EC_{50}: 0.035 mg/L for 5-14 d *Daphnia*, LC_{50}: >100 mg/L for 48 h Fish, LC_{50}: >100 mg/L for 96 h	Extremely toxic to algae.
Trichlorfon (OP)	1-27 d (soil) (Extoxnet 1993b)	0.43 Very low	120 g/L	na	na	Rat: 250 mg/kg	Birds: na Bees: Low toxicity	Algae, EC_{50}: >10 mg/L *Daphnia*, LC_{50}: 0.0009 mg/L for 48 h Fish, LC_{50}: 0.7 mg/L for 96 h	Extremely toxic to *Daphnia* and fish.
Trifloxystrobin (F)	4.2-9.5 d (soil); 0.3-1 d (water)	4.5 High	610 µg/L	na	na	Rat: >5,000 mg/kg	Birds: >2,000 mg/kg Bees: >200 µg/bee	Algae, EC_{50}: 0.0053 mg/L *Daphnia*, LC_{50}: 0.016 mg/L for 48 h Fish, LC_{50}: 0.015 mg/L for 96 h	Highly bioaccumulative, extremely toxic to fish, *Daphnia* and algae.
Triflumuron (I)	Quickly degrade (soil)	4.91 High	0.025 mg/L	na	na	Rat: >5,000 mg/kg	Birds: 561 mg/kg Bees: Toxic to bees	Algae, EC_{50}: >25 mg/L for 96 h *Daphnia*, LC_{50}: 0.225 mg/L for 48 h Fish, LC_{50}: >320 mg/L for 96 h	Bioaccumulative, extremely toxic to *Daphnia* and toxic to bees.
Trifluralin (H)	57-126 d (soil)	4.83 High	0.184-0.39 mg/L (pH related)	3	ED	Rat: >5,000 mg/kg	Birds: >2,000 mg/kg Bees: >100 µg/bee	Algae, EC_{50}: 12.2 mg/L for 7 d *Daphnia*, LC_{50}: 0.245 mg/L for 48 h Fish, LC_{50}: 0.088 mg/L for 96 h	Persistent, bioaccumulative, extremely toxic to *Daphnia* and fish.

Pesticides	Chemistry (half-life, log K_{ow}, solubility) (Tomlin 2000 and 2006)			Toxicity (Cancer, Endocrine disruption, LD_{50} of mammals)			Ecotoxicity (LD_{50} of birds, bees and EC_{50}/ LC_{50} of algae, *Daphnia*, fish)		Remarks
	Half-life *(d = days, h = hours, mo = months, w = weeks, Y = year)*	**log K_{ow}** *(bioaccumulation potential)*	**Water solubility**	**Cancer group** (IARC 2019)	**Endocrine Disruptor (ED)** (McKinlay et al 2008, Mnif et al 2011)	**Toxicity Mammal** LD_{50} (Rats) oral (Tomlin 2000 and 2006)	**Toxicity birds and bees** LD_{50} birds- mallard ducks/ quail; LD_{50} bees oral/ contact (Tomlin 2000 and 2006)	**Ecotoxicity to aquatic organisms** Algae, *Selenastrum capricornutum;* Daphnia, *Daphnia magna*; Fish, rainbow trout (Tomlin 2000 and 2006)	
Triticonazole (F)	224-260 d (soil)	3.29 Medium	7 mg/L	na	na	Rat: >2,000 mg/kg	Birds: >2,000 mg/kg Bees: Non-toxic	Algae, EC_{50}: >1.0 mg/L for 96 h *Daphnia*, LC_{50}: >9.3 mg/L for 48 h Fish, LC_{50}: 10 mg/L for 96 h	Persistent, bioaccumulative, very toxic to algae.
Ziram (F)	42 h (soil)	1.23 low	1.58-18.3 mg/L	3	ED	Rat: 2,068 mg/kg	Birds: >5,250 mg/kg Bees: > 100 µg/bee	*Daphnia*, LC_{50}: 0.048 mg/L for 48 h Fish, LC_{50}: 0.16 mg/L for 96 h	Extremely toxic to fish and *Daphnia*.

- **Organophosphates:** Chlorpyrifos (log K_{ow}: 4.7), Coumaphos (log K_{ow}: 4.13), Ethion (log K_{ow}: 4.28), Fenthion (log K_{ow}: 4.84), Primiphos-methyl (log K_{ow}: 4.2), Profenopos (log K_{ow}: 4.4), Prothiofos (log K_{ow}: 5.6), Temephos (log K_{ow}: 4.91),
- **Synthetic pyrethroids:** Bifenthrin (log K_{ow}: >6.0), Cyfluthrin (log K_{ow}: 6.0), Cyhalothrin (log K_{ow}: 6.6), Cypermethrin (log K_{ow}: 6.6), Deltamethrin (log K_{ow}: 4.6), Esfenvalerate (log K_{ow}: 6.22), Fenvalerate (log K_{ow}: 5.01), Permethrin (log K_{ow}: 6.1), Phenothrin (log K_{ow}: 6.01).

7.2.3. Toxicity (mammals)

All pesticides are toxic to some forms of life. There is a standard toxicity classification schemes for rating the relative toxicity of pesticides to different groups of organisms (mammals, birds, bees, fish, daphnia and algae). The classification helps in interpreting toxicity data and pesticide risk assessment as given in Table 7.5. Several pesticides are **extremely** and **acutely toxic to Mammals** (LD_{50}: <50 mg/kg) including the following: *Acrolein* (LD_{50}: 29 mg/kg); ***Aldicarb*** (LD_{50}: 0.93 mg/kg); *Aldrin* (LD_{50}: 38 mg/kg); *Azinphos-methyl* (LD_{50}: 9 mg/kg); *Carbofuran* (LD_{50}: 8 mg/kg); *Chlorfenvinphos* (LD_{50}: 10 mg/kg), *Coumaphos* (LD_{50}: 16 mg/kg); *Demeton* (LD_{50}: 30 mg/kg); *Dichlorvos* ((LD_{50}: 50 mg/kg)*; Dieldrin* (LD_{50}: 38 mg/kg); *Endosulfan* (LD_{50}: 44.9 mg/kg); *Methidathion* (LD_{50}: 26 mg/kg); *Methomyl* (LD_{50}: 30 mg/kg); *Monocrotophos* (LD_{50}: 18 mg/kg); *Omethoate* (LD_{50}: 25 mg/kg); *Parathion* (LD_{50}: 2 mg/kg); *Parathion-methyl* (LD_{50}: 3 mg/kg); *Phosphine* (LD_{50}: 8.7 mg/kg)*; Terbufos* (LD_{50}: 1.6 mg/kg); *Toxaphene* (LD_{50}: 12.27 mg/kg) (see Table 7.4 for details). Based on LD_{50} values, Aldicarb (a carbamate pesticide) is highly toxic to mammals. In addition, a number of pesticides are also **carcinogenic** (cancer-causing) and are as follows: *IARC*[49] *group 1 - carcinogenic to humans* (Lindane); *IARC group 2A - probably carcinogenic to humans* (Aldrin, DDT, Diazinon, Dieldrin, Glyphosate, Malathion); *IARC group 2B - possibly carcinogenic to humans* (2,4-D; Chlordane, Chlorothalonil, HCB (hexachlorobenzene), Heptachlor, Mirex and Toxaphene). A total of 109 pesticides have been identified as **endocrine disruptors**[50] including herbicides, insecticides and fungicides (Table 7.6).

[49]***IARC = International Agency for Research on Cancer.*** *IARC is an inter-governmental agency forming part of the World Health Organization (WHO) of the United Nations. Its role is to conduct and coordinate research into the causes of cancer. It also collects and publishes surveillance data regarding the occurrence of cancer worldwide.*

[50]***Endocrine-disrupting*** *(ED) chemicals are compounds that alter the normal functioning of the endocrine system, potentially causing disease or deformity in organisms and their offsprings.*

Table 7.5. Toxicity classification of pesticides (Kamrin 1997).

Mammals	Birds	Aquatic organisms
High toxicity Oral LD_{50}: 0-50 mg/kg	***High toxicity*** Oral LD_{50}[51]: 10-50 mg/kg	***Very high toxicity*** Acute[52] LC_{50}: 0.1-1 mg/L; Chronic[53] LC_{50}: <0.01-0.1 mg/L
Moderate toxicity Oral LD_{50}: >50-500 mg/kg	***Moderate toxicity*** Oral LD_{50}: >50-500 mg/kg	***Moderate toxicity*** Acute LC_{50}: >1-10 mg/L; Chronic LC_{50}: >0.1-1 mg/L
Slight toxicity Oral LD_{50}: >500-5,000 mg/kg	***Slight toxicity*** Oral LD_{50}: >500-5,000 mg/kg	***Slight-toxicity*** Acute LC_{50}: > 10-100 mg/L; Chronic LC_{50}: >1-10 mg/L

Table 7.6. List of pesticides considered as an endocrine disruptor

[based on McKinlay et al 2008, Mnif et al 2011; A = acaricide; F = fungicide; H = herbicide; I = insecticide; M = molluscicide; N = nematicide; T = termiticide].

Sl. No.	Pesticides	Sl. No.	Pesticides	Sl. No.	Pesticides
1	2,4-D (H)	38	Epoxyconazole (F) (preventive and curative)	75	Phenylphenol (F) (disinfectant)
2	Acephate (I) *(systemic[54] insecticides)*	39	Fenarimol (F)	76	Primicarb (I)
3	Acetochlor (H)	40	Fenbuconazole (F)	77	Maneb (F)
4	Alachlor (H)	41	Fenitrothion (I)	78	Metriam (F)

[51]***LD_{50} or LC_{50}*** *is the median lethal dose (LD_{50}) or median lethal concentration (LC_{50}), which is the dose or the concentration of the test substance that kills 50% of test organisms within a given time period.*

[52]***Acute toxicity*** *is the adverse effects of a substance that result either from a single exposure or from multiple exposures in a short period of time (usually less than 24 h).*

[53]***Chronic toxicity*** *is the development of adverse effects as the result of long-term exposure to a toxicant or other stressor.*

[54]***Systemic pesticide*** *is water-soluble pesticide, is absorbed by a plant when applied to seeds, soil, or leaves (https://www.gardeningknowhow.com/plant-problems/pests/pesticides/systemic-pesticide-in-gardens.htm).*

Sl. No.	Pesticides	Sl. No.	Pesticides	Sl. No.	Pesticides
5	Aldicarb (I)	42	Fenoxycarb (I)	79	Methomyl (I, A)
6	Aldrin (I) *(non-systemics)*	43	Fenvalerate (I, A)	80	Methoxychlor (I)
7	Amitrole (H)	44	Fipronil (I)	81	Prochloraz (F)
8	Atrazine (H)	45	Fluvalinate (I, A) *(contact insecticides)*	82	Procymidone (F)
9	Bendiocarb (I, A, N)	46	Flusilazole (F)	83	Prodiamine (I)
10	Benomyl (F)	47	Flutriafol (F)	84	Propamocarb (F)
11	Bifenthrin (I, A)	48	Glyphosate (H)	85	Propanil (H)
12	Bioallethrin (I)	49	HCB (F)	86	Propazine (H)
13	Bitertanol (F)	50	HCH (Lindane) (I)	87	Propiconazole (F)
14	Bupirimate (F)	51	Heptachlor (I)	88	Propoxur (I)
15	Captan (F)	52	Hexaconazole (F)	89	Prothiophos (I)
16	Carbaryl (I)	53	Isoproturon (H)	90	Pyridate (H)
17	Carbofuran (I)	54	Iprodione (F)	91	Pyrifenox (F)
18	Chlorothalonil (F)	55	Kepone (I)	92	Pyrimethanil (F)
19	Chlordane (I)	56	Linuron (H)	93	Pyripyroxifen (I)
20	Chlordecone (I)	57	Malathion (I)	94	Resmethrin (I)
21	Chlorfenviphos (I)	58	Methiocarb (H)	95	Simazine (H)
22	Chlorpyrifos methyl (I)	59	Methomyl (I)	96	Sumithrin (I)
23	Cyhalothrin (I, A)	60	Methoxychlor (I)	97	Tebuconazole (F)
24	Cypermethrin (I)	61	Metolachlor	98	Tetrabutylin *(antifouling agents)*
25	Cyproconazole (F)	62	Metribuzin (H)	99	Tetramethrin (I)
26	DDT and metabolites (I)	63	Mirex (I)	100	Thiazopyr (H)
27	Deltamethrin (I)	64	Molinate (H)	101	Tolchlofos-methyl (I)
28	Diazinon (I, A)	65	Myclobutanil (F)	102	Toxaphene (I)
29	Dibutyltin dichloride *(antifouling agents)*	66	Nitrofen (H) *(selective herbicides)*	103	Triadimefon (F)
30	Dichlorvos (I, A)	67	Nonachlor (T)	104	Tribenuron-methyl (H)

Sl. No.	Pesticides	Sl. No.	Pesticides	Sl. No.	Pesticides
31	Dicofol (I, A)	68	Nonylphenols (M)	105	Trichlorfon (I)
32	Dieldrin (I)	69	Oxamyl (I, A, N)	106	Trifluralin (H)
33	Diflubenzuron (I)	70	Oxychlordane (I)	107	Triphenyltin (F)
34	Dimethoate (I)	71	Parathion (I, A)	108	Vinclozolin (F)
35	Diuron (H)	72	Penconazole (F)	109	Zineb (F)
36	Endosulfan (I, A)	73	Pentachlorophenol (H, F, I)		
37	Endrin (I) *(foliar insecticides)*	74	Permethrin (I)		

7.2.4. Ecotoxicity[55] to birds, bees, algae, daphnia and fish

7.2.4.1. Birds

Several pesticides are **extremely** and **acutely toxic to Birds** (LD_{50}: < 50 mg/kg) including the following: *Acrolein* (LD_{50}: 19 mg/kg); *Aldrin* (LD_{50}: 50 mg/kg); *Azamethiphos* (LD_{50}: 48.4 mg/kg), *Azinphos-methyl* (LD_{50}: 32 mg/kg); ***Carbofuran*** (LD_{50}: 2.5 mg/kg); *Chlorfenapyr* (LD_{50}: 10 mg/kg); *Chlorfenvinphos* (LD_{50}: 16 mg/kg); *Coumaphos* (LD_{50}: 29.8 mg/kg); *Diazinon* (LD_{50}: 2.7 mg/kg); *Dimethoate* (LD_{50}: 10.5 mg/kg); *Methidathion* (LD_{50}: 23.6 mg/kg); *Methomyl* (LD_{50}: 15.9 mg/kg); *Monocrotophos* (LD_{50}: 4.8 mg/kg); *Phosphine* (LD_{50}: 37.5 mg/kg); *Primicarb* (LD_{50}: 28.5 mg/kg). Based on LD_{50} values, carbofuran (a carbamate pesticide) is very highly toxic to birds.

7.2.4.2. Bees

Several pesticides are **extremely** and **acutely toxic to Bees** (LD_{50}: < 1 μg/bee) including the following: *Azamethiphos* (LD_{50}: <0.1 μg/bee), *Bifenthrin* (LD_{50}: 0.1μg/bee); *Bupirimate* (LD_{50}: 0.2 μg/bee); *Carbaryl* (LD_{50}: 1μg/bee); *Chlorfenapyr* (LD_{50}: 0.2 μg/kg); *Chlorfenvinphos* (LD_{50}: 0.55 μg/bee); *Chlorpyrifos* (LD_{50}: 0.38 μg/bee); *Cypermethrin* (LD_{50}: 0.035 μg/bee); *Cyproconazole* (LD_{50}: >0.1 mg/bee); *Deltamethrin* (LD_{50}: 0.079 μg/bee); *Dimethoate* (LD_{50}: 0.1 mg/kg); *Esfenvalerate* (LD_{50}: 0.017 μg/bee); *Fenthion* (LD_{50}: 0.436 μg/bee); *Iprodione* (LD_{50}: 0.4 μg/bee); *Lindane* (LD_{50}: 0.011 μg/bee); *Malathion* (LD_{50}: 0.71 μg/bee); *MCPA* (LD_{50}: 0.104 μg/bee); *Methomyl* (LD_{50}: 0.1 μg/bee); *Phosmet* (1 μg/bee); *Profenopos* (LD_{50}: 0.102 μg/bee); ***Spinosad*** (LD_{50}: 0.0029

[55]***Ecotoxicology** is the study of the toxic effects of chemicals (pollutants/ toxicants/ stressors) on the aquatic and terrestrial environment including individual organism, species in food webs, food resources such as birds, bees, algae and fish.*

μg/bee); *Tebuconazole* (LD_{50}: 0.6 μg/bee). Based on LD_{50} values, Spinosad (an insecticide found in the bacterial species *Saccharopolyspora spinosa*) is very highly toxic to birds.

7.2.4.3. Aquatic organisms (algae, daphnia and fish)

Several pesticides are **extremely and acutely toxic to aquatic organisms** (Algae, Daphnia and Fish (LC_{50}/ EC_{50} <1 mg/L) including the following:

- *Acrolein* (Algae, EC_{50}[56]: 0.050 mg/L); *Aldicarb* (*Daphnia*, LC_{50}: 0.18 mg/L); *Aldrin* (Fish, LC_{50}: 0.0037 mg/L); *Amitraz* (*Daphnia*, LC_{50}: 0.035 mg/L); *atrazine* (Algae, EC_{50}: 0.13 mg/L); ***Azamethiphos*** (*Daphnia*, LC_{50}: 0.00067 mg/L); *azinphos-methyl* (*Daphnia*, LC_{50}: 0.0011 mg/L); *azoxystrobin* (*Daphnia*, EC_{50}: 0.08 mg/L);
- *Benomyl* (Fish, LC_{50}: 0.27 mg/L); ***Bifenthrin***(Fish, LC_{50}: 0.00015 mg/L); *Bromoxynil* (Fish, LC_{50}: 0.65 mg/);
- *Captan* (Fish, LC_{50}: 0.034 mg/L); *Carbendazim* (*Daphnia*, LC_{50}: 0.13 mg/L); *Carbofuran* (*Daphnia*, LC_{50}: 0.038 mg/L); *Carfentrazone-ethyl* (Algae, EC_{50}: 0.012 mg/L); *Chlordane* (Fish, LC_{50}: 0.09 mg/L); *Chlorfenapyr* (Algae, EC_{50}: 0.0013 mg/L); *Chlorfenvinphos* (*Daphnia*, EC_{50}: 0.0003 mg/L); *Chlorfluazuron* (*Daphnia*, LC_{50}: 0.908 μg/L); *Chlorothalonil* (Fish, LC_{50}: 0.047 mg/L); *Chlorpyrifos* (*Daphnia*, LC_{50}: 0.0017 mg/L); *Chlorsulfuron* (Algae, EC_{50}: 0.05 mg/L); *Clodinafop-propargyl* (Fish, LC_{50}: 0.39 mg/L); *Coumaphos* (*Daphnia*, LC_{50}: 0.001 mg/L); ***Cyfluthrin*** (*Daphnia*, LC_{50}: 0.00016 mg/L); *Cyhalothrin* (*Daphnia*, LC_{50}: 0.00038); ***Cypermethrin*** (*Daphnia*, LC_{50}: 0.00015 mg/L); *Cyprodinil* (*Daphnia*, LC_{50}: 0.033 mg/L);
- *DDT* (*Daphnia*, LC_{50}: 0.0011 mg/L); *Deltamethrin* (*Daphnia*, LC_{50}: 0.0035 mg/L); *Demeton* (*Daphnia*, LC_{50}: 0.023 mg/L); ***Diazinon*** (*Daphnia*, LC_{50}: 0.00096 mg/L); *Diclofop-methyl* (*Daphnia*, LC_{50}: 0.23 mg/L); *Dicofol* (Algae, EC_{50}: 0.075 mg/L); *Dieldrin* (Fish, LC_{50}: 0.0037); *Difenoconazole* (*Daphnia*, LC_{50}: 0.77 mg/L); *Diflubenzuron* (*Daphnia*, LC_{50}: 0.0071 mg/L); *Diquat dibromide* (*Daphnia*, LC_{50}: 0.0022 mg/L); *Dodine* (Algae, EC_{50}: 0.0051 mg/L);
- *Endosulfan* (Fish, LC_{50}: 0.002 mg/L); ***Endrin*** (*Daphnia*, LC_{50}: 0.00004 mg/L); ***Esfenvalerate*** (*Daphnia*, LC_{50}: 0.00024 mg/L); *Ethion* (*Daphnia*, LC_{50}: 0.056 mg/L);
- *Fenitrothion* (*Daphnia*, EC_{50}: 0.0086 mg/L); *Fenoxaprop-p-ethyl* (*Daphnia*, LC_{50}: 0.56 mg/L); *Fenoxycarb* (*Daphnia*, LC_{50}: 0.4 mg/L); *Fenpyroximate* (*Daphnia*, LC_{50}: 0.0032 mg/L); *Fenthion* (*Daphnia*, LC_{50}: 0.0057 mg/L); *Fenvalerate* (Fish, LC_{50}: 0.0036 mg/L); *Fipronil* (Algae, EC_{50}: 0.0048 mg/L); *Fluazinam* (*Daphnia*, LC_{50}: 0.22 mg/L);

[56] ***EC_{50}** = median effective concentration, the concentration of test substance which results in a 50 percent reduction in either algae growth/ death.*

Fluazuron (*Daphnia*, LC_{50}: 0.0006 mg/L); *Fluquinconazole* (Algae, EC_{50}: 0.014 mg/L); *HCB* (Fish, LC_{50}: 0.05-0.2 mg/L);

- *Heptachlor* (Fish, LC_{50}: 0.007 mg/L); *Indosulfuron-methyl -sodium* (Algae, EC_{50}: 0.02 mg/L); *Indoxacarb* (*Daphnia*, LC_{50}: 0.60 mg/L); *Iprodione* (*Daphnia*, LC_{50}: 0.25 mg/L); *Kreoxsim-methyl- sodium* (*Daphnia*, LC_{50}: 0.186 mg/L);
- *Lindane* (Fish, LC_{50}: 0.022 mg/L); *Linuron* (*Daphnia*, LC_{50}: 0.75 mg/L);
- *Malathion* (*Daphnia*, LC_{50}: 0.001 mg/L); *Methidathion* (Fish, LC_{50}: 0.01 mg/L); *Methomyl* (*Daphnia*, LC_{50}: 0.0317 mg/L); *Methoprene* (*Daphnia*, LC_{50}: 0.036 mg/L); ***Methoxychlor*** (*Daphnia*, LC_{50}: 0.00078 mg/L); *Metiram* (Algae, EC_{50}: 0.3 mg/L); *Metolachlor* (Algae, EC_{50}: 0.1 mg/L); *Metosulam* (Algae, EC_{50}: 0.075 mg/L); *Mirex* (Fish, LC_{50}: 0.023 mg/L); ***Monocrotophos*** (*Daphnia*, LC_{50}: 0.00024 mg/L); *Omethoate* (*Daphnia*, LC_{50}: 0.022 mg/L); *Paraquat dichloride* (Algae, EC_{50}: 0.28 mg/L);
- *Parathion* (*Daphnia*, LC_{50}: 0.0025 mg/L); *Parathion-methyl* (*Daphnia*, LC_{50}: 0.0073 mg/L); *Pendimethalin* (Fish, LC_{50}: 0.14 mg/L); ***Permethrin*** (*Daphnia*, LC_{50}: 0.0006 mg/L); *Phenothrin* (Fish, LC_{50}: 0.0027 mg/L); *Phosmet* (*Daphnia*, LC_{50}: 0.0085 mg/L); *Phosphine* (*Daphnia*, LC_{50}: 0.2 mg/L); *Primicarb* (*Daphnia*, LC_{50}: 0.017 mg/L; ***Primiphos-methyl*** (*Daphnia*, LC_{50}: 0.00021 mg/L); *Profenopos* (*Daphnia*, LC_{50}: 0.001 mg/L); *Propachlor* (Algae, EC_{50}: 0.015 mg/L); *Propargite* (*Daphnia*, LC_{50}: 0.092 mg/L); *Propiconazole* (Algae, EC_{50}: 0.02-13.6 mg/L); *Prothiofos* (*Daphnia*, LC_{50}: 0.014 mg/L); *Pyraclostrobin* (Fish, LC_{50}: 0.006 mg/L);
- *Quintozene* (Algae, EC_{50}: 0.058 mg/L);
- *Simazine* (Algae, EC_{50}: 0.042 mg/L);
- *Tebufenozide* (Algae, EC_{50}: 0.64 mg/L); *Tebufenpyrad* (Fish, LC_{50}: 0.03 mg/L); *Tebuthiuron* (Algae, EC_{50}: 0.05 mg/L); *Terbufos* (Fish, LC_{50}: 0.01 mg/L); *Thiabendazole* (Fish, LC_{50}: 0.055 mg/L); *Thiobencarb* (Algae, EC_{50}: 0.017 mg/L); *Thiodicarb* (*Daphnia*, LC_{50}: 0.053 mg/L); *Thiram* (Algae, EC_{50}: 0.065 mg/L); *Toxaphene* (Fish, LC_{50}: 0.56 mg/L); *Triasulfuron* (algae, EC_{50}: 0.035 mg/L); ***Trichlorfon*** (*Daphnia*, LC_{50}: 0.0009 mg/L); *Trifloxystrobin* (Fish, LC_{50}: 0.015 mg/L); *Triflumuron* (*Daphnia*, LC_{50}: 0.225 mg/L); *Trifluralin* (fish, LC_{50}: 0.088 mg/L);
- *Ziram* (*Daphnia*, LC_{50}: 0.048 mg/L).

Based on LC_{50} values, **Organochlorine** (endrin), synthetic **pyrethroids** (bifenthrin, cypermethrin, cyfluthrin, esfenvalerate) and **organophosphates** (primiphos-methyl, monocrotophos) are very highly toxic to aquatic organisms (Fig. 7.3).

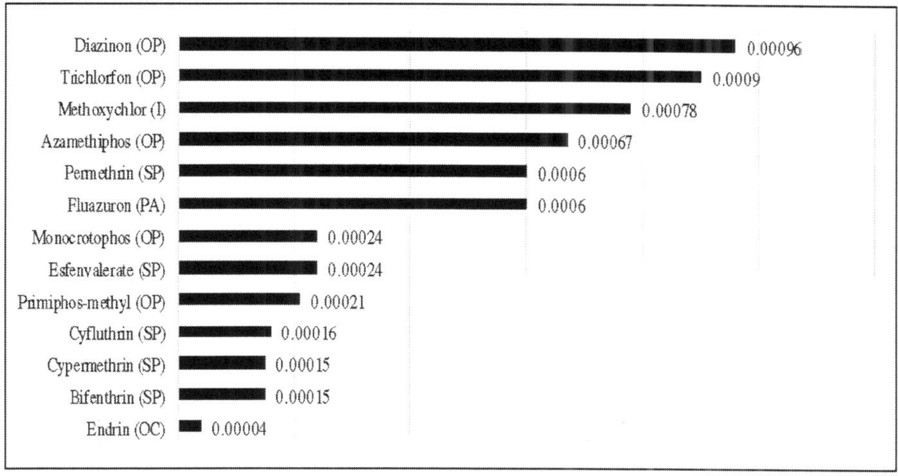

Fig. 7.3. The lowest acutely toxic concentrations of pesticides (LC_{50} in mg/L) to aquatic organisms *[the Figure is produced based on data in Table 7.4; I = insecticides; OC = organochlorines; OP = organophosphates; PA = parasiticides; SP = synthetic pyrethroids]*

7.3. Impacts of pesticides on social, economic and environmental sectors

7.3.1. Pesticide residues

Residues are the remains (small amount) of either the chemical itself or breakdown products that remain in or on a substance (food, product or water). The pesticides found in or on foods or in water are called **residues**. Residues may enter the environment via surface or groundwater as a result of:

- spray drift
- surface run-off
- leaching
- drainage discharges
- vaporisation
- spills
- unlawful acts,
- soil dust (Kookana et al 2003, Kibria et al 2010) (Fig. 7.4).

The surface run-off is known to be one of the important and major pathways of movement of pesticide residues into the environment (Kibria 2019). Residues in food and agricultural produce may result from the application of pesticides. For example, fungicides used to control fungal diseases in plants and plant products; insecticides and acaricides used to control insect and mite pests and herbicides used to control weeds in the crop. The other paths of residue contamination is an accidental contamination or intentional treatment of animals or feed with pesticides or from environmental contamination or an unintended exposure of plants and animals to pesticides that are no longer registered in Australia and other countries, such as Organochlorines - dieldrin, DDT but remain in the soil for long periods (Poolpak et al 2008).

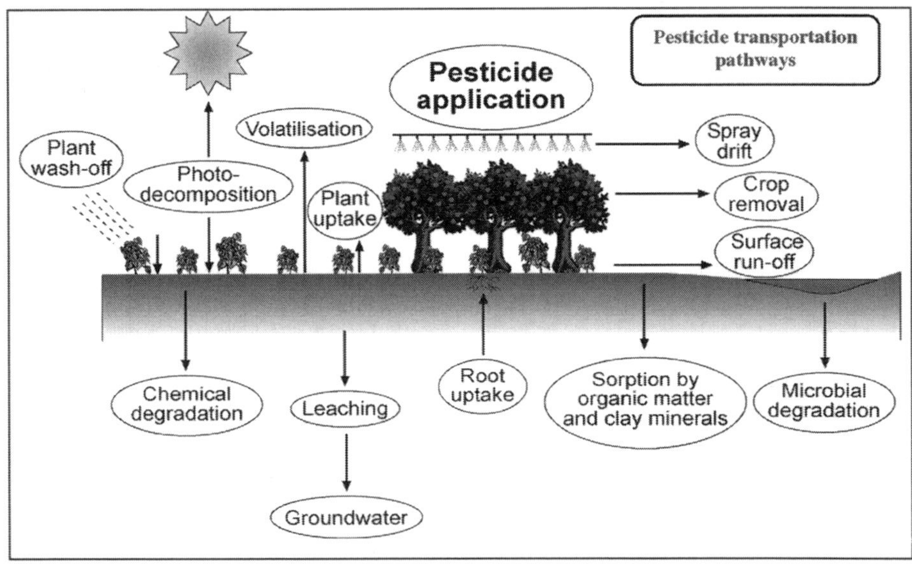

Fig. 7.4. Fate and transport of agrochemicals (pesticides) from the site of application to off-site environments including non-target organisms
(Source: Sarmah et al 2004; page 131).

Humans can be exposed to pesticide residues via contaminated water and food or non-target drift or application. Most importantly, pesticide residues may enter the human body through food chains and by legal application of pesticides. For example, livestock can be exposed to pesticides through intake of grain, fodder, forage as well as feeding of processed by-products and from ingestion of contaminated soil. These pesticides can then be accumulated in the animals because of the lipophilic (fat-loving) nature of some of the pesticides that may cause them to be incorporated and accumulated in the tissues (e.g. fat, milk) of living organisms. However, the tendency of a pesticide to accumulate in fat and milk of animals is generally related to the physico-chemical properties of the pesticide such as log K_{ow}. Therefore, the risk of chemical residues to human is in the risk of residues remaining on the food consumed (such as meat, milk and milk products). Pesticide chemicals, as residue, if consumed by humans more than the recommended maximum residue limits (**MRL**) can be hazardous. It is therefore advisable that, a national pesticide residue monitoring program be conducted on a regular basis to assess pesticide risk. Because monitoring pesticide residues are the only effective way to control the concentrations of pesticides in foods or water. The purpose of such monitoring is as follows:

- an estimate of the occurrence (frequency) of pesticide residue in agriculture products or in water;
- confirm that residues are below the guideline levels for beneficial water usage;
- confirm that residues in agriculture produce are below the set limit of MRL and
- take up corrective measures or action.

7.3.1.1. Residues in food (fruits, vegetables, grain, seafood)

Residue monitoring surveys conducted in different countries reveal that food (fruit, vegetables, crops, and seafood) are contaminated with pesticides (Table 7.7). The surveys found that MRL compliance rate was higher in developed countries (Australia, Europe, USA) compared to developing countries (Bangladesh, China, Pakistan, Saudi Arabia, Thailand). For, example, pesticide residue in stone fruit, aquaculture and seafood compliance rate in Australia was 100%. In developed countries, a significant number of imported/ exotic fruits (USA) exceeded recommended MRL for pesticides (Table 7.7).

Table 7.7. Pesticide residue detected in fruits and vegetables in Australia, Bangladesh, Belgium, China, Croatia, Europe, India, Nigeria, Pakistan, Saudi Arabia and the USA and respective MRL compliance rate
[*MRL = maximum residue limits*].

Country and Survey period	Pesticide residues monitored	MRL compliance	References
Australia (2017-18)	A total of 9,729 animal food samples and 6,798 plant food samples were monitored. Food samples analysed for a range of pesticides including fungicides, organophosphates, Organochlorines, synthetic pyrethroids, herbicides and dithio-carbamates as part of Australian National Residue Survey (ANRS).	The overall compliance rate against Australian standard was: Beef (99.9%); hen eggs (99.4%); honey (98.4%); poultry (100%) Aquaculture (100%); wild-caught seafood (100%) Grains (99.2%); pome fruit (94.3%); stone fruit (100%); citrus fruit (99.5%); Macadamia nuts (100%).	Australian Government 2018
Bangladesh (2011)	Two common vegetables (egg plants and tomatoes; a total of 140 samples) were analysed for five organophosphates	Egg plants (50%) and tomatoes (60%) were contaminated with pesticides and were above the maximum	Alam et al 2015

Country and Survey period	Pesticide residues monitored	MRL compliance	References
	(diazinon, dimethoate, fenitrothion, parathion, and phosphamidon), one carbamate (carbofuran), and one phenyl urea herbicide (linuron).	residual limit (MRL) proposed by the EC. Diazinon was the most common (35%) pesticide detected in the vegetable samples at the concentration of 45–450 times higher than the MRL.	
Bangladesh (2011-12)	Vegetables such as brinjal (egg plant), yard long bean, bitter gourd, snake gourd, pointed gourd, okra (ladys finger), tomatoes, hyacinth bean and cabbage collected from local markets of eight districts. 170 samples analysed for residues of eight commonly used insecticides (fenvalerate, diazinon, quinalphos, fenitrothion, acephate, chlorpyriphos, cypermethrin and malathion).	21.78% of samples were found contaminated with four insecticides: chlorpyriphos, quinalphos, acephate and cypermethrin. Most of the samples contained chlorpyrifos (13.53%) and quinalphos (8.4%) residue. 18.26% of samples exceeded MRL.	Ahmed et al 2016
Belgium (2008)	1,413 samples of fruit and vegetables analysed for 349 different pesticides.	6% of the samples exceeded MRL; approximately 94% of the samples were compliant. The main fruit groups for which residue levels exceeded the MRL were exotic fruits (passion fruit), citrus fruit (oranges), berries and small fruits (currants and strawberries).	Claeys et al 2010
China (2006 to March 2009)	3,009 samples (fruits, vegetables) were analysed for 22 targeted pesticides.	11.7% (353 samples) exceeded MRLs.	Chen et al 2011
Croatia (2007–2009)	866 samples of 28 different kinds of fresh fruits and vegetables were analysed for a range of pesticides.	5.3% (46 samples) contained pesticides above the MRL.	Knezevic et al 2012

Country and Survey period	Pesticide residues monitored	MRL compliance	References
European Union (2016)	12,168 samples analysed covering 165 pesticides in 11 food products (2016) under the EU coordinated control programme (EUCP).	98.3% of the samples analysed in 2016 of EUCP fell within the legal limits (1,965 samples). 6,367 samples (52.3%) were free of quantifiable residues (residues below the limit of quantification (LOQ). Only 1.7% of the samples analysed exceeded the MRL (203 samples)	EFSA 2018
India (2012-2013)	Vegetables, fruit, spices, rice, wheat and other food items (16,790 samples) were analysed for pesticides.	509 samples out of 16,790 were above pesticides MRL including: 221 (2.29%) vegetable samples (out of 7,591 samples); 36 (1.61%) fruit samples (out of 2,235 samples); 128 (11.02%) spices samples (out of 1,161 samples); 73 (8.23%) rice sample (out of 886 samples); 39 (4.74%) wheat samples (out of 823 samples); 06 (0.77%) fish samples (out of 776 samples); The pesticides commonly found above MRL in the samples were chlorpyrifos, ethion, acetamiprid, dichlorvos and cypermethrin.	* see weblink underneath of this Table.
Nigeria	Cereal grains (millets, Guinea corn, maize) analysed for 10 Organochlorine pesticide residues	Lindane and Aldrin residues were above the MRLs set by WHO/ FAO.	Anzene et al 2014
Pakistan (2002-03)	27 vegetables (206 samples) analysed for 24 pesticides.	63% of the samples were contaminated with pesticides. 46% of the contaminated samples exceeded the MRL.	Parveen et al 2005

Country and Survey period	Pesticide residues monitored	MRL compliance	References
Saudi Arabia (2008-09)	160 different vegetable samples analysed for 23 targeted pesticides (13 insecticides, 3 fungicides, 3 herbicides, 2 acaricides, 1 rodenticide and 1 nematicide).	34% (54 samples) exceeded MRLs. MRL values exceeded most often in cabbage followed by squash, green pepper and carrot, and also followed by cucumber and lettuce, egg-plant and then tomatoes.	Osman et al 2011
Thailand (2013-14)	Three commonly consumed vegetables bought from local kitchen and supermarket (Chinese kale, pak choi and water morning glory) were tested for 28 pesticides residue.	The pesticide residues exceeding the MRL in those three vegetables studied were remarkably high. MRL exceedance rate: 35%-48% (for Chinese kale); 55%-71% (for pak choi), and 42% -49% (for morning glory).	Wanwimolruk et al 2016
USA (2010)	12,028 samples tested from contributing states.	0.25% (30 samples) exceeded MRLs (17 of them imported). 41% of the samples tested had no detectable pesticides.	Fothergill and Abdelghani 2013, Shipman 2012

**//economictimes.indiatimes.com/articleshow/46108223.cms?utm_source=contentofinterest&utm_medium=text&utm_campaign=cppst*

7.3.1.2. Residues in Australian waterways (surface, groundwater and sediments)

Water contaminated with pesticides can be unsuitable for a wide range of beneficial water usages (irrigation, drinking, recreation, agriculture, industries, stock and domestic supply, aquaculture and aquatic ecosystems protection). Pesticides residues have been detected in surface water, groundwater and sediments (Table 7.8). The detected pesticides comprised herbicides, insecticides and fungicides are summarised below and detailed in Table 7.8:

7.3.1.2.1. Herbicides

2,4-D (surface water: <0.005-2.50 μg/L, sediment: <0.01-1.8 mg/kg); **ametryn** (surface water 0.31-2.1 μg/L); **amitrole** (groundwater: 20-2,300 μg/L); **atrazine** (surface water: 0.01-58 μg/L, groundwater: 0.02-0.45 μg/L, sediment: <5-<40 μg/kg); **atrazine-desethyl** (surface water: 0.02-1.3 μg/L); **atrazine desisopropyl**

(<0.005-0.714 µg/L); **atrazine-2-hydroxy** (0.5-1.2 µg/L); **bromacil** (surface water: 0.005-0.03 µg/L, groundwater: 2.5 µg/L); **chlorproham** (groundwater: 0.02 µg/L); **chlorsulfuron** (surface water: 0.001-0.001 µg/L); **clopyralid** (surface water: <0.4-0.41 µg/L); **dicamba** (surface water: 0.036-0.120 µg/L); **diuron** (surface water: 0.3-8.2 µg/L, sediment: <2.5-5316 µg/kg); **fluometuron** (surface water: 0.20 µg/L); **fluroxypyr** (surface water: 0.021-0.044 µg/L); **glyphosate** (groundwater: 380 µg/L); **hexazinone** (surface water <0.005-123 µg/L, groundwater: 0.01 µg/L); **linuron** (surface water: 0.04-19 µg/L, groundwater: 0.41 µg/L, sediment: <20-<100 µg/kg); **MCPA** (surface water: <0.01- 0.513 µg/L, sediment: <0.01-trace µg/kg); **metasulfuronethyl** (surface water: 0.002-0.057 µg/L); **metolachlor** (surface water: 0.001-421 µg/L, groundwater: 0.41 µg/L, sediment: <2-51.9 µg/kg); **metribuzin** (surface water: 0.009-0.28 µg/L, groundwater: 0.13 µg/L); **molinate** (surface water: trace-µg/L); **pendimethalin** (surface water: <0.05-trace µg/L); **picloram** (surface water: 0.079-0.129 µg/L); **prometryn** (surface water: 0.069-300 µg/L, sediment: <2-374 µg/kg); **propazine** (surface water: 0.004-0.40 µg/L); **simazine** (surface water: 0.002-6 µg/L, groundwater: 0.02-3.1 µg/L, sediment: <5-260 µg/kg); **sulfometuron methyl** (surface water: 0.005-0.073 µg/L); **tebuthiuron** (surface water: 0.02-0.03 µg/L); **terburtyn** (surface water: 0.02-0.03 µg/L); **terburtyn** (0.002-0.309 µg/L); **triasulfuron** (surface water: 0.001-0.003 µg/L); **triclopyr** (surface water: <0.002-2.03 µg/L) (see Table 7.8 for details)

7.3.1.2.2. Insecticides

Aldrin (groundwater: 0.06 µg/L); **azinphos methyl** (surface water: 0.02-0.97 µg/L); **BHC** (groundwater: 0.01-0.08 µg/L); **bifenthrin** (surface water: 100 µg/L, sediment: <2-330 µg/kg); **carbaryl** (surface water: 0.039-0.3 µg/L, sediment: 2-276 µg/kg); **chlordane** (groundwater: 0.01-0.07 µg/L); **chlorpyrifos** (surface water: 0.002-110 µg/L, groundwater: 0.01-2.4 µg/L, sediment: 0.07-42 mg/kg); **cypermethrin** (sediment: 10-<400 µg/kg); **DDE** (p,p'- DDE) (sediment: <3-16 µg/kg dw); **DDT** (surface water: 0.00073-0.34 µg/L, groundwater: 0.01-0.04 µg/L, sediment: trace-80 mg/kg); **diazinon** (surface water: <0.002-17 µg/L, sediment: <1-17 µg/kg); **dieldrin** (surface water: 0.0006-0.005 µg/L, groundwater: 0.03-22.5 µg/L, sediment: 0.005-8 µg/kg); **dimethoate** (surface water: 0.12 µg/L, sediment: <25-69.1 µg/kg); **endosulfan** (surface water: 0.007-0.029 µg/L); **endrin** (groundwater: 0.02 µg/L); **esfenvalerate** (surface water: 65 µg/L); **fenamiphos** (sediment: <10-69.6 µg/kg); **fipronil** (sediment: <5-<30 µg/kg); **HCB** (groundwater: 0.01-0.05 µg/L); **heptachlor** (groundwater: 0.01-1.7 µg//L); **imidacloprid** (surface water: 0.02-0.06 µg/L, sediment: <5-5 µg/kg); **indoxacarb** (sediment: <10-<125 µg/kg); **malathion** (sediment: <5-<20 µg/kg); **methiocarb** (surface water: 1.2-153 µg/L); **oxychlordane** (sediment: <4-6 µg/kg); **parathion methyl** (surface water: 0.23-0.92 µg/L); **permethrin** (sediment: <3-930.5 µg/kg); **pirmicarb** (surface water: <0.002-7.3 µg/L, sediment: 26 µg/kg); **taufluvalinate** (75 µg/L); **tebufenozide** (surface water: <0.002-0.002 µg/L) (see Table7.8 for details).

7.3.1.2.3 Fungicides

Azoxystrobin (surface water: <0.002-0.016 µg/L, sediment: <1-<150 µg/kg); **boscalid** (sediment: <2-13.6 µg/kg); **chlorothalonil** (sediments: 118-460 µg/kg); **cyproconazole** (surface water: 0.39 µg/L); **difenoconazole** (surface water: 0.03 µg/L, sediment: <50-<150 µg/kg); **dimethomorph** (sediment: <15-<150 µg/kg); **fenarimol** (surface water: 0.20 µg/L); **iprodione** (surface water: 2-170 µg/L, sediment: <20-120 µg/kg); **metaxyl** (surface water: <0.002-0.03 µg/L, sediment: <10-<125 µg/kg); **myclobutanil** (surface water: 0.24-26 µg/L, sediment: <2-120 µg/kg); **oxadixyl** (surface water: <0.001-0.4 µg/L, sediment: 2 µg/kg); **penconazole** (surface water: 0.001-0.02 µg/L); **prochloraz** (surface water: 0.02 µg/L, sediment: <50-<400 µg/kg); **pyraclostrobin** (surface water: 0.005 µg/L, sediment: 1-<80 µg/kg); **pyrimethanil** (surface water: 0.27-1,390 µg/L, groundwater: 0.02-0.95 µg/L, sediment: <1-272 µg/kg); **tebuconazole** (surface water: 0.03 µg/L, sediment: <4-trace µg/kg); **tetraconazole** (surface water: 0.02 µg/L); **triadimenol** (<0.002-<0.002 µg/L); **trifloxystrobin** (surface water: 0.008-793 µg/L, sediment: <2-68 µg/kg) (see Table 7.8 for details).

Table 7.8. Some common pesticides detected in the surface, groundwater and sediments in Australia

[results are based on both grab water samples and passive samples; ***dw*** *= dry weight;* ***F*** *= fungicides;* ***H*** *= herbicides;* ***I*** *= insecticides,* ***max.*** *= maximum;* ***nd*** *= not detected or data is not available;* ***OC*** *= Organochlorines;* ***OP*** *= organophosphates;* ***SP*** *= synthetic pyrethroids;* ***TRIMPS*** *= 2,2,4- trimethyl pentane containing passive samplers; see Box 7.1].*

Pesticides	Surface water	Groundwater	Sediment
2-4-D (H)	2.20-2.50 µg/L (Davis et al 2008); <0.005-0.237 µg/L (Mossop et al 2013); 0.111 (0.0022-1.320 µg/L) (Zhang et al 2014)	nd	1.8 mg/kg dw (SKM 2002); <0.01- 0.011 µg/kg dw (Mossop et al 2013).
Aldrin (I)	nd	0.06 µg/L (SKM 1995)	nd
Ametryn (H)	0.31-2.10 µg/L (Davis et al 2008)	nd	nd
Amitrole (H)	nd	20-2,300 µg/L (Wenig and Lawrence 1998)	nd

Pesticides	Surface water	Groundwater	Sediment
Atrazine (H)	0.14 – 3.2 μg/L (Chapman and Stranger 1994); 0.24–1.9 μg/L; (Chapman and Stranger 1994); 0.02 – 0.92 μg/L (Chapman and Stranger 1993); 0.01-0.062 μg/L (water) (Rose and Kibria 2007); 6.5-8.80 μg/L (Davis et al 2008); <0.003-0.003 μg/L (water) (Mossop et al 2013); <0.5-0.8 μg/L (**Chemcatcher**[57]); 0.58 (0.002-0.783 μg/L) (Zhang et al 2014)	0.02–0.25 μg/L (Watkins et al 1999a); 0.06 μg/L (Chapman and Stranger 1994); 0.022 – 0.06 μg/L (Bauld 1994, Bauld et al 1998) 0.02–0.45 μg/L (Ivkovic et al 2001)	<5-<40 μg/kg dw (Marshall et al 2016).
Atrazine-Desethyl (H) (metabolite)	1.2-1.3 μg/L (Davis et al 2008); 1.3 μg/L (max. in water) (Schafer et al 2011); 0.021 (0.02-0.040 μg/L) (Zhang et al 2014)	nd	nd
Atrazine desisopropyl (H) (metabolite)	0.06 μg/L (Chapman and Stranger 1994; SKM 1995); 0.29-0.41 μg/L (Davis et al 2008); <0.005 -<0.005 μg/L (water) (Mossop et al 2013); >0.5- 0.5 μg/L (in **Chemcatcher**) (Mossop et al 2013); 0.098 (0.022-0.714 μg/L) (Zhang et al 2014)	nd	nd
Atrazine-2-hydroxy (H) (metabolite)	<0.5-1.2 μg/L (in **Chemcatcher**) *(La Trobe Catchment)* (Mossop et al 2013)	nd	nd
Azinphos methyl (I)	0.02-0.97 μg/L (water) (Rose and Kibria 2006, Rose and Kibria 2007)	nd	nd
Azoxystrobin (F)	0.02 μg/L) (mean) (Wightwick et al 2012); <0.002-0.016 μg/L (Mossop et al 2013)	nd	<1-5 μg/kg dw (Mossop et al 2013); <30-<150 μg/kg dw (Marshall et al 2016).

[57] ***Chemcatcher*** *is a passive water quality sampling device for monitoring a variety of pollutants in water.*

Pesticides	Surface water	Groundwater	Sediment
BHC (Benzene hexachloride) (I- OC)	nd	0.03-0.08 µg/L (SKM 1995); 0.01 – 0.02 µg/L (SKM 1995)	nd
Bifenthrin (I-SP)	100 µg/L (**TRIMPS**[58] + **Dodecanol**) (one sample) (Rose and Kibria 2006) (see **Box 7.1** for explanations on **TRIMPS**)	nd	<2-330 µg/kg dw (36.4 mean) (Marshall et al 2016).
Boscalid (F)	nd	nd	<2-13.6 µg/kg dw (Marshall et al 2016).
Bromacil (H)	0.03 µg/L (Davis et al 2008); 0.005 (0.005-0.0055 µg/L) (Zhang et al 2014)	2.5 µg/L (Ivkovic et al 2001)	nd
Carbaryl (I)	0.039 (µg/) (max. in water) (Schafer et al 2011); 0.3 µg/L (max. in **TRIMPS**) (Schafer et al 2011)	nd	2 µg/kg dw (max.) (Schafer et al 2011); <5-276 µg/kg dw (Marshall et al 2016).
Chlordane (1)	nd	0.03 µg/L (SKM 1995); 0.01- 0.07 µg/L (SKM 1995)	nd
Chloroth-alonil (F)	nd	nd	460 µg/kg dw (max.) (Schafer et al 2011); 118 µg/kg dw (mean) (Wightwick et al 2012).
Chlorpropham (H)	nd	0.02 µg/L (Watkins et al 1999a)	nd
Chlorpyrifos (I-OP)	0.002 µg/L (Chapman and Stranger 1994); 0.01-0.089µg/L (in **TRIMPS**) (Rose and Kibria 2007); 0.04 µg/L (max in water) (Schafer et al 2011); 110 µg/L	0.19 µg/L (SKM 1995); 0.09 µg/L (SKM 1995); 2.4 µg/L (Wenig and Lawrence 1998);	0.07 mg/kg dw (Chapman and Stranger 1993); 42 µg/kg dw (max.) (Schafer et al 2011); <10-25.2 µg/kg

[58] ***TRIMPS*** *= trimethylpentane containing passive samplers.*

Pesticides	Surface water	Groundwater	Sediment
	(max. in **TRIMPS**) (Schafer et al 2011) (see **Box 7.1** for explanations on **TRIMPS**)	0.01 µg/L (SKM 1995); 0.2 µg/L (SKM 1995)	dw (Marshall et al 2016).
Chlorsul-furon (H)	(0.001-0.001) µg/L (Zhang et al 2014)	nd	nd
Clopyralid (H)	<0.4-trace (Mossop et al 2013); 0.085 (0.010-0.401 ng/L) (Zhang et al 2014)	nd	nd
Cypermethrin (I)	nd	nd	<10-<400 µg/kg dw (Marshall et al 2016).
Cyproco-nazole (F)	0.39 µg/L (mean) (Wightwick et al 2012)	nd	nd
DDE (p,p'-DDE[59]) (I-OC)	nd	nd	<3-16 µg/kg dw (Mossop et al 2013).
DDE (p,p'-DDE (1% carbon) (I-OC)	nd	nd	1-4 µg/kg dw (Mossop et al 2013).
DDT (total) (I)	0.00073–0.340 µg/L (McKenzie-Smith et al 1994); 0.17 µg/L (Chapman and Stranger 1994)	0.03 µg/L (SKM 1995); 0.01 – 0.04 µg/L (SKM 1995); 0.02 µg/L (SKM 1995)	6.3–>80 mg/kg dw (McKenzie-Smith et al 1994); 0.0005 – 0.0099 mg/kg dw (Moore et al 1996); 0.01-0.04 mg/kg dw (SKM 2002); Trace-4 µg/kg dw (Mossop et al 2013).
Diazinon (I-OP)	17 µg/L (max. in **TRIMPS**) (Schafer et al 2011); <0.002-0.03 µg/L(Mossop et al 2013)	nd	17 µg/kg dw (max.) (Schafer et al 2011); <1-trace (Mossop et al 2013).
Dicamba (H)	0.078 (0.036-0.120 µg//L) (Zhang et al 2014)	nd	nd

[59] ***p,p'- DDE** = main metabolite of pesticide dichlorodiphenyltrichloroethane (DDT).*

Pesticides	Surface water	Groundwater	Sediment
Dieldrin (I)	0.0006–0.00266 µg/L (McKenzie-Smith et al 1994); 0.005 µg/L (Chapman and Stranger 1994)	0.04 µg/L (SKM 1995); 0.03–22.5 µg/L (Wenig and Lawrence 1998)	0.005–0.02 µg/kg dw (Chapman and Stranger 1994); 3-42 mg/kg dw (McKenzie-Smith et al 1994); 0.0007–0.010 mg/kg dw (Moore et al 1996); <4-8 µg/kg dw (Mossop et al 2013).
Dieldrin (1% carbon) (I-OC)	nd	nd	3 µg/kg dw (Mossop et al 2013).
Difenocona-zole (F)	0.03 µg/L (mean-water) (Wightwick et al 2012)	nd	<50-<150 µg/kg dw (Marshall et al 2016).
Dimethoate (I-OP)	0.12 µg/L (Chapman and Stranger 1994)	nd	<25-69.1 µg/kg (Marshall et al 2016)
Dimetho-morph (F)	nd	nd	<15-<150 µg/kg dw (Marshall et al 2016).
Diuron (H)	0.3- 4.8 µg/L (Chapman and Stranger 1994); 3.8-8.20 µg/L (Davis et al 2008); 0.084 (0.007-0.450 µg/L) (Zhang et al 2014)	nd	<2.5-5316 µg/kg (84.6 mean) (Marshall et al 2016).
Endosulfan (I)	0.0007-0.029 (µg/L in **TRIMPS**) (Rose and Kibria 2007)	nd	nd
Endosulfan (α) Endosulfan (β) Endosulfan (SO_4) (I)	Endosulfan (α) (0.01–0.035); Endosulfan (β) (0.012–0.014) (Chapman and Stranger 1993); Endosulfan (α) (0.004-0.005 µg/L); Endosulfan (β) (0.004 µg/L); Endosulfan (SO4) (0.016–0.031 µg/L) (Chapman and Stranger 1994)	nd	Endosulfan(α) (0.021 mg/kg); Endosulfan (β) (0.061 mg/kg dw); Endosulfan (SO_4) (0.015 mg/kg dw) (Chapman and Stranger 1993).
Endrin (I)	nd	0.02 µg/L (SKM 1995)	nd

Pesticides	Surface water	Groundwater	Sediment
Esfenvalerate (I-SP)	65 µg/L (in **TRIMPS + Dodecanol**) (one sample) (Rose and Kibria 2006)	nd	nd
Fenamiphos (I)	nd	nd	<10-69.6 µg/kg dw (12.4 mean) (Marshall et al 2016).
Fenarimol (F)	0.20 µg/L (mean) (Wightwick et al 2012)	nd	nd
Fipronil (I)	nd	nd	<5-<30 µg/kg dw (Marshall et al 2016).
Fluometuron (H)	0.04µg/L (Davis et al 2008)	nd	nd
Fluroxypyr (H)	0.033 (0.021-0.044 µg//L) (Zhang et al 2014)	nd	nd
Glyphosate (H)	nd	380 µg/L (Larsen et al 1998)	nd
HCB (I) *(Hexachlorobe nzene)*	nd	0.03 µg/L (SKM 1995); 0.01–0.02 µg/L (SKM 1995); 0.03–0.05 µg/L (SKM 1995)	nd
Heptachlor (I)	nd	0.03-1.7 µg/L (SKM 1995); 0.01–0.04 µg/L (SKM 1995)	nd
Heptachlor-epoxide (I)	nd	0.05 µg/L (SKM 1995); 1.3 µg/L (SKM 1995)	nd
Hexazinone (H)	0.09-0.038 µg/L (Davis et al 2008); 0.96 µg/L (max. in water) (Schafer et al 2011); <0.05-0.027 µg/L (water) (Mossop et al 2013); <0.5-20.6 µg/L (in **Chemcatche**r) (Mossop et al 2013); 0.015 (0.006-0.030 µg//L) (Zhang et al 2014).	0.01 µg/L (Hydrotechnology 1994)	123 µg/kg dw *(Yarra catchment* (Schafer et al 2011).

Pesticides	Surface water	Groundwater	Sediment
Imidacloprid (I)	0.02-0.06 µg/L (Davis et al 2008)	nd	<5-5 µg/kg dw (Mossop et al 2013).
Indoxacarb (I)	nd	nd	<10-<125 µg/kg dw (Marshall et al 2016).
Iprodione (F)	3 µg/L (max. in water) (Schafer et al 2011); 120 µg/L (max. in **TRIMPS**) (Schafer et al 2011); 2.0 µg/L (mean-water); (Wightwick et al 2012); 3.7 µg/L (mean-in **TRIMPS**) (Wightwick et al 2012)	nd	170 µg/kg dw (max.) (Schafer et al 2011); <20-84.5 µg/kg dw (Marshall et al 2016).
Linuron (H)	0.6 µg/ L (max. in water (Schafer et al 2011); 19 µg/L (max. in **TRIMPS**) (Schafer et al 2011); 0.040 (0.040-0.040 ng/L) (Zhang et al 2014)	0.41 µg/L (Chapman and Stranger 1994)	18 µg/kg dw (max.) (Schafer et al 2011); <20-<100 µg/kg dw (Marshall et al 2016).
Malathion (I)	nd	nd	<5-<20 µg/kg dw (Marshall et al 2016).
MCPA (H) *[2-methyl-4-chloroph-enoxyacetic acid]*	0.20 µg/L (Davis et al 2008); <0.01-0.076 µg/L (Mossop et al 2013); 0.087 (0.002-0.513 ng/L) (Zhang et al 2014)	nd	<0.01-trace µg/kg dw (Mossop et al 2013).
Metalaxyl (F)	0.004 µg/L (mean, in water) (Wightwick et al 2012); 0.03 µg/L (mean, in **TRIMPS**) (Wightwick et al 2012); <0.002-0.021 µg/L (Mossop et al 2013)	nd	1.0 µg/kg dw (mean) (Wightwick et al 2012); <10-<125 µg/kg dw (Marshall et al 2016).
Metasulfu-ronmethyl (H)	0.017 (0.002-0.057 µg/L) (Zhang et al 2014)	nd	nd
Methiocarb (I)	1.2 µg/L(max. in water) (Schafer et al 2011); 153 µg/L (max. in **TRIMPS**) (Schafer et al 2011)	nd	nd

Pesticides	Surface water	Groundwater	Sediment
Metolachlor (H)	<0.005-0.011 µg/L (water) (Mossop et al 2013); <0.3-0.4 µg/L (in **Chemcatcher**) (Mossop et al 2013); 0.082 (0.001-0.421 µg/L) (Zhang et al 2014)	0.04 µg/L (Watkins et al 1999a)	<10-trace µg/kg dw (Mossop et al 2013); <2-51.9 µg/kg dw (22.6 mean) (Marshall et al 2016).
Metribuzin (H)	0.15–0.28 µg/L (Chapman and Stranger 1994); 0.07 µg/L (Chapman and Stranger 1994); <0.05-0.162 µg/L (Mossop et al 2013); 0.009 (0.009-0.009 µg/L) (Zhang et al 2014)	0.13 µg/L (Chapman and Stranger 1994)	nd
Molinate (H)	Trace (in **TRIMPS + Dodecanol**) (Rose and Kibria 2006 and 2007)	nd	nd
Myclobutanil (F)	2.9 µg/L(max. in water) (Schafer et al 2011); 26 µg/L (max. in **TRIMPS**) (Schafer et al 2011); 0.24 µg/L (mean-water) (Wightwick et al 2012); 0.39 µg/L (mean in **TRIMPS**) (Wightwick et al 2012)	nd	120 µg/kg dw (max.) (Schafer et al 2011); 18 µg/kg dw (mean) (Wightwick et al 2012); <2-trace µg/kg dw (Mossop et al 2013); <10-15.2 µg/kg dw (Marshall et al 2016).
Oxadixyl (F)	0.4 µg/L (max. in water) (Schafer et al 2011); 0.2 µg/L (max. in **TRIMPS**) (Schafer et al 2011); <0.001 µg/L (Wightwick et al 2012)	nd	2 µg/kg dw (max.) (Schafer et al 2011).
Oxychlor-dane (I-OC)	nd	nd	<4-6 µg/kg dw (mean) (Mossop et al 2013).
Parathion methyl (I-OP)	0.23-0.92 µg/L (in **TRIMPS + dodecanol**) (Rose and Kibria 2007)	nd	nd
Penconazole (F)	0.02 µg/L (mean in water) (Wightwick et al 2012); 0.001 µg/L (mean in TRIMPS) (Wightwick et al 2012)	nd	nd

Pesticides	Surface water	Groundwater	Sediment
Pendimethalin (H)	<0.05-trace (Mossop et al 2013)	nd	nd
Permethrin (I)	nd	nd	80 µg/kg dw (max.) (Schafer et al 2011); <3-930.5 µg/kg (26.1 mean) (Marshall et al 2016).
Picloram	<0.8-1.02 µg/L (Mossop et al 2013); 0.140 (0.079-0.129 µg/L) (Zhang et al 2014)	nd	nd
Pirmicarb (I-C)	1.4 µg/L (max. in water) (Schafer et al 2011); 7.3 µg/L (max. in TRIMPS) (Schafer et al 2011); <0.002-0.005 µg/L (Mossop et al 2013)	nd	26 µg/kg dw (max.) (Schafer et al 2011).
Prochloraz (F)	0.02 µg/L (mean) (Wightwick et al 2012)	nd	<50-<400 µg/kg dw (Marshall et al 2016)
Prometryn (H)	0.03 µg/L (Davis et al 2008); 21 µg/L (max. in water) (Schafer et al 2011); 300 µg/L (Schafer et al 2011); 0.069 (0.011-0.168 µg/L) (Zhang et al 2014)	nd	374 µg/kg dw (max.) (Schafer et al 2011); <2-34.7 µg/kg dw (15.4 mean) (Marshall et al 2016).
Propazine (H)	0.40 µg/L (Davis et al 2008); 0.005 (0.004-0.007 µg/L) (Zhang et al 2014)	nd	nd
Pyraclostrobin (H)	0.0.05 µg/L (mean) (Wightwick et al 2012)	nd	1.0 µg/kg dw (mean) (Wightwick et al 2012); <10-<80 µg/kg dw (Marshall et al 2016).
Pyrimethanil (F)	70 µg/L (max. in water) (Schafer et al 2011); 1,390 µg/L (max. in **TRIMPS**) (Schafer et al 2011); 0.27 µg/L (mean-water) (Wightwick et al 2012); 2.3 µg/L (mean-**TRIMPS)** (Wightwick et al 2012)	nd	272 µg/kg dw (max.) (Schafer et al 2011); 27 µg/kg dw (mean) (Wightwick et al 2012); <1-7.6 µg/kg dw (2.3 mean) (Marshall et al 2016).

Pesticides	Surface water	Groundwater	Sediment
Simazine (H)	0.61 µg/L (Chapman and Stranger 1994); 0.43–1.4 µg/L (Chapman and Stranger 1994); 0.014-0.18 µg/L (Davis et al 2008); 15 µg/L (max. in water) (Schafer et al 2011); 6 µg/L (max. in **TRIMPS** (Schafer et al 2011); <0.05-0.062 µg/L (water) (Mossop et al 2013); <0.3-36.1 µg/L (in **Chemcatcher**) (Mossop et al 2013); 0.452 (0.002-4.450 µg/L) (Zhang et al 2014)	0.02–0.45 µg/L) (Ivkovic et al 2001); 0.04–0.08 µg/L (Watkins et al 1999a); 0.022-0.255 µg/L (Bauld 1994; Bauld et al 1998) 0.95 µg/L (Watkins et al 1999b); 3.1 µg/L (Wenig and Lawrence 1998)	260 µg/kg dw (max.) (Schafer et al 2011; <5-50 (Mossop et al 2013); <10-trace µg/kg dw (Mossop et al. 2013); <5-16 µg/kg dw (14 mean) (Marshall et al 2016).
Sulfometuron methyl (H)	0.041 (0.005-0.073 µg/L) (Zhang et al 2014)	nd	nd
Taufluva-linate (I)	75 µg/L (in **TRIMPS + Dodecanol**) (one sample) (Rose and Kibria 2006)	nd	nd
Tebuconazole (F)	0.03 µg/L (mean) (Wightwick et al 2012)	nd	<4-trace µg/kg dw (Mossop et al 2013).
Tebufenozide (I)	<0.002- 0.002 µg/L (Mossop et al 2013)	nd	nd
Tebuthiuron (H)	0.02-0.03 µg/L (Davis et al 2008)	nd	nd
Terbutryn (H)	0.035 (0.002-0.309 µg/L) (Zhang et al 2014)	nd	nd
Tetraconazole (F)	0.02 µg/L (mean) (Wightwick et al 2012)	nd	nd
Triadimenol (F)	<0.002-trace (Mossop et al 2013)	nd	nd
Triasulfuron (H)	0.002 (0.001-0.003 µg/L) (Zhang et al 2014)	nd	nd

Pesticides	Surface water	Groundwater	Sediment
Triclopyr (H)	<0.02-2.03 µg/L (Mossop et al 2013); 0.102 (0.002-1.450 µg/L) (Zhang et al 2014)	nd	nd
Triflox-ystrobin (F)	0.73 g/L (max.in water) (Schafer et al 2011); 793 µg/L (max. in **TRIMPS)** (Schafer et al 2011); 0.08 µg/L (mean in water) (Wightwick et al 2012); 0.23 µg/L (mean in **TRIMPS**) (Wightwick et al 2012)	na	8 µg/kg dw (max.) (Schafer et al 2011); 3.2 µg/kg dw (Wightwick et al 2012); <2-68 µg/kg dw (10.8 mean) (Marshall et al 2016).

Box 7.1. Whats is a TRIMPS technology?

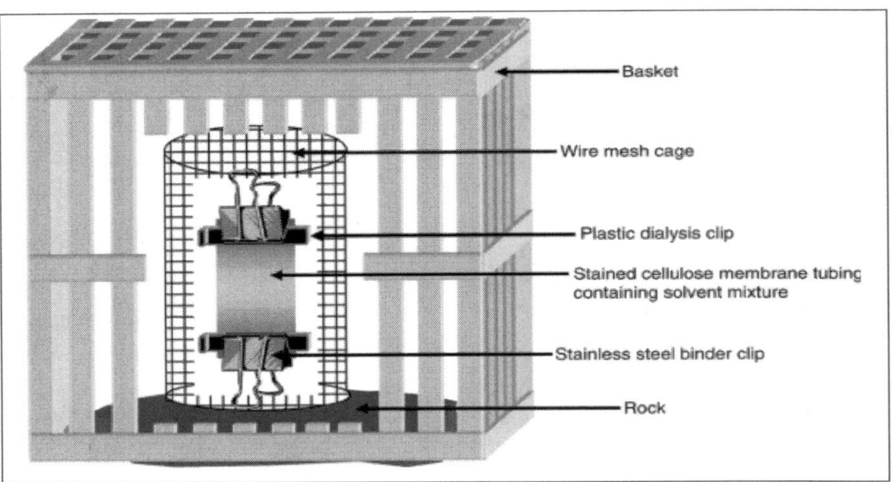

Fig. 7.5. Schematic diagram of the passive sampling device "TRIMPS"

[The device was deployed in the GMWs (G-MW = Goulburn-Murray) six irrigation districts, Victoria, Australia as part of one of the several collaborative pesticides research carried out during 2001-2012 in Australia and led by Golam Kibria (see Kibria et al 2019) (image source Hyne and Aistrope 2008, page 613)].

TRIMPS (trimethylpentane solvent/ chemical containing passive sampler bags) is an innovative pollution (pesticide) monitoring device/ tool. TRIMPS were used in GMW's (GMW= Goulburn-Murray Rural Water Authority) six irrigation districts, North Victoria, Australia (GMW is the largest waste authority in Australia, *https://www.g-mwater.com.au/*) to monitor and assess the risks of pesticides to water quality, aquatic biota (fish) and human health. The monitoring (using

TRIMPS) covered irrigation channels, rivers, creeks, lakes, lagoon, channel outfalls, stock and domestic supply, town supply, mixed farming areas close by waterways including pome and stone fruit, tomatoes, olive, vineyards within 68,000 square kilometres of G-MW. The monitoring targeted both hydrophobic[60] pesticides (log K_{ow}: > 3) and hydrophilic[61] pesticides (log K_{ow}: <3). Two types of TRIMPS were used **i.** passive samplers with solvent 2,2,4-trimethylpentane (for hydrophobic pesticides), **ii.** passive samplers with a solvent mixture of 1-dodecanol: 2,2,4–trimethylpentane (3:2 ratio) (for hydrophilic pesticides). The monitoring found three pesticides on a regular basis across the six irrigation areas: endosulfan, an Organochlorine insecticide (log K_{ow}: 4.79), atrazine, an herbicide (log K_{ow}: 2.55) and copper, a fungicide (inorganic chemical, no log K_{ow} values). These three pesticides were associated with intensive orchards (stone, pome fruit) and vineyards, vegetable (tomatoes) and pastures (legume plants). Other pesticides that were irregularly detected at some sites were bifenthrin (log K_{ow}: 6), chlorpyrifos (log K_{ow}: 4.77), molinate (log K_{ow}: 2.88), parathion methyl (log K_{ow}: 3) and taufluvalinate (log K_{ow}: 4.26). The G-MW collaborated with local chemical retailers, agriculture growers/irrigators, CSIRO (Federal, Commonwealth Scientific and Industrial Research Organisation), DPI (Department of Primary Industries, Victoria) and the Centre for Ecotoxicology, New South Wales to pull together a wide range of information, expertise and funding to carryout out this innovative pesticide research works. Dr Golam Kibria, as Senior Environmental Scientist with G-MW (the lead author of this book) planned, developed and managed the collaborative pesticide research during 2001-2012 periods (Rose and Kibria 2007, Kibria et al 2019).

Fig. 7.6. Passive samplers (TRIMPS) in triplicate deployed at G-MWs risk sites, Australia (Photo: Golam Kibria, the site is at Shepparton)

[60]***Hydrophobic pesticides*** *are those pesticides that do not mix with water or water-repelling (hydrophobic pesticides have a log K_{ow}: >3). Such pesticides do accumulate in biota such as fish.*

[61]***Hydrophilic pesticides*** *are those pesticides which interact well with water or water-attracting (hydrophilic pesticides have a log K_{ow}: <3). Such pesticides in principle donot accumulate in biota such as fish.*

7.3.1.3. DDT Residues in human (women) breast milk

Residue surveys conducted in developed and developing countries revealed DDT residues in human breast milk across the globe (as highlighted in Table 7.9). Though DDT levels are decreasing in women's breast milk in **developed countries** (Australia, Japan, Germany, Sweden, the UK, the USA) (Table 7.9; Fig. 7.7) but it is still at elevated levels in **developing countries**. It appears that DDT is still being used in some developing countries (India, Jordan, South Africa, Vietnam and Zimbabwe) to control malaria and pests in agriculture which may have attributed to the recent detection of elevated levels of DDT in those developing countries (Table 7.9). For example, recent research (2007) reported very high DDT levels in women's breast milk in some **South African** villages (9,500-18,000 µg/g in milk fat) (Bouwman et al 2012) where DDT is still being used to control malaria. In **China**, DDTs were predominant in environmental samples collected from coastal areas and animal feed, which may indicate that DDTs are still being used in China, though it is officially banned. Furthermore, in **Vietnam** higher levels of DDTs in sediments from populated locations were also reported indicating the recent application of DDTs for public health purposes, such as malaria control. Over 20,000 tonnes of DDTs are currently used annually in Vietnam and current results of DDTs in women's breast milk support recent usage of DDTs in Vietnam (Kibria et al 2010). The Asia-Pacific Mussel Watch Programme indicated that DDT concentrations in mussels and fish from Vietnamese coastal waters are among the highest values (Monirith et al 2003), therefore, food chain transfer of DDTs to women in Vietnam is another possibility.

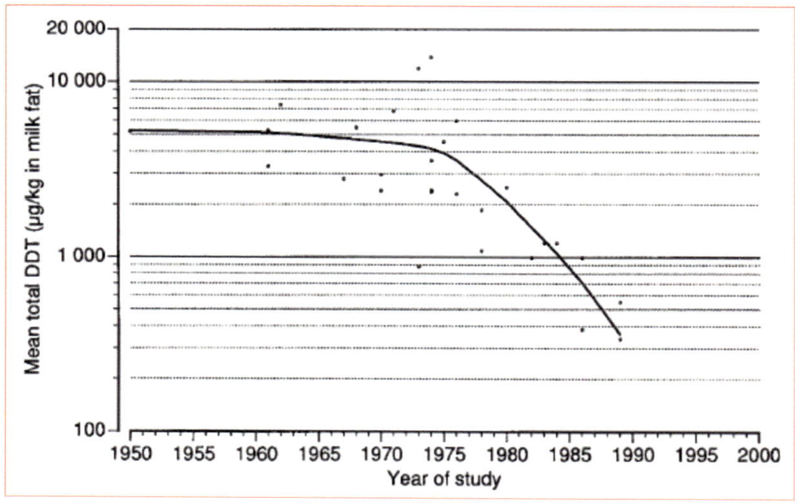

Fig. 7.7. Mean total DDT compounds in women's breast milk (fat) (by year) reported in studies from the USA and Canada *[(Smith 1999, page 184); the figure shows that there was an average of 4,000-5,000 µg/kg in women's breast milk (fat) during 1950-1975 and then a steady decline from 1975 onwards].*

Table 7.9. Reported concentration of total DDT compounds in women's breast milk from some selected developed and developing countries.

Location	Year	Number of samples	Total DDT (µg/g in milk fat)	Remarks	References
Australia (Rural)	1970	36	4,630	none	Newton and Greene 1972
Australia (Urban)	1970	26	3,730	none	Newton and Greene 1972
Australia (Rural)	1979	45	1,281	none	Stacey et al 1985
Australia (Urban)	1980	14	1,277	none	Stacey et al 1985
Australia (Victoria)	1995	60	1,200	DDT was used extensively in Victoria prior to early 1980s; gradually reduced and finally prohibited in 1987.	Quinsey et al 1995
Australia (Western Australia)	2013-2015	99	79.9	Maternal age, parity and body mass index have been associated with pesticide residues in women's milk.	Du et al 2016
Belgium	1969	20	4,800	none	Siyali 1973
Brazil	1992		1,700	none	Paumgartten et al 2000
Canada	1968	147	5,500	none	Ritcey et al 1972
Canada	1983	21	1,200	none	Mes et al 1993
Canada	1986	412	385	none	Mes et al 1993
China	1982		6,200	none	Slorach and Vaz 1983

Location	Year	Number of samples	Total DDT (μg/g in milk fat)	Remarks	References
China (Guangzhou)	2000	54	3,550	none	Wong et al 2002
China	2002		2,100	none	Kunisue et al 2004
Germany	1980	836	2,040	none	Somogyi and Beck 1993
Germany	1990	286	561	none	Furst et al 1994
Germany	1995-1997	246	240	none	Schade and Heinzow 1998
Hong Kong	1976	21	18,870	none	Ip 1983
Hong Kong	1985	25	13,840	Probable routes of uptake of DDT include high consumption of seafood (mussels) (*Perna viridis*) (significantly contaminated by a range of Organochlorines).	Ip and Phillips 1989
Hong Kong	1999	132	2,870	none	Wong et al 2002
India	1980	25	4,800	none	Siddiqui et al 1981
India	1984	6	4,750	none	Ramakrishnan et al 1985
India	1988	25	1,200	DDT was used to control agricultural pests and for public health purposes in earlier years; 85% used for mosquito control and the major route of DDT exposure is through inhalation and through contamination of food during spraying at homes.	Tanabe et al 1990
India	2000	8	420	none	Kunisue et al 2002

Location	Year	Number of samples	Total DDT (µg/g in milk fat)	Remarks	References
Iran	1991		2,000	none	Cok et al 1999
Japan	1977	100-141	1,900	none	Yakushiji et al 1979
Japan	1988	45	287.8	none	Konishi et al 2001
Japan	1998	49	290	none	Konishi et al 2001
Jordan (05 districts)	2012-2013	100	400-1,500	Elevated levels were found in areas where there was intensive vegetable cultivation.	Al Antary et al 2015
Jordan	1992	59	2,740	none	Alawi et al 1992
Kenya	1986	8	4,860	A significant correlation was found between the levels of DDT in sub-cutaneous fat and milk fat of the mothers.	Kanja et al 1992
Mexico	1976	15	13,180	none	Albert et al 1981
Mexico	1997–1998	60	4,648	none	Waliszewski et al 1999
New Zealand (North)	1988	21	1,051	DDE increased linearly with increasing age of the breast-feeding women.	Bates et al 1994
New Zealand (South)	1988	17	3,022	DDE increased linearly with increasing age of the breast-feeding women.	Bates et al 1994
Norway	1976	44	1,758	none	Brevik and Bjerk 1978
Norway	1982	14	1,000	none	Skaare and Polder 1990

Location	Year	Number of samples	Total DDT (μg/g in milk fat)	Remarks	References
Russia	1996		2,000	none	Hansen 1998
South Africa	1987	129	15,830	Primiparous[62] mothers from the malaria's area had significantly higher DDT and metabolites than multiparous[63] mothers from the same area.	Bouwman et al 1990
South Africa (villages)	2007	163	9,500-18,000	Villages where malaria is controlled with DDT (sprayed indoors)	Bouwman et al 2012
Spain	1979	45	7,200	none	Hernandez et al 1993
Spain	1991	51	660	none	Hernandez et al 1993
Spain (Canary Islands)	2014	72	0.08-16.96 ng/g	Positive association found between DDT levels and vegetable growing (intensive agriculture).	Vall et al 2014
Sweden	1972	75	2,930	none	Nore´n et al 1996
Sweden	1992	20	283	none	Nore´n et al 1996
Sweden	1997		170	none	Nore´n and Meironyte´ 2000
Thailand	1989	3	43,403	none	Schecter et al 1989
Turkey	1997	104	2,161	none	Schinas et al 2000

[62]***Primiparous*** *= A woman who has given birth to one child or who is giving birth for the first time.*

[63]***Multiparous*** *= A woman who has given birth more than once.*

Location	Year	Number of samples	Total DDT (µg/g in milk fat)	Remarks	References
Turkey	1995-1996		2,400	none	Cok et al 1997
UK	1963	19	5,040	none	Egan et al 1965
UK	1980	102	1,710	none	Collins et al 1982
UK	1997-1998	168	470	none	Harris et al 1999
Ukraine	1993–1994	197	2,793	Unusual contamination due to excessive use of pesticides DDT in the past.	Gladen et al 1999
USA (Hawaii)	1979-1980	102	2,200	none	Takei et al 1983
USA (Washington)	1961	5	3,300	none	Quinby et al 1965
USA (Seven cities)	1971	101	6,800	none	Wilson et al 1973
USA (Arizona)	1973	6	11,900	none	Hagyard et al 1973
USA (Arkansas)	1986	990	942	none	Mattison et al 1992
USA	1994-1996		408.85	none	Bagga et al 2000
Vietnam	1989	12	1,1400	none	Schecter et al 1989
Vietnam	2001-2001	48	2,200	Elevated concentrations due to recent applications of DDTs for sanitation and malaria control; The Asia-Pacific Mussel Watch Program indicated that DDT concentrations in mussels and fish from Vietnamese coastal waters are among the highest values in the world.	Minh et al 2004

Location	Year	Number of samples	Total DDT (μg/g in milk fat)	Remarks	References
Zimbabwe	1990	40	6,000	Continued use of DDT in agriculture and malaria control program caused elevated levels.	Chikuini et al 1991

7.3.2. Impacts on non-target organisms (biota)

Pesticides are toxic chemicals and if not handled and used properly (i.e. incorrectly) can lead to adverse effects on the environment, non-target organisms, livestock and humans. The harmful or unwanted effects that may happen from the application or use of pesticides on non-target organisms are summarised below:

- Fish kills due to accidental spills or uncontrolled washing of pesticides from treatment tanks and equipment.
- Alteration of wildlife due to erroneous use of pesticides (sub-lethal and lethal effects).
- Accumulation of hydrophobic pesticides (log K_{OW}: >3) residues in the food crops, vegetables, fruits and fatty tissues of animal organisms (seafood) (section 7.3.1.1).
- Contamination of surface and groundwater by water-soluble or hydrophilic pesticides (log K_{OW}: <3).
- Water contaminated with pesticides can be unfit for human consumption (drinking), irrigation, food processing, livestock drinking, recreation use, aquatic ecosystems protection and aquaculture.

The following section highlights some pesticide-related effects on birds, earthworms, fish, frogs and bees.

7.3.2.1. Birds

Birds can be poisoned by pesticide residues that remain in/ on food after spraying, for example, when wild animals enter fields for feeding that were sprayed with pesticides. The Organochlorine pesticides are toxic to a wide range of biota. A reduction in the reproductive success of carnivorous birds such as, the peregrine[64] falcon has been related to DDE[65] (a metabolite of DDT). In addition,

[64]**Peregrine** = *migratory birds; a powerful falcon (birds of prey) found on most continents, breeding chiefly on mountains and coastal cliffs.*

[65]**DDE** = *dichlorodiphenyldichloroethylene*

the accumulation of DDT has also been linked to the thinning of eggshells (Lundholm 1997, *http://www.ecosmagazine.com/?act=view_file&file_id=EC20p8.pdf*) causing them to break during incubation and dramatically reducing the population of predatory birds, such as falcons which are at the top of the food chain. In the farmland of Britain (1979-1999), birds' population have declined due to loss of plant and invertebrate species on which the birds feed because of pesticides use (Sotherton and Holland 2003) (see Table 7.4 for LD_{50} for birds).

7.3.2.2. Earthworms

Earthworms decompose the organic matter and create channels in the soil; however, they are threatened by carbamate pesticides which are very toxic to earthworms. In addition, some organophosphate pesticides are reported to reduce earthworm populations (Edwards 1987).

7.3.2.3. Fish and other aquatic biotas

Fish and other aquatic biotas (invertebrates) can be harmed by water contaminated with pesticides. For example, application of herbicides can cause fish kills when dead plants rot and use up most of the dissolved oxygen in the water. Copper sulphate ($CuSO_4$, an herbicide) is used to kill aquatic plants, however, the chemical is very toxic to fish and other aquatic organisms. Repeated exposures of lethal doses of pesticides can cause physiological and behavioural changes in fish, therefore, may reduce the populations in the wild. Monocrotophos (an organophosphorus pesticide) caused 100% mortality of *Labeo rohita* fish at 0.0044 mg/L (96 h) and 50% mortality at 0.0036 mg/L (96 h) (Muthukumaravel et al 2013) (see Table 7.4 for LC_{50} for fish and invertebrates).

7.3.2.4. Frog (tadpoles)

Frog (tadpoles) chronically exposed to endosulfan at 0.01-0.05 mg/L suffered a high mortality rate, decreased larval growth, delayed development, an increased incidence of malformations (Lavorato et al 2013). According to a recent review (Mann et al 2009), pesticides may cause the following effects on frogs and toads:

- impaired behaviour and growth (e.g. by pyrethroids and fungicides);
- limb deformities (e.g. by mixtures of atrazine and carbaryl);
- altered sex ratios (e.g. by mancozeb);
- acute and chronic toxicity (e.g. by azinphos-methyl, glyphosate-based products, paraquat, permethrin);
- endocrine disruption (some controversy exists whether atrazine is a potential endocrine disruptor since some of the research suggests it may adversely affect the reproductive capacity of frogs and toads; whereas other and more recent research did not find links between atrazine

exposure and the development of abnormal gonads in *Xenopus laevis, Rana pipienis* and *Hyla versicolr*; and

- delayed metamorphosis, survival and growth (e.g. by mixtures of pesticides).

7.3.2.5. Pollinators (bees)

Bees are essential for pollination of many plants and agricultural crops in natural ecosystems. For example, bumble bees (*Bombus* spp.), honey-bees (*Apis* spp.) are efficient pollinators. Bees can be exposed to pesticides via spray drifts during the foliar[66] spraying of crops, dust from seed drilling at planting, inhalation of volatile pesticides during or after application to the crops and by exposure to pesticide residues present in pollen, wax, nectar, honey etc. Bees can also be exposed during drinking of water contaminated with pesticides (Sanchez-Bayo and Goka 2014). Pesticide residues found in pollen and nectar of crop plants could pose a major risk to bee pollinators, worker bees and ultimately in the collapse of bees in a hive. Neonicotinoid insecticides (acetamiprid, clothianidin, dinotefuran, imidacloprid, nitenpyram, thiamethoxam, thiacloprid) caused the lethal and sub-lethal effects on foraging behaviour, learning and memory abilities of bees (Blacquie`re et al 2012). Gill et al (2012) showed that chronic exposure of bumble bees to two pesticides (neonicotinoid and pyrethroid) impaired natural foraging behaviour, increased worker bees' mortality, reduced pollen collecting efficiency (see also Table 7.4 for LD_{50} for bees and the Case study 3: neonicotinoid pesticide - a threat to efficient pollinators).

7.3.3. Impacts of pesticide on humans health

Humans can be exposed to pesticides via oral ingestion, dermal absorption and inhalation and through water and food contamination. The immediate health effects from pesticide exposure include irritation of nose, throat and skin causing burning, stinging and itching as well as formation of the skin rashes and blisters. Workers who handle pesticides may experience nausea, abdominal pain, headaches, vomiting, dizziness, and diarrhoea as well as skin and eye problems. In many cases, symptoms of pesticide poisoning are similar to common colds or the flu. The long-term health effects include cancer and other tumours; brain and nervous system damage; birth defects; miscarriages, infertility, male sterility and other reproductive problems. It may cause damage to the liver, kidneys, lungs and other body organs (Agnihorti 1999, USEPA 2006, Abhilash and Singh 2009) (see Table 7.10 for a list of chronic disease that can be caused by exposure to pesticides). Investigations revealed a positive relationship between exposure of pesticides and different types of cancers including kidney, pancreatic, prostate (atrazine related) and stomach cancer. However, the chronic effects may not

[66]**Foliar** = *relating to leaves.*

appear even after years of exposure. Some pesticides act as endocrine disruptors (e.g. atrazine, simazine, endosulfan, DDT, see also Table 7.6) and have been reported to cause serious harm to animals, including cancer, sterility and developmental problems. Other pesticides related effects are disruption of the endocrine system and suppression of immune systems. In Australia, during 1989-1992, at least two fatalities were related to pesticides poisoning (a rate of one fatality every two years). The WHO and UNEP estimate that about 3 million workers (in the agriculture sector) in the developing world experience severe poisoning from pesticides in each year of which about 18,000 would ultimately die (WHO/ UNEP 1990).

Table. 7.10. List of chronic disease that is linked to pesticides
(original reviewed by Gill and Garg 2014).

Diseases linked to pesticides	Remarks	References
Alzheimer disease; Parkinson disease	Professional/ occupational exposure to pesticides (Organochlorines insecticides) linked to Alzheimer disease[67] and dementia[68]; **ii.** paraquat is associated with Parkinson's disease[69].	Elbaz et al 2009, Hayden et al 2010, Tanner et al 2011
Birth defects	**i.** Atrazine maybe related to birth defects; **ii.** multiple pesticide exposure (occupational and contaminated food) (carbendazim, 2,4-dichlor ophenoxyacetic acid, glyphosate, ioxynil, linuron, trifluralin and vinclozolin) may put farmers and their offspring at risk of reproductive disorder.	Winchester et al 2009, Mesnage et al 2010
Cancer	There are associations between **i.** some specific agricultural pesticide exposures (bufencarb, chlorpyrifos, coumaphos metribuzin, paraquat) and the risk of glioma (tumour) among male farmers; **ii.** between parental exposure to	Lee et al 2005, Shim et al 2009, Heck et al 2010, Xu et al 2010, Band et al 2011,

[67]***Alzheimer's disease*** *is a type of dementia that causes problems with memory, thinking and behaviour. Symptoms usually develop slowly and get worse over time, becoming severe enough to interfere with daily tasks. (https://www.alz.org/alzheimers-dementia/what-is-alzheimers).*

[68]***Dementia*** *is a broad category of brain diseases that cause a long-term and often gradual decrease in the ability to think and remember that is severe enough to affect daily functioning. The most common type of dementia is Alzheimer's disease, which makes up 50% to 70% of cases (https://en.wikipedia.org/wiki/Dementia).*

[69]***Parkinson's disease*** *is a progressive, degenerative neurological condition that affects the control of body movements. It causes trembling (shaking) in the hands, arms, legs, jaw, and face; rigidity or stiffness of the limbs or trunk; slowness of body movements and unstable posture and difficulty in walking. Early symptoms are subtle and occur gradually (https://brainfoundation.org.au/disorders/parkinsons-disease/).*

Diseases linked to pesticides	Remarks	References
	pesticides (herbicides) and risk of astrocytoma[70] (cancer of the brain) in offspring; **iii.** between renal cell cancer and agricultural work; **iv.** OC (Organochlorines) pesticide (β-hexachloro-cyclohexane or HCH and dieldrin) exposures and cancer risk; **v.** between prostate cancer risk and exposure to DDT; **vi.** an increase in the risk of specific lymphoma[71] sub-types associated with exposure to pesticides (organophosphates).	Cocco et al 2013
Cardio-vascular disease; artery disease	There is link **i.** between chronic organophosphate pesticide exposure and coronary artery disease; **ii.** pre-natal (before birth) pesticide exposure and adverse cardiovascular risks.	Abdullah et al 2011, Andersen et al 2012
Diabetes (Type 2 Diabetes)	There is strong association **i.** between low-dose Organochlorine pesticides and type 2 diabetes; **ii.** environmental exposure to some POPs (Organochlorines OCs[72], PCBs) can substantially increase the risk of future type-2 diabetes in elderly people.	Son et al 2010, Lee et al 2011
Reproductive disorders	Exposure to environmental pesticides **i.** could damage the prenatal[73], perinatal[74], and postnatal[75] reproductive system in humans including spontaneous abortion; **ii.** may impair female fertility.	Petrelli and Mantovani 2002, Greenlee et al 2003
Respiratory diseases (Asthma, chronic obstructive pulmonary disease - COPD)	Long-term exposure to cholinesterase-inhibiting agricultural pesticides (organophosphates OP, carbamates) is associated with **i.** a reduction in lung function, COPD and rise in respiratory symptoms; **ii.** a link between pesticide use and adult-onset asthma among male farmers was found.	Chakraborty et al 2009, Hoppin et al 2009

[70]***Astrocytoma*** *is a type of cancer that can form in the brain or spinal cord. Astrocytoma begins in cells called astrocytes that support nerve cells. Astrocytoma signs and symptoms depend on the location of your tumour (https://www.mayoclinic.org/diseases-conditions/astrocytoma/cdc-20350132).*

[71]***Lymphomas*** *refers to types of cancer that begin in the lymphatic system (part of the circulatory system and the immune system) (lymph = liquid/ water). Lymphomas are the sixth most common form of cancer overall (excluding non-melanoma skin cancer) (https://www.cancer.org.au/about-cancer/types-of-cancer/lymphoma.html).*

[72]***OCs*** *= Organochlorines.*

[73]***Pre-natal*** *= before birth; during or relating to pregnancy.*

[74]***Peri-natal*** *= pertaining to the period immediately before and after birth.*

[75]***Post-natal*** *= begins immediately after the birth of a child.*

7.4 Case studies

Case study 1: Glyphosate herbicide - the new threat to public health?

Glyphosate (also known as a **Round-Up** or Weed killer or and Monsanto) currently has the highest global production of all herbicides. It is the most widely used pesticide on the planet. It has the largest use worldwide in agriculture, followed by forestry, urban, and home applications (Gill et al 2018). A new joint R&D - published in the "Science of the Total Environment" (Silva et al 2019) shows that almost **half of European soils** (11 Countries including the UK, Denmark, Portugal, Italy, Greece, Spain, Hungary, Poland, the Netherlands, France and Germany) are contaminated with glyphosate. The studies revealed that 45% of agricultural land in Europe contains glyphosate and its metabolite aminomethylphosphonic acid (**AMPA**). The analysis is based on 300 top soils (15-20 cm). It is known that glyphosate and AMPA are highly persistent (strongly adsorb[76] and accumulate in the top few centimetres of agriculture soils) once they are attached to soil particles. There is a risk of spreading soil-bound glyphosate particles (as dust) through wind erosion or water erosion (surface run-off) and exposure to human. Humans can be exposed to glyphosate by inhalation of contaminated **i. dust particles**; **ii. dermal contact** and **iii. ingestion of contaminated surface water**. Glyphosate was detected in the urine of farmers, forest workers, pregnant women, fishermen, people living close by glyphosate sprayed areas in Colombia, Finland, France, Ireland, Mexico, Sri Lanka, Thailand, USA (Table 7.11, Gillezeau et al 2019). A wide variety of human diseases, including various forms of cancer, kidney damage and mental conditions such as ADHD[77], autism[78], Alzheimer's and Parkinson's disease have been correlated with Glyphosate (Séralini et al 2014, Samsel and Seneff 2016; *https://www.ecowatch.com/15-health-problems-linked-to-monsantos-roundup-1882002128.html*). The International Agency for Research on Cancer (IARC) of the WHO reclassified glyphosate as probably carcinogenic (2A) (IARC 2019).

---

*[76]**Adsorption** is the adhesion of atoms, ions, molecules, or pollutants from gas, water or dissolved solids to a surface of an adsorbent. Adsorption plays an important role in the transport of contaminants (see Chapter 4, section 4.3.2 for Adsorption of contaminants by plastic).*

*[77]**ADHD** stands for 'attention deficit hyperactivity disorder'. It is a medical condition. A person with ADHD has differences in brain development and brain activity that affect attention, the ability to sit still, and self-control. ADHD can affect a child at school, at home, and in friendships (https://kidshealth.org/en/parents/adhd.html).*

*[78]**Autism** spectrum disorder (ASD) is a broad term used to describe a group of neuro-developmental disorders. These disorders are characterized by problems with communication and social interaction. People with ASD often demonstrate restricted, repetitive, and stereo-typed interests or patterns of behaviour (https://www.healthline.com/health/autism).*

Table 7.11. Glyphosate concentrations in humans' urine in different countries

[LOD = Limit of detection; NR= not reported].

Country *(NR=not reported)*	Urine sample	Glyphosate (G) or aminomethylphosphonic acid metabolite (AMPA) in urine	References
Colombia (2005-06)	112 individuals living in the sprayed areas (aerially administered glyphosate).	G concentration: 7.6 ± 18.6 μg/L (range: 0–130 μg/L) AMPA: 1.6 ± 8.4 μg/L (range: 0– 56 μg/L) A statistically significant relationship was found between the use of terrestrial glyphosate (manual) and the levels of herbicide in the urine.	Varona et al 2009
Europe (18 cou-ntries) (2001)	Volunteers (lived in cities, and none had handled or used glyphosate products in the run-up to the tests) (Austria, Belgium, Bulgaria, Croatia, Cyprus, Czech Republic, France, Georgia, Germany, Hungary, Latvia, Macedonia, Malta, Poland, Spain, Switzerland, the Netherlands, and the UK)	Highest G concentration: 1.8 μg/L (Latvia) Highest AMPA concentration: 2.6 μg/L (Croatia) Mean G concentration in 12 countries: 0.21 μg/L (0.09-0.82) Mean AMPA in 12 countries: 0.18 μg/L (0.08-0.29) On an average 44% urine samples contained glyphosate. *https://www.foeeurope.org/weed-killer-glyphosate-found-human-urine-across-Europe-130613*	Hoppe 2013
Finland (1988)	Forest workers sprayed round-up containing the solution for 6 h/ d for 1 w.	Urine samples remained < LOD for G and AMPA. G concentration: One urine sample had 85 μg/L glyphosate.	Jauhiainen et al 1991
France (NR)	Farmer and his family, using glyphosate-based herbicide.	G concentration: 9.5 μg/L after spraying in case of the farmer. G concentration: 2 μg/L 2 days later in case of the farmer and one of the children.	Mesnage et al 2012
Ireland (2015)	Amenity horticul-turalists, before and after spraying (18 workers).	Pre-spraying mean G concentration: 0.71±0.92 μg/L Post-spraying G concentration: 1.35 ± 2.18 μg/L	Connolly et al 2017

Country *(NR=not reported)*	**Urine sample**	**Glyphosate (G) or aminomethylphosphonic acid metabolite (AMPA) in urine**	**References**
Mexico (NR)	Farmers (subsistence farmers from the Francisco J. Mújica communities).	Mean G concentration in farming areas: 0.26 ± 0.23 μg/L Highest concentration of G in urine: 0.47 μg/L	Rendón-von Osten et al 2017
Mexico (NR)	Fishermen in the urban area.	Mean G concentration: 0.16±0.1 μg/L (median: 0.20 μg/L) The highest concentration of G in urine was 0.47 μg/L and of subsistence farmers (0.2 μg/L, reference group).	Rendónvon Osten and Dzul-Caamal 2017
Sri Lanka (NR)	Healthy farmers from areas with chronic endemic kidney disease.	Endemic group median G concentration: 73.5 (40.2- >80) μg/L Non-endemic group G concentration: 3.3 (1.2-5.5) μg/L	Jayasumana et al 2015
Thailand (2011)	Pregnant women, age 19–35 years who delivered a baby.	Maternal serum median: 17.5 (0.2–189.1) μg/L Umbilical cord serum median: 0.2 (0.2–94.9) μg/L Pregnant women who work in agriculture or live in families that work in agriculture can have higher exposures to the herbicide glyphosate.	Kongtip et al 2017
USA (South Carolina, Minnesota) (NR)	Farm's families on application day and 3 days later (48 farmers, 48 spouses, 79 children).	G concentration: 3.2 ± 6.4 μg/L 67% of farmers had a detectable level of glyphosate in the urine on the day of application. Farmers who did not use gloves had higher G concentrations in urine than other farmers who used gloves (10 μg/L vs 2 μg/L)	Acquavella et al 2004
USA (Indiana) (2015-16)	Pregnant women, age 18–39 years.	G concentration in Urine: mean ± SD: 3.40±1.24μg/L Higher G levels were found in women who lived in rural areas or having higher caffeine intake. G concentration in water: not detected in drinking water.	Parvez et al 2018

Case study 2: Agricultural insecticides – a threat to surface waters and aquatic biodiversity at the global scale.

A comprehensive meta-analysis of 838 peer-reviewed research studies (in 73 countries) found that the world surface waters are threatened by 28 toxic insecticides[79] comprising of Organochlorines, organophosphates, carbamates, pyrethroid and neonicotinoid (as shown in Table 7.2) (Stehle and Schulz 2015). About 52.4% of the measured insecticide concentrations (MICs) (n = 11,300) exceeded the regulatory threshold levels (RTLs) for either surface water or sediments. The study revealed that the surface waters and aquatic biodiversity (water-based and sediment-based) located in and around cultivated areas are at risk from exposure of insecticides. The study found that RTLs exceedance[80] rate was significantly higher for newer-generation insecticides (i.e. pyrethroids) (*the RTLs exceedance rate was 65.8%, 43.7%, 24.4%* and *6.1% for pyrethroids, organophosphates, Organochlorines* and *neonicotinoids* respectively). The results demonstrated that agricultural insecticides are a threat to surface waters at the global scale. It found that the entry route of insecticide into surface waterways were non-point (65.2%), run-off (25.2%), rice-field effluent (3.8%), spray drift (3.1%), aerial application (1.6%) and drainage discharge (1.1%). The study revealed that in the future, there would be an increase in the application of pesticides in many developing countries due to the demand for growing more food in line with the projected increase in population which will force farmers to switch from traditionally extensive agricultural practices to more intensive agriculture that requires significant use of pesticides. Furthermore, the proliferation of insects in a warming world (global warming and climate change) will require more use of chemical pesticides to control pests of animals and plants (see also section 7.5 for linkages between pesticides and climate change).

Case study 3: Neonicotinoid pesticide - a threat to efficient pollinators (Bees).

Neonicotinoids (NNI) are a common class of insecticides derived from nicotine (*NNIs included clothianidin, thiamethoxam, imidacloprid* and *acetamiprid).* NNIs are highly toxic to insects including bees. A 2-years research in Canada quantified effects of NNI dust in 55 bee colonies on corn-growing areas (Indiana, Ontario, and Québec). Honey bees (*Apis mellifera*) experienced high colony mortality. NNI residues were found in dead bees and hives. NNI dust generated during seeding (corn planting) is believed to be the main route of acute exposure to bees. NNI increased worker-bees' mortality and was associated with declines in social immunity and increased queen-lessness (devoid of queen bee) over time. It may be mentioned here that bee colonies near corn fields were naturally exposed to neonicotinoids for up to 4 months — the majority of the honey bee's active season (Tsvetkov et al 2017). It can be mentioned that Registered insecticide labels in Australia have required the user to identify bee mortality risks.

[79]***28 toxic insecticides*** = *see Table 7.2 for the list of 28 toxic insecticides, those have been identified as a threat to global surface waters (after Stehle and Schulz 2015).*

[80]***Exceedance*** = *exceeding the regulatory threshold levels or guidelines standard.*

7.5. Linkages between pesticides and climate change

The warmer climates and climate extreme events could be more favourable to the proliferation of insect pests and plant diseases resulting in increased use of chemical pesticides. Climate-change-related extreme events, such as floods can lead to increased wash-off of pesticides and herbicides from soils to water bodies and increase erosion of pesticide-rich soil particles from fields to drains and surface waters and thus may lead to water contamination in rivers. More frequent and intense rainfall events may lead to more by-pass flow increasing the likelihood of rapid pesticide movement to drains, surface waters and vertically below the soil layer. Moreover, higher temperatures will increase the volatilisation[81] of pesticides and degradation of pesticide residues in the soil and surface waters (Kibria 2014) (see also the case studies 1 and 2).

7.6. Management of pesticides risk

Major management options to reduce risks of pesticides are summarised below:

- Monitoring of pesticide residues in soil, water, food, and human breast milk and comparing with recommended guideline thresholds values (WQ guidelines, MRL) for the protection of the environment, food and water and public health (Table 7.12);
- Carrying out ecotoxicological studies with native fish with the aim to assess the lethal and sublethal effects of locally used pesticides for the protection of aquatic species (Box 7.2, Table 7.13);
- Assessing threats and risks of legacy pesticides as well as new pesticides in local and new farming areas;
- Encourage and enforce acts, regulations and good agricultural practices (GAP) for the management of pesticide residues;
- Researching, educating and training communities, through field days, focus group discussions (FGDs), field farmers schools, etc. regarding the proper use of pesticides and their risk and impact and use of pesticides according to the material safety data sheet (MSDS) prescription if applicable and approved by the authority;
- Adopting and popularising Integrated Pest Management (IPM) System;
- Banned pesticides should never be allowed to enter in the country (through import, trade and supply) for use in pest control;
- Use of light-traps, biopesticides (biofungicides e.g., *Trichoderma* spp.), bioherbicides e.g. *Phytopthora* spp.), bioinsecticides (spore-forming

---

[81]***Volatilisation*** *= The process of converting a chemical substance from a liquid or solid state to a gaseous or vapour state.*

Table 7.12. Australian food and water quality guidelines for pesticides

*[**A** = Acaricide; **C** = carbamate; **F** = fungicide; **H** = herbicide; **I** = insecticide; **M** = Molluscicide; **N** = nematicide; **Neo** = neonicotinoid; **OC** = Organochlorine; **OP** = organophosphate; **R** = Rodenticide; **SP** = synthetic pyrethroid; **MRL** = maximum residue limits; **-** = data is not available].*

Pesticides	Food Australian standard (MRL)				Water Australian guidelines				
	Cereal grains (maize, wheat) (mg/kg) (NRS 2010/11)	**Apple** (mg/kg) (NRS 2010/11)	**Cattle** (mg/kg) (NRS 2010/11)	**Eggs** (mg/kg) (NRS 2010/11)	**Irrigation** (mg/L) (ANZECC and ARMCANZ 2000)	**Aquaculture** (µg/L) (freshwater = Fw; marine = Mr; ANZECC and ARMCANZ 2000)	**Drinking water** Human health- (mg/L) (NHMRC and NRMMC 2011)	**Recreation** (µg/L; ANZECC and ARMCANZ 2000) (max. conc.)	**Aquatic ecosystems protection** (µg/L); (freshwater =Fw; marine = Mr; ANZECC and ARMCANZ 2000)
2,2-DPA (Dalapon)	-	-	-	-	0.004	-	-	-	-
2,4,5-T (H)	-	-	-	-	-	-	0.1	2.0	36.0 (Fw)
2,4-D (H)	0.2	-	-	-	-	<4.0 (Fw)	0.03	100	280.0 (Fw)
Acephate (OP)	-	-	-	-		-	0.008	20.0	-
Acrolein (H)					0.1	-	-	-	-
Aldicarb (C)	-	-	-	-	-	-	0.004	1.0	-
Aldrin and Dieldrin (OC)	0.02	0.05	0.2	0.1	-	<0.01 (Fw)	0.0003	-	-
Amitraz (I, A)	-	-	-	-	-	-	0.009	-	-
Amitrole (H)	0.01	-	-	-	0.002	-	0.009	1.0	-
Atrazine (H)	0.1	-		--	-	-	0.02	-	13.0 (Fw)

Pesticides	Food Australian standard (MRL)				Water Australian guidelines				
	Cereal grains (maize, wheat) (mg/kg) (NRS 2010/11)	**Apple** (mg/kg) (NRS 2010/11)	**Cattle** (mg/kg) (NRS 2010/11)	**Eggs** (mg/kg) (NRS 2010/11)	**Irrigation** (mg/L) (ANZECC and ARMCANZ 2000)	**Aquaculture** (µg/L) (freshwater = Fw; marine = Mr; ANZECC and ARMCANZ 2000)	**Drinking water** Human health- (mg/L) (NHMRC and NRMMC 2011)	**Recreation** (µg/L; ANZECC and ARMCANZ 2000) (max. conc.)	**Aquatic ecosystems protection** (µg/L); (freshwater =Fw; marine = Mr; ANZECC and ARMCANZ 2000)
Azamethiphos (OP, I)	0.1	-	-	-	-	-	-	-	-
Azinphos-methyl (OP)	-	2.0	-	-	-	<0.01 (Fw)	0.03	10.0	0.01(Fw)
Azoxystrobin (F)	0.02		-	-	-	-	-	-	-
Benomyl (F)	-	-	-	-	-	-	0.09	-	-
Bifenazate	-	2.0	-	-	-	-	-	-	-
Bifenthrin (SP)	2.0	0.05	2.0	-	-	-	-	-	-
Boscalid (F)	-	2.0	-	-	-	-	-	-	-
Bromacil (H)	-	-	-	-	-	-	0.4	-	-
Bromoxynil (H)	0.2	-	-	-	-	-	0.01	30.0	-
Bupirimate (F)	-	1.0	-	-	-	-	-	-	-
Captan (F)	-	10.0	-	-	-	-	0.4	-	-
Carbaryl (C)	5.0	5.0	0.2	-	-	-	0.03	60.0	-
Carbendazim (F)	0.05	5.0	-	-	-	-	0.09	200.0	-

Pesticides	Food Australian standard (MRL)				Water Australian guidelines				
	Cereal grains (maize, wheat) (mg/kg) (NRS 2010/11)	**Apple** (mg/kg) (NRS 2010/11)	**Cattle** (mg/kg) (NRS 2010/11)	**Eggs** (mg/kg) (NRS 2010/11)	**Irrigation** (mg/L) (ANZECC and ARMCANZ 2000)	**Aquaculture** (µg/L) (freshwater = Fw; marine = Mr; ANZECC and ARMCANZ 2000)	**Drinking water** Human health- (mg/L) (NHMRC and NRMMC 2011)	**Recreation** (µg/L; ANZECC and ARMCANZ 2000) (max. conc.)	**Aquatic ecosystems protection** (µg/L); (freshwater =Fw; marine = Mr; ANZECC and ARMCANZ 2000)
Carbofuran (C)	-	-	-	-	-	-	0.01	30.0	0.06 (Fw)
Carfentrazone-ethyl (H)	0.05	-	-	-	-	-	0.1	-	-
Chlorantraniliprole (I)	-	0.3	-	-	-	-	6.0	-	-
Chlordane (OC)	0.02	-	0.2	0.02	-	<0.01 (Fw); 0.004 (Mr)	0.002	6.0	0.03 (Fw)
Chlorfenapyr (I, A)	-	0.5	0.05	-	-	-	-	-	-
Chlorfenvinphos (OP)	0.05	-	0.2	-	-	-	0.002	10.0	-
Chlorfluazuron (I)	-	-	1.0	-	-	-	-	-	-
Chlorothalonil (F)	-	-	-	-	-	-	0.05	-	-
Chlorpyrifos (OP)	0.1	0.5	0.5	-	-	<0.001 (Fw)	0.01	2.0	0.01 (Fw); 0.009 (Mr)
Chlorpyrifos-methyl (OP)	10.0	-	0.05	-	-	-	-	-	-
Chlorsulfuron (H)	0.05	-	-	-	-	-	0.2	-	-
Clethodim (H)	0.1	-	-	-	-	-	-	-	-

Pesticides	Food Australian standard (MRL)				Water Australian guidelines				
	Cereal grains (maize, wheat) (mg/kg) (NRS 2010/11)	**Apple** (mg/kg) (NRS 2010/11)	**Cattle** (mg/kg) (NRS 2010/11)	**Eggs** (mg/kg) (NRS 2010/11)	**Irrigation** (mg/L) (ANZECC and ARMCANZ 2000)	**Aquaculture** (µg/L) (freshwater = Fw; marine = Mr; ANZECC and ARMCANZ 2000)	**Drinking water** Human health- (mg/L) (NHMRC and NRMMC 2011)	**Recreation** (µg/L; ANZECC and ARMCANZ 2000) (max. conc.)	**Aquatic ecosystems protection** (µg/L); (freshwater =Fw; marine = Mr; ANZECC and ARMCANZ 2000)
Clodinafop-propargyl (H)	0.05	-	-	-	-	-	-	-	-
Clopyralid (H)	2.0	-	-	-	-	-	-	-	-
Clothianidin (I, Neo)	-	0.5	-	-	-	-	-	-	-
Coumaphos (OP)	-	-	0.2	-	-	-	-	-	-
Cyfluthrin (SP)	2.0	-	0.5	-	-	-	0.05	-	-
Cyhalothrin (SP)	0.05	-	0.5	-	-	-	-	-	-
Cypermethrin (SP)	0.2	1.0	0.5	-	-	-	0.2	-	-
Cyproconazole (F)	0.02	-	0.03	-	-	-	-	-	-
Cyprodinil (F)	-	0.05	-	-	-	-	-	-	-
DDT (OC)	0.1	1.0	5.0	0.5	-	<0.0015 (Fw)	0.009	3.0	0.006 (Fw)
Deltamethrin (SP)	2.0	-	0.5	-	-	-	0.04	-	-
Demeton (I,A)	-	-	-	-	-	<0.01 (Fw)	-	-	-
Diazinon (OP)	0.1	0.5	0.7	-	-	-	0.004	10.0	0.01 (Fw)

Pesticides	Food Australian standard (MRL)				Water Australian guidelines				
	Cereal grains (maize, wheat) (mg/kg) (NRS 2010/11)	**Apple** (mg/kg) (NRS 2010/11)	**Cattle** (mg/kg) (NRS 2010/11)	**Eggs** (mg/kg) (NRS 2010/11)	**Irrigation** (mg/L) (ANZECC and ARMCANZ 2000)	**Aquaculture** (µg/L) (freshwater = Fw; marine = Mr; ANZECC and ARMCANZ 2000)	**Drinking water** Human health- (mg/L) (NHMRC and NRMMC 2011)	**Recreation** (µg/L; ANZECC and ARMCANZ 2000) (max. conc.)	**Aquatic ecosystems protection** (µg/L); (freshwater =Fw; marine = Mr; ANZECC and ARMCANZ 2000)
Dicamba (H)	0.05	-	-	-	-	-	0.1	300.0	-
Dichlorvos (OP)	5.0	-	-	-	-	-	0.005	20.0	-
Diclofop-methyl (H)	0.1	-	-	-	-	-	0.005	3.0	-
Dicofol (OC, A)	-	5.0	-	-	-	-	0.004	100.0	-
Dieldrin (OC)	-	-	-	-	-	<0.005 (Fw)	-	1.0	-
Difenoconazole (F)	0.01	0.3	-	-	-	-	-	-	-
Diflubenzuron (I)	2.0	-	-	-	-	-	0.07	-	-
Dimethoate (OP, I, A)	0.05	5.0	-	-	-	-	0.007	100.0	0.15 (Fw)
Diquat dibromide (H)	0.1	-	-	-	-	-	0.007	10.0	1.4 (Fw)
Dithiocarbamates (F)	0.5	3.0	-	-	-	-	-	-	-
Diuron (H)	0.1	0.5	-	-	0.002	-	0.02	40.0	-
Dodine (F)	-	5.0	-	-	-	-	-	-	-
Endosulfan (OC)	0.1	1.0	0.2	0.05	-	<0.003 (Fw);	0.02	40.0	0.03 (Fw);

Pesticides	Food Australian standard (MRL)				Water Australian guidelines				
	Cereal grains (maize, wheat) (mg/kg) (NRS 2010/11)	**Apple** (mg/kg) (NRS 2010/11)	**Cattle** (mg/kg) (NRS 2010/11)	**Eggs** (mg/kg) (NRS 2010/11)	**Irrigation** (mg/L) (ANZECC and ARMCANZ 2000)	**Aquaculture** (µg/L) (freshwater = Fw; marine = Mr; ANZECC and ARMCANZ 2000)	**Drinking water** Human health- (mg/L) (NHMRC and NRMMC 2011)	**Recreation** (µg/L; ANZECC and ARMCANZ 2000) (max. conc.)	**Aquatic ecosystems protection** (µg/L); (freshwater =Fw; marine = Mr; ANZECC and ARMCANZ 2000)
						0.001 (Mr)			0.005 (Mr)
Endrin (OC)	-	-	-	-	-	<0.002 (Fw)	-	-	0.01 (Fw); 0.004 (Mr)
Epoxiconazole (F)	0.05	-	-	-	-	-	-	-	
Esfenvalerate (SP)	-	-	-	-	-	-	0.03	-	0.001 (Fw)
Ethion (I, A)	-	-	2.5	-	-	-	0.004	6.0	-
Ethofumesate (H)	-	-	0.5	-	-	-	-	-	-
Ethoprofos (I, N)	0.005	-	-	-	-	-	-	-	-
Etoxazole (A)	-	0.2	-	-	-	-	-	-	-
Fenarimol (F)	-	0.2	-	-	-	-	-	-	-
Fenitrothion (OP)	10.0	1.0	-	-	-	-	0.007	-	0.2 (Fw)
Fenoxaprop-ethyl (H)	0.01		-	-	-	-	-	-	-
Fenoxycarb (I)	-	2.0	-	-	-	-	-	-	-

Pesticides	Food Australian standard (MRL)				Water Australian guidelines				
	Cereal grains (maize, wheat) (mg/kg) (NRS 2010/11)	**Apple** (mg/kg) (NRS 2010/11)	**Cattle** (mg/kg) (NRS 2010/11)	**Eggs** (mg/kg) (NRS 2010/11)	**Irrigation** (mg/L) (ANZECC and ARMCANZ 2000)	**Aquaculture** (µg/L) (freshwater = Fw; marine = Mr; ANZECC and ARMCANZ 2000)	**Drinking water** Human health- (mg/L) (NHMRC and NRMMC 2011)	**Recreation** (µg/L; ANZECC and ARMCANZ 2000) (max. conc.)	**Aquatic ecosystems protection** (µg/L); (freshwater =Fw; marine = Mr; ANZECC and ARMCANZ 2000)
Fenpyroximate (A)	-	0.3	-	-	-	-	-	-	-
Fenthion (OP)	-	2.0	1.0	-	-	-	0.007	-	-
Fenvalerate (SP)	2.0	1.0	1.0	-	-	-	0.06	40.0	-
Fipronil (I)	-	-	-	-	-	-	0.007	-	-
Flamprop-m-methyl (H)	0.05	-	-	-	-	-	-	-	-
Fluazinam (F)	-	0.01	-	-	-	-	-	-	-
Fluazuron	-	-	7.0	-	-	-	-	-	-
Flumethrin (SP)	-	-	0.2	-	-	-	-	-	-
Fluquinconazole (F)	0.02	0.3	0.5	-	-	-	-	-	-
Flusilazole (F)	-	0.2	-	-	-	-	-	-	-
Flutriafol (F)	0.02	-	0.05	-	-	-	-	-	-
Glufosinate-ammonium (H)	0.2	-	-	-	-	-	-	-	-
Glyphosate (H)	5.0	-	-	-	-	-	-	200.0	370.0 (Fw)

Pesticides	Food Australian standard (MRL)				Water Australian guidelines				
	Cereal grains (maize, wheat) (mg/kg) (NRS 2010/11)	**Apple** (mg/kg) (NRS 2010/11)	**Cattle** (mg/kg) (NRS 2010/11)	**Eggs** (mg/kg) (NRS 2010/11)	**Irrigation** (mg/L) (ANZECC and ARMCANZ 2000)	**Aquaculture** (µg/L) (freshwater = Fw; marine = Mr; ANZECC and ARMCANZ 2000)	**Drinking water** Human health- (mg/L) (NHMRC and NRMMC 2011)	**Recreation** (µg/L; ANZECC and ARMCANZ 2000) (max. conc.)	**Aquatic ecosystems protection** (µg/L); (freshwater =Fw; marine = Mr; ANZECC and ARMCANZ 2000)
HCB: Hexachlorobenzene) (F)	0.05	-	1.0	1.0	-	<0.00001 (Fw)	-	-	-
HCH: Hexachlorocyclohexane (OC)	0.1	-	0.3	0.1	-	-	-	-	-
Heptachlor (OC)	0.02	-	0.2	0.05	-	<0.005 (Fw)	0.0003	3.0	0.01 (Fw)
Hexythiazox (I, A)	-	1.0	-	-	-	-	-	-	-
Imazalil (F)	-	5.0	-	-	-	-	-	-	-
Imazapyr (H)	-	-	-	-	-	-	9.0	-	-
Imidacloprid (I, N)	0.05	0.3	-	-	-	-	-	-	-
Indosulfuron-methyl-sodium (H)	0.01	-	-	-	-	-	-	-	-
Indoxacarb (I, A)	-	2.0	1.0	-	-	-	-	-	-
Iprodione	-	3.0	-	-	-	-	-	-	-
Kreoxsim-methyl (F)	-	0.1	-	-	-	-	-	-	-
Lindane (gamma-HCH)	0.5	2.0	2.0	0.1	-	<0.1 (Fw); 0.004	0.01	10.0	0.07 (Fw)

Pesticides	Food Australian standard (MRL)				Water Australian guidelines				
	Cereal grains (maize, wheat) (mg/kg) (NRS 2010/11)	**Apple** (mg/kg) (NRS 2010/11)	**Cattle** (mg/kg) (NRS 2010/11)	**Eggs** (mg/kg) (NRS 2010/11)	**Irrigation** (mg/L) (ANZECC and ARMCANZ 2000)	**Aquaculture** (µg/L) (freshwater = Fw; marine = Mr; ANZECC and ARMCANZ 2000)	**Drinking water** Human health- (mg/L) (NHMRC and NRMMC 2011)	**Recreation** (µg/L; ANZECC and ARMCANZ 2000) (max. conc.)	**Aquatic ecosystems protection** (µg/L); (freshwater =Fw; marine = Mr; ANZECC and ARMCANZ 2000)
Hexachlorocyclohexanes (OC)						(Mr)			
Malathion (OP)	8.0	2.0	1.0	-		<0.1 (Fw)	0.07	-	0.05 (Fw)
Mancozeb (F)	-	-	-	-	-	-	0.009	-	-
MCPA (H)	0.02	-	-	-	-	-	0.04	-	-
Methidathion (OP)	0.01	0.2	0.5	-	-	-	0.006	60.0	-
Methomyl (C)	0.1	1.0	-	-	-	-	0.02	60.0	3.5 (Fw)
Methoprene (I)	2.0	-	-	-	-	-	-	-	-
Methoxychlor (I)	-	-	-	-	-	<0.03 (Fw)	0.3	-	-
Metiram (F)	-	-	-	-	-	-	0.009	-	-
Metolachlor (H)	0.1		0.05		-		0.3	800.0	
Metosulam (H)	0.02	-	-	-	-	-	-	-	-
Metsulfuron-methyl (H)	0.02	-	-	-	-	-	0.04	-	-
Mirex (OC)	-	-	-	-	-	<0.001 (Fw)	-	-	-

Pesticides	Food				Water				
	Australian standard (MRL)				Australian guidelines				
	Cereal grains (maize, wheat) (mg/kg) (NRS 2010/11)	**Apple** (mg/kg) (NRS 2010/11)	**Cattle** (mg/kg) (NRS 2010/11)	**Eggs** (mg/kg) (NRS 2010/11)	**Irrigation** (mg/L) (ANZECC and ARMCANZ 2000)	**Aquaculture** (µg/L) (freshwater = Fw; marine = Mr; ANZECC and ARMCANZ 2000)	**Drinking water** Human health- (mg/L) (NHMRC and NRMMC 2011)	**Recreation** (µg/L; ANZECC and ARMCANZ 2000) (max. conc.)	**Aquatic ecosystems protection** (µg/L); (freshwater =Fw; marine = Mr; ANZECC and ARMCANZ 2000)
Molinate (H)	-	-	-	-	-	-	0.004	1.0	3.4 (Fw)
Monocrotophos (OP)	-	-	-	-	-	-	0.002	2.0	-
Myclobutanil (F)	-	0.5	-	-	-	-	-	-	-
Omethoate (OP)	0.05	2.0	-	-	-	-	0.001	0.4	-
Oxychlordane	0.02	-	-	-	-	-	-	-	-
Paraquat (H)	0.1	-	-	-	-	-	0.02	40.0	-
Parathion (OP)	-	-	-	-	-	<0.04 (Fw)	0.02	30.0	0.004 (Fw)
Parathion-methyl (OP)	-	0.5	0.05	-	-	-	0.0007	6.0	-
Penconazole (F)	-	0.1	-	-	-	-	-	-	-
Pendimethalin (H)	0.05	-	-	-	-	-	0.4	600.0	-
Permethrin (SP)	2.0	-	1.0	-	-	-	0.2	300.0	-
Phenothrin (SP)	2.0	-	-	-	-	-	-	-	-
Phosmet (OP)	0.05	-	1.0	-	-	-	-	-	-

Pesticides	Food Australian standard (MRL)				Water Australian guidelines				
	Cereal grains (maize, wheat) (mg/kg) (NRS 2010/11)	**Apple** (mg/kg) (NRS 2010/11)	**Cattle** (mg/kg) (NRS 2010/11)	**Eggs** (mg/kg) (NRS 2010/11)	**Irrigation** (mg/L) (ANZECC and ARMCANZ 2000)	**Aquaculture** (µg/L) (freshwater = Fw; marine = Mr; ANZECC and ARMCANZ 2000)	**Drinking water** Human health- (mg/L) (NHMRC and NRMMC 2011)	**Recreation** (µg/L; ANZECC and ARMCANZ 2000) (max. conc.)	**Aquatic ecosystems protection** (µg/L); (freshwater =Fw; marine = Mr; ANZECC and ARMCANZ 2000)
Phosphine (I, R)	0.01	-	-	-	-	-	-	-	-
Picloram (H)	0.2	-	-	-	-	-	0.3	30.0	-
Primicarb (C)	0.02	0.5	-	-	-	-	0.007	100.0	-
Primiphos-methyl (OP)	10.0	-	0.05	-	-	-	0.09	-	-
Procymidone (F)	-	-	0.2	-	-	-	-	-	-
Profenopos (OP)	-	-	-	-	-	-	0.0003	0.6	-
Propachlor (H)	-	-	0.02	-	-	-	0.07	-	-
Propanil (H)	-	-	-	-	-	-	0.7	1,000.0	-
Propargite (A)	-	3.0	-	-	-	-	0.007	1,000.0	-
Propiconazole (F)	0.05	-	0.1	-	-	-		-	-
Prothioconazole (F)	0.3	-	-	-	-	-	0.1	-	-
Prothiofos (OP)	-	0.05	-	-	-	-	-	-	-
Pyraclostrobin (F)	0.01	-	-	-	-	-	-	-	-

Pesticides	Food				Water				
	Australian standard (MRL)				Australian guidelines				
	Cereal grains (maize, wheat) (mg/kg) (NRS 2010/11)	**Apple** (mg/kg) (NRS 2010/11)	**Cattle** (mg/kg) (NRS 2010/11)	**Eggs** (mg/kg) (NRS 2010/11)	**Irrigation** (mg/L) (ANZECC and ARMCANZ 2000)	**Aquaculture** (µg/L) (freshwater = Fw; marine = Mr; ANZECC and ARMCANZ 2000)	**Drinking water** Human health- (mg/L) (NHMRC and NRMMC 2011)	**Recreation** (µg/L; ANZECC and ARMCANZ 2000) (max. conc.)	**Aquatic ecosystems protection** (µg/L); (freshwater =Fw; marine = Mr; ANZECC and ARMCANZ 2000)
Quintozene (F)	-	-	-	-	-	-	-	-	-
Simazine (H)	-	0.1	-	-	-	-	0.02	-	3.2 (Fw)
Spinosad (I)	1.0	0.5	-	-	-	-	-	-	-
Tebuconazole (F)	0.2	-	-	-		-	-	-	-
Tebufenozideb (I)	-	1.0	-	-	-	-	-	-	-
Tebufenpyrad (A)	-	1.0	-	-	-	-	-	-	-
Tebuthiuron (H)	-	-	-	-	-	-	-	-	2.2 (Fw)
Temephos (OP, I)	-	-	5.0	-	-	-	0.4	30.0	0.05 (Mr)
Terbufos (OP, I, N)	0.01	-	-	-	-	-	0.0009	-	-
Tetradifon (A)	-	5.0	-	-	-	-	-	-	-
Thiabendazole (F)	-	10.0	-	-	-	-	-	-	-
Thiacloprid (I, Neo)		1.0			-			-	-
Thiobencarb (H)	-	-	-	-	-	-	0.04	40.0	2.8 (Fw)

Pesticides	Food Australian standard (MRL)				Water Australian guidelines				
	Cereal grains (maize, wheat) (mg/kg) (NRS 2010/11)	**Apple** (mg/kg) (NRS 2010/11)	**Cattle** (mg/kg) (NRS 2010/11)	**Eggs** (mg/kg) (NRS 2010/11)	**Irrigation** (mg/L) (ANZECC and ARMCANZ 2000)	**Aquaculture** (µg/L) (freshwater = Fw; marine = Mr; ANZECC and ARMCANZ 2000)	**Drinking water** Human health- (mg/L) (NHMRC and NRMMC 2011)	**Recreation** (µg/L; ANZECC and ARMCANZ 2000) (max. conc.)	**Aquatic ecosystems protection** (µg/L); (freshwater =Fw; marine = Mr; ANZECC and ARMCANZ 2000)
Thiodicarb (C, I, M)	0.1	-	-	-	-	-	-	-	
Thiram (F)	-	-	-	-	-	<0.002 (Fw)	0.007	30.0	0.01 (Fw)
Toxaphone (I)	-	-	-	-	-		-	-	-
Tralkoxydim (H)	0.02	-	-	-	-	-	-	-	-
Triadimenol (F)	0.01	-	-	-	-	-	-	-	-
Triasulfuron (H)	0.02		-	-	-	-	-	-	-
Trichlorfon (OP, I)	0.1	0.1	-	-	-	-	0.007	-	-
Trifloxystrobin (F)	-	0.3	-	-	-	-	-	-	-
Triflumuron (I)	0.05		0.05		-	-	-	-	-
Triflumuron (I)	0.05	-	-	-	-	-	-	-	-
Trifluralin (H)	0.05	-	-	-	-	-	0.09	500.0	2.6 (Fw)
Triticonazole (F)	0.05	-	-	-	-	-	-	-	-

bacteria, *Bacillus thuringiensis*, and *B. popilliae*, Actinomycetes), naturally occurring fungi (*Beauveria bassiana*), microscopic roundworms (*Entomopathogenic nematodes*), Spinosad, insect hormones (e.g. pheromone hormone traps) and insect growth regulators (Gupta and Dikshit 2010, Singh and Mandal 2013) should be popularised.

Box 7.2. The effects of pulse exposure of pesticides on the early life stages of selected native fish from the Goulburn-Murray River Region, Victoria, Australia (Raymond et al 2006).

A series of ecotoxicological experiments were carried out to assess the lethal and sublethal effects of six pesticides including four aquatic herbicides and two other pesticides (used in the G-MW region) on two species of selected Australian native fish (the iconic Australian species Murray Cod (*Maccullochella peellii peelii*) and the sensitive native fish Murray River Rainbowfish (*Melanotaenia fluviatilis)*. These herbicides were glyphosate, amitrole, 2,4-D amine and acrolein which are commonly used by Goulburn -Murray Rural Water Authority (G-MW) to control aquatic plants in channels and drains. Two other pesticides that were regularly detected in G-MW irrigation channels (2004) i.e. were endosulfan and copper and those were also tested. There were **no previous data on the toxicity of herbicides used by G-MW on native fish**. Early life stage (< 2 days old larvae) of two native fish including the Murray Cod and the Murray River Rainbowfish were selected for a series of ecotoxicological experiments. All experiments were conducted at the RMIT University purpose-built ecotoxicological laboratory following the American Society of Testing Materials (ASTM) protocols for independent treatments. **Larval** fish were exposed to a variety of concentrations (1, 10, 100, 1,000 and 10,000 µg/L) of **six pesticides** (glyphosate, amitrole, 2,4-D amine and acrolein, endosulfan and copper referred to as "toxicants") for 24 hours (h) and then placed in a toxicant free medium (i.e. pulse-exposed). **The main objectives** of this study were to observe and record the effects of these toxicants on the early life stages of Murray cod and Murray River rainbow fish in relation to some selected endpoints[82] including survival and mortality, growth and development (total length, yolk sac length, heart rate) and other behavioural responses at 24 h, 48 h, 72 h and 96 h.

The study found that, acrolein can significantly affect the survival of Murray cod larvae, as 100% mortality was recorded when fish were exposed to acrolein at 10,000 µg/L (12 h). However, the study also recorded insignificant numbers of cod deaths with acrolein at other concentrations. The other effects due to acrolein were a

---

[82]***Toxicity endpoint:*** *A toxic endpoint is the result of a study conducted to determine* ***how dangerous a substance is****. The data collected from such studies are used to report the* ***relative toxicity*** *of the compound to various* ***regulatory agencies*** *and environmental compliance groups. Toxic endpoints can include* ***mortality, behaviour, reproductive status or physiological*** *and biochemical changes. Toxic endpoints are acute (short, few hours) or chronic (long few days). With acute studies, a common endpoint is an* LD_{50}/LC_{50}, *which is the dose or concentration of a compound required to kill half the organisms in the study. Chronic studies are longer in duration* ***(more than a week)*** *and include endpoints such as* ***reproduction, long-term survival and growth*** *(https://sciencing.com/definition-toxic-endpoint-6603458.html).*

reduction in fish length, yolk-sac length, and heart rate. On the other hand, there was virtually no effect on Murray cod larvae from exposure to other herbicides such as glyphosate, 2,4-D amicide and amitrole at the concentrations tested. All Murray cod larvae died from exposure to copper at 10,000 µg/L but no other behavioural or morphological variations were observed following exposure to other concentrations. Endosulfan did not result in any deaths in Murray cod larvae but caused a significant reduction in cod growth due to spinal curvature. In contrast, Murray River rainbowfish larvae were found to be very sensitive to pulse exposed endosulfan and acrolein at very low concentrations with mortality recorded as low as 1 µg/L. Rainbowfish exposed to endosulfan and acrolein at the tested concentrations were also lethargic, motionless, comatose, or twitching.

Adult Murray river rainbowfish *Melanotaenia fluviatilis*

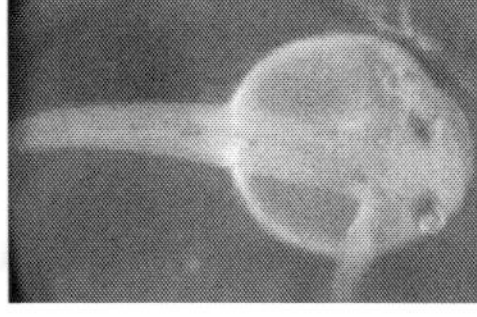

A one-day-old Murray cod, M*accullochella peelii peelii*, larvae.

All experiments were conducted at the RMIT University purpose built ecotoxicological laboratory (Flow through glass aquaria)

Based on these preliminary experiments, the 96-h LC_{50} on fish larvae was estimated, With acrolein following a 24-h pulse exposure for Murray cod, the 96 h LC_{50} was >1000 and <10,000 µg/L and for Murray River rainbowfish it was <10 µg/L while for endosulfan 96 h LC50 was >10,000 µg/L for Murray cod and for Murray River rainbowfish it was <10 µg/L (Table 7.13). **The research** results therefore indicate that the Murray River rainbowfish is very sensitive to pesticides, while Murray cod is a moderately hardy fish (in the larval stage) and less sensitive to pesticides. The 96-h LC_{50} values for Northern hemisphere fish, such as rainbow trout in continuous exposure were reported to be 150 µg/L for acrolein and 2 µg/L for endosulfan respectively (Table 7.13). The larger yolk-sac and thicker yolk membrane in Murray cod larvae may play a key role in minimising the toxic effects of herbicides and pesticides on this species whereas Murray River rainbow fish have a smaller yolk sac and much thinner yolk membrane which may have provided less protection from the toxicants. This study **generated for the first time ecotoxicological data** for the iconic Australian native fish Murray cod i.e. lethal and sub-lethal effects of a range of pesticides on Murray cod. The data shows that larval Murray cod are less sensitive to the tested chemicals (in particular to acrolein, endosulfan and copper) than alien fish species (like rainbow trout). **This preliminary** study demonstrates the effects of individual herbicides and pesticides on Murray cod and Murray River rainbowfish. However, in natural waterways, they are generally exposed to a mixture of chemicals simultaneously. Therefore, it is recommended that further research should be conducted using mixtures of G-MW herbicides and pesticides identified in G-MW irrigation areas to evaluate their effects on sensitive native fish such as rainbowfish.

Dr Golam Kibria, as the Senior Environmental Scientist with the G-MW (the lead author of this book) planned, lead and managed this expensive and extensive research during 2001-2012 periods. The research was carried out in collaboration with the researchers from RMIT University, Australia.

Table 7.13. A comparison of LC_{50} (96 h) values of Murray cod, Murray River Rainbow fish (based on current experiments) and alien fish species (Rainbow trout) and guideline trigger value for the protection of aquatic species for each tested chemical (Raymond et al 2006)

Toxicant	LC_{50} (96 h) for Murray cod (**current study**)	LC_{50} (96 h) for Murray River Rainbow Fish (**current study**)	LC_{50} (96 h) for **Alien fish** (Tomlin 2000; ANZECC & ARMCANZ 2000)	**Guideline trigger value** for the protection of freshwater species (95% or 99%) (ANZECC & ARMCANZ 2000)	**Remarks**
Acrolein (950 g/L)	>1000 <10,000 µg/L	<10 µg/L	150 µg/L (rainbow trout) 14-125 µg/L (4 spp.)	0.01 µg/L Interim working levels	**Murray cod** are **less sensitive** than native rainbow fish and alien fish to acrolein **Murray Rainbow fish** are **very sensitive** to acrolein
2,4-D amicide (625 g/L)	>10,000 µg/L	not done	100,000 mg/L (rainbow trout) 1400-4800,000 µg/L (23 spp)	280 µg/L (95% protection) 140 µg/L (99% protection)	2,4-D affects Murray cod larvae **at very high concentration**
Amitrole T (250 g/L)	>10,000 µg/L	not done	>1000,000 mg/L (rainbow trout) 65,000-410,000 µg/L 4 spp.)	22 µg/L (not specified)	Amitrole affects Murray cod larvae at **very high concentration**
Glyphosate (360 g/L)	>100,000 µg/L	not done	86,000 µg/L (rainbow trout) 11,000-4290,000 µg/L (10 spp)	1200 µg/L (95% protection) 370 µg/L (99% protection)	Glyphosate may affect Murray cod larvae at **very high concentration**

Copper	>100 µg/L	not done	40-80 µg/L (salmonid including rainbow trout)	1.4 µg/L (95% protection) 1.0 µg/L (99% protection)	Copper is **highly toxic** to Murray cod larvae at low concentrations
Endosulfan (350 g/L)	>10,000 µg/L	10 µg/L	2 µg/L (golden orfe) 0.1-63µg/L (42 spp)	0.2 µg/L (95% protection) 0.03 µg/L (99% protection)	**Murray cod** are **less sensitive** than Murray rainbow fish and alien fish to endosulfan **Rainbow fish** are **very sensitive** to endosulfan

7.7. Conclusion

Pesticides use has greatly helped to increase and improve world food production. Despite this, there are remaining concerns about the adverse effects of pesticides on non-target organisms including humans, fish and wildlife. Intensive agriculture is contributing to increased pesticide inputs into surface waters (rivers, creeks) via spray drift, drainage discharge, run-off and volatilisation. Pesticide residues in irrigation tail water, the water discharged from irrigated crops and water from sub-surface drains can be a major source of pesticides in the environment. As hydrophobic pesticide residues can be transferred via water and food chains to humans, a residue monitoring program of agri-produces (e.g. vegetables, fruits and grains, fish/ seafood) would help to protect the health of people and the production of high quality, pesticide/chemical-free food for selling in domestic and export markets at premium prices. Pesticide residue monitoring is the only tool to control the use of the quantity of currently registered or even unregistered pesticides on/in food and ensures whether pesticides are used according to GAPs. Residue monitoring is also recognised as a significant aspect of initiatives to reduce potential hazards on the environment (FAO 2005) and human health. Furthermore, monitoring of surface and groundwater for high-risk pesticide residue and comparing those results with guideline reference threshold values (Table 7.12) would help to clearly understand and define pesticides predicted impacts/ effects.

7.8 References

Abhilash PC and Singh N 2009 Review–Pesticide use and application: An Indian scenario. *Journal of Hazardous Materials* **165** 1-12

Abdullah NZ, Ishaka A, Samsuddin N, Mohd Rus R and Mohamed AH 2011 *Chronic organophosphate pesticide exposure and coronary artery disease*: finding a bridge. IIUM Research, Invention and Innovation Exhibition (IRIIE) 2011 223; *http://irep.iium.edu.my/13579/*

Acquavella JF, Alexander BH, Mandel JS et al 2004 Glyphosate biomonitoring for farmers and their families: results from the farm family exposure Study. *Environmental Health Perspectives* **112(3)** 321–326

Agnihorti NP 1999 *Pesticide safety and monitoring.* All India Coordinate Research Project on Pesticides Residues. Indian Council of Agricultural Research, New Delhi, India

Ahmed MS, Begum A, Rahman MA, Akon MW and Chowdhury MAZ 2016 Extent of insecticide residue load in vegetables grown under conventional farming in Bangladesh. *The Agriculturists* **14(2)** 38-47

https://pdfs.semanticscholar.org/9324/a2ea654946ee25daa86f5b04f1696d945e9d.pdf

Alam MN, Chowdhury MAZ, Hossain MS et al 2015 Detection of residual levels and associated health risk of seven pesticides in fresh eggplant and tomato Samples from Narayanganj District. *Journal of Chemistry* 2015, Article ID 243574, 7 pages; *http://dx.doi.org/10.1155/2015/243574*

Alawi MA, Ammari N, al-Shuraiki Y 1992 Organochlorine pesticide contaminations in human milk samples from women living in Amman, Jordan. *Archives of Environmental Contamination and Toxicology* **23** 235–239

Albert L, Vega P and Portales A 1981 Organochlorine pesticide residues in human milk samples from Comarca Lagunera, Mexico, 1976. *Pesticide Monitoring* **15** 135-138

Al Antary TM, Alawi MA, Estityah H, Haddad N, and Al-Antary 2015 Chlorinated pesticide residues in human breast milk collected from southern Jordan in 2012/2013. *Toxin Reviews* **34(4)** 190–194

Andersen HR, Wohlfahrt-Veje C, Dalgård C, Christiansen L et al 2012 1 polymorphism Pesticides: and prenatal pesticide exposure associated with adverse cardiovascular risk profiles at school age. *PloS One* **7**: e36830.

ANZECC (Australian and New Zealand Environment and Conservation Council) and ARMCANZ (Agriculture and Resource Management Council of Australia and New Zealand) 2000 (Australian and New Zealand guidelines for fresh and marine water quality, Volume 1, The Guidelines Chapters 1-7).

http://www.mincos.gov.au/publications/australian_and_new_zealand_guidelines_for_fresh and_marine_water_quality

Anzene JS, Tyohemba RL, Ahile UJ and Emezi KSA 2014 Organochlorine pesticide residues analysis of postharvest cereal grains in Nasarawa State, Nigeria. *International Journal of Agronomy and Agricultural Research* **5(5)** 59-64

Arnoult MH and Lukac M 2018 *A brief history of pesticide use DOI: 10.13140/RG.2.2.34405.24806 https://www.researchgate.net/publication/330760976_Food_and_Pesticides-A_Brief_History_of_Pesticides*

ATSDR (Agency for Toxic Substances and Disease Registry) 2002 ToxGuideTM for Aldrin/Dieldrin

U.S. Department of Health and Human Services Public Health Service Agency for Toxic Substances and Disease Registry. *https://www.atsdr.cdc.gov/toxguides/toxguide-1.pdf*

Australian Government 2018 *National Residue Survey 2017–18*. Department of Agriculture and Water Resources. 4 pages *http://www.agriculture.gov.au/ag-farm-food/food/nrs/nrs-results-publications/industry-brochures/summary*

Bagga D, Anders KH, Wang HJ, Roberts E and Glaspy JA 2000 Organochlorine pesticides content or breast adipose tissue from women with breast cancer and control subjects. *Journal of the National Cancer Institute* **92 (9)** 750–753

Band PR, Abanto Z, Bert J, Lang B, Fang R, Gallagher RP and Le ND 2011 Prostate cancer risk and exposure to pesticides in British Columbia farmers. *Prostate* **71** 168-183

Bates MN, Hannah DJ, Buckland SJ, Taucher JA and van Maanen T 1994 Chlorinated organic contaminants in breast milk of New Zealand women. *Environment Health Perspectives* **102** 211-217

Bauld J 1994 *Groundwater quality in irrigation areas of Australia*: interactions of agriculture and hydrogeology. In Water Down Under 94, Volume 2 Part B - Groundwater Papers, Adelaide Convention Centre, Adelaide South Australia, 21-25 November 1994. p 423-428

Bauld J, Ivkovic KM, Please PM and Watkins KL 1998 *Groundwater quality benchmarks for areas of irrigated agriculture*, in *Proceedings of the International Groundwater Conference 1998, Groundwater: Sustainable Solutions*, (eds) TR Weaver and CR Lawrence, University of Melbourne, Melbourne, Australia, 8-13 February 1998. *International Association of Hydro-geologists* p 395-399

Blacquie`re T, Smagghe G, Cornelis AM, Gestel van and Mommaerts V 2012 Neonicotinoids in bees: a review on concentrations, side-effects and risk assessment. *Ecotoxicology* **21** 973-992

Bouwman H, Kylin H, Sereda B and Riana Bornman R 2012 High levels of DDT in breast milk: Intake, risk, lactation duration, and involvement of gender. *Environmental Pollution* **170** 63-70

Bouwman H, Reinecke AJ, Cooppan RM and Becker PJ 1990 Factors affecting levels of DDT and metabolites in human breast milk from KwaZulu. *Journal of Toxicology and Environmental Health* **31** 93-115

Brevik EM and Bjerk JE 1978 Organochlorine compounds in Norwegian human fat and milk. *Acta Pharmacologica et Toxicologica* **43** 59-63

CDC (Centre for Disease Control and Prevention) 2017 Mirex. Biomonitoring summary. Organochlorine pesticides overview. CAS No. 2385-85-5

https://www.cdc.gov/biomonitoring/Mirex_BiomonitoringSummary.html

Chapman RN and Stranger JW 1994 *Horticultural Pesticide Residues in Water* - A Survey of Pesticide Residues Conducted in Gippsland Victoria, 1994 Victorian Department of Agriculture, Victoria, Australia.

Chapman RN and Stranger JW 1993 Horticultural Pesticide Residues in Water: A Survey of Pesticides Conducted in the Mitchell Valley, Victoria, 1992/93. Victorian Department of Agriculture, Victoria, Australia.

Chakraborty S, Mukherjee S, Roychoudhury S, Siddique S, Lahiri T and Ray MR 2009 Chronic exposures to cholinesterase-inhibiting pesticides adversely affect respiratory health of agricultural workers in India. *Journal of Occupational Health* **51** 488-497

Chikuni O, Skaare JU, Nyazema N, Polder A 1991 Residues of Organochlorine pesticides in human milk from others living in the greater Harare area of Zimbabwe. *Central African Journal of Medicine* **37** 136-141

Chen C, Qia Y, Chen Q, Tao C, Li C, and Li Y 2011 Evaluation of pesticide residues in fruits and vegetables from Xiamen, China. *Food Control* **22** 1114–1120

Claeys WL, Schmit J-F, Bragard C, Maghuin-Rogister G et al 2010 Exposure of several Belgian consumer groups to pesticide residues through fresh fruit and vegetable consumption. *Food Control* **22** 508–516

Cocco P, Satta G, Dubois S, Pili C, Pilleri M et al 2013 Lymphoma risk and occupational exposure to pesticides: results of the Epilymph study. *Occupational and Environmental Medicine* **70** 91-98

Cok I, Karakaya AE, Afkham BL, and Burgaz S 1999 Organochlorine pesticide contaminants in human milk samples collected in Tebriz (Iran). *Bulletin of Environmental Contamination and Toxicology* **63** 444-450

Cok I, Bilgili A, O¨ zdemir M, et al 1997 Organochlorine pesticide residues in human breast milk from agricultural regions of Turkey, 1995e1996. *Bulletin of Environmental Contamination and Toxicology* **59** 577-582

Collins GB, Holmes DC and Hoodless RA 1982 Organochlorine pesticide residues in human milk in Great Britain, 1979- 1980. *Human Toxicology* **1** 425-431

Connolly A, Jones K, Galea KS et al 2017 Exposure assessment using human biomonitoring for glyphosate and fluroxypyr users in amenity horticulture. *International Journal of Hygiene and Environmental Health* **220(6)** 1064–73

Cumming H and Ru¨cker C 2017 Octanol–Water Partition coefficient measurement by a simple[1]H NMR method *American Chemical Society Omega* **2** 6244–6249 *DOI: 10.1021/acsomega.7b01102*

Davis A, Lewis S, Bainbridge Z, Brodie J and Shannon E 2008 Pesticide residues in waterways of the lower Brurdekin region: Challenges in ecotoxicological interpretation of monitoring data. *Australasian Society of Ecotoxicology* **14** 89-108

Du J, Gridneva Z, Gay MC, Lai CT, Trengove RD, Hartmann PE, Geddes DT 2016 *Longitudinal study of pesticide residue levels in human milk from Western Australia during 12 months of lactation:* Exposure assessment for infants. *Scientific Reports* **6** 38355. *doi: 10.1038/srep38355.*

Edwards CA 1987 The environmental impact of insecticides. In: Delucchi V (ed.) Integrated pest management, Protection Integŕee Quo vadis? An International Perspective. *Parasitis* **86** Geneva, Switzerland; p 309-329

Egan H, Goulding R, Roburn J and Tatton JO 1965 Organochlorine pesticide residues in human fat and human milk. *British Medical Journal* **2** 66-69

Elbaz A, Clavel J, Rathouz PJ, Moisan F, Galanaud JP, Delemotte B, Alperovitch A, Tzourio C 2009 Professional exposure to pesticides and Parkinson disease. *Annals of Neurology* **66** 494-504.

EFSA (European Food Safety Authority) 2018 The 2016 European Union report on pesticide residues in food. *EFSA Journal* **16(7)** 5348, 139 p. *https://doi.org/10.2903/j.efsa. 2018.5348 ISSN: 1831-4732*

https://efsa.onlinelibrary.wiley.com/doi/pdf/10.2903/j.efsa.2018.5348

EPA 2012 Persistent, bioaccumulative, and toxic profiles estimated for organic chemicals on-line. *http:// www.pbtprofiler.net*

EXTOXNET (Extension Toxicology Network) 1993a Aldicarb

http://webcache.googleusercontent.com/search?q=cache:g3H9xSmtdtsJ:pmep.cce.cornell.edu/profiles/extoxnet/24d-captan/aldicarb-ext.html+&cd=2&hl=en&ct=clnk&gl=au

EXTOXNET (Extension Toxicology Network) 1993b Trichlorfon

http://pmep.cce.cornell.edu/profiles/extoxnet/pyrethrins-ziram/trichlorfon-ext.html

EXTOXNET (Extension Toxicology Network) 1993c Imazalil

http://pmep.cce.cornell.edu/profiles/extoxnet/haloxyfop-methylparathion/imazalil-ext.html

EXTOXNET (Extension Toxicology Network) 1993d Lindane

http://pmep.cce.cornell.edu/profiles/extoxnet/haloxyfop-methylparathion/lindane-ext.html

EXTOXNET (Extension Toxicology Network) 1993e Parathion

http://pmep.cce.cornell.edu/profiles/extoxnet/metiram-propoxur/parathion-ext.html

FAO (Food and Agriculture Organization of the UN) 2005 Proceedings of the Asia Regional Workshop on the implementation, monitoring and observance. International Code of Conduct on the distribution and use of pesticides. Food and Agriculture Organization of the UN, Office for Asia and the Pacific, RAP, Bangkok, Thailand; RAP publication 2005/29.

Fothergill A and Abdelghani A 2013 A review of pesticide residue levels and their related health exposure risks. *WIT Transactions on Ecology and The Environment* ***10*** *2495/FENV130181; doi:10.2495/FENV130181*

Furst P, Furst C, Wilmers K 1994 Human milk as a bioindicator for body burden of PCDDs, PCDFs, Organochlorine pesticides and PCBs. *Environmental Health Perspectives* **102** 187 193

Gillezeau C, Maaike van Gerwen, Rachel M et al 2019 The evidence of human exposure to glyphosate: a review *Environmental Health* **18** 2

Gill K and Garg H 2014 chapter 8. Pesticides: Environmental Impacts and Management Strategies, in *Pesticides - Toxic Effects*, Chapter: Pesticide: Environmental Impacts and Management Strategies. S Solenski, and M. Larramenday (eds) Publisher: Intech. *http://dx.doi.org/10.5772/57399*

Gill JPK, Sethi N, Mohan A, Datta S and Girdhar M 2018 Glyphosate toxicity for animals. Environmental Chemistry Letters **16** 401 https://doi.org/10.1007/s10311-017-0689-0;

https://www.intechopen.com/books/pesticides-toxic-aspects/pesticides-environmental-impacts-and-management-strategies

Gill RJ, Ramos-Rodriguez O, Raine NE 2012 Combined pesticide exposure severely affects individual- and colony-level traits in bees. *Nature* **491** 105-108

Gladen BC, Monaghan SC, Lukyanova EM et al 1999 Organochlorines in breast milk from two cities in Ukraine. Environmental Health Perspectives **107** 459–62

Greenlee AR, Arbuckle TE and Chyou PH 2003 Risk factors for female infertility in an agricultural region. *Epidemiology* **14** 429-436

Gupta S, and Dikshit AK 2010 Biopesticides: An eco-friendly approach for pest control. *Journal of Biopesticides* **3**186-188

Hagyard SB, Brown WH, Stull JW and Whiting FM 1973 DDT and DDE content of human milk in Arizona. *Bulletin Environmental & Toxicology* **9** 169-172

Hansen JC 1998 Assessment report: arctic pollution issues. ***AMAP***, pp. 775-837 (Chapter 12)

Harris CA, O'Hagan S, Merson GHJ 1999 Organochlorine pesticide residues in human milk in the United Kingdom 1997-98. *Human and Experimental Toxicology* **18** 602-606

Hatfield Consultant 1995a Toxaphene. *http://www.popstoolkit.com/about/chemical/toxaphene.aspx*

Hatfield Consultant 1995b Endrin. *http://www.popstoolkit.com/about/chemical/endrin.aspx*

Hayden KM, Norton MC, Darcey D, Østbye T, Zandi PP, Breitner JCS, Welsh-Bohmer KA 2010 Occupational exposure to pesticides increases the risk of incident AD The Cache County Study. *Neurology* **74** 1524-1530

Heck JE, Charbotel B, Moore LE, Karami S et al 2010 Occupation and renal cell cancer in Central and Eastern Europe. *Occupational and Environmental Medicine* **67** 47-53

Hernandez LM, Fernandez MA, Hoyas E, Gonzales MJ, Garca JF 1993 Organochlorine insecticide and polychlorinated biphenyl residues in human breast milk in Madrid (Spain). *Bulletin Environmental Contamination and Toxicology* **50** 308-315

Hoppe 2013 Determination of Glyphosate residues in human urine samples from 18 European countries.

https://www.foeeurope.org/sites/default/files/glyphosate_studyresults_june12.pdf

https://www.foeeurope.org/weed-killer-glyphosate-found-human-urine-across-Europe-130613

Hydrotechnology 1994 Summary of Pesticide Sampling Program, August 1994. Department of Conservation and Natural Resources, Victoria, Australia.

Hyne RV and Aistrope M 2008 Calibration and field application of a solvent-based cellulose membrane passive sampling device for the monitoring of polar herbicides. *Chemosphere* **71** 611–620

Hoppin JA, Umbach DM, London SJ, Henneberger PK et al 2009 Pesticide use and adult-onset asthma among male farmers in the agricultural health study *European Respiratory Journal* **341** 296-1303

IARC 2019 IARC monographs on the identification of carcinogenic hazards to humans. *https://monographs.iarc.fr/agents-classified-by-the-iarc/*

IPCS (International Programme on Chemical Safety) 1986 Tetradifon, Environmental health criteria 67

http://www.inchem.org/documents/ehc/ehc/ehc67.htm

Ip HMH 1983 Breast milk contaminants in Hong Kong. Bulletin Hong Kong Medical Association **35** 1-16

Ip HMH and Phillips DJH 1989 Organochlorine chemicals in human breast milk in Hong Kong. *Archives Environmental Contamination and Toxicology* **18** 490-494

Ivkovic KM, Watkins KL, Cresswell, RG and Bauld J 2001 *A Groundwater Quality Assessment of the Upper Shepparton Formation Aquifers*: Cobram Region, Victoria. Bureau of Rural Sciences, Department of Agriculture, Fisheries and Forestry, ACT Australia.

Jauhiainen A, Rasanen K, Sarantila R, Nuutinen J and Kangas J 1991 Occupational exposure of forest workers to glyphosate during brush saw spraying work. *American Industrial Hygiene Association Journal* **52(2)** 61–64

Jayasumana C, Gunatilake S, and Siribaddana S 2015 Simultaneous exposure to multiple heavy metals and glyphosate may contribute to Sri Lankan agricultural nephropathy. *BMC Nephrology* **16** 103

Kamrin MA 1997 *Pesticide profile: Toxicity, environmental impact, and fate*. CRC, Lewis Publishers, Boca Raton, Florida.

Kanja LW, Skaare JU, Ojwang SBO, Maitai CK 1992 A comparison of Organochlorine pesticide residues in maternal adipose tissue maternal blood, cord blood and human milk from mother/ infant pairs. *Archives Environmental Contamination and Toxicology* **22** 21-24

Kibria G 2019 Presentation: Pesticide Pollution- Risk and Impact. *DOI: 10.13140/RG.2.2.32666.29129/1*

Kibria G 2014 Climate Change and Chemical Contaminants. *DOI: 10.13140/RG.2.1.2733.6169*

https://www.researchgate.net/publication/261216780_Climate_Change_and_Chemical_Contaminants

Kibria G, Kookana R, Correll R et al 2019 Pesticides in channels, drains and natural Carriers in Australia: Summary of research findings (2001- 2012).DOI: 10.13140/RG.2.2.34574.51520

https://www.researchgate.net/publication/335013136_Pesticides_in_Channels_Drains_and_Natural_Carriers_in_Australia-_Summary_of_Research_Findings_2001-2012

Kibria G, Haroon AKY, Nugegoda D and Rose G 2010 *Climate change and chemicals: Environmental Biological aspects.* New India Publishing Agency. New Delhi. 470 p. ISBN: 9789380235301.

https://www.researchgate.net/publication/261216635_Climate_Change_and_Chemicals_Environmental_and_Biological_Aspects

Knezevic Z, Serdar M and Ahel M 2012 Risk assessment of the intake of pesticides in Croatian diet. *Food Control* **23** 59-65

Konishi Y, Kuwabara K and Hori S 2001 Continuous surveillance of Organochlorine compounds in human breast milk from 1972 to 1998 in Osaka Japan. *Archives of Environmental Contamination and Toxicology* **40** 571-578

Kookana R, Kibria G, Barnes M B and Correll, R L 2003 *First-Tier Assessment of the Risks Associated with Pesticides Used in Goulburn-Murray Irrigation Areas* - A Pesticide Risk Reduction Program for G-MW Channels. Report prepared under a research collaboration agreement between GMW & CSIRO. Goulburn-Murray Rural Water Authority (G-MW), Tatura and CSIRO, Adelaide. *https://www.g-mwater.com.au/downloads/gmw/R_D/pesticides_riskassessment.pdf*

Kongtip P, Nankongnab N, Phupancharoensuk R et al 2017 Glyphosate and Paraquat in maternal and fetal serums in Thai women. *Journal of Agromedicine* **22(3)** 282–289

Kunisue T, Someya M, Kayama F, Jin Y, Tanabe S 2004 Persistent Organochlorines in human breast milk collected from primiparae in Dalian and Shenyang, China. *Environmental Pollution* **131** 381-392

Kunisue T, Watanabe M, Someya M et al 2002 PCDDs, PCDFs, PCBs and Organochlorine insecticides in human breast milk collected from Asian developing countries: risk assessment for infants. *Organohalogen Compounds* **58** 285–287

Larsen RM, Watkins KL, Steel NA, Appleyard SJ and Bauld J 1998 *A Groundwater Quality Assessment of the Jandakot Mound, Swan Coastal Plain, Western Australia.* ASGSO record 1998/18. Australian Geological Survey Organisation, Canberra, Australia.

Lavorato M, Bernabò I, Crescente A, Denoël M, Tripepi S and Brunelli E 2013 Endosulfan effects on *Rana dalmatina* tadpoles: quantitative developmental and behavioural analysis. *Archives of Environmental Contamination & Toxicology* **64(2)** 253-62 *doi: 10.1007/s00244-012-9819-7. Epub 2012 Oct 13.*

Lee DH, Lind PM, Jacobs DR, Salihovic S, van Bavel B and Lind L 2011 Polychlorinated biphenyls and Organochlorine pesticides in plasma predict development of type 2 diabetes in the elderly. The Prospective Investigation of the Vasculature in Uppsala Seniors (PIVUS) study. *Diabetes Care* **34** 1778-1784

Lee WJ, Colt JS, Heineman EF et al 2005 Agricultural pesticide use and risk of glioma in Nebraska, United States. *Occupational & Environmental Medicine* **62** 786-792

Lundholm CD 1997 "DDE-induced eggshell thinning in birds: effects of p, p'-DDE on the calcium and prostaglandin metabolism of the eggshell gland". *Comparative Biochemistry and Physiology C.* **118(2)** 113–128 *doi: 10.1016/S0742-8413(97)00105-9. PMID9490182*

Mann RM, Hyne RV, Choung CB and Wilson SP 2009 Amphibians and agriculture chemicals: Review of the risks in a complex environment. *Environmental Pollution* **157(11)** 2903-2927 *doi: 10.1016/j.envpol.2009.05.015*

Marshall S, Sharley D, Jeppe K et al 2016 Potentially toxic concentrations of synthetic pyrethroids associated with low density residential land use. *Frontiers Environmental Science* **4** article 75 *https://doi.org/10.3389/fenvs.2016.00075*

Mattison DR, Wohlleb J, To et al 1992 Pesticide contamination in Arkansas breast milk. *Journal of Arkansas Medical Society* **88** 445-454

McKinlay R, Plant JA, Bell J.N, Voulvoulis N 2008 Endocrine disrupting pesticides, Implications for risk assessment. *Environmental International* **34** 168-183

McKenzie-Smith F, Tiller D and Allen D 1994 Organochlorine biocide residues in water and sediment from the Ovens and King Rivers, North-East Victoria, Australia. *Archives of Environmental Contamination and Toxicology* **26** 483-490

Mes J, Davies DJ, Doucet J et al 1993 Levels of chlorinated hydrocarbon residues in Canadian human breast milk and their relationship to some characteristics of the donors. *Food Additives and Contaminants* **10** 429-441

Mesnage R, Moesch C, Grand R et al 2012 Glyphosate exposure in a farmer's family. *Journal of Environmental Protection* **3(9)** 1001

Mesnage R, Clair E, de Vendômois JS and Seralini GE 2010 Two cases of birth defects overlapping Stratton-Parker syndrome after multiple pesticide exposure. *Occupational and Environmental Medicine* **67** 359-359

Minh NH, Someya M, Minh TB et al 2004 Persistent Organochlorine residues in human breast milk from Hanoi and Hochiminh City, Vietnam: contamination, accumulation kinetics and risk assessment for infants. *Environmental Pollution* **129** 431–441 *doi: 10.1016/j.envpol.2003.11.012*

Mnif W, Hassine AIH, Bouaziz A, Bartegi A, Thomas O, Roig B 2011 Effect of endocrine disruptor pesticides: a review. *Int. J. Environ. Res. Public Health* **8** 2265–2203 *doi: 10.3390/ijerph8062265*

Monirith I, Ueno D, Takahashi S, Nakata H et al 2003 Asia-Pacific mussel watch: monitoring of persistent Organochlorine compounds in coastal waters of Asian countries. *Marine Pollution Bulletin* **46** 281-300

Moore S, Lewin K, Talman S and O'Connor N 1996 *Effects of past and present biocide Usage on aquatic fauna in the Ovens catchment. Water Ecoscience* Victoria, Australia.

Mossop D, Kellar C, Jeppe K, Myers J, Rose G, Weatherman K et al 2013 Impacts of intensive agriculture and plantation forestry on water quality in the Latrobe catchment, Victoria. Publication number 1528 April 2013 Authorised and published by EPA Victoria, 200 Victoria Street, Carlton EPA Victoria, Monitoring and Assessment Unit Centre for Aquatic Pollution Investigation and Management

MSDS (Material Safety Data Sheet) Pubchem 2009 (updated 2020). Boscalid.

https://pubchem.ncbi.nlm.nih.gov/compound/Boscalid

MSDS (Material Safety Data Sheet) Pubchem 2005a (updated 2020) Thiacloprid

https://pubchem.ncbi.nlm.nih.gov/compound/Thiacloprid

MSDS (Material Safety Data Sheet) Pubchem 2005b (updated 2020) Chlorfenapyr

https://pubchem.ncbi.nlm.nih.gov/compound/Chlorfenapyr

MSDS (Material Safety Data Sheet) Pubchem 2005c (updated 2020) Chlorfenvinphos

https://pubchem.ncbi.nlm.nih.gov/compound/Chlorfenvinphos

MSDS (Material Safety Data Sheet) Pubchem 2005d (updated 2020) Primicarb

https://pubchem.ncbi.nlm.nih.gov/compound/Pirimicarb

MSDS-Dunrite Toxaphene 2000 Material Safety Data Sheet 8 p

https://www.dunrite.com.au/wp-content/uploads/2018/06/011.-GM2000-Current-May-2016.pdf

MSDS-Sigma-Mirex 2012 Material Safety Data Sheet 8 p

https://www.nwmissouri.edu/naturalsciences/sds/m/Mirex.pdf

MSDS (Material Safety Data Sheet) Sunitomo-Chem 2015 Phenothrin 5 p

https://sumitomo-chem.com.au/sites/default/files/sds-label/d-phenothrin_tg_msds_2015.pdf`

Muthukumaravel K, Sukumaran M and Sathick O 2013 Studies on the acute toxicity of pesticides on the freshwater fish *Lebo rohita. Journal of Pure and Applied Zoology* **1(2)** 185-192

Newton and Greene 1972 Organochlorine pesticide residue levels in human milk-Victoria, Australia-1970-71 *Pest Monitoring Journal* **6** 4-8

NHMRC and NRMMC 2011 *Australian Drinking Water Guidelines. Paper 6 National Water Quality Management Strategy*. National Health and Medical Research Council, National Resource Management Ministerial Council, Commonwealth of Australia, Canberra. National Resource Management Ministerial Council, Commonwealth of Australia, Canberra.

Nicolopoulou-Stamati P, Maipas S, Kotampasi C, Stamatis P and Hens L 2016 Chemical Pesticides and Human Health: The Urgent Need for a New Concept in Agriculture. *Frontier in Public Health* **4** 148. *doi: 10.3389/fpubh.2016.00148*

Nore´n K and Meironyte´D 2000 Certain Organochlorine and organobromine contaminants in Swedish human milk in perspective of past 20-30 years. *Chemosphere* **40** 1111-1123

Nore´n K, Lunden A, Pettersson E and Bergman A 1996 Methyl sulfonyl metabolites of PCBs and DDE in human milk in Sweden, 1972-1992. *Environmental Health Perspectives* **104** 766-773

NRS 2010/11 *National Residue Survey Annual Report 2010-11*. Australian Government. Department of Agriculture, Fisheries and Forestry. Commonwealth of Australia. 271p. GPO Box 858, Canberra ACT 2601. ISSN 1323-4722

Osman, KA, Al-Humid AI, Al-Rehiayani SM and Al-Redhaiman, KN 2011 Estimated Daily intake of pesticide residues exposure by vegetables grown in greenhouses in Al-Qassim region, Saudi Arabia. *Food Control* **22** 947–953

Parvez S, Gerona RR, Proctor C et al 2018 Glyphosate exposure in pregnancy and shortened gestational length: a prospective Indiana birth cohort study. *Environmental Health* **17(1)** 23

Pariona, A 2017 "*Top pesticide using countries*." World Atlas, Apr. 25, 2017, worldatlas.com/articles/top-pesticide-consuming-countries-of-the-world.html.

https://www.worldatlas.com/articles/top-pesticide-consuming-countries-of-the-world.html

Parveen Z, Khuhro MI and Rafiq N 2005 Monitoring of pesticide residues in vegetables (2000-2003) in Karachi, Pakistan. *Bulletin of Environmental Contamination and Toxicology* **74(1)**170-176

Paumgartten FJR, Cruz CM, Chahoud I, Palavinskas R and Mathar W 2000 PCDDs, PCDFs, PCB, and other Organochlorine compounds in human milk from Rio de Janeiro, Brazil. *Environmental Research* **83** 293-297

Petrelli G and Mantovani A 2002 Environmental risk factors and male fertility and reproduction. *Contraception* **65** 297-300.

Pretty J and Bharucha ZP 2015 Integrated Pest Management for Sustainable Intensification of Agriculture in Asia and Africa. *Insects* **6** 152-182 *doi: 10.3390/insects6010152*

file:///C:/Users/User1/Downloads/insects-06-00152.pdf

Poolpak T, Pokethitiyook P, Kruatrchue M, Arjarasirikoon U and Thanwaniwat N 2008 Residue analysis of Organochlorine pesticides in the Mae Klong River of central Thailand. *Journal of Hazardous Materials* **156** 230-239

Quinby GE, Armstrong JF, Durham WF 1965 DDT in human milk. *Nature* **207** 726-728

Quinsey PM, Donohue DC and Ahokas JT 1995 Persistence of Organochlorines in breast milk of women in Victoria-Australia. *Food and Chemical Toxicology* **33** 49–56

Ramakrishnan N, Kaphalia BS, Seith TD and Roy NK 1985 Organochlorine pesticide residues in mother milk: a source of toxic chemicals in suckling infants. *Human Toxicology* **4** 7-12

Raymond S, Nugegoda D and Kibria, G. 2006 The effects of pulse exposure of six agricultural chemicals (including four herbicides used by G-MW) on the early life stages of selected native fish from the Goulburn-Murray River regions. Report prepared under a research collaboration agreement between Goulburn- Murray Rural Water Authority, Tatura, Victoria, Australia and RMIT University, Melbourne, Australia. *DOI: 10.13140/2.1.2320.7364.*

https://www.researchgate.net/publication/264789845_The_Effects_of_Pulse_Exposure_of_Six_Agricultural_Chemicals_including_four_herbicides_on_the_Early_Life_Stages_of_Selected_Native_Fish_from_the_Goulburn-Murray_River_Region_Victoria_Australia_research

Rendón-von Osten J and Dzul-Caamal R 2017 Glyphosate residues in groundwater, drinking water and urine of subsistence farmers from intensive agriculture localities: a survey in Hopelchén, Campeche, Mexico. *International Journal of Environmental Research & Public Health* **14** 595

Ritcey WR, Savary G, Mc Mullen E 1972 Organochlorine insecticide residues in human milk, evaporated milk, and some milk substitutes in Canada. *Canadian Journal of Public Health* **63:** 125-132

Rose G and Kibria G 2007 Pesticide and heavy metal residues in Goulburn-Murray Irrigation Water 2004-2006. *Australasian Journal of Ecotoxicology* **13** 65-79

Rose G and Kibria G 2006 *Pesticide Monitoring in Goulburn-Murray Water's Irrigation Supply Channels covering the Six Irrigation Areas* [2004-2006 Irrigation Season Study Report]. Report Prepared under a research collaboration agreement between G-MW and PIRVic. Goulburn Murray Rural Water Authority (G-MW), Tatura and Primary Industries Research, Vic, Werribee.

https://www.researchgate.net/publication/264899276_Pesticide_Monitoring_in_Irrigation_Water_Using_Innovative_Passive_Sampling_Technology_Victoria_Australia_%28research_collaboration_between_GM_Water_Australia_and_Department_of_Primary_Industries_Victoria_Australia%29

Sanchez-Bayo F and Goka K 2014 Pesticide Residues and Bees – A Risk Assessment. *PLoS ONE* **9(4)** e94482 *doi:10.1371/journal.pone.0094482*

Sarmah AK, Müller K and Ahmad R 2004 Fate and behaviour of pesticides in the agroecosystem—a review with a New Zealand perspective. *Australian Journal of Soil Research* **42(2)** 125-154

Samsel A and Seneff S 2016 Glyphosate pathways to modern diseases V: Amino acid analogue of glycine in diverse proteins. *Journal of Biological Physics and Chemistry* **16** 9–46

file:///C:/Users/User1/Downloads/Glyphosate_pathwaystomoderndiseasesV-Aminoacidanalogueofglycineindiverseproteins_FNL_Published.pdf

Schade G and Heinzow B 1998 Organochlorine pesticides and polychlorinated biphenyls in human milk of mothers living in northern Germany: current extent of contamination, time trend from 1986 to 1997 and factors that influence the levels of contamination. *Science of the Total Environment* **215** 31–39

Schafer RB, Pettigrove V, Rose G, Allinson G, Wightwick A et al 2011 Effects of pesticides monitored with three sampling methods in 24 Sites on macro-invertebrates and microorganisms. *Environmental Science and Technology* **45** 1665–1672

Schecter AJ, Furst P, Kruger C, Meemken H-A and Grobel W and Constable D 1989 Levels of polychlorinated dibenzofurans, dibenzodioxins, PCBs, DDT and DDE, hexachlorobenzene, dieldrin, hexachlorocyclohexane and oxychlordane in human breast milk from the United States, Thailand, Vietnam and Germany. *Chemosphere* **18** 445-454

Schinas V, Leotsinidis M, Alexopoulos A et al 2000 Organochlorine pesticide residue in human breast milk from Southwest Greece: associations with weekly food consumption patterns of mothers. *Archives of Environmental Health* **55** 411–417

Séralini, G-E, Clair E, Mesnage R et al 2014 Republished study: long-term toxicity of a Roundup herbicide and a Roundup-tolerant genetically modified maize *Environmental Sciences Europe* **26(1)** 14. Epub 2014 Jun 24.

Shipman D 2012 *Pesticide Data Program, Annual Summary, Calendar Year 2010*, U.S. D. O. Agriculture, Editor 2012 USDA.

https://www.ams.usda.gov/sites/default/files/media/2010%20PDP%20Annual%20Summary.pdf

Shim YK, Mlynarek SP, and van Wijngaarden E 2009 Parental exposure to pesticides and childhood brain cancer: US Atlantic coast childhood brain cancer study. *Environmental Health Perspectives* **117** 1002-1006

Siddiqui MKJ, Saxena MC, Bhargava AK, Seth TD et al 1981 Agrochemicals in the maternal blood, milk, and cord blood: a source of toxicants for parents and neonates. *Environmental Research* **24** 24-32

Singh B and Mandal K 2013 Environmental impact of pesticides belonging to newer chemistry. In: Dhawan AK, Singh B, Brar-Bhullar M, Arora R (eds.) *Integrated pest management*. Scientific Publishers, Jodhpur, India, 2013. p. 152-190

Silva V, Mol HGJ, Zomer P,Tienstra M et al 2019 Pesticide residues in European agricultural soils – A hidden reality unfolded. *Science of the Total Environment* **653** 1532–1545

Siyali DS 1973 Polychlorinated biphenyls, hexachlorobenzene and other Organochlorine pesticides in human milk. *Medical Journal Australia* **1** 815-818

SKM (Sinclair Knight Merz) 2002 Goulburn-Murray Water sediment Toxicants – biocide and metal levels in drain sediments of the Shepparton irrigation region 2002. Sinclair Knight Merz, Goulburn-Murray Water, Victoria, Australia.

SKM (Sinclair Knight Merz) 1995 *Summary of 1994/95 Pesticide Sampling Program,* Groundwater Report No. 37. Sinclair Knight Merz, Department of Natural Resources and Environment, Victoria, Australia.

Skaare JU and Polder A 1990 Polychlorinated biphenyls and Organochlorine pesticides in milk of Norwegian women during lactation. *Archives Environmental Contamination and Toxicology* **19** 640-645

Slorach AS and Vaz R 1983 *Assessment of human exposure to selected Organochlorine compounds through biological monitoring*. Global Environment Monitoring System

(GEMS) UNEP/ WHO. Prepared by the Swedish National Food Administration, Uppsala, Sweden, 1983, 134 p.

Smith D 1999 Worldwide trends in DDT in human breast milk. *International Journal of Epidemiology* **28(2)** 179-88

Sotherton, N and Holland, J 2003 *Indirect effects of pesticides on farmland wildlife*, in *Handbook of Ecotoxicology,* p. 1173-1195, 2nd edition, (eds) DJ Hoffman, BA Rattner, GA Burton Jr and J Cairns Jr, Lewis Publishers, Boca Raton, Florida

Somogyi A and Beck H 1993 Nurturing and breast feeding: exposure to chemicals in breast milk. *Environmental Health Perspectives* **101** 45-52

Son HK, Kim SA, Kang JH, Chang YS et al 2010 Strong associations between low-dose Organochlorine pesticides and type 2 diabetes in Korea. *Environment International* **36** 410-414

Stacey CI, Perrian WS, Whitney S 1985 Organochlorine pesticide residue levels in human milk: Western Australia. 1979-1980. *Archives Environment Health* **40** 102-108

Stehle S and Schulz R 2015 Agricultural insecticides threaten surface waters at the global scale. *PNAS.* **112** (18) 5750–5755

Strempel S, Scheringer M, Ng CA and Hungerbühler K 2012 Screening for PBT chemicals among the "Existing" and "New" Chemicals of the EU. *Environmental Science & Technology* **46(11)** 5680–5687; *DOI: https://doi.org/10. 1021/es3002713*

Takei GH, Kauahikauua SM and Leong GH 1983 Analyses of human milk samples collected in Hawaii for residues of Organochlorine pesticides and polychloro-biphenyls. *Bulletin Environmental Contamination and Toxicology* **30** 606-613

Tanabe S, Gondaira F, Subramanian A et al 1990 Specific pattern of persistent Organo-chlorine residues in human breast milk from south India. *Journal of agricultural Food Chemistry* **38** 899 903

Tanner CM, Kamel F, Ross GW, Hoppin JA et al 2011 Rotenone, paraquat, and Parkinson's disease. *Environmental Health Perspectives* **119** 866-872

Tsvetkov N, Samson-Robert O, Sood K et al 2017Chronic exposure to neonicotinoids reduces honey bee health near corn crops. *Science* **356** 1395–1397

Tomlin CDS 2006 A world compendium – The Pesticide Manual. 14th edition. Editor: C. D. S. Tomlin. British Crop Production Council. 7 Omni Business Centre, Omega Park, Alton, Hampshire, GU34 2QD, London. 1349 p

Tomlin CDS 2000 A world compendium – *The Pesticide Manual.* 12th edition. Editor: CDS Tomlin. British Crop Production Council. 49 Downing Street, Farnham, Surrey GU9 7 PH, I UK. 1250 p

USEPA (United States Environment Protection Agency) 2006 *List of drinking water contaminants and MCLS. http://www.epa.gov/safewater/mcl.html.*

USEPA (United States Environment Protection Agency) 1980 Ambient Water Quality Criteria for Endrin.

https://www.epa.gov/sites/production/files/2019-03/documents/ambient-wqc-endrin-1980.pdf

UTZ 2015 List of banned pesticides and pesticides watch list. UTZ Standard and Certification Department De Ruyterkade 6 bg 1013 AA Amsterdam The Netherlands *https://utz.org/wp-content/uploads/2015/12/EN_UTZ_List-of-Banned-PesticidesWatchlist_v1.0_2015.pdf*

Vall O, Gomez-Culebras M, Puig C, Rodriguez-Carrasco E, Gomez Baltazar A et al 2014 Prenatal and Postnatal Exposure to DDT by Breast Milk Analysis in Canary Islands. *PLoS ONE* **9(1)** e83831. *doi:10.1371/journal.pone.0083831*

Varona M, Henao GL, Diaz S, Lancheros A, Murcia A, Rodriguez N and Alvarez VH 2009 Effects of aerial applications of the herbicide glyphosate and insecticides on human health. *Biomedica* **29(3)** 456–75

Waliszewski SM, Aguire AA, Infanzon RM, Benitez A and Rivera J 1999 Comparison of Organochlorine pesticide levels in adipose tissue and human milk of mothers living in Veracruz, Mexico. *Bulletin of Environmental Contamination and Toxicology* **62** 685–690

Wanwimolruk S, Phopin K, Boonpangrak S, and Prachayasittikul V 2016 Food safety in Thailand 4: comparison of pesticide residues found in three commonly consumed vegetables purchased from local markets and supermarkets in Thailand. *Peer J* **4** e2432 *.DOI: 10.7717/peerj.2432*; *https://www.ncbi.nlm.nih.gov/pmc/articles/PMC5012412/*

Watkins KL, Ivkovic KM and Bauld J 1999a A Groundwater Quality Assessment of the Goulburn Catchment, Victoria: Kyabram-Tongala, Australian Groundwater Quality Assessment Project Report No. 7. Bureau of Rural Sciences, Department of Agriculture, Fisheries & Forestry – Australia, ACT, Australia.

Watkins KL, Ivkovic KM and Bauld J. 1999b A Groundwater Quality Assessment of the Goulburn Catchment, Victoria: Nagambie – Mangalore, Australian Groundwater Quality Assessment Project Report no. 8. Bureau of Rural Sciences, Department of Agriculture, Fisheries & Forestry – Australia, ACT, Australia.

Wenig D and Lawrence C 1998 Pesticide contamination of shallow aquifers at Ardmona, Girgarre and Kyvalley, Shepparton Irrigation Region, Victoria, in *International Groundwater Conference 1998, Proceedings: Groundwater Sustainable Solutions* (eds) TR Weaver, CR Lawrence, University of Melbourne, Melbourne, Australia, 8-13 February 1998. International Association of Hydrogeologists pp 586-590

WHO (World Health Organization) /UNEP (United Nations Environment Program) 1990 Working group. Public health impact of pesticides used in agriculture. Geneva: World Health Organisation.

Wilson DJ, Locker DJ, Ritzen CA, Watson JT and Schaffner W 1973 DDT concentrations in human milk. *American Journal of Diseases in Children* **125** 814-817

Winchester PD, Huskins J and Ying J 2009 Agrichemicals in surface water and birth defects in the United States. *Acta Paediatrica* **98** 664-669 DOI:0.1111/j.1651-2227.2008.01207.x

Wightwick A, Bui, AD, Zhang P, Rose, G, Allinson M et al 2012 Environmental fate of fungicides in surface waters of a horticultural-production catchment in South-eastern Australia. *Archives of Environmental Contamination and Toxicology* **62** 380–390

Wikoff D, Fitzgerald L and Birnbaum L 2012 Persistent organic pollutants: An overview (pages: 1-35). *In*: *Dioxins and Health*: *Including Other Persistent Organic Pollutants and Endocrine Disruptors*. 3rd Edition, Arnold Schecter (ed). John Wiley & Sons, Inc.

Wong CKC, Leung KM, Poon BHT, Lan CY and Wong MH 2002 Organochlorine hydrocarbons in human breast milk collected in Hong Kong and Guangzhou. *Archives of Environmental Contamination and Toxicology* **43** 364–372

Xu X, Dailey AB, Talbott EO et al 2010 Associations of serum concentrations of organochlorine pesticides with breast cancer and prostate cancer in US adults. *Environmental Health Perspectives* **118** 60-66

Yakushiji T, Watanabe I, Kuwabara K, Yoshida S, Koyama K and Kunita N 1979 Levels of polychlorinated biphenyls (PCBs) and Organochlorine pesticides in human milk and blood collected in Osaka Prefecture from 1972 to 1977. *International Archives of Occupational and Environmental Health* **43** 1-15

Zhang P, Bui A, Rose G and Allinson G 2014 Mixed-mode solid-phase extraction coupled with liquid chromatography tandem mass spectrometry to determine phenoxy acid, sulfonylurea, triazine and other selected herbicides at nano gram per litre levels in environmental waters. *Journal of Chromatography* **1325** 56–64

Web

https://foodprint.org/issues/pesticides/?cid=263

https://www.foodstandards.gov.au/consumer/chemicals/maxresidue/pages/default.aspx

https://www.google.com/search

https://en.wikipedia.org/wiki/

https://www.biologyonline.com/dictionary/cuticle

https://www.atsdr.cdc.gov/substances/toxchemicallisting.asp?sysid=39; https://www.ncbi.nlm.nih.gov/pmc/articles/PMC5464684/pdf/ITX-9-90.pdf; https://www.cdc.gov/nceh/clusters/fallon/organophosfaq.htm)

https://www.lsuagcenter.com/articles/page1508333080784

https://www.sciencedirect.com/topics/pharmacology-toxicology-and-pharmaceutical-science/thiacloprid

https://www.epa.gov/ingredients-used-pesticide-products/what-are-biopesticides#classes

https://ipmworld.umn.edu/ware-intro-herbicides

https://www.who.int/water_sanitation_health/dwq/chemicals/chlorophenoxyherb.pdf

https://link.springer.com/referenceworkentry/10.1007%2F978-3-319-17900-1_6

file:///C:/Users/User1/Downloads/Fact%20Sheet%20Weeds%20Management%20and%20Acrolein%20(3).pdf

file:///C:/Users/User1/Downloads/Azcarateetal2015.pdf

https://link.springer.com/article/10.2165/00139709-200524030-00008

https://en.wikipedia.org/wiki/Amid

https://simple.wikipedia.org/wiki/Paris_green

https://www.worldatlas.com/articles/top-pesticide-consuming-countries-of-the-world.html

https://www.worldatlas.com/articles/top-pesticide-consuming-countries-of-the-world.html

http://vro.agriculture.vic.gov.au/dpi/vro/vrosite.nsf/pages/lwm_invasive-plants_common-name https://www.sciencedirect.com/topics/pharmacology-toxicology-and-pharmaceutical-science/bioaccumulation

https://www.nicnas.gov.au/notify-your-chemical/types-of-assessments/permit-categories/controlled-use-permit/controlled-use-eligibility-criteria/Persistent-organic-pollutants-criteria]

https://economictimes.indiatimes.com/articleshow/46108223.cms?utm_source=contentofinterest&utm_medium=text&utm_campaign=cppst

http://www.ecosmagazine.com/?act=view_file&file_id=EC20p8.pdf

https://www.alz.org/alzheimers-dementia/what-is-alzheimers

https://en.wikipedia.org/wiki/Dementia

https://brainfoundation.org.au/disorders/parkinsons-disease/

https://www.mayoclinic.org/diseases-conditions/astrocytoma/cdc-20350132

https://www.ecowatch.com/15-health-problems-linked-to-monsantos-roundup-1882002128.html

https://www.foeeurope.org/weed-killer-glyphosate-found-human-urine-across-Europe-130613

https://kidshealth.org/en/parents/adhd.html

https://www.healthline.com/health/autism

■■■■

8 Nutrient Pollution, Algal Toxins, Hypoxia Risks, Impacts and Management

Summary

Nutrient pollution due to nitrogen (N) and phosphorus (P) are now considered as the major pollution problem in freshwater and coastal areas across the globe. Such pollution can cause a facet of environmental problems including eutrophication, algal blooms, algal toxins and low levels of dissolved oxygen (DO) in waterways causing "hypoxia" (low oxygen). Globally, the hypoxic areas are expanding rapidly in coastal areas and since 1950, more than 500 sites in coastal waters have reported been having oxygen concentrations of ≤2 mg/L, a threshold level often used to define hypoxia. Hypoxia can cause mass mortality of demersal fishes and benthos resulting in the loss of fisheries resources and impacting on livelihoods of fishers. Water with blue-green algal/cyanobacterial bloom could be unsuitable for agriculture, human drinking, livestock drinking, irrigation, wildlife and recreation. Algal toxins can accumulate in crops and seafood and can be transferred to humans via the food and water. This chapter reviews: **i.** chemistry, toxicity and ecotoxicity of algal toxins and hypoxia; **ii**. Impacts of algal toxins on social, economic and environmental sectors including water quality, crops, seafood, fish kill, livestock, wildlife, recreation and humans; **iii**. impacts of hypoxia on social, economic and environmental sectors including marine life, corals, fish reproduction, habitat, tourism, fisheries, seafood security and livelihoods, ocean acidification, and chemical release and **iv**. linkages among algal blooms, hypoxia and climate change; **v.** management of nutrient pollution, algal blooms, algal toxins and hypoxia.

Contents

List of Tables

List of Figures

8.1. Introduction

Nutrient pollution[1] is the pollution caused by nutrients, mainly nitrogen (N) and phosphorus (P). Excess of nutrients are now considered as the major pollution problem in freshwater and coastal areas in many countries (NRC 2000, Kibria et al 2010, Kibria 2016, Kibria et al 2016). The main sources of nutrients (N, P) in the environment are: **i.** effluents discharged from untreated sewage/ wastewater; **ii.** Run-off from agriculture land (having N, P fertilizers that was used in crops); **iii.** fossil fuel[2] combustion (automobiles) and N depositions; **iv**. effluents discharged from livestock and poultry farms; **v**. stormwater run-off from urban areas (lawn fertilizers); **vi**. effluents[3] discharged from fish and poultry farms (Fig. 8.1). The other sources are pet and wildlife wastes; detergents; forest fires (ash), etc. In freshwater systems, the limiting nutrient is phosphate, whereas in marine systems the limiting nutrient is nitrogen (Kibria et al 1997a, Kibria et al 1997b, Kibria et al 1998). The largest contribution to the global P load into freshwater is domestic sewage (54%), agriculture (38%) and Industry (8%) (Mekonnen et al 2018). Sources of N in the marine environment are sewage treatment plant

---

[1] ***Nutrient pollution** is the process when too much nutrients, mainly nitrogen and phosphorus, get added to water bodies and can act like fertiliser, causing excessive growth of algae. Nutrients can run-off from land in urban areas where lawn and garden fertilisers are used or runoff from rural agricultural farms (https://oceanservice.noaa.gov/facts/nutpollution.html).*

[2] ***Fossil fuels** are hydrocarbons, primarily coal, fuel oil or natural gas, formed from the remains of dead plants and animals; (https://www.sciencedaily.com/terms/fossil_fuel.htm).*

[3] ***Effluents** are liquid waste or sewage discharged into the river or the sea. For example, effluents discharged from aquaculture facilities contain elevated concentrations of nutrients (N, P), organic matter and suspended solids and can pollute natural waters or promote algal blooms.*

discharges, fertilisers associated with agriculture, other activities (e.g. lawns, gardens, golf courses), marine aquaculture, increased atmospheric N deposition, increased stormwater from urban areas, and fossil fuels (Boyer et al 2006, Boyer and Howarth 2008, Paerl and Piehler 2009). Nutrient pollution can cause several environmental problems including **eutrophication** (increased load of nutrients to lakes, rivers, estuaries and coastal waters), **algal blooms** (excessive growth of algae such as blue-green algae/ cyanobacteria[4] in freshwater and diatoms and dinoflagellates in marine environments due to the oversupply of nutrients), and low levels of **dissolved oxygen** (DO) in waterways causing **hypoxia[5]/ anoxia[6]** (dead zones[7]) (Kibria et al 1997a, Diaz and Rosenberg 2008, Chislock et al 2013, Kibria 2016, Breitburg et al 2018) (Table 8.1).

Table. 8.1. Flowchart showing mechanisms of eutrophication, algal blooms and hypoxia.

[There are eight steps in nutrient pollution leading to eutrophication, algal blooms, hypoxia. Excess nutrients promote eutrophication and then stimulate algal bloom and ultimately deplete oxygen (when algal blooms die and sink). The bacterial degradation of dead algae consumes the oxygen in the waterways creating either hypoxic (when dissolved oxygen level <2 mg/L) or anoxic (no oxygen) environment. Kibria et al 2019; https://en.wikipedia.org/wiki/Eutrophication#/media/File:Eutrophicationmodel.svg https://gulfhypoxia.net/about-hypoxia/].

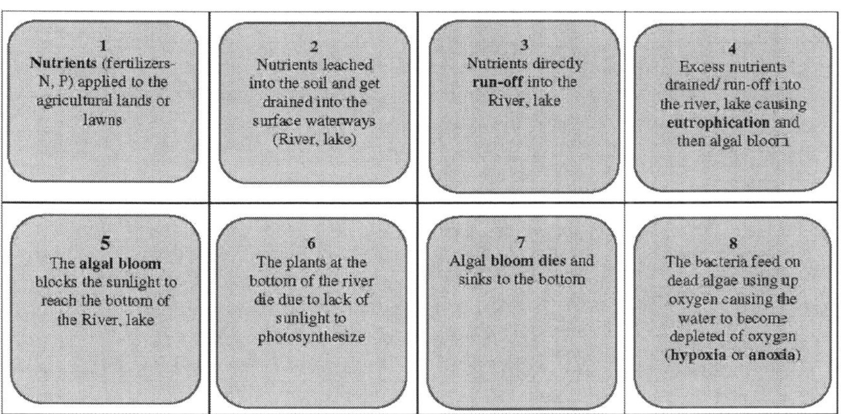

1 **Nutrients** (fertilizers-N, P) applied to the agricultural lands or lawns	2 Nutrients leached into the soil and get drained into the surface waterways (River, lake)	3 Nutrients direcdy **run-off** into the River, lake	4 Excess nutrients drained/ run-off into the river, lake causing **eutrophication** and then algal bloom
5 The **algal bloom** blocks the sunlight to reach the bottom of the River, lake	6 The plants at the bottom of the river die due to lack of sunlight to photosynthesize	7 Algal **bloom dies** and sinks to the bottom	8 The bacteria feed on dead algae using up oxygen causing the water to become depleted of oxygen (**hypoxia** or **anoxia**)

---

[4] ***Cyanobacteria*** *are aquatic bacteria (cyano = blue; also referred to as blue-green algae). These water-dwelling bacteria do photosynthesize and can be found in many different environments. There are many different types of cyanobacteria, but not all produce toxins. Microcystin and Anatoxin are two of the more common toxins that are produced by cyanobacteria.*

[5] ***Hypoxia*** *= "low oxygen". In aquatic ecosystems, low oxygen usually means a concentration of less than 2 mg/L of dissolved oxygen per litre of water.*

[6] ***Anoxia*** *= A complete lack of oxygen (0 mg/L).*

[7] ***Dead zones*** *are hypoxic (low-oxygen) areas in the world's oceans and large lakes, caused by mainly "excessive nutrient pollution" from human activities coupled with other factors that deplete the oxygen required to support most marine life in bottom and near-bottom water (https://en.wikipedia.org/wiki/Dead_zone_(ecology)#cite_note-2).*

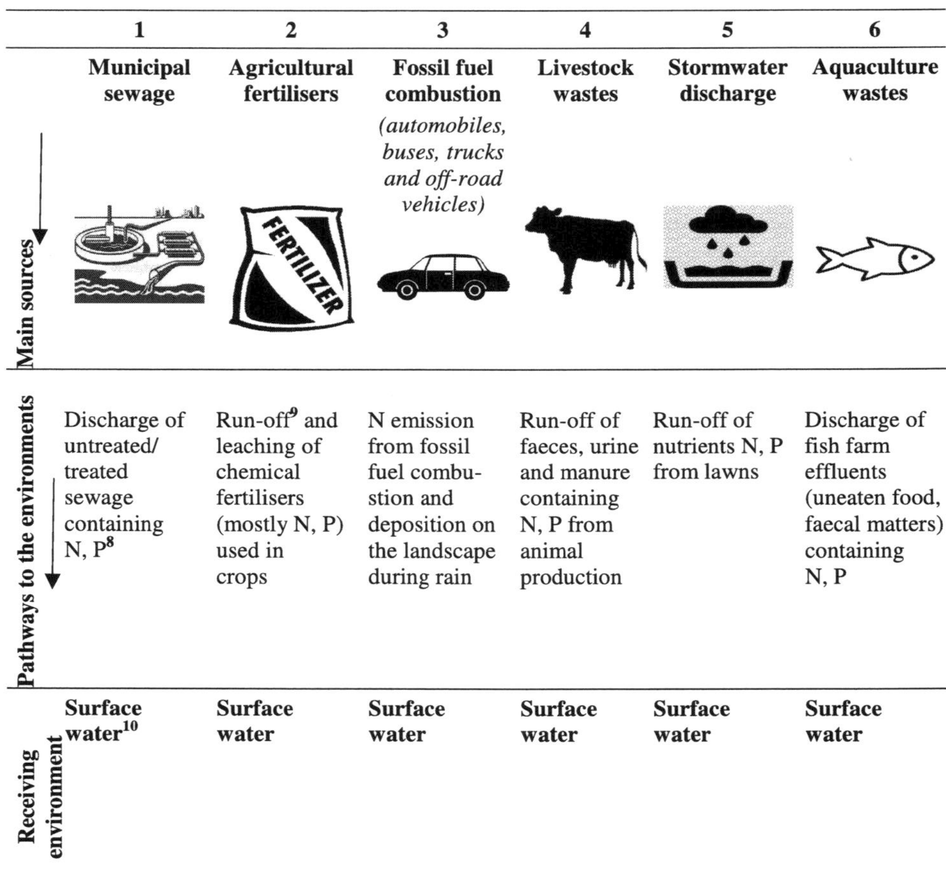

	1	2	3	4	5	6
Main sources	**Municipal sewage**	**Agricultural fertilisers**	**Fossil fuel combustion** *(automobiles, buses, trucks and off-road vehicles)*	**Livestock wastes**	**Stormwater discharge**	**Aquaculture wastes**
Pathways to the environments	Discharge of untreated/ treated sewage containing N, P[8]	Run-off[9] and leaching of chemical fertilisers (mostly N, P) used in crops	N emission from fossil fuel combu-stion and deposition on the landscape during rain	Run-off of faeces, urine and manure containing N, P from animal production	Run-off of nutrients N, P from lawns	Discharge of fish farm effluents (uneaten food, faecal matters) containing N, P
Receiving environment	**Surface water**[10]	**Surface water**	**Surface water**	**Surface water**	**Surface water**	**Surface water**

Fig. 8.1. Main sources of nutrient (N, P) pollution in the freshwater and marine environment

[N, P = Nitrogen, phosphorus; Kibria et al 1997a, Mekonnen and Hoekstra 2018, Nagatia et al 2019; http://www.oceanhealthindex.org/methodology/components/nutrient-pollution].

---

[8] ***NP** = Nitrogen (N), phosphorus (P). In freshwater systems, the "limiting nutrient" is phosphate (P), whereas in marine systems the "limiting" nutrient is nitrogen (N). An abundance or excess of N, P in both marine and freshwater ecosystems can promote the growth of algae, aquatic plant, algal blooms and algal toxin production. The situation may create hypoxic or anoxic conditions leading to dead zones, fish kill, alteration of plant species diversity, food web disruption, tourism disruption and health issues etc. (Nagatia et al 2019).*

[9] ***Run-off** = the draining away of water (or substances carried with it) from the surface of an area of land, a building or structure, etc. Surface run-off is the flow of water that occurs when excess stormwater, meltwater, or other sources flow over the Earth's surface. This can occur when the soil is saturated to full capacity, and rain arrives more quickly than the soil can absorb it.*

[10] ***Surface water** is water on the surface of continents such as in river, lake, or wetland.*

8.2. Chemistry, toxicity and ecotoxicity of algal toxins

8.2.1. Algal blooms[11] and algal toxins

An abundance or oversupply of nutrients can cause a rapid increase in the population of algae (algal blooms) either in freshwater (by blue-green algae/ cyanobacteria) or in marine waters (by diatoms and dinoflagellates) causing the discolouration of the water (due to their pigments), reduce water clarity and degrade the water quality (Chislock et al 2013, Kibria 2016). Some algae (also called harmful algal blooms) produce toxins (e.g. Microcystins by the freshwater *Microcystis*[12] spp. and Paralytic shell-fish poisoning (PSP) by dinoflagellates[13], *Alexandrium* spp.) (Table 8.2). Human poisoning due to freshwater algae/ cyanobacterial toxins are mainly caused by microcystins, cylindrospermopsin and anatoxin-a (Hudnell 2010). Cyanotoxins such as Anatoxin (a), Anatoxin a(s) and Saxitoxin are hepatotoxin (damages/ affects the liver) and microcystin, nodularin and cylindrospermopsin are neurotoxins (damages/ affects the nervous system) (Chorus and Bartram 1999, Duy et al 2000). Human poisoning due to marine algal toxins are caused by domoic acid, okadaic acid, brevetoxins, saxitoxins, and ciguatoxins (Kibria et al 2010; Table 8.2).

8.2.1.1 Persistence of algal toxins

Most algal toxins are highly soluble in water with a short **half-life** (few hours to few days) with the exception of microcystins and saxitoxins which have a long half-life (more than 40 days) (Table 8.2). The half-life of microcystins and saxitoxins are reported to be up to 12 weeks (Harada et al 1996) and 10 weeks (Jones and Negri 1997) respectively (Table 8.2).

8.2.1.2. Bioaccumulation of algal toxins

Chemicals having a **log K_{ow}** >3 have the potential or tendency for accumulation in biota (OECD 2012). Most algal toxins have a log K_{ow} values <3. However, a few algal toxins have moderate to high bioaccumulation potential such as

---

[11] ***Algal bloom*** *is the accumulation or rapid increase in the number of algae in a water body. Algal blooms are usually visible as brightly coloured water or colourful scum on the water's surface.*

[12] ***Microcystis*** *is a genus of freshwater cyanobacteria (blue-green algae) which includes the harmful algal bloom of Microcystis aeruginosa. The cyanobacteria can produce neurotoxins and hepatotoxins, such as microcystin and cyanopeptolin.*

[13] ***Dinoflagellates*** *= one-celled aquatic organisms bearing two dissimilar flagella and most are marine and is an important component of phytoplankton and food chain. Under certain conditions, several species can reproduce rapidly to form water blooms or red tides that discolour the water and may poison fish and other animals. Some dinoflagellates produce toxins that are among the most poisonous known (Ciguatera fish poisoning, Paralytic shell-fish poisoning).*

microcystins (log K_{ow} 2.3) (MSDS Merck 2006); ciguatoxins (log K_{ow} 3.4) (MSDS PubChem 2006b) and brevetoxins (log K_{ow} 5.3) (MSDS PubChem 2006a) (Table 8.2).

8.2.1.3. Toxicity of algal toxins

The LD_{50} values reveal that most of the algal toxins are **extremely toxic** to mammals and wildlife (LD_{50}: <1 mg/kg) [**freshwater** - *anatoxin (a)*, LD_{50}: 0.20-0.25 mg/kg (Hudnell 2010); *anatoxin a(s),* LD_{50}: 0.02 mg/kg (Hudnell 2010); *saxitoxins,* LD_{50}: 0.01 mg/kg (Hudnell 2010); *microcystins,* LD_{50}: 0.05 mg/kg (Hudnell 2010); *nodularin,* LD_{50}: 0.05 mg/kg (MSDS Caymanchem 2015); *cylindrospermopsin,* LD_{50}: 0.18-0.30 mg/kg (Hudnell 2010); **marine** - *domoic acid,* LD_{50}: 0.12 mg/kg (Garthwaite 2000); *okadaic acid,* LD_{50}: 0.2 mg/kg (Garthwaite 2000); *brevetoxins,* LD_{50}: 0.2 mg/kg (Garthwaite 2000); *ciguatoxins,* LD_{50}: 0.00025-0.0009 mg/kg (Garthwaite 2000) (Table 8.2)]. Apart from toxicity, **microcystins** is also carcinogenic (IARC group 2B), endocrine disrupting and bioaccumulative (BCF is up to 2,409) (see Table 8.2). In 1996, microcystins caused the most serious human poisoning resulting in the death from liver failure of more than 50 patients who received contaminated dialysis water (Azevedo et al 2002). Among the marine algal toxins, **ciguatoxins** (CTX) is one of the most lethal natural substances known with LD_{50} in mice in the range of 0.00025-0.0009 mg/kg (IP, intraperitoneal) (Garthwaite 2000) (Table 8.2). It is caused by consumption of tropical fish contaminated with toxins produced by Dinoflagellates *(Gambierdiscus toxicus, Ostreopsis* spp., *Prorocentrum* spp.).

8.2.1.4. Ecotoxicity of algal toxins

The ecotoxicity data (EC_{50}/ LC_{50} for algae, daphnia and fish) for almost all algal toxins are not available. It is only available for microcystins and nodularin which shows that these two toxins are **extremely toxic** to biota (LC_{50} <1 mg/kg) and are as follows: **i.** *microcystins: Daphnia,* LC_{50}: 0.00045-0.0214 mg/L (DeMott et al 1991, Herrera et al 2015), Fish, LC_{50}: 0.02-1.5 mg/L (Malbrouck and Kestemont 2006, Butler et al 2009); **ii.** nodularin, *Daphnia,* LC_{50}: 0.00052-0.0141 mg/L (De Mott et al 1991) (Table 8.2).

Table 8.2. Chemistry, toxicity and ecotoxicity of freshwater and marine algal toxins

[Kibria et al 2010, Kibria et al 2013; ***BCF***[14] *= Bioconcentration factor,* ***na*** *= data is not available].*

Toxins and their causative species	**Chemistry** log K_{ow}[15], half-life[16]	**Toxicity** (LD_{50}[17]; BCF)	**Ecotoxicity** (EC_{50}[18]/ LC_{50}[19] algae, daphnia, fish
Fresh- and brackish water algal toxins: Cyanobacterial toxins/ blue-green algal toxins			
Anatoxin-a $C_{10}H_{15}NO$ (Neurotoxins); caused by *Anabaena* spp., *Oscillatoria (Planktothrix)* spp., *Aphanizomenon* spp.	**log K_{ow}:** 1.12 (MSDS Pubchem 2008a) **Half-life:** 1-2 h (photochemical[20]); 14 d (normal day and night at pH 8-10) (Smith and Sutton 1993, MSDS Pubchem 2008a)	Rat, LD_{50}: 0.20-0.25 mg/kg (IP[21]) (Carmichael et al 1990, MSDS Pubchem 2008a, Hudnell 2010) **BCF:** 30-47 (fish, rainbow trout) (Osswald et al 2011)	na
Anatoxin-a(s) $C_7H_{17}N_4O_4P$ *(Neurotoxins);* caused by *Anabaena* spp., *Aphanizomenon* spp., *Lyngbya* spp., *Cylindrospermopsis* spp.	**log K_{ow}**: -1.3 to -1.6 (T3DB 2014), -4.4 (MSDS PubChem 2008b)	Rat, LD_{50}: 0.02 mg/kg (IP) (Carmichael et al 1990, Hudnell 2010)	na

[14] ***BCF = bioconcentration factor (BCF)*** *of a chemical, it is the ratio of its concentrations in the organism and in water during steady state or equilibrium (Opperhuizen and Sijm 1990, van der Oost et al 2003) or concentration in fish ÷ concentration in water medium.*

[15] ***log K_{ow}*** *= Octanol–water partition coefficient. It is a measure of hydrophilicity/ lipophilicity of a compound. log K_{ow} is also called log P. The biological properties such as bioaccumulation and toxicity are largely determined by log K_{ow} (Cumming and Rücker 2017).*

[16] ***Half-life*** *= the time it takes for one-half of the toxin to degrade.*

[17] ***LD_{50}*** *= (median lethal dose), the dose of a material that is estimated to be lethal to 50% of the test organisms (rats, mice, etc.).*

[18] ***EC_{50}*** *= effective concentration of the test substance that results in 50% effect or the concentration of test substance which results in a 50 percent reduction in either algal growth or algae growth rate or Daphina immobilization etc.*

[19] ***LC_{50}*** *= (median lethal concentration), the concentration of a material in the water that is estimated to be lethal to 50% of the test organisms (fish, daphnia, etc.). The LC_{50} is usually expressed as a time-dependent value, (24 h or 96 h).*

[20] ***Photochemical*** *= relating to or caused by the chemical action of light.*

[21] ***IP*** *= intraperitoneal.*

Toxins and their causative species	Chemistry log K_{ow}[15], half-life[16]	Toxicity (LD_{50}[17]; BCF)	Ecotoxicity (EC_{50}[18/] LC_{50}[19] algae, daphnia, fish
Saxitoxins (STX) $C_{10}H_{17}N_7O_4$ (Neurotoxins); caused by *Anabaena* spp., *Aphanizomenon* spp., *Lyngbya* spp., *Cylindrospermopsis* spp.	**log K_{ow}:** -4.6 (MSDS PubChem 2012) **Half-life:** 1-10 weeks (Jones and Negri 1997)	Rat, LD_{50}: 0.01 mg/kg (IP) (Hudnell 2010) Bird (pigeon), LD_{50}: 0.091 mg/kg (oral) (Chemol and Toan 1969, MSDS PubChem 2012)	na
Microcystins (MC-LR) $C_{49}H_{74}N_{10}O_{12}$ (Hepatotoxins); caused by *Microcystis* spp., *Anabaena* spp., *Nostoc* spp., *Oscillatoria (Planktothrix)* spp., *Hapalosiphon* spp.	**log K_{ow}**: 2.3 (MSDS Merck 2006) **Half-life**: 3 weeks (pH 1 and 40°C); 10 weeks (ambient conditions) (Tsuji et al 1995, Butler et al 2009); 12 weeks (pH 9) (Harada et al 1996) **Water solubility:** Highly soluble in water (MSDS Merck 2006)	Mouse, MC-LR[22], LD_{50}: 0.05 mg/kg (IP) (Hudnell 2010) Rat, MC-LR, LD_{50} (rat): >5 mg/kg (Oral) (MSDS Caymanchem 2017) Bird (quail), LD_{50}: 0.256 mg/kg (MC-RR[23]) (IP) (Takahashi and Kaya 1993) **Carcinogenicity:** MC-LR: Carcinogen, IARC group 2B (IARC 2019) **Endocrine disruption[24]:** Caused endocrine disruption in male frogs, *Rana nigromaculata*	Zooplankton, LC_{50}: 0.45-21.4 mg/L for 48 h (Chorous and Bartram 1999) *Daphnia* (four species, *D. hyaline, D. pulex, D. pulcaria, D. birgie*), LC_{50}: 0.00045-0.0214 mg/L for 48 h (DeMott et al 1991, Herrera et al 2015) Fish, LC_{50}: 0.02-1.5 mg/L (Malbrouck and Kestemont 2006, Butler et al

[22]***MC-LR** = [MC = microcystin; L = Leucine amino acids; R = Arginine amino acids or MC lysine, arginine) is one of the groups of toxins produced and released by cyanobacteria is called microcystin, named from the species Microcystis aeruginosa. The **microcystins** are a group of cyclic heptapeptide hepatotoxins produced by a number of cyanobacterial genera. The peptide ring is made up of **two protein** amino acids and **five non-protein** amino acids. It is the two protein amino acids that distinguish microcystin types from each other, while the other amino acids are relatively constant. By using amino acid single letter code classification, each microcystin is designated a name depending on the variable amino acids which complete their structure. For instance, one of the most common toxins found in water supplies around the world, microcystin-LR contains the amino acids Leucine (L) and Arginine (R) in these variable positions].*

[23] ***MC-RR** = [MC = microcystin; R = Arginine amino acids; R = Arginine amino acids] or (MC arginine, arginine) (https://journals.sagepub.com/doi/pdf/10.1177/0192623308315356).*

[24] ***Endocrine disruptors** are chemicals that can interfere with endocrine systems at certain doses. These disruptions can cause cancerous tumours, birth defects, and other developmental disorders (https://www.niehs.nih.gov/health/topics/agents/endocrine/).*

Toxins and their causative species	Chemistry log K_{ow}[15], half-life[16]	Toxicity (LD_{50}[17]; BCF)	Ecotoxicity (EC_{50}[18]/ LC_{50}[19] algae, daphnia, fish
		(MSDS PubChem 2005a, Jia et al 2014) **BCF:**14 to 2,409 (fish) (do bioaccumulate) (Franke et al 1994, MSDS PubChem 2005a, Berry et al 2011)	2009)
Nodularin $C_{41}H_{60}N_8O_{10}$ (Hepatotoxins); caused by *Nodularia* spp.,	**log K_{ow}:** 1.7 (MSDS PubChem 2009a) **Half-life:** 24 h (lake water with algal bloom) and 54 h (lake water without alga bloom) (Heresztyn and Nicholson 1997)	Mouse, LD_{50}: 0.06 mg/kg (IP) Rat, LD_{50}: 0.05 mg/kg (IP) (MSDS Caymanchem 2015) **BCF:** 12-18 (three species of zooplankton) (Karjalainen et al 2003)	*Daphnia* (same four species as mentioned above), LC_{50}: 0.00052-0.0141 mg/L (De Mott et al 1991)
Cylindrosp-ermopsin $C_{15}H_{21}N_5O_7S$ *(Hepatotoxins);* caused by *Cylindros-permopsis* spp., *Aphanizomenon* spp., *Umezakia* spp.	**log K_{ow}:** -2.7 (MSDS PubChem 2009b) **Half-life:** 1.5-18 h (high levels of pigment), 11-15 days (natural water-low levels of pigment) (Chiswell et al 1999, ILS 2000) **Solubility:** High solubility in water (ILS 2000)	Rat, LD_{50}: 0.18-0.30 mg/kg (IP) (Hudnell 2010) **BCF**: 0.045 (*Hydrilla verticillata*) (White et al 2005, Kinnear 2010); BCF: 1.08 (duckweed, *Lemna punctate*) (Seifert 2007)	na
Marine algal toxins: ASP, DSP, NSP, PSP, CFP			
Domoic acid $C_{15}H_{21}NO_6$ (Amnesic shell-fish poisoning (**ASP**); caused by diatoms, *Pseudonitzschia* spp.	**Log K_{ow}:** -1.3 to 0.89 (MSDS PubChem 2005b, USEPA 2012) **Half-life:** 12-36 h (sunlight) (Fisher et al 2006)	Mouse, LD_{50}: 0.12 µg/kg (IP)(Garthwaite 2000) Mouse, LD_{50}: 3.6 mg/kg (IP) (Grimmelt et al 1990, MSDS PubChem 2005b) **BCF**: 3 (fish) (MSDS PubChem 2005b, USEPA 2012) (Low bioaccumulation)	na

Toxins and their causative species	**Chemistry log K_{ow}[15], half-life[16]**	**Toxicity (LD_{50}[17]; BCF)**	**Ecotoxicity (EC_{50}[18]/ LC_{50}[19] algae, daphnia, fish**
Okadaic acid $C_{44}H_{68}O_{13}$ (Diarrhoetic shell-fish poisoning (**DSP**); caused by dinoflagellates, *Dinophysis* spp. and *Prorocentrum* spp.	**Log K_{ow}**: 3.4 (MSDS PubChem 2005c)	Mouse, LD_{50}: 0.2 mg/kg (IP) (Garthwaite 2000)	na
Brevetoxins $C_{49}H_{70}O_{13}$ (Neurotoxic shell-fish poisoning (**NSP**); caused by dinoflagellates, *Kareniabrevis* (formerly *Gymnodinium breve)*	**Log K_{ow}:** 5.3 (MSDS PubChem 2006a)	Mouse, LD_{50}: 0.2 mg/kg (IP) (Garthwaite 2000)	na
Saxitoxins $C_{10}H_{17}N_7O_4$, Paralytic shell-fish poisoning (**PSP**); caused by Dinoflagellates (*Alexandrium* spp., *Gymno-dinium* spp. and *Pyrodinium* spp.)	**Log K_{ow}:** -4.6 (MSDS PubChem 2012) **Half-life:** 1-10 weeks (Jones and Negri 1997)	Rat, LD_{50}: 0.01 mg/kg (IP) (Hudnell 2010) Bird (pigeon), LD_{50}: 0.091 mg/kg (oral) (Chemol and Toan 1969, MSDS PubChem 2012)	na
Ciguatoxins $C_{60}H_{86}O_{18}$, Ciguatera fish poisoning (**CFP**); caused by Dinoflagellates *(Gambierdiscus toxicus, Ostreo-psis* spp. and *Prorocentrum* spp.)	**Log K_{ow}:** 3.4 (MSDS PubChem 2006b)	Mouse, LD_{50}:0.00025-0.0009 mg/kg (IP) (Garthwaite 2000) Mouse, LD_{50}: 0.00052 to 0.0141 mg/L (De Mott et al 1991)	na

8.2.2. Hypoxia

Hypoxia (DO <2 mg/L) are associated with excess nutrient inputs from both point- and non-point sources (Diaz and Rosenberg 2008; *https://gulfhypoxia.net/about-hypoxia/*). DO <2 mg/L is considered as a useful threshold of hypoxia (Vaquer-Sunyer and Duarte 2008, Breitburg et al 2018). When algae/ algal blooms die and sink to the bottom, they are decomposed by bacteria, using the available dissolved oxygen (Diaz and Rosenberg 2008). This causes the hypoxic environment and creates so-called "**dead zones**" [i.e. lack of sufficient oxygen or the oxygen concentration is so low that animals suffocate and die, and as a result, are often called "dead zones"] (Diaz and Rosenberg 2008; *https://oceanservice.noaa.gov/podcast/feb18/nop13-hypoxia.html*). The hypoxic areas are expanding rapidly across the globe in both freshwater (Jenny et al 2016; Fig. 8.2), and coastal areas (Breitburg et al 2018; Fig. 8.3) and the number of dead zones doubled in each decade since the 1950s (Diaz and Rosenberg 2008, Vaquer-Sunyer and Duarte 2008). Since 1950, more than 500 sites in coastal waters have reported oxygen concentrations ≤2 mg/L (Breitburg et al 2018). The Baltic Sea and the Gulf of Mexico is the world's largest dead zones" ranking 1st and 2nd respectively (BNI 2004, Rabalais et al 2002).

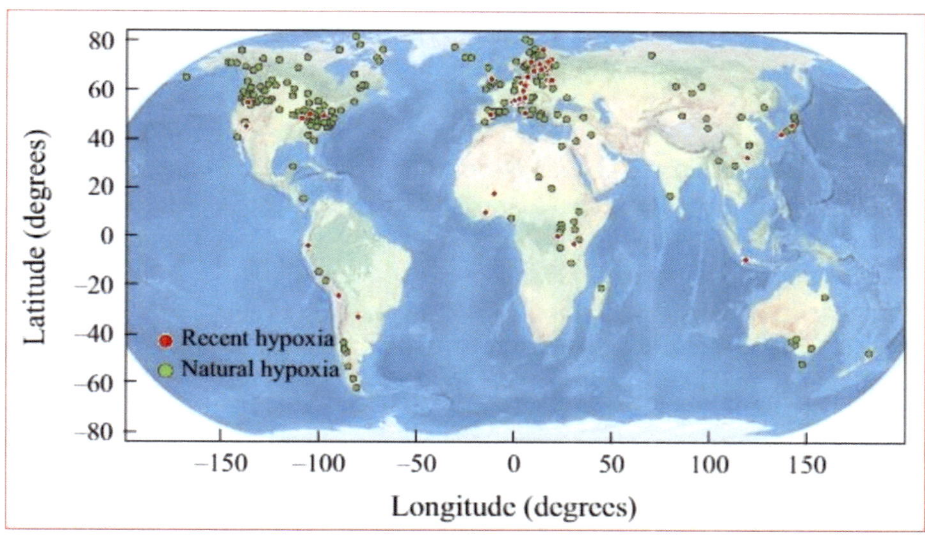

Fig. 8.2. Global map showing freshwater hypoxic areas (lakes) [***Red dots*** *= Recent hypoxia after AD 1700;* ***Green dots*** *= Natural hypoxia for the last 300 years (Jenny et al 2016, page 1482). This study compiles the onset and duration of hypoxia recorded in sediments of 365 lakes worldwide since AD 1700].*

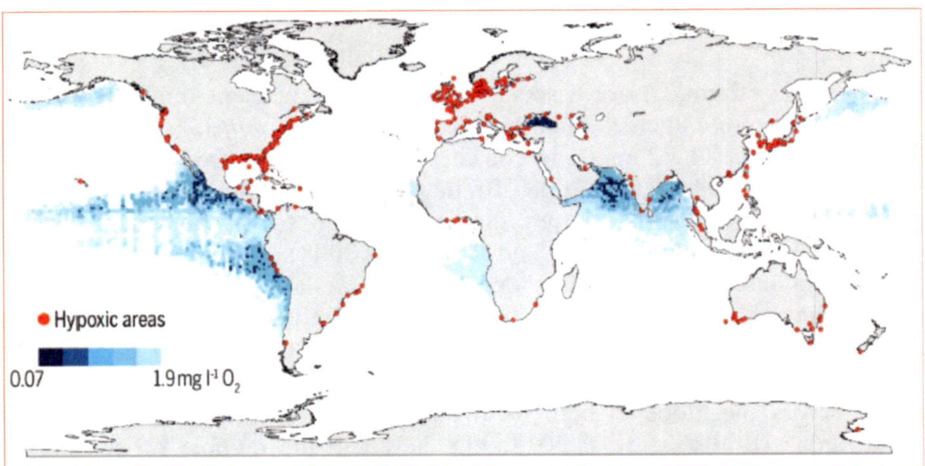

Fig. 8.3. Global map showing coastal and open ocean where nutrients inputs caused a decline in dissolved oxygen (DO <2 mg/L) resulting in hypoxia *[**Red dots:** coastal areas; **Blue shades:** open oceans, Brietburg et al 2018; note: DO is fundamental to aquatic biota (fish, crustaceans, molluscs) for survival, growth and reproduction].*

8.2.2.1. Ecotoxicity of hypoxia (LC_{50} of DO)

Prolonged low-oxygen/ hypoxic events can lead to sub-lethal effects (reduced growth, reproduction etc.) and lethal effects (mortalities) of fish, prawns and bivalves (Gehrke et al 1993, Whitworth et al 2012). LC_{50} (the toxicity endpoint of DO causing 50% mortality of the test aquatic organisms) ranged between 0.25 mg/L and 3.7 mg/L (Table 8.3). Some fish and crustaceans are **most sensitive to hypoxia** (LC_{50}: > 2 mg/) including New Zealand, whitebait, *Galaxias maculatus*, LC_{50}: 2.65 mg/L (Landman et al 2005); the USA striped bass, *Morone saxatilis* post-larvae, LC_{50}: 2.4 mg/L, (Miller et al 2002); the USA American lobster larvae, *Homarus americanus*, LC_{50}: 3.1 mg/L and the USA Say mud crab, *Dyspanopeus sayi,* LC_{50}: 3.7 mg/L (Miller et al 2002) (Table 8.3). On the other hand, some species are **most tolerant to hypoxia** (LC_{50}: <1.0 mg/L) including Australian eel-tailed catfish, *Tandanus tandanus*, LC_{50}: 0.25 mg/L (Small et al 2014); New Zealand short-fin eel elvers[25], *Anguilla australis,* LC_{50}: 0.54 mg/L (Landman et al 2005) and the USA bivalvia[26], *Spisula solidissima,* LC_{50}: 0.7 mg/L (Miller et al 2002). Dissolved oxygen tolerance of fish and invertebrates of the Patuxent River (a seasonally hypoxic estuary, Maryland, the USA) shows that gelatinous zooplankton (jellyfish), some benthic fish and invertebrates (polychaetes, bivalves) are the most tolerant species, while larval fish and larval

---

[25] ***Elvers*** *= baby eel.*

[26] ***Bivalvia*** *= an aquatic mollusc which has a compressed body enclosed within a hinged shell, such as oysters, mussels, and scallops; having a shell composed of two valves.*

crustaceans tend to be the most sensitive (Breitburg et al 2003). A comprehensive review (based on 872 published experimental results on oxygen thresholds of 206 species) reveals that crustaceans are most sensitive to hypoxia and gastropods are most tolerant to hypoxia (Vaquer-Sunyer and Duarte 2008).

Table. 8.3. Concentrations of dissolved oxygen (DO) causing 50% mortality (LC_{50}) of some selected fish, crustaceans and molluscs from Australia, New Zealand, the USA and across the globe.

Country	Species	LC_{50} (DO)	Remarks
Australia *Four native freshwater fish from the Murray-Darling Basin (Small et al 2014)*	Murray cod, *Maccullochella peelii*	1.58 mg/L	
	Silver perch, *Bidyanus bidyanus*	1.04 mg/L	
	Golden perch, *Macquaria ambigua*	0.85 mg/L	Most tolerant
	Eel-tailed catfish, *Tandanus*	0.25 mg/L	Most tolerant
New Zealand *Seven native freshwater fish, shrimp and crayfish from the Rangitaiki, Waikato and Kaituna Rivers, Lake Tarawera, Turangi Trout Centre (Landman et al 2005)*	White bait (*Galaxias maculatus*)	2.65 mg/L	Most sensitive
	Common smelt (*Retropinna retropinna*)	1.83 mg/L	
	Rainbow trout (*Oncorhynchus mykiss*)	1.61 mg/L	
	Common bully (*Gobiomorphus cotidianus*)	0.91 mg/L	Most tolerant
	Shrimp (*Paratya curvirostris*)	0.82 mg /L	Most tolerant
	Freshwater crayfish (*Paranephrops planifrons*)	0.77 mg/L	Most tolerant
	Shortfin eel elvers (*Anguilla australis*)	0.54 mg/L	Most tolerant
USA *35 fish species from small streams in Missouri (Smale and Rabeni 1995)*	35 fish species from the mid-western United States	0.49 mg/L - 1.59 mg/L	

Country	Species	LC_{50} (DO)	Remarks
USA	**Fish,** larva	1.4-1.8 mg/L	
22 marine and estuarine species *including 12 fish, 9 crustaceans, 1 bivalve, all from the Virginian Province, East and Gulf coasts, western Atlantic (Miller et al 2002)*	Fish, post-larva	2.4 mg/L	Most sensitive
	Fish, juvenile	0.6-1.6 mg/L	
	Crustacea, larva	1.6-3.7 mg/L	Most sensitive
	Crustacea, post-larva	1.0-2.2 mg/L	
	Crustacea, juvenile	0.7-1.2 mg/L	
	Bivalve, juvenile	0.5 mg/L	Most tolerant
USA *Patuxent River* (Chesney and Houde 1989)	Planktonic eggs of the bay anchovy *Anchoa mitchilli*	2.7 mg/L	Most sensitive
USA *Patuxent River* (Brietburg et al 2003)	Sea nettles	0.5 mg/L	Most tolerant
Global *based on 872 experimental assessments of 206 species of benthic metazoans (Vaquer-Sunyer and Duarte 2008)*	Crustaceans (n = 168)	2.45 mg/L	Most sensitive
	Fish (n = 77)	1.54 mg/L	
	Bivalves (n =12)	1.42 mg/L	
	Gastropods (n = 19)	mg/L	Most tolerant

8.3. Impacts of algal toxins and hypoxia on social, economic and environmental sectors

8.3.1. Impacts of algal toxins

Some freshwater and marine algae produce toxins (Table 8.2), which can contaminate water and seafood. Human poisoning due to cyanobacteria is mainly caused by microcystins, cylindrospermopsin and anatoxin-a (Hudnell 2010). Cyanotoxins are intra-cellular toxins (i.e. contained within the cyanobacterial cells) and released to the surrounding waters during cell senescence[27], death and

[27]***Cell senescence*** *= the cessation of cell division.*

lysis[28] or water treatment processes (e.g. per-chlorination or algicide application). Once released into the water, microcystins can persist for a relatively long period before being removed by biodegradation or photolysis (Chorus and Bartram 1999, van Apeldoorn et al 2007). Cyanotoxins are among the most known potent toxins; only dioxin group are among the few toxic substances that are known to be more potent than cyanotoxins (Table 8.4). Algal blooms and algal toxins can cause significant impacts on various sectors including crops, drinking water, fish, humans, irrigation, livestock, recreation, seafood, water quality and wildlife and are highlighted below (***note**: Blue-green algae (BGA) and cyanobacteria are interchangeably used in this chapter*).

Table. 8.4. Toxicity scale: Consumption thresholds for toxic effects

[based on daily ingestion in milligrams of toxins per kilograms of body weight; mg/kg-d = mg per kg per day; https://www.flickr.com/photos/fondriestenvironmental/32533628442].

Cyanotoxins/BGA toxins	Dioxin groups	Other groups (metals, pesticides etc.)
Anatoxin (a): 0.0005 mg/kg-d *(covered in Chapter 8)*	PCBs: 0.00002 mg/kg-d *(covered in Chapter 5)*	Aluminium: 1 mg/kg-d *(covered in Chapter 2)*
Cylindrospermopsin: 0.00003 mg/kg-d *(covered in Chapter 8)*	**Dioxin: 0.000001 mg/kg-d** *(covered in Chapter 5)*	Atrazine: 0.04 mg/kg-d *(covered in Chapter 7)*
		Cyanide: 0.02 mg/kg-d
Saxitoxin: 0.000005 mg/kg-d *(covered in Chapter 8)*		Botulinum toxin: 0.001 mg/kg-d
Microcystin L-R: 0.000003 mg/kg-d *(covered in Chapter 8)*		DDT: 0.0005 mg/kg-d *(covered in Chapter 7)*
		Methylmercury: 0.0001mg/kg-d *(covered in Chapter 2)*

8.3.1.1. Water quality

Kibria et al (2010) compiled measured concentrations of BGA/ cyanotoxins in surface waters, the summary of which is highlighted below: **Anatoxin-a**: range 0.02-4,400 μg/g dw (dry weight); highest concentrations (4,400 μg/g dw)

[28]***Lysis*** *= the breaking down of the membrane of a cell.*

reported from Finland; **Anatoxin-a** (S): range 4-3,300 µg/g dw; highest concentrations (3,300 µg/g dw) reported from the USA; **Saxitoxins**: range 5-3,400 µg/g dw; highest concentrations (3,400 µg/g dw) reported from Australia; **Microcystin:** range 0.004-25,000 µg/g dw; highest concentrations (25,000 µg/g dw) reported from Germany; **Nodularin**: range 3,000-18,000 µg/g dw; highest concentrations (18,000 µg/g dw) reported from the Baltic Sea; **Cylindrospermopsin**: 5,500 µg/g dw reported from Australia. Algal blooms of toxic BGA can affect the water quality by causing peculiar tastes and odours, discolouration and unsightly scums. Cyanobacteria have the ability to produce earthy and musty smelling compounds or odour metabolites (compounds that are not essential for primary cell metabolism). Over 200 compounds have been identified as being produced by algae, of which filamentous forms produce more than 25% of all known off-flavour compounds. **Geosmin**[29] (E1, 10-dimethyl-E-9-decalol) and 2-methyl-Isoborneol (**MIB**) accounts for the majority off-flavour odour. The cyanobacterial genera that are known to produce geosmin are *Anabaena, Aphanizomenon, Lyngbya, Microcystis, Oscillatoria (Planktothrix), Phormidium, Schizothrix* and *Symploca* (Smith et al 2008). The presence of objectionable odour has caused an estimated 30% of the potential revenue loss in the US channel catfish industry alone (Smith et al 2008, Burr et al 2012). Off-flavours have also impacted Nile tilapia, Atlantic salmon, rainbow trout, and shrimp aquaculture industry. Moreover, the presence of cyanobacteria in water can cause clogging of irrigation and water delivery systems including filters, meters, valves and sprinklers, thereby increases the costs of water treatment. Some microcystins (such as MC-LR and other toxin variants) may be persistent and in certain cases leach and caused groundwater contamination (Chen et al 2006).

8.3.1.2. Crops/ edible plants

Water with cyanobacterial bloom could be unsuitable for agriculture use since (**BGA**)/ cyanobacterial toxins can be accumulated on the external surfaces of edible plant material. For example, the toxins are deposited when the water dries on the plant surface between irrigation periods or when water becomes trapped in the centres of, for example, salad plants (*http://agriculture.vic.gov.au/__data/assets/pdf_file/0007/319453/Blue-green-algal-poisoning-of-stock-InfoNote-.pdf*; Chorus and Bartram 1999, Kibria et al 2010). The cyanotoxins such as microcystins can be taken up by plants and when present in sufficient concentrations can induce various metabolic and morphological changes such as inhibition of photosynthesis, germination, growth inhibition, a decrease in roots

[29] ***Geosmin and Methyl-Isoborneol (MIB)*** *are naturally occurring compounds that have a very strong, earthy taste and odour. Some kinds of algae and bacteria present in dams naturally produce Geosmin and MIB. Geosmin and MIB can not be removed from water using normal treatment processes; https://www.hunterwater.com.au/Water-and-Sewer/Water-Supply/Water-Quality/Geosmin-and-MIB.aspx.*

development and leaf necrosis (Babica et al 2006). In a laboratory bioassay experiment (McElhiney et al 2001), microcystin-LR (1.12 μg/mL) showed necrosis (death of cells) in leaves of runner bean (*Phaseolis vulgaris*) after 25 days of exposure, the effects followed by the desiccation and loss of older leaves in exposed plants. In addition, MC-LR impaired the development of the roots of the exposed plants. When potato (*Solanum tuberosum*) plants were exposed to solutions containing 1-5,000 μg/L MC-LR, no significant growth effect was seen at 1 μg/L, but at concentrations of 5, 10, 50, 100 and 500 μg/L significant growth inhibitions were recorded. Studies conducted by Chen et al (2004), found that microcystin significantly inhibited the elongation of primary roots of rape and rice seedlings. Rape seed and rice plants exposed to different levels of microcystins showed an accumulation of toxins that were correlated to the toxin levels in the water (Chen et al 2004). This is an area of great concern, especially in countries with high reliance on the irrigation for high-valued crops, and with a high incidence of eutrophication and blooms, like Australia (Kibria et al 2013).

8.3.1.3. Bioaccumulation in seafood organisms

Freshwater cyanobacterial/ **blue-green algal toxins** (microcystins- produced by *Microcystis* spp.) can accumulate in the gut, liver, kidney, blood, gills and muscle of fish and other freshwater seafood (mussels and shrimps) (de Figueiredo et al 2004, Ibelings and Chorus 2007, van Apeldoorn et al 2007, Ibelings and Havens 2008, Kibria et al 2013). Investigations carried out in Brazil, China, Egypt, Portugal and Uganda reveals that microcystin levels in edible parts of freshwater seafood (e.g. fish/ shrimp/ mussel) have exceeded the WHO recommended limit for the tolerable daily intake (TDI) on many occasions for human consumption of 0.04 μg/kg per day of MC-LR (Ibelings and Chorus 2007, Poste et al 2011). **Marine algal toxins** (domoic acid, okadaic acid, brevetoxins, saxitoxins, ciguatoxins, produced by diatoms and dinoflagellates) can accumulate in filter feeders such as shell-fish (clams, oysters, mussels, scallops and crabs) and tropical fishes (Lehane and Lewis 2000, FAO 2004, FSANZ 2019; *http://www.foodstandards.gov.au/consumer/safety/foodborne-illness/Pages/ Toxins-in-seafood.aspx*). Therefore, the transfer of the freshwater and marine algal toxins to humans via the food chain is easily possible.

8.3.1.4. Fish kill

Fish can be exposed to cyanotoxins during normal feeding, drinking or gill respiration; the toxins can be accumulated in the gut, liver, kidney, blood, gills and muscle causing damage to the vital organs leading to death (de Figueiredo et al 2004, van Apeldoorn et al 2007, Ibelings and Havens 2008). Blooms of marine dinoflagellates- *Alexandrium* sp. (causing PSP) can cause fish kills (FAO 2004). Most of the global fish kills are frequently linked to algal blooms/ algal toxins, and low dissolved oxygen (hypoxia or anoxia) related to blooms etc. (Kibria 2016).

8.3.1.5. Livestock

As some cyanobacterial toxins could be toxic in dry form, therefore, continued application of heavily affected waters on pastures may lead to significant toxin build-up on foliages. Long-term feeding of livestock with contaminated pastures may cause poisoning of stocks (cattle, sheep, poultry) (McElhiney et al 2001, Chen et al 2004, Babica et al 2006, van Apeldoorn et al 2007, Crush et al 2008; *http://agriculture.vic.gov.au/__data/assets/pdf_file/0007/319453/Blue-green-algal-poisoning- of-stock-InfoNote-.pdf*).

8.3.1.6. Wildlife

Microcystins are responsible for repeated poisoning in wild and domesticated animals, including the death of cattle, geese, sheep, pigs, horses, dogs, cats, squirrels, poultry, waterfowl, fish and birds (reviewed by Kibria et al 2010, WHO 2014). Animals killed due to cyanobacterial toxins are birds, cats, cattle, chicken, dogs, ducks, fish/ shrimp, horses, lambs, monkeys, sheep, squirrels, pigs, rhinoceros, rodents and invertebrates (de Figueiredo et al 2004, Stewart et al 2008). Marine biotoxin PSP (paralytics shell-fish poisoning) causes the deaths of fish, seabirds and marine mammals (FAO 2004, Kibria et al 2010).

8.3.1.7. Recreation

Water bodies affected by BGA blooms having a pungent smell can make the water unsuitable for recreational activities, such as swimming, boating or fishing. People can be exposed to BGA toxins via accidental ingestion or direct contact during recreational water sports. Swimmers in water containing cyanobacterial toxins may suffer allergic reactions, such as asthma, eye irritation, rashes, and blisters around the mouth and nose (Chorus and Bartram 1999, Hudnell 2005 and 2010, WHO 2014).

8.3.1.8. Humans

Humans maybe exposed to algal toxins (freshwater and marine algal toxins) through several paths including **i.** oral ingestion of untreated infested drinking water; **ii**. body surface contact during recreational and bathing activities; **iii**. through consumption of toxins contaminated fin-fish, shell-fish, crops and vegetables; and **iv**. eating food supplements such as BGAS (blue-green algae supplements) (Manganelli et al 2012). Microcystins are highly toxic to humans (Falconer and Humpage 2005, Zegura et al 2011, Kibria et al 2013), bioaccumulative in edible plants, pastures and seafood (Falconer and Humpage 2005, Zegura et al 2011, Kibria et al 2013, Kibria 2016), and is a possible human carcinogen (IARC 2B) (Falconer and Humpage 2005, Zegura et al 2011, IARC (2019); *https://monographs.iarc.fr/wp-content/uploads/2018/09/Classifications AlphaOrder.pdf*). Cyanobacterial toxins (microcystins) caused the most serious human poisoning resulting in the death from liver failure of more than 50 patients who received contaminated dialysis water in Brazil in 1996 (Azevedo et al 2002). PSP is poisonous, bioaccumulative and causes harm to humans (Kibria et al 2010). The consumption of PSP (saxitoxins) contaminated shell-fish (mussels, clams, oysters) (raw or cooked) can cause paralytic phenomena such as cramps, signs of paralysis and blocking of respiration and even death. (Garthwaite 2000; *https://www.whoi.edu/redtide/page.do?pid=15159&tid=523&cid=27690*). Ciguatoxins (CTX-1) is one of the most lethal natural substances known with LD_{50} in mice around 0.25-0.90 μg/kg (0.00025-0.0009 mg/kg IP (Garthwaite 2000). Oral intake of as little as 0.1μg ciguatoxin can cause illness in the adult human (FAO 2004).

8.3.2. Impacts of hypoxia

Hypoxia is among the most widespread deleterious anthropogenic influences on estuarine and marine environments (Diaz and Rosenberg 2008). Since 1950, more than 500 sites in coastal waters have reported oxygen concentrations ≤2 mg/L, a threshold level often used to define "hypoxia" (Diaz and Rosenberg 2008, Isensee et al 2016). The number of coastal sites (where hypoxia has been reported) has increased with the exponential growth rate of 5.54% per year over time (Vaquer-Sunyer and Duarte 2008). Coastal hypoxia resulted in so-called ''**dead zone** (low oxygen areas)'' (Rabalais et al 2002, Renaud 1986) [("dead zone" because of most marine life either dies or if they are mobile such as fish, leave the area)]. Dead zones have developed in continental seas, such as the Baltic, Kattegat, Black Sea, Gulf of Mexico, and East China, all of which are major fishery areas (Diaz and Rosenberg 2008). The Baltic Sea and The Gulf of Mexico is the world's largest dead zones" ranking 1st (BNI 2004) and 2nd (Rabalais et al 2002) respectively. The Baltic Sea dead zones have increased from 5,000 km^2 in 1900 to more than 60,000 km^2 in recent years (BNI 2004) and in the Gulf of Mexico dead zones have increased from 39 km^2 in 1988 to 22,730 km^2 in 2017 (NOAA 2017). The anthropogenic nutrient discharges and effluents from municipal wastewater treatment plants are the primary driving factor creating widespread hypoxic conditions in both The Gulf of Mexico and the Baltic Sea (Carstensen et al 2014).The following section briefly discusses the effects of hypoxia on the following sectors:

- marine life,
- corals,
- fish reproduction,
- habitat and ecosystems compression,
- tourism,
- fisheries, ocean acidification, seafood security,
- chemicals release

8.3.2.1. Marine life

Coastal hypoxia is a major threat to coastal ecosystems and coastal biodiversity across the globe, resulting in so-called ''dead zones'' and triggering both lethal (mortality) and sub-lethal stresses (reduced growth and reproduction) and causing damage/ loss of fisheries resources (Rabalais et al 2002, Renaud 1986). Hypoxia can cause forced migration of biota, reduction of suitable habitat, increased vulnerability to predation and disruption of life cycles (Service 2004, Rabalais et al 2002). **Mobile** organisms such as fish can leave/ flee the hypoxic area (DO <2 mg/L), whereas **benthic**[30] and **sessile**[31] organisms (mussels, echinoderms) which are very slow to move out would generally die or experience stress (at **DO <1.0 mg/L**) due to lack of sufficient oxygen (either hypoxia or anoxia; Fig. 8.4; *https://gulfhypoxia.net/about-hypoxia/*). Among the aquatic biota, benthic organisms are most vulnerable to coastal hypoxia because they live farthest from contact with atmospheric oxygen supply (Vaquer-Sunyer and Duarte 2008). Many benthic organisms (polychaetes, annelids, crustaceans, bivalves, priapulids, and anemones) leave their burrows or tubes to move to the sediment surface or reduce their burial depth (Pihl et al 1992, Nilsson and Rosenberg 1994, Diaz and Rosenberg 1995) in the presence of hypoxia. Hypoxia reduces the abundance of large, slow-growing, deeper-dwelling animals, which were replaced by smaller, fast-growing species that can colonize the surface sediments rapidly following hypoxia (Díaz and Rosenberg 1995). On the contrary, the jellyfish population may thrive in low-oxygen waters due to their high tolerance of low oxygen conditions (de Mutsert et al 2016) and being an efficient predator, predate and compete in feeding plankton and fish larvae at low DO (Breitburg et al 1997, Shoji et al 2005) [***note***: jellyfish can store oxygen in their jelly (mesoglea) (*https://gulfhypoxia.net/are-jellyfish-a-harbinger-of-dying-seas/*)].

8.3.2.2. Corals[32]

The unprecedented hypoxic event in the Caribbean coast of Panama caused coral bleaching and massive mortality of corals and other reef-associated organisms (Altieri et al 2017). It is known that coral reefs are associated with more than half of the known tropical dead zones worldwide, with >10% of all coral reefs are at elevated risk for hypoxia based on local and global risk factors (Altieri et al 2017; Fig. 8.5).

---

[30] ***Benthic*** *= Organisms that live on, or near the seabed, river, lake, or stream bottom.*

[31] ***Sessile*** *= Animals that attach to a surface and stay in the same place for most of their time.*

[32]***Corals*** *are marine invertebrates within the class Anthozoa of the Phylum Cnidaria. They typically live in compact colonies of many identical individual polyps. Corals species include the important reef builders that inhabit tropical oceans and secrete calcium carbonate to form a hard skeleton; https://en.wikipedia.org/wiki/Coral*

8.3.2.3. Fish reproduction

Hypoxia can cause major reproductive impairments by inhibiting testicular and ovarian development, affecting the production and quality of sperm and egg, reducing fertilization and hatching success, affecting larval survival as well as the quality and fitness of juveniles (Wu 2009, Wang et al 2016).

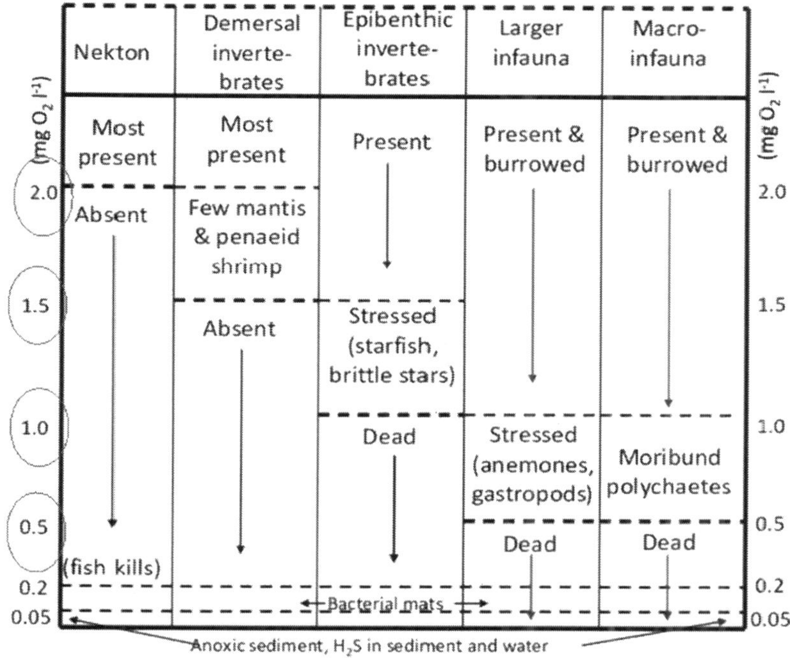

Figure 8.4. Progressive changes in fish and invertebrate fauna as oxygen concentration decreases from 2 mg/L to anoxia (0 mg/L) on the Louisiana continental shelf (Rabalais et al 2001).

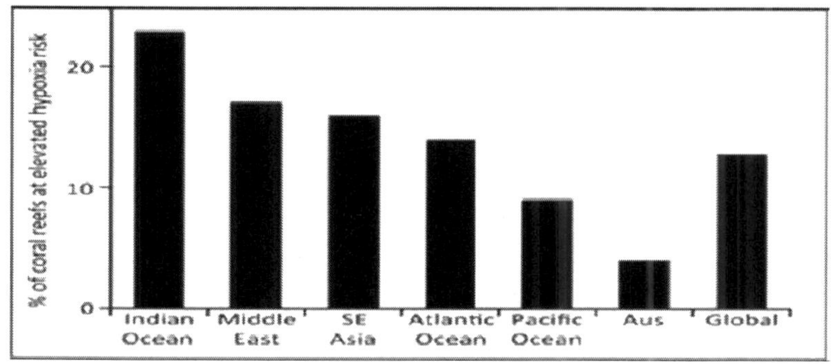

Fig. 8.5. Proportion of coral reefs at elevated risk of hypoxia (Altieri et al 2017, page 5).

8.3.2.4. Habitat/ ecosystem compression

Hypoxia can cause habitat compression and reduction of suitable habitat (Eby and Crowder 2002, Stramma et al 2012, UNESCO 2018). Hypoxia-induced habitat compression can restrict **pelagic**[33] fish species (e.g. billfishes and tunas) to shallower waters, therefore, can increase vulnerability to fishing (Stramma et al 2012, UNESCO 2018) and may be associated with a 10–50% worldwide decline of pelagic predator diversity (Worm et al 2005). For **demersal**[34] fish species, habitat compression can lead to crowding and increased competition for food (prey), and fish may be forced to occupy poorer habitat which may increase bioenergetic costs, potentially resulting in sub-lethal effects (decreased growth and condition) (Eby and Crowder 2002). The food limitations due to hypoxia potentially resulted in decreased body condition of important fishery species such as the Baltic cod (*Gadus morhua*) in the Baltic Sea ecosystem during the past two decades where there has been the fivefold increase in the extent of hypoxic and anoxic areas (Casini et al 2016).

8.3.2.5. Tourism

Inland lakes and coastal areas are hotspots for tourism and recreational activities. They are an important economic source for tourism. Threats posed by nutrient pollution, algal blooms, algal toxins and hypoxia can discourage economic investment and damage tourism and hospitality industry (Le et al 2010). This is due to poorer water quality (odours, discolouration and unsightly scums making the water unsuitable for recreational activities, such as swimming, boating or fishing). People can be exposed to BGA toxins via accidental ingestion or direct contact during recreational water sport (Kibria 2016).

8.3.2.6. Fisheries, seafood security and livelihoods

Hypoxia is a threat to lucrative commercial and recreational fisheries (Diaz and Rosenberg 2008), because it causes large-scale mortality of fin-fish and shell-fish (mussels, bivalves) worldwide (UNESCO 2018) used as food (Zaitsev and Mamaev 1997). Chronic exposure to milder hypoxia can reduce fish and bivalve populations by interfering with reproduction (Thomas et al 2007, Long et al 2014). In addition, hypoxia can cause fish and crustaceans to migrate away, therefore, can reduce the abundance and diversity of species that are important to food security and local economies (UNESCO 2018). Hypoxia can damage local fishers' catch and communities livelihood as well. The catch per unit effort of the targeted species can either decrease (when the abundance of target species is reduced by hypoxia) or increase (due to aggregation of target species at the edge of the hypoxic zone) (Craig 2012). Nevertheless, hypoxia can reduce fish

[33] ***Pelagic fish*** *= Fish living near the ocean surface or in middle depths.*

[34] ***Demersal fish*** *= Fish living on or near the bottom*

landings and biomass (Wu 2009, de Mutsert et al 2016) impacting local fishers' catch and their livelihoods. For example, the bottom water DO was found significantly correlated with fish biomass and the number of brown and white shrimps abundance in a study carried out in Louisiana coastal waters, USA during 1983 hypoxia. Shrimp (Brown shrimp, *Penaeus aztecus;* white shrimp, *P. setiferus*) and bottom fish (Atlantic croaker, *Micropogonias undulatus*) were generally absent in hypoxic stations (Renaud 1986). The overall economic consequences of hypoxia include reduced production of commercially and recreationally important fin-fish and shell-fish species (Diaz and Rosenberg 1995, Breitburg 2002).

8.3.2.7. Ocean acidification

Hypoxic and anoxic waters are more acidic than average healthy ocean waters. This is due to the fact that the consumption of oxygen by microbes for respiration (that drives hypoxia; see Table. 8.1 Flowchart for hypoxia) results in the production of very high levels of dissolved inorganic carbon (DIC), including CO_2 gas (CO_2 is acidic in nature) (Borges and Gypens 2010). The elevated carbon dioxide would enhance ocean acidification and increase the vulnerability of estuaries and coastal systems and result in to be brittle[35] of calcifying organisms such as molluscs and crustaceans (Howarth et al 2011).

8.3.2.8. Chemicals release and production

Hydrogen sulphide (H_2S) (Wright et al 2012), nitrous oxide (N_2O), and methane (CH_4) production enhanced in hypoxic waters/ sediments (Naqvi et al 2010, Brietburg et al 2018) *[**note:** both CH_4 and N_2O are potent greenhouse gases that cause climate change (GWP^{100} for CH_4 and N_2O are 34 and 298 respectively) (Myhre et al 2013) (**GWP^{100}** =Global Warming Potential for 100 years; CO_2 has a GWP of 1)]*. Hypoxia increases conversion of As(V) (arsenate) to the more toxic As(III) (arsenite) (Li et al 2018). Cadmium (Cd), copper (Cu), and zinc (Zn) form solid sulfide precipitates in the ocean under conditions of anoxic or extremely oxygen-deficient waters where sulphide is present (Janssen et al 2014) (***note:** Cd is a trace nutrient for marine algae*). The phosphorus (P) and iron (Fe) released from the sediments are generally enhanced under anoxic conditions/ O_2 depleted bottom waters (Ingall and Jahnke 1994, Scholz et al 2014) *(**note:** these nutrients (P and Fe) have the potential to further stimulate biological production (of algae) and in marine productivity)*.

---

[35]***Brittle** = liable to break easily.*

8.4. Linkages among algal blooms, hypoxia and climate change

i. Climate change is anticipated to lead to enhanced algal blooms in the future, in particular, global warming would allow the algal bloom to occur sooner and collapse later. For example, cyanobacterial blooms frequency may increase in a warming world since there is a competitive advantage for cyanobacteria to grow better in warmer temperatures (Kibria et al 2013, Kibria 2016). The rising temperatures would favour cyanobacteria since cyanobacteria generally exhibit optimal growth rates at relatively high temperatures, usually in excess of 25°C (Paerl and Huisman 2008). At these elevated temperatures, cyanobacteria compete most effectively with eukaryotic[36] algae, such as diatoms, chlorophytes, and dinoflagellates. Thus, the frequency and intensity of cyanobacterial blooms may enhance in the future due to global warming. This may increase potential exposure of cyanotoxins to aquatic organisms (fish) and humans via water or food.

ii. Global warming would worsen hypoxia/ deoxygenation in the open oceans and coastal waters. There is a link between climate change/ projected rise in temperatures and hypoxia. This is mainly due to three factors: **a.** oxygen is less soluble in higher temperatures so less of it dissolves into the ocean (oxygen solubility is inversely related to temperatures); **b.** aquatic life (fish) consumes more oxygen at higher temperatures because higher temperatures contribute to higher metabolic rates; and **c.** higher temperatures lead to more stratification, which means the more oxygenated surface water doesn't mix well with more hypoxic bottom waters. Therefore, global warming is expected to reduce overall oxygen levels in the oceans resulting in increasing hypoxia (Kibria et al 2013, Breitburg et al 2018). A recent study (Pauly and Cheung 2017) demonstrated that warmer waters would cause fish to shrink in size (may shrink by 20-30% if ocean temperatures continue to climb) due to lack of sufficient oxygen and more demand for oxygen in a warming world.

iii. Global warming may enhance ocean acidification due to the production of very high levels of dissolved inorganic carbon (DIC), including CO_2 gas in a hypoxic environment. The elevated carbon dioxide would enhance ocean acidification (see also section 8.3.2.7).

---

[36]***Eukaryotic*** *= A eukaryote is an organism whose cells contain a nucleus within a membrane. Eukaryotes vary from single-celled organisms to complex* ***multicellular*** *animals and plants. Most living things are eukaryotes, made up of cells with distinct nuclei and chromosomes that contain their DNA.* ***(contrast) Prokaryote*** *= a* ***unicellular organism*** *having cells lacking membrane-bound nuclei; bacteria are the prime example but also included are blue-green algae and actinomycetes and mycoplasma https://www.vocabulary.com/dictionary/procaryote*

8.5. Management of nutrient pollution, algal blooms, algal toxins and hypoxia

There are several steps, strategies, actions and policies that can be undertaken/ implemented to slow down and reverse nutrient pollution and its impacts. Some of the measures are listed below but not limited to:

8.5.1. Managing algal blooms and algal toxins

- Avoid over fertilisation of croplands (Kibria et al 2019);
- Minimise erosion and nutrient run-off (Kibria et al 2019);
- Improve wastewater treatment plants to filter out nutrients and to reduce river and coastal nutrient load (Kibria et al 2019);
- Apply fertilisers (N, P) when it is needed most during the crops growing season (Nagatia et al 2019);
- Apply the right quantity of fertilisers (Nagatia et al 2019);
- Construct artificial wetlands to receive and contain stormwater run-off into the natural waterways and to act as N, P fertilisers sink (Nagatia et al 2019);
- Construct lagoons/ponds to hold animal waste (urine, faces) and fertilisers (Nagatia et al 2019);
- Control livestock (manure) and other animal wastes, septic systems and sewage treatment effluents (Kibria et al 2019);
- Control/limit/ or contain effluents discharge from agriculture and aquaculture facilities into natural waterways (Kibria et al 2019);
- Educate and build the capacity of the community about the risks and impacts of nutrient pollution, algal blooms, algal toxins (Kibria et al 2019); and
- Monitor cyanobacterial and marine algal blooms and algal toxins levels in water and foods and compare with the guideline threshold values (Kibria et al 2019) (Table 8.5).

Table 8.5. Regulatory limits/ action levels of freshwater and marine algal toxins in drinking water and seafood *[na = data is not available].*

Toxins	**Drinking water** (Burch 2008)	**Seafood**
Freshwater algal toxins		
Anatoxin-a	6.0 µg/L (Burch 2008)	na
Anatoxin-a(s)	1.0 µg/L	na

Toxins	Drinking water (Burch 2008)	Seafood
Microcystin (MC-LR)	1.0 µg/L	Tolerance level: 80 µg STX eq/100 g mussel meat (IPCS 1984, WHO 2006).
Saxitoxin	3 µg /L	Na
Nodularin	1.0 µg/L	Health alert level: Fin-fish: 214 µg/kg food; Prawn: 943 µg/kg food; Mussels: 1,286 µg/kg food (van Buynder et al 2001).
Cylindrospermopsin	15 µg/L	Health alert level: Fin-fish: 158 µg/kg food; Prawn: 720 µg/kg food; Mussels: 933 µg/kg food (Saker and Eaglesham 1999).
Marine algal toxins		
Amnesic shell-fish poisoning (ASP)	na	2,000 µg/100 g shell-fish flesh (Garthwaite 2000).
Diarrhoetic shell-fish poisoning (DSP)	na	16-20 µg/100 g shell-fish flesh (Garthwaite 2000).
Neurotoxic shell-fish poisoning (NSP)	na	80 µg/100 g shell-fish flesh (Garthwaite 2000).
Paralytic shell-fish poisoning (PSP)	na	80 µg/100 g shell-fish flesh (Garthwaite 2000).
Ciguatera fish poisoning (CFP)	na	3.5 µg/100 g fish flesh (Garthwaite 2000).

8.5.2. Managing hypoxia (to reduce impacts of hypoxia on aquatic biodiversity, the following strategies can be implemented)

- Reduce greenhouse gas emissions to reduce deoxygenation (Kibria et al 2019);
- Reduce fishing pressure on hypoxia-intolerant species (Breitburg et al 2018, UNESCO 2018);
- Create marine protected areas (MPAs) and no-catch zones in well-oxygenated areas that can serve as refugia to protect populations when oxygen is low (Breitburg et al 2018, UNESCO 2018);

- Carryout research on the effects of low dissolved oxygen on fisheries production, seafood industries and livelihoods;
- Use those fishing gears, that minimise additional stress on oxygen-impacted fish stocks and ecosystems (Breitburg et al 2018, UNESCO 2018);
- Improve the management of fisheries and marine habitats that are sensitive to low oxygen (hypoxia) conditions to protect economies, livelihoods, and food security (Breitburg et al 2018, UNESCO 2018);
- Create protected areas (viz. MPAs) that can serve as refugia in hypoxic areas or during hypoxic events (Breitburg et al 2018, UNESCO 2018);
- Exert local, national, and global efforts to limit further oxygen declines, restore oxygen levels to previously well-oxygenated values and enhance the resilience of ecosystems affected by deoxygenation (Breitburg et al 2018, UNESCO 2018);
- Educate and build the capacity of the community about the risks and impacts of hypoxia on fisheries and seafood security (Kibria et al 2019); and
- Monitor DO levels in waterways (coastal waters, estuaries, open ocean) (Kibria et al 2019) and take consequent proper actions (Fig. 8.6).

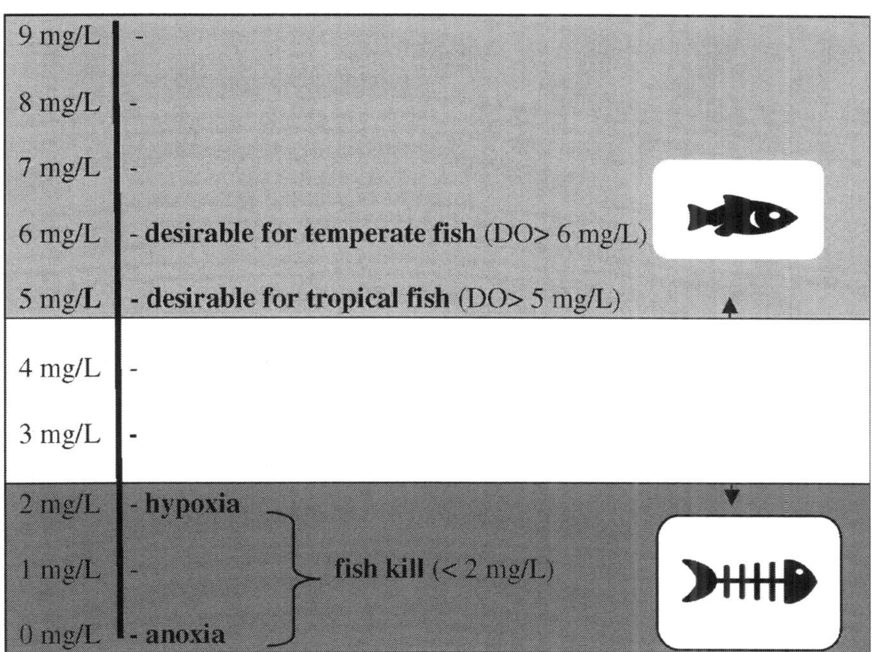

Fig 8.6. Dissolved oxygen (DO) level, ecological hypoxia and anoxia and fish kill (Kibria 2014).

8.6. Conclusion

Nutrient pollution causes or stimulates to form algal blooms, algal toxins and hypoxia in freshwater lakes, coastal and open oceans. Both algal bloom and hypoxia are significant threats to ecosystems, water quality, irrigation, biodiversity, seafood security, public health, and local economies. Water with blue-green algae/ cyanobacterial bloom is unsuitable for agriculture, drinking water, livestock drinking, irrigation, wildlife and recreational activities. Algal toxins can be accumulated in crops and seafood and can be transferred to humans via the food and water. Hypoxia can cause habitat compression and reduce survival and growth, impair reproduction and mass mortality of demersal fishes and benthos, other seafood organisms resulting in loss of fisheries resources. Livelihoods of people associated with fisheries, food industries and tourism can be threatened (due to poorer water quality, mass mortality of fish, shell-fish and corals and forced movement of fish into other areas as a consequence of algal blooms and hypoxia). Apart from nutrients, climate change/global warming is predicted to exacerbate oxygen depletion in many nutrient-enriched coastal systems through decreased oxygen solubility, increased intensity and duration of stratification and accelerated biological respiration in a warming world. Therefore, all-out efforts are required to reduce and minimise the causes of nutrient pollution and inclusion of climate change as an additional stressor on hypoxia. Avoiding over-fertilisation of croplands; minimising erosion and nutrient run-off; controlling livestock (manure) and other animal wastes, septic systems and sewage treatment effluents entering into waterways; and monitoring DO levels in waterways, monitoring cyanobacterial and marine (PSP) toxins levels in water and foods, and educating the community about the risk and impact of hypoxia and algal toxins are some of the strategies that can be implemented to reduce threats and risks of nutrient pollution on the environment, biodiversity and public health.

8.7. References

Altieri AH, Harrison SB, Seemann J et al 2017 Tropical dead zones and mass mortalities on coral reefs. *Proceedings National Academy of Sciences of the U.S.A.* **114** 3660–3665. *doi: 10.1073/pnas.1621517114; pmid: 28320966*

Azevedo SMFO, Carmichael WW, Jochimsen EM et al 2002 Human intoxication by microcystins during renal dialysis treatment in Caruaru-Brazil. *Toxicology* **181-182** 441-446

Babica P, Bláha L and Maršálek B 2006 Exploring the natural role of microcystins - A review. *Journal of Phycology* **42** 9-20

Berry J, Lee E, Walton K et al 2011 Bioaccumulation of microcystins by fish associated with a persistent cyanobacterial bloom in Lago de Patzcuaro (Michoacan, Mexico). *Environmental Toxicology & Chemistry* **30(7)** 1621-1628

BNI (Baltic Nest Institute) 2014 Dead zones have increased by more than 10-fold in the last century.

http://www.balticnest.org/balticnest/activities/news/news/deadzoneshaveincreasedbymorethan10foldinthelastcentury.5.3186f824143d05551ad52e7.html

Borges AV and Gypens N 2010 Carbonate chemistry in the coastal zone responds more strongly to eutrophication than to ocean acidification. *Limnology & Oceanography* **55** 3463–53

Boyer EW, Howarth RW, Galloway JN et al 2006 Riverine nitrogen export from the continents to the coasts. *Global Biogeochemical Cycles* **20(1)** 1-91

doi: 10.1029/2005GB002537. https://agupubs.onlinelibrary.wiley.com/doi/10.1029/2005GB002537

Boyer EW and Howarth RW 2008 Nitrogen Fluxes from Rivers to the Coastal Oceans. Chapter 36. p. 1565-1587. *In*: Capone DG, Bronk DA, Mulholland MR, Carpenter, EJ (Eds.) Nitrogen in the Marine Environment (2nd edition). Elsevier. ISBN: 978-0-12-372522-6

Breitburg DL 2002 Effects of hypoxia, and the balance between hypoxia and enrichment, on coastal fishes and fisheries. *Estuaries* **25(4)** 767–781

Breitburg D, Loher T, Pacey CA and Gerstein A 1997 Varying Effects of Low Dissolved Oxygen on Trophic Interactions in an Estuarine Food Web. *Ecological Monographs* **67(4)**489-507;

https://doi.org/10.1890/0012-9615(1997)067[0489:VEOLDO]2.0.CO;2

Breitburg D, Adamack A, Rose KA et al 2003 "The pattern and influence of low dissolved oxygen in the Patuxent River, a seasonally hypoxic estuary". *Estuaries* **26(2A)** 280–297

Breitburg D, Levin LA, Oschlies A et al 2018 Declining oxygen in the global ocean and coastal waters. *Science* **359(6371)** 1-11 *DOI: 10.1126/science.aam7240*

http://science.sciencemag.org/content/sci/359/6371/eaam7240.full.pdf

Vol. 359, Issue 6371,

Burch MD 2008 Effective doses, guidelines and regulations. In: Hudnell, H.K (editor). Cyanobacterial harmful algal blooms: State of the science and research needs. *Advances in Experimental Medicine and Biology* **619** chapter 36. Springer Press, New York, 831-853 p.

Burr GS, Wolters WR, Schrader KK, Summerfelt ST 2012 Impact of depuration of earthy-musty off-flavors on fillet quality of Atlantic salmon, *Salmo salar,* cultured in a recirculating aquaculture system. *Aquacultural Engineering* **50** 28–36

Butler N, Carlisle JC, Linville R, Barbara Washburn B 2009 Microcystins: A brief overview of their toxicity and effects, with special reference to fish, wildlife, and livestock. Office of Environmental Health Hazard Assessment. Ecotoxicology Program, Integrated Risk Assessment Branch. California Environmental Protection Agency.

https://oehha.ca.gov/media/downloads/ecotoxicology/document/microcystin031209.pdf

Carmichael WW, Mahmood NA and Hyde EG 1990 Natural toxins from cyanobacteria (blue-green algae). *In*: Hall S and Strichartz G (Eds.) Marine Toxins, Origin, Structure and Molecular Pharmacology, vol. 418. *American Chemical Society, Washington, D.C*, **418** 87–106

Carstensen, J, Andersen, JH, Gustafsson et al 2014 Deoxygenation of the Baltic Sea during the last century. *PNAS* **111(15)** 5628-5633

https://doi.org/10.1073/pnas.1323156111

Casini M, Käll F, Hansson M et al 2016 Hypoxic areas, density-dependence and food limitation drive the body condition of a heavily exploited marine fish predator. *Royal Society Open Science* **3** 160416 *doi: 10.1098/rsos.160416; pmid: 27853557*

Chemol J and Toan T 1969 Paralyzing neuromuscular activity of saxitoxin in rats. *Therapie* **24(1)** 191-198

Chen JZ, Song LR, Dai E et al 2004 Effects of microcystins on the growth and the activity of superoxide dismutase and peroxidase of rape (*Brassica napus* L.) and rice (*Oryza sativa* L.). *Toxicon* **43(4)** 393-400

Chen W, Song L, Gan N and Li L 2006 Sorption, Degradation and mobility of microcystins in Chinese agriculture soils: risk assessment for groundwater protection, *Environmental Pollution* **144** 752–758

Chesney EJ and House ED 1989 Laboratory studies on the effect of hypoxic waters on the survival of eggs and yolksac larvae of the bay anchovy, *Anchoa mitchilli,* p. 184—191. *In:* Houde ED, Chesney EJ, Newberger TA, Vazquez AV, Zastrow CE, Morin LG, Harvey HR, and Gooch JW, (eds.) Population Biology of Bay Anchovy in Mid-Chesapeake Bay. Final Report to Maryland Sea Grant Ref. No. (UMCEES) CBL 89-141. Solomons, Maryland.

Chislock MF, Doster E, Zitomer RA and Wilson AE 2013 Eutrophication: Causes, Consequences, and Controls in Aquatic Ecosystems. *Nature Education Knowledge* **4(4)** 10.

https://www.nature.com/scitable/knowledge/library/eutrophication-causes-consequences-and-controls-in-aquatic-102364466

Chiswell RK, Shaw GR, Eaglesham G, Smith MJ et al 1999 Stability of Cylindrospermopsin, the Toxin from the Cyanobacterium, *Cylindrospermopsis raciborskii:* Effect of pH, Temperature, and Sunlight on Decomposition. *Environmental Toxicology* **14(1)** 155-161

Chorus I and Bartram J (eds.) 1999 Toxic cyanobacteria in water. A guide to their public health consequences, monitoring and management. WHO, E and FN Spon Press, London. 416 p.

Craig J K 2012 Aggregation on the edge: Effects of hypoxia avoidance on the spatial distribution of shrimp and demersal fishes in the northern Gulf of Mexico. *Marine Ecology Progress Series* **445** 75–95

Crush JR, Briggs R and Sprosen SN 2008. Effect of irrigation with lake water containing microcystin content and growth of rye grass, clover, rape, and lettuce. *Environmental Toxicology* **23(2)** 246-252

Cumming H and Ru¨cker C 2017 Octanol–Water partition coefficient measurement by a simple HNMR method.

American Chemical Society Omega **2** 6244–6249 *DOI: 10.1021/acsomega.7b01102*

de Figueiredo DR, Azeiteiro UM, Estves SM et al 2004 Microcystin –producing blooms - a serious global public health issue. *Ecotoxicology and Environmental Safety* **59** 151-163

DeMott WR, Zhang QX and Carmichael WW 1991 Effects of toxic cyanobacteria and purified toxins on the survival and feeding of a copepod and three species of Daphnia. *Limnology & Oceanography* **36(7)** 1346-1357

de Mutsert K, Steenbeek J, Lewis K et al 2016 Exploring effects of hypoxia on fish and fisheries in the northern Gulf of Mexico using a dynamic spatially explicit ecosystem model. *Ecological Modelling* **331** 142-150

Diaz RJ and Rosenberg R 1995 Marine benthic hypoxia: A review of its ecological effects and the behavioural responses of benthic macrofauna. *Oceanography & Marine Biology: Annual Review* **33** 245-303

Diaz RJ and Rosenberg R 2008 Spreading Dead Zones and Consequences for Marine Ecosystems. *Science* **321** 926-929. *DOI: 10.1126/science.1156401*

Duy TN, Lam PKS, Shaw GR and Connell DW 2000 Toxicology and risk assessment of freshwater cyanobacteria (blue-green algal) toxins in water. *Reviews of Environmental Contamination and Toxicology* **163** 113-186

Eby LA and Crowder LB 2002 Hypoxia-based habitat compression in the Neuse River Estuary: Context-dependent shifts in behavioral avoidance thresholds. *Canadian Journal of Fisheries and Aquatic Science* **59** 952–965. *doi: 10.1139/f02-067*

Falconer IR and Humpage AR 2005 Health Risk Assessment of Cyanobacterial (Blue-green Algal) Toxins in Drinking Water. *International Journal of Environmental Research and Public Health* **2(1)** 43–50

FAO (Food and Agricultural Organization of the United Nations) 2004. The marine biotoxins. Food and Agriculture Organization of the United Nations. FAO, Food and Nutrition Paper 80. *http://www.fao.org/3/a-y5486e.pdf*

Fisher JM, Reese JG, Pellechia PJ et al 2006 Role of Fe(III), phosphate, dissolved organic matter, and nitrate during the photodegradation of domoic acid in the marine environment. *Environmental Science & Technology* **40** 2200-2205

Franke C, Studinger G, Berger G et al 1994 The assessment of bioaccumulation. *Chemosphere* **29** 1501-1514

FSANZ (Food Standard Australia and New Zealand) 2019 Toxins in seafood.

http://www.foodstandards.gov.au/consumer/safety/foodborne-illness/Pages/Toxins-in-seafood.aspx

Garthwaite I 2000 Keeping shellfish safe to eat: brief review of shellfish toxins, and methods for their detection. *Trends in Food Science and Technology* **11** 235-244

Gehrke PC, Revell MB, Philbey AW 1993 Effects of river red gum, *Eucalyptus camaldulensis*, litter on golden perch, Macquaria ambigua. *Journal of Fish Biology* **43** 265–279

Grimmelt B, Nijjar MS, Brown J et al 1990 Relationship between domoic acid levels in the blue mussel (*Mytilus edulis*) and toxicity in mice. *Toxicon* **28(5)** 501-508

Harada K-I, Tsuji K, Watanabe MF 1996 Stability of microcystins from cyanobacteria. III. Effect of pH and temperature *Phycologia* **35(6)** 83-88

Heresztyn T and Nicholson BC 1997 Nodularin concentrations in Lakes Alexandrina and Albert, South Australia, during a bloom of the cyanobacterium (blue-green alga) *Nodularia spumigena* and degradation of the toxin *Environmental Toxicology and Water Quality* **12(4)** 273-282

Herrera NA, Echeverri LF, Ferr~ao-Filho AS 2015 Effects of phytoplankton extracts containing the toxin microcystin-LR on the survival and reproduction of cladocerans. *Toxicon* **95** 38-45

Howarth R, Chan F, Conley DJ et al 2011 Coupled biogeochemical cycles: Eutrophication and hypoxia in temperate estuaries and coastal marine ecosystems. *Frontiers in Ecology and the Environment* **9(1)** 18-26

Hudnell HK 2005 Chronic biotoxin-associated illness; Multiple-system symptoms, a vision deficit and effective treatment. *Neurotoxicology and Teratology* **27** 733-743

Hudnell HK 2010 The State of US freshwater harmful algal blooms assessments, policy and legislation. *Toxicon* **55** 1024–1034. *doi:10.1016/j.toxicon.2009.07.021*

IARC (International Agency for Cancer Research) 2019 Agents Classified by the IARC Monographs, Volumes 1–123. *https://monographs.iarc.fr/wp-content/uploads/2018/09/ClassificationsAlphaOrder.pdf*

Ibelings BW and Chorus I 2007 Accumulation of cyanobacterial toxins I. freshwater "seafood" and its consequences for public health: A review. *Environmental Pollution* **150** 177-192

Ibelings BW and Havens KE 2008 Cyanobacterial toxins: a qualitative meta-analysis of concentrations, dosage and effects in freshwater, estuarine and marine biota. *In*: Hudnell, HK (ed.). Cyanobacterial harmful algal blooms: State of the science and research needs. *Advances in Experimental Medicine & Biology* **619** chapter 12. Springer Press, New York, 675-732 p.

ILS (Integrated Laboratory Systems) 2000 Cylindrospermopsin [CASRN 143545-90-8] Review of toxicological literature.

https://ntp.niehs.nih.gov/ntp/htdocs/chem_background/exsumpdf/cylindrospermopsin_508.pdf

IPCS (International Programme on Chemical Safety) 1984 Aquatic Marine and Freshwater Biotoxins. Environmental Health Criteria 37, International Programme on Chemical Safety, World Health Organization, Geneva

Isensee K, Levin LA, Breitburg D et al 2016 p. 20-32; *https://ocean-climate.org/wp-content/uploads/2020/02/4.-The-Ocean-is-losing-its-breath-Scientific-notes-2016.pdf*

Ingall E and Jahnke R 1994 Evidence for enhanced phosphorus regeneration from marine sediments overlain by oxygen depleted waters. *Geochimica* et *Cosmochimica Acta* **58** 2571–2575. *doi: 10.1016/0016-7037(94)90033-7*

Janssen DJ, Conway TM, John SG et al 2014 Undocumented water column sink for cadmium in open ocean oxygen-deficient zones. *Proceedings of the National Academy of Sciences of the USA* **111** 6888–6893. *doi: 10.1073/ pnas.1402388111; pmid: 24778239*

Jenny JP, Francus P, Normandeau A et al 2016 Global spread of hypoxia in freshwater ecosystems during the last three centuries is caused by rising local human pressure. *Global Change Biology* **22** 1481–1489. *doi: 10.1111/gcb.13193*

Jia X, Cai C, Wang J and Zhang H 2014 Endocrine-disrupting effects and reproductive toxicity of low dose MCLR on male frogs (*Rana nigromaculata*) in vivo. *Aquatic Toxicology* **155** 24-31

Jones GJ and Negri AP 1997 Persistence and degradation of cyanobacterial paralytic shellfish poisons (PSPs) in freshwaters. *Water Research* **31** 525-533

Karjalainen M, Reinikainen M, Lindvall F et al 2003 Uptake and accumulation of dissolved, radiolabelled nodularin in Baltic Sea zooplankton. *Environmental Toxicology* **18(1)** 52-60

Kinnear S 2010 Cylindrospermopsin: A Decade of Progress on Bioaccumulation. Research. *Marine Drugs* **8** 542-564 *doi: 10.3390/md8030542*

Kibria G, Nugegoda D, Fairclough R and Lam P 1997a Pollution from aquaculture. *Chemistry in Australia*. Jan/Feb. p. 19-20 *DOI: 10.13140/RG.2.1.1193.5206*

https://www.researchgate.net/publication/261216729_Pollution_from_Aquaculture

Kibria G, Nugegoda D, Fairclough R and Lam P 1997b Aspects of phosphorus pollution from aquaculture. *Naga-ICLARM Quarterly*. p. 20-24 *DOI: 10.13140/2.1.3762.5284*

Kibria G, Nugegoda D, Fairclough R and Lam P 1998 Can nitrogen pollution from aquaculture be reduced? *Naga-ICLARM Quarterly*. p. 17-25. *https://www.researchgate.net/publication/309571115_Can_nitrogen_pollution_from_aquaculture_be_reduced*

Kibria G, Haroon AKY, Nugegoda D and Rose G 2010 Climate change and chemicals: Environmental Biological aspects. New India Publishing Agency. New Delhi. 470 p. ISBN: 9789380235301 *DOI: 10.13140/2.1.4384.0963;*

https://www.researchgate.net/publication/261216635_Climate_Change_and_Chemicals_Environmental_and_Biological_Aspects;
https://www.researchgate.net/publication/331275314_Nutrient_Pollution_Causing_Algal_Blooms_Hypoxia_Dead_Zones_Across_the_Globe_Impacting_Ecosystems_Biodiversity_Public_Health_Livelihoods

Kibria G, Haroon AKY and Nugegoda D 2013 Climate change and Agricultural food production: Impacts, vulnerabilities and remedies. New India Publishing Agency (NIPA), New Delhi, India, 300 p. ISBN: 9789381450512 *DOI: 10.13140/2.1.3245.4081*

https://www.researchgate.net/publication/261178054_Climate_Change_and_Agricultural_Food_Production_Impacts_Vulnerabil ities_and_Remedies

Kibria G 2014 Global fish Kills: Causes and Consequences *DOI: 10.13140/RG.2.1.1422.8965*

https://www.researchgate.net/publication/261216309_Global_fish_Kills_Causes_and_Consequences

Kibria G 2016 Blue-green algal/ cyanobacterial blooms (BGA), climate change and BGA impacts on water quality, fish kills, recreation, crops, seafood, livestock, wild animals and humans. 7 p. *DOI: 10.13140/RG.2.1.1306.9765/2*; *http://www.worldfishcenter.org/Naga/na_2242.pdf*

Kibria G, Haroon AKY and Nugegoda D 2016 Climate Change and Water security. Impacts, future projections, adaptations and mitigations. New India Publishing Agency (NIPA), New Delhi, India, 330 p, ISBN: 978-93-85516-26-9. *https://www.researchgate.net/publication/303218799_Climate_Change_Water_Security*

https://www.researchgate.net/publication/267864673_Blue-green_algalcyanobacterial_blooms_BGA_climate_change_and_BGA_impacts_on_water_quality_fish_kills_recreation_crops_seafood_livestock_wild_animals_and_humans/stats

Kibria G, Haroon AKY, Nugegoda D and Hossain MMM 2019 Nutrient pollution causing algal blooms, hypoxia (Dead Zones) across the globe impacting ecosystems, biodiversity, public health and livelihoods *DOI: 10.13140/RG.2.2.28654.97609*

Landman MJ, Van den Heuvel MR and Ling N 2005 Relative sensitivities of common freshwater fish and invertebrates to acute hypoxia. *New Zealand Journal of Marine and Freshwater Research* **39** 1061–1067

Le C, Zha Y, Li Y et al 2010 Eutrophication of Lake waters in China: Cost, causes, and control. *Environmental Management* **45** 662-668

Lehane L and Lewis R 2000 Ciguatera: recent advances but the risk remains-review. *International Journal of Food Microbiology* **61** 91-125

Li L, Pohl C, Ren JL et al 2018 Revisiting the biogeochemistry of arsenic in the Baltic Sea: Impact of anthropogenic activity. *Science of the Total Environment* **613-614** 557-568 *doi: 10.1016/j.scitotenv.2017.09.029. Epub 2017 Sep 26*

Long WC, Seitz RD, Brylawski BJ and Lipcius RN 2014 Individual, population, and ecosystem effects of hypoxia on a dominant benthic bivalve in Chesapeake Bay. *Ecological Monographs* **84** 303-327

Malbrouck C and Kestemont P 2006 Effects of microcystins on fish. *Environmental Toxicology & Chemistry* **25(1)** 72-86

Manganelli M, Scardala S, Stefanelli M, Palazzo F, Funari E, Vichi S et al. 2012 Emerging health issues of cyanobacterial blooms. *Ann. Ist Super Sanita* **48(4)** 415–28

McElhiney J, Lawton LA and Leifert C 2001 Investigations into the inhibitory effects of microcystins on plant growth, and the toxicity of plant tissues following exposure. *Toxicon* **39** 1411-1420

Miller DC, Poucher S and Coiro L 2002 Determination of lethal dissolved oxygen levels for selected marine and estuarine fishes, crustaceans, and a bivalve. *Marine Biology* **140(2)** 287-296

Mekonnen MM and Hoekstra AY 2018 Global anthropogenic phosphorus loads to freshwater and associated grey water footprints and water pollution levels: A high-resolution global study. *Water Resources Research* **54** 345–358

MSDS (Material Safety Data Sheet) Caymanchem 2015 Nodularin.

https://www.caymanchem.com/msdss/10007190m.pdf

MSDS (Material Safety Data Sheet) Caymanchem 2017 Microcystin L-R.

https://www.caymanchem.com/msdss/10007188m.pdf

MSDS (Material Safety Data Sheet) Merck 2006 Microcystin-LR, *Microcystis aeruginosa. file:///C:/Users/User1/Downloads/475815_SDS_AU_EN%20(1).PDF*

MSDS (Material Safety Data Sheet) PubChem 2005a (updated 2020) Microcystins-LR. U.S. National Library of Medicine. National Center for Biotechnology Information.

https://pubchem.ncbi.nlm.nih.gov/compound/Microcystin-LR

MSDS (Material Safety Data Sheet) PubChem 2005b (updated 2020). Domoic acid. U.S. National Library of Medicine. National Center for Biotechnology Information.

https://pubchem.ncbi.nlm.nih.gov/compound/Domoic-acid

MSDS (Material Safety Data Sheet) PubChem 2005c (updated 2020). Okadaic acid. U.S. National Library of Medicine. National Center for Biotechnology Information.

https://pubchem.ncbi.nlm.nih.gov/compound/Okadaic-acid

MSDS (Material Safety Data Sheet) PubChem 2006a (updated 2020). Brevetoxin-A. U.S. National Library of Medicine. National Center for Biotechnology Information.

https://pubchem.ncbi.nlm.nih.gov/compound/Brevetoxin-A

MSDS (Material Safety Data Sheet) PubChem 2006b (updated 2020). Ciguatoxin. U.S. National Library of Medicine. National Center for Biotechnology Information.

https://pubchem.ncbi.nlm.nih.gov/compound/Domoic-acid

MSDS (Material Safety Data Sheet) PubChem 2008a Anatoxin-A. U.S. National Library of Medicine. National Center for Biotechnology Information.

https://pubchem.ncbi.nlm.nih.gov/compound/Anatoxin-A

MSDS (Material Safety Data Sheet) PubChem 2008b Anatoxin-A(s) U.S. National Library of Medicine. National Center for Biotechnology Information.

https://pubchem.ncbi.nlm.nih.gov/compound/Anatoxin-a_s

MSDS (Material Safety Data Sheet) PubChem 2009a (updated 2015) Nodularin. U.S. National Library of Medicine. National Center for Biotechnology Information.

https://pubchem.ncbi.nlm.nih.gov/compound/Nodularin

MSDS (Material Safety Data Sheet) PubChem 2009b (updated 2015) Cylindrospermopsin. U.S. National Library of Medicine. National Center for Biotechnology Information.

https://pubchem.ncbi.nlm.nih.gov/compound/7-Epi-cylindrospermopsin

MSDS (Material Safety Data Sheet) PubChem 2012 Saxitoxin. U.S. National Library of Medicine. National Center for Biotechnology Information.

https://pubchem.ncbi.nlm.nih.gov/compound/Saxitoxin

Myhre G, Shindell D, Bréon G-M et al 2013 Anthropogenic and natural radiative forcing. In: Climate Change 2013: The Physical Science Basis. Contribution of Working Group I to the Fifth Assessment Report of the Intergovernmental Panel on Climate Change. *In*: Stocker TF, Qin D, Plattner GK, Tignor M, Allen SK, Boschung J, Nauels A, Xia Y, Bex V and Midgley PM (eds.). Cambridge University Press, Cambridge, United Kingdom and New York, NY, USA.

Ngatia L, Grace JM, Moriasi D and Taylor R 2019 Nitrogen and phosphorus eutrophication in marine Ecosystems, monitoring of marine pollution, Houma Bachari Fouzia, IntechOpen, *DOI: 10.5772/intechopen.81869.* Available from *https://www.intechopen.com/books/monitoring-of-marine-pollution/nitrogen-and-phosphorus-eutrophication-in-marine-ecosystems*

Nilsson HC and Rosenberg R 1994 Hypoxic response of two marine benthic communities. *Marine Ecology Progress Series* **115(3)** 209-217 *DOI: 10.3354/meps115209*

NRC (National Research Council) 2000 Clean Coastal Waters: Understanding and Reducing the Effects of Nutrient Pollution. National Academies Press, Washington, DC.

Naqvi SWA, Bange HW, Farías L et al 2010 Marine hypoxia/anoxia as a source of CH_4 and N_2O. *Biogeosciences* **7** 2159–2190 *doi: 10.5194/ bg-7-2159-2010*

NOAA (National Oceanic and Atmospheric Administration) 2017 Gulf of Mexico 'dead zone' is the largest ever measured. *http://www.noaa.gov/media-release/gulf-of-mexico-dead-zone-is-largest-ever-measured*

OECD 2012 Test No. 305: Bioaccumulation in Fish: Aqueous and Dietary Exposure, OECD Guidelines for the Testing of Chemicals, Section 3, O.P. 2012. OECD guidelines for testing of chemicals. Bioaccumulation in Fish: Aqueous and Dietary Exposure. *doi: 10. 1787/2074577x; https://www.oecd.org/env/test-no-305-bioaccumulation-in-fish-aqueous-and-dietary-exposure-9789264185296-en.htm*

Opperhuizen A and Sijm DTHM 1990 Bioconcentration and bioaccumulation of polychlorinated dibenzo-p-dioxins and dibenzofurans in fish. *Environmental Toxicology and Chemistry* **9** 175-186

Osswald J, Azevedo J, Vasconcelos V and Guilhermino L 2011 Experimental determination of the bioconcentration factors for anatoxin-a in juvenile rainbow trout (*Oncorhynchus mykiss*). *Proceedings of the International Academy of Ecology and Environmental Sciences* **1(2)** 77-86

Paerl HW and Huisman J 2008 Blooms like it hot. *Science* **320** 57–58 *DOI: 10.1126/science.1155398*

Paerl HW and Piehler MF 2009 Nitrogen and Marine Eutrophication. Chapter 11, p. 529-567. *In*: Capone DG, Bronk DA, Mulholland MR, Carpenter EJ (Eds.) Nitrogen in the Marine Environment (2nd edition). Elsevier. ISBN: 978-0-12-372522-6

Pauly D and Cheung WWL 2017 Sound physiological knowledge and principles in modeling shrinking of fishes under climate change. *Global Change Biology* **24** e15–e26 *https://doi.org/10.1111/gcb.13831*

Pihl L, Baden SP, Diaz RJ and Schaffner LC 1992 Hypoxia-induced structural changes in the diet of bottom-feeding fish and crustacea. *Marine Biology* **112** 349–361

Poste AE, Hecky RE and Guildford SJ 2011 Evaluating Microcystin Exposure Risk through Fish Consumption. *Environmental Science & Technology* **45** 5806–5811

Rabalais NN, Harper Jr., DE, Turner RE 2001 Responses of nekton and demersal and benthic fauna to decreasing oxygen concentrations. *In*: Rabalais NN and Turner RE (eds.), *Coastal and estuarine studies: coastal hypoxia consequences for living resources and ecosystems*. American Geophysical Union, Washington D C, 115–128

Rabalais NN, Turner RE and Wiseman WJ 2002 Gulf of Mexico hypoxia, aka "The dead zone". *Annual Review of Ecology, Evolution & Systematics* **33** 235–263

Renaud ML 1986 Hypoxia in Louisiana coastal waters during 1983: Implications for fisheries. *Fishery Bulletin* **84** 9–26

Saker ML and Eaglesham GK 1999 The accumulation of cylindrospermopsin from the cyanobacterium *Cylindrospermopsis raciborskii* in tissue of the Redclaw crayfish, *Cherax quadricarinatus*. *Toxicon* **37** 1065-1077

Scholz F, McManus J, Mix AC et al 2014 The impact of ocean deoxygenation on iron release from continental margin sediments. *Nature Geoscience* **7** 433–437 *doi: 10.1038/ngeo2162*

Seifert M 2007 The ecological effects of the cyanobacterial toxin cylindrospermopsin. Ph.D Thesis. School of Medicine. The University of Queensland; Brisbane, Australia *https://espace.library.uq.edu.au/view/UQ:158799*

Service RF 2004 Oceanography. New dead zone off Oregon coast hints at sea change in currents. *Science* **305** 1099–1099

Shoji U, Masuda R, Yamashita Y and Tanaka, M 2005 Predation on fish larvae by moon jellyfish, *Aurelia auritaunder* low dissolved oxygen concentrations. *Fisheries Science* **71(4)** 748–753

Small K, Kopf RK, Watts RJ, and Howitt J 2014 Hypoxia, Blackwater and Fish Kills: Experimental Lethal Oxygen thresholds in Juvenile Predatory Lowland River Fishes. *PLoS ONE* **9(4)** e94524

Doi:10.1371/journal.pone.0094524

Smale MA and Rabeni CF 1995 Hypoxia and hyperthermia tolerances of headwater stream fishes. *Transactions of the American Fisheries Society* **124** 698–710

Small K, Kopf RK, Watts RJ and Howitt J 2014 Hypoxia, blackwater and fish Kills: Experimental lethal oxygen thresholds in juvenile predatory lowland River fishes. *PLoS ONE* 9(**4**) e94524 *doi:10.1371/journal.pone.0094524*

Smith C and Shutton A 1993 The persistence of anatoxin-a in reservoir water. Foundation for Water Research, UK Report No.FR0427

Smith JL, Boyer GL, Zimba PV 2008 A review of cyanobacterial odours and bioactive metabolites: impacts and management alternatives in aquaculture. *Aquaculture* **280** 5-20

Stewart I, Seawright AA and Shaw GW 2008 Cyanobacterial poisoning in livestock, wild mammals and birds-an overview. In: Hudnell, HK (editor). Cyanobacterial harmful algal blooms: State of the science and research needs. *Advances in Experimental Medicine & Biology* **619**: chapter **28**. Springer Press, New York, 613-637 p.

Stramma L, Prince ED, Schmidtko S et al 2012 Expansion of oxygen minimum zones may reduce available habitat for tropical pelagic fishes. *Nature Climate Change* 2 33-37

T3DB (The Toxin and Toxin Target Database) 2014 Anatoxin-a(s) (T3D4138). *http://www.t3db.ca/toxins/T3D4138*

Takahashi S. and Kaya K 1993 Quail spleen is enlarged by microcystin RR as a blue-green algal hepatotoxin. *Natural Toxins* **1(5)** 283-285

Tsuji K, Watanuki T, Kondo F et al 1995 Stability of microcystins from cyanobacteria--II. Effect of UV light on decomposition and isomerization. *Toxicon* **33(12)** 1619-1631

Thomas P, Rahman MS, Khan IA and Kummer JA 2007 Widespread endocrine disruption and reproductive impairment in an estuarine fish population exposed to seasonal hypoxia. *Proceedings of the Royal Society of London B: Biological Sciences* **274** 2693-2702

US EPA 2012 Estimation Program Interface (EPI) Suite. Ver. 4.11. November 2012.

https://www.epa.gov/tsca-screening-tools/epi-suitetm-estimation-program-interface

UNESCO (United Nations Educational, Scientific and Cultural Organization) 2018 Global Ocean Oxygen Network 2018. The ocean is losing its breath: Declining oxygen in the world's ocean and coastal waters. *In*: Breitburg D, Grégoire M and Isensee K (eds.) IOC-UNESCO, IOC Technical Series, No. 137 40 p. *https://unesdoc.unesco.org/ark:/48223/pf0000265196*

van Apeldoorn ME, van Egmond HP, Gerrit JAS and Bakker GJI 2007 Review. Toxins in cyanobacteria. *Molecular Nutrition & Food research* **51** 7-60

van Buynder PG, Oughtred T, Kirkby B et al 2001 Nodularin uptake by seafood during a cyanobacterial bloom. *Environmental Toxicology* **16** 468–471

Van der Oost R, Beyer J and Vermeulen NPE 2003 Fish bioaccumulation and biomarkers in environmental risk assessment: a review. *Environmental Toxicology and Pharmacology* **13** 57-149

Vaquer-Sunyer R and Duarte CM 2008 Thresholds of hypoxia for marine biodiversity. *PNAS* **105(40)** 15452–15457.

Wang SY, Lau K, Lai KP et al 2016 Hypoxia causes transgenerational impairments in reproduction of fish. *Nature Communications* Article number **# 712114** *doi: 10.1038/ncomms12114*

White SH, Duivenvoorden LJ and Fabbro LD 2005 Absence of Free-Cylindrospermopsin bioconcentration in Water Thyme (*Hydrilla verticillata*). *Bulletin of Environmental Contamination & Toxicology* **75** 574–583

Whitworth KL, Baldwin DS and Kerr JL 2012 Droughts, floods and water quality: Drivers of a severe hypoxic blackwater event in a major river system (the southern Murray-Darling Basin, Australia). *Journal of Hydrology* **450-451** 190–198

http://www.climatechange2013.org/images/report/WG1AR5_Chapter08_FINAL.pdf

WHO (World Health Organisation) 2006 Guidelines for drinking water quality, incorporating first addendum. Chapter 5. Safe level and safe practices.

https://www.who.int/water_sanitation_health/resourcesquality/toxcyanchap5.pdf

WHO (World Health Organisation) 2014 Gender, climate change and health. ISBN 978 92 4 150818 6.

https://apps.who.int/iris/bitstream/handle/10665/144781/9789241508186_eng.pdf;jses

Worm B, Sandow M, Oschlies A, Lotze et al 2005 Global patterns of predator diversity in the open oceans. *Science* **308** 1365–1369

Wright J, Konwar KM and Hallam SJ 2012 Microbial ecology of expanding oxygen minimum zones (2012). *Nature Review Microbiology* **10** 381–394 *https://doi.org/10.1038/nrmicro2778*

Wu RSS 2009 Effects of hypoxia on fish reproduction and development. Chapter 3. *In*: Jeffrey GR, Farrell AP, and Brauner CJ (Eds.): Fish Physiology, Vol 27, 79-141. Burlington: Academic Press, 2009, p. 79-141 *DOI: 10.1016/S1546-5098(08)00003-4*

Zaitsev Y and Mamaev V 1997 Marine biological diversity in the Black Sea. A study of change and decline. United Nations Publications New York. 208 p.

Zegura B, Straser AS, and Metka Filipic M 2011 Genotoxicity and potential carcinogenicity of cyanobacterial toxins – a review. *Mutation Research* **727** 16–41

Web

https://oceanservice.noaa.gov/facts/nutpollution.html

https://www.sciencedaily.com/terms/fossil_fuel.htm

https://en.wikipedia.org/wiki/Dead_zone_(ecology)#cite_note-2

http://www.oceanhealthindex.org/methodology/components/nutrient-pollution

https://en.wikipedia.org/wiki/Eutrophication#/media/File:Eutrophicationmodel.svg

https://gulfhypoxia.net/about-hypoxia/

https://oceanservice.noaa.gov/podcast/feb18/nop13-hypoxia.html

https://www.flickr.com/photos/fondriestenvironmental/32533628442

https://www.hunterwater.com.au/Water-and-Sewer/Water-Supply/Water-Quality/Geosmin-and-MIB.aspx

http://agriculture.vic.gov.au/__data/assets/pdf_file/0007/319453/Blue-green-algal-poisoning-of-stock-InfoNote-.pdf

http://www.foodstandards.gov.au/consumer/safety/foodborne-illness/Pages/Toxins-in-seafood.aspx

https://monographs.iarc.fr/wp-content/uploads/2018/09/ClassificationsAlphaOrder.pdf

https://www.whoi.edu/redtide/page.do?pid=15159&tid=523&cid=27690

https://gulfhypoxia.net/about-hypoxia/

https://gulfhypoxia.net/are-jellyfish-a-harbinger-of-dying-seas/

https://en.wikipedia.org/wiki/Coral

■■■■

9 Pharmaceuticals Risks, Impacts and Management

Summary

Pharmaceuticals are synthetic or natural chemicals that can be found in prescription medicines, over-the-counter therapeutic and veterinary drugs. The occurrence of pharmaceuticals in the environment (surface water, groundwater, drinking water, sediments, and sewage treatment effluents) at trace levels (ng/L to μg/L) has been of great concern because of threats it poses to the environment and human health. A total of 16 pharmaceuticals have been detected in all the environmental matrices (surface, ground, tap or drinking water sources) across the globe. They are: acetylsalicylic acid/ aspirin (analgesic); carbamazepine (antiepileptic/analgesic); ciprofloxacin (antibiotic); clofibric acid (lipid-lowering); diclofenac (Non-steroidal anti-inflammatory drugs (NSAID)/analgesic); estrone-E1 (estrogen); 17β-estradiol-E2 (estrogen); estriol-E3 (estrogen); 17α-ethinylestradiol-EE2 (estrogen/ contraceptive pill); ibuprofen (analgesics/ anti-inflammatory); naproxen (antibiotic/antimicrobial); norfloxacin (antibiotic/ antimicrobial); ofloxacin (antibiotic/antimicrobial); paracetamol (analgesics/NSAID); sulfamethoxazole (antibiotic/antimicrobial) and trimethoprim (antibiotic/ antimicrobial). This chapter reviews **i**. classification of pharmaceuticals and environmental sources of pharmaceuticals; **ii**. chemistry, toxicity and ecotoxicity of pharmaceuticals; **iii.** environmental risk assessment of pharmaceuticals; **iv.** impacts of pharmaceuticals on social, economic and environmental sectors including environmental matrices, such as aquatic biota (bacteria, algae, daphnia, molluscs, fish), wildlife (frogs, birds), crops and human health; **v.** linkages between pharmaceuticals' use and climate change; and **vi.** management of pharmaceuticals (including drinking water quality guideline threshold values).

Contents

List of Tables

List of Figures

List of Box

9.1. Introduction

Pharmaceuticals are synthetic or natural chemicals that can be found in prescription medicines, over-the-counter therapeutic and veterinary drugs. Pharmaceuticals are used in the treatment, cure, prevention or diagnosis of diseases or used to otherwise enhance the physical or mental well-being of humans. Tonnes of pharmacologically active substances are used annually in human and veterinary medicines. Some of these substances are excreted (via human and animal excretion) as non-metabolised or active metabolites and enter into the environment directly via different pathways (e.g. wastewater treatment effluents, landfill leachate, stormwater run-off, livestock and aquaculture effluents) (Fig. 9.1). The occurrence of pharmaceuticals in the environment (surface water, groundwater, drinking water, sediments, sewage treatment effluents) at trace levels (ng/L to µg/L) has been of great concern (Farŕe et al 2008, Richman and Castensson 2008, WHO 2012).

Pharmaceuticals have been identified as "**contaminants of emerging concern** (CECs) (also called emerging contaminants[1] or emerging pollutants) (Farŕe et al 2001, Beausse 2004, Rosenfeld and Feng 2011). According to OECD (2019), in the future, the use of pharmaceuticals will eventually increase due to the following reasons: **i.** increase in populations' age and lifespan, **ii.** growing economies - particularly in emerging and developing economies requiring improved lifestyle and hygiene; **iii.** increasing ability and expectation to treat ageing-related chronic diseases; **iv.** new clinical practices and facilities for earlier treatment and prolonged treatment, **v.** on-going intensification of livestock and aquaculture practices; and **vi**. innovation and production of new pharmaceuticals.

9.1.1. Classification of pharmaceuticals

Pharmaceuticals can be classified in various ways, e.g. by chemical properties, mode of administration, the biological system those are affected or therapeutic effects (beneficial effects). However, most acceptable classifications are according to their anatomical and therapeutical (treatment) purposes. Based on the therapeutical application, pharmaceuticals are classified into antacid, antianginal, antiasthmatic, antibiotic, anticoagulant, anticonvulsant/ anti-epileptics, antidepressant; antianxiety-antipsychotic, anti-diabetic, antifungal, antihelminthic, antihistamine, antihypertensive, antineoplastic, beta-blocker, birth control/ contraceptive pill, bronchodilator, cardiac stimulant, contrast media, diuretic, lipid regulator, NSAID (Non-steroidal anti-inflammatory drugs), opioid analgesic, psychostimulant, steroid hormones, and vasodilator (Table 9.1).

[1]***Emerging contaminants*** *(or* ***contaminants of emerging concern****) are synthetic or naturally occurring chemicals that are not commonly monitored in the environment but have the potential to enter the environment and cause known or suspected adverse ecological and/ or human health effects. They consist of pharmaceuticals, industrial chemicals, surfactants, plasticisers, fire retardants, and personal care products that are consistently being found in groundwater, surface water, municipal wastewater, drinking water, and food sources. They also include endocrine-disrupting compounds, analgesics, antibiotics, hormones, and a whole range of other pharmaceutical compounds including anti-inflammatory, anti-diabetic, and anti-epileptic drugs (Rosenfeld and Feng 2011).*

9.1.2. Environmental sources of pharmaceuticals

Pharmaceuticals can enter into the environment (surface water, groundwater) from the treatment of humans via different routes, for example, from households, hospitals, schools, retirement facilities (Fig. 9.1). Improper disposal[2] of unused or expired drugs thrown in garbage, toilets or sewerage may end up in landfills, which are the local points of contaminations (audits have shown that a significant percent of waste medicines (unused or expired) are disposed-off in dustbins/ domestic garbage (>45%) and 10-28% are disposed directly to sewerage; Kotchen et al 2009).

	1	2	3	4	5	6	7	8
Main sources	**Human medicine** *(households, retirement facilities)*	**Hospital wastewater**	**Storm-water/ urban run-off**	**Pharm-aceutical production factories**	**Landfill disposal**	**Waste-water Treatment plants**	**Livestock** *(antibi-otics)*	**Aquacult ure** *(antib-iotics)*
Pathways to the environment	**Human excretion** (urine, faeces) *[excretion rate of pharm-aceuticals - 40% to 70%]*	**Drainage discharge**	**Drainage discharge**	**Drainage discharge**	**Landfill leachate**	**Leakage, treated/ untreated effluent discharge**	**Animal excretion, manure, bio-solids use**	**Effluent discharge**
Receiving environment	**Surface water**	**Surface water**	**Surface water**	**Surface water**	**Groundwa ter**	**Ground water, Surface water**	**Ground water, Surface water**	**Surface water**
	Drinking water							

Fig. 9.1. Sources and pathways of pharmaceutical products entering the environment.

[2] *The local community pharmacy in Australia provides a free and convenient way to dispose of unwanted medicines for free, safe disposal. Most medicines can be placed directly in the disposal bin provided by the Return Unwanted Medicines (RUM) Project (https://www.tga.gov.au/safe-disposal-unwanted-medicines).*

Table 9.1. Classification of pharmaceuticals (human and veterinary) based on the therapeutic uses *[Mompelat et al 2009; note: pharmaceuticals highlighted in bold fonts (column 2) are those detected in the environment including sewage outfall, river water, seawater, groundwater, drinking water as cited in Hernando et al 2006, Richman and Castensson 2008, WHO 2012, Gaw et al 2014, Aus der Beek et al 2016, Pereira et al 2017); compounds in italic (column 2) are a metabolite of parent compound].*

Therapeutic class	**Examples** *(pharmaceuticals in bold are those detected in the water environment)*	**References**
Antacid; histamine 2 blockers *(to prevent or correct acidity in the stomach)*	**cimetidine**; **ranitidine**	Mompelat et al 2009
Anti-anginal *(to relief from symptoms and prevent angina i.e. chest pain attacks)*	*dehydronifedipine* (nifedipine metabolite)	Mompelat et al 2009
Anti-asthmatic *(to treat or prevent asthma attacks)*	albuterol; **terbutaline; theophylline**	Mompelat et al 2009
Anti-biotic/ antimicrobial *(a medicine, such as penicillin that inhibits the growth of or destroys microorganisms)*	amoxicillin; **azithromycin**; ceftriaxone; cefuroxime; cephalexin; **chloramphenicol; chlortetracycline; clarithromycin**; clindamycin; danofloxacin; **ciprofloxacin;** difloxacin; doxycycline, enoxacin, enrofloxacin, **erythromycin**; *erythromycin H_2O* (erythromycin metabolite); **flumequine; ivermectin;** lincomycin; lomefloxacin; monensin; nalidixic acid; **norfloxacin**; **ofloxacin**; oleandomycin; ornidazole; oxolonic acid; **oxytetracycline**; penicillin G1/2 – benzathine salt; penicillin V - potassium salt; demeclocycline; **roxithromycin;** sarafloxacin; spiramycin; **sulfadiazine**; sulfadimethoxine; sulfamethazine; **sulfamethoxazole;** sulfapyridine, **sulfasalazine;** sulfathiazole; **tetracycline**; **triclosan**; **trimethoprim; tylosin**, virginiamycin	Mompelat et al 2009
Anti-coagulant *(blood thinners, that prevent or reduce coagulation of blood, prolonging the clotting time)*	**warfarin**	Mompelat et al 2009
Anti-convulsant/anti-epileptics *(drugs used to prevent or reduce the severity of epileptic fits or other convulsions)*	**carbamazepine;** dilantin; gabapentin; primidone; *10,11-dihydro-10-hydroxycarbamazepine* (Carbamazepine metabolite)	Mompelat et al 2009

Therapeutic class	**Examples** *(pharmaceuticals in bold are those detected in the water environment)*	**References**
Anti-depressant; antianxiety-anti-psychotic/ Selective serotonin reuptake inhibitors (SSRI) *(medications to help relieve symptoms of depression, social anxiety disorder, anxiety disorders)*	amitryptiline; **citalopram; diazepam; fluoxetine**; gabapentin; primidone; **sertraline**	Mompelat et al 2009
Anti-diabetic *(a substance that helps a person with diabetes to control their level of glucose (sugar) in the blood such as insulin and the oral hypoglycaemic agents)*	glibenclamide; **metformin**	Mompelat et al 2009
Anti-fungal *(a pharmaceutical fungicide used to treat and prevent mycosis such as athlete's foot, ringworm, candidacies, serious systemic infections such as cryptococcal meningitis, and others)*	**clotrimazole**	Mompelat et al 2009
Anti-helminthic *(antiparasitic drugs that expel parasitic worms and other internal parasites from the body by either stunning or killing them and without causing significant damage to the host)*	ivermectin	Mompelat et al 2009
Anti-histamine (drugs that can relieve allergic symptoms such as the runny nose, sneezing, and congestion) .	diphenhydramine, cetirizine	Mompelat et al 2009
Anti-hypertensive *(drugs used to treat hypertension i.e. to prevent the complications of high blood pressure, such as stroke and myocardial infarction)*	**diltiazem; enalapril**; *enalaprilat* (= enalapril maleate metabolite); **hydrochlorothiazide**; valsartan	Mompelat et al 2009
Anti-inflammatory substances	See NSAID	Mompelat et al 2009

Therapeutic class	**Examples** *(pharmaceuticals in bold are those detected in the water environment)*	**References**
Antineoplastic *(medications used to treat cancer, also called anticancer, chemotherapy, chemo, cytotoxic, or hazardous drug)*	**bleomycin; cyclophosphamide**; epirubicin/ doxorubicin; 5-fluoroacil; **ifosafamide**; methotrexate; **tamoxifen**	Mompelat et al 2009
Beta-blocker/antihypertensor *(medications that are predominantly used to manage abnormal heart rhythms, and to protect the heart from a second heart attack after a first heart attack)*	acebutolol; **atenolol; betaxolol; bisoprolol**; celiprolol; **metoprolol**; nadolol; pindolol; **propanolol; sotalol; timolol**	Mompelat et al 2009
Birth control/ contraceptive *(estrogenic medication which is used widely in birth control pills in combination with progestin)*	**17α-ethinylestradiol** (synthetic steroid hormone or EE2); 6α-methyl-17α-hydroxyprogesterone acetate	Mompelat et al 2009
Bronchodilator (medication that relaxes and opens the airways, or bronchi, in the lungs)	clenbuterol; **fenoterol**; **salbutamol**; terbutaline	Mompelat et al 2009
Cardiac stimulant *(a substance which acts as a stimulant of the heart)*	digoxin; *digoxigenin* (digoxin metabolite)	Mompelat et al 2009
Contrast media *(substances used to increase the contrast of structures or fluids within the body in medical imaging)*	diatrizoate; gadolinium (Gd-DTPA: GD-diethylenetriamine pentaacetic acid); iopromide-X-ray; **lomeprol;** lohexol; **lopamidol**	Mompelat et al 2009
Diuretic *(any substance that promotes diuresis[3], the increased production of urine. All diuretics increase the excretion of water from bodies)*	bendroflumethiazide; **furosemide**	Mompelat et al 2009

[3] ***Diuresis*** *= increased or excessive production of urine.*

Therapeutic class	**Examples** *(pharmaceuticals in bold are those detected in the water environment)*	**References**
Lipid regulator *(lipid-regulating drugs are used to treat dyslipidaemias (abnormal amounts of lipids, primarily raised cholesterol)*	atorvastatin; **benzafibrate**; clofibrate; ***clofibric acid*** (active metabolite of clofibrate); **fenofibrate;** *fenofibric acid* (active metabolite of fenofibrate); **gemfibrozil**; *o-and p-hydroxyatorvastatin* (atorvastain metabolite); pravastain; **simvastain**; *simvastain hydroxy acid* (simvastain metabolite)	Mompelat et al 2009
Non-steroidal anti-inflammatory drugs (NSAID) and analgesic *(drug class that reduces pain, decreases fever, prevents blood clots, and in higher doses, decreases inflammation)*	5-aminosalicyclic acid; aminopyrine; **aspirin (acetyl salicylic acid); diclofenac**; dimethylaminophenazone (DMAA); fenoprofen; flufenamic acid; flurbiprofen; **ibuprofen; indomethacin**; **ketoprofen**; meclofenamic acid; **mefenamic acid**; **naproxen;** nimesulide; **paracetamol** (acetaminophen); **phenazone**; **propyphenazone**; *salicylic acid* (ASA active metabolite); sulfapyridine; sulfasalazine; tolfenamic acid; DMAA metabolites/ or *degradation products* [AAA *(acetaminoantipyrine),* ADMOPH *(1-acetyl-1-methyl-2-dimethyloxamoyl-2-phenylhydrazine);* AMPH *(1-acetyl-1-methyl-2 – phenylhydrazine);* FAA *(formylaminoantipyrine)*] Phenazone *degradation products* [DP *(1,5-dimethyl-1,2-dehydro-3-pyrazolone);* DMADP *(4-(N,N-dimethyl)-amino-1,5-dimethyl-1,2-dehydro-3-pyrazolone)*]. Prophyphenazone *degradation products* [PDP *(4-(2-methylethyl)-1,5-dimethyl-1,2-dehydro-3-pyrazolone)*]	Mompelat et al 2009
Opioid analgesic *(Opioids, sometimes called narcotics, are medications prescribed by doctors to treat persistent or severe pain. such as chronic headaches and backaches by patients recovering from surgery or experiencing severe pain associated with cancer, and by adults and children who have gotten hurt playing sports or who have been injured)*	**codeine**; fentanyl (synthetic); **dextropropoxyphene**; morphine; methadone (synthetic); tramadol; *6-acetylmorphine* (6-AM, morphine metabolite); EDDP, *norcodeine* (codeine metabolite); *normorphine* (morphine metabolite)	Mompelat et al 2009

Therapeutic class	**Examples** *(pharmaceuticals in bold are those detected in the water environment)*	**References**
Psychostimulant *(a substance that has mood-enhancing and stimulant properties used to treat patients with ADHD (attention deficit hyperactivity disorder)*	amphetamine; **caffeine;** *1,7-dimethylxanthine* (caffeine metabolite)	Mompelat et al 2009
Steroid hormones *(The term steroid describes both hormones produced by the body and externally given medications that duplicate the action for the naturally occurring steroids; Steroid hormones are required for survival of stress, injury (and illness), metabolism, inflammation, salt and water balance, immune functions, and development of sexual characteristics)*	androsterone, **diethylstilbestrol; 17β- estradiol** (E2); **estriol** (E3); **estrone** (E1); **norethisterone**; progesterone, **testosterone**	Mompelat et al 2009
Vasodilator *(a group of medicines that dilate (open) blood vessels, which allows blood to flow more easily)*	pentoxyfilline	Mompelat et al 2009

The pharmaceutical ingredients, metabolites and conjugates[4] are mainly excreted in urine or faeces (excretion rate of some pharmaceuticals is very high, ranging between 40% and 70%; Table 9.2), from where they enter municipal sewage treatment plants (STPs) where they can be degraded, adsorbed to sewage sludge or eventually diluted into surface water. One of the most important sources of pharmaceutical products (PPs) in aquatic environments is the effluents from STPs since PPs are not completely removed during treatment. For example, Hernando et al (2006) reported that STPs partially removed carbamazepine, atenolol, metoprolol, trimethoprim or diclofenac (<10% for most of them and 10-39% for diclofenac). Therefore, those compounds that cannot be removed by sewage treatment can be found in surface water.

[4] ***Conjugate*** *= a compound formed by the joining of two or more chemical compounds.*

9.1.2. Environmental sources of pharmaceuticals

Pharmaceuticals can enter into the environment (surface water, groundwater) from the treatment of humans via different routes, for example, from households, hospitals, schools, retirement facilities (Fig. 9.1). Improper disposal of unused or expired drugs thrown in garbage, toilets or sewerage may end up in landfills, which are the local points of contaminations. An audit has shown that a significant percent of waste medicines (unused or expired) are disposed-off in dustbins/domestic garbage (>45%) and 10-28% are disposed directly to sewerage (Kotchen et al 2009). The pharmaceutical ingredients, metabolites and conjugates are mainly excreted in urine or faeces (excretion rate of some pharmaceuticals is very high, ranging between 40% and 70%; Table 9.2), from where they enter municipal sewage treatment plants (STPs) where they can be degraded, adsorbed to sewage sludge or eventually diluted into surface water. One of the most important sources of pharmaceutical products (PPs) in aquatic environments is the effluents from STPs since PPs are not completely removed during treatment. For example, Hernando et al (2006) reported that STPs partially removed carbamazepine, atenolol, metoprolol, trimethoprim or diclofenac (<10% for most of them and 10-39% for diclofenac). Therefore, those compounds that cannot be removed by sewage treatment can be found in surface water.

Table 9.2. Class of pharmaceutical products based on their excretion rates (proportion of parent compounds excreted)
[compiled by Jjemba 2006, Mompelat et al 2009; pharmaceuticals in bold fonts are those detected in the water environment].

Excretion rate (%)	**Compound (excretion rate %)** *Pharmaceuticals in bold were detected in the water environment; based on Table 9.1)*	**Therapeutic use**	**References**
Low (≤5% excretion)	amitryptilline (little)	antidepressants-anti-anxiety	Jjemba 2006, Mompelat et al 2009
	amphetamine (1-74)	psycho-stimulants	Jjemba 2006, Mompelat et al 2009
	carbamazepine (1-2)	anticonvulsant	Jjemba 2006, Mompelat et al 2009
	cyclophosphamide (≤5-20)	antineoplastics	Jjemba 2006, Mompelat et al 2009
	diazepam (1)	antidepressants-anti-anxiety	Jjemba 2006, Mompelat et al 2009
	diltiazem (2-4)	antidepressants-anti-anxiety	Jjemba 2006, Mompelat et al 2009

Excretion rate (%)	**Compound (excretion rate %)** *Pharmaceuticals in bold were detected in the water environment; based on Table 9.1)*	**Therapeutic use**	**References**
	erythromycin (5-10)	antibiotics	Jjemba 2006, Mompelat et al 2009
	ibuprofen (1-10)	NSAIDs and analgesics	Jjemba 2006, Mompelat et al 2009
	ifosafamide (upto 15)	antineoplastics	Jjemba 2006, Mompelat et al 2009
	metoprolol (3-30)	beta-blockers	Jjemba 2006, Mompelat et al 2009
	paracetamol (acetaminophen) (<5)	NSAIDs and analgesics	Jjemba 2006, Mompelat et al 2009
	propranolol (<0.5)	beta-blockers	Jjemba 2006 Mompelat et al 2009
Moderately low (6-39% excretion)	5-aminosalicyclic acid (<12)	NSAIDs and analgesics	Jjemba 2006. Mompelat et al 2009
	5-fluorouracil (11-20)	antineoplastics	Jjemba 2006, Mompelat et al 2009
	acebutolol (39)	beta-blockers	Jjemba 2006, Mompelat et al 2009
	amphetamine (1-74)	psycho-stimulants	Jjemba 2006, Mompelat et al 2009
	chloramphenicol (5-10 or 8-12)	Antibiotics	Jjemba 2006, Mompelat et al 2009
	clarithromycin (25)	Antibiotics	Jjemba 2006, Mompelat et al 2009
	cyclophosphamide (≤5-20)	antineoplastics	Jjemba 2006, Mompelat et al 2009
	diclofenac (15)	NSAIDs and analgesics	Jjemba 2006, Mompelat et al 2009
	enalapril (30-36)	anti-hypertensives	Jjemba 2006, Mompelat et al 2009

Excretion rate (%)	Compound (excretion rate %) *Pharmaceuticals in bold were detected in the water environment; based on Table 9.1)*	Therapeutic use	References
	epirubicin/doxorubicin (biliary excretion: 9-11 within 48 h)	antineoplastics	Jjemba 2006, Mompelat et al 2009
	erythromycin (5-10)	Antibiotics	Jjemba 2006, Mompelat et al 2009
	fluorouracil	antineoplastics	Jjemba 2006, Mompelat et al 2009
	ifosfamide (upto 15)	antineoplastics	Jjemba 2006, Mompelat et al 2009
	metronidazole (40 or 20)	Antibiotics	Jjemba 2006, Mompelat et al 2009
	metoprolol (3-30)	beta-blockers	Jjemba 2006, Mompelat et al 2009
	norfloxacin (30)	antibiotics	Jjemba 2006, Mompelat et al 2009
	primidone (15-40)	anticonvulsant	Jjemba 2006, Mompelat et al 2009
	salbutamol (30)	bronchodilatators	Jjemba 2006, Mompelat et al 2009
	simvastatin (10-15)	lipid regulator	Jjemba 2006, Mompelat et al 2009
	spiramycin (20)	antibiotics	Jjemba 2006, Mompelat et al 2009
	sulfamethoxazole (15 or 30)	antibiotics	Jjemba 2006, Mompelat et al 2009
	tramadol (15-35)	opioanalgesics	Jjemba 2006, Mompelat et al 2009

Excretion rate (%)	**Compound (excretion rate %)** *Pharmaceuticals in bold were detected in the water environment; based on Table 9.1)*	**Therapeutic use**	**References**
Relatively high (40-69% excretion)	amphetamine (1-74)	psycho-stimulants	Jjemba 2006, Mompelat et al 2009
	atenolol (50-90)	beta-blockers	Jjemba 2006, Mompelat et al 2009
	benzafibrate (45-50)	lipid regulators	Jjemba 2006, Mompelat et al 2009
	cimetidine (48-75)	antacids	Jjemba 2006, Mompelat et al 2009
	ciprofloxacin (83.7 or 50-70)	antibiotics	Jjemba 2006, Mompelat et al 2009
	digoxin (50-70)	cardiac stimulants	Jjemba 2006, Mompelat et al 2009
	EE2 (57)	contraceptive pills	Johnson and Williams 2004
	epirubicin/ doxorubicin (40 biliary excretion within 72 h)	antineoplastics	Jjemba 2006, Mompelat et al 2009
	furosemide (40-90)	diuretics	Jjemba 2006, Mompelat et al 2009
	ranitidine (30-70)	antacids	Jjemba 2006, Mompelat et al 2009
	testosterone (60)	steroid hormones	Jjemba 2006, Mompelat et al 2009
High (≥70% excretion)	amoxicillin (80-90 or 60-80)	antibiotics	Jjemba 2006, Mompelat et al 2009
	amphetamine (1-74)	psycho-stimulants	Jjemba 2006, Mompelat et al 2009
	atenolol (50-90)	beta-blockers	Jjemba 2006, Mompelat et al 2009
	ceftriaxone (70)	antibiotics	Jjemba 2006, Mompelat et al 2009

Excretion rate (%)	Compound (excretion rate %) *Pharmaceuticals in bold were detected in the water environment; based on Table 9.1)*	Therapeutic use	References
	cimetidine (48-75)	antacids	Jjemba 2006, Mompelat et al 2009
	ciprofloxacin (83.7 or 50-70)	antibiotics	Jjemba 2006, Mompelat et al 2009
	codeine (70 free or as conjugates)	opioanalgesics	Jjemba 2006, Mompelat et al 2009
	doxycycline (70)	antibiotics	Jjemba 2006, Mompelat et al 2009
	furosemide (40-90)	diuretics	Jjemba 2006, Mompelat et al 2009
	gabapentin (100)	antibiotics	Jjemba 2006, Mompelat et al 2009
	morphine (71.6)	opioanalgesics	Jjemba 2006, Mompelat et al 2009
	sotalol (>75)	beta-blockers	Jjemba 2006, Mompelat et al 2009
	tetracycline (80-90)	antibiotics	Jjemba 2006, Mompelat et al 2009
	trimethoprim (60 or 80)	antibiotics	Jjemba 2006, Mompelat et al 2009
	valsartan (80)	anti-hypertensives	Jjemba 2006, Mompelat et al 2009

Veterinary pharmaceuticals can enter into the environment via different pathways. These include emissions during manufacture, formulation, treatment processes, and/or disposal of unused medicines and disposal of containers. The most important routes of entry into the environment are direct discharge of fish/ aquaculture products (via the use of medicines or as fish farm effluents), excretion of substances in urine and faeces of livestock animals, and the wash-off of topical treatments from livestock animals. It can enter indirectly via the application of manure containing excreted products (manure, slurry) to land. The main by-products from cattle feedlots are manure and liquid effluents. Run-off from animal feedlots could also increase the amount of livestock chemicals input into surface waters and groundwater (from leaching). Livestock's direct access to open land and water would also contribute additional chemicals to the environment by urine and faeces. Aquaculture is an increasingly significant contribution to veterinary antibiotic transfer to the environment.

9.2. Chemistry, toxicity and ecotoxicity of pharmaceuticals

Pharmaceutical products are generally polar (water-soluble) (Daughton and Ternes 1999), with low volatility and have high mobility in the environment. In addition, a number of pharmaceuticals are persistent, bioaccumulative and toxic (PBT) as well as endocrine-disrupting and carcinogenic (Table 9.3). Persistent, bioaccumulative and toxic criteria can be used to assess risks and harmful effects posed by organic contaminants on aquatic and terrestrial biota (EPA 2012, Strempel et al 2012). Such PBT based risk assessment strategy (half-life, log K_{ow}, bioconcentration factors (BCF), toxicity (LD_{50} for rats, birds)) and ecotoxicity (EC_{50}/ LC_{50} of bacteria, algae, daphnia and fish) has been applied throughout this book (see section 9. 2.5).

9.2.1. Persistence

Several pharmaceuticals are persistent (i.e half-life in freshwater **>40 days**; FASS 2012) including aspirin; caffeine; carbamazepine; ciprofloxacin; clofibric acid; diclofenac; E2; EE2; fluoxetine; gemfibrozil; ibuprofen; norfloxacin; ofloxacin; paracetamol; roxithromycin; salicylic acid; tetracyclines; triclosan; trimethoprim (see Table 9.3). The half-life of some of the pharmaceuticals are very long [e.g. (*caffeine* ≈10 years in surface water (Buerge et al 2006, Edwards et al 2015); *carbamazepine:* 780-5,700 days in lake water (Zou et al 2015); *salicylic acid:* 30,142 days (82 years) in aqueous solution (MSDS-GlaxoSmithKline 2014); *ciprofloxacin:* 120-2,310 days in biosolids (Walters et al 2010); *ofloxacin:* 360-1,386 days in biosolids (Walters et al 2010); *fluoxetine:* 120-1,000 days in biosolids (Walters et al 2010) and *ibuprofen:* 17.8-247.6 days in lake water (Araujo et al 2014).

9.2.2. Bioaccumulation

Several pharmaceuticals are bioaccumulative (**log K_{ow} >3**) including *benzafibrate* (log K_{ow} 4.25), *diclofenac* (log K_{ow} 4.51), *E1* (log K_{ow} 3.43); *E2* (log K_{ow} 3.94); *EE2* (log K_{ow} 4.15); *erythromycin* (log K_{ow} 3.06); *fluoxetine* (log K_{ow} 4.6); *gemfibrozil* (log K_{ow} 4.77); *ibuprofen* (log K_{ow} 3.97); *naproxen* (log K_{ow} 3.18); *triclosan* (log K_{ow} 4.76) (Table 9.3). Chemicals having a log K_{ow} > 3 have the potential or tendency for accumulation in biota (OECD 2012). Another measure of bioaccumulation potential is the bioconcentration factor (**BCF**) (concentration in fish ÷ concentration in water medium). BCF has been the most widely accepted measure of bioaccumulation potential. Pharmaceuticals having a BCF >500 have high bioaccumulation potential (FASS 2012). A number of pharmaceuticals have BCF >500 including *diclofenac* (BCF: 12-2,700; Schwaiger et al 2004); *EE2* (BCF: 600-610; Al-Ansari et al 2013); *fluoxetine* (BCF: 1,000; Bringolf et al 2010, Brooks 2014); *gemfibrozil* (BCF: 92-500; Mimeault et al 2005); *naproxen* (BCF: 75–1,285; Brozinski et al 2013); *triclosan* (BCF: 2,532-4,175 (in Zebra fish) (WFD UKTAG 2012) (see Table 9.3) *[**note:** the modern chemical legislation requires measuring the BCF of large numbers of chemicals in fish (*Adolfsson-Erici et al 2012).

Table 9.3. Chemistry, toxicity and ecotoxicity of pharmaceuticals.

*[Classification of pharmaceuticals based on their half-life, log K_{ow}, BCF and ecotoxicity: **i. Half-life**[5] ($t_{1/2}$) (biotic/ abiotic degradation); **persistent**: freshwater >40 days; marine water >60 days; freshwater sediments >120; marine sediment >180 days; **very persistent** (freshwater >60 days; freshwater sediment >180 days (FASS 2012); **ii. Log K_{ow}** >3 = bioaccumulation[6] potential or tendency for bioaccumulation (OECD 2012); **iii. BCF (bioconcentration factor** is the most widely accepted measure of bioaccumulation potential ; BCF <500 = low bioaccumulation potential; BCF >500 = high bioaccumulation potential (FASS 2012); **iv. ecotoxicity** (EC_{50}/ LC_{50} of bacteria, algae, daphnia, fish): extremely toxic = <0.1 mg/L; very toxic = 0.1 mg/L to 1.0 mg/L; toxic = 1.0 mg/L to 10.0 mg/L; non-toxic = > 100 mg/L (CEC 1996, Kim et al 2006, 2008); **na** = data is not available)].*

Pharmac-euticals	**Chemistry**		**Toxicity**		**Ecotoxicity**
*(*pharmaceuticals in **bold** fonts with an asterisk were detected in environ-mental water worldwide (Aus der Beek et al 2016)*	**Log K_{ow}**	**Solubility and Half-life ($t_{1/2}$) in water/ soil/ sediment**	**Endocrine disruptor[7] (ED), Biocon-centration factor (BCF[8])**	**Acute toxicity to mammal, bird (LD_{50}[9]), IARC Cancer group**	**Acute aquatic toxicity (bacteria, algae, daphnia and fish, EC_{50}[10]/ LC_{50}[11])**

[5] ***Half-life ($t½$)*** *= defined as the time it takes for an amount of a compound to be reduced by half through degradation in an environmental compartment (water, soil or sediment etc). It is used to measure the persistence of a substance.*

https://www.chemsafetypro.com/Topics/CRA/Toxicology_Dose_Descriptors.html.

[6] ***Bioaccumulation*** *= the gradual accumulation of substances/ chemicals in an organism (algae, daphnia, fish). Bioaccumulation refers to uptake from all sources combined (e.g. water, food, air, etc.) whereas bioconcentration is referred to uptake and accumulation of a substance from water alone (Alexander 1999).*

[7] ***Endocrine disruptors*** *are chemicals that can interfere with endocrine systems at certain doses. These disruptions can cause cancerous tumours, birth defects, and other developmental disorders. https://www.niehs.nih.gov/health/topics/agents/endocrine/*

[8] ***BCF*** *= bioconcentration factor is the ability of an organism to accumulate a chemical from the ambient environment. BCF (concentration in fish ÷ concentration in medium); Bioconcentration is the specific bioaccumulation process by which the concentration of a chemical in an organism becomes higher than its concentration in the air or water around the organism. https://www.sciencedirect.com/topics/chemistry/bioconcentration*

[9] ***LD_{50}*** *= (median lethal dose), the dose of a material that is estimated to be lethal to 50% of the test organisms (rats, mice, etc.).*

[10] ***EC_{50}*** *= effective concentration of the test substance that results in 50% effect or the concentration of test substance which results in a 50 percent reduction in either algae growth or algae growth rate or Daphina immobilization etc.*

[11] ***LC_{50}*** *= = (median lethal concentration), the concentration of a material in the water that is estimated to be lethal to 50% of the test organisms (fish, daphnia, etc.). The LC_{50} is usually expressed as a time-dependent value, (24 h or 96 h).*

Pharmaceuticals	**Chemistry**		**Toxicity**		**Ecotoxicity**
*(*pharmaceuticals in **bold** fonts with an asterisk were detected in environmental water worldwide (Aus der Beek et al 2016)*	**Log K_{ow}**	**Solubility and Half-life ($t_{1/2}$) in water/ soil/ sediment**	**Endocrine disruptor[7] (ED), Bioconcentration factor (BCF[8])**	**Acute toxicity to mammal, bird (LD_{50}[9]), IARC Cancer group**	**Acute aquatic toxicity (bacteria, algae, daphnia and fish, EC_{50}[10]/ LC_{50}[11])**
***Acetylsalicylic acid/ Aspirin** *(analgesics/ NSAID)* $HC_9H_7O_4$	1.19 (Beausse 2004)	Solubility: 4,600 mg/L (MSDS-Pubchem 2004a) Half-life: 40 days *(https://www.researchgate.net/post/Aspirin_dissolution_or_hydrolysis)*	ED (Caliman and Gaurilescu 2009) Low/ no potential for bioaccumulation(FASS-no date-Aspirin)	Rat, LD_{50}: 200 mg/kg (MSDS-Pubchem 2004a)	Algae, LC_{50}: 106.7 mg/L for 24 h *Daphnia*, LC_{50}: 88 mg/L for 48 h Fish, LC_{50}: 150 mg/L for 96 h (Cemagref 2007, Sanderson and Thomsen 2009).
Benzafibrate *(lipid regulators)* $C_{19}H_{20}ClNO_4$	4.25 (Beausse 2004)	Solubility: na Half-life: 12-17 days (lake) (Zou et al 2015)	na	Rat, LD_{50}: 1,082 mg/kg (MSDS-Pubchem 2005a)	Algae, LC_{50}: >100 mg/L for 24 h *Daphnia*, LC_{50}:>100 mg/L for 48 h Fish, LC_{50}: >100 mg/L for 96 h (Park 2005, Sanderson and Thomsen 2009).
Caffeine *(stimulant)* $C_8H_{10}N_4O_2$	-0.07 (Kibuye et al 2019)	Solubility: 13,000 mg/L (highly soluble) Half-life: ≈ 10 yrs (surface water) (Buerge et al 2006, Edwards et al 2015)	Low bioaccumulation BCF: 29 (mosquito fish) (Wang and Gardinali 2012)	Rat, LD_{50}: 367 mg/kg (Adamson 2016)	Algae, LC_{50} : na *Daphnia*, LC_{50} : 182 mg/L for 48 h Fish, LC_{50}: 37 mg/L for 96 h (NOAA 2006, Sanderson and Thomsen 2009).
***Carbamazepine** *(analgesics/ antiepileptics)* $C_{15}H_{12}N_2O$	2.45 (Beausse 2004)	Solubility: 17.7 mg/L (MSDS-Pubchem 2005b) Half-life: 780-5,700 days (lake) (Zou et al 2015)	Low bioaccumulation BCF: 2.2-12.6 (algae and crustaceans) (Vernouillet et al 2010); BCF: 108 (mosquito fish) (Wang and Gardinali 2012)	Rat, LD_{50}: > 1,500 mg/kg (MSDS-Pubchem 2005b)	Algae, LC_{50}: 33.6 mg/L for 24 h *Daphnia*, LC_{50}: 13.8 mg/L for 48 h Fish, LC_{50}: na (Park 2005, Cunningham et al 2006, Sanderson and Thomsen 2009).

Pharmaceuticals	**Chemistry**		**Toxicity**		**Ecotoxicity**
*(*pharmaceuticals in **bold** fonts with an asterisk were detected in environmental water worldwide (Aus der Beek et al 2016)*	**Log K_{ow}**	**Solubility and Half-life ($t_{1/2}$) in water/ soil/ sediment**	**Endocrine disruptor[7] (ED), Bioconcentration factor (BCF[8])**	**Acute toxicity to mammal, bird (LD_{50}[9]), IARC Cancer group**	**Acute aquatic toxicity (bacteria, algae, daphnia and fish, EC_{50}[10]/ LC_{50}[11])**
***Ciprofloxacin** *(antibiotics/ antibacterial)* $C_{17}H_{18}FN_3O_3$	0.4 (Beausse 2004)	Solubility: 30,000 mg/L (MSDS-Pubchem 2005c) Half-life: 120-2,310 days (biosolid) (Walters et al 2010)	Low bioaccumulation BCF: 3 (fish)) (MSDS-Pubchem 2005c)	Rat, LD_{50}: 2,000 mg/kg (MSDS-Pubchem 2005c)	Bacteria: 0.006 mg/L to 0.61 mg/L (sewage sludge) (Halling-Sørensen et al 2000 and 2003) Algae, LC_{50}: 3 mg/L for 24 h *Daphnia*, LC_{50}: 60 mg/L for 48 h Fish, LC_{50}: na (Cunningham et al 2006, Cemagref 2007, Sanderson and Thomsen 2009).
***Clofibric acid** *(lipid regulators)* $C_{10}H_{11}ClO_3$	2.57 (Beausse 2004)	na Half-life: 131-495 days (lake water) (Araujo et al 2014)	ED (Caliman and Gaurilescu 2009) Low bioaccumulation BCF: 2.3-7.4 (Lumpfish - *Eleuthero embryo*) (Molina-Fernandez et al 2017); BCF: 40.5 (Arnot and Gobas 2006)	Rat, LD_{50}: 897 mg/kg (MSDS-Pubchem 2005d)	Bacteria, *Vibrio fischeri*, EC_{50}: 100 mg/L for 30 minutes (Henschel et al 1997) Algae, LC_{50}: 72 mg/L for 24 h Daphnia, LC_{50}: 86 mg/L for 48 h Fish, LC_{50}: 89 mg/L for 96 h (Roche 2007, Sanderson and Thomsen 2009).
***Diclofenac** *(analgesics/ NSAID)* $C_{14}H_{11}Cl_2NO_2$	4.51 (Beausse 2004)	Solubility: 2.37 mg/L (MSDS-Pubchem 2005e) Half-life: 8-35 days (lake) (Zou et al 2015); 5.76-60 days (lake water) (Araujo et al 2014)	ED (Caliman and Gaurilescu 2009) Moderate to high bioaccumulation BCF: 12-2,700 (rainbow trout liver, kidney, gills and muscles) (Schwaiger et al 2004); BCF: 113 (Fish) (Brown et al 2007)	Rat, LD_{50}: 55-240 mg/kg (MSDS-Pubchem 2005e) Birds, *Gyps bengalensis* LD_{50}: 0.1–0.2 mg /kg ((Oaks et al 2004, Swan et al 2006) (extremely toxic)	Algae, LC_{50}: 14.5 mg/L for 24 h *Daphnia*, LC_{50}: 22.4 mg/L for 48 h Fish, LC_{50}: na (Cemagref 2007, Sanderson and Thomsen 2009).

Pharmaceuticals	**Chemistry**		**Toxicity**		**Ecotoxicity**
*(*pharmaceuticals in **bold** fonts with an asterisk were detected in environmental water worldwide (Aus der Beek et al 2016)*	**Log K_{ow}**	**Solubility and Half-life ($t_{1/2}$) in water/ soil/ sediment**	**Endocrine disruptor[7] (ED), Bioconcentration factor (BCF[8])**	**Acute toxicity to mammal, bird (LD_{50}[9]), IARC Cancer group**	**Acute aquatic toxicity (bacteria, algae, daphnia and fish, EC_{50}[10]/ LC_{50}[11])**
***E1 (estrone)** *(steroid hormones)* $C_{18}H_{22}O_2$	3.43 (Beausse 2004)	Solubility: 5 mg/L (Zorita et al 2009) Half-life: 0.1-10.9 days (River water) (Jürgens et al 2002)	ED (Caliman and Gaurilescu 2009) Low bioaccumulation BCF: 104-106 (rainbow trout) (Larsson et al 1999)	Rat, LD_{50}: na (known to be human carcinogens, IARC group 1) (MSDS-Pubchem 2004b)	Fish, Zebra fish, *Danio rerio,* EC_{50}: 0.0000078 mg/L (or 0.078) µg/L for 18 days (vitellogenin) (MSDS-Pubchem 2004b, Holbech et al 2006).
***E2 (17β-estradiol)** *(steroid hormones)* $C_{18}H_{24}O_2$	3.94 (Beausse 2004)	Solubility: 3 mg/L (Zorita et al 2009), 3.9 mg/L (MSDS-Pubchem 2004c) Half-life: 56 days (seawater) (Ying and Kookana 2003	ED (Caliman and Gaurilescu 2009, Kumar and Xagoraraki 2010) Low bioaccumulation BCF: 104-106 (rainbow trout) (Larsson et al 1999)	Rat, LD_{50}: na (known to be human carcinogens, IARC group 1) (MSDS-Pubchem 2004c)	*Daphnia*, EC_{50}: 2.87 mg/L for 48 h (Brennan et al 2006, MSDS-Pubchem 2004c) Fish, fathead minnow, EC_{50}: 0.00012 mg/L (0.12 µg/L) for 19 days (Kramer et al 1998).
***E3 (estriol)** *(steroid hormones)* $C_{18}H_{24}O_3$	2.5 (Beausse 2004)	Solubility: 27.34 mg/L (MSDS-Pubchem 2005f)	ED (Kumar and Xagoraraki 2010) Low bioaccumulation BCF: 19-50 (fish) (MSDS-Pubchem 2005f)	Rat, LD_{50}: >2,000 mg/kg (known to be human carcinogens-IARC group 1) (MSDS-Pubchem 2005f)	Sea urchin, *Strongylocentrotus purpuratu,* EC_{50}: 1.51 mg/L (1515.4 µg/L) for 95 h (MSDS-Pubchem 2005f , Roepke et al 2005).

Pharmaceuticals	**Chemistry**		**Toxicity**		**Ecotoxicity**
*(*pharmaceuticals in **bold** fonts with an asterisk were detected in environmental water worldwide (Aus der Beek et al 2016)*	**Log K_{ow}**	**Solubility and Half-life ($t_{1/2}$) in water/ soil/ sediment**	**Endocrine disruptor[7] (ED), Bioconcentration factor (BCF[8])**	**Acute toxicity to mammal, bird (LD_{50}[9]), IARC Cancer group**	**Acute aquatic toxicity (bacteria, algae, daphnia and fish, EC_{50}[10]/ LC_{50}[11])**
***EE2 (17α-ethinylestradiol)** *(oral contraceptive/ steroid hormones)* $C_{20}H_{24}O_2$	4.15 (Beausse 2004)	Solubility: 2 mg/L (Zorita et al 2009) 13 mg/L (MSDS-Pubchem 2005g) Half-life: >52 days (seawater) (Ying and Kookana 2003)	ED (Purdom et al 1994, Caliman and Gaurilescu 2009) High bioaccumulation BCF: 337 (goldfish) (Al-Ansari et al 2013) BCF: 600–610 (fathead minnow) (Länge et al 2001)	Rat, LD_{50}: 960 mg/kg (known to be human carcinogens, IARC group 1) (MSDS-Pubchem 2005g)	Algae, LC_{50}: 0.84 mg/L for 24 h *Daphnia*, LC_{50}: 5.7 mg/L for 48 h Fish, LC_{50}: 1.6 mg/L for 96 h (Park 2005, Cunningham et al 2006, Sanderson and Thomsen 2009).
Erythromycin *(antibiotics)* $C_{37}H_{67}NO_{13}$	3.06 (Beausse 2004)	Solubility: 2,000 mg/L (MSDS-Pubchem 2005h) Half-life: 6.8 to 37.9 days (fresh and seawater) (Kwon 2016)	Low bioaccumulation BCF: 72.2 (carp) (Liu et al 2014); BCF: 49 (fish) (McFarland et al 1997, MSDS-Pubchem 2005h)	Rat, LD_{50}: 4,600 mg/kg (MSDS-Pubchem 2005h)	Bacteria, *Photorhabdus luminescens*, EC_{50}: 0.022 mg/L (González et al 2016) Algae, LC_{50}: 0.02 mg/L for 24 h *Daphnia*, LC_{50}: 0.94 mg/L for 48 h Fish, LC_{50}: 80 mg/L for 96 h (Cunningham et al 2006, Sanderson and Thomsen 2009).

Pharmaceuticals	**Chemistry**		**Toxicity**		**Ecotoxicity**
*(*pharmaceuticals in **bold** fonts with an asterisk were detected in environmental water worldwide (Aus der Beek et al 2016)*	**Log K_{ow}**	**Solubility and Half-life ($t_{1/2}$) in water/ soil/ sediment**	**Endocrine disruptor[7] (ED), Bioconcentration factor (BCF[8])**	**Acute toxicity to mammal, bird (LD_{50}[9]), IARC Cancer group**	**Acute aquatic toxicity (bacteria, algae, daphnia and fish, EC_{50}[10]/ LC_{50}[11])**
Fluoxetine *(antidepressants/ psychiatric drugs)* $C_{17}H_{18}F_3NO$	1.57 (pH 7) (Brooks et al 2003); 4.6 (MSDS-Pubchem 2005r)	Solubility: na Half-life: 112-133 days (sediment free) (Kwon and Armbrust 2006); 120-1,000 days (biosolid) (Walters et al 2010); highly persistent (Peake et al 2016)	ED (Brooks et al 2003, Sanchez-Arguello et al 2009, El-Bassat et al 2012, Mennigen et al 2017) High bioaccumulation (Bringolf et al 2010, Brooks 2014); BCF: 1,000 (FW mussels) (Bringolf et al 2010, Brooks 2014)	Rat, LD_{50}: 825 mg/kg (MSDS-Pubchem 2005r)	Algae, *Pseudokirchneriella subcapitata,* EC_{50}: 0.024 mg/L *Daphnia*, EC_{50}: 0.82 mg/L Fish, *Pimephales promelas,* EC_{50}: 0.705 mg/L (Brooks et al 2003).
Gemfibrozil *(lipid regulators)* $C_{15}H_{22}O_3$	4.77 (Beausse 2004)	Solubility: 11 mg/L (MSDS-Pubchem 2005i) Half-life: 40-57 days (lake) (Zou et al 2015); 75-231 days (biosolid) (Walters et al 2010)	Moderate bioaccumulation BCF: 92-500 (plasma goldfish) (Mimeault et al 2005) BCF: 421 (fish) (Brown et al 2007)	Rat, LD_{50}: 1,414 mg/kg (MSDS-Pubchem 2005i)	Bacteria, *V. fischeri* (bioluminescence), EC_{50}: 64.6 mg/L for 24 h) (Zurita et al 2007) Algae, *Chlorella vulgaris* (growth) EC_{50}: 195 mg/L for 24 h (Zurita et al 2007) *Daphnia*, LC_{50}: >100 mg/L for 48 h Fish, LC_{50}: na (Cunningham et al 2006, Sanderson and Thomsen 2009).

Pharmaceuticals	**Chemistry**		**Toxicity**		**Ecotoxicity**
*(*pharmaceuticals in **bold** fonts with an asterisk were detected in environmental water worldwide (Aus der Beek et al 2016)*	**Log K_{ow}**	**Solubility and Half-life ($t_{1/2}$) in water/ soil/ sediment**	**Endocrine disruptor[7] (ED), Bioconcentration factor (BCF[8])**	**Acute toxicity to mammal, bird (LD_{50}[9]), IARC Cancer group**	**Acute aquatic toxicity (bacteria, algae, daphnia and fish, EC_{50}[10]/ LC_{50}[11])**
***Ibuprofen** *(analgesics/ NSAID)* $C_{13}H_{18}O_2$	3.97 (Beausse 2004)	Solubility: 21 mg/L (MSDS-Pubchem 2004d) Half-life: 17.8-247.6 days (lake water) (Araujo et al 2014)	ED (Caliman and Gaurilescu 2009, Han et al 2010) Low bioaccumulation BCF: 0.08-1.4 (fathead minnow, channel catfish) (Nallani et al 2011), BCF: 1.3 (bluntnose minnow-gill) (Bhandari and Venables 2011); BCF: 78 (fish) (Brown et al 2007)	Rat, LD_{50}: 636 mg/kg (MSDS-Pubchem 2004d)	Algae, LC_{50}: 7.1 mg/L for 24 h *Daphnia*, LC_{50}: 9.1 mg/L for 48 h Fish, LC_{50}: 173 mg/L for 96 h (Park 2005, Cunningham et al 2006, Sanderson and Thomsen 2009).

Pharmaceuticals	**Chemistry**		**Toxicity**		**Ecotoxicity**
*(*pharmaceuticals in **bold** fonts with an asterisk were detected in environmental water worldwide (Aus der Beek et al 2016)*	**Log K_{ow}**	**Solubility and Half-life ($t_{1/2}$) in water/ soil/ sediment**	**Endocrine disruptor[7] (ED), Bioconcentration factor (BCF[8])**	**Acute toxicity to mammal, bird (LD_{50}[9]), IARC Cancer group**	**Acute aquatic toxicity (bacteria, algae, daphnia and fish, EC_{50}[10]/ LC_{50}[11])**
Ketoprofen *(analgesics/ NSAID)* $C_{16}H_{14}O_3$	3.12-3.16 (Beausse 2004)	Solubility: na Half-life: <1–2 days (lake) (Zou et al 2015); 5.76-14.6 days (lake water) (Araujo et al 2014)	Low bioconcentration BCF: 20 (fish plasma) (Brown et al 2007)	Rat, LD_{50}: 62.4 mg/kg (MSDS-Pubchem 2005j)	Algae, LC_{50}: na *Daphnia*, LC_{50}: 64 mg/L for 48 h Fish, LC_{50} na (Cemagref 2007, Sanderson and Thomsen 2009).
***Naproxen** *(analgesics-NSAID)* $C_{14}H_{14}O_3$	3.18 (Beausse 2004)	Solubility: 15.9 mg/L (MSDS-Pubchem 2005k) Half-life: 14 days (Lake) (Tixier et al 2003)	ED (Caliman and Gaurilescu 2009) Low bioconcentration BCF: 1.9–9.3 (Lumpfish - Eleuthero[12] embryo) (Molina-Fernandez et al 2017); BCF: 75–1,285 (wild fish bile) (Brozinski et al 2013)	Rat, LD_{50}: 248 mg/kg (MSDS-Pubchem 2005k)	Algae, LC_{50}: 21 mg/L for 24 h *Daphnia*, LC_{50}: 37 mg/L for 48 h Fish, LC_{50}: 560 mg/L for 96 h (Cemagref 2007, Sanderson and Thomsen 2009).
***Ofloxacin** *(antibiotics)* $C_{18}H_{20}FN_3O_4$	0.35 (Beausse 2004) -0.39 (MSDS-Pubchem 2005m)	Solubility: 230 mg/L (Zorita et al 2009) Half-life: 360-1,386 days (biosolids) (Walters et al 2010)	BCF: 3 (fish) (Hansch et al 1995)	Rat, LD_{50}: 3,590 mg/kg (MSDS-Pubchem 2005m)	Algae, LC_{50}: 1.44 mg/L for 24 h *Daphnia*, LC_{50}: 17.44 mg/L for 48 h Fish, LC_{50}: na (Cunningham et al 2006, Sanderson and Thomsen 2009).

[12] ***Eleuthero embryos****= fish hatched but not yet free*

Pharmaceuticals (*pharmaceuticals in **bold** fonts with an asterisk were detected in environmental water worldwide (Aus der Beek et al 2016)	Chemistry		Toxicity		Ecotoxicity
	Log K_{ow}	**Solubility and Half-life ($t_{1/2}$) in water/ soil/ sediment**	**Endocrine disruptor[7] (ED), Bioconcentration factor (BCF[8])**	**Acute toxicity to mammal, bird (LD_{50}[9]), IARC Cancer group**	**Acute aquatic toxicity (bacteria, algae, daphnia and fish, EC_{50}[10]/ LC_{50}[11])**
***Paracetamol** (acetaminophen) *(analgesics/ NSAID)* $C_8H_9NO_2$	0.46 (MSDS-Pharmacopedia 2006)	Solubility: 14,200 mg/L (MSDS-Pharmacopedia 2006) Half-life: 72 days (persistent) (Dusing et al 2009, Peake et al 2016)	ED (Caliman and Gaurilescu 2009) Donot bioconcentrate (Peake et al 2016)	Rat, LD_{50}: 2,404 mg/kg (MSDS-Pharmacopedia 2006) (non-toxic (Peake et al 2016)	Bacteria, *V. fischri,* EC_{50}: 650 mg/L for 30 minutes (Henschel et al 1997) Algae, LC_{50}: 134 mg/L for 24 h Daphnia, LC_{50}: 9.2 mg/L for 48 h Fish, LC_{50}: 378 mg/L for 96 h (Park 2005, Cemagref 2007, Sanderson and Thomsen 2009). Non-toxic (Peake et al 2016).
Roxithromycin *(antibiotics)* $C_{41}H_{76}N_2O_{15}$	2.75 (Beausse 2004) 3.1 (MSDS-Pubchem 2005n)	Half-life: 180 days (water) (Choi et al 2008)	Low bioaccumulation BCF: 13.4-93.5 *(Daphnia)* (Ding et al 2008)	Rat, LD_{50}: 830 mg/kg (MSDS-Pubchem 2005n)	Bacteria, *V. fischeri,* IC_{50}[13]: 1,000 mg/L for 5-15 minutes *Daphnia*, EC_{50}: 74.3 mg/L for 48 h Fish, *Oryzias latipes,* LC_{50}: 288.3 mg/L for 96 h (Choi et al 2008).

[13] ***IC_{50}*** *= The half-maximal inhibitory concentration (IC_{50}) is a measure of the potency of a substance in inhibiting a specific biological or biochemical function. IC_{50} is a quantitative measure that indicates how much of a particular inhibitory substance (e.g. drug) is needed to inhibit, in* ***vitro (in the lab)****, a given biological process or biological component by 50%. The biological component could be an enzyme, cell, cell receptor or microorganism. https://en.wikipedia.org/wiki/IC50#cite_note-fda-1*

Pharmaceuticals	**Chemistry**		**Toxicity**		**Ecotoxicity**
*(*pharmaceuticals in **bold** fonts with an asterisk were detected in environmental water worldwide (Aus der Beek et al 2016)*	**Log K_{ow}**	**Solubility and Half-life ($t_{1/2}$) in water/ soil/ sediment**	**Endocrine disruptor[7] (ED), Bioconcentration factor (BCF[8])**	**Acute toxicity to mammal, bird (LD_{50}[9]), IARC Cancer group**	**Acute aquatic toxicity (bacteria, algae, daphnia and fish, EC_{50}[10]/ LC_{50}[11])**
Salicyclic acid (*Acetyl salicyclic acid metabolite*) $C_7H_6O_3$	2.24 (Sanderson and Thomsen 2009)	Solubility: 2,240 mg/L (MSDS-Pubchem 2004e) Half-life: 30,142 days (photolysis aqueous) (MSDS-GlaxoSmithKline 2014)	Low bioaccumulation BCF: 8.32-30.9 (MSDS-GlaxoSmithKline 2014)	Rat, LD_{50}: 891 mg/kg (MSDS-Pubchem 2004e)	Bacteria, *V fischeri,* EC_{50}: 90 mg/L for 30 minutes (Henschel et al 1997) Algae, LC_{50}: >100 mg/L for 24 h *Daphnia*, LC_{50}: 118 mg/L for 48 h Fish, LC_{50}: 37 mg/L for 96 h (Park 2005, Cemagref 2007, Cunningham et al 2006, Sanderson and Thomsen 2009).
***Sulfamethoxazole** *(antibiotics)* $C_{10}H_{11}N_3O_3S$	0.89 (Beausse 2004)	Solubility: 610 mg/L (MSDS-Pubchem 2005o) Half-life: 20-35 days (lake) (Zou et al 2015), >50 days (anaerobic conditions-soil) (moderately persistent) (Lin and Gan 2011)	Low bioaccumulation BCF: 3 (fish) (Hansch et al 1995)	Rat, LD_{50}: 6,200 mg/kg (MSDS-Pubchem 2005o)	Algae, LC_{50}: 10 mg/L for 24 h *Daphnia*, LC_{50}: 0.03 mg/L for 48 h Fish, LC_{50}: na (Cunningham et al 2006, NOAA 2006, Sanderson and Thomsen 2009).
Tetracyclines (*antibiotics*) $C_{22}H_{24}N_2O_8$	-1.19 (Beausse 2004)	Solubility : 2,310- 52,000 mg/L (Thiele-Bruhn 2003) Half-life: 120- 578 days (biosolids) (Walters et al 2010)	ED (Caliman and Gaurilescu 2009)	Rat, LD_{50}: 807 mg/kg (MSDS-Pubchem 2011)	Algae, LC_{50}: 2.2 mg/L for 24 h *Daphnia*, LC_{50}: 0.165 mg/L for 48 h Fish, LC_{50}: 220 mg/L for 96 h (Park 2005, Cunningham et al 2006, Sanderson and Thomsen 2009).

Pharmaceuticals *(*pharmaceuticals in **bold** fonts with an asterisk were detected in environmental water worldwide (Aus der Beek et al 2016)*	**Chemistry**		**Toxicity**		**Ecotoxicity**
	Log K_{ow}	**Solubility and Half-life ($t_{1/2}$) in water/ soil/ sediment**	**Endocrine disruptor[7] (ED), Bioconcentration factor (BCF[8])**	**Acute toxicity to mammal, bird (LD_{50}[9]), IARC Cancer group**	**Acute aquatic toxicity (bacteria, algae, daphnia and fish, EC_{50}[10]/ LC_{50}[11])**
Triclosan *(antibiotics/ antiseptic agent)*	4.76 (MSDS-Pubchem 2005q)	Solubility: 10 mg/L (water) (MSDS-Pubchem 2005q) Half-life: 120-187 days (biosolids) (Walters et al 2010)	ED (Kumar and Xagoraraki 2010) Low to high bioaccumulation BCF: 2.7–90 (carp, *Cyprinus carpio)* (WFD UKTAG 2012); BCF: 2,532-4,175 (Zebra fish) (Orvos et al 2002, WFD UKTAG 2012)	Rat, LD_{50}: 3,700 mg/kg (MSDS-Pubchem 2005q)	Bacteria, *V. fischeri,* EC_{50}: 0.52 mg/L for 15 minutes Algae, *P. Subcapitata,* EC_{50}: 0.012 mg/L for 96 h *Daphnia*, EC_{50}: 0.0014-3 mg/L Fish, *Pimephales promelas,* LC_{50}: 0.26 mg/L Amphibians, *Xenopus laevis*, LC_{50}: 0.82 mg/L for 96 h (Harada et al 2008, Dhilon et al 2015).
***Trimethoprim** *(antibiotics)* $C_{14}H_{18}N_4O_3$	0.64 (Sanderson and Thomsen 2009)	Solubility: 400 mg/L (MSDS-Pubchem 2005p) Half-life: > 42 days (freshwater); 20-100 days (freshwater sediment) (Boxall et al 2002, Straub 2013)	Low bioaccumulation BCF: ~ 0.32 (marine fish liver); BCF: ~ 0.16 freshwater fish liver) (Bergsjø and Søgnen 1980)	Rat, LD_{50}: 5,300 mg/kg (MSDS-Pubchem 2005p)	Bacteria, *V. fischeri,* IC_{50}: 165.1 mg/L for 5 minutes (Choi et al 2008) Algae, LC_{50}: 110 mg/L for 24 h *Daphnia*, LC_{50}: >100 mg/L for 48 h Fish, LC_{50}: >100 mg/L for 96 h (Park 2005, Cunningham et al 2006, Sanderson and Thomsen 2009).

9.2.3. Toxicity

Several pharmaceuticals are **endocrine disruptors** including aspirin; clofibric acid; E1; E2; EE2; fluoxetine; ibuprofen; naproxen; tetracyclines (see Table 9.3). Pharmaceuticals such as E1, E2 and EE2 are also **carcinogenic** to humans (IARC group 1) (Table 9.3). The LD_{50} values reveal that most of the pharmaceuticals have low to moderately acutely toxicity to mammals (LD_{50}> 500 mg/kg). However, diclofenac has been reported to be extremely toxic to birds (*Gyps bengalensis*) having LD_{50}: 0.1–0.2 mg /kg (Oaks et al 2004, Swan et al 2006). The exposure of diclofenac has caused a decline of vulture population in the entire Indian sub-continent (see section 9.3).

9.2.4. Ecotoxicity

Several pharmaceuticals (ciprofloxacin, E1, E2, EE2, erythromycin, fluoxetine, sulfamethoxazole, tetracyclines, triclosan) are **extremely toxic** (EC_{50}/ LC_{50} <1 mg/L) to aquatic organisms including bacteria, algae, daphnia and fish. These pharmaceuticals include: **i.** *Ciprofloxacin* [(bacteria, EC_{50}: 0.006 mg/L (Halling-Sørensen et al 2000 and 2003)], **ii.** *E1-estrone* [(zebra fish, EC_{50}: 0.0000078 mg/L (MSDS-Pubchem 2004b, Holbech et al 2006)], **iii.** *E2-17β estradiol* [(fish-fathead minnow, EC_{50}: 0.00012 mg/L (Kramer et al 1998)], **iv.** *EE2-17α ethinylestradiol* [(algae, LC_{50}: 0.84 mg/L; fish LC_{50}: 1.6 mg/L (Park 2005, Cunningham et al 2006)], **v.** *erythromycin* [(bacteria, EC_{50}: 0.022 mg/L; algae, LC_{50}: 0.02 mg/L; daphnia, LC_{50}: 0.94 mg/L (Cunningham et al 2006, Sanderson and Thomsen 2009)], **vi.** *fluoxetine* [(algae, EC_{50}: 0.024 mg/L; daphnia, EC_{50}: 0.82 mg/L; fish fathead minnow, EC_{50}: 0.705 mg/L (Brooks et al 2003)], **vii.** *sulfamethoxazole* [(daphnia, LC_{50}: 0.03 mg/L (Cunningham et al 2006, NOAA 2006)], **viii.** *Tetracyclines* [(daphnia LC_{50}: 0.165 mg/L) (Park 2005, Cunningham et al 2006)]; **ix.** *Triclosan* [(bacteria, EC_{50}: 0.52 mg/L; algae, EC_{50}: 0.012 mg/L; daphnia, EC_{50}: 0.26 mg/L; amphibians, LC_{50}: 0.82 mg/L (Harada et al 2008)] (see also Table 9.3).

9.2.5. Environmental risk[14] assessment of pharmaceuticals

9.2.5.1. Risk assessment based on PEC (Predicted environmental concentration) and PNEC (Predicted no-effect concentration)

The environmental risk (Risk = Exposure x Hazard) of a pharmaceutical can be calculated from the ratio of predicted environmental concentration (**PEC**) and predicted no-effect concentration (**PNEC**) which is called **Risk quotient** (RQ). **RQ = PEC/ PNEC** (Richman and Castensson 2008, Peake et al 2016). The PEC is the calculated/ or the measured concentration of a chemical in the environment. PNEC is the concentration of a pharmaceutical at which no pharmacological effect is expected to occur for a specific organism/ biota (usually derived using ecotoxicity/ecotoxicology[15] testing data - EC_{50} or LC_{50} values of the most sensitive species such as algae, daphnia and fish) (Richman

---

[14] ***Risk** = exposure x hazard*

[15] ***Ecotoxicology** is the study of the toxic effects of chemicals (pollutants/ toxicants/ stressors) on the aquatic and terrestrial environment including individual organism, species in food webs, food resources such as fish and shell-fish. Ecotoxicology assesses acute and chronic toxicity. Acute (short-term) and chronic (long-term) toxicity tests are designed to assess the short- and long-term effects of chemicals on key biological parameters, including survival, reproduction, physiological and biochemical responses (Newman 2010).*

and Castensson 2008). A PEC/ PNEC ratio of < 1 suggests that there is no significant risk associated with the environmental presence of the pharmaceutical, whereas a ratio ≥ 1 suggests that further environmental risk characterisation or management of the pharmaceutical is necessary (see Table 9.4 for risk assessment of pharmaceuticals based on PEC/PNEC; Table 9.5 provides an example of risk assessment of selected pharmaceuticals based on measured environmental concentrations) [***note:*** The US EPA has developed a software programme which is known as **ECOSAR** (Ecological Structure Activity Relationships) which can be freely downloaded from the US EPA website *https://www.epa.gov/tsca-screening-tools/ecological-structure-activity-relationships-ecosar-predictive-model*. This programme can be used to estimate acute (short-term) and chronic (long-term or delayed) toxicity of a chemical to aquatic organisms such as fish, aquatic invertebrates, and aquatic plants by using computerized structure-activity relationships (SARs)].

Table 9.4. Risk assessment of pharmaceuticals based on PEC/PNEC or Risk quotient (RQ)
(Richman and Castensson 2008, Wennmalm and Gunnarsson 2009).

Risk quotient (RQ)/ (PNC/ PNEC)	**Environmental risk level**	**Remarks**
<0.1	Insignificant / no risk	Use of medicine will result in insignificant environmental risk.
0.1-1.0	Low risk	Use of the medicine will result in low or minimal environmental risk.
1.0-10	Medium/moderate risk	Use of the medicine will result in moderate environmental risk.
>10	High risk	Use of medicine will result in high environmental risk.

Table 9.5. Examples of risk assessment of selected pharmaceuticals based on measured environmental concentrations in influent[16] (MEC[17]-influent) and effluent[18] (MEC-effluent) samples, predicted no-effect concentrations (PNEC) and risk quotients (PEC/PNEC) of each pharmaceutical compound (Santos et al 2007).

Compound	MEC-influent (µg/l)	MEC-effluent (µg/l)	PNEC (µg/l)	MEC influent/ PNEC	MEC effluent/ PNEC	Comments
Caffeine	11.44	3.20	182[a]	0.06	0.02	**No risk or low risk** (risk quotient in influent and effluent wastewaters is <0.1).
Carbamazepine	2.15	1.29	6.36[a]	0.34	0.20	**Low risk** (risk quotient in influent and effluent wastewaters is <1).
Ibuprofen	373.11	48.24	9.1[c]	41	5.3	**An ecological risk** is present since risk quotient is >1 (risk ratios were 41 and 5.32 for influent and effluent respectively (columns 4, 5).
Ketoprofen	3.59	1.50	15.6[b]	0.23	0.10	**No risk or low risk** (risk quotient in influent and effluent wastewaters is <1).
Naproxen	27.04	4.28	21.2[b]	1.3	0.20	**Reduced ecological risk** (since the risk quotient in effluent water is 0.20 (column 5).

[a] *PNEC data estimated using ECOSAR*

[b] *PNEC data from Farré et al (2001)*

[c] *PNEC data from Stuer-Lauridsen et al (2000)*

---

[16] ***Influent=*** *the untreated wastewater or raw sewage coming into a wastewater treatment plant.*

[17] ***MEC=*** *measured environmental concentrations.*

[18] ***Effluent*** *= "wastewater (treated or untreated) that flows out of a treatment plant, sewer, or industrial outfall.*

9.2.5.2. Risk assessment based on Persistence (P), Bioaccumulation (B) and Toxicity (T)

Organic contaminants can be screened and classified for potential harmful effects using **PBT** (persistence, bioaccumulation, and toxicity) criteria (Klečka et al 2009, EPA 2012, Strempel et al 2012). Use of PBT criteria is a global standard approach to assess potential risks posed by a pharmaceutical ingredient in the environment. The **"PBT" hazard assessment** *strategy/ scheme was* devised by the Stockholm County Council in Sweden (Castensson et al 2009). In this scheme, a pharmaceutical is given a ranking score of 0–3 under each of the PBT categories (Table 9.6). The score derived from the sum of these categories provide an overall measure of the environmental hazard of the pharmaceutical (Deblonde and Hartemann 2013). The total of these values constitutes the PBT index for the substance, which ranges from **zero to nine.** The higher the PBT index of a pharmaceutical, the greater the hazard to the environment. Most pharmaceuticals have an insignificant or low acute toxic risk for the environment. However, some pharmaceuticals with a PBT index of 9 are antifungals (ketoconazole, miconazole, terbinafine), anti-infectives (ofloxacin, efavirenz), antineoplastic agents (dasatinib, docetaxel, tamoxifen, megestrol) and drugs for the nervous system (propofol, bromocriptine, clozapine, citalopram, etc.) (Deblonde and Hartemann 2013). Estradiol and ethinyl estradiol have also high environmental risk (Deblonde and Hartemann 2013).

Table 9.6 Principles for the assignment of numerical values in the environmental hazard classification developed by Stockholm County Council and Apoteket AB [(National Corporation of Swedish Pharmacies) (Castensson et al 2009).

Parameters		**Score**
Persistence (P)- *(readily biodegradable)*	Yes	0
	No	3
Bioaccumulation (B) *(potentially bioaccumulating in adipose tissue of aquatic organisms)*	Yes	3
	No	0
Toxicity (T) *(potential to poison aquatic organisms)*	Very high toxicity	3
	High toxicity	2
	Moderate toxicity	1
	Low toxicity	0

9.2.5.3. Risk assessment based on bioaccumulation (log K_{ow}) and ecotoxicity

Kumar and Xagoraraki (2010) ranked 100 pharmaceuticals, personal care products and endocrine-disrupting chemicals found in U.S. stream water/ source water and finished drinking water. The ecological risks/ effects of these 100 chemicals were based on bioaccumulation potential (log K_{ow}) and ecotoxicity data (lethal concentrations with respect to EC_{50} and LC_{50} of algae, daphnids and fish). Out of 100 chemicals, 59 were pharmaceuticals. The pharmaceuticals that are within the top 50 chemicals (and could cause most harm to aquatic biota) include *fluoxetine* (antidepressant); *triclosan* (antibiotic); *17α-ethynyl estradiol (EE2)* (birth control pill/ hormone); *17β-estradiol* (E2) (steroid hormone); *estrone* (E1) (steroid hormone); *azithromycin* (antibiotic); *gemfibrozil* (lipid regulator); *erythromycin* (antibiotic); *bezafibrate* (lipid regulator), *testosterone* (steroid hormone); *ibuprofen* (NSAID), *naproxen* (antibiotic), *diclofenac* (NSAID) (Table 9.7).

Table 9.7. Ranking of 100 chemicals, found in US streams/ source water and finished drinking water, based on ecological effects (ecotoxicity on algae, daphnia and fish and log K_{ow}/ bioaccumulation potential)

*[Kumar and Xagoraraki 2010; **EDCs**= endocrine-disrupting chemicals; **na** = not available; **shaded rows** = pharmaceuticals; **Teratogenic** = an agent that can disturb the development of the embryo or fetus]*

Prioritised chemicals	Chemical class	Remarks
1. Cis-chlordane	Industrial chemicals	EDC
2. Bis(2-ethylhexyl) phthalate	Industrial chemicals	EDC
3. AHTN *(7-acetyl-1,13,4,4,6-hexamethyl-1,2,3,4-tetrahydronaphthalene)*	Personal care products	EDC
4. HHCB *(1,3,4,6,7,8-hexahydro- 4,6,6,7,8,8-hexamethylcyclopenta-(γ)-2-benzopyran)*	Personal care products	EDC
5. Benzo(a)pyrene	Industrial chemicals	EDC
6.. 4-nonylphenol	Industrial chemicals	EDC, Teratogenic
7. Celestolide	Personal care products	Teratogenic

Prioritised chemicals	Chemical class	Remarks
8. Fluoxetine	Pharmaceuticals – antidepressants	Teratogenic
9. 2,6-di-tert-butyl-1,4- benzoquinone	Industrial chemicals	EDC, Teratogenic
10. Pentachlorophenol	Industrial chemicals	EDC
11. 2,6-di-tert-butylphenol	Industrial chemicals	Teratogenic
12. Triclosan	Pharmaceuticals- antibiotics	EDC, Teratogenic
13. Dieldrin	Industrial chemicals	EDC
14. Fipronil	Industrial chemicals	na
15. Ethylhexyl methoxycinnamate	Personal care products	na
16. Chlorpyrifos	Industrial chemicals	EDC, Teratogenic
17. 17α-ethynyl estradiol (EE2)	Pharmaceuticals - contraceptive pill	EDC
18. Metolachlor	Industrial chemicals	na
19. Equilenin	Pharmaceuticals - steroidal hormone	EDC
20. 17α-estradiol	Pharmaceuticals - steroidal hormone	EDC
21. 17β-Estradiol (E2)	Pharmaceuticals - steroidal hormone	EDC, Teratogenic
22. Diphenhydramine	Pharmaceuticals – antihistamine	Teratogenic
23. Musk xylene	Personal care products	na

Prioritised chemicals	Chemical class	Remarks
24. 3-tert-butyl-4-hydroxy anisole	Industrial chemicals	EDC, Teratogenic
25. Musk Ketone	Personal care products	na
26. Lindane	Industrial chemicals	EDC, Teratogenic
27. Musk ambrette	Personal care products	na
28. Mestranol	Pharmaceuticals - birth control	EDC, Teratogenic
29 Diazinon	Industrial chemicals	EDC, Teratogenic
30. Estrone (E1)	Pharmaceuticals - steroidal hormone	EDC, Teratogenic
31. Equilin	Pharmaceuticals - steroidal hormone	EDC
32. Azithromycin	Pharmaceuticals - antibiotics	Teratogenic
33. Risperidone	Pharmaceuticals - antipsychotic/ neuroleptics	Teratogenic
34. Linuron	Industrial chemicals	na
35. Bisphenol A	Industrial chemicals	na
36. Gemfibrozil	Pharmaceuticals - lipid regulator	Teratogenic
37. Erythromycin	Pharmaceuticals - antibiotics	Teratogenic
38. Progesterone	Pharmaceuticals - steroidal hormone	EDC, Teratogenic
39. Bezafibrate	Pharmaceuticals - lipid regulator	na

Prioritised chemicals	Chemical class	Remarks
40. TDIP *(tris (dichloroisopropyl) Phosphate)*	Industrial chemicals	EDC
41. Propylparaben	Personal care products	na
42. Testosterone	Pharmaceuticals - steroidal hormone	EDC, Teratogenic
43. Cis-androsterone	Pharmaceuticals - steroidal hormone	EDC
44. Tri(2-butoxyethyl) phosphate	Industrial chemicals	na
45. Ibuprofen	Pharmaceuticals - analgesics/ NSAID	Teratogenic
46 Benzophenone	Personal care products	na
47. Naproxen	Pharmaceuticals - antibiotics	Teratogenic
48. Diclofenac	Pharmaceuticals - NSAID/ analgesics	Teratogenic
49. Dehydronifedipine	Pharmaceuticals- hypertension	na
50. Norfluoxetine	Pharmaceuticals - selective serotonin reuptake inhibitor	na
51. Diethoxyoctylphenol	Industrial chemicals	EDC
52. Musk moskene	Personal care products	na
53. Octocrylene	Personal care products	na
54. Atorvastatin	Pharmaceuticals - cholesterol-lowering	Teratogenic

Prioritised chemicals	Chemical class	Remarks
55. Demeclocycline	Pharmaceuticals – antibiotics	Teratogenic
56. Desulfinyl fipronil	Industrial chemicals	na
57. 3-Methyl-1H-indole (skatole)	Personal care products	Teratogenic
58. Carbaryl	Industrial chemicals	EDC
59. Dilitiazem	Pharmaceuticals- blood pressure and angina	Teratogenic
60. Estriol (E3)	Pharmaceuticals- steroidal hormone	EDC, Teratogenic
61. Sulfadimethoxine	Pharmaceuticals – antibiotics	na
62. Methyl parathion	Industrial chemicals	EDC
63. Roxithromycin	Pharmaceuticals – antibiotics	na
64. Sulfathiazole	Pharmaceuticals – antibiotics	Teratogenic
65. Trimethoprim	Pharmaceuticals – antibiotics	Teratogenic
66. Methylbenzyldene camphor	Personal care products	na
67. N,N-diethyltoluamid	Personal care products	na
68. Warfarin	Pharmaceuticals - anti-coagulant (blood thinner)	Teratogenic
69. Atrazine	Industrial chemicals	EDC
70. Sulfamethoxazole	Pharmaceuticals – antibiotics	Teratogenic

Prioritised chemicals	Chemical class	Remarks
71. Codeine	Pharmaceuticals - opioid analgesic	Teratogenic
72. Sulfamethizole	Pharmaceuticals – antibiotics	na
73. Digoxigenin	Pharmaceuticals - steroid found in the flowers	Teratogenic
74. Cimetidine	Pharmaceuticals - Antacid/ histamine	Teratogenic
75. Diethyl-phthalate	Industrial chemicals	EDC
76. 19-norethisterone	Pharmaceuticals - birth control	EDC, Teratogenic
77. Atenolol	Pharmaceuticals - beta-blocker/ high blood pressure	Teratogenic
78. Lincomycin	Pharmaceuticals – antibiotics	EDC
79. Methylparaben	Personal care products	na
80. Phenytoin	Pharmaceuticals - anti-epileptic	Teratogenic
81. Carbamazepine	Pharmaceuticals-analgesics/ antiepileptics	Teratogenic
82. Acetophenone	Personal care products	na
83. Acetaminophen/ paracetamol	Pharmaceuticals - analgesics/ NSAID	Teratogenic
84. Cotinine	Pharmaceuticals - antidepressant	na

Prioritised chemicals	Chemical class	Remarks
85. Clofibric acid	Pharmaceuticals - lipid regulators	Teratogenic
86. Sarafloxacin	Pharmaceuticals – antibiotics	na
87.Chlortetracycline	Pharmaceuticals – antibiotics	na
88. TCEP *(tris (2-chloroethyl)phosphate)*	Industrial chemicals	EDC
89. Enrofloxacin	Pharmaceuticals – antibiotics	Teratogenic
90. Tylosin	Pharmaceuticals – antibiotics	Teratogenic
91 Tetracycline	Pharmaceuticals – antibiotics	Teratogenic
92. Flumequine	Pharmaceuticals – antibiotics	EDC, Teratogenic
93. Ethylparaben	Personal care products	na
94. Ciprofloxacin	Pharmaceuticals – antibiotics	Teratogenic
95. Meprobamate	Pharmaceuticals - anxiety disorders	na
96. Norfloxacin	Pharmaceuticals – antibiotics	Teratogenic
97. Diazepam	Pharmaceuticals – antidepressant	Teratogenic
98. Oxytetracycline	Pharmaceuticals - antibiotics	Teratogenic

Prioritised chemicals	Chemical class	Remarks
99. 1,7-dimethyl-xanthine	Industrial chemicals	na
100. Triethyl-citrate	Personal care products	EDC

9.3. Impacts of pharmaceuticals on social, economic and environmental sectors

9.3.1 Impacts on the environmental matrices

Pharmaceutical ingredients have been reported to be ubiquitously present in *sewage treatment effluents* (Hernando et al 2006, Fent et al 2006, Spongberg and Witter 2008, Zhou et al 2008, Kümmerer 2009, Miége et al 2009, Aus Der Beek et al 2016), *surface waters* (Boxall 2004, Hernando et al 2006, Mompelat et al 2009, Kümmerer 2009, Watkinson et al 2009, Gaw et al 2014, Aus Der Beek et al 2016), *groundwaters* (Barnes et al 2008, Richman and Castensson 2008, Kümmerer 2009) and *drinking water* reservoirs (Jones et al 2005, Mompelat et al 2009, New Scientist 2009, Aus Der Beek et al 2016) in the ng/L to µg/L range (Kibria et al 2010). There are approximately 4,000 types of drugs worldwide for human and veterinary use of which, 600 have been shown to have been spread to terrestrial and aquatic habitats at a global scale (Ne´stor and Mariana 2017).

Pharmaceutical residues (therapeutic classes detected in the environment, expressed in relative percentage) that have been detected in different environmental compartments include *NSAID/ analgesics* (pain reliver) (16%); *antibiotics* (15%); *lipids regulators* (12%); *estrogens/ hormones* (9%); *beta/ β-blockers* (heart protection) (8%); *antiepileptics/seizure prevention* (8%); *antidepressants* (4%); *anxiolytic (anxiety prevention)* (4%); *antineoplastic/ cytostatic (cancer)* (4%); *antihypertensives* (high blood pressure prevention) (4%); *antidiabetics* (3%); antacids (3%); *x-ray contrast media* (3%); *veterinary products* (3%); *β2-sympathomimetics* (stimulant) (3%) and *antipsychotic* (1%) (Santos et al 2010). This is based on data collected from 134 articles published between 1997 and 2010.

A recent survey shows that 16 pharmaceuticals have been detected in all the environmental water matrices (surface water, groundwater, tap or drinking water sources) across the **globe** (Africa, Asia, Europe, America). These pharmaceuticals are **i. *acetylsalicylic acid/aspirin*** (analgesic); **ii. *carbamazepine*** (antiepileptic/analgesic); **iii. *Ciprofloxacin*** (antibiotic); **iv. *clofibric acid*** (lipid-lowering); **v. *diclofenac*** (NSAID/ analgesic); **vi. *estrone-E1*** (estrogen); **vii. *estradiol-E2*** (estrogen); **viii. *estriol-E3*** (estrogen); **ix. *17α-ethinylestradiol-EE2*** (estrogen); **x. *ibuprofen*** (analgesics/ anti-inflammatory); **xi. *naproxen*** (antibiotic); **xii. *norfloxacin*** (antibiotic); **xiii. *ofloxacin*** (antibiotic); **xiv.**

paracetamol (analgesics/ NSAID); **xv.** ***sulfamethoxazole*** (antibiotic); **xvi.** ***trimethoprim*** (antibiotic) (Aus Der Beek et al 2016) (see Tables 9.8 and 9.9 for pharmaceuticals detected in surface water and drinking water and Fig. 9.2 for worldwide distribution of estrogens (E1, E2, E3 and EE2 in rivers and surface water sites).

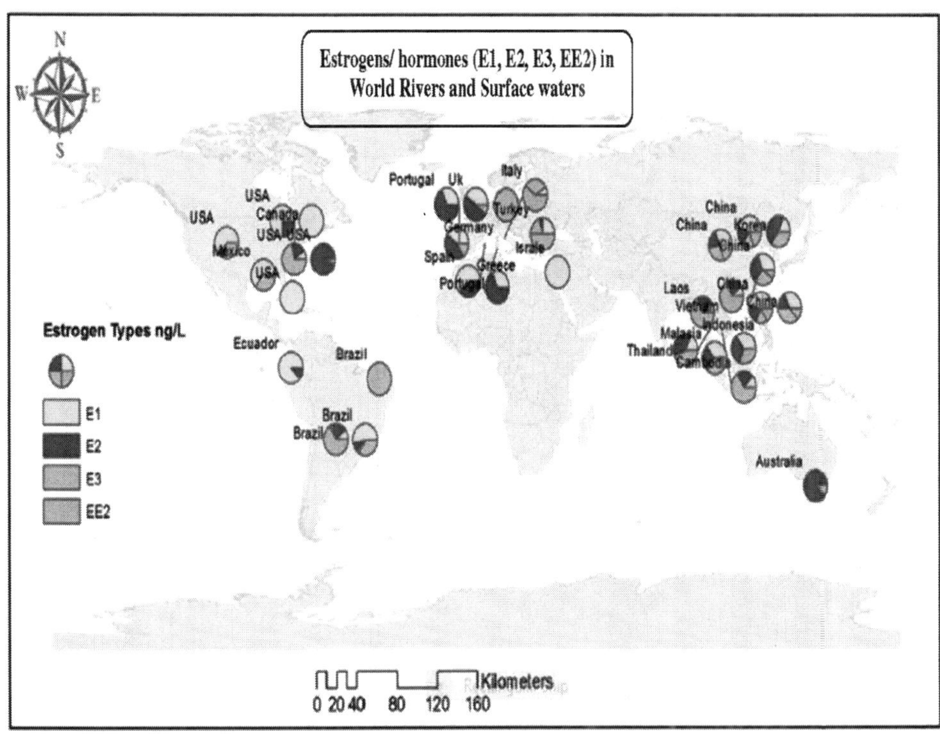

Fig. 9.2. Worldwide distribution of estrogens/ hormones[19] **(E1, E2, E3, EE2) in rivers and surface water sites** (Adeel et al 2017, page 111)

[Each pie chart comprises the natural estrogens: E1, E2, E3 and the synthetic EE2 as percentages of the total at each site; the Figure shows that there has been a major occurrence of estrogens in Rivers in the American mid-west, to the eastern sea board of North America (Canada, USA), to South America (Mexico, Ecuador, Brazil), and to countries bordering or close to the Mediterranean basin of Europe (Greece, Israel, Italy, Germany, Portugal, Spain, Turkey, UK), and to Asia (China, Cambodia, Indonesia, Korea, Laos, Thailand, Malaysia), and Australia (Southern Australia). The intensive farming in the mid-West of the USA and Southern Australia may have caused to the rise in estrogen levels in rivers in those areas. Elevated levels of EE2 occurred in Brazil, China, Cambodia, Indonesia, Thailand, and Vietnam].

[19] ***Estrogens*** *are hormones that are important for sexual and reproductive development, mainly in women. They are also referred to as female sex hormones.*

Table 9.8 Average concentrations of the 16 pharmaceuticals those found globally in surface water compartments (µg/L) (Aus der Beek et al 2016).

[Values in parentheses describe the number of samples; na = no data is available].

Pharmaceutical substance	Africa	Asia-Pacific	Eastern Europe	Latin America and Caribbean	Western Europe and others	Global
Acetylsalicylic acid/ aspirin	0.043 (6)	na	0.062 (29)	2.07 (112)	0.002 (107)	0.922 (254)
Carbamazepine	0.868 (14)	0.026 (18)	0.131 (302)	0.0 (10)	0.188 (24776)	0.187 (25115)
Ciprofloxacin	0.017 (6)	61.94 (206)	0.002 (22)	0.001 (95)	0.008 (343)	18.99 (672)
Clofibric acid	0.007 (6)	0.003 (3)	0.009 (48)	0.0007 (94)	0.023 (2796)	0.022 (2947)
Diclofenac	0.273 (14)	0.090 (152)	0.111 (420)	0.239 (130)	0.020 (6301)	0.032 (7017)
E1 (Estrone)	0.004 (96)	0.012 (159)	0.001 (262)	0.347 (79)	0.004 (1632)	0.016 (2228)
E2 (Estradiol)	0.0001 (46)	0.0003 (48)	0.0005 (20)	na	0.005 (183)	0.003 (297)
E3 (Estriol)	0.0 (34)	0.008 (66)	0.002 (131)	0.0002 (21)	0.002 (536)	0.009 (790)
EE2 (Ethinylestradiol)	0.0 (32)	0.008 (159)	0.002 (57)	1.175 (50)	0.005 (1238)	0.043 (1530)
Ibuprofen	3.181 (13)	0.059 (23)	0.183 (529)	0.0 (121)	0.097 (6264)	0.108 (6950)
Naproxen	0.012 (12)	0.008 (139)	0.023 (403)	0.017 (9)	0.057 (2666)	0.050 (3229)
Norfloxacin	0.076 (6)	9.11 (238)	0.004 (15)	0.0006 (95)	0.009 (274)	3.457 (628)

Pharmaceutical substance	Africa	Asia-Pacific	Eastern Europe	Latin America and Caribbean	Western Europe and others	Global
Ofloxacin	0.012 (6)	0.617 (206)	0.0003 (11)	0.335 (86)	0.123 (451)	0.278 (760)
Paracetamol	5.667 (6)	0.023 (132)	0.028 (68)	0.74 (112)	0.046 (629)	0.161 (937)
Sulfamethoxazole	2.53 (58)	0.258 (518)	0.033 (139)	0.001 (95)	0.068 (7789)	0.095 (8599)
Trimethoprim	0.985 (12)	0.128 (333)	0.012 (22)	0.005 (95)	0.022 (2598)	0.037 (3060)

Table 9.9. Concentrations of pharmaceuticals detected in drinking water in different countries.

*[Jones et al 2005; UL 2011; **AMDOPH** = (1-acetyl-1-methyl-2-dimethyloxamoyl-2-phenylhydrazide; **DP** = 1,5-dimethyl-1,2-dehydro-3-pyrazolone; **PDP** = 4-(2-methylethyl)-1,5-dimethyl-1,2-dehydro-3-pyrazolone)].*

Pharmaceuticals (range detected)	Therapeutic class	Drinking water (ng/L)	References
Acetaminophen/ paracetamol (210-219 ng/L)	NSAID and analgesics	210.1 ng/L (France) 219 mg/L (USA)	Togola and Budzinski 2008 UL 2011
AMDOPH (900-1,250 ng/L)	NSAIDs and analgesics	900-1,250 ng/L (Germany)	Heberer et al 2004, Reddersen et al 2002
Amitriptyline (1.4 ng/L)	Antidepressant, anti-anxiety	1.4 ng/L (France)	Togola and Budzinski 2008
Bezafibrate (27 ng/L)	Lipid regulator	27 ng/L (Germany)	Stumf et al 1996
Bleomycin (Cytotoxic drugs) (13 ng/L)	Antineoplastic	13 ng/L (UK)	Aherne et al 1990

Pharmaceuticals (range detected)	Therapeutic class	Drinking water (ng/L)	References
Caffeine (22.9-119 ng/L)	Psychostimulants	60–119 ng/L (USA)	Stackelberg et al 2004, Stackelberg et al 2007
		22.9 ng/L (France)	Togola and Budzinski 2008
Carbamazepine (10.3-258 ng/L)	Antiepileptic/ Analgesics	24 ng/L (Canada)	Tauber 2003
		43.2 ng/L (France)	Togola and Budzinski 2008
		60 ng/L (Germany)	Heberer et al 2004
		140–258 ng/L (USA)	Stackelberg et al 2004, Stackelberg et al 2007
		10.3 mg/L (USA)	UL 2011
Clofibric acid (5.3-270 ng/L)	Metabolite of lipid regulators	50-270 ng/L (Germany)	Heberer et al 1997, Stumpf et al 1996, Heberer et al 2004
		5.3 ng/L (Italy)	Zuccato et al 2000
Codeine (30 ng/L)	Opioidanalgesics	30 ng/L (USA)	Stackelberg et al 2007
Diatrizoate (1200 ng/L)	X ray contrast media	1200 ng/L (Germany)	Pérez and Barceló 2007
Diazepam (10-23.5 ng/L)	Antidepressant/ Psychiatric drug	10 ng/L (Italy)	Waggot 1981
		23.5 ng/L (Italy)	Zuccato et al 2000
Diclofenac (2.5-35 ng/L)	NSAIDs and analgesics	6-35 ng/L (Germany)	Stumpf et al 1996, Heberer et al 2004
	Analgesic anti-inflammatory	2.5 ng/L (France)	Togola and Budzinski 2008
Dilantin (1.3 ng/L)	Antiepileptics/ anticonvulsants	1.3 ng/L (USA)	Vanderford and Snyder 2006
DP (1.1 ng/L)	NSAIDs and analgesics	1.10 (Germany)	Zühlke et al 2004

Pharmaceuticals (range detected)	Therapeutic class	Drinking water (ng/L)	References
17α-Ethinyl estradiol (EE2) (0.65-1.36 ng/L)	Oral contraceptive/ estrogens	0.65 to 1.36 ng/L (USA)	MPCA 2013, MDH 2016
Gemfibrozil (70-264 ng/L)	NSAIDs and analgesics/ Lipid regulators	70 ng/L (Canada)	Tauber 2003
		264 mg/L (USA)	UL 2011
Ibuprofen (0.6-1350 ng/L)	NSAIDs and analgesics	3 ng/l (Germany)	Stumpf et al 1996
		0.6 ng/L (France)	Togola and Budzinski 2008
		8.5 ng/L (Finland)	Vieno et al 2005
		1350 ng/L (USA)	Loraine and Pettigrove 2006
Iopromide (<50 ng/L)	Iodinated X-ray contrast media	<50 ng/L (Germany)	Pérez and Barceló 2007
Ketoprofen (3-8 ng/L)	NSAIDs and analgesics	8 ng/L (Finland)	Vieno et al 2005
		3 ng/L (France)	Togola and Budzinski 2008
Meprobamate (5.9-42 ng/L)	Antidepressant, anti-anxiety	5.9 ng/L (USA)	Vanderford and Snyder 2006
		42 ng/L (USA)	Benotti et al 2009
Mestranol (0.7 ng/L)	Estrogens	0.7 mg/L	Erickson et al 2014, MDH 2016
PDP (0.24 ng/L)	Metabolites	0.24 ng/L (Germany)	Zühlke et al 2004
Phenazone (250-400 ng/L)	NSAIDs and analgesics	250-400 ng/L (Germany)	Zühlke et al 2004, Reddersen et al 2002
Primidone (40 ng/L)	antiepileptics/ anticonvulsants	40 ng/L (Germany)	Heberer et al 2004
Propyl phenazone (80-240 ng/L)	NSAID and analgesics	80-240 ng/L (Germany)	Zühlke et al 2004, Reddersen et al 2002, Heberer et al 2004

Pharmaceuticals (range detected)	Therapeutic class	Drinking water (ng/L)	References
Sulfamethoxazole (219-1200 ng/L)	Antibiotics	1200 ng/L (USA) 219 mg/L (USA	Jones et al 2005 UL 2011
Triclosan (734 ng/L)	Antibiotics	734 ng/L (USA)	Loraine and Pettigrove 2006
Tylosin (1.7 ng/L)	Antibiotic (growth promoter in livestock)	1.7 (Italy)	Zuccato et al 2000

9.3.2. Impacts on aquatic biota (bacteria, algae, daphnia, molluscs, fish), wildlife (frogs, birds) and crops

Pharmaceuticals or active pharmaceutical ingredients can enter aquatic organisms through their diet and exchange surfaces (gills, cuticles and epidermis). The uptake of soluble, hydrophilic compounds occurs mainly via the gills (fish, molluscs) or dermally via cuticles (invertebrates) and via epidermis (larvae), whereas uptake of less soluble, hydrophobic compounds more often occurs via a dietary route (Arnold et al 2014, Ne´stor and Mariana 2017). Wildlife (bats, birds and other insectivores) can be exposed to pharmaceuticals when they forage in wastewater treatment plants. Pharmaceuticals can also be taken up into plants or remain bound to the external surfaces of leaves (e.g. from use of contaminated biosolids as a natural fertiliser or contaminated wastewater for irrigation). Scavengers (such as vultures) can potentially be exposed to veterinary drugs from eating carcases of cattle treated with the drug-diclofenac (Arnold et al 2014).

Exposure to pharmaceuticals via different paths affected biota (bacteria, algae, daphnia, invertebrates, molluscs, fish), wildlife (frogs, birds), crops/ plants, biofilms and stream communities and ecosystems function etc. The following sections (9.3.2.1- 9.3.2.9) **summarise**s the impact of most frequently detected and most harmful pharmaceuticals to biota and wildlife. Table 9.10 provides a **detailed account** of the possible impact of individual pharmaceutical including acetaminophen/paracetamol, amphetamine, antibiotics, anticancer, anticonvulsant, antifungal, antihistamine, antiparasitic, atenolol, beta-blockers, caffeine, carbamazepine, cimetidine, ciprofloxacin, citalopram, clofibrate/ clofibric acid, diclofenac, EE2 (17α-ethinylestradiol), erythromycin, fenofibrate, fluoxetine, ibuprofen, indomethacin, ivermectin, ketoprofen, metformin, naproxen, oxazepam, pharmaceutical mixtures, psychiatric drugs, progestogens, propranolol, tetracycline, triclosan.

9.3.2.1. Antibiotics (drugs to kill microorganism)

Antibiotics (ciprofloxacin, erythromycin, tetracycline, triclosan) are toxic to biota and impacted bacteria, algae, molluscs, fish and plants/ crops and caused a diverse range of effects: **i.** ***Ciprofloxacin*** inhibited active growing microorganisms (Girardi et al 2011) and caused a decline in algal richness (Wilson et al 2003); **ii.** ***Erythromycin*** enhanced antimicrobial resistance in waterbodies and reduced growth of bacteria (OECD 2019)' inhibited photosynthesis-related processes (e.g. photochemistry, carbon assimilation) in microalgae, *Selenastrum capricornutum* (Liu et al 2011); reduced growth of algae and aquatic plants (OECD 2019); inhibited root elongation in lettuce plants (Pan and Chu 2016); **iii.** ***Tetracycline*** enhanced antibacterial resistance in soil bacteria (Sengelov et al 2003) and aquatic bacteria (Quinlan et al 2011); suppressed the immune system in rainbow trout (Grondel et al 1985) and changed the aquatic biotic community; abundance and productivity (bacteria, algae, cyanobacteria, nematodes and organic biomass) (Quinlan et al 2011). It (tetracycline) was the most toxic among the five veterinary antibiotics evaluated causing effects on seed germination and root elongation in lettuce, tomato, carrot and cucumber (Pan and Chu 2016); **iv.** ***Triclosan*** caused an increase of bacterial mortality (Ricart et al 2010) and antimicrobial-resistant bacteria (Proia et al 2011, Drury et al 2013, Nietch et al 2013); it affected photosynthesis and reproduction in green algae, *Scenedesmus vacuolatus* (Yang et al 2008); caused reduction in algal biomass (Lawrence et al 2009); stimulated the glutathione transferase and glutathione reductase enzymes activity in the Mediterranean mussel, *Mytilus galloprovincialis* (Canesi et al 2007); delayed embryonic development; caused spinal curvature and increased heartbeat rate in medaka fish (Nassef et al 2010b); induced edema in Zebra fish (Oliveira et al 2009, Saley et al 2016) and delayed hatching, and increased ChE (cholinesterase enzyme) and GSH (glutathione S-transferase enzyme) in zebra fish (Oliveira et al 2009).

9.3.2.2. Caffeine (stimulants)

Caffeine impacted algae and corals. **Algae:** Caffeine decreased biofilms biomass (algae) (Lawrence et al 2012). **Corals**: in the seawater, it may exacerbate the effects of other environmental factors that aid in coral bleaching (Edwards et al 2015).

9.3.2.3. Carbamazepine (anti-convulsant drugs)

Carbamazepine impacted algae, daphnia, mollusc, frogs, and fish. **Algae:** carbamazepine altered antioxidant enzymes [superoxide dismutase (SOD) and catalase (CAT)] activity in algae (Zhang et al 2012). **Daphnia:** it decreased growth (Dietrich et al 2010, Lürling et al 2006, Lamichhane et al 2013), fecundity in *Daphnia* (Lürling et al 2006, Lamichhane et al 2013), and inhibited acetylcholinesterase (the enzyme responsible for the hydrolytic metabolism) activity in rotifers (Rhee et al 2013). **Molluscs:** it affected gills of bivalve (Chen et al 2014), caused oxidative stress and protein damage in zebra mussel (Contardo-Jara et al 2011) and enhanced bioaccumulation in clam (Almeida et al 2015). **Fish:** it altered the brain physiology in Atlantic salmon (Hampel et al 2014), increased hatching time, altered feeding and swimming in medaka fish (Nassef et al 2010a,b), increased antioxidant enzyme (SOD, CAT, GR (glutathione reductase) activities in rainbow trout (Li et al 2011) and inhibited basal EROD (ethoxy resorufin-o-deethylase) activity in cultures of rainbow hepatocytes (Laville et al 2004). EROD is a highly sensitive indicator of contaminant uptake in fish.

9.3.2.4. Clofibrate/ clofibric acid (lipid regulators)

Clofibrate/ clofibric acid impacted fish. **Fish:** *Clofibrate* is toxic to mosquito fish. It induced a decrease in liver CAT activity and an increase in muscle lactate dehydrogenase activity (LDH) (Nunes et al 2004), caused oxidative stress, and inhibited EROD) activity in rainbow trout fish (Lawrence et al 2012). *Clofibric* acid lowered sperm count and plasma androgen concentration in fathead minnow fish (Runnells et al 2007).

9.3.2.5. Diclofenac (NSAID-Non-steroidal anti-inflammatory drugs)

Diclofenac impacted algae, mollusc, frogs, fish and birds. **Algae:** diclofenac decreased biofilm/ algal biomass (Lawrence et al 2012). **Molluscs:** it caused genotoxicity, neurotoxicity and oxidative stress in molluscs (OECD 2019) as well as bioaccumulated in the Baltic Sea blue mussels (Ericson et al 2010). **Frogs:** it caused hormones disruption in frogs and damaged organs (OECD 2019). **Fish:** it bioaccumulated in trout, bream and roach fishes (Schwaiger et al 2004, Kallio et al 2010, Brozinski et al 2013), hindered lipid synthesis (Hoeger et al 2005) and damaged gills, liver, and kidney of brown trout fish (Hoeger et al 2005, Brozinski et al 2013), caused necrosis (tissue death) and hyperplasia (enlargement of tissue) in fish (Triebskorn et al 2007, Mehinto et al 2010) e.g. rainbow trout (Triebskorn et al 2007), reduced hatching success in fish (OECD 2019), inhibited cytochrome P450 enzyme[20] (CYP2M[21]) activity in carp (Thibaut

---

[20]***Cytochrome P450*** *enzyme are called because they are bound to membranes within a* ***cell (cyto)*** *and contain a* ***heme pigment*** *(chrome and P) that absorbs light at a wavelength* ***of 450 nm*** *when exposed to carbon monoxide; cytochrome P45 enzyme is primarily found in liver cells but are also located in cells throughout the body; this enzyme is essential for the production of numerous agents including cholesterol and steroids, moreover, these enzymes are necessary for the detoxification of foreign chemicals and the metabolism of drugs (https://www.drugguide.com/ddo/view/Davis-Drug-Guide/109519/all/The_Cytochrome_P450_System:_What_Is_It_and_Why_Should_I_Care). https://ghr.nlm.nih.gov/primer/genefamily/cytochromep450*

et al 2006), inhibited basal EROD activity in cultures of rainbow hepatocytes (Laville et al 2004). **Birds:** it caused a decline of vulture (scavenging bird of prey) population in the Indian sub-continent due to the birds' feeding on the carcasses of cattle treated with the drug diclofenac and dying due to kidney failure (Oaks et al 2004) (LD_{50} of diclofenac to vultures is 0.1 mg/kg (Swan et al 2006).

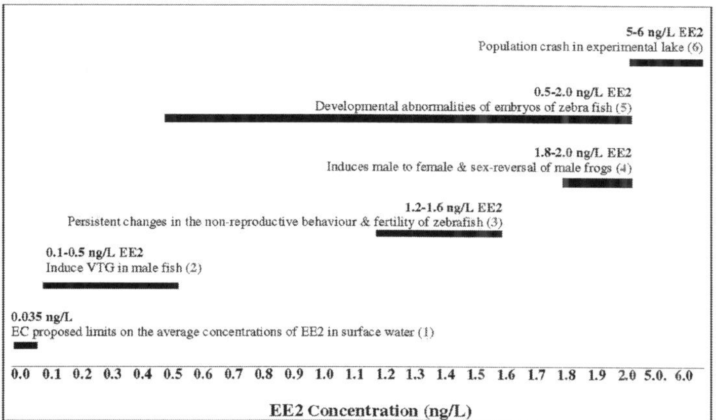

Fig. 9.3. Harmful effects in fish associated with different EE2 concentrations in surface waters

*(**1.** Gilbert 2012, **2.** Purdom et al 1994, **3.** Volkova et al 2015, **4.** Gyllenhammar et al 2009, **5.** Soares et al 2009, **6.** Kidd et al 2007; **EC** = European Commission; **EE2** = ethinyl estradiol; **VTG**= Vitellogenin[22]).*

***Note about EE2:** More than 100 million women worldwide use contraceptive pills. The active ingredient of contraceptive pills in most formulations is ethinyl estradiol or EE2. About 57% of EE2 is excreted in faeces and urine ((Johnson and Williams 2004). EE2 and other estrogens cause widespread damage in the aquatic environment by disrupting endocrine systems in wildlife (intersex, feminisation of male fish, lower sperm counts, and irreversible development of eggs in the testes of male fish). The European Commission (**EC**) proposed in January 2012 that the EU's member states limit the drug's annual average concentrations in surface waters to no more than **0.035 nanograms per litre** (ng/L). The commission has also proposed that lawmakers to make law regarding the use of **diclofenac** that disrupts cell function in the liver, kidneys and gills of fish and devastated vulture populations in Asia (Gilbert 2012).*

---

[21] ***CYP2M*** *= There are approximately 60 cytochrome P450 genes in humans. Each cytochrome P450 gene is named with CYP, indicating that it is part of the cytochrome P450 gene group. The gene is also given a number associated with a specific group within the gene group, a letter representing the gene's subgroup, and a number assigned to the specific gene within the subgroup. For example, the cytochrome P450 gene that is in the group 27, subgroup A, gene 1 is written as CYP27A1 (https://ghr.nlm.nih.gov/primer/genefamily/cytochromep450).*

[22] ***Vitellogenin*** *= Vitellogenin is an egg yolk precursor protein expressed in the females of nearly all oviparous (egg-laying) species including fish, amphibians, reptiles, birds, and most invertebrates. They are a source of nutrients during early development of egg-laying vertebrates and invertebrates. It is used as a biomarker in vertebrates of exposure to environmental estrogens which stimulate elevated levels in males as well as females (https://en.wikipedia.org/wiki/Vitellogenin).*

9.3.2.6. EE2 (17α-ethinylestradiol) (birth control pills/ hormones)

EE2 impacted frogs and fish, crops/ plants. **Frogs**: it caused sex reversal and reduced fertility in male frogs (Gyllenhammar et al 2009). **Fish:** it caused developmental abnormalities and induced VTG (vitellogenin; VTG is egg yolk precursor protein) gene in male zebra fish (Schmid et al 2002, Soares et al 2009) and male rainbow trout (Purdom et al 1994). Artificial exposure of EE2 caused a complete collapse of the fathead minnow's fish population in a Canadian lake (Kidd et al 2007). Brain and head kidney StAR (steroidogenic acute regulatory) protein and cytochrome P450 enzyme expression in salmon was modulated/ modified from waterborne exposure (Lyssimachou and Arukwe 2007). **Plants/ crops**: Plants can accumulate estrogen hormones (E1, E2, and EE2) in both roots and shoots and hormones or their precursors can affect root and shoot development, flowering and germination (Adeel et al 2017) (see Fig. 9.3 for the harmful effects in fish associated with EE2).

9.3.2.7. Fluoxetine (antidepressants/selective serotonin reuptake inhibitor)

Fluoxetine impacted algae, invertebrates, molluscs, frogs, and fish. **Algae:** it suppressed primary productivity (Richmond et al 2016). **Invertebrates:** it decreased activity of amphipod (amphipods are crustaceans and important food for fish, bird and amphibian species) (De Lange et al 2006). **Molluscs:** it affected reproduction in mollusc (Sanchez-Arguello et al 2009), accumulated in the tissue of freshwater mussel, *Elliptio complanata* (Bringolf et al 2010). **Frogs:** exposure to fluoxetine caused tadpoles of *Bufo arabicus* to be more susceptible and vulnerable to predation (Barry 2014). **Fish:** it decreased feeding and growth rate (Stanley et al 2007) and bioaccumulated in fathead minnows (Brooks 2014) and caused endocrine disruption in medaka fish (Brooks et al 2003) and goldfish (Mennigen et al 2017).

9.3.2.8. Ibuprofen (NSAIDs and analgesics)

Ibuprofen impacted invertebrates, molluscs, frogs, and fish. **Invertebrates**: it caused significant decrease in the activity of amphipod (De Lange et al 2006), reduced fecundity (Heckmann et al 2007) and reduced amino acids levels in *Daphnia* (Kovacevic et al 2016). **Molluscs:** it induced apoptosis (death of cells) in bivalve mussel (Parolini et al 2009, Rocco et al 2010), altered antioxidant and detoxifying enzyme activities in the zebra mussel, *Dreissena polymorpha* (Parolini et al 2011). **Frogs:** it can cause hormones disruption in frogs (OECD 2019). **Fish:** it reduced hatching success in fish (OECD 2019), decreased spawning rates in medaka fish (Flippin et al 2007, Han et al 2010), caused endocrine disruption [e.g. induced VTG in male medaka fish, altered testosterone and 17β-estradiol (E2) concentrations] in medaka fish (Han et al 2010), impaired/

disturbed ion regulation (Cl^-, K^+), metabolic and cortisol (steroid hormone) responses in rainbow trout (Gravel et al 2009) and inhibited CYP2M (cytochrome P450 enzyme) catalysed activity in carp (Thibaut et al 2006).

9.3.2.9. Pharmaceutical mixtures

Pharmaceuticals do not occur alone in the environment but as a mixture of different active substances, their metabolites and transformation products. Experimental results reveal that mixed pharmaceuticals exhibited much more toxic effects (additive effects) on biota (algae, daphnia and fish) compared to single pharmaceutical. Some examples are highlighted below: **i.** exposure of pharmaceuticals mixture (acetaminophen, carbamazepine, gemfibrozil, venlafaxine) significantly decreased embryo production (developmental abnormalities and embryo mortality) in zebra fish. Embryo mortality was elevated/ higher with exposure to the high concentration of mixed pharmaceuticals (Galus et al 2013); **ii**. a mixture of three pharmaceuticals (erythromycin, triclosan and trimethoprim) caused a shift in sex ratio in *Daphnia* population (where the individual components did not provoke any significant effects) (Flaherty and Dodson 2005); **iii. a** mixture of two pharmaceuticals (fluoxetine and clofibric acid) killed more than 50% of a *Daphnia* population after exposure of six days (while there are no apparent effects for the same concentrations of individual pharmaceutical) (Flaherty and Dodson 2005); **iv.** A mixture of diclofenac and ibuprofen had strong toxic effects on *Daphnia magna*, while diclofenac alone did not exhibit any effects (Cluvers 2003) (see also Table 9.10).

Table 9.10. Impacts of individual pharmaceuticals on aquatic biota (bacteria, algae, daphnia, molluscs), wildlife (frogs, birds) and crops.

Pharmaceuticals compound	Impacts	References
Amphetamine *(illicit drugs[23])*	An environmentally relevant concentration (at 1 μg/L) of amphetamine suppressed biofilm[24]	Lee et al 2016

---

[23]***Illicit drugs*** *are substances that either stimulate (such as cocaine or amphetamines) or inhibit (such as heroin or sedative-hypnotics) the central nervous system or cause hallucinogenic effects (such as marijuana or LSD), hence their use has been prohibited globally.*

[24]***Biofilm*** *is defined as an assemblage of microbial cells that is irreversibly associated with a surface and enclosed in a matrix of primarily polysaccharide material. Biofilms also termed as micro-fouling, are composed of different species of bacteria, diatoms and protozoa surrounded by a matrix of extra polymeric substances (EPS). Biofilms in aquatic ecosystems colonize various surfaces (sand, rocks, leaves) and play a key role in the environment. Aquatic biofilms supply energy and organic matter to the food chain; they are important in recycling organic matter and contribute to water quality (Romani et al 2016). Stream biofilms are the base of the food web for streams, and negative impacts on these microbial communities by exposure to pharmaceuticals may affect both nutrient cycling in an ecosystem and higher trophic levels that depend on biofilms.*

Pharmaceuticals compound	Impacts	References
	primary production (lowered chlorophyll-*a* and seston[25] communities).	
Antibiotic mixture	Exposure to low concentrations of an antibiotic mixture (sulfamethoxazole, trimethoprim, erythromycin, roxithromycin, and clarithromycin at 2 μg/L) led to shifts in leaf microbial communities, resulting in an increased amphipod (*Gammarus* spp.) body mass.	Bundschuh et al 2017
Antibiotics[26]	Discharge of antibiotics to water bodies can exacerbate the problem of antimicrobial resistance[27].	OECD 2019
	Caused reduced growth of environmental bacteria, algae and aquatic plants.	OECD 2019
	Antibiotics (*erythromycin, ciprofloxacin and sulfamethoxazole*) could inhibit the primary photochemistry, electron transport, photophosphorylation[28] and carbon assimilation in algae, *Selenastrum capricornutum;* photosynthesis of green algae can be affected by exposure of antibiotics, in particular, erythromycin was found more toxic than ciprofloxacin and sulfamethoxazole.	Liu et al 2011
Anti-cancer drugs	Cyclophosphamide, mitomycin-C, fluorouracil etc. can cause genotoxic[29] effects on biota.	OECD 2019
Anti-convulsant[30] drugs	Carbamazepine, phenytoin, valproic acid can cause a delay in fish development, reproduction toxicity in invertebrates.	OECD 2019

[25] ***Seston*** *= are composed of both living organisms (such as plankton and nekton) and non-living matter (such as plant debris or suspended soil particles).*

[26] ***Antibiotic*** *= drugs that inhibit the growth of or destroy desired/ targeted microorganisms.*

[27] ***Antimicrobial resistance*** *(AMR) = is the ability of a microbe to resist the effects of medication that could once successfully destroy or inhibit the microbe. AMR is a global health crisis with the potential for enormous health, food security and economic consequences (OECD 2019).*

[28] ***Photophosphorylation*** *= the process of utilising light energy during photosynthesis to convert ADP (Adenosine diphosphate) to ATP (Adenosine triphosphate).*

[29] ***Genotoxic*** *= chemical agents that damages the genetic information within a cell causing mutations, which may lead to cancer.*

[30] ***Anti-convulsant*** *= also commonly known as antiepileptic drugs or as antiseizure drugs, are a diverse group of pharmacological agents used in the treatment of epileptic seizures.*

Pharmaceuticals compound	Impacts	References
Antifungals drugs	Clotrimazole, ketoconazole, triclosan etc. can cause reduced growth in algae and fish.	OECD 2019
Antihistamines	Fexofenadine, hydroxyzine, diphenhydramine, etc. can affect behaviour, growth and feeding rate in fish and behaviour and reproduction in invertebrates.	OECD 2019
Antiparasitic	Ivermectin reduces growth and reproduction in invertebrates.	OECD 2019
Atenolol *(beta-blockers)*[31]	Impaired growth in fathead minnow fish (*Pimephales promlas*).	Winter et al 2006
Beta-blockers *(propranolol)*	Affected behaviour (in fish) and reproduction (invertebrates)	OECD 2019
Caffeine *(stimulants)*	Caused increased an abundance of organisms (protozoa, micrometazoa) and feeding activity and decreased biofilm (algae) biomass at 5 μg/L.	Lawrence et al 2012
Carbamazepine *(analgesics/ antiepileptics)*	Carbamazepine altered the brain physiology of juvenile Atlantic salmon (*Salmo salar).*	Hampel et al 2014
	Carbamazepine altered the survival of medaka fish embryos with an EC_{50} at 13.1 ng/egg. Hatching time significantly increased under carbamazepine exposure.	Nassef et al 2010b
	Carbamazepine at 100 or 1,000 μg/ L inhibited AChE[32] (acetylcholinesterase) activity in the monogonont rotifer (*Brachionus koreanus*), as well as decreased AChE mRNA expression under the concentration and time-dependent carbamazepine exposure.	Rhee et al 2013

*[31] **Beta blockers** are a class of medications that are predominantly used to manage abnormal heart rhythms, and to protect the heart from a second attack after a first heart attack.*

*32 **Acetylcholinesterase (AChE)** is the primary **enzyme** responsible for the hydrolytic metabolism of the neurotransmitter acetylcholine (ACh) into choline and acetate.*

Pharmaceuticals compound	Impacts	References
	Carbamazepine exposure was shown to alter the feeding behaviour and swimming speed in Japanese medaka fish (*Oryzias latipes*).	Nassef et al 2010a
	Carbamazepine is relatively hydrophobic/ bioaccumulative.	Liu et al 2015
	Caused developmental abnormalities in the zebra fish, *Danio rerio*.	van den Brandhof and Montforts 2010
	Caused growth retardation in the water flea, *Daphnia magna*.	Dietrich et al 2010
	Fish exposed to carbamazepine have displayed increased antioxidant[33] enzyme activities, such as SOD (superoxide-dismutase[34]), CAT (catalase[35]), GR (glutathione reductase[36]) and GPx (glutathione peroxide[37]) in the liver, muscle and intestine of rainbow trout (*Oncorhynchus mykiss*).	Li et al 2011
	Gills of freshwater bivalve, *Corbicula fluminea* has been shown to be significantly affected by carbamazepine.	Chen et al 2014

[33] ***Antioxidant*** *= A substance that* ***reduces the damage due to oxygen*** *such as that caused by free radicals (an atom, molecule, or ion that has an unpaired valence electron). Well-known antioxidants include* ***enzymes****, which are capable of counteracting the damaging effects of oxidation.* ***Antioxidant enzymes*** *are proteins involved in the catalytic transformation of reactive oxygen species and their by-products into stable non-toxic molecules, therefore, representing the most important defence mechanism against oxidative stress-induced cell damage.*

[34] ***Superoxide dismutase*** *(SOD) is a group of* ***enzymes*** *that catalyse the dismutation (partitioning) of superoxide radicals (O_2^-) to molecular oxygen (O_2) and hydrogen peroxide (H_2O_2), providing cellular defence against reactive oxygen species.*

[35] ***Catalase (CAT)*** *is a common* ***enzyme*** *found in nearly all living organisms exposed to oxygen (such as bacteria, plants, and animals). It catalyses the decomposition of hydrogen peroxide to water and oxygen. It is a very important enzyme in protecting the cell from oxidative damage by reactive oxygen species (ROS).*

[36] ***Glutathione reductase*** *(GR) is an* ***enzyme*** *that catalyses the reduction of glutathione disulfide (GSSG) to the sulfhydryl form glutathione (GSH), which is a critical molecule in resisting oxidative stress and maintaining the reducing environment of the cell.*

[37] ***Glutathione peroxide*** *is the general name of an* ***enzyme*** *family with peroxidase activity whose main biological role is to protect the organism from oxidative damage.*

Pharmaceuticals compound	Impacts	References
	High carbamazepine concentrations have been shown to promote decreasing rates of *Daphnia* population growth and delayed maturity and significantly decreased fecundity[38].	Lürling et al 2006, Lamichhane et al 2013
	High carbamazepine levels can promote both bioaccumulation and LPO (lipid peroxidation[39]) because this compound is able to inhibit SOD and CYP3A4[40]. The accumulation of carbamazepine in clam tissues, *Ruditapes philippinarum* increased with the increase of exposure concentration.	Almeida et al 2015
	Inhibited basal EROD (ethoxyresorufin-O-deethylase) activity in the cultures of rainbow trout hepatocytes. EROD is a highly sensitive indicator of contaminant uptake in fish.	Laville et al 2004
	Rat embryos exposed to carbamazepine suffered pericardial edema[41] and delayed heartbeat as the main sensitive endpoints in the FET (*Fish Embryo Acute Toxicity)* test.	van den Brandhof and Montforts 2010
	Carbamazepine exposure to aquatic organisms caused oxidative stress and protein damage in the zebra mussel *Dreissena polymorpha*.	Contardo-Jara et al 2011
	Several algae have displayed altered antioxidant enzyme activity after carbamazepine exposure. Green algae, *Scenedesmus obliquus* and *Chlorella pyrenoidosa*, showed increases in both SOD (superoxide-dismutase) and CAT (catalase) activity when exposed to carbamazepine.	Zhang et al 2012

[38] ***Fecundity*** *is the natural capability to produce offspring, measured by the number of gametes (eggs).*

[39] ***Lipid peroxidation*** *is the oxidative degradation of lipids. It is the process in which free radicals "steal" electrons from the lipids in cell membranes, resulting in cell damage*

[40] ***CYP3A4*** *(or Cytochrome P450 3A4) is an important* ***enzyme*** *in the body, mainly found in the liver and in the intestine. It oxidizes small foreign organic molecules (xenobiotics), such as toxins or drugs, so that they can be removed from the body system.*

[41] ***Edema*** *= swelling of a cell with excessive fluids*

Pharmaceuticals compound	Impacts	References
Cimetidine *(antihistamines)*	Reduced growth and biomass of invertebrates (*Gammarus fasciatus*) at 0.07 µg/L and caused low survival of invertebrates (*Psephenus herricki*) when exposed to high concentrations (at 70 µg/L).	Hoppe et al 2012
Ciprofloxacin *(antibiotics)*	Caused marked shifts in the community structure of suspended and attached algae at both upstream and downstream sites of wastewater treatment plants and caused a decline in algal genus richness.	Wilson et al 2003
	Highly toxic and inhibited active and growing microorganisms (EC_{50} to bacteria, *Pseudomonas putida,* is 0.25 mg/L).	Girardi et al 2011
Citalopram *(antidepressant)*	Caused suppression of primary productivity and community respiration (algae).	Richmond et al 2016
Clofibrate *(lipid regulator)*	Acutely toxic to mosquito fish, *Gambusia holbrooki*. Acute exposure of *G. holbrooki* to both clofibrate and clofibric acid induced a decrease in liver catalase (CAT) activity and an increase in muscle lactate dehydrogenase[42] activity (LDH)[43].	Nunes et al 2004
	Induced oxidative stress[44], inhibited EROD[45] activity (EROD = ethoxy resorufin-o-deethylase) (CYP1a) in rainbow trout fish. Clofibrate was found cytotoxic (toxic to cells) to fish.	Laville et al 2004

[42] ***Lactate dehydrogenase*** *(also called lactic acid dehydrogenase or LDH) is* ***an enzyme*** *found in almost all body tissues.*

[43] ***LDH*** *is an enzyme found in almost all body tissues. It plays an important role in cellular respiration, the process by which glucose (sugar) from food is converted into usable energy for the cell.*

[44] ***Oxidative stress*** *results from an imbalance between the production of reactive oxygen species and the antioxidants defences, in favour of the former. Oxidative stress indicators are increasingly being used as biomarkers to understand the mechanisms of various human-induced stressors (Birnie-Gauvin et al 2017).*

[45] ***EROD*** *(Ethoxyresorufin-O-deethylase) is a highly sensitive indicator of contaminant uptake in fish. EROD activity in fish is a well-established in* ***vivo*** *(within the living organism) biomarker of exposure to certain planar halogenated and polycyclic aromatic hydrocarbons (https://www.cerc.usgs.gov/pubs/BEST/EROD.pdf).*

Pharmaceuticals compound	Impacts	References
Clofibric acid (*metabolite of clofibrate)*	Lowered sperm count and plasma androgen reported at 10–1,000 μg/L (0.01–1 mg/L) concentration in fathead minnow (*Pimephales promelas*).	Runnells et al 2007
Diclofenac *(NSAID)*	Diclofenac and its metabolites have been found to accumulate in the bile of rainbow trout (*O. mykiss*) (Kallio et al 2010), and in the wild fish species (bream, *Abramis brama*), roach (*Rutilus rutilus)* sampled downstream of a Finnish wastewater treatment facility (Brozinski et al 2013).	Kallio et al 2010, Brozinski et al 2013
	Diclofenac has been linked with a major ecological disaster leading to the decline of vulture population in the Indian subcontinent.	Oaks et al 2004
	Can affect the expression of multiple genes (at concentrations of 1–500 μg/L in surface water) in exposed fish (rainbow trout, *O. mykiss*).	Cuklev et al 2011
	Inhibited basal EROD (ethoxyresorufin-O-deethylase) activity in cultures of rainbow trout hepatocytes. EROD is a highly sensitive indicator of contaminant uptake in fish.	Laville et al, 2004
	Caused renal lesions[46] and alterations of rainbow trout, *Oncorhynchus mykiss* gills at 5 μg/L; bioaccumulated in all fish organs examined (Liver> kidney> gills> muscle tissue).	Schwaiger et al 2004
	Hindered prostaglandin (lipid) synthesis, damage to gill, liver, and kidney in brown trout (*Salmo trutta*) at 0.50–50 μg/L (0.0005–0.05 mg/L).	Hoeger et al 2005
	Caused tubular necrosis[47] in the kidney and hyperplasia[48] and fusion of the villi[49] in the intestine of fish rainbow trout, *Oncorhynchus mykiss* at 1 μg/L; exposure to environmental concentrations of diclofenac can interfere with the biochemical functions of fish and lead to tissue damage.	Mehinto et al 2010

[46] ***Lesion** = damage.*

[47] ***Necrosis** = tissue death*

[48] ***Hyperplasia** = the enlargement of an organ or tissue caused by an increase in the duplication/ reproduction rate of its cells, often as an initial stage in the development of cancer.*

[49] ***Villi** = small, finger-like projections that extend into the lumen of the small intestine.*

Pharmaceuticals compound	Impacts	References
	Inhibited CYP2M catalysed activity in carp (*Cyprinus carpio*) at 296.1500 (1 mM) (CYP = cytochrome P450 enzymes; mM = milli molar).	Thibaut et al 2006
	Increased abundance of organisms (protozoa, micrometazoa) and feeding activity (direct) and decreased biofilm (algae) biomass (indirect) at 5 μg/L.	Lawrence et al 2012
	Caused damage to organs of fish, frogs and molluscs; reduced hatching success in fish; caused genotoxicity[50], neurotoxicity and oxidative stress in molluscs; caused hormones disruption in frogs.	OECD 2019
	Diclofenac is suspected of causing damage to the inner organs (liver, kidney and gills) in rainbow trout, *O. mykiss* (e.g. collapse of cellular compartmentation of the liver, necrosis of endothelial cells of kidney and necrosis, hyperplasia and hypertrophy[51] of chloride cells of gills).	Triebskorn et al 2007
	The consumption of carcases of livestock (that were medicated with diclofenac) resulted in the death of over 95% of *Gyps bengalensis* vultures due to kidney failure. This has resulted in their widespread decline and even localised population extinctions (Oaks et al 2004, Cuthbert et al 2014) *[note: LD_{50} of diclofenac to vultures is 0.1 mg/kg (Swan et al 2006); the* Oriental white-backed vulture, *G. bengalensis was once one of the most common raptors*[52] *in the Indian subcontinent].*	Oaks et al 2004

[50] ***Genotoxicity*** *= describes the property of chemical agents that damages the genetic information within a cell causing mutations, which may lead to cancer.*

[51] ***Hypertrophy*** *= the enlargement of an organ or tissue from the increase in size of its cells.*

[52] ***Raptors*** *= Birds of prey, or raptors, include species of bird that primarily hunt and feed on vertebrates that are large relative to the hunter. They have keen eyesight for detecting food at a distance or during flight at high altitudes, strong feet equipped with talons for grasping or killing prey, and powerful, curved beaks for tearing flesh.*

Pharmaceuticals compound	Impacts	References
	Diclofenac (D) and propranolol (P) bioaccumulated in the Baltic Sea blue mussels, *Mytilus edulis trossulus* (have bioconcentration factors (BCFs[53]) of 10–180 in blue mussels); exposure of D and P caused lower byssus[54] strength, lower scope of growth.	Ericson et al 2010
EE2 (*17α-ethinylestradiol*) *(birth control pills/ hormones)*	Caused developmental abnormalities of eggs and embryos in zebra fish, increased embryo mortality and induction of the VTG (vitellogenin) gene in male zebra fish.	Soares et al 2009
	EE2 in the water (at 5 ng/ L) has been shown to cause reproductive failure in colonies of laboratory-maintained zebra fish.	Nash et al 2004
	Exposure of roach (*Rutilus rutilus*), living in the UK rivers (downstream of WWTW (wastewater treatment works discharges to 4 ng/L of EE2 for a 3-year period) resulted in feminisation of the exposed populations.	Lange et al 2009
	Larval exposure to environmentally relevant concentrations of EE2 (at 1.8 ng/L) induced sex-reversal of *Xenopus tropicalis* frogs leading to reduced fertility in the adult frogs.	Gyllenhammar et al 2009
	Low doses (at 1.2-1.6 ng/L) of EE2 resulted in persistent changes in the behaviour and fertility of zebra fish.	Volkova et al 2015
	The dosing of a lake in Canada (Lake 206) with 5–6 ng/L of EE2 resulted in the collapse of the population of the fathead minnows (*Pimephales promelas*).	Kidd et al 2007

[53] ***Bioconcentration*** *is the accumulation of a chemical in or on an organism when the source of chemical is solely water. Bioconcentration factor is the ratio of the concentration of a chemical in an organism/ biota to the concentration of the chemical in the surrounding environment (BCF = concentration in biota/ concentration in water) (Gobas and Morrison 2000).*

[54] ***Byssus*** *is a bundle of filaments secreted by many species of bivalve mollusc that function to attach the mollusc to a solid surface.*

Pharmaceuticals compound	Impacts	References
	Fish (*P. promelas*) exposed to 50 ng/L of EE2 for 35-days caused VTG induction in males and 100% mortality.	Schmid et al 2002
	EE2 has been shown to induce VTG production in male rainbow trouts at 0.1 to 0.5 ng/L.	Purdom et al 1994
	EE2 is extremely potent in fish and environmentally relevant concentrations (at 1 ng/L) have been shown to induce feminisation in fish, including induction of the female yolk precursor[55] vitellogenin (VTG) in males.	Thorpe et al 2007, Corcoran et al 2010, Caldwell et al 2012
	Exposure of endocrine disruptor (EE2) at 5 or 50 ng/L to Atlantic salmon impaired a time-dependent steroid production in Atlantic salmon.	Lyssimachou and Arukwe 2007
	EE2 at concentrations of 5 ng/L to 6 ng/L has been demonstrated to cause population collapse of fathead minnow, *Pimephales promelas* (as a result of the feminisation of male fish in a Canadian whole-lake experiment).	Kidd et al 2007
	Plants can accumulate estrogens in both roots and shoots; treatment of plants with steroid estrogen hormones or their precursors can affect root and shoot development, flowering and germination.	Adeel et al 2017
Erythromycin *(antibiotics)*	Caused phytotoxicity in lettuce (inhibited root elongation). The median effect concentration (EC_{50}) of erythromycin to lettuce was 68.8 mg/L.	Pan and Chu 2016
	Erythromycin caused toxic effects at environmentally relevant levels for the luminescent terrestrial bacterium, *Photorhabdus luminescens* (EC_{50}:= 0.022 mg/L) and for the algae *P. subcapitata* (EC_{50}:= 0.35 mg/L).	González et al 2016

[55] ***Precursor*** *= a substance, cell or cellular component from which another substance, cell or cellular component is formed.*

Pharmaceuticals compound	Impacts	References
	Erythromycin was found to be the most harmful to sensitive algae group (algae >daphnids >fish).	Liu et al 2015
	Many photosynthesis-related processes (photochemistry, electron transport, and photophosphorylation and carbon assimilation) were inhibited in *Selenastrum capricornutum* by low concentrations (0.06 mg/L).	Liu et al 2011
Estrogens, progestogens	Estrogens and progestogens can impair the reproductive functions of amphibians, through their effects on vitellogenesis and reproductive behaviour.	Sa¨fholm et al 2014
Fenofibrate *(lipid regulator)*	Inhibited EROD activity (CYP1a) in rainbow trout fish; Fenofibrate was also found cytotoxic (toxic to cells) to fish.	Laville et al 2004
Florfenicol *(veterinary medicine)*	Florfenicol was found to be more toxic to *Daphnia* at an ambient temperature of 25°C compared to 20°C.	Martins et al 2013
Fluoxetine *(psychiatric drugs/ antidepressants)*	Affected reproduction of freshwater mollusc, *Physa acuta.*	Sanchez-Arguello et al 2009
	Bioaccumulated in the fatty tissue in freshwater fish, specifically in the brain, and caused effects such as anxiety behaviour in adult male fathead minnow, *P. promelas.*	Brooks 2014
	Decreased ability to catch prey hybrid striped bass (*Morone saxatilis* x *M. chrysops*) at 23.20–100 µg/L (0.0232–0.1 mg/L).	Gaworecki and Klaine 2008
	Decreased feeding rate and growth rate in fathead minnow, *P. promelas.*	Stanley et al 2007
	Decreased territorial aggression in Bluehead wrasse, *Thalassoma bifasciatum* (coral reef fish) at 6 µg/L (0.006 mg/L).	Perreault et al 2003

Pharmaceuticals compound	Impacts	References
	Exposure (10-100 ng/L) to amphipod, *Gammarus pulex*, resulted in a significant decrease in their activity	De Lange et al 2006
	Increased estradiol levels and developmental abnormalities in medaka fish, *Oryzias latipes* at 0.10–5.00 μg/L (0.0001–0.005 mg/L).	Brooks et al 2003
	Waterborne exposure to environmentally relevant levels of fluoxetine caused endocrine disruption[56] in goldfish (a significant decrease in circulating E2, estradiol) exposed to 0.54 μg/L fluoxetine).	Mennigen et al 2017
	Fluoxetine accumulated in the freshwater mussel, *Elliptio complanata* tissue. It has a bioconcentration factor of over 1,000 in freshwater mussels. It has the potential to disrupt several aspects of reproduction in freshwater mussels.	Bringolf et al 2010
	Caused suppression of primary productivity by 23% and community respiration by 43% on biofilms (algae) at low concentrations.	Richmond et al 2016
	Exposure to fluoxetine (*selective serotonin re-uptake inhibitor)* caused tadpoles of *Bufo arabicus* to be more susceptible and vulnerable to predation from predatory dragonfly larvae (*Anax imperator*).	Barry 2014
	Altered fish behaviour making them vulnerable to predators.	OECD 2019
Gemfibrozil *(lipid regulator)*	Inhibited CYP2M catalysed activity in carp, *Cyprinus carpio* at 296.1500 (1 mM) (CYP = cytochrome P450 enzymes; mM = milli molar).	Thibaut et al 2006

[56] (See Table 9.3)

Pharmaceuticals compound	Impacts	References
Ibuprofen *(NSAIDs and analgesics/ pain killer)*	Can damage organs of fish, frog and molluscs; reduced hatching success in fish; can cause genotoxicity, neurotoxicity and oxidative stress in molluscs; and can also cause hormones disruption in frogs.	OECD 2019
	Induced apoptosis[57] in bivalve zebra mussel haemocytes (Parolini et al 2009) and zebrafish erythrocytes (Rocco et al 2010) (*note: ibuprofen concentrations may cause DNA damage, leading to apoptosis*).	Parolini et al 2011, Rocco et al 2010
	Exposure of ibuprofen (10-100 ng/L) to amphipod[58], *Gammarus pulex,* resulted in a significant decrease in their activity.	De Lange et al 2006
	Exposure of ibuprofen affected growth rate and reproduction in *Daphnia magna* (reduced fecundity).	Heckmann et al 2007
	Exposure to ibuprofen altered antioxidant and detoxifying enzyme activities in the zebra mussel, *Dreissena polymorpha*.	Parolini et al 2011
	Ibuprofen caused endocrine disruption (reproductive disorders) including altered testosterone and 17β-estradiol (E2) concentrations and induced VTG in male medaka fish.	Han et al 2010
	Ibuprofen exposure at low concentrations induced a significant decrease in certain essential amino acids, such as leucine, arginine and lysine in crustaceans (*D. magna*).	Kovacevic et al 2016

[57] ***Apoptosis*** = *the death of cells.*

[58] ***Amphipods** are crustaceans. They are often found in high densities in headstreams where they are an important reserve of food for fish, bird and amphibians. They play a major role in the leaf litter breakdown process and consequently in the entire food web (Xuereb et al 2009).*

Pharmaceuticals compound	Impacts	References
	The chronic (long-term) exposure of Ibuprofen significantly increased egg numbers per reproductive event while decreasing spawning rates in Japanese medaka fish, *Oryzias latipes*. Changed reproduction pattern in *O. latipes* occurred at 1.00–100.00 µg/L (0.001-0.1 mg/L).	Flippin et al 2007, Han et al 2010
	Impaired/ disturbed ion regulation (Cl^-, K^+), metabolic and cortisol[59] responses in rainbow trout, *Oncorhynchus mykiss* at 1,000 µg/L (1 mg/L).	Gravel et al 2009
	Inhibited CYP2M catalysed activity in carp, *C. carpio* at 296.1500 (1 mM) (CYP = cytochrome P450 enzymes; mM = milli molar).	Thibaut et al 2006
Indomethacin *(NSAID)*	Disrupted oocyte[60] maturation/ ovulation in Zebra fish at 100,000 µg/L (100 mg/L).	Lister and van der Kraak 2008
Ivermectin *(antiparasitic)*	Can affect growth and reproduction in invertebrates.	OECD 2019
Ketoprofen	Inhibited CYP2M catalysed activity in carp, *C. carpio* at 296.1500 (1 mM) (CYP = cytochrome P450 enzymes; mM = milli molar).	Thibaut et al 2006
Metformin *(antidiabetics)*	Potential for endocrine-disrupting effects in fish.	OECD 2019
Naproxen (NSAID)	Inhibited CYP2M catalysed activity in carp at 296.1500 (1 mM) (CYP = cytochrome P450 enzymes; mM = milli molar).	Thibaut et al 2006

[59] ***Cortisol** = a **steroid hormone** that regulates a wide range of processes throughout the body, including metabolism and the immune response. It also has a very important role in helping the body to respond to stresses.*

[60] ***Oocyte** = a germ cell involved in reproduction. It is an immature ovum or egg cell. An oocyte is produced in the ovary during female gametogenesis.*

Pharmaceuticals compound	Impacts	References
Oral contraceptives	Caused feminisation of fish and amphibians.	OECD 2019
Oxazepam *(antidepressants)*	Altered behaviour of fish, European perch (*Perca fluviatilis)* including increased predator activity/ feeding rates) and increased bioaccumulation via consumption of exposed prey.	Brodin et al 2014
Pharmaceutical mixtures	A mixture of diclofenac and ibuprofen exposed to *D. magna* had stronger toxicity, while single diclofenac exposure didnot not exhibit effects.	Cleuvers 2003
	Adult zebrafish, *D. rerio* showed significantly decreased embryo production (developmental abnormalities and embryo mortality) after a 6-week exposure to a pharmaceutical mixture (MIX; 0.5 and 10 μg/L) of acetaminophen, carbamazepine, gemfibrozil and venlafaxine to diluted wastewater effluent (WWE; 5% and 25%).	Galus et al 2013
	A mixture of three pharmaceuticals (erythromycin, triclosan and trimethoprim) caused a shift in sex ratio in *Daphnia* population (the individual components did not provoke significant effects).	Flaherty and Dodson 2005
	A mixture of two pharmaceuticals (fluoxetine and clofibric acid) killed more than 50% of a *Daphnia* population after an exposure of 6 days, while there are no apparent effects for the same concentrations of any individual pharmaceuticals.	Flaherty and Dodson 2005
	Exposure to pharmaceuticals alone or in combination/mix treatment (including caffeine, ciprofloxacin, cimetidine and diphenhydramine) suppressed algal growth (82% reduction of chlorophyll-*a* in mix treatment) and microbial respiration and altered bacterial communities in stream biofilms.	Rosi-Marshall et al 2013

Pharmaceuticals compound	Impacts	References
	Exposure of a mixture of 13 pharmaceuticals found in Italian wastewaters and rivers (atenolol, benzafibrate, carbamazepine, ciprofloxacin, lincomycin, ofloxacin, sulfamethoxazole, cyclophosphamide, furosemide, hydrichlorothiazide, ibuprofen, ranitidine, salbutamol) were found to be very sensitive to unicellular algae, *Pseudokirchneriella subcapitata,* in particular, the chloroplasts.	Vannini et al 2011
Propranolol *(beta blockers)*	Altered blood flow in gills in rainbow trout fish.	Payan and Girard 1977
	Altered heart rate in zebra fish.	Fraysse et al 2006
	Can affect reproduction behaviour (in fish), reproduction toxicity (in invertebrates).	OECD 2019
	Caused a drop in glucose production in rainbow trout fish.	McKinley and Hazel 1993
	Impaired growth in rainbow trout fish.	Owen et al 2007
	Growth, eggs production and viable egg hatching in Medaka fish was decreased at concentrations as low as 0.5 μg/L.	Huggett et al 2002
Psychiatric drugs	Psychiatric drugs chlorpromazine, citalopram, fluoxetine, sertraline, oxazepam can cause a change in behaviour, activity, feeding and hormones disruption in fish and invertebrates	OECD 2019
Sulphonamides *(six sulfonamide related antibiotics)*[61]	Sulphonamides were found acutely toxic to medaka fish at > 100 mg/L.	Kim et al 2006, 2008
Tetracyclines *(antibiotics)*	Tetracycline was associated with the highest level of toxicity among the five major veterinary	Pan and Chu 2016

[61]***Sulfonamide related antibiotics*** *(sulfamethoxazole, sulfachlorpyridazine, sulfathiazole, sulfamethazine, sulfadimethoxine, and trimethoprim).*

Pharmaceuticals compound	Impacts	References
	antibiotics evaluated (tetracycline> norfloxacin> erythromycin> sulfamethazine> chloramphenicol) on seed germination and root elongation in lettuce, tomato, carrot and cucumber. Lettuce was found to be sensitive to most of the veterinary antibiotics. The median effect concentration (EC_{50}) of tetracycline to lettuce was 14.4 mg/L.	
	Tetracycline exposure at concentrations of 0.5 μg/L caused significant changes in the stream biotic community including significant changes in antibiotic resistance, bacterial abundance and productivity, and algal biomass, cyanobacteria, organic biomass, and nematodes.	Quinlan et al 2011
	Antibacterial resistance was reported in soil bacteria (sites treated with pig slurry).	Sengelov et al 2003
	Suppressed the immune system in rainbow trout fish.	Grondel et al 1985
Triclosan *(antibiotic/ antibacterial)*	Affected on the mitochondria of zebra fish embryos including an increase in oxygen consumption rate, and inhibited ATP-linked respiration etc.	Shim et al 2016, Raftery et al 2017
	Caused delays in embryonic development (eye, internal organs) in medaka fish, spinal curvature in the newly hatched larvae with significantly increased haemorrhage, heartbeat rate and hatching time.	Nassef et al 2010b
	Found to be immunotoxic to the Manila clam, *Ruditapes philippinarum*.	Matozzo et al 2012
	Low triclosan concentrations had significant effects on the swimming and feeding behaviours of Japanese medaka fish.	Nassef et al 2010b
	Affected the photosynthesis (EC_{50}: 3.7 μg/L) and the reproduction system (EC_{50}: 1.9 μg/L) of the green algae, *Scenedesmus vacuolatus* (chlorophyte).	Yang et al 2008

Pharmaceuticals compound	Impacts	References
	Triclosan exposure can cause cytotoxic and genotoxic impacts on the zebra mussel, *Dreissena polymorpha.*	Binelli et al 2009, Kovacevic et al 2016
	Caused cardiac problems (pericardial edema and alterations in heart, structure) in zebra fish at environmentally relevant concentrations.	Saley et al 2016
	Induced morphological malformations in *Bufo gargarizans* (an endemic frog species in China) embryos, including hyperplasia, abdominal edema, and axial flexures resulting in delayed growth and development and also caused teratogenic effects (substances that can interfere with normal embryonic development).	Chai et al 2016
	Caused deleterious effects in both adults and early stages of zebra fish including embryo toxicity, hatching delay, delays in otolith formation, eye and body pigmentation, spine malformations, pericardial edema and under-size and increased levels of cholinesterase[62] (ChE), lactate dehydrogenase (LDH) and glutathione S-transferase (GSH) in the early stages.	Oliveira et al 2009
	Most toxic to benthic invertebrates, *Chironomus tentans* (LC_{50}: 0.4 mg/L) and *Hyalella azteca* (LC_{50}: 0.2 mg/L).	Dussault et al 2008
	Highly toxic and affected different aquatic web levels including bacteria (EC_{50}: 0.52 mg/L), algae (EC_{50}: 0.012 mg/L), crustaceans (EC_{50}: 0.26 mg/L), amphibians (LC_{50}: 0.82 mg/L) and protozoa (EC_{50}: 0.21 mg/L).	Harada et al 2008
	Wheat seedlings (*Triticum aestivum*) exposed to triclosan was damaged (due to the accumulation of chlorophyll, synthesis of soluble protein, and activity of peroxidase (POD) and superoxide dismutases (SOD).	An et al 2009

[62] ***Cholinesterase*** *= an* ***enzyme****, found especially in the heart, brain, and blood, that hydrolyses acetylcholine to acetic acid and choline.*

Pharmaceuticals compound	Impacts	References
	Triclosan altered the structure of bacterial communities (decreased benthic bacterial diversity and shifts in bacterial community composition) and increased antimicrobial-resistant bacteria both *in situ* and in artificial streams amended with triclosan.	Drury et al 2013, Proia et al 2011
	Environmentally relevant concentrations caused an increase of bacterial mortality; inhibition of photosynthetic efficiency in biofilm algae with increasing triclosan concentrations.	Ricart et al 2010
	Caused significant reduction in algal biomass as a result of exposure to triclosan.	Lawrence et al 2009
	Exposure of triclosan caused an increase in the abundance of triclosan resistant bacteria and stimulated periphyton growth.	Nietch et al 2013
	Stimulated the activity of glutathione transferase[63] and glutathione reductase in the Mediterranean mussel, *Mytilus galloprovincialis*.	Canesi et al 2007

9.3.3. Impacts on human health

Pregnant women and their babies, as well as children, are at particular risk if drinking water is contaminated with pharmaceuticals. The detection of drug residues in drinking water can expose newborn human babies from a very low age before the blood-brain barrier and the detoxifying systems are fully developed. The possible impacts of life-long exposure of humans are not yet available. However, the exposure of the pregnant mother or her foetus to cytotoxic drugs and chemotherapy drugs via drinking water could be very harmful. As shown in Table 9.9, 28 drugs have been detected in drinking water, of which, several drugs are not recommended for children. If children are exposed to these pharmaceuticals, the relative risk is up to eight-fold higher than adults according to Collier (2007). Several drugs are **teratogenic**[64] which are not

---

[63] ***Glutathione transferase*** *(GHS) = are a family of* ***enzymes*** *that catalyse the conjugation of reduced glutathione to a wide range of substrates, usually resulting in detoxification. They also function as transport proteins. GSH conjugation activity plays a critical role in cellular detoxification against xenobiotics and noxious compounds as well as against oxidative stress.*

[64] ***Teratogenic*** *= substances or agents that can interfere with normal embryonic development.*

recommended during breast feeding but detected in drinking water (acetaminophen/paracetamol, carbamazepine, codeine, diazepam, diclofenac, gemfibrozil, mestranol, sulfamethoxazole (Kumar and Xagoraraki 2010; Table 9.9). Antibiotics present in the environment can produce resistance in microbial assemblages, which can have potentially drastic effects on human health. The transfer of resistant bacteria to humans may occur via water or food if plants are watered with surface water or sewage sludge, or if manure is used as a fertiliser or if resistant bacteria are present in the meat of food animals.

9.4. Linkages between pharmaceuticals use and climate change

Climate change could exacerbate existing human diseases, which would require more pharmaceuticals. For example, non-communicable diseases (e.g. cardiovascular disease and mental illness) and respiratory, water-borne, vector-borne and food-borne toxicants and infections are expected to become more common as climate change intensifies (OECD 2019). In addition, warming (rise in temperatures) would cause the proliferation of diseases in livestock and aquaculture, requiring more use of antibiotics (Kibria et al 2013).

9.5. Management of pharmaceuticals

The following approaches, measures and strategies (e.g. targeted/ prioritised monitoring, source control, green innovation, enforcement of laws, acts and regulations, education of communities and upgrading wastewater treatment plants to improve efficiencies) can be implemented to reduce/ limit amounts of pharmaceuticals released to the environment (Boxall 2004, Kummerer 2009, Toma and Crisan 2018, OECD 2019):

9.5.1. Monitoring: Introduce a regular but targeted and prioritised monitoring of high-risk pharmaceutical residues in waterways (rivers, channels, drains, lakes, estuaries, coastal areas) and drinking water and compare with guideline threshold values (Table 9.11) to minimise environmental risks and risks to human health *[Two pharmaceuticals such as hormones (17α-ethinylestradiol and 17-β estradiol) and diclofenac have been placed on a watch list for "**contaminants of emerging concern**" under the WFD (Water Framework Directive) of the European Union (Gaw et al 2014)].*

9.5.2. Source control: Provide information on possible environmental impacts on medicine labels/ packages, so that physicians and veterinary surgeons can consider the potential environmental impact when prescribing the particular drugs. There are several sources and pathways (wastewater treatment effluents, landfill leachate, stormwater run-off, livestock and aquaculture effluents) of pharmaceuticals entering the environment (see Fig. 9.1), hence, minimisations of discharge from such sources would be essential.

9.5.3. Medical doctors: Medical, livestock and aquaculture health officials/ managers should prescribe medicines according to environmental criteria if alternatives are available.

9.5.4. Pharmacists: Pharmaceutical companies and pharmacies can participate in take-back systems of expired or unused medicines if appropriate.

9.5.5. Patients: Patients should be educated not to dispose of expired medications down the drain; instead either return to the pharmacy if the take-back system is established or into the household waste if appropriate (check with the local council regulations).

9.5.6. Pharmaceutical companies: Pharmaceutical companies should publish data relevant to environmental impacts of pharmaceutical drugs, participate in green pharmacy strategy and establish and accept take-back systems of expired or unused medicines (***note***: after collection of unused or expired medicines, they are disposed of by incineration; Sanofi 2018).

9.5.7. Hospitals: Hospitals should integrate/ negotiate with pharmacy/ wholesaler to handle expired medicines/ pharmaceuticals/ medications.

9.5.8. Disposal: There is a need to implement an appropriate and safe disposal of unused, expired medicines and containers both locally, regionally and globally.

9.5.9. Green innovation: Encourage and invest to promote innovative green pharmacy concepts (biological therapies, production of environmentally friendly pharmaceuticals, eco-labelling on pharmaceutical products to improve consumer choice selection).

9.5.10. Stringent conditions (laws, acts and regulations): **i**. Implement laws, acts and regulations for more stringent conditions for putting a pharmaceutical on the market that is of high-risk to the environment; **ii.** Implement laws, acts and regulations to prevent open discharge of pharmaceutical effluents in natural waterways (without treatment) or limiting discharge pharmaceutical wastewater into the environment; **iii**. Restrict and ban using unnecessary high-risk pharmaceuticals (e.g. antibiotics and growth hormones in livestock and aquaculture) (see Box 9.1 for policy instruments to control pharmaceuticals in the environment in OECD countries).

9.5.11. Collection programs: Promote community pharmacies or local Councils or waste disposal companies to collect unused or expired medicines in order to mitigate the environmental impacts of medicinal products.

9.5.12. Separation: Separate hospital wastes from other wastes; separate treated animals from untreated animals to allow targeted treatment; separate urine from solid materials.

9.5.13. Good farming practice: Sludge (mud) and slurry (a watery mixture of insoluble matter) can be applied to land as good farming practice (non-food plants) to reduce/ limit the potential for drugs to contaminate surface and groundwaters.

9.5.14. Education: **i.** Educate the community to avoid the disposal of pharmaceuticals via sink or toilet; **ii.** Encourage community for safe disposal of unused drugs and expired drugs (such as to join in public pharmaceutical collection schemes); **iii**. Provide adequate information on medicine labels for possible environmental impacts (to consider potential environmental impacts when prescribing).

9.5.15. Manure management: Improve manure management by passive storage or anaerobic fermentation in biogas plants.

9.5.16. Wastewater treatment plants: Upgrade wastewater treatment plants to improve removal efficiencies (novel and more effective techniques) [***note:*** the removal rates for pharmaceuticals in wastewater treatment plants (WWTPs) range from <10% to almost 100%. Pharmaceuticals such as aspirin, 17β-estradiol, estriol, norfloxacin, triclosan have highly removal efficiency (>80%), whereas atenolol, carbamazepine, clofibric acid, diclofenac, ketoprofen, metoprolol, roxithromycin, trimethoprim, mefenamic acid, sulfamethazine have low removal efficiency (<40%) in WWTPs; Miége et al 2009; page 1724).

***9.5.17. Phytoremediation*:** Phytoremediation of pollutants from waterbodies is an economical option. Wetland macrophytes, leafy vegetables and algae are capable of accumulating animal or synthetic estrogens (Kumar and Xagoraraki 2010). In an experiment, Bircher (2011) showed that *Scirpus validus* (emergent wetland plant) and *Populus deltoides nigra* (wetland plant) reduced the concentration of estrone (E1), estriol (E3), 17β-estradiol (E2), and 17α-ethinylestradiol (EE2) from the solution through root tissues (Bircher 2011).

Box 9.1. Selected examples of policy instruments to control pharmaceuticals in the environment in OECD countries (OECD 2019).

Australia: *National pharmaceutical collection and disposal programme, with retail pharmacies commonly acting as collection sites.*

Germany: *An environmental check-list for veterinarians and farmers to reduce the use and release of veterinary pharmaceuticals to the environment.*

Korea: *Uses suspect and non-target screening to identify and prioritise pharmaceuticals for water quality monitoring.*

Sweden: *Public procurement system in which pollution control during manufacturing is considered.*

Sweden: *Consider medications that are equally safe and equally suitable for the medical purpose but causes less environmental impacts.*

UK: *Voluntary antibiotic stewardship/ management programme in the poultry industry.*

USA: *National regulations on the disposal of hazardous pharmaceutical wastes in the health sector.*

Table 9.11. Australian recommended drinking water quality guideline values for pharmaceuticals *[NWQMS 2008; NSAID = Nonsteroidal anti-inflammatory drugs].*

Pharma-ceuticals	µg/L	Pharma-ceuticals	µg/L	Pharma-ceuticals	µg/L
i. Antibiotics		***ii. NSAIDs***		***iv. General pharmaceuticals***	
Amoxicillin	1.5 µg/L	Diclofenac	1.8 µg/L	Cyclophosphamide	3.5 µg/L
Azithromycin	3.9 µg/L	Dipyrone (vet)	525 µg/L	Dehydronifedipine	20 µg/L
Chloramphenicol	175 µg/L	Fenoprofen	450 µg/L	Diltiazem	60 µg/L
Chlortetracycline	105 µg/L	Ibuprofen	400 µg/L	Enalaprilat	1.3 µg/L
Ciprofloxacin	250 µg/L	Indomethacin	25 µg/L	Fluoxetine	10 µg/L
Clarithromycin	250 µg/L	Ketoprofen	3.5 µg/L	Gemfibrozil	600 µg/L
Clindamycin	300 µg/L	Tolfenamic acid (vet)	17.5 µg/L	Iohexol	720 µg/L
Doxycycline	10.5 µg/L	***iii. β-adrenergic blockers***		Iopamidol	400 µg/L
Enrofloxacin	22 µg/L	Betaxolol	10 µg/L	Iopromide	750 µg/L
Erythromycin	17.5 µg/L	Bisoprolol	0.63 µg/L	Isophosphamide	3.5 µg/L
Monensin	35 µg/L	Carazolol	0.35 µg/L	Metformin	250 µg/L
Nalidixic acid	1,000 µg/L	Nadolol	20 µg/L	Paracetamol	175 µg/L
Oxytetracycline	105 µg/L	Propranolol	40 µg/L	Salbutamol	3 µg/L
Penicillin G	1.5 µg/L	Timolol	10 µg/L	Salicyclic acid	105 µg/L
Penicillin V	1.5 µg/L	***iv. General pharmaceuticals***		Sulfasalazine	500 µg/L
Roxithromycin	150 µg/L	Antipyrine	1,000 µg/L	Temazepam	5 µg/L
Penicillin G	1.5 µg/L	Atorvastatin	5 µg/L	Valium	2.5 µg/L
Sulfamethazine	35 µg/L	Benzafibrate	300 µg/L	***v. Hormones***	

Pharma-ceuticals	µg/L	Pharma-ceuticals	µg/L	Pharma-ceuticals	µg/L
Sulfamethizole	35 µg/L	Carbamazepine	100 µg/L	17α-ethinyl estradiol	1.5 ng/L
Sulfamethoxazole	35 µg/L	Cimetidine	200 µg/L	17β-estradiol	175 ng/L
Tetracycline	105µg/L	Clenbuterol	15 µg/L	Equilin	30 ng/L
Trimethoprim	70 µg/L	Clofibric acid	750 µg/L	Estriol	50 ng/L
Tylosin	1,050 µg/L	Codeine	50 µg/L	Estrone	30 ng/L
Aspirin (Acetylsalicylic acid)	29 µg/L	Cotinine	10 µg/L	Mestranol	2.5 ng/L
				vi. others	
				Caffeine	0.35 µg/L
				Triclosan	0.35 µg/L

9.6. Conclusion

Pharmaceuticals are **"contaminants of emerging concern"**. Several thousand tonnes of medicinally active pharmaceutical substances are used every year in the world for human, livestock and aquaculture purposes. The use of pharmaceuticals will obviously increase over the years across the globe due to a projected increase in populations, and a need for improved lifestyle and hygiene such as 2020 scare of Coronavirus (COVID-19) and on-going intensification of livestock and aquaculture practices. One of the most important sources of pharmaceutical products (PP) in aquatic environments is the effluents from wastewater treatment plants (WWTPs) since PPs are not completely removed during treatment (removal efficiency of many pharmaceuticals is less than 40%). Pharmaceuticals have been detected in all the environmental matrices (WWTP effluents, surface water, groundwater, tap or drinking water sources).

Many pharmaceuticals are *persistent* (half-life >40 days); *bioaccumulative* (log K_{ow} >3); extremely *toxic* (EC_{50}/ LC_{50} <1 mg/L; e.g. ciprofloxacin, E1, E2, EE2, erythromycin, fluoxetine, sulfamethoxazole, tetracyclines, triclosan), and some are also *endocrine disrupting* (e.g. aspirin, clofibric acid; E1, E2, EE2, fluoxetine; ibuprofen; naproxen; tetracyclines) and *carcinogenic* (E1, E2, EE2). The presence of human and veterinary pharmaceuticals in the environment is of increasing global concern due to their effects on ecological receptors (bacteria, algae, daphnia, fish, frogs, birds, plants/ crops). Pharmaceuticals or active pharmaceutical ingredients can enter aquatic organisms through their diet and exchange surfaces (gills, cuticles and epidermis) whereas humans can be exposed from eating pharmaceuticals contaminated food and water. *Antibiotics* have caused a diverse range of effects (e.g. enhanced antimicrobial resistance; reduced algal richness; inhibited photosynthesis-related processes, delayed embryonic development, induced edema in fish and effects on seed germination and root elongation in crops/vegetables). Ecological disasters such as the decline of vultures in the Indian sub-continent (Bangladesh, Pakistan and India) due to *diclofenac* toxicity is one of the worst ecological effects of pharmaceuticals.

Residues of *EE2 (17α-ethinylestradiol)* cause sex reversal, reduced fertility and vitellogenesis (egg yolk) in male frogs and fish. Experimental results reveal that *mixed pharmaceuticals* exhibited much more toxic effects (additive effects) on biota (algae, daphnia and fish) compared to a single pharmaceutical. Pregnant women and their babies, as well as children, are at higher risk if drinking water is contaminated with pharmaceuticals (28 pharmaceuticals have been detected in drinking water). Pharmaceuticals such as hormones (*17α-ethinylestradiol and 17β-estradiol*) and diclofenac have been placed on a watch list for **"contaminants of emerging concern"** under the WFD (Water Framework Directive) of the European Union.

Monitoring (prioritised and targeted), source control, green innovation, enforcement of laws, acts and regulations, education of communities for improving awareness and upgrading wastewater treatment plants are some of the approaches, measures and strategies that can be implemented to reduce and limit amounts of pharmaceuticals released to the environment and safeguard environmental and human health.

9.7. References

Adamson R.H 2016 The acute lethal dose 50 (LD_{50}) of caffeine in albino rats, Regulatory Toxicology & Pharmacology **80** 274-276 *doi: 10.1016/j.yrtph.2016.07.011. Epub 2016 Jul 25*

Adeel M, Song X, Yuanyuan Wang Y et al 2017 Environmental impact of estrogens on human, animal and plant life: A critical review. *Environment International* **99** 107–119

Adolfsson-Erici M1, Åkerman G and McLachlan MS 2012 Measuring bioconcentration factors in fish using exposure to multiple chemicals and internal benchmarking to correct for growth dilution. *Environmental Toxicology & Chemistry* **31(8)** 1853-60 *doi: 10.1002/etc.1897. Epub 2012 Jul 3*

Aherne GW, Hardcastle A and Nield AH 1990 Cytotoxic drugs and the aquatic environment – estimation of Bleomycin in river and water samples. *Journal of Pharmacy & Pharmacology* **42** 741–742

Al-Ansari AM, Atkinson SK, Doyle JR et al 2013 Dynamics of uptake and elimination of 17α-ethinylestradiol in male goldfish (*Carassius auratus*). *Aquatic Toxicology* **132-133** 134-140 *doi: 10.1016/j.aquatox.2013.02.006. Epub 2013 Feb 16*

Almeida AFR, Calisto V, Esteves, VI et al 2015 Chronic toxicity of the antiepileptic carbamazepine on the clam *Ruditapes philippinarum. Comparative Biochemistry &*

Physiology Part C Toxicology & Pharmacology **172-173** 26–35 *Doi:10.1016/j.cbpc.2015.04.00*

An J, Zhou Q, Sun Y and Xu Z 2009 Ecotoxicological effects of typical personal care products on seed germination and seedling development of wheat (*Triticum aestivum* L.). *Chemosphere* **76** 1428–1434 *Doi:10.1016/j.chemosphere.2009.06.004*

Araujo L, Troconis ME, Espina MB and Prieto A 2014 Persistence of ibuprofen, ketoprofen, diclofenac and clofibric acid in natural waters. *Journal of Environment and Human* **1(2)** 32-38

Arnot JA and Gobas FA 2006 A review of bioconcentration factor (BCF) and bioaccumulation factor (BAF) assessments for organic chemicals in aquatic organisms. *Environmental Reviews* **14** 257–297

Arnold KE, Brown AR, Ankley GT and Sumpter JP 2014 Medicating the environment: assessing risks of pharmaceuticals to wildlife and ecosystems. *Philosophical Transaction of the Royal Society B* **369** 20130569 *http://dx.doi.org/10.1098/rstb.2013.0569*

Aus der Beek T, Weber RA, Hickmann S et al 2016 "Pharmaceuticals in the environment. Global occurrences and perspectives". *Environmental Toxicology & Chemistry* **35(4)** 823-835 *http://dx.doi.org/10.1002/ etc.3339*

Barnes KK, Kolpin DW, Furlong ET et al 2008 A national reconnaissance of pharmaceuticals and other organic wastewater contaminants in the United States. 1 Groundwater. *Science of the Total Environment* **402** 192-200

Barry MJ 2014 Fluoxetine inhibits predator avoidance behaviour in tadpoles. Toxicological & Environmental Chemistry **96(4)** 641–649 *DOI: https://doi.org/10. 1080/02772248.2014.966713*

Beausse J 2004 Selected drugs in solid matrices: a review of environmental determination, occurrence and properties of principal substances. *Trends in Analytical Chem*istry **23(10-11)** 753-761

Benotti MJ, Trenholm RA, Vanderford BJ et al 2009 Pharmaceuticals and endocrine disrupting compounds in U.S. drinking water. *Environmental Science & Technology* **43(3)** 597-603

Bergsjø T and Søgnen E 1980 Plasma and tissue levels of trimethoprim in the rainbow trout, *Salmo gairdneri,* after absorption from fresh and salt water. *Acta Vet. Scand.* **21(1)** 18-25

Bhandari K and Venables B 2011 Ibuprofen bioconcentration and prostaglandin E2 levels in the bluntnose minnow, *Pimephales notatus. Comparative Biochemistry & Physiology Part C: Toxicology & Pharmacology* **153(2)** 251-257

Binelli A, Cogni D, Parolini M et al 2009 In vivo experiments for the evaluation of genotoxic and cytotoxic effects of triclosan in zebra mussel hemocytes. *Aquatic Toxicology* **91** 238–244

Bircher S 2011. "Phytoremediation of natural and synthetic steroid growth promoters used in livestock production by riparian buffer zone plants". M.Sc. Thesis. University of Iowa, U.S.A

https://www.semanticscholar.org/paper/Phytoremediation-of-natural-and-synthetic-steroid-Bircher/ee6aedc01f9ae3dc55d1deb64aba64be3a7f1051

Birnie-Gauvin K, Costantini D, Cooke Sj and Willmore WG 2017 A comparative and evolutionary approach to oxidative stress in fish: A review. *Fish & Fisheries* **18** 928–942

Boxall ABA, Fogg L, Blackwell PA et al 2002 Review of veterinary medicines in the environment. R&D Technical Report P6–012/8/TR, Environment Agency, Bristol, UK, p. 187; Available online: *http://a0768b4a8a31e106d8b0-50dc802554eb38a24458b98ff72d550b.r19.cf3.rackcdn.com/sp6-012-8-tr-e-e.pdf*

Boxall ABA 2004 The environmental side effects of medication. *EMBO Reports* **5(12)** 1111-1116

Brennan SJ, Brougham CA, Roche JJ and Fogarty AM 2006 Multi-generational effects of four selected environmental oestrogens on *Daphnia magna*. *Chemosphere* **64(1)** 49-55

Bringolf RB, Heltsley RM, Newton TJ et al 2010 Environmental occurrence and reproductive effects of the pharmaceutical fluoxetine in native freshwater mussels. *Environmental Toxicology & Chem*istry **29** 1311-318; (doi:10.1002/etc.157)

Brodin T, Piovano S, Fick J et al 2014 Ecological effects of pharmaceuticals in aquatic systems—impacts through behavioural alterations. *Philosophical Transactions of the Royal Society B* **369(1656)** 20130580 *DOI: https://doi.org/10.1098/rstb.2013.0580*

Brooks BW, Foran CM, Richards SM 2003 Aquatic ecotoxicology of fluoxetine. *Toxicology Letters* **142** 169-183

Brooks BW 2014 Fish on Prozac (and Zoloft): ten years later. *Aquatic Toxicology* **151** 61–67

Brown JN, Paxeus N, Forlin L et al 2007 Variations in bioconcentration of human pharmaceuticals from sewage effluents into fish blood plasma *Environmental Toxicology & Pharmacology* **24** 67–274

Brozinski JM, Lahti M, Meierjohann A et al 2013 The anti-inflammatory drugs Diclofenac, Naproxen and Ibuprofen are found in the bile of wild fish caught downstream of a wastewater treatment plant. *Environmental Science & Technology* **47** 342–348

Bundschuh M, Hahn T, Gessner MO et al 2017 Antibiotic mixture effects on growth of the leaf-shredding stream detritivore, *Gammarus fossarum*. *Ecotoxicology* **26(4)** 547–554 *DOI: https://doi.org/10.1007/s10646-017-1787-2*

Buerge IJ, Poigner T, Müller MD and Muser HR 2006 Combined sewer overflow to surface waters detected by the anthropogenic marker caffeine. *Environ. Sci. Technol.* **40(13)** 4096–4102

Caldwell DJ, Mastrocco F, Anderson PD, La¨nge R and Sumpter JP 2012 Predicted-no-effect concentrations for the steroid estrogens estrone, 17β-estradiol, estriol, and 17α-ethinylestradiol. *Environmental Toxicology & Chemistry* **31** 1396–1406 *doi:10.1002/etc.1825*

Caliman FA and Gaurilescu M 2009 Pharmaceuticals, personal care products and endocrine disrupting agents in the environment – a review. *Clean* **37(4–5)** 277–303

Canesi L, Ciacci C, Lorusso LC et al 2007 Effects of triclosan on *Mytilus galloprovincialis* hemocyte function and digestive gland enzyme activities: possible modes of action on non-target organisms. *Comparative Biochemistry & Physiology C: Toxicology & Pharmacology* **145** 464–472

Castensson S, Eriksson V, Lindborg K and Wettermark B 2009 A method to include the environmental hazard in drug prescribing. *Pharmacy World & Science* **31** 24–31

CEC (Commission of the European Communities) 1996 Technical guidance document in support of commission directive 93/67/EEC on risk assessment for new modified substances. Part II. Environmental risk assessment. Luxembourg: Office for official publication of the European Communities.

Cemagref 2007 Environmental database for pharmaceuticals *http://pharmaecobase.lyon.cemagref.fr/*

Chai L, Wang H, Zhao H and Deng H 2016 Chronic effects of triclosan on embryonic development of Chinese toad, *Bufo gargarizans. Ecotoxicology* **25** 1600–1608 *Doi: 10.1007/s10646-016-1715-x*

Chen H, Zha J, Liang X et al 2014 Effects of the human antiepileptic drug carbamazepine on the behavior, biomarkers, and heat shock proteins in the Asian clam *Corbicula fluminea. Aquatic Toxicology* **155** 1–8 *Doi:10.1016/j.aquatox.2014.06.00*

Choi K, Kim Y, Jung J et al 2008 Occurrences and ecological risks of roxithromycin, trimethoprim, and chloramphenicol in the Han River, Korea. *Environmental Toxicology & Chemistry* **27(3)** 711–719

Cleuvers M 2003 Aquatic ecotoxicity of pharmaceuticals including the assessment of combination effects. *Toxicology Letters* **142** 185–194

Collier AC 2007 Pharmaceutical contaminants in potable water: potential concerns for pregnant women and children. *EcoHelath* **4** 164-171

Contardo-Jara V, Lorenz C, Pflugmacher S et al 2011 Exposure to human pharmaceuticals carbamazepine, ibuprofen and bezafibrate causes molecular effects in *Dreissena polymorpha. Aquatic Toxicology* **105** 428–437

Corcoran J, Winter MJ and Tyler CR. 2010 Pharmaceuticals in the aquatic environment: a critical review of the evidence for health effects in fish. *Crit. Rev. Toxicol.* **40** 287–304 *doi: 10.3109/10408440903373590*

Cuklev F, Kristiansson E, Fick J et al 2011 Diclofenac in fish: blood plasma levels similar to human therapeutic levels affect global hepatic gene expression. *Environmental Toxicology & Chemistry* **30(9)** 2126–2134 *doi: 10.1002/etc.599*

Cunningham L, Buzby M, Huthcinson T et al 2006. Effects of human pharmaceuticals on aquatic life: next steps. *Environmental Science & Technology* **40** 3456–3462

Cuthbert RJ, Taggart MA, Prakash V et al 2014 Avian scavengers and the threat from veterinary pharmaceuticals. *Philosophical Transaction of the Royal Society B* **369** 20130574 *doi: 10.1098/rstb.2013.0574*

Daughton CG and Ternes TA 1999 Pharmaceuticals and Personal Care Products in the Environment: Agents of Subtle Change? *Environmental Health Perspectives* **107** 907

Deblonde T and Hartemann P 2013 Environmental impact of medical prescriptions: assessing the risks and hazards of persistence, bioaccumulation and toxicity of pharmaceuticals. *Public Health* **127** 312–317

De Lange HJ, Noordoven W, Murk AJ et al 2006 Behavioural responses of *Gammarus pulex* (Crustacea, Amphipoda) to low concentrations of pharmaceuticals. *Aquatic Toxicology* **78** 209–216 *Doi: 10.1016/j.aquatox.2006.03.002*

Dhilon GS, Kaur S, Pulicharla R al 2015 Triclosan: current status, occurrence, environmental risks and bioaccumulation potential 2015 *International Journal of*

Environmental Research & Public Health **12** 5657-5684 *doi: 10.3390/ijerph120505657*

Dietrich S, Ploessl F, Bracher F and Laforsch C 2010 Single and combined toxicity of pharmaceuticals at environmentally relevant concentrations in *Daphnia magna* – a multigenerational study. *Chemosphere* **79** 60–66

Ding J, Lu G, Liu J et al 2008 Uptake, depuration, and bioconcentration of two pharmaceuticals, roxithromycin and propranolol, in *Daphnia magna. Ecotoxicology & Environmental Chemistry* **126** 85-93

Drury B, Scott J, Rosi-Marshall EJ and Kelly JJ 2013 Triclosan exposure increases triclosan resistance and influences taxonomic composition of benthic bacterial communities. *Environmental Science & Technology* **47(15)** 8923–8930 *DOI: https://doi.org/10.1021/es401919k*

Dusing R, Handrock R, Klebs S et al 2009 Impact of supportive measures on drug adherence in patients with essential hypertension treated with valsartan: the randomized, open-label, parallel group study VALIDATE. *Journal of Hypertension* **27** 894–901

Dussault ÈB, Balakrishnan VK, Sverko E et al 2008. Toxicity of human pharmaceuticals and personal care products to benthic invertebrates. *Environmental Toxicology & Chemistry* **27** 425 *Doi:10.1897/07-354R.1*

Edwards QA, Kulikov SM, and Garner-O'Neal LD 2015 Caffeine in surface and wastewaters in Barbados, West Indies. *Springer Plus* **4** 57 *DOI 10.1186/s40064-015-0809-x*

El-Bassat RA, Touliabah HE and Harisa GI 2012 Toxicity of four pharmaceuticals from different classes to isolated plankton species. *African Journal of Aquatic Science* **37** 71–80

EPA 2012 Persistent, bioaccumulative, and toxic profiles estimated for organic chemicals on-line. *http:// www.pbtprofiler.net*

Ericson H, Thorsen G and Kumblad L 2010 Physiological effects of diclofenac, ibuprofen and propranolol on Baltic Sea blue mussels. *Aquatic Toxicology* **99** 223 – 231 *doi:10.1016/j.aquatox.2010.04.017*

Erickson ML, Langer SK, Roth JL and Kroening SE. 2014 Contaminants of emerging concern in ambient groundwater in urbanized areas of Minnesota, 2009-12. US Geological Survey Scientific Investigations Report 2014-5096

Farré M, Ferrer I, Ginebreda A et al 2001 Determination of drugs in surface water and wastewater samples by liquid chromatography–mass spectrometry: methods and preliminary results including toxicity studies with *Vibrio fischeri. Journal of Chromatography* A **938** 187–97

FASS (Swedish medicines information portal) -no date Aspirin. *file:///C:/Users/User1/Downloads/19350131000010%20(1).pdf*

FASS (Swedish medicines information portal) 2012 Environmental classification of pharmaceuticals. Guidance for pharmaceutical companies.

https://www.fass.se/pdf/Environmental_classification_of_pharmaceuticals-120816.pdf

Flaherty CM and Dodson SI 2005 Effects of pharmaceuticals on *Daphnia* survival, growth, and reproduction. *Chemosphere* **61** 200-207 *doi:10. 1016/j.chemosphere.2005.02.016*

Flippin JL, Huggett D, Foran CM 2007 Changes in the timing of reproduction following chronic exposure to ibuprofen in Japanese medaka (*Oryzias latipes*). *Aquatic Toxicology* **81** 73–83

Fent K., Weston AA and Caminada D 2006 Review- Ecotoxicology of human pharmaceuticals. *Aquatic Ecotoxicology* **76** 122-159

Fraysse B, Mons R, Garric J 2006 Development of a zebrafish 4-day embryo larval bioassay to assess toxicity of chemicals. *Ecotoxicology & Environmental Safety* **63** 253–267

Galus M, Jeyaranjaan J, Smith E et al 2013 Chronic effects of exposure to a pharmaceutical mixture and municipal wastewater in zebra fish. *Aquatic Toxicology* **132-133** 212-222 *doi:10.1016/j.aquatox.2012.12.016*

Gaw S, Thomas KV and Hutchinson TH. 2014 Sources, impacts and trends of pharmaceuticals in the marine and coastal environment. *Philosophical Transactions of the Royal Society B* **369** 20130572 *http://dx.doi.org/10.1098/rstb.2013.0572*

Gaworecki KMand Klaine SJ 2008 Behavioral and biochemical responses of hybrid striped bass during and after fluoxetine exposure. *Aquatic Toxicology* **88(4)** 207-213 *doi: 10.1016/j.aquatox.2008.04.011*

Gilbert N 2012 Drug-pollution law all washed up. *Nature* **491** 503–504 *Doi: 10.1038/491503a*

Girardi C, Greve J, Lamshöft M et al 2011 Biodegradation of ciprofloxacin in water and soil and its effects on the microbial communities. *Journal of Hazardous Materials* **198** 22–30

Gobas FAPC and Morrison HA 2000 "Bioconcentration and biomagnification in the aquatic environment". *In*: Boethling RS; Mackay D (eds.) Handbook of Property Estimation Methods for Chemicals: Environmental and Health Sciences. Boca Raton, FL, USA: Lewis. p. 189–23

González O, Bayarri B, Aceña J, Pérez S and Barceló D 2016 Treatment technologies for wastewater reuse: fate of contaminants of emerging concerns. In: Fatta-Kassinos D, Dionysiou D and Kümmerer K (eds.). *The Handbook of Environmental Chemistry* **45** Advanced Treatment Technologies for Urban Wastewater Reuse. Springer, Cham-Heidelberg-New York, Dordrecht-London, p. 305.

Gravel A, Wilson JM, Pedro DFN and Vijayan MM 2009 Non-steroidal antiinflammatory drugs disturb the osmoregulatory, metabolic and cortisol responses associated with seawater exposure in rainbow trout. *Comparative Biochemistry & Physiology C* **149** 481–490

Grondel JL, Gloudemans AG and Van Muiswinkle WB 1985 The influence of antibiotics on the immune system II. Modulation of fish leukocyte responses in culture. *Veterinary Immunology & Immunopathology* **9** 251–260

Gyllenhammar I, Holm L, Eklund R and Berg C 2009 Reproductive toxicity in *Xenopus tropicalis* after developmental exposure to environmental concentrations of ethynylestradiol. *Aquatic Toxicology* **91** 171–178 *Doi: 10.1016/j.aquatox.2008.06.019*

Halling-Sørensen B, Lützhøft HCH, Andersen HR and Ingerslev F 2000 Environmental risk assessment of antibiotics: comparison of mecillinam, trimethoprim and ciprofloxacin. *Journal of Antimicrobial Chemotherapy* **46** 53–58

Halling-Sørensen B, Sengeløv G, Ingerslev F and Jensen LB 2003 Reduced antimicrobial potencies of oxytetracycline, tylosin, sulfadiazin, streptomycin, ciprofloxacin, and olaquindox due to environmental processes. *Archives of Environmental Contamination & Toxicology* **44** 0007–0016

Han S, Choi K, Kim J et al 2010 Endocrine disruption and consequences of chronic exposure to ibuprofen in Japanese medaka (*Oryzias latipes*) and freshwater cladocerans *Daphnia magna* and *Moina macrocopa*. *Aquatic Toxicology* **98** 256–264; Doi:10.1016/j.aquatox.2010.02.013.

Hampel M, Bron JE, Taggart JB and Leaver MJ 2014 The antidepressant drug Carbamazepine induces differential transcriptome expression in the brain of Atlantic salmon, *Salmo salar*. *Aquatic Toxicology* **151** 114–23 *Doi: 10.1016/j.aquatox.2013.12.018*

Hansch C, Leo A and Hoekman D 1995 Exploring QSAR. Fundamentals and applications in chemistry and biology p. 141. In: Hydrophobic, electronic and steric Constants. American Chemical Society, Washington, DC. 348 p; ISBN 0-8412-2993-7.

Harada A, Komori K, Nakada, N et al 2008 Biological effects of PPCPs on aquatic lives and evaluation of river waters affected by different wastewater treatment levels. *Water Science & Technology* **58** 1541 *Doi: 10.2166/wst.2008.742*

Heberer T, Dünnbier U, Reilich C and Stan H-J 1997 Detection of drugs and drug metabolites in ground water samples of a drinking water treatment plant. *Fresenius Environmental Bulletin* **6** 438–443

Heberer T, Mechlinski A, Fanck B et al 2004 Field studies on the fate and transport of pharmaceutical residues in bank filtration. *Ground Water Monitoring & Remediation* **24(2)** 70–77

Heckmann LH, Callaghan A, Hooper HL et al 2007 Chronic toxicity of ibuprofen to *Daphnia magna*: Effects on life history traits and population dynamics. *Toxicology Letters* **172** 137–145 *Doi:10.1016/j.toxlet.2007.06.001*

Hernando MD, Mezcua M, Fernández-Alba AR and Barceló D 2006 Environmental risk assessment of pharmaceutical residues in wastewater effluents, surface waters and sediments. *Talanta* **69** 334-342

Henschel K-P, Wenzel A, Diedrich M, and Fliedner A 1997 Environmental hazard assessment of pharmaceuticals *Regulatory Toxicology & Pharmacology* **25 (3)** 220–225

Hoeger BA, Köllner B, Dietrich DR and Hitzfeld B 2005 Water-borne diclofenac affects kidney and gill integrity and selected immune parameters in brown trout (*Salmo trutta* f. fario). *Aquatic Toxicology* **75** 53–64

Holbech H, Kinnberg K, Petersen GI, et al 2006.Detection of endocrine disrupters: evaluation of a fish sexual development test (FSDT). *Comparative Biochemistry & Physiology C Toxicology & Pharmacology* **144(1)** 57-66

Hoppe, PD, Rosi-Marshall, EJ and Bechtold, HA 2012 The antihistamine cimetidine alters invertebrate growth and population dynamics in artificial streams. *Freshwater Science* **31(2)** 379–388 *DOI: https://doi.org/10.1899/11-089*

Huggett DB, Brooks BW, Peterson B et al 2002 Toxicity of select beta-adrenergic receptor blocking pharmaceuticals (β blockers) on aquatic organisms. *Archives of Environmental Contamination & Toxicology* **43** 229–235

Jjemba PK 2006 Excretion and ecotoxicity of pharmaceutical and personal care products in the environment. *Ecotoxicology & Environmental Safety* **63** 113-130

Johnson, AC and Williams RJ 2004 A model to estimate influent and effluent concentrations of estradiol sewage treatment works. *Environmental Science and Technology* **38** 3649–3658; *DOI:10.1021/es035342u*

Jones OA, Lester JN and Voulvoulis N 2005 Pharmaceuticals: a threat to drinking water? *Trends in Biotechnology* **23(4)** 163-167

Jürgens MD, Holthaus KIE, Johnson AC et al 2002 The Potential for estradiol and ethinylestradiol degradation in English rivers. *Environmental Toxicology & Chemistry* **21(3)** 480-488; *DOI: 10.1002/etc.5620210302*

Kallio JM, Lahti M, Oikari A and Kronberg L 2010 Metabolites of the aquatic pollutant diclofenac in fish bile. *Environmental Science & Technology* **44** 7213–7219

https://www.researchgate.net/publication/261216635_Climate_Change_and_Chemicals_Environmental_and_Biological_Aspects

Kibria G, Haroon AKY, Nugegoda D and Rose G 2010 *Climate change and chemicals: Environmental Biological aspects.* New India Publishing Agency (NIPA), New Delhi. 470 p. ISBN: 9789380235301

https://www.researchgate.net/publication/261216635_Climate_Change_and_Chemicals_Environmental_and_Biological_Aspects

Kibria G, Haroon AKY, Nugegoda D 2013 Climate change and Agricultural food production: Impacts, vulnerabilities and remedies. 300 p, New India Publishing Agency (NIPA), New Delhi. ISBN: 9789381450512

https://www.researchgate.net/publication/261178054_Climate_Change_and_Agricultural_Food_Production_Impacts_Vulnerabilities_and_Remedies.

Kibuye FA, Gall HE, Kyle R. Elkin KR et al 2019 Occurrence, concentrations, and risks of pharmaceutical compounds in private wells in Central Pennsylvania. *Journal of Environmental Quality* **48** 1057–1066 *doi:10.2134/jeq2018.08.0301*

Kidd KA, Blanchfield PJ, Mills KH et al 2007 Collapse of a fish population after exposure to a synthetic estrogen. *Proceedings of the National Academy of Science, USA* **104** 8897–8901 *doi: 10. 1073/pnas.0609568104*

Klečka GM, Muir DCG, Dohmen P et al 2009 Introduction to special series: Science-based guidance and framework for the evaluation and identification of PBTs and POPs. Integrated *Environmental Assessment & Management* **5(4)** 535–538 *DOI: https://doi. org/10.1897/IEAM_2009-045.1*

Kim Y, Choi K, Jung J, Park S, Kim PG and Park J 2006 Aquatic toxicity of acetaminophen, carbamazepine, cimetidine, diltiazem, and six major sulphonamides and their potential ecological risks in Korea. *Environment International* **33** 370–375

Kim Y, Jung J, Kim M et al 2008 Prioritizing veterinary pharmaceuticals for aquatic environment in Korea. *Environmental Toxicology & Pharmacology* **26** 167-176

Kotchen M, Kallaos J, Wheeler K et al 2009 Pharmaceuticals in wastewater: Behaviour, preferences, and willingness to pay for a disposal program. *Journal of Environmental Management* **90** 1476-1482 *doi: 10.1016/j.envman.2008.10.002*

Kovacevic V, Simpson AJ, Simpson MJ 2016 1H NMR-based metabolomics of *Daphnia magna* responses after sub-lethal exposure to triclosan, carbamazepine and ibuprofen. *Comparative Biochemistry and Physiology Part D Genomics Proteomics* **19** 199–210 *Doi:10.1016/j.cbd.2016.01.004*

Kramer VJ, Miles-Richardson S, Pierens SL and Giesy JP 1998 Reproductive impairment and induction of alkaline-labile phosphate, a biomarker of estrogen exposure, in fathead minnows *Pimephales promelas* exposed to water borne 17 beta-estradiol. *Aquatic Toxicology* **40** 335-360

Kumar A and Xagoraraki I 2010 Pharmaceuticals, personal care products and endocrine-disrupting chemicals in U.S. surface and finished drinking waters: A proposed ranking system. *Science of the Total Environment* **408** 5972–5989

Kümmerer K 2009 Review The presence of pharmaceuticals in the environment due to human use – present knowledge and future challenges. *Journal of Environmental Management* **90** 2354–2366

Kwon J-W and Armbrust KL 2006 Laboratory persistence and fate of fluoxetine in aquatic environments. *Environmental Toxicology & Chemistry* **25(10)** 2561-2568 *DOI: 10.1897/05-613R.1*

Kwon J-W 2016 Environmental impact assessment of veterinary drug on fish aquaculture for food safety. *Drug Testing Analysis* **8** 556–564

Lamichhane K, Garcia SN, Huggett DB et al 2013 Chronic effects of carbamazepine on life-history strategies of *Ceriodaphnia dubia* in three successive generations. *Archives of Environmental Contamination & Toxicology* **64** 427–438 *Doi: 10.1007/s00244-012-9845-5*

Länge R, Hutchinson TH, Croudace CP et al 2001 "Effects of the synthetic estrogen 17α-ethinylestradiol on the life-cycle of the fathead minnow (*Pimephales promelas*)." *Environmental Toxicology & Chemistry* **20** 1216-1227

Lange A, Paull GC, Coe TS et al 2009 Sexual reprogramming and estrogenic sensitization in wild fish exposed to ethinylestradiol. *Environmental Science Technology* **43(4)** 1219-25

Larsson DGJ, Adolfsson-Erici M, Parkkonen J et al 1999 Ethinyloestradiol—An undesired fish contraceptive? *Aquatic Toxicology* **45** 91–97

Laville N, Aït-Aïssa S, Gomez E, Casellas C and Porcher J.M 2004 Effects of human pharmaceuticals on cytotoxicity, EROD activity and ROS production in fish hepatocytes. *Toxicology* **196** 41–55

Lawrence JR, Zhu B, Swerhone GDW et al 2009 Comparative microscale analysis of the effects of triclosan and triclocarban on the structure and function of river biofilm communities. *Science of the Total Environment* **407** 3307–3316

Lawrence, JR, Zhu, B, Swerhone, GDW et al 2012 Molecular and microscopic assessment of the effects of caffeine, acetaminophen, diclofenac, and their mixtures on river biofilm communities. *Environmental Toxicology & Chemistry* **31(3)** 508–517 *DOI: https://doi.org/10.1002/etc.1723*

Lee SS, Paspalof AM, Snow D et al 2016 Occurrence and potential biological effects of amphetamine on stream communities. *Environmental Science & Technology* **50(17)** 9727–9735 *DOI: https://doi.org/10.1021/acs.est.6b03717*

Li ZH, Zlabek V, Velisek J et al 2011 Acute toxicity of carbamazepine to juvenile rainbow trout (*Oncorhynchus mykiss*): Effects on antioxidant responses, hematological parameters and hepatic EROD. *Ecotoxicology & Environmental Safety* **74** 319–327 *Doi: 10.1016/j. ecoenv.2010.09.008*

Lin KD and Gan J 2011 Sorption and degradation of wastewater-associated non-steroidal anti-inflammatory drugs and antibiotics in soils. *Chemosphere* **83** 240–246

Lister A, Van der Kraak G 2008 An investigation into the role of prostaglandins in zebrafish oocyte maturation and ovulation. *General & Comparative Endocrinology* **159** 46–57

Liu BY, Nien XP, Liu WQ et al 2011 Toxic effects of erythromycin, ciprofloxacin and sulfamethoxazole on photosynthetic apparatus in *Selenastrum capricornutum. Ecotoxicology & Environmental Safety* **74(4)** 1027–1035 *doi: 10.1016/j.ecoenv.2011.01.022*

Liu J, Lu G, Ding J et al 2014 Tissue distribution, bioconcentration, metabolism, and effects of erythromycin in crucian carp (*Carassius auratus*). *Science of the Total Environment* **490** 914-20 *doi: 10.1016/j.scitotenv.2014.05.055. Epub 2014 Jun 6*

Liu J, Lu G, Xie, Z et al 2015 Occurrence, bioaccumulation and risk assessment of lipophilic pharmaceutically active compounds in the downstream rivers of sewage treatment plants. *Science of the Total Environment* **511** 54–62 *http://dx.doi.org/10. 1016/j.scitotenv.2014.12.03*3

Loraine G and Pettigrove M 2006 Seasonal variations in concentrations of pharmaceuticals and personal care products in drinking water and reclaimed wastewater in southern California. *Environmental Science & Technology* **40** 687–95

Lürling M, Sargant E and Roessink I 2006 Life-history consequences for *Daphnia pulex* exposed to pharmaceutical carbamazepine. *Environmental Toxicology* **21** 172–180 *Doi:10.1002/tox.20171*

Lyssimachou A and Arukwe A 2007 Alteration of brain and interrenal StAR protein, P450scc, and Cyp11βmRNA levels in Atlantic salmon after nominal waterborne exposure to the synthetic pharmaceutical estrogen ethinylestradiol. *Journal of Toxicology & Environmental Health Part A* **70(7)** 606–613 *doi: 10.1080/10937400600882905*

Martins A, Guimaraes L and Guilhermino L 2013 Chronic toxicity of the veterinary antibiotic florfenicol to *Daphnia magna* assessed at two temperatures. *Environ. Toxicol. Pharmacol.* **36** 1022–1032

Matozzo V, Devoti AC and Marin MG 2012 Immunotoxic effects of triclosan in the clam *Ruditapes philippinarum. Ecotoxicology* **21** 66–74

McFarland JW, Berger CM, Froshauer SA et al 1997 Quantitative structure-activity relationships among macrolide antibacterial agents: in vitro and in vivo potency against *Pasteurella multocida. Journal of Medicinal Chemistry* **40** 1340-1346

McKinley SJ and Hazel JR 1993 Epinephrine stimulation of glucose release from perfused trout liver: Effects of assay and acclimation temperature. *Journal of Experimental Biology* **177** 51–62

MDH (Minnesota Department of Health) 2016 Contaminants of emerging concern program. 17α-Ethinylestradiol and mestranol and drinking water.

https://www.health.state.mn.us/communities/environment/risk/docs/guidance/gw/mestraethinyleinfo.pdf

Mehinto AC, Hill EM and Tyler CR 2010 Uptake and biological effects of environmentally relevant concentrations of the nonsteroidal anti-inflammatory pharmaceutical diclofenac in rainbow trout (*Oncorhynchus mykiss*). *Environmental Science & Technology* **44(6)** 2176–2182 *https://doi.org/10.1021/es903702m*

Mennigen JA, Zamora JM, Chang JP and Trudeau VL 2017 Endocrine disrupting effects of waterborne fluoxetine exposure on the reproductive axis of female goldfish, *Carassius auratus. Comparative Biochemistry & Physiology C Toxicology & Pharmacology* **202** 70-78 *doi: 10.1016/j.cbpc.2017.08.003. Epub 2017 Aug 16, 10.1021/es903702m*

Miége C, Choubert JM, Ribeiro L et al 2009 Fate of pharmaceuticals and personal care products in wastewater treatment plants-conception of a database and first results. *Environmental Pollution* **157(5)** 1721-1726 *doi.10.1016/j.envpol.2008.11.045.*

Mimeault C, Woodhouse AJ, X.-S, Miao X-S et al 2005 The human lipid regulator, gemfibrozil bioconcentrates and reduces testosterone in the goldfish, *Carassius auratus. Aquatic Toxicology* **73** 44–54

Molina-Fernandez N, Perez-Conde C, Rainieri S and Sanz-Landaluze J 2017 Method for quantifying NSAIDs and clofibric acid in aqueous samples, lumpfish (Cyclopteruslumpus) roe, and zebrafish (*Danio rerio*) eleutheroembryos and evaluation of their bioconcentration in zebrafish eleuthero embryos. *Environmental Science & Pollution Research* **24** 10907–10918 *DOI 10.1007/s11356-016-6671-8*

Mompelat S, LeBot B and Thomas O 2009 Occurrence and fate of pharmaceutical products and by–products, from resource to drinking water. *Environment International* **35** 803–814

MPCA (Minnesota Pollution Control Authority) 2013 Pharmaceuticals and endocrine active chemicals in Minnesota lakes

http://www.pca.state.mn.us/index.php/viewdocument.html?gid=19427

MSDS (Material Safety Data Sheet) GalxoSmithKline 2014 Salicylic acid. https://www.briggatemedical.com/pdfs/msds_duofilm.pdf

MSDS (Material Safety Data Sheet) Pharmacopedia 2006 Paracetamol.

https://www.pharmacopoeia.com/Catalogue/Preview?uri=%2Fcontent%2Ffile%2Fproducts%2Fhealthandsafety%2FCat_371_GB.pdf

MSDS (Material Safety Data Sheet) Pubchem 2004a Aspirin. US National Library of Medicine. National Center for Biotechnology Information; *https://pubchem.ncbi.nlm.nih.gov/compound/Aspirin*

MSDS (Material Safety Data Sheet) Pubchem 2004b Estrone. US National Library of Medicine. National Center for Biotechnology Information; *https://pubchem.ncbi.nlm.nih.gov/compound/Estrone*

MSDS (Material Safety Data Sheet) Pubchem 2004c Estradiol. US National Library of Medicine. National Center for Biotechnology Information; *https://pubchem.ncbi.nlm.nih.gov/compound/Estradiol*

MSDS (Material Safety Data Sheet) Pubchem 2004d Ibuprofen US National Library of Medicine. National Center for Biotechnology Information; *https://pubchem.ncbi.nlm.nih.gov/compound/Ibuprofen*

MSDS (Material Safety Data Sheet) Pubchem 2004e. Salicyclic acid Library of Medicine. US National Center for Biotechnology Information; *https://pubchem.ncbi.nlm.nih.gov/compound/Salicylic-acid*

MSDS (Material Safety Data Sheet) Pubchem 2005a Benzafibrate. US National Library of Medicine. National Center for Biotechnology Information; *https://pubchem.ncbi.nlm.nih.gov/compound/Bezafibrate*

MSDS (Material Safety Data Sheet) Pubchem 2005b Carbamazepine. US National Library of Medicine. National Center for Biotechnology Information; *https://pubchem.ncbi.nlm.nih.gov/compound/Carbamazepine*

MSDS (Material Safety Data Sheet) Pubchem 2005c Ciprofloxacin. US. National Library of Medicine. National Center for Biotechnology Information; *https://pubchem.ncbi.nlm.nih.gov/compound/Ciprofloxacin*

MSDS (Material Safety Data Sheet) Pubchem 2005d Clofibric acid. US National Library of Medicine. National Center for Biotechnology Information; *https://pubchem.ncbi.nlm.nih.gov/compound/Clofibric-acid*

MSDS (Material Safety Data Sheet) Pubchem 2005e Diclofenac. US National Library of Medicine. National Center for Biotechnology Information; *https://pubchem.ncbi.nlm.nih.gov/compound/Diclofenac*

MSDS (Material Safety Data Sheet) Pubchem 2005f Estriol. US National Library of Medicine. National Center for Biotechnology Information; *https://pubchem.ncbi.nlm.nih.gov/compound/Estriol*

MSDS (Material Safety Data Sheet) Pubchem 2005g Ethinylestradiol. US National Library of Medicine. National Center for Biotechnology Information; *https://pubchem.ncbi.nlm.nih.gov/compound/Ethinyl-estradiol*

MSDS (Material Safety Data Sheet) Pubchem 2005h Erythromycin. US National Library of Medicine. National Center for Biotechnology Information; *https://pubchem.ncbi.nlm.nih.gov/compound/Erythromycin*

MSDS (Material Safety Data Sheet) Pubchem 2005i Gemfibrozil. US National Library of Medicine. National Center for Biotechnology Information; *https://pubchem.ncbi.nlm.nih.gov/compound/Gemfibrozil*

MSDS (Material Safety Data Sheet) Pubchem 2005j Ketoprofen. US National Library of Medicine. National Center for Biotechnology Information; *https://pubchem.ncbi.nlm.nih.gov/compound/Ketoprofen*

MSDS (Material Safety Data Sheet) Pubchem 2005k Naproxen. US National Library of Medicine. National Center for Biotechnology Information; *https://pubchem.ncbi.nlm.nih.gov/compound/Naproxen*

MSDS (Material Safety Data Sheet) Pubchem 2005m Ofloxacin. US National Library of Medicine. National Center for Biotechnology Information; *https://pubchem.ncbi.nlm.nih.gov/compound/ofloxacin*

MSDS (Material Safety Data Sheet) Pubchem 2005n Roxithromycin. US National Library of Medicine. National Center for Biotechnology Information; *https://pubchem.ncbi.nlm.nih.gov/compound/Z_-Roxithromycin*

MSDS (Material Safety Data Sheet) Pubchem 2005o Sulfamethoxazole. US National Library of Medicine. National Center for Biotechnology Information; *https://pubchem.ncbi.nlm.nih.gov/compound/Sulfamethoxazole*

MSDS (Material Safety Data Sheet) Pubchem 2005p Trimethoprim. US National Library of Medicine. National Center for Biotechnology Information; *https://pubchem.ncbi.nlm.nih.gov/compound/Trimethoprim*

MSDS (Material Safety Data Sheet) Pubchem 2005q Triclosan. US National Library of Medicine. National Center for Biotechnology Information; *https://pubchem.ncbi.nlm.nih.gov/compound/Triclosan*

MSDS (Material Safety Data Sheet) Pubchem 2005r. Fluoxetine US National Library of Medicine. National Center for Biotechnology Information;

https://pubchem.ncbi.nlm.nih.gov/compound/Fluoxetine

MSDS (Material Safety Data Sheet) Pubchem 2011 Tetracyclines. US National Library of Medicine. National Center for Biotechnology Information;

https://pubchem.ncbi.nlm.nih.gov/compound/Tetracycline

Nallani GC1, Paulos PM, Constantine LA, Venables BJ, Huggett DB 2011 Bioconcentration of ibuprofen in fathead minnow (*Pimephales promelas*) and channel catfish (*Ictalurus punctatus). Chemosphere.* **84(10)** 1371-1377 *doi: 10.1016/j.chemosphere.2011.05.008. Epub 2011 Jun 11*

Nash JP, Kime DE, Van der Ven LTM et al 2004 Long-term exposure to environmental concentrations of the pharmaceutical ethynylestradiol causes reproductive failure in fish. *Environmental Health Perspec*tives **112** 1725–1733

Nassef M, Matsumoto S, Seki M et al 2010a Acute effects of triclosan, diclofenac and carbamazepine on feeding performance of Japanese medaka fish (*Oryzias latipes*). *Chemosphere* **80** 1095–1100 *Doi:10.1016/j.chemosphere.2010.04.073.*

Nassef M, Kim SG, Seki M et al 2010b In Ovonano injection of triclosan, diclofenac and carbamazepine affects embryonic development of medaka fish (*Oryzias latipes*). *Chemosphere* **79** 966–973 *Doi: 10.1016/j.chemosphere.2010.02.002.*

Newman MC 2010 Fundamentals of Ecotoxicology, third ed. CRC Press, Boca Raton, United States.

New Scientist 2009 Top 11 compounds in US drinking water. 12 January. *http://www.newscientist.com/article/dn16397-top-11-compounds-in-us-drinking-water.html*

Ne´stor MC and Mariana C 2017 Impact of pharmaceutical waste on biodiversity. In: LM Go´mez-Oliva´n (ed.), Eco-pharmaco-vigilance: Multidisciplinary Approaches to Environmental Safety of Medicines, Hdb Env Chem, © Springer International Publishing AG 2017 *DOI: 10.1007/698_2017_151*

Nietch CT, Quinlan EL, Lazorchak JM et al 2013 Effects of a chronic lower range of triclosan exposure on a stream mesocosm community. *Environmental Toxicology & Chemistry* **32(12)** 2874–2887 *DOI: https://doi.org/10.1002/ etc.2385*

NOAA 2006 United States National Oceanic and Atmospheric Administration (NOAA), Pharmaceuticals in the environment. *http://www. chbr.noaa.gov/peiar/*

Nunes B, Carvalho F, Guilhermino L 2004 Acute and chronic effects of clofibrate and clofibric acid on the enzymes acetylcholinesterase, lactate dehydrogenase and catalase of the mosquitofish *Gambusia holbrooki. Chemosphere* **57** 1581–1589

National Water Quality Management Strategy (NWQMS) 2008 Australian Guidelines for Water Recycling: Managing Health and Environmental Risks (Phase 2) — Augmentation of Drinking Water Supplies. Natural Resource Management Ministerial Council. Environment Protection and Heritage Council. Australian Health Ministries Conference. Biotex Pty Ltd, Canberra (140 p.).

https://www.nhmrc.gov.au/about-us/publications/australian-guidelines-water-recycling

Oaks JL, Gilbert M, Virani MZ et al 2004 Diclofenac residues as the cause of vulture population decline in Pakistan. *Nature* **427** 630–633

OECD 2012. Test No. 305: Bioaccumulation in Fish: Aqueous and Dietary Exposure, OECD Guidelines for the Testing of Chemicals, Section 3, O.P., 2012. OECD guidelines for testing of chemicals. Bioaccumulation in Fish : Aqueous and Dietary Exposure *doi:10. 1787/2074577x*

OECD 2019. Pharmaceutical residues in freshwater. Hazards and Policy responses. *https://www.oecd.org/environment/resources/pharmaceutical-residues-in-freshwater-policy-highlights.pdf*

Oliveira R, Domingues I, Koppe C et al 2009 Effects of triclosan on zebrafish early- life stages and adults *Environmental Science & Pollution Research* **16** 679–688 *Doi:10.1007/s11356-009-0119-3*

Orvos DR, Versteeg DJ, Inauen J et al 2002 Aquatic Toxicity of Triclosan. *Environmental Toxicology & Chemistry* **21(7)** 1338-1349

Owen SF, Giltrow E, Huggett DB et al 2007 Comparative physiology, pharmacology and toxicology of β-blockers: Mammals versus fish. *Aquatic Toxicol*ogy **82** 145–162

Pan M and Chu LM 2016 Phytotoxicity of veterinary antibiotics to seed germination and root elongation of crops. *Ecotoxicology & Environmental Safety* **126** 228–237 *Doi:10.1016/j.ecoenv.2015.12.027*

Peake BM, Braund R, Tremblay LA and Tong AYC 2016 Impact of pharmaceuticals on the environment, chapter 5, p. 109-152, In: The Life-Cycle of Pharmaceuticals in the Environment. Elsevier Ltd. *http://dx.doi.org/10.1016/B978-1-907568-25-1.00005-0*

Park J 2005 Pharmaceuticals in the environment and management approaches in Korea. KEI, 2005, RE-12. *http://www.kei.re.kr/04 publ/pdf/report/ 05 RE12.pdf*

Parolini M, Binelli A, Cogni D et al 2009 An in-vitro biomarker approach for the evaluation of the ecotoxicity of non-steroidal anti-inflammatory drugs (NSAIDs). *Toxicology in Vitro* **23** 935–942 *Doi:10.1016/j.tiv.2009.04.014*

Parolini M, Binelli A and Provini A 2011 Chronic effects induced by ibuprofen on the freshwater bivalve *Dreissena polymorpha*. *Ecotoxicology & Environmental Safety* **74(6)** 1586-1594 *doi: 10.1016/j.ecoenv.2011.04.025*

Payan P and Girard J-P 1977 Adrenergic receptors regulating patterns of blood flow through the gills of trout. *Amr. Jour. Physiol. Heart Circ. Physiol.* **232** H18–H23

Peake BM, Braund R, Tremblay LA and Tong AYC 2016 Impact of pharmaceuticals on the environment, chapter 5 Pages 109-152 In: The Life-Cycle of Pharmaceuticals in the Environment. *http://dx.doi.org/10.1016/B978-1-907568-25-1.00005-0*

Pereira AMPT, Silva LJG, Laranjeiro CSM et al 2017 Human pharmaceuticals in Portuguese rivers: The impact of water scarcity in the environmental risk. *Science of the Total Environment* **609** 1182–1191

Perreault HAN, Semsar K and Godwin J 2003 Fluoxetine treatment decreases territorial aggression in a coral reef fish. *Physiology & Behaviour* **79** 719–724

Pérez S and Barceló D 2007 Fate and occurrence of X-ray contrast media in the environment. *Analytical & Bioanalytical Chemistry* **387** 1235–1246

Proia L, Morin S, Peipoch M et al 2011 Resistance and recovery of river biofilms receiving short pulses of triclosan and diuron. *Science of The Total Environment* **409(17)** 3129–3137 *DOI: https://doi.org/10.1016/j. scitotenv.2011.05.013*

Purdom CE, Hardiman PA, Bye VVJ et al 1994 Estrogenic effects of effluents from sewage treatment works. *Chemical Ecology* **8** 275–285 *Doi: 10.1080/02757549408038554*

Quinlan EL, Nietch CT, Blocksom K et al 2011 Temporal dynamics of periphyton exposed to tetracycline in stream mesocosms. *Environmental Science & Technology* **45** 10684–10690

Raftery TD, N. Jayasundara and RT Di Giulio 2017 A bioenergetics assay for studying the effects of environmental stressors on mitochondrial function *in vivo* in zebra fish larvae. *Comparative Biochemistry & Physiology Part C Toxicology & Pharmacology* **192** 23–32 *Doi: 10.1016/j.cbpc.2016.12.001*

Reddersen K, Heberer T and Dünnbier U 2002 Identification and significance of phenazone drugs and their metabolites in ground and drinking water. *Chemosphere* **49** 539–544

Rhee JS, Kim BM, Jeong CB et al 2013 Effect of pharmaceuticals exposure on acetylcholinesterase (AchE) activity and on the expression of AchE gene in the monogonont rotifer, *Brachionus koreanus. Comparatve Biochemistry & Physiology C Toxicology & Pharmacology* **158** 216–224 *Doi: 10.1016/j.cbpc.2013.08.005*

Richman C and Castensson S 2008 Impact of waste pharmaceuticals: an environmental hazard or green wash. *The Pharmaceutical Journal* **280** 335-342

Ricart M, Guasch H, Alberch M et al 2010 Triclosan persistence through wastewater treatment plants and its potential toxic effects on river biofilms. *Aquatic Toxicology* **100** 346–353

Richmond EK, Rosi-Marshall EJ, Lee SS et al 2016 Antidepressants in stream ecosystems: Influence of selective serotonin reuptake inhibitors (SSRIs) on algal production and insect emergence. *Freshwater Science* **35(3)** 845–855 *DOI: https://doi.org/10.1086/687841*

Roche 2007 Pharmaceuticals sustainability database.

<http://www.roche.com/home/sustainability/sus_env/sus_sds_search/sus_sds_search_result.htm?sdssearch_name+&sdsearch_sum=%25&sdsearch.cas=&submit_Show (SDS)>

Rocco L, Frenzilli G, Fusco D et al 2010 Evaluation of zebra fish DNA integrity after exposure to pharmacological agents present in aquatic environments. *Ecotoxicology & Environmental Safety* **73** 1530–1536; *Doi: 10.1016/j.ecoenv.2010.07.032*

Roepke TA, Snyder MJ and Cherr GN 2005 Estradiol and endocrine disrupting compounds adversely affect development of sea urchin embryos at environmentally relevant concentrations. *Aquatic Toxicology* **71(2)** 155-173

Romaní AM, Guasch H, Balaguer MD and de Montilivi C (eds.) 2016 *Aquatic Biofilms: Ecology, Water Quality and Wastewater Treatment.* Caister Academic Press, Spain. ISBN: 978-1-910190-17-3

Rosi-Marshall, EJ, Kincaid, DW, Bechtold et al 2013 Pharmaceuticals suppress algal growth and microbial respiration and alter bacterial communities in stream biofilms. *Ecological Applications* **23(3)** 583–593 *DOI: https://doi. org/10.1890/12-0491.1*

Rosenfeld PE and Feng LGH 2011 Chapter 16. Emerging Contaminants. P. 215-221. In: Risks of Hazardous Wastes. ISBN 978-1-4377-7842-7

Runnells TJ, Hala DN and Sumpter JP 2007 Preliminary studies into the effects of the human pharmaceutical clofibric acid on sperm parameters in adult fathead minnow. *Aquatic Toxicology* **84** 111–118

Sa¨fholm M, Ribbenstedt A, Fick J, and Berg C 2014 Risks of hormonally active pharmaceuticals to amphibians: a growing concern regarding progestagens. *Philosophical Transactions of the Royal Society B* **369** 20130577 *doi: 10.1098/rstb.2013.0577*

Saley A, Hess M, Miller K Howard D and King-Heiden TC 2016 Cardiac toxicity of triclosan in developing zebra fish. *Zebrafish* **13** 399–404 *Doi: 10.1089/zeb.2016.1257*

Sanderson H and Thomsen M 2009 Comparative analysis of pharmaceuticals versus industrial chemicals acute toxicity classification according to the United Nations classification system for chemicals. *Toxicology Letters* **187** 84–93 *doi: 10.1016/j.toxlet.2009.02.003*

Sanchez-Arguello P, Fernandez C and Tarazona JV 2009 Assessing the effects of fluoxetine on *Physa acuta* (Gastropoda, Pulmonata) and *Chironomus riparius* (Insecta, Diptera) using a two-species water-sediment test. *Science of the Total Environment* **407** 1937–1946

Sanofi 2018 Disposal of unused medicines and user recommendations. Corporate social responsibility.

https://www.sanofi.com/-/media/project/one-sanofi-web/websites/global/sanofi-com/home/common/docs/download-center/disposal_of_unused_medicines_and_user_recommendations_2018.pdf/

Santos JL, Aparicio I and Alonso E 2007 Occurrence and risk assessment of pharmaceutically active compounds in wastewater treatment plants: a case study: Seville city (Spain). *Environment International* **33** 596-601

Santos L, Araujo AN, Fachini A et al 2010 Ecotoxicological aspects related to the presence of pharmaceuticals in the aquatic environment. *Journal of Hazardous Materials* **175** 45–95

Schmid T, Gonzalez-Valero J, Rufli H and Dietrich DR 2002 Determination of vitellogenin kinetics in male fathead minnows (*Pimephales promelas*). *Toxicology Letters* **131** 65–74 *Doi: 10.1016/S0378-4274(02)00043-7*

Schwaiger J, Ferling H, Mallow U et al 2004 Toxic effects of the non-steroidal anti-inflammatory drug diclofenac: Part I: Histopathological alterations and bioaccumulation in rainbow trout. *Aquatic Toxicology* **68**(2) 141–150 *doi: 10.1016/j.aquatox.2004.03.014*

Sengelov G, Agerso Y, Halling-Sorensen B et al 2003 Bacterial antibiotic resistance levels in Danish farmland as a result of treatment with pig manure slurry. *Environment International* **28** 587–595

Shim J, Weatherly LM, Luc RH et al 2016 Triclosan is a mitochondrial uncoupler in live zebra fish. *Journal of Applied Toxicology* **36** 1662–1667 *Doi: 10.1002/jat.3311*

Soares J, Coimbra AM, Reis-Henriques MA et al 2009 Disruption of zebrafish (*Danio rerio*) embryonic development after full life-cycle parental exposure to low levels of ethinylestradiol. *Aquatic Toxicology* **95** 330–338 *Doi: 10.1016/j.aquatox.2009.07.021*

Spongberg AL and Witter JD 2008 Pharmaceutical compounds in the wastewater process stream in Northwest Ohio. *Science of the total Environment* **397** 148-157

Stackelberg PE, Furlong ET, Meyer MT et al 2004 Persistence of pharmaceutical compounds and other organic wastewater contaminants in a conventional drinking-water-treatment plant. *Science of the Total Environment* **329** 99-113

Stackelberg PE, Gibs J, Furlong ET et al 2007 Efficiency of conventional drinking-water-treatment processes in removal of pharmaceuticals and other organic compounds. *Science of the Total Environment* **377** 255–72

Stanley JK, Ramirez AJ, Chambliss CK and Brooks BW 2007 Enantiospecific sublethal effects of the antidepressant fluoxetine to a model aquatic vertebrate and invertebrate. *Chemosphere* **69** 9–16

Straub JO 2013 An environmental risk assessment for human-use trimethoprim in European surface waters *Antibiotics* **2** 115-162 *doi: 10.3390/antibiotics2010115*

Strempel S, Scheringer M, Ng CA and Hungerbühler K 2012 Screening for PBT chemicals among the "Existing" and "New" Chemicals of the EU. *Environmental Science & Technology* **46(11)** 5680–5687 *DOI: https://doi.org/10. 1021/es3002713*

Stumpf M, Ternes TA, Haberer K et al 1996 Determination of drugs in sewage treatment plants and river water (in German language). *Vom Wasser* **86** 291–303

Stuer-Lauridsen F, Birkved M, Hansen LP et al 2000 Environmental risk assessment of human pharmaceuticals in Denmark after normal therapeutic use. *Chemosphere* **40** 783–93

Swan GE, Cuthbert R, Quevedo M et al 2006 Toxicity of diclofenac to *Gyps* vultures. *Biology Letters* **2** 279– 282 *doi: 10.1098/rsbl.2005.0425*

Tauber R 2003 Quantitative analysis of pharmaceuticals in drinking water from ten Canadian cities, Enviro-Test Laboratories, Xenos Division, Ontario, Canada

Thibaut R, Schnell S and Porte C 2006 The interference of pharmaceuticals with endogenous and xenobiotic metabolizing enzymes in carp liver: an in-vitro study. *Environmental Science & Technology* **40(16)** 5154-5160

Thiele-Bruhn S 2003 Pharmaceutical antibiotic compounds in soils- a review. *Journal of Plant Nutrition & Soil Science* **166** 145-167

Thorpe KL, Benstead R, Hutchinson TH, Tyler CR. 2007 Associations between altered vitellogenin concentrations and adverse health effects in fathead minnow (*Pimephales promelas*). *Aquatic Toxicology* **85** 176–183 *doi: 10.1016/j.aquatox. 2007.08.012*

Tixier C, Singer HP, Oellers S et al 2003 Occurrence and fate of carbamazepine, clofibric acid, diclofenac, ibuprofen, ketoprofen, and naproxen in surface waters. *Environmental Science & Technology* **37(6)** 1061-1068.

Togola A and Budzinski H 2008 Multi-residue analysis of pharmaceutical compounds in aqueous samples. *Journal of Chromatography* **1177** 150–158

Toma A and Crisan O 2018 Green pharmacy- A narrative review. Clujul Medical **91(4)** 391-398 *DOI: 10.15386/cjmed-1129;*

https://www.ncbi.nlm.nih.gov/pmc/articles/PMC6296717/pdf/cm-91-391.pdf

Triebskorn R, Casper H, Scheil V and Schwaiger J 2007 Ultrastructural effects of pharmaceuticals (carbamazepine, clofibric acid, metoprolol, diclofenac) in rainbow trout (*Oncorhynchus mykiss*) and common carp (*Cyprinus carpio*). *Analytical & Bioanalytical Chemistry* **387** 1405–1416

UL (Underwriters Laboratories) 2011 Pharmaceuticals and personal care products in drinking water;

https://legacy-uploads.ul.com/wp-content/uploads/sites/40/2015/02/UL_WP_Final_Pharmaceuticals-and-Personal-Care-Products-in-Drinking-Water_v2_HR1.pdf

van den Brandhof E and Montforts M 2010 Fish embryo toxicity of carbamazepine, diclofenac and metoprolol. *Ecotoxicology & Environmental Safety* **73** 1862–1866

Vanderford BJ and Snyder SA 2006 Analysis of pharmaceuticals in water by isotope dilution liquid chromatography/ tandem mass spectrometry. *Environmental Science & Technol*ogy **40** 7312–7320

Vannini C, Domingo G, Marsoni M et al 2011 Effects of a complex mixture of therapeutic drugs on unicellular algae *Pseudokirchneriella subcapitata. Aquatic Toxicology* **101(2)** 459–465 *doi: 10.1016/j.aquatox.2010.10.011*

Vernouillet G, Eullaffroy P, Lajeunesse A et al 2010 Toxic effects and bioaccumulation of carbamazepine evaluated by biomarkers measured in organisms of different trophic levels. *Chemosphere* **80** 1062–1068

Vieno NM, Tuhkanen T and Kronberg L 2005 Seasonal variation in the occurrence of pharmaceuticals in effluents from a sewage treatment plant and in the recipient water. *Environmental Science & Technology* **39** 8220–8226

Volkova K, Reyhanian Caspillo N, Porseryd T et al 2015 Developmental exposure of zebrafish (*Danio rerio*) to 17α-ethinylestradiol affects non-reproductive behavior and fertility as adults and increases anxiety in unexposed progeny. *Hormones & Behaviour* **73** 30–38 *Doi: 10.1016/j.yhbeh.2015.05.014*

Waggot A 1981 Trace organic substances in the River Lee (Great Britain). In Chemistry in Water Reuse, 1st edn., (Cooper, WJ, ed.), p. 55–99, Ann Arbour Science.

Walters E, McClellan K, and Halden RU 2010 Occurrence and loss over three years of 72 pharmaceuticals and personal care products from biosolids soil mixtures in outdoor mesocosms. *Water Research* **44(20)** 6011-6020

Wang J and Gardinali PR 2012 Analysis of selected pharmaceuticals in fish and the freshwater bodies directly affected by reclaimed water using liquid chromatography-tandem mass spectrometry. *Analytical and Bioanalytical Chemistry* **404** 2711–2720

Watkinson AJ, Murby EJ, Kolpin DW and Costanzo S D 2009 The occurrence of antibiotics in an urban watershed: from wastewater to drinking water. *Science of the Total Environment* **8(1)** 775-777

Wennmalm A and Gunnarsson B 2009 Pharmaceutical management through environmental product labelling in Sweden. *Environment International* **35** 775–777 *doi:10.1016/j.scitoenv.2008.11.059*

WFD UKTAG (Water Framework Directive, United Kingdom Technical Advisory Group) 2012 Proposed EQS for Water Framework Directive Annex VIII substances: triclosan (For consultation). SNIFFER/ ENVIRONMENT AGENCY 2012 *file:///C:/Users/User1/Downloads/Triclosan-UKTAG.pdf*

WHO (World Health Organization) 2012 Pharmaceuticals in drinking-water. Technical report. 30 p. ISBN 978 92 4 150208 5 *https://apps.who.int/iris/bitstream/handle/10665/44630/9789241502085_eng.pdf;jsessionid=6F4B71C131B7F4E0A4B3728EDB83D0EF?sequence=1*

Wilson BA, Smith VH, deNoyelles F Jr and Larive CK 2003 Effects of three pharmaceutical and personal care products on natural freshwater algal assemblages. *Environmental Science & Technology* **37(9)** 1713-1719

Winter MJ, Caunter JE, Glennon Y and Hutchinson TH 2006 Atenolol: 28-day assessment of survival and growth in fathead minnow (*Pimephales promlas*) embryo-larvae. *Astra Zeneca* Brixham Environmental Laboratory, Report Number BL8269A.

Xuereb B, Lefèvre E, Garric J and Geffard O 2009 Acetylcholinesterase activity in *Gammarus fossarum* (Crustacea Amphipoda): Linking AChE inhibition and behavioural alteration. *Aquatic. Toxicology* **94** 114–122 *Doi: 10.1016/j.aquatox.2009.06.010*

Yang L-H, Ying G-G, Su H-C et al 2008 Growth-inhibiting effects of 12 antibacterial agents and their mixtures on the freshwater microalga *Pseudokirchneriella subcapitata. Environmental Toxicology & Chem*istry **27** 1201 *Doi: 10.1897/07-471.1.*

Ying GG and Kookana RS 2003 Degradation of five selected endocrine-disrupting chemicals in seawater and marine sediment. *Environmental Science & Technology* **37** 1256–1260

Zhang, W, Zhang M, Lin K et al 2012 Eco-toxicological effect of Carbamazepine on Scenedesmus obliquus and Chlorella pyrenoidosa. Environmental Toxicology & Pharmacology **33** 344–352 *Doi:10.1016/j.etap.2011.12.024*

Zhou JL, Zhang ZL, Banks E et al 2008 Pharmaceutical residues in wastewater treatment work effluents and their impact on receiving river water. *Journal of Hazardous Materials* **166(2-3)** 655-661 *doi: 10.1016/j.hazmat.2008.11.070*

Zorita S, Mårtensson L and Mathiasson L 2009 Occurrence and removal of pharmaceuticals in a municipal sewage treatment system in the south of Sweden. *Science of the Total Environment* **407** 2760-2770

Zou H, Radke M, Kierkegaard A et al 2015 Using chemical benchmarking to determine the persistence of chemicals in a Swedish Lake *Environmental Science & Technology* **49** 1646-1653

Zuccato E, Calamari D, Natangelo M and Fanelli R 2000 Presence of therapeutic drugs in the environment. *Lancet* **355** 1789–1790

Zühlke S, Dünnbier U and Heberer T 2004 Detection and identification of phenazone-type drugs and their microbial metabolites in ground and drinking water applying solid-phase extraction and gas chromatography with mass spectrometric detection. *Journal of Chromatography A* **1050** 201–209

Zurita JL, Repetto G, Jos A et al 2007 Toxicological effects of the lipid regulator gemfibrozil in four aquatic systems *Aquatic Toxicology* **81** 06–115

Web

https://www.researchgate.net/post/Aspirin_dissolution_or_hydrolysis

https://www.niehs.nih.gov/health/topics/agents/endocrine/

https://www.sciencedirect.com/topics/chemistry/bioconcentration

https://en.wikipedia.org/wiki/IC50#cite_note-fda-1

https://www.epa.gov/tsca-screening-tools/ecological-structure-activity-relationships-ecosar-predictive-model

https://www.cerc.usgs.gov/pubs/BEST/EROD.pdf

https://en.wikipedia.org/wiki/Vitellogenin

https://en.wikipedia.org/wiki/STAR_(gene)

https://ghr.nlm.nih.gov/primer/genefamily/cytochromep450

https://www.youtube.com/watch?v=PZRnF2a56RQ&ab_channel=ProfessorG

■■■■

10 Surfactants Pollution Risks, Impacts and Management

Summary

Surfactants are principally used in household cleaning, detergents, common soaps and personal care products. They are also used in many industrial applications such as additives in paints, textile softeners, antistatic agents, metal processing, oil drilling operations, biocides/ antimicrobial, pharmaceuticals, agrochemicals, food industry, petroleum industry, soil remediation, contaminated aquifer remediation, gas and coal recovery, microelectronics and pulp and paper industries. Based on the charge of the hydrophilic group of surfactants, they are classified as: *anionic* (alkyl ethoxy sulphates [AES], alkyl sulphates, linear alkylbenzene sulfonates); *cationic* (quaternary ammonium compounds); *non-ionic* (alkylphenol ethoxylates, nonylphenol, alcohol ethoxylates) and *amphoteric/ zwitterionic* (Betaines). The discharge of sewage effluents (treated or untreated containing surfactants) into the surface waters and application of sewage sludge (containing surfactants) on the land has the potential to impact water quality, aquatic biodiversity, seafood, agricultural food and human health. This chapter reviews **i.** Chemistry, toxicity and ecotoxicity of surfactants; **ii.** Impacts of surfactants on social, economic and environmental sectors including foam pollution, algal blooms, aquatic biota, seafood, crops, vegetables, fruits and human health; and **iii**. management of surfactants.

Contents

List of Tables

List of Figures

10.1. Introduction

Surfactants[1] (Surf-Act-Ants/ surface-active agents) are a diverse group of organic chemicals used in household cleaning, detergents, personal care products, textiles, paints, polymers, pesticide formulations, pharmaceuticals, mining, oil recovery and pulp and paper industries etc. Surfactants may act as detergents[2], wetting agents[3], emulsifiers[4], foaming[5] agents, and dispersants[6]. They are also used in many industrial applications such as additives in paints, textile softeners, antistatic[7] agents, metal processing, oil drilling operations, biocides/ anti-microbial, pharmaceuticals, agrochemicals, food industry, petroleum industry, soil remediation, contaminated aquifer remediation, gas and coal recovery, microelectronics and pulp and paper industries (Ying 2006, Rebello et al 2013, Calvo-Flores 2018). Please refer to Table 10.1 for industrial applications of surfactants.

---

[1] ***Surfactants*** *(surface-active agents) are chemically synthesised compounds made of molecules with a hydrophobic (water-repelling) section that is attracted to oils and fat and a hydrophilic (water-attracting) part which has an affinity for water (https://www.uts.edu.au/sites/default/files/2019-12/UTS_Consumer_Products_Report_v5.pdf).*

[2] ***Detergents*** *= It is a chemical substance, usually in the form of a* **powder or liquid**, *which is used for washing clothes or dishes. Detergent is similar to soap, but it is stronger and dissolves more completely in water. Detergents are special, powerful cleansers that can break up dirt, oils, and grease in clothing or on dishes. The Latin root of detergent is "detergere", means to wipe away or cleanse (https://www.vocabulary.com/dictionary/detergent).*

[3] ***Wetting agents*** *are substances that reduce the* ***surface tension*** *(the tendency of liquid surfaces to shrink to the minimum* **surface** *area possible) of water to allow it to spread drops onto a surface, increasing the spreading abilities of a liquid.* ***Dish soap*** *is a great example of a wetting agent. The soap dissolves all unwanted particles, exposing a clean surface. The soap also lowers the surface tension of water, allowing it to spread evenly across the entire surface.*

(https://chem.libretexts.org/Bookshelves/Physical_and_Theoretical_Chemistry_Textbook_Maps/Supplemental_Modules_(Physical_and_Theoretical_Chemistry)/Physical_Properties_of_Matter/States_of_Matter/Properties_of_Liquids/Wetting_Agents).

[4] ***Emulsion*** *= It is a mixture of two or more liquids that are normally immiscible (do not mix). Emulsions are part of a more general class of two-phase systems of matter called colloids.*

[5] ***A foaming agent*** *is a material that facilitates the formation of foam such as a surfactant. A surfactant, when present in small amounts, reduces the surface tension of a liquid (reduces the work needed to create the foam) or increases its colloidal stability by inhibiting coalescence of bubbles (https://en.wikipedia.org/wiki/Foaming_agent).*

[6] ***Dispersant*** *= a liquid or gas used to disperse small particles in a medium.*

[7] ***An antistatic agent*** *is a compound used for the treatment of materials or their surfaces in order to reduce or eliminate the build-up of static electricity.*

Table 10.1. Typical industrial applications of surfactants.

[https://infogram.com/different-types-of-surfactants-and-their-applications-1gjk92ejx86rp16].

Industries	Examples
Construction	To use as plasticisers for fluid concrete; to act as air-entraining agents in lightweight concrete; as concrete compaction agents and grouting aids; for processing bitumen in road construction.
Fibres and textile processing	To act as detergents and cleaning agents for the preparation and treatment of textiles such as silk and wool.
Food processing and production	To help water and fat phases of margarine separation; help to make ice creams smooth and creamy.
Leather processing	To shorten the soaking time of raw hides/ skins; accelerate the effect of liming and de-liming agent; remove the unwanted natural fat; facilitate uniform dyeing; and soften the leather.
Metal processing	To use as coolants, cleaners, and wetting agents.
Mining	To use in the extraction of minerals, coal and ores from the rock.
Oil production	To help keep the decomposition products dispersed and prevent them from becoming deposited on metal components.
Paints, inks and coatings	To act as dispersing agents for colour pigments, stabilise colours and simplify stirring and homogenisation.
Petroleum products	To keep the decomposition products dispersed and prevent them from becoming deposited on the metal components.
Plastics production	To use in both emulsion and suspension polymerisation production processes.
Pulp and paper processing	To help increase the softness and absorbency of paper and keep the screens and suction rollers used in paper production clean.

Surfactants generally consist of a polar/hydrophilic head group (either charged or uncharged), which is easily soluble in water, and a non-polar/ hydrophobic hydrocarbon tail, which does not dissolve in water easily (Ying 2006) (Fig. 10.1a and 10.1b). The hydrophobic group in a surfactant consists of an eight to 18 carbon hydrocarbon, which can be aliphatic (open chain), aromatic (ring-shaped), or a mixture of both (Cowan-Ellsberry et al 2014). Based on the charge of the hydrophilic group of surfactants, they are classified as: **i.** ***anionic*** (negatively charged); **ii.** ***cationic*** (positively charged)**; iii.** ***non-ionic*** (without any charge or uncharged) and **iv.** ***amphoteric/ zwitterionic*** (having both charges) (Myers 2006, Ivankovic and Hrenovic 2010, Prieto-Blanco et al 2019) (see Table 10.2 for classification of surfactants).

Household detergents and personal care applications are the most important application sector for surfactants, accounting for approximately 46% and 14% respectively of global consumption (*https://ihsmarkit.com/products/chemical-surfactants-scup.html*). World production of surfactants is estimated at 15 million tonnes per year (Mton/yr) (15 x 10^6 tonnes/year), of which about half are **soaps** (8 million tonnes per year) (8 x 10^6 tons/annum). Other surfactants produced on a particularly large scale are **LAS** (linear alkylbenzene sulfonates) (2 million tonnes per year), fatty alcohol **ethoxylates** (700 kton/yr), lignin sulfonates (600 kton/yr), alkylphenol ethoxylates (500 kton/yr) (Kosswig et al 2012, page 494).

In Europe, anionic (945 million tons/yr) and non-ionic (ethoxylates) (784 million tons/yr) surfactants are the major users (based on surfactants sales in 2014; Calvo-Flores et al 2018). Alkyl ethoxy sulphates, alkyl sulphates, linear alkylbenzene sulphonates (*anionic surfactants*), quaternary ammonium compounds (*cationic surfactants*) and alkylphenol ethoxylates, alcohol ethoxylates (*non-ionic surfactants*) are commonly used commercial surfactants (Ying 2006, Cowan-Ellsberry et al 2014) for which data and information is extensively available. The chemistry, toxicity and ecotoxicity of these surfactants are given in Table 10.3.

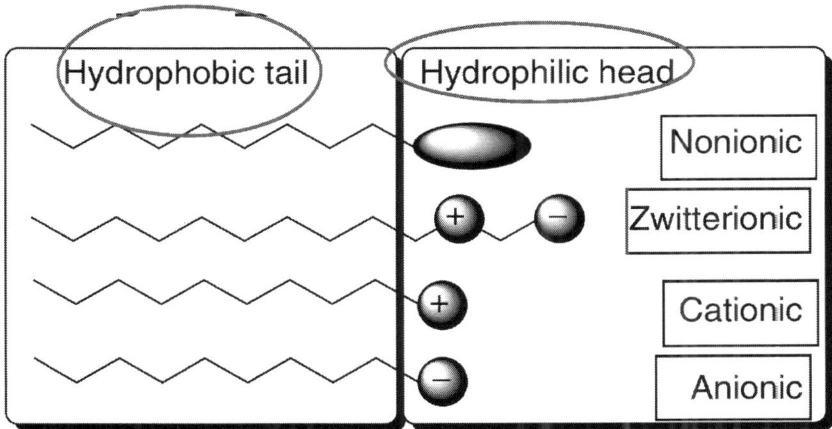

Fig. 10.1a. Generalized Structure of surfactants (Calvo-Flores et al 2018; page 342) *[surfactant molecular structures consist of two parts. One is called the* ***hydrophilic group****, which is soluble in water, the other part is called the* ***hydrophobic group*** *which is insoluble in water but is soluble in fat].*

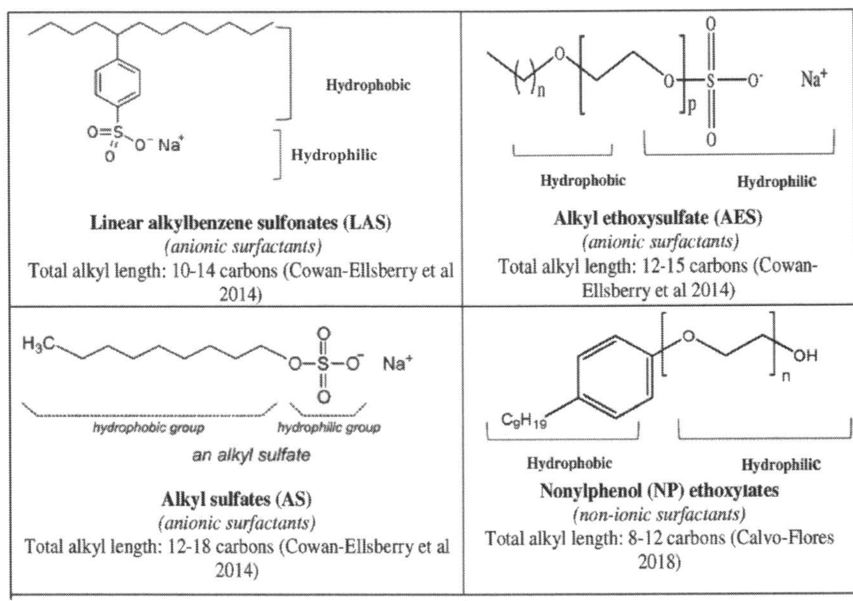

Figure 10.1b. Chemical structure of some common surfactants.

Table 10.2. Classification of surfactants
[Ivankovic and Herneovic 2010, Rebello et al 2013, Yuan et al 2014, Zhang et al 2015, Calvo-Flores et al 2018].

Surfactants category		Uses	Examples
Anionic surfactants *(negatively charged)*	Sulphates	Acne (skin) treatment, anti-ageing, bleaching, **bubble bath**, cleanser, dandruff/ scalp treatment, exfoliant[8]/ scrub, foot treatment, **hair colour**, liquid **hand soaps**, **shampoos**, **shaving creams**.	**Alkyl ethoxy sulphates** (AES), **Alkyl sulphates** (AS), Ammonium lauryl sulphate (ALS), Sodium dodecyl sulphate (SDS), Sodium Laurel Sulphates (SLS), Sodium lauryl ether sulphate (SLES) (Rebello et al 2013, Chong et al 2019). SLSs are mostly oleochemical[9] origin (palm or coconut oil), whereas SLES are mostly oleochemical/ natural or synthetic/ petrochemical (Chong et al 2019).
	Sulfonates	Fabric protectors, **fire-fighting foams**, metal plating, **stain repellents**.	**Linear alkylbenzene sulfonates** (LAS), Perfluorooctanesulfonic acid (PFOS), Perfluorobutanesulfonate. LAS is petrochemical-derived (Chong 2019). *note: LAS is the most popularly used synthetic anionic surfactants; about 82-*

[8] ***Exfoliation** involves the removal of the old, dead skin cells on the skin's outermost surface.*

[9] ***Oleochemistry** is the study of **vegetable oils and animal oils** and fats, and oleochemicals are derived from these fats and oils or from petrochemical feed stocks through physico-chemical modifications or transformation. The resulting product is called oleochemicals (from Latin: oleum "olive oil") (https://en.wikipedia.org/wiki/Oleochemistry).*

Surfactants category		Uses	Examples
			*87% of North American consumption is LAS (*Cowan-Ellsberry et al 2014*).* *note: see Chapter 5, sections 5.2.3, 5.3.3. for chemistry and impacts of PFOS).*
	Phosphates	**Detergents** (powders or liquids).	Sodium tripolyphosphate *(note: since surfactants contain phosphate, therefore, the discharge of surfactants contaminated water into waterways may cause algal blooms, note: see section 10.3.2.2. for Algal blooms)*
	Carboxy-lates	**Soaps**, fluorocarboxylates (in teflon[10], wire insulators, fire-fighting foam, wax papers, water and oil repellent **fabrics**)	Perfluorooctanoate (PFO), ***Perfluorononanoate*** (PFOA), Sodium stearate.
Cationic surfactants *(positively charged)*		Antiseptics, breath **sprays**, corrosion inhibitors, **cosmetics**, **fabric softeners**, **hair conditioners**, lozenges, **mouth washes**, nasal sprays, **tooth paste,** throat sprays (Ying 2006, Ivankovic and Herneovic 2010, Rebello et al 2013, Zhang et al 2015)	**Quaternary ammonium compounds** (QAC) include Benzalkonium chloride (BAC), Cetylpyridinium chloride (CPC), Cetyl trimethyl ammonium bromide (CTAB). QAC is oleochemical derived (Chong et al 2019)
Non-ionic surfactants *(without any charge/ uncharged)*		**Creams**, dye stuff and pigment preparations, **foam** stabilising agents, laundry detergents, **leather** industries (as degreasing), **lotions, paper** processing, **pesticide** formulation, shampoos, **textile** industries (wet processing) (Ying 2006, Rebello et al 2013, Calvo-Flores et al 2018).	**Alkylphenol ethoxylates (APE)**, Alcohol ethoxylates[11] (AE), **Nonyl phenol ethoxylates**[12]**,** Octyl phenol ethoxylates. APEs or AEs are either petrochemical or oleochemical derived (Chong et al 2019). *note: the degradation products of APEs are nonylphenol (NP) and Octylphenol (**OP**); the worldwide production of APEs consisted of 80-85% NP ethoxylates and 20% OP ethoxylates (Ying 2006, USEPA 2010); NP are toxic to both freshwater and marine species and identified as an endocrine disruptor to aquatic species (fish) (Ying 2006, Calvo-Flores et al*

[10] ***Teflon*** *(also polytetrafluoroethylene or PTFE) is a synthetic chemical made of carbon and fluorine. Surfaces coated with teflon become non-sticky and frictionless, making it ideal for cookware like pans and pots. Teflon-coated pans are easier to clean, use less oil to cook with, and last longer in the kitchen (https://www.bhg.com.au/the-truth-about-teflon-are-non-stick-pans-safe).*

[11] ***Alcohol ethoxylates*** *(AE) are a class of non-ionic surfactants that have been used for many years in formulated products, including laundry detergents and other household, institutional, and industrial cleaning products.*

[12] ***Nonylphenol*** *arises from the environmental degradation of nonylphenol ethoxylates (NPEs - the metabolites of commercial detergents called alkylphenol ethoxylates. NPEs are used as detergents, cleaners, emulsifiers, and a variety of other applications).*

Surfactants category	Uses	Examples
		2018).
Amphoteric/ zwitter-ionic *(both charges)*	**Cosmetics**, industrial softeners, non-stick cooking spray, protein purification, **shampoo, shower gel** (Ivankovic and Herneovic 2010, Rebello et al 2013, Yuan et al 2014).	Amino oxide, Betaines, Lecithin, Imidazolines.

10.1.1. Environmental occurrences of the surfactants

After their use as household and industrial detergents (laundry products, and cleaners), surfactants and their metabolites[13] are released into the surface water/ aquatic environment as treated or untreated waste waters and these enter various environmental compartments such as receiving waters and sediments (Fig. 10.2). A range of anionic, cationic, and non-ionic surfactants are commonly found in wastewater (see section 10.3.1). Surfactants can also enter aquatic ecosystems by stormwater discharge. In addition, spills from industrial processes are another way of surfactants entering in the environment. Recent studies have suggested that up to 60% of the total surfactant produced globally eventually enters the aquatic environment (Chong et al 2019). Several commercial industries also use surfactants for various needs (Table 10.1), therefore, industries could also be a potential source of surfactants pollution in the environment (in addition to household uses).

10.2. Chemistry, toxicity and ecotoxicity of surfactants

10.2.1 Persistence

Most surfactants are **non-persistent** in the environment having a short half-life, which ranges from a few hours to a few days. For example, the quaternary ammonium compounds (*cationic surfactants*) has a half-life of 2.5 h in sludge (Games and King 1982), whereas the half-life of linear alkylbenzene sulfonates (*anionic surfactants*) is reported to be three days in the river water (Larson and Payne 1981) (Table 10.3). Alkyl ethoxy sulphates such as nonylphenol (*non-ionic surfactants)* is the only surfactant that persists little bit longer (compared to other surfactants, Table 10.3) with a half-life of 50-70 d in seawater (Groshart et al 2001).

[13] ***Metabolites**: a substance produced during or taking part in metabolism.*

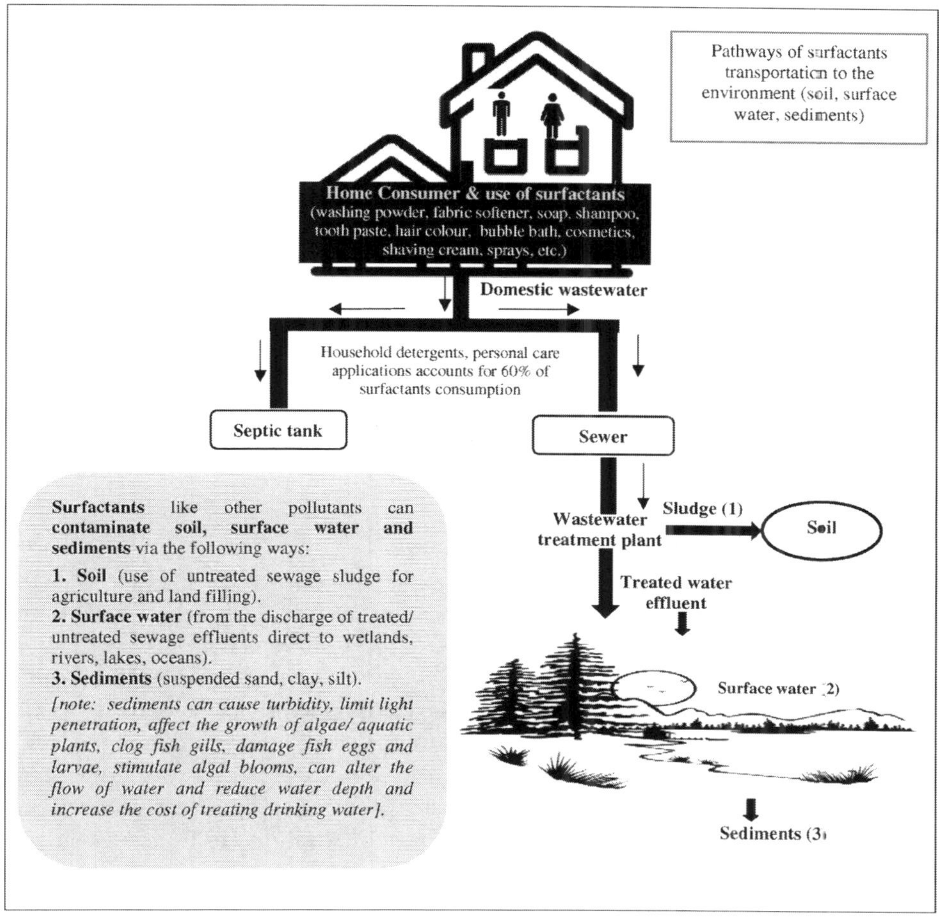

Fig. 10.2. Transportation of surfactants from household use to the environment (soil, surface water and sediments).

10.2.2 Biodegradation

Most surfactants are **readily biodegradable** in both aerobic and anaerobic conditions with a half-life of a few hours to few days (Table 10.3). However, under anaerobic conditions some surfactants degrade very slowly, causing it to exist in high concentrations in sewage sludge. For example, LAS (a group of anionic surfactants and a major ingredient of synthetic detergents), degrade very slowly under anaerobic conditions. Therefore, treatment plants do not completely remove LAS, and it can be found in very high concentrations in most sewage sludge and enter the soil compartment as a result of sludge application. The half-life of LAS in sediment is estimated to be 90 d. The USEPA proposed a standard **surfactant criterion** based on the rate of biodegradation and acute aquatic toxicity. In general, surfactants with a slower rate of biodegradation (>28 days) and the higher aquatic toxicity (LC_{50}/ EC_{50}/ IC_{50} <10 mg/L) is the surfactants of concern (see *https://www.epa.gov/ saferchoice/safer-choice-criteria-surfactants*). In this regard, nonylphenol could be of concern having a half-life of more than 70 d (seawater) and acutely toxic to algae, *Daphnia* and fish (EC_{50}/LC_{50}: 0.02 mg/L to 0.3 mg/L) (see Table 10.3).

10.2.3. Bioaccumulation

Most surfactants have a **low bioaccumulation potential** with a log K_{ow} values generally < 3 (*note: Log K_{ow} values of surfactants varies with their carbon chain lengths;* Cowan-Ellsberry et al 2014). Alkylphenol ethoxylates (APE) and their degradation products (nonylphenol and octylphenol) have high bioaccumulation potential of log K_{ow} values > 4 (Table 10.3). Nonylphenol residues have been detected in seafood organisms such as anchovies, hakes, mullets, octopus, swordfish and tuna (Ferrara et al 2008) (see also Table 10.5).

10.2.4. Toxicity

Most anionic, cationic and non-ionic surfactants are **nontoxic**, having LD_{50} (the lethal dose that kills 50% of a test sample) is generally more than 2,000 (range 2,000–8,000 mg/kg (Table 10.3) *[note: LD_{50} of surfactants are comparable to common salt or sodium chloride; LD_{50} of salt is 3,000 mg/kg; MSDS ChemOx 2015*]. Surfactants are reported to be **non-carcinogenic** to humans (Swisher 1966) but a recent report revealed that certain quaternary ammonium compounds (anionic surfactants) can contribute to the formation of a class of carcinogenic compounds called nitrosamines in the presence of certain chemicals (Greenwood 2011, EU 2012). Nonylphenol causes endocrine disruption (ED) effects on fish (Jobling and Sumpter 1993, Purdom et al 1994, Jobling et al 1996), crabs and mussels (Ricciardi et al 2008). Nonylphenol is reported to induce vitellogenin[14] in male rainbow trout at 20.3 mg/L (Jobling et al 1996) and medaka fish at 0.1 mg/L (Tabata et al 2001). It has been reported that nonylphenol has a three times higher endocrine disruption or estrogenic activity[15] than that of DDT (Soto et al 1991). Some other surfactants (LAS, AES and QAS) are moderate to slightly toxic to aquatic organisms.

---

[14]**Vitellogenin** = *Vitellogenin is a precursor protein of egg yolk normally in the blood or hemolymph only of* ***females*** *that are used as a biomarker in vertebrates of exposure to environmental estrogens which stimulate elevated levels in males as well as females (https://en.wikipedia.org/wiki/Vitellogenin).*

[15]**Estrogenic activity (EA)** *mimics human estrogen, by binding to the cell's normal estrogen receptor location. Chemicals with EA are the most common subset of endocrine-disrupting chemicals (EDC). http://www.plastipure.com/safer-products/what-is-estrogenic-activity/*

10.2.5 Ecotoxicity

Surfactants such as alkyl ethoxy sulphates, alkyl sulphates, linear alkylbenzene sulfonates (*anionic*) are slight to moderately toxic to aquatic organisms (algae, *Daphnia* and fish) with EC_{50}/ LC_{50} (EC/ LC lethal concentration that effects or kills 50% of a test sample) is > 1 mg/L (Table 10.3). Quaternary ammonium compounds (*cationic*), alkyl/alcohol ethoxylates and alkylphenol ethoxylates (nonylphenol) surfactants are extremely toxic to algae, *Daphnia* and fish with EC_{50}/ LC_{50} of < 1.0 mg/L (Table 10.3). Nonylphenol ethoxylate is extremely toxic to aquatic organisms (toxicity range 0.02 mg/L to 0.3 mg/L) which could cause lethal (mortality) and sub-lethal effects (reduced growth and reproduction) in aquatic organisms (algae, invertebrates and fish) (Table 10.3).

Table 10.3. Chemistry, toxicity and ecotoxicity of most common surfactants (anionic, cationic, non-ionic)

[For definitions of EC_{50}, endocrine disruptor, LC_{50}, LD_{50}, Log K_{ow}, please consult footnotes given in Chapters 7 and 9; d = days; h = hours; na = data is not available; NP = nonylphenol].

Surfactants type/ category	**Chemistry**			**Toxicity** *(Mammal)*	**Ecotoxicity** *(Aquatic organisms)*	**Remarks**
	Half-life *(d = days; h = hours)*	**Biodeg-radation**	**Log K_{ow}** *(**note:** Log K_{ow} values of surfactants vary with carbon chain lengths)*	**LD_{50} (Rat) ED (Endocrine disruptor)**	**Acute aquatic toxicity (EC_{50} / LC_{50})** *(algae, invertebrate/ Daphnia and fish)*	
i. Anionic surfactants (negatively charged)						
Alkyl ethoxy sulphates (AES)/ Alcohol ether sulphates Carbon chain lengths: 12-18 (Cowan-Ellsberry et al 2014).	1.4 d (river water), 2.4 d (soil) (HERA 2000); < 12 h (surface water (Guckert et al 1996)	Readily degradable in aerobic (Scott and Jones 2000) and anaerobic condition (Scott and Jones 2000)	log K_{ow}: 0.95-3.9 (Cowan-Ellsberry et al 2014).	**LD_{50}:** 2,000 mg/kg (MSDS Meguiars 2018)	Algae (*Raphidocelis subcapitata*), IC_{50}[16]: 2.18 mg/L (cell density) (Liwarska-Bizukojc et al 2005); *Artemia franciscana*, LC_{50}: 23.92 mg/L (nauplii mortality) (Sibila et al 2008); Fish, Fathead minnow (*Pimephales promelas*) LC_{50}: 0.8-80 mg/L (Painter 1992)	AES is readily degradable in both aerobic and anaerobic conditions and are slightly to moderately toxic to aquatic organisms.

[16] ***IC_{50}** = The half-maximal inhibitory concentration (IC_{50}). It is a measure of the effectiveness of a substance in inhibiting a specific biological or biochemical function. (https://en.wikipedia.org/wiki/IC50).*

	Chemistry			Toxicity (Mammal)	Ecotoxicity (Aquatic organisms)	
Alkyl sulphates (AS)/ Alcohol sulphates Carbon chain lengths: 12-18; (Cowan-Ellsberry et al 2014).	160-460 h (coastal waters) (George 2002)	Readily degradable in aerobic (George 2002; Margesin and Schinner 1998) and anaerobic condition (Feitkenhauer and Meyer 2002a, b).	log K_{ow}: 1.60-4.64 (Cowan-Ellsberry et al 2014).	**LD_{50}:** 1,400-8,000 mg/kg (HERA 2002)	Algae (*Scenedesmus subspicatus*), EC_{50}: 27 mg/L (Verge and Moreno 1996); *Daphnia magna,* EC_{50}: 42- 2,300 mg/L (Lundahl and Cabridenc 1978, Sanchez Leal et al 1991); Fish (*Leuciscus idus*), LC_{50}: 172->270 mg/L (Cowan-Ellsberry et al 2014)	AS is readily degradable in both aerobic and anaerobic conditions and are slightly toxic to aquatic organisms.
Linear alkylbenzene sulfonates[17] (LAS) *(anionic-surfactants)* Carbon chain lengths: 10-14; (Cowan-Ellsberry et al 2014).	3 d (river water) (Larson and Payne 1981); 90 d (sediment) (Lara-Mart´ın et al 2007); 7-33 d (soil) (Ying 2006); 12 h (river water (HERA 2004)	Degradable in aerobic condition (De Wolf and Feijtel 1998, Takada et al 1994, Van Ginkel 1996); persistent in anaerobic condition (Krueger et al 1998, Scott and Jones 2000)	log K_{ow}: 1.94 -3.42 (Ying 2006, Cowan-Ellsberry et al 2014).	LD_{50}: 401-1900 mg/kg (IPCS 1996)	Algae, IC_{50}: 9.1 mg/L *D. magna*, EC_{50}: 4.1 mg/L Fish *(Lepomis macrochirus*), LC_{50}: 4.1 mg/L (HERA 2013)	LAS is readily degradable in aerobic condition but persistent in anaerobic conditions; it is moderately toxic to aquatic organisms.
ii. Cationic surfactants (positively charged)						
Quaternary	2.5 h	QAC	log K_{ow}: 1.81-	LD_{50}: 3190	Algae (*Dunaliella* sp),	QAC is

[17] ***Linear alkylbenzene sulphonate (LAS)*** *is an anionic surfactant. It was introduced in 1964 as the readily biodegradable replacement for highly branched alkylbenzene sulphonates (ABS). LAS is one of the major anionic surfactants in the market*

Chemistry				Toxicity (*Mammal*)	Ecotoxicity (*Aquatic organisms*)	
ammonium compounds[18] (QAC)	(sludge (octadecyl trimethyl ammonium chloride) (Games and King 1982)	(DTDMAC) is readily degradable in aerobic condition (Games and King 1982, Garcia et al 2001) but persistent in anaerobic condition (Garcia et al 1999).	2.63 (Ying 2006).	mg/kg (ECHA 2011) ED (e.g. benzalkonium chloride (Greenwood 2011)	EC_{50}: 0.79 mg/L for 24 h (Ying 2006); *D. magna*, EC_{50}: 0.38 mg/L (immobilisation) (Garcia et al 2001); Fish (*Salmo gairdneri)*, EC_{50}: 1.21 mg/L (immobilisation) (Ying 2006).	degradable in aerobic condition but persistent in anaerobic condition; they are **extremely toxic** to aquatic organisms.
iii. Non-ionic surfactants (without any charge)						
Alkyl/ Alcohol ethoxylates[19] (AE) Carbon chain lengths: C_8-C_{18}; Ethylene oxide chain lengths: 3-12 (Cowan-Ellsberry et al 2014).	< 12 h (water) (Guckert et al 1996)	Readily degradable in aerobic condition (Marcomini and Pojana 1997, Reznickova et al 2002, Szymanski et al 2002) and degradable in anaerobic condition (Manzano et al 1999, Huber et al 2000)	log K_{ow}: 3.03-5.19 (Cowan-Ellsberry et al 2014).	LD_{50}: > 3000 mg/kg (ECHA 2011)	Algae (*Lemna minor*). EC_{10}: 0.10 mg/L (frond count) (Belanger et al 2006); Invertebrate (*Ceriodaphnia dubia),* EC_{50}: 0.39 mg/L (immobilisation) (Warne et al 1999); Fish, Fathead minnow (*P. promelas*), LC_{50}: 0.63-1.65 mg/L (Lewis and Suprenant 1983).	AE is degradable in both aerobic and anaerobic conditions and are **extremely toxic** to aquatic organisms.

[18] ***Quaternary ammonium compounds*** *(QACs) are the major class of cationic surfactants used as the ingredients in fabric softeners, antistatics, disinfectants, biocides, detergents, phase transfer agent and numerous personal care products, such as hair care products. The three most frequently detected QACs in natural environments are* ***dialkyldimethyl ammonium*** *compounds (DADMACs) (with alkyl chain lengths from C8 to C18),* ***alkyltrimethyl ammonium compounds*** *(ATMACs) (with alkyl chain lengths from C12 to C18) and* ***benzylalkyldimethyl ammonium*** *compounds (BACs) (with alkyl chain lengths from C12 to C18) (Zhang et al 2015). Other QAC includes benzalkonium chloride, benzethonium chloride, methylbenzethonium chloride, cetalkonium chloride, cetylpyridinium chloride, cetrimonium, cetrimide, dofanium chloride, tetraethylammonium bromide, didecyldimethylammonium chloride and domiphen bromide (https://en.wikipedia.org/wiki/Quaternary_ammonium_cation).*

[19] ***Ethoxylation*** *is a chemical reaction in which ethylene oxide adds to a substrate.*

	Chemistry			Toxicity (Mammal)	Ecotoxicity (Aquatic organisms)	
Alkylphenol ethoxylates (APE) *NP (nonyl-phenol), OP (octylph-enol) – (the degradation products of APEs and widely detected in the enviro-nment)*	**NP:** 15- 20 d (fresh-water), 50-70 d (seawater) (Groshart et al 2001)	Readily degradable in aerobic condition (Mann and Boddy 2000, Manzano et al 1999), partially degradable in anaerobic condition (Charles et al 1996).	NP: log K_{ow}: 4.48; OP Log K_{ow}: 4.12 (Ahel and Giger 1993 a, b).	LD_{50}: 2,000-5,000 mg/kg (APE) (MSDS Farnell 1992); LD_{50} : 3,700-5,400 (OP) (Swisher 1966) ED: fish (rainbow trout) (Jobling and Sumpter 1993, Purdom et al 1994, Jobling et al 1996), crabs and mussels (Ricciardi et al 2008).	Algae EC_{50}: 0.025–0.75 mg/L (NP); *D. magna*, LC_{50}: 0.19 mg/L for 48 h (NP); Fish, fathead minnow (*P. promelas*), LC_{50}: 0.3 mg/L for 96 h (NP) (Naylor 1995).	APEs (NP) are **bio-accumulative** (Log K_{ow} >4), **extremely toxic** to aquatic organisms and a well-known **endocrine disruptor.**

10.3. Impacts of surfactants on social, economic and environmental sectors

A range of anionic, cationic, and non-ionic surfactants are commonly found in wastewater (Table 10.4). The discharge of sewage effluents (treated or untreated containing surfactants) into surface waters and application of sewage sludge (containing surfactants) on the land have the potential to impact water quality, aquatic ecosystems, aquatic biodiversity, and terrestrial plants (Yin 2006, Rebello et al 2013, Calvo- Flores et al 2018) via foaming, toxicity, bioaccumulation, fish kill, and algal blooms.

10.3.1. Impacts on the environmental matrices

Anionic, cationic and non-ionic surfactants have been detected in the wastewater treatment plant (WWTP) effluents, sludge[20], surface water and sediments[21] in a number of countries (Australia, Canada, China, Denmark, Germany, Japan, Holland, Korea, Italy, Netherlands, Norway, Portugal, Spain, Switzerland, UK, the USA). However, surfactants monitoring data from developing countries are

[20] **Sludge** = *thick, soft, wet mud.*

[21] **Sediments**= *suspended sand, clay, silt.*

scanty or lacking. Table 10.4 provides an account of surfactants detected in different environmental matrices. The following sections (10.3.1.1 through 10.3.1.5) provide a synopsis of Table 10.4.

Table 10.4. Detected concentrations of surfactants in environmental compartments from different countries

[environmental compartments include effluent, sludge, surface water and sediments; na = data is not available; Soares et al 2008*].*

Surfactants type	**Wastewater treatment plants (WWTP)** (effluent[22], influent[23])	**Sludge**	**Surface water**	**Sediments**
i. Anionic surfactants				
Alkyl ethoxy sulphates (AES)	***Denmark*: Effluent 3-1.64 mg/L (AE/AES) (DK-EPA 2001, Matthijs et al 1999) ***The Netherlands***: Effluent 0.003-0.012 mg/L (Matthijs et al 1999, Sibila et al 2008) ***USA*: Effluent 0.24-2.85 mg/L (Sanderson et al 2006, Sibila et al 2008)	na	na	na
Alkyl/ alcohol sulphates (AS)	***Netherlands:*** Effluent 0.0012-0.012 mg/L) (Matthijs et al 1999) ***USA:*** Influent 0.401-0.755 mg/L; effluent 0.036-0.046 mg/L (Fendinger et al 1992)	na	na	na

[22] ***Effluent:*** *wastewater (treated or untreated) that flows out of a treatment plant, sewer, or industrial outfall.*

[23] ***Influent****: the untreated wastewater or raw sewage coming into a wastewater treatment plant.*

Surfactants type	**Wastewater treatment plants (WWTP)** (effluent[22], influent[23])	**Sludge**	**Surface water**	**Sediments**
Linear alkylbenzene sulfonates (LAS)	***The Netherlands***: Effluent 0.019-0.071 mg/L (DK-EPA 2001, Matthijs et al 1999) ***UK***: Effluent 1,090 mg/L (Holt et al 1998)	***European countries***: 16-66.4 mg/kg (Ying 2006) ***Germany:*** 30,200 mg/kg (Berna et al 1989, Ivankovic and Herneovic 2010)	***Japan***: 0.004-0.081 mg/L (Miura et al 2008) ***UK***: 0.416 mg/L (Fox et al 2000)	***Japan:*** 0.5-45.1 µg/g(Takada et al 1992a, b) ***Italy***: 0.03-17.76 mg/kg (Hampel et al 2009) ***Spain***: 0.44-19 µg/g (González-Mazo et al 1997) ***USA:*** 0.01-20 mg/kg (Tabor and Barber 1996)
ii. Cationic-surfactants				
Quaternary ammonium compounds (QAC) (cationic-surfactants)	***USA:*** Effluent 0.062 mg/L (Versteeg et al 1997)	***China:*** Alkyltrimethyl ammonium (ATMACs): 0.38-293 µg/g); Benzylakyl dimethylethylammonium (BACs): 0.09-191 µg/g); Dialkyl dimethylammonium (DADMACs): 0.64-344 µg/g (Ruan et al 2014) ***Switzerland***: 5,870 mg/kg (Fernandez et al 1996)	na	***Austria***: 1,079 µg/kg (Martinez-Carballo et al 2007) ***USA***: 29 µg/g (Li and Brownawell 2010) ***USA***: Benzalkonium chlorides: 22-206 µg/kg (Ferrer and Furlong 2002)
iii. Non-ionic surfactants				
Alkyl/ Alcohol ethoxylates (AE)	***Canada***: Effluent 0.001-0.022 mg/L (mean 0.0068) (Belanger et al 2006) ***Europe***: Effluent 0.001-0.017 mg/L (mean 0.0050) (Belanger et al 2006) ***Netherlands***: Effluent 0.0022-	na	***Japan:*** 0.00002-0.031 mg/L (Miura et al 2008)	na

Surfactants type	**Wastewater treatment plants (WWTP)** (effluent[22], influent[23])	**Sludge**	**Surface water**	**Sediments**
	0.013 mg/L) (Matthijs et al 1999) ***USA*:** Effluent 0.0015- 0.021 mg/L (mean 0.0036) (Belanger et al 2006) ***USA*:** 0.002-0.017 mg/L (Dunphy et al 2001)			
Alkylphenol ethoxylates (APE) (nonylphenol[24])	***Australia*:** 0.000287-0.002058 mg/L (Ying et al 2009) ***Germany*:** Effluent 0.00025–0.002.3 µg/L (Spengler et al 2001) ***Japan*:** Effluent 0.0005–0.001.1 mg/L (Nakada et al 2006) ***Italy:*** 0.0003–0.208 mg/L (Olkowska et al 2014) ***Norway*:** Effluent: 0.00005–0.001.31 mg/L (Johnson et al 2005) ***Switzerland*:** Influent: 0.0435 mg/L; Effluent: 0.015 mg/L (Ahel et al 1994a)	***Germany:*** 80–500 mg/kg (aerobically stabilised) (Giger et al 1984) ***Italy:*** 308 mg/kg (anaerobically stabilised) (Bruno et al 2002) ***Switzerland*:** 450–2,530 mg/kg (anaerobically stabilised) (Giger et al 1984) ***Switzerland:*** 640–2,200 mg/kg (anaerobically stabilised) (Brunner et al 1988) ***USA:*** 754 mg/kg (anaerobically stabilised); 496 mg/kg (heat treated); 470 mg/kg (limed treated); 64 mg/kg (composted) (La Guardia et al 2001)	***Canada:*** 0.01–0.92 µg/L, (lake and river) (Bennie et al 1997); <0.092 µg/L, river (Sabik et al 2003) ***China*:** 1.9–32.8 µg/L (lake) (Wu et al 2007) ***Germany:*** 0.0007–0.0044 µg/L (river); <0.03 µg/L, estuaries) (Bester et al 2001); 0.0006–0.135 µg/L (river) (Kuch and Ballschmiter 2001); 0.028–1.22 µg/L (river) (Fries and Puttmann 2003) ***Holland:*** 0.031–0.934 µg/ L (estuary) (Jonkers et al 2003) ***Italy:*** 158 µg/L (river) (1994); 8.8 µg/ L (river) (1996) (Davi and Gnudi 1999)	***Canada :*** 0.17–72 mg/kg (river) (Bennie et al 1997); <37 mg/kg (lake) (Bennett and Metcalfe 1998); 0.0403–0.293 mg/kg (river (Sabik et al 2003) ***China:*** 3.5–32.4 mg/kg (lake) (Wu et al 2007) ***Germany:*** 0.01–0.153 mg/kg (marine); <0.01–0.055 mg/kg (sea) (Bester et al 2001) ***Holland:*** 0.0004–1.08 mg/kg (estuary) (Jonkers et al 2003) ***Korea:*** 0.0254–0.932 mg/kg (river) (Li et al 2004) ***Spain:*** River sediments: 0.022–0.645 mg/kg (Petrovic et al 2002b); 0.008–1.05 mg/kg (harbour)

[24] ***Nonylphenol** has been referred to as the list of priority substances in the Water Frame Directive of the EU. Nonylphenol is a toxic xenobiotic compound classified as an endocrine disrupter capable of interfering with the hormonal system of numerous organisms (Soares et al 2008).*

Surfactants type	**Wastewater treatment plants (WWTP)** (effluent[22], influent[23])	**Sludge**	**Surface water**	**Sediments**
			Japan: 0.051–1.08 µg/L, (river) (Isobe et al 2001) ***Korea:*** 0.0232–0.1876 µg/L, (river) (Li et al 2004) ***Portugal:*** <10 µg/ L (river and ocean) (Azevedo et al 2001) ***Singapore:*** 0.02–2.76 mg/m^3 (Basheer et al 2004) ***Spain*** 15 µg/L (river) (Petrovic et al 2002b); 0.15–4.1 µg/L (ocean) (Petrovic et al 2002a) ***USA:*** < 1 µg/L (river) (Naylor et al 1992); 0.1–0.5 µg/L (river) (Rice et al 2003)	(Petrovic et al 2002a) ***Switzerland:*** 3,520 mg/kg (river) (Ahel et al 1994b) ***USA:*** < 3,000 mg/kg (river) (Naylor et al 1992); < 1,240 mg/kg (river) (Hale et al 2000); 0.01–60 mg/kg (river) (Kannan et al 2001); 0.075–0.340 mg/kg (river) (Rice et al 2003)

10.3.1.1. WWTP effluent

- **Anionic surfactants detected: i.** Alkyl ethoxy sulphates (AES): 0.003 mg/L to 2.85 mg/L (Denmark, the Netherlands, and USA); **ii**. Alkyl/ alcohol sulphates (AS): 0.0012 to 0.075 mg/L (the Netherlands, USA); **iii**. Linear alkylbenzene sulfonates (LAS): 0.019 to 1090 mg/L (the Netherlands, UK) (see Table 10.4 for details).
- **Cationic surfactants detected: i.** Quaternary ammonium compounds: 0.062 mg/L (USA) (see Table 10.4 for details).
- **Non-ionic surfactants detected: i.** Alkyl/ Alcohol ethoxylates (AE): 0.001 to 0.022 mg/L (Canada, Europe, the Netherlands, and the USA) (see Table 10.4 for details); **ii.** Alkylphenol ethoxylates (APE) (**nonylphenol**): 0.00025 to 0.208 mg/L (Australia, Germany, Japan, Italy, Norway, and Switzerland) (see Table 10.4 for details).

10.3.1.2. Sludge

- **Anionic surfactants detected: i**. Alkyl ethoxy sulphates (AES): data is not available/not detected; **ii**. Alkyl/ alcohol sulphates (AS): data is not available/not detected; **iii**. Linear alkylbenzene sulfonates (LAS): 16-30,200 mg/kg (Europe, Germany) (see Table 10.4 for details).
- **Cationic surfactants detected:** i. Quaternary ammonium compounds: 0.09-5,870 µg/g (China, Switzerland) (see Table 10.4 for details).
- **Non-ionic surfactants detected: i.** Alkyl/ Alcohol ethoxylates (AE): data is not available/ not detected; **ii**. Alkylphenol ethoxylates (APE) (**nonylphenol**): 80-2,530 mg/kg (Germany, Italy, Switzerland, and USA) (see Table 10.4 for details).

10.3.1.3. Surface water

- **Anionic surfactants detected: i**. Alkyl ethoxy sulphates (AES): data is not available/not detected; **ii**. data is not available/ not detected; **iii**. Linear alkylbenzene sulfonates (LAS): 0.004-0.416 mg/L (Japan, UK) (see Table 10.4 for details).
- **Cationic surfactants detected: i.** Quaternary ammonium compounds: data is not available/ not detected.
- **Non-ionic surfactants detected: i.** Alkyl/Alcohol ethoxylates (AE): 0.00002-0.031 mg/L (Japan); **ii. alkylphenol ethoxylates (APE)** (**nonylphenol**): 0.0006-13.8 mg/L (Canada, China, Germany, Holland, Italy, Japan, Korea, Portugal, Spain, USA) (see Table 10.4 for details).

10.3.1.4. Sediments

- **Anionic surfactants detected: i**. Alkyl ethoxy sulphates (AES): data is not available/ not detected; **ii**. Alkyl/ alcohol sulphates (AS): data is not available/ not detected; **iii**. Linear alkylbenzene sulfonates (LAS):0.01-20 mg/kg (Italy, Japan, and the USA) (see Table 10.4 for details).
- **Cationic surfactants detected:** i. Quaternary ammonium compounds: 22- 1079 µg/kg (Austria, the USA).
- **Non-ionic surfactants detected: i.** Alkyl/ Alcohol ethoxylates (AE): data is not available/ not detected; **ii.** Alkylphenol ethoxylates (APE) (**nonylphenol**): 0.0004-3,520 mg/kg (Canada, China, Germany, Holland, Korea, Spain, Switzerland, the USA) (see Table 10.4 for details)

10.3.1.5. Nonylphenol

Apart from the detection in the effluent, sludge, surface water, and sediment (Table 10.4), the most persistence surfactant, nonylphenol (APEs degradation product) has also been detected in the air, drinking water, groundwater, rainwater, snows, soft drinks, seafood, agricultural food and agricultural fields (Table 10.5).

Table 10.5. Nonylphenol concentrations in the air, drinking water, groundwater, rainwater, snows, soft drinks, seafood, agricultural food and agricultural fields *[fw = fresh weight; dw = dry weight; dm = cubic decimetre (1 dm = 1 litre); ng = nanogram; pg = picogram].*

Medium	Concentrations	References
Air	air (<0.001-81 ng/m^3)	van Ry et al 2000, Ying et al 2002, Cincinelli et al 2003
Drinking water	*Italy:* <0.0077 to 0.084 mg/m^3 (drinking **fountains**)	Maggioni et al 2013
	Italy: <0.0077 mg/m^3 (PET **- bottled water**)	Maggioni et al 2013
	USA: 0.015 to 0.300 mg/m^3 (**Spring water** bottled in HDPE and PVC, sampled from supermarkets)	Loyo-Rosales et al 2004
	China: 0.1 to 2.7 µg/L	Shao et al 2005
	Germany: 300 pg/L to 2 ng/L	Kuch and Ballschmiter 2001
Ground-water	*Switzerland:* 200-760 ng/L	Ahel et al 1996
Rainwater	*Belgium, Germany:* 0.030 µg/dm^3 to 0.599 µg/dm^3	Fries and Puttmann 2004
Snows	*Belgium, Germany:* 0.025µg/dm^3 to 0.802 µg/dm^3	Fries and Puttmann 2004
Soft drinks	<0.03–0.22 mg/m^3	Li et al 2013
Seafood	*Cambodia, India, Indonesia, Malaysia, the Philippines, Singapore, Thailand, Vietnam:* 18	Isobe et al 2007

Medium	Concentrations	References
	to 663 μg/kg dw (**green mussel** during 1994–1999)	
	China: <0.05 to 55.98 μg/kg fw (meat/ **seafood** from supermarkets in Beijing)	Shao et al 2007
	The Netherlands: <10 to 10 μg/kg fw (edible marine species, **flounder** from North Sea Canal)	Vethaak et al 2005
	The Netherlands: <10 to 160 μg/kg fw (edible freshwater species, **bream**)	Vethaak et al 2005
	Singapore: 46.6 to 197 μg/kg fw (**seafood** from supermarkets)	Basheer et al 2004
	USA: 96 to 3000 μg/kg dw (**mussels** along the California coast)	Dodder et al 2014
Agriculture food	*Sweden:*<10 to 71 μg/kg fw (**fruits, cereal products**, and **vegetables** commercially available)	Gyllenhammar et al 2012
	USA:<0.3 to 1.0 μg/kg fw (**vegetables and fruits**, from local commercial sources, Florida)	Lu et al 2013
	339.2 to 926.9 μg/kg fw (**lettuce and collards**[25], roots)	Dodgen et al 2013
Agricultural fields	*Mexico:*<25 to 299 μg/kg dw (irrigated **wastewater**)	Gibson et al 2010

10.3.2. Impacts on water quality and aquatic biodiversity

10.3.2.1. Foam pollution

Surfactants-containing wastewater (e.g. detergents, shampoos, hair care products) discharged into the environment can cause water pollution problems in the form

[25]***Collards*** *= members of the cabbage family but are also close relatives to kale.*

of persistent foams (a thin film floating on the water surface which is bright white in colour) (Fig. 10.3). When this film is agitated by waves, strong currents or wind, the surfactant molecules trap small bubbles, producing foam or "**Foam pollution**" (Schilling and Zessner 2011, NSW-EPA 2015).

Figure 10.3. Foam spill in Blackburn Lake, Victoria, Australia (January 2019) caused by hair care products (shampoo)
[https://ref.epa.vic.gov.au/about-us/news-centre/news-and-updates/news/2019/january/24/blackburn-foam].

In general, when the concentration of the surfactant reaches to 0.1 mg/L, the foam can be formed (Yuan et al 2014). It is reported that anionic surfactants such as **LAS, AES, AE** can cause much foam in rivers (Kikuchi and Wakabayashi 1992), in particular, LAS surfactants are not biodegradable in anaerobic condition and can result in extremely persistent foam (Ramachandra et al 2017). Chemical analysis of field foam samples (originating from the detergents, oil and grease) reveal that foams are enriched with particulate organic and inorganic compounds including **nutrients** (nitrogen, phosphorus and carbon) and cations (sodium, potassium, calcium and magnesium) and have high **biological oxygen demand** (BOD) and chemical oxygen demand (COD) (Table 10.6). Contaminants such as **heavy metals, pesticides, lipids** can also be accumulated in 'freshwater foams' (Schilling and Zessner 2011).

Nevertheless, foam can act as an insulating layer, thus prevents/ weakens exchange between the water body and atmospheric oxygen. Therefore, the dissolved oxygen level in water bodies can be depleted which may cause fish kills (*see chapter 8, Fig 8.6 for the relationships among dissolved oxygen level, ecological hypoxia and anoxia and fish kill*). Research carried out in the UK found that, foam in the Rivers of Exe and the Creedy (due to the discharge of alkylphenol ethyoxylates) changed skin morphology of fish and caused "exceptional incidences" of **fungal infections** to salmon fish (House 1993). Foaming can reduce the **aesthetic values** of the water bodies impeding and increasing recreational and navigational hazards in waterways. Nevertheless, detergents can destroy the external mucus layers that protect the fish from bacteria and parasites and can damage **fish gills** (Lenntech 2019). Gills are the vital organ used by fish for respiration.

Table 10.6. Physico-chemical analysis of water and foam samples *(Ramachandra et al 2017).*

Parameters	Water samples	Foam samples
TDS (mg/L)	451.0	7,000.0
EC (μS)	756.5	17,000.0
pH	7.40	6.98
BOD (mg/L)	42.7	650.41
COD (mg/L)	66.0	1,140.0
Alkalinity (mg/L)	336.0	12,000.0
Chloride (mg/L)	120.0	3,195.0
Total hardness (mg/L)	215.0	13,000.0
Ca hardness (mg/L)	609.25	3,607.2
Mg hardness (mg/L)	37.44	2,282.45
Phosphate (mg/L)	1.07	74.59
Nitrate (mg/L)	0.451	129.72
Sodium (mg/L)	165.25	770.0

10.3.2.2. Algal blooms

When the untreated sewage/wastewater mostly having phosphorus-containing surfactants (Table 10.7), are discharged into the waterways it could easily lead to enhancement of nutrients and proliferation of blue-green algae/ cyanobacterial blooms in waterways. Many surfactants contain polyphosphate, one of the ingredients for a freshwater algal bloom. For example, laundry detergents itself contain 35% to 75% sodium triphosphate ($Na_5P_3O_{10}$) (Choudhury 2018). Water with elevated levels of blue-green/ cyanobacterial bloom could be unsuitable for agriculture, drinking, livestock drinking, irrigation, and wildlife since some algae such as *Microcystis* spp. produce toxins. Cyanobacterial blooms of toxic blue-green algae can affect the water quality by causing peculiar tastes and odours, discolouration and unsightly scum (Kibria 2016) (see Chapter 8 on nutrient pollution, algal blooms, algal toxins and hypoxia).

Table 10.7. Typical values for phosphorus in surfactants related consumer products – Australia, 2010 [Tjandraatmadja et al 2010, Chong et al 2019]

Surfactants product	Median concentration of phosphorus	Maximum concentration of phosphorus
Laundry products		
Laundry powder	25 g/kg	55 g/kg
Laundry liquid	8 g/kg	46 g/kg
Fabric softener	0.04 g/kg	0.2 g/kg
Dishwashing products		
Dishwasher tablet	29 g/kg	98 g/kg
Dishwasher powder	22 g/kg	53 g/kg
Dishwashing liquid	0.01 g/kg	1.5 g/kg
Personal care products		
Toothpaste	10 g/kg	79 g/kg
Soaps	0.9 g/kg	2.5 g/kg
Shampoo	0.15 g/kg	1.7 g/kg

10.3.2.3. Aquatic biota

Surfactants such as AES, LAS is slight to moderately toxic to aquatic organisms (EC_{50}/ LC_{50} > 1 mg/L). On the other hand, surfactants like QAC, AE and APEs are extremely toxic (EC_{50}/ EC_{50} <1 mg/L) to algae, daphnia and fish (see Table 10.3). In particular, nonylphenol (APEs degradation product) is extremely toxic to aquatic organisms with EC_{50} for algae 0.02 mg/L, and LC_{50} for *Daphnia* and fathead minnow fish are 0.19 mg/L and 0.3 mg/L respectively. Surfactants can affect the growth, motility and photosynthetic ability of algae (Lewis 1990). For example, anionic Sodium dodecyl sulfate (SDS) at 0.1 mg/mL causes inhibition of asexual and sexual reproduction of freshwater green algae, *Closterium ehrenbergii* (Matsui and Park 2000) whereas higher concentrations of SDS (1 mg/mL) hindered the growth and motility of alga *Plagioselmis prolonga* (Aizdaicher and Markina 2006), diatomaceous alga *Thalassiosira pseudonana* (Aizdaicher and Reunova 2002) and duckweeds (Diriligen and Ince 1995). In fact, the photosynthetic ability of algae was found to decrease exponentially with increasing surfactant concentration (Maksimov and Parshikova 2006), with cationic surfactants causing the most potent inhibition.

10.3.2.4. Seafood, crops, vegetables, fruits

Water bodies (creeks, dams, ponds, lakes and rivers) are the common source of water for agricultural crops and aquaculture farms which can be contaminated with surfactants. Sewage sludge can contain high concentrations of surfactants (Ying et al 2006). Irrigation of crops and vegetables with surfactant contaminated water or growing of crops using contaminated sludge may cause surfactant residues (such as nonylphenol) to accumulate in food and possible transfer of such residues to humans via the food chain. Due to a high degree of hydrophobicity of nonylphenol (Log $K_{ow} > 4$), it is likely to adsorb to biosolids, sediments and soils and can accumulate in biota. Nonylphenol residues have been detected in mussels from California, edible marine species from Asia and vegetables, fruits, pumpkin, carrot, citrus and apple in Florida, USA (Careghini 2014, USDA 2015) (See also Table 10.5 for accumulation of nonylphenol in seafood (green mussel, flounder, bream) and agricultural produce (lettuce, collards). Water contaminated with detergents are reported to reduce the photosynthetic rate and chlorophyll content in bean plants (Jovanic et al 2010) and sunflower leaves (Gadallah 1996).

10.4. Impacts on human health

The long-term use of surfactants can cause skin irritation and lead to some degree of skin damage (Stepanets et al 2001, Yuan et al 2014). QAC is known to cause skin and respiratory irritation and some people are highly allergic to QAC (Greenwood 2011). Nonylphenol (NP) is a well-known endocrine-disrupting chemical (EDC), which can enhance the progression of cancer (colorectal cancer) by functioning as an estrogen-like factor (Yang et al 2017).

10.5. Management of surfactants

To reduce the risks and impacts of surfactants on the environment and public health the following steps can be undertaken:

- Routine monitoring of surfactants residue in industrial effluents, sludge, surface water, sediments and comparing those with the guidelines threshold values should be the top priority (see Table 10.8 for guidelines on surfactants for the protection of aquatic species and water quality) *[A rapid screening test called "methylene blue active substances assay" (MBAS assay) can be used for monitoring of surfactants in water. It is a colorimetric test that uses methylene blue to detect the presence of anionic surfactants (such as a detergent or foaming agent) in a sample of water. An anionic surfactant is detected by the colour reaction) (Hennes-Morgan and de Oude 1994, Baird et al 2017)].*
- Communities and industries should be educated about the threats and risks posed by surfactants on the environment so that they can shoulder responsibility and act for wise and limited use of detergents to reduce surfactant pollution.
- Use of low-foam surfactants, green-surfactants, bio-surfactants, phosphate-free surfactants, phasing out/banning most harmful surfactants to the environment, wise and limited usage of surfactants

and proper physical, chemical and bioremediation strategies would significantly help to reduce the toxicity of surfactants and surfactants pollution.

Table 10.8. Guidelines for the protection of aquatic species and water quality (surface water or sewage water)

[[MAC = maximum allowed concentrations; MPC = maximum permissible concentrations; TV = trigger values; these are the concentrations of the key performance indicators measured for the ecosystem, below which there exists a low risk or low adverse biological/ ecological effects].

Surfactants	**Trigger values** (ANZECC and ARMCANZ 2000)	**Maximum permissible concentrations** (Britton 1998)	**Maximum allowed concentrations** (Ivankovic and Herneovic 2010)
Alcohol ethoxylated (AE)	50 μg/L (99% protection) 140 μg/L (95% protection) 220 μg/L (90% protection) 360 μg/L (80% protection)	110 μg/L	na
Alcohol ethoxyolated sulfate (AES)	340 μg/L (99% protection) 650 μg/L (95% protection) 850 μg/L (90% protection) 1,100 μg/L (80% protection)	400 μg/L	na
Linear alkylbenzene sulfonates (LAS)	65 μg/L (99% protection) 280 μg/L (95% protection) 520 μg/L (90% protection) 1,000 μg/L (80% protection)	250 μg/L	na
Nonylphenol (NP)	0.1 μg/L (fresh water) 1.0 μg/L (marine water)	na	na
Soap	na	27	na
Anionic	na	na	1.0 mg/L surface water 10.0 mg/L sewage water
Cationic	na	na	0.2 mg/L surface water 2.0 mg/L sewage water
Non-ionic	na	na	1.0 mg/L surface water 10.0 mg/L sewage water

10.6. Conclusion

It is known that up to 60% of globally produced surfactants enter the environment. The most visible potential ecosystem impact from surfactants is foaming (bright white in colour) on waterways, algal blooms and aquatic toxicity. Nonylphenol (NP), the biodegradation product of APEs detected in drinking waters, is persistent, most toxic to aquatic organisms (algae, invertebrates and fish), bioaccumulate in seafood and agricultural foods and cause endocrine disruption in fish. We cannot totally avoid using surfactants or detergents as they are part of our household cleaning and personal care products. They are also needed for many industries for the processing of products (textile, leather, paper, oil). Hence, communities and industries should be educated about the threats and risks posed by surfactants, that enter into the environment, and the public be encouraged to purchase surfactants as minimum as possible and use those which are environmentally friendly and phosphate-free.

10.7. References

Ahel M and Giger W 1993a Aqueous solubility of alkylphenols and alkylphenol polyethoxylates. *Chemosphere* **26** 1461–1470

Ahel M and Giger W 1993b Partitioning of alkylphenols and alkylphenol polyethoxylates between water and organic solvents. *Chemosphere* **26** 1471–1478

Ahel M, Giger W and Koch M. 1994a Behaviour of alkylphenol polyethoxylate surfactants in the aquatic environment 1. Occurrence and transformation in sewage treatment. *Water Research* **28** 1131–1142

Ahel M, Giger W and Schaffner C 1994b Behavior of alkylphenol polyethoxylated surfactants in the aquatic environment 2. Occurrence and transformation in rivers. *Water Research* **28** 1143–1152

Ahel M, Schaffner C and Giger W 1996 Behavior of alkyphenol polyethoxylate surfactants in the aquatic environment III: Occurrence and elimination of their persistent metabolites during infiltration of river water to groundwater. *Water Resources* **30** 37–46

ANZECC and ARMCANZ 2000 Australian and New Zealand guidelines for fresh and marine water quality, Volume 1, The Guidelines Chapters 1-7.

https://www.waterquality.gov.au/sites/default/files/documents/anzecc-armcanz-2000-guidelines-vol1.pdf; https://www.waterquality.gov.au/anz-guidelines/guideline-values/default/water-quality-toxicants/toxicants/detergents-2000

Aizdaicher NA and Markina ZV 2006 Toxic effects of detergents on the alga *Plagioselmis prolonga* (Cryptophyta). *Russian Journal of Marine Biology* **32** 45–49 *doi: 10.1134/S1063074006010068*

Aizdaicher NA and Reunova YA 2002 Effects of detergents on in vitro growth of diatom alga *Thalassiosira pseudonana. Russian Journal of Marine Biology* **28** 324–328 *doi: 10.1023/A:1020907501713*

Azevedo DA, Lacorte S, Viana P and Barcelo D 2001 Occurrence of nonylphenol and bisphenol A in surface waters from Portugal. *Journal of the Brazilian Chemical Society* **12** 532–537

Baird RB, Eaton AD and Rice EW 2017 *Standard Methods For the Examination of Water and Wastewater* 23rd edition. American Public Health Association, American Water Works Association and Water Environment Federation. Washington DC *DOI: 10.2105/SMWW.2882.030*

Basheer C, Lee HK and Tan KS 2004 Endocrine-disrupting alkylphenols and bisphenol-A in coastal waters and supermarket seafood from Singapore. *Marine Pollution Bulletin* **48** 1145–1167

Belanger SE, Dorn PB, Toy R, Boeije G, Marshall SJ, Wind T, Van Compernolle R. Zeller D 2006 Aquatic risk assessment of alcohol ethoxylates in North America and Europe. *Ecotoxicology and Environmental Safety* **64** 85-99

Bennie DT, Sullivan CA, Lee HB et al 1997 Occurrence of alkylphenols and alkylphenol mono-and diethoxylates in natural waters of the Laurentian Great Lakes basin and the upper St Lawrence River. *Science of the Total Environmen*t **193** 263–275

Bennett ER and Metcalfe CD 1998 Distribution of alkylphenol compounds in Great Lakes sediments, United States and Canada. *Environmental Toxicology and Chemistry* **17** 1230–1255

Berna JL, Ferrer J, Moreno A, Prats D and Bevia FR 1989 The fate of LAS in the environment. *Tenside Surfactants Detergents* **26** 101-107

Berna JL, Ferrer J, Moreno A, Prats D and Bevia FR 1989 The fate of LAS in the environment. *Tenside* Surfactants *Detergents* **26** 101-107

Bester K, Theobald N and Schroder HF 2001 Nonylphenols, nonylphenol-ethoxylates, linear alkylbenzenesulfonates (LAS) and bis(4-chlorophenyl)-sulfone in the German Bight of the North Sea. *Chemosphere* **45** 817–826

Britton LN 1998 Surfactants and the environment *https://doi.org/10.1007/s11743-998-00*

Brunner PH, Capri S, Marcomini A and Giger W 1988 Occurrence and behaviour of linear alkylbenzenesulphonates, nonylphenol, nonylphenol mono-and nonylphenol diethoxylates in sewage and sewage sludge treatment. *Water Research* **22** 1465–1472

Bruno F, Curini R, Di Corcia A et al 2002 Determination of surfactants and some of their metabolites in untreated and anaerobically digested sewage sludge by subcritical water extraction followed by liquid chromatography-mass spectrometry. *Environmental Science & Technology* **36** 4156–4161

Calvo-Flores FG, Joaquín Isac-García and José A. Dobado 2018 Surfactants in the Environment, Chapter 10, in *Emerging Pollutants: Origin, Structure, and Properties* (eds) Wiley-VCH Verlag GmbH & Co KGaA, John Wiley & Sons, Incorporated, 341-384 pages

Careghini A, Mastorgio AF, Saponaro S and Sezenna E 2014 Bisphenol A, nonylphenols, benzophenones, and benzotriazoles in soils, groundwater, surface water, sediments, and food: a review. *Environmental Science and Pollution Research* **22(8)** 5711-5744 *doi: 10.1007/s11356-014-3974-5. Epub 2014 Dec 30*

Charles W, Ho G and Cord-Ruwisch R 1996 Anaerobic bioflocculation of wool scouring effluent: the influence of non-ionic surfactant on efficiency. *Water Science and Technology* **34(11)** 1–8

Chong J, Dominish E, Tjandraatmadja G et al 2019 Ecosystem impacts of phosphorus and surfactants in consumer products. Report prepared by the Institute for Sustainable Futures, University of Technology Sydney for Stewart Investors.

https://www.uts.edu.au/sites/default/files/2019-12/UTS_Consumer_Products_Report_v5.pdf

Chowdhury AKR 2018 Eco-friendly dyes and dyeing, You may download the paper. *Advanced Materials and* Technologies for Environmental Applications **2(1)** 145-176

Cincinelli A, Mandorlok S, Dickhut RM and Lepra L 2003 Particulate organic compounds in the atmosphere surrounding an industrialised area of Prato (Italy). *Atmospheric Environment* **37** 3125–3133

Cowan-Ellsberry C, Belanger S, Dorn P, Dyer S, McAvoy D et al 2014 Environmental Safety of the Use of Major Surfactant Classes in North America, *Critical Reviews in Environmental Science and Technology* **44(17)** 1893-1993 *DOI: 10.1080/10739149.2013.803777*

Davi ML and Gnudi F 1999 Phenolic compounds in surface water. *Water Research* **33** 3213–3219

De Wolf W and Feijtel T 1998 Terrestrial risk assessment for linear alkylbenzene sulfonate (LAS) in sludge-amended soils. *Chemosphere* **36** 1319– 1343

Diriligen N and Ince N 1995 Inhibition effect of the anionic surfactant SDS on duckweed, *Lemna minor* with considerations of growth and accumulation. *Chemosphere* **31** 4185–4196

DK-EPA (Danish Environmental Protection Agency) 2001 Environmental Project 615: Environmental and health assessment of substances in household detergents and cosmetic detergent products. *http://www.mst.dk/udgiv/Publications/ 2001/87-7944-596-9/html/helepubl_eng.htm*

Dodder NG, Maruya KA, Ferguson PL et al 2014 Occurrence of emerging contaminants of emerging concern in mussels (*Mytilus* spp.) along the California coast, and the influence of land use, storm water discharge, and treated wastewater effluent. *Marine Pollution Bulletin* **81** 340–346

Dodgen LK, Li J, Parker D and Gan JJ 2013 Uptake and accumulation of four PPCP/EDCs in two leafy vegetables. *Environment Pollution* **182** 150-156 *doi: 10.1016/j.envpol.2013.06.038. Epub 2013 Jul 31*

Dunphy JC, Pessler DG, Morall SW, Evans KA, Robaugh DA, Fujimoto G. Negahban A 2001 Derivatization LC/MS for the simultaneous determination of fatty alcohol and alcohol ethoxylate surfactants in water and wastewater samples. *Environmental Science and Technology* **35** 1223-30

ECHA (European Chemical Agency) 2011 Quaternary ammonium compounds, C20-22-alkyltrimethyl, chlorides. Endpoint summary. EC number: 271-756-9

https://echa.europa.eu/registration-dossier/-/registered-dossier/14334/7/3/1

EU (European Union) 2012 Opinion on nitrosamines and secondary Amines in Cosmetic Products. Scientific Committee on Consumer Safety SCCS. SCCS/1458/11; *https://ec.europa.eu/health/scientific_committees/consumer_safety/docs/sccs_o_090.pdf*

Feitkenhauer H and Meyer U 2002a Anaerobic digestion of alcohol sulfate (anionic surfactant) rich wastewater-batch experiments: Part I. Influence of the surfactant concentration. *Bioresource Technology* **82** 115–21

Feitkenhauer H and Meyer U 2002b Anaerobic digestion of alcohol sulfate (anionic surfactant) rich wastewater-batch experiments: Part II. Influence of the hydrophobic chain length. *Bioresource Technology* **82** 123–9

Fendinger NJ, Begley WM, McAvoy DC and Eckhoff WS 1992 Determination of alkyl sulfate surfactants in natural waters. *Environmental Science and Technology* **26(12)** 2493–2498 *doi: 10.1021/es00036a024*

Fernandez P, Alder AC, Suter MJF and Giger W 1996 Determination of the quaternary ammonium surfactant ditallowdimethylammonium in digested sludges and marine sediments by supercritical fluid extraction and liquid chromatography with post-column ion-pair formation. *Analytical Chemistry* **68** 921-929

Ferrer I and Furlong ET 2002 Accelerated solvent extraction followed by on-line solid-phase extraction coupled to ion trap LC/ MS/MS for analysis of benzalkonium chlorides in sediment samples. *Analytical Chemistry* **74** 1275-1280

Fries E and Puttmann W 2003 Occurrence and behaviour of 4-nonylphenol in river water of Germany. *Journal of Environmental Monitoring* **5** 598–603

Fries E and Puttmann W 2004 Occurrence of 4-Nonylphenol in rain and snow. *Atmospheric Environment* **38** 2013–2016

Ferrara F, Ademollo N, Delise M, Fabietti F and Funari E 2008 Alkylphenols and their ethoxylates in seafood from the Tyrrhenian Sea. *Chemosphere* **72** 1279-1285

Fox K, Holt M, Daniel M, Buckland H, Guymer 2000 Removal of linear alkylbenzene sulphonate from a small Yorkshire stream: contribution to GREAT-ER project. *Science of the Total Environment* **251/252** 265-275

Gadallah MAA 1996 Phytotoxic effects of industrial and sewage waste waters on growth, chlorophyll content, transpiration rate and relative water content of potted sunflower plants. *Water Air and Soil Pollution* **89** 33–47

Games LM and King JE 1982 Fate and distribution of a quarternary ammonium surfactant, octadecyl trimethyl ammonium chloride (OTAC) in wastewater treatment. *Environmental Science and Technology* **16** 483–488 *http://dx.doi.org/10.1021/es00102a011*

García MT, Campos E, Sanchez-Leal J and Ribosa I 1999 Effect of the alkyl chain length on the anaerobic biodegradability and toxicity of quaternary ammonium-based surfactants *Chemosphere* **38** 3473–3483

García MT, Ribosa, I, Guindulain T Sánchez-Leal J and Vives-Rego J 2001 Fate and effect of monoalkyl quaternary ammonium surfactants in the aquatic environment. *Environmental Pollution* **111** 169–175

George AL 2002 Seasonal factors affecting surfactant biodegradation in Antarctic coastal waters: comparison of a polluted and pristine site. *Marine Environmental Research* **53** 403-415

Gibson R, Durán-Álvarez J, León Estrada K et al 2010 Accumulation and leaching potential of some pharmaceuticals and potential endocrine disruptors in soils irrigated with wastewater in the Tula Valley, Mexico. *Chemosphere* **81** 1437–1445

Giger W, Brunner PH and Schaffner C 1984 4-Nonylphenol in sewage-sludge: accumulation of toxic metabolites from nonionic surfactants. *Science* **225** 623–625

González-Mazo E, Honing M, Barceló D and Gómez-Parra A 1997 Monitoring long-chain intermediate products from the degradation of linear alkylbenzene sulfonates in the marine environment by solid-phase extraction followed by liquid chromatography/isospray mass spectrometry. *Environmental Science and Technology* **31(2)** 504–510

Greenwood S 2011 (updated 2013) Quaternary Ammonium Compounds; *http://chemicaloftheday.squarespace.com/most-controversial/2011/8/31/quaternary-ammonium-compounds.html*

Groshart CP, Okkerman PC, Wassenberg WBA et al 2001 Chemical study on alkylphenols; *https://repository.tudelft.nl/islandora/object/uuid%3A20647de5-7a88-416a-b78a-e5e8ab650b3b*

Guckert JB, Walker DD and Belanger S.E 1996 "Environmental chemistry for a surfactant ecotoxicology study supports rapid degradation of C12 alkyl sulfate in a continuous-flow stream mesocosm". *Environmental Toxicology and Chemistry* **15(3)** 262–269 *doi*: *10.1002/etc.5620150306*

Gyllenhammar I, Glynn A, Darnerud PO, Lignell S, van Delft R, Aune M 2012 4-Nonylphenol and bisphenol A in Swedish food and exposure in Swedish nursing women. *Environmental International* **43** 21–28

Hale RC, Smith CL, de Fur PO et al 2000 Nonylphenols in sediments and effluents associated with diverse wastewater outfalls. *Environmental Toxicology and Chemistry* **19** 946–952

Hampel M, Canário J, Branco V et al 2009 Environmental levels of Linear alkylbenzene Sulfonates (LAS) in sediments from the Tagus estuary (Portugal): environmental implications. *Environmental Monitoring and Assess*ment **149** 151–161 *DOI 10.1007/s10661-008-0190-0*

Hennes-Morgan EC and de Oude NT 1994 Detergents. In *Handbook of ecotoxicology*, vol 2, ed P Calow, Blackwell Scientific, London.

HERA (Human and Environmental Risk Assessment) 2000 Human and Environmental Risk Assessment on ingredients of European household cleaning products Alcohol ethoxysulphates (AES). Environmental Risk Assessment.

https://www.heraproject.com/files/1-E-04-HERA%20AES%20ENV%20%20web%20wd.pdf

HERA (Human and Environmental Risk Assessment) 2002 Human and Environmental Risk Assessment on ingredients of European household cleaning products Alcohol Sulphates Human Health Risk Assessment;

https://www.heraproject.com/files/3-HH-04-%20HERA%20AS%20HH%20web%20wd.pdf

HERA (Human and Environmental Risk Assessment) 2004 LAS Linear Alkylbenzene Sulphonate (Cas No. 68411-30-3); 2004.

http://www.heraproject.com/files/4-F-690FDB82-0387- 49B5-92889661008359DD.pdf

HERA (Human and Environmental Risk Assessment) 2013 Environmental Risk Assessment of LAS (Linear Alkylbenzene Sulphonate) (CAS No. 68411-30-3). *https://www.heraproject.com/files/HERA-LAS%20revised%20February%202013.pdf*

Holt MS, Fox KK, Burford M, Daniel M, Buckland H. 1998 UK monitoring study on the removal of linear alkylbenzene sulphonate in trickling filter type sewage treatment plants Contribution to GREAT-ER project #2. *Science of the Total Environment* **210-211** 255-69

House WA 1993 The effects of surfactants in the Rivers Exe and Creedy. Natural Environment Research Council Institute of Freshwater Ecology, River Laboratory,

East Stoke, Wareham, Dorset BH20 6BB IFE Report Ref: RL/T11053n1/1TFS Project No: T11053n1; *http://nora.nerc.ac.uk/id/eprint/502109/1/N502109CR.pdf*

Huber M, Meer U, Rys P 2000 Biodegradation mechanisms of linear alcohol ethoxylates under anaerobic conditions. *Environmental Science and Technology* **34** 1737– 41

IPCS (International Programme on Chemical Safety) 1996 Linear alkylbenzene sulfonates and related compounds. Environmental Health Criteria 169. World Health Organization, Geneva, Switzerland.

Isobe T, Takada H, Kanai M et al 2007 Distribution of polycyclic aromatic hydrocarbons (PAHs) and phenolic endocrine disrupting chemicals in South and Southeast Asian mussels. *Environmental Monitoring and Assessment* **135** 423–440

Isobe T, Nishiyama H, Nakashima A and Takada H 2001 Distribution and behavior of nonylphenol, octylphenol and nonylphenol monoethoxylate in Tokyo metropolitan area: their association with aquatic particles and sedimentary distributions. *Environmental Science and Technology* **35** 1041–1049

Ivankovic T and Herneovic J 2010 Surfactants in the environment. *Arh Hig Rada Toksikol (Archives of Industrial Hygiene and Toxicology)* **61** 95-110

Jobling SJ and Sumpter JP 1993 Detergent components in sewage effluent are weakly oestrogenic to fish: an in vitro study suing rainbow trout hepatocytes. *Aquatic Toxicology* **27** 361–72

Jobling S, Sheahan D, Osborne A, Mathiessen P and Sumpter JP 1996 Inhibition of testicular growth in rainbow trout (*Oncorhynchus mykiss*) exposed to estrogenic alkylphenolic chemicals. *Environmental Toxicology and Chemistry* **15** 194-202

Johnson AC, Aerni H-R, Gerritsen A et al 2005 Comparing steroid estrogen, and nonylphenol content across a range of European sewage plants with different treatment and management practices. *Water Research* **39** 47–58

Jonkers N, Laane R and De Voogt P 2003 Fate of nonylphenol ethoxylates and their metabolites in two Dutch estuaries: evidence of biodegradation in the field. *Environmental Science and Technology* **37** 321–327

Jovanic BR, Bojovic S, Panic B, Radenkovic B, Despotovic M 2010 The effect of detergent as polluting agent on the photosynthetic activity and chlorophyll content in bean leaves. *Health* **2** 395–399

Kannan K, Kober JL, Kang YS et al 2001 Polychlorinated naphthalenes, biphenyls, dibenzo-p-dioxins, and dibenzofurans as well as polycyclic aromatic hydrocarbons and alkylphenols in sediment from the Detroit and Rouge Rivers, Michigan, USA. *Environmental Toxicology and Chemistry* **20** 1878–1889

Kibria G 2016 Blue-green algal/cyanobacterial blooms (BGA), climate change and BGA impacts on water quality, fish kills, recreation, crops, seafood, livestock, wild animals and humans *DOI: 10.13140/RG.2.1.1306.9765/2*

Kikuchi M and Wakabayashi M 1992 Foam property of selected surfactants. *Journal of Japan Society on Water Environment.* **15(10)** 734-738 *DOI: 10.2965/jswe.15.734*

Kosswig K 2012 "Surfactants" in Ullmann's Encyclopedia of Industrial chemistry. *DOI: 10.1002/14356007.a25_747*

Krueger CJ, Radakovich KM, Sawyer TE et al 1998 Biodegradation of the surfactant linear alkylbenzenesulfonate in sewage-contaminated groundwater: a comparison of column experiments and field tracer tests. *Environmental Science and Technology* **32** 3954–3961

http://www.ugr.es/~tep028/pqi/descargas/Industria%20quimica%20organica/tema_5/tensioactivos_a25_747.pdf

Kuch HM and Ballschmiter K 2001 Determination of endocrine-disrupting phenolic compounds and estrogens in surface and drinking water by HRGC-(NCI)-MS in the picogram per liter range. *Environmental Science and Technology* **35** 3201–3206

La Guardia MJ, Hale RC, Harvey EP and Mainor TM 2001 Alkylphenol ethoxylate degradation products in land-applied sewage sludge (biosolids). *Environmental Science and Technology* **35** 4798–804

Lara-Mart´ın PA, G´omez-Parra A, K¨ochling T et al 2007 Anaerobic degradation of linear alkylbenzene sulfonates in coastal marine sediments. *Environmental Science and Technology* **41** 3573–3579

Larson R and Payne A 1981 Fate of the benzene ring of linear alkylbenzene sulfonate in natural waters. *Applied and Environmental Microbiology* **41** 621– 627

Lenntech 2019 Detergents occurring in freshwater. What happens when detergents get into freshwater Ecosystems *https://www.lenntech.com/aquatic/detergents.htm* (accessed 29 August 2019)

Lewis MA. and Suprenant D 1983 Comparative acute toxicity of surfactants to aquatic invertebrates. *Ecotoxicology and Environmental Safety* **7** 313-322

Lewis MA 1990 Chronic toxicities of surfactants and detergent builders to algae: a review and risk assessment. *Ecotoxicology and Environmental Safety* **20** 123–140

Li X and Brownawell BJ 2010 L Quaternary ammonium compounds in urban estuarine sediment environments – A class of contaminants in need of increased attention? *Environmental Science and Technology* **46** 7561-7568

Li Y, Zhang S, Song C and You J 2013 Determination of bisphenol A and alkylphenols in soft Drinks by high-performance liquid chromatography with fluorescence detection. *Food Analytical Methods* **6** 1284–1290

Li D, Kim M, Shim WJ et al 2004 Seasonal flux of nonylphenol in Han River, Korea. *Chemosphere* **56** 1–6

Liwarska-Bizukojc E, Miksch K, Malachowska-Jutsz A, Kalka J 2005 Acute toxicity of five selected anionic and nonionic surfactants. *Chemosphere* **58** 1249-1253

Loyo-Rosales JE, Rosales-Rivera GC, Lynch AM et al 2004 Migration of nonylphenol from plastic containers to water and a milk surrogate. *Journal of Agricultural and Food Chemistry* **52** 2016–2020

Lu J, Wu J, Stoffella PJ, Wilson PC 2013 Analysis of bisphenol A, nonylphenol, and natural estrogens in vegetables and fruits using gas chromatography–tandem mass spectrometry. *Journal of Agricultural and Food Chemistry* **61** 84–89

Lundahl P and Cabridenc R 1978 Molecular structure-biological properties relationships in anionic surface-active agents. *Water Research* **12** 25–30

Maggioni S, Balaguer P, Chiozzotto C and Benfenati E 2013 Screening of endocrine-disrupting phenols, herbicides, steroid estrogens, and estrogenicity in drinking water from the waterworks of Italian cities and from PET-bottled mineral water. *Environmental Science and Pollution Research* **20** 1649–1660

Maksimov VN and Parshikova TV 2006 Influence of surfactants on the photosynthetic activity of algae. *Hydrobiology Journal* **42** 67–76

Mann RM and Boddy MR 2000 Biodegradation of a nonylphenol ethoxylate by the autochthonous microflora in lake water with observations on the influence of light. *Chemosphere* **41** 1361– 1369

Manzano MA, Perales JA, Sales D, Quiroga JM 1999 The effect of temperature on the biodegradation of a nonylphenol polyethoxylate in river water. *Water Research* **33** 2593–2600

Marcomini A and Pojana G 1997 Surfactant metabolites in environmental aqueous samples: a liquid chromatographic approach to the routine analysis of alcohol polyethoxylae biointermediates. *Analusis* **25(7)** 35– 37

Margesin R and Schinner F 1998 Biodegradation of the anionic surfactant sodium dodecyl sulfate at low temperatures. *International Biodeterioration and Biodegradation* **41** 139–43

Martinez-Carballo E, Gonzalez-Barreiro C, Sitka A et al 2007 Determination of selected quaternary ammonium compounds by liquid chromatography with mass spectrometry. Part II. Application to sediment and sludge samples in Austria. *Environmental Pollution* **146** 543–547 *http://dx.doi.org/10.1016/j. envpol.2006.07.016.*

Matsui S and Park H 2000 Morphological effects and ecotoxicity of nonionic and anionic surfactants to *Closterium ehrenbergii* using AGZI (algal growth and zygospore inhibition) test. *Environmental Engineering Research* **5(2)** 63–69

Matthijs E, Holt MS, Kiewiet AT and Rijs GB 1999 Environmental monitoring for linear alkylbenzene sulfonate, alcohol ethoxylate, alcohol ethoxy sulfate and soap. *Environmental Toxicology and Chemistry* **18** 2634–2644

Mezzanotte V, Bolzacchini E, Orlandi M, Rozzi A and Rullo S 2002 Anaerobic removal of linear alcohol ethoxylates. *Bioresource Technology* **82** 151– 156

Miura K, Nishiyama N and Yamamoto A 2008 Aquatic environmental monitoring of detergent surfactants. *Journal of Oleo Science* **57(3)** 161-70 *DOI: 10.5650/jos.57.161*

MSDS ChemOx 2015 Safety data for sodium chloride

https://web.archive.org/web/20110607224738/http://msds.chem.ox.ac.uk/SO/sodium _chloride.html

MSDS Farnell 1992 (revised 2004) Alkylphenol Ethoxylate;

http://www.farnell.com/datasheets/ACL/ACL4100U.pdf

MSDS Meguiars 2018 Alkyl ethoxy sulphates;

https://www.meguiars.com/sites/default/files/pdf/M65%20SDS.pdf

Myers D 2006 Surfactant science and technology, 3rth edition. John Wiley & Sons, Inc., Hoboken, New Jersey. 379 p. ISBN-13 978-0-471-68024-6

Nakada N, Tanishima T, Shinohara H et al 2006 Pharmaceutical chemicals and endocrine disrupters in municipal wastewater in Tokyo and their removal during activated sludge treatment. *Water Research* **40** 3297–3303

Naylor CG 1995 Environmental fate and safety of nonylphenol ethoxylates. *Text Chem Color* **27** 29–33

Naylor CG, Mieure JP, Adams WJ et al 1992 Alkylphenol ethoxylates in the environment. *Journal of the American Oil Chemists' Society* **69** 695–703

NSW EPA (New South Wales Environment Protection Agency) 2015 Foam & chemical contamination in waterways;

https://www.epa.nsw.gov.au/~/media/EPA/Corporate%20Site/resources/epa/Foam-chemical-contamination-in-waterway.ashx

Olkowska E, Ruman M and Polkowska Z 2014 Occurrence of surface-active agents in the environment. Volume **2014,** Article ID 769708. *Journal of Analytical Methods in Chemistry http://dx.doi.org/10.1155/2014/769708*

Painter HA 1992 Anionic surfactants. In: Oude NT, Huntzinger O, editors. Anthropogenic compounds. Detergents, 3 (Part F). Berlin, Germany: Springer.

Petrovic M, Fernandez-Alba AR, Borrull F et al 2002a Occurrence and distribution of nonionic surfactants, their degradation products, and linear alkylbenzene sulfonates in coastal waters and sediments in Spain. *Environmental Toxicology and Chemistry* **21** 37–46

Petrovic M, Sole M, de Alda MJL and Barcelo D 2002b Endocrine disruptors in sewage treatment plants, receiving river waters, and sediments: integration of chemical analysis and biological effects on feral carp. *Environmental Toxicology and Chemistry* **21** 2146–2156

Prieto-Blanco MC, Fernández-Amado M, López-Mahía P et al 2019 *Surfactants*, in *Encyclopedia of Analytical Science, (3rd ed)* (eds) P Worsfold C Poole A Townshend M Miró vol. 9, Elsevier, ISBN: 9780081019832, 410–417 pages

Purdom CE, Hardiman PA, Bye VJ et al 1994. Estrogenic effects from sewage treatment works. *Journal of Chemical Ecology* **8** 275–285

Ramachandra TV, Mahapatra DM, KS Asulaba and Varghese 2017 Foaming or algal bloom in water bodies of India. Remedial measures-Restrict phosphate (P) detergents. ENVIS Technical Report 108. Environmental Information System. CES, Indian Institute of Science, Bangalore-560012.

Rebello S, Asok AK Sathish Mundayoor and Jisha MS 2013 Chapter 5. Surfactants: *Chemistry, Toxicity and Remediation.* Pollutant Diseases, Remediation and Recycling. p 277-320

Reznickova I, Hoffmann J and Komarek K 2002 Biodegradation of technical mixtures of oxyethylenated aliphatic alcohols in an aqueous environment. *Chemosphere* **48** 83–87

Rice CP, Schmitz-Afonso I, Loyo-Rosales JE et al 2003 Alkylphenol and alkylphenol-ethoxylates in carp, water, and sediment from the Cuyahoga River, Ohio. *Environmental Science and Technology* **37** 3747–3754

Ricciardi F, Matozzo V and Marin MB 2008 Effects of 4-nonylphenol exposure in mussels (*Mytilus galloprovincialis)* and crabs (*Carcinus aestuarii*) with particular emphasis on vitellogenin induction. *Marine Pollution Bulletin* **57** 365-372

Ruan T, Song S, Wang T et al 2014 Identification and composition of emerging quaternary ammonium compounds in municipal sewage sludge in China. *Environmental Science and Technology* **48** 4289–4297 *http://dx.doi.org/10.1021/es4050314.*

Sabik H, Gagne F, Blaise C, Marcogliese DJ and Jeannot R 2003 Occurrence of alkylphenol polyethoxylates in the St. Lawrence River and their bioconcentration by mussels (*Elliptio complanata*). *Chemosphere* **51(5)**349-356

Sanchez Leal J, Gonzalez JJ, Compelles F, Campos E, and Cigana T 1991 Biodegradability and toxicity of anionic surfactants. *Acta Hydrochimica et Hydrobiologica* **19** 703-709

Sanderson H, Dyer SD, Price B et al 2006 Occurrence and weight-of-evidence risk assessment of alkyl sulfates, alkyl ethoxysulfates, and linear alkylbenzene sulfonates (LAS) in river water and sediments. *Science of the Total Environment* **368** 695–712 *https://www.cleaninginstitute.org/sites/default/files/research-pdfs/Surfactant_effects_on_humans_and_other_mammals.pdf*

Schilling K and Matthias Zessner M 2011 Foam in the aquatic environment. *Water Research* **45** 4355-4366

Scott MJ and Jones MN 2000 The biodegradation of surfactants in the environment. *Biochimica et Biophyica Acta* **1508** 235–251

Shao B, Han H, Li D et al 2007 Analysis of alkylphenol and bisphenol A in meat by accelerated solvent extraction and liquid chromatography with tandem mass spectrometry. *Food Chemistry* **105** 1236–1241

Shao B, Hu J, Yang M et al 2005 Nonylphenol and nonylphenol ethoxylates in river water, drinking water, and fish tissues in the area of Chongqing, China. *Archives of Environmental Contamination and Toxicology* **48** 467–473

Sibila MA, Garrido MC, Perales JA and Quiroga JM 2008 Ecotoxicity and biodegradability of an alkyl ethoxysulphate surfactant in coastal waters. *Science of Total Environment* **394** 265-274

Soares, A., Guieysse, B., Jefferson, B et al 2008 Nonylphenol in the environment: A critical review and treatment in wastewaters. *Environmental International* **34** 1033-1049

Soto AM, Justica H, Wray JW and Sonnenschein C 1991 P Nonylphenol: an estrogenic xenobiotic release from ''modified'' polystyrene. *Environmental Health Perspectives* **92** 167–173

Spengler P, Korner W and Metzger JW 2001 Substances with estrogenic activity in effluents of sewage treatment plants in southwestern Germany. 1. Chemical analysis. *Environmental Toxicology and Chemistry* **20** 2133–2141

Stepanets OV, Solov'eva GY, Mikhailova AM and Kulapin IL 2001 Rapid determination of anionic surfactants in seawater under shipboard conditions *Journal of Analytical Chemistry* **56(3)** 290

Swisher RD 1966 Surfactant effects on humans and other mammals. The Soap and Detergent Association. Scientific and Technical Report Number #4, New York. 25pp.

https://www.aciscience.org/docs/Surfactant%20effects%20on%20humans%20and%20other%20mammals.pdf

Szymanski A, Wyrwas B, Swit Z, Jaroszynski T, Lukaszewski Z 2000 Biodegradation of fatty alcohol ethoxylates in the continuous flow activated sludge test. *Water Research* **34** 4101– 4109

Tabata A, Kashiwada S, Ohnishi Y, Ishikawa H, Miyamoto N, Itoh M 2001 Estrogenic influences of estradiol-17β, *p*-nonylphenol and bis-phenol-A on Japanese medaka (*Oryzias lattipes*) at detected environmental concentrations. *Water Science and Technology* **43** 109-116

Tabor CF and Barber LB 1996 Fate of linear alkylbenzene sulphonate in the Mississippi River. *Environmental Science and Technology* **30** 161-171

Takada H, Ogura N and Ishiwatari R 1992a Seasonal variations and modes of riverine input of organic pollutants to Tokio Bay. *Environmental Science and Technology* **26(12)** 2517–2523

Takada H, Ishiwatari R and Ogura N 1992b Distribution of lineal alkylbenzenes (LABs) and linear alkylbenzenesulphonates (LAS) in Tokyo Bay sediments. *Estuarine Coastal and Shelf Science* **35** 141–156

Takada H, Mutoh K, Tomita N et al 1994 Rapid removal of linear alkylbenzenesulfonates (LAS) by attached biofilm in an urban shallow stream. *Water Research* **28** 1953–60

Tjandraatmadja G, Pollard C, Sheedy C and Gozukara Y 2010 Sources of contaminants in domestic wastewater: nutrients and additional elements from household products. CSIRO: Water for a Healthy Country National Research Flagship.

http://www.clw.csiro.au/publications/waterforahealthycountry/2010/wfhc-contaminants-domestic-wastewater.pdf

Tsuda T, Takino A, Muraki K et al 2001 Evaluation of 4- nonylphenols and 4-tert-octylphenol contamination of fish in rivers by laboratory accumulation and excretion experiments. *Water Research* **35** 1786– 1792

USDA (United States Department of Agriculture) 2015 Nonylphenol ethoxylates in crops. Technical evaluation report. Compiled by Pesticide Research Institute for the USDA National Organic Program.

https://www.ams.usda.gov/sites/default/files/media/NPE%20Technical%20Evaluation%20Report%20%282015%29.pdf

USEPA (United States Environment Protection Agency) 2010 Nonylphenol (NP) and Nonylphenol Ethoxylates (NPEs) Action Plan [RIN 2070-ZA09]. *https://www.epa.gov/sites/production/files/2015-09/documents/rin2070-za09_np-npes_action_plan_final_2010-08-09.pdf*

Van Ginkel CG 1996 Complete degradation of xenobiotics surfactants by consortia of aerobic microorganisms. *Biodegradation* **2** 151– 64

van Ry DA., Dachs J, Gigliotti CL et al 2000 Atmospheric seasonal trends and environmental fate of alkylphenols in the Lower Hudson River Estuary. *Environmental Science and Technology* **34** 2410–2417

Verge C and Moreno A 1996 Toxicity of anionic surfactants to green microalgae "*Scenedesmus Subspicatus*" and "*Selenastrum Capricornutum*". *Tenside Surfactants Detergents* **33** 166–169

Vethaak AD, Lahr J, Schrap SM et al 2005 An integrated assessment of estrogenic contamination and biological effects in the aquatic environment of The Netherlands. *Chemosphere* **59** 511–524

Versteeg DJ, Stanton DT, Pence MA, Cowan C. 1997 Effects of surfactants on the rotifer, *Brachionus calyciflorus*, in a chronic toxicity test and in the development of QSARs. *Environmental Toxicology and Chemistry* **16** 1051-1059

Warne MS and Schifko AD 1999 Toxicity of laundry detergent components to a freshwater cladoceran and their contribution to detergent toxicity. *Ecotoxicology and Environmental Safety* **44(2)** 196-206

Wu ZB, Zhang Z, Chen SP et al 2007 Nonylphenol and octylphenol in urban eutrophic lakes of the subtropical China. *Fresenius Environmental Bulletin* **16** 227–234

Yang X, Huang H, Wang M et al 2017 Effect of nonylphenol on the regulation of cell growth in colorectal cancer cells. *Molecular Medical Reports* **16** 2211-2216 *https://doi.org/10.3892/mmr.2017.6817*

Ying G-G 2006 Fate, behavior and effects of surfactants and their degradation products in the environment. *Environment International* **32** 417–431

Ying G-G, Williams B and Kookana R 2002 Environmental fate of alkyphenols and alkyphenol ethoxylates – a review. *Environment International* **28** 215–226

Ying G-G, Kookana RS, Kumar A and Mortimer M 2009 Occurrence and implications of estrogens and xenoestrogens in sewage effluents and receiving waters from southeast Queensland. *Science of the Total Environment* **407** 5147-5155 *Doi:10.1016/j.scitotenv.2009.002*

Yuan CL, Xu ZZ, Fan MX et al 2014 Study on characteristics and harm of surfactants. *Journal of Chemical and Pharmaceutical Research* **6(7)** 2233-2237

Zhang C, Cui, F, Zeng G-M et al 2015 Quaternary ammonium compounds (QACs): A review on occurrence, fate and toxicity in the environment. *Science of the Total Environment* **518–519** 352–362

Web

https://infogram.com/different-types-of-surfactants-and-their-applications-1gjk92ejx86rp16

https://www.uts.edu.au/sites/default/files/2019-12/UTS_Consumer_ Products_ Report_ v5.pdf

https://www.vocabulary.com/dictionary/detergent

https://chem.libretexts.org/Bookshelves/Physical_and_Theoretical_Chemistry_Textbook_Maps/Supplemental_Modules_(Physical_and_Theoretical_Chemistry)/Physical_Properties_of_Matter/States_of_Matter/Properties_of_Liquids/Wetting_Agents

https://en.wikipedia.org/wiki/Foaming_agent

https://www.bhg.com.au/the-truth-about-teflon-are-non-stick-pans-safe

https://ihsmarkit.com/products/chemical-surfactants-scup.html

https://www.epa.gov/saferchoice/safer-choice-criteria-surfactants

https://www.epa.gov/saferchoice/safer-choice-criteria-surfactants

https://en.wikipedia.org/wiki/Vitellogenin

http://www.plastipure.com/safer-products/what-is-estrogenic-activity/

https://en.wikipedia.org/wiki/Quaternary_ammonium_cation

■■■■

11 Synthetic Dyes and Textile Effluents Risks, Impacts and Management

Summary

Synthetic dyes can be considered as "**contaminants of emerging concern**" or emerging **contaminant**. These dyes are made from chemicals, petroleum by-products, and earth minerals. They include dyes namely, acid, azo, basic, direct, disperse, fluorescent, mordant, reactive, solvent and vat etc. The discharge of dye residues into water bodies are ecological and human health risks. The effluents discharged from textile factories are high in colour, pH, suspended solids, dissolved solids, biochemical oxygen demand and chemical oxygen demand, temperature and salts and heavy metals. The most frequently utilised textile dye is Azo dye (60-70% of all dye used), the residue of which can remain in the environment for a long time (~50 years), is highly bioaccumulative and carcinogenic. This chapter reviews **i.** Classification of dyes; **ii**. Chemistry, toxicity and ecotoxicity of synthetic dyes/textile dyes; **iii.** Impacts of synthetic dyes/ textile effluents on social, economic and environmental sectors including water quality, the aquatic environment and aquatic biota, crops and vegetables, human's health; and **iv.** Management of synthetic dyes/textile effluents.

Contents

List of Tables

List of Figures

List of Boxes

11.1. Introduction

Synthetic dyes are coloured organic compounds that are used to colour various substrates (Katheresan et al 2018), including textile, pulp and paper, leather, food processing, fur, hair, drugs/ pharmaceuticals, cosmetics, waxes, fuel, plastics, colour photography and printing and rubber (Table 11.1). The first synthetic dye "mauveine" was discovered in 1856, and it was the beginning of the era for manufacturing dyes synthetically and in large scales (Gupta and Suhas 2009). Worldwide, approximately 10^6 tons (10 million tons) of synthetic dyes are produced annually (Zollinger 1987, Pereira and Alves 2012). Among the industries, the textile industry[1] utilises the highest amount of dyestuff (2/3rd of dyes) per year worldwide (O'Neill et al 1999, Rodríguez-Couto et al 2009). In fact, the textile industry plays a major role in the economy of China, EU (European Union), India, the USA (Pereira and Alves 2012, Ghaly et al 2013; Fig. 11.1). China, the EU (EU28), Bangladesh, and Vietnam unshakably remained the world's top four largest apparel (clothing, readymade garments) exporter in 2018 (Fig. 11.1). The Indian subcontinent, China, and the EU accounts for 70% of worlds dyeing industries (Ghaly et al 2013). Though the textile and clothing industries are contributing significant economic benefits to the aforementioned countries, but those industries are causing major pollution problems in waterways, threatening biodiversity, ecosystems and public health. Five major industries (textile, dyeing, paper and pulp, tannery and paint and dye manufacture) are responsible for the presence of dye effluents in the environment (waterways), of which, the textile industry alone accounts for 54%, followed by dyeing (21%), paper and pulp (10%), tannery and paint (8%) and dye manufactures (7%) (Katheresan et al 2018). Besides, in aquaculture, some chemicals are often used as a disinfectant treatment for fungal and external parasites (such as malachite green) (Verdon and Andersen 2017). Textile industries are regarded as the most polluting industries by the Ministry of Environment and Forests, Government of India (Garg and Kaushik 2007). Synthetic dyes can be considered as "**contaminants of emerging concern**" or **emerging contaminan**ts because they are potentially toxic and have been found in the aquatic environment and there are no regulations stating the maximum allowable concentrations in water to ensure the protection of aquatic biota and human health (Ribeiro and de Aragão Umbuzeiro 2014).

11.1.1. Classification of dyes

Dyes are classified according to chemical structure or by their usage or application method (Hunger 2003). The synthetic dyes include acid, azo, basic, direct, disperse, fluorescent, mordant, reactive, solvent and vat etc (Table 11.1). Acid, basic, direct, mordant and reactive dyes are examples of **soluble dyes**; whereas azo, disperse, sulphur and vat dyes are examples of **insoluble dyes** (Rauf and Ashraf 2012). Dyes are also classified based ionic charges: **anionic**

---

[1]*The **textile industry** is a group of related industries which uses a variety of natural (cotton, wool, etc.) and/ or synthetic fibres to produce the fabric. It is a significant contributor to many national economies. The textile industry is very **energy, water, and chemical-intensive**. About 60% of the energy is used by dyeing and finishing operations (Kim et al 1983). Environmental problems associated with the textile industry are typically associated with water pollution. **Effluents** are generally hot, alkaline, having a strong obnoxious smell and coloured by chemicals used in dyeing processes. Some of the chemicals discharged are also **toxic** (https://p2infohouse.org/ref/11/10489/sectors71.html).*

dyes/ negatively (⁻) charged (acid, direct and reactive dyes), **cationic dyes/** positively (⁺) charged (azo, basic) and **non-ionic dyes (no charge)** (disperse, solvent, vat) (Pereira and Alves 2012, Gupta et al 2013, Yagub et al 2014, Yazdani 2018). Among all dye types, **azo dyes** are the highest produced dye type globally (Balapure et al 2015), it is the most frequently utilised dye worldwide (60-70% of all organic dye production) (Vijaykumar et al 2007, Cotillas et al 2018) and over 2,000 different azo dyes were in use as of 2007 (Vijaykumar et al 2007).

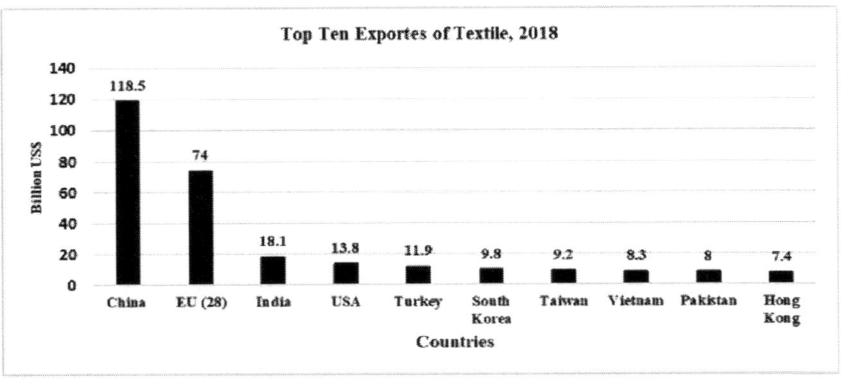

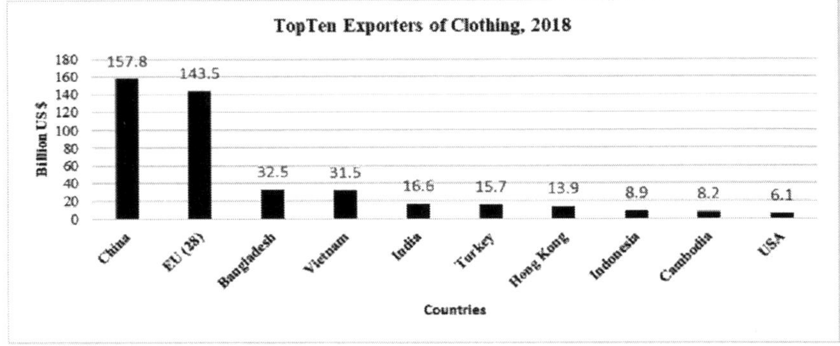

Fig. 11.1. World textile and apparel/ clothing trade in 2018.
[note: textile (top figure) and apparel (bottom figure); data are based on World Trade Organization (WTO) report https://shenglufashion.com/2019/08/16/wto-reports-world-textile-and-apparel-trade-in-2018/) The graph is plotted based on data of World Trade Organization (WTO)].

Dyes are also classified by their **Colour Index** (CI) (Hunger 2003) (CI was developed by the Society of Dyes and Colourist). CI system is made up of three parts, **i.** the first is the class of dye/generic name based on its application, **ii.** the second is the colour, and **iii.** the third is the CI number (5 digits) based on its chemical structure if known (O-Neill et al 1999, Hunger 2003). Some examples are given below: Acid-Blue-80 (26400), Acid-yellow-36 (13065), Basic-violet-2 (42520), Basic-blue-11 (44040), Direct-blue-150 (35110), Direct-brown-44

(35005), Disperse-yellow-3 (11855), Disperse-blue-19 (61110), Reactive-Red-5 (18105), Reactive-yellow-39 (18971), Mordant-Yellow-10 (14010), Pigment-blue-15 (74160) (copper phthalocyanine) Vat-orange 9 (59700), Vat-yellow-3 (61725) (Hao et al 2000, Pereira and Alves 2012). The most common letters used to designate hue/colour are **R** for reddish, **B** for bluish, and **G** for greenish shades.

Table 11.1. Classification of synthetic and textile dyes by use or application method

[Hao et al 2000, Hunger 2003, Pereira and Alves 2012, Croce et al 2017, Katheresan et al 2018, https://www.essentialchemicalindustry.org/materials-and-applications/colorants.html, https://www.sciencedirect.com/topics/chemistry/metal-complex-dye;

*[**note:** anionic= $^-$ charged; cationic= $^+$ charged; non-ionic= no charge; 85% of disperse dyes are azo or anthraquinone dyes (Hamprecht and Westerkamp 2000; approximately 95% of reactive dyes are azo dyes (ESC 2013)].*

Dye class (chemistry)	**Principal applications** *(bold items are major usage)*	**Chemical class** *(bold chemical classes are most frequently used)*	**Examples (Colour index or CI name)**
Acid *(water-soluble; **anionic**)*	Cosmetics, colouring **tissue proteins**, **food**, ink-jet printing, leather, modified **acrylics**, **nylon**, paper, printing ink, **silk**, and **wool**	**Azo** (including pre-metallized), **anthraquinone**, azine, nitro, nitroso, triphenylmethane and **xanthene**	Acid Black 1, Acid Black 24 Acid Blue 80, Acid Blue 113 Acid Brown 235, Acid Brown 354, Acid Green 68, Acid Green 111, Acid Orange 7, Acid Orange 33/95, Acid Red 119, Acid Red 131, Acid Yellow 42, Acid Yellow 49, Acid Violet 48

Dye class (chemistry)	Principal applications *(bold items are major usage)*	Chemical class *(bold chemical classes are most frequently used)*	Examples (Colour index or CI name)
Azo[2]/azoic *(water-insoluble; 60–70% of all organic dye production, mostly **anionic and some cationic**)*	Acetate, cellulose, cotton, rayon and polyester	**Azo** (-N=N-) (depending upon the number of azo groups in the molecules, azo dyes can be classified as **Monoazo** (Acid orange 7, Orange G, Methyl red); **Diazo** (Congo red), Trypan blue); **Traiazo** (Direct Blue 71) (Rauf and Ashraf 2012)	Acid Orange 5, Acid Orange 7 Acid Red 13, Acid Red 97, Basic Yellow 15, Basic Blue 41 **Congo Red[3]**, Direct Blue 1, Direct Blue 15, Direct Blue 160, Disperse Orange 1, Disperse Red 1, Methyl Orange, Methyl Red, Mordant Brown 33, Mordant Red 19 Orange B, Orange G, Orange II Reactive Red 2, Reactive Orange 16, Reactive Black 5, Solvent Red 1, Solvent Black 3, Solvent Red 26, Sudan Red 7B, Sudan Red G (Rauf and Ashraf 2012, *https://en.wikipedia.org/wiki/Category:Azo_dyes*)

[2]***Azo*** *= are aromatic compounds with one or more nitrogen groups (–N=N– groups), constitute the largest class of synthetic dyes used in commercial applications, constituting* ***60–70% of all dyestuffs*** *produced. or utilised dye worldwide. The linkage* ***(–N=N–)*** *may be present more than once (mono azo dyes have one azo linkage; diazo dyes have two linkages; triazo dyes have three linkages). Azo dyes have one or more azo groups (R1-N=N-R2) having* ***aromatic rings*** *mostly substituted by* ***sulfonate groups****. These complex aromatic substituted structures make a conjugated system and are responsible for the intense colour, high water solubility, and resistance to degradation of azo dyes under natural conditions (reviewed by Khan and Malik 2014). Approximately* ***95% of reactive dyes*** *are* ***azo dyes*** *(ESC 2013). They are cost-effective and easy to use, which make them the most popular synthetic dye (Jamee and Siddique 2019).*

[3]***Congo red*** *(CR) is the most commonly used azo dyes worldwide. CR is a diazo based anionic dye containing double azo (-N=N-) linkage. It is toxic to many organisms and can be metabolized to* ***benzidine****, a known human* ***carcinogen****. Due to health concerns, CR dye has been banned by various countries, but it is still being used in many countries (Xia et al 2013, Ismail et al 2019).*

Dye class (chemistry)	**Principal applications** *(bold items are major usage)*	**Chemical class** *(bold chemical classes are most frequently used)*	**Examples (Colour index or CI name)**
Basic/Cationic *(water-soluble; **cationic**)*	Inks, medicine, modified **nylon**, modified polyester, **paper,** polyacrylonitrile, **polyester**, **silk,** tannin-mordanted cotton and **wool**	Acridine, **anthraquinone**, azine, **azo**, cyanine, diazahemicyanine, **diphenylmethane**, hemicyanine, thiazine, **triarylmethane**, oxazine and **xanthene**	Basic Blue 3, Basic Blue 11 Basic Violet 1, Basic Violet 6 Basic Orange 2, Basic Orange 21 Basic Red 2, Basic Red 18, Basic Yellow 11, Basic Yellow 13 Basic Green 4 (**malachite green**[4])
Direct dyes[5] *(water-soluble; **anionic**; polar, high molecular weight)*	**Cotton**, **leather**, **nylon**, rayon, silk and **paper**; **pH indicators** and as **biological stains**	**Azo**, **phthalocyanine**, polyazo, stilbene, and oxazine	Direct Blue 44, Direct Blue 71, Direct Blue 293, Direct Green 26 Direct Red 23, Direct Red 81 Direct Red 227, Direct Orange 26 Direct Yellow 11, Direct Yellow 28
Disperse dyes[6] *(water-insoluble; **non-ionic**; nonpolar/ hydrophobic)*	Acetate, **acrylic fibres**, **cellulose,** cellulose acetate, **nylon**, polyamide, **polyester**, **polyester**-cotton and plastic	**Anthraquinone**, **azo**, benzodifuranone, nitro and styryl	Disperse Blue 3, Disperse Blue 27 Disperse Blue 79, Disperse Orange 30, Disperse Orange 61, Disperse Red 4, Disperse Yellow 3, Disperse Yellow 54

[4] ***Malachite green*** *(MG) is a **benzidine based** dye used in the dyeing process and has extensively been used in fish farms to prevent fungal infections. Due to its toxicological and carcinogenic effects, the uses of MG have been banned worldwide since 2002, although, in some environments, these dyes may persist (Chen et al 2010). It is hydrophobic (log K_{ow} 5.7) and have been detected in eel, seafood imported in the USA from China (https://en.wikipedia.org/wiki/Malachite_green).*

[5] ***Direct dye**, also called substantive dye which has an affinity for fibre and is taken up **directly**, such as the **benzidine derivatives** (https://www.britannica.com/technology/direct-dye).*

[6]***85%** of **disperse dyes** are azos or anthraquinone dyes (Hamprecht and Westerkamp 2000, https://en.wikipedia.org/wiki/Disperse_dye).*

Dye class (chemistry)	**Principal applications** *(bold items are major usage)*	**Chemical class** *(bold chemical classes are most frequently used)*	**Examples (Colour index or CI name)**
			*(**85%** of disperse dyes are **azos** or anthraquinone dyes; Hamprecht and Westerkamp 2000)*
Fluorescent brighteners *(a fluorescent tracer)*	All fibres, oils, paints, plastics and soaps as well as detergents	Coumarin, naphthylamides, pyrazolos and stilbene	4,4′-bis (ethoxycarbonylvinyl) stilbene
Food, drug, and Cosmetic	Food, drug, and cosmetics	**Anthraquinone**, **azo**, carotenoid and **triarylmethane**	Food Yellow 4 and Tartrazine
Metal-complex dyes[7]	Leather finishing, wood stains, printing inks, inks, colouring of metals, plastic and wool	**Azo, anthraquinone**	Acid Violet 78
Mordant[8] *(water-soluble; **anionic**)*	Anodized aluminium, natural fibres, **leather** and **wool**	**Anthraquinone** and **azo**	Mordant Black 11/ Eriochrome black Mordant Red 11 Mordant Yellow 10
Oxidation bases	Cotton, fur and hair	Aniline black and indeterminate structures	Direct Blue
Reactive[9]	**Natural fibres**	**Anthraquinone,**	Reactive Blue 5,

[7]***Metal-complex dyes****= where a metal atom within the dye structure are often used to form a complex. Such metals can be* ***Cu, Cr, Co, Ni.*** *They are mainly applied to wool, silk and nylon to achieve better wash fastness for dyed fabrics, compared to those obtained with the parent acid dye (Gregory 2007, https://www.sciencedirect.com/science/article/pii/B9781845696955500131)*

[8]***A mordant*** *or dye fixative is a substance used to set dyes on fabrics by forming a coordination complex with the dye, which then attaches to the fabric. It may be used for dyeing fabrics or for intensifying stains in cell or tissue preparations (https://www.yourdictionary.com/mordant). Alum, tin and iron are typical mordants used in natural dyeing. (https://organiclifestyle.com/articles/fibre-reactive-low-impact-dyes#mordants)*

[9] ***Reactive dye*** *is the dye that can react with a fibre to form a covalent link (contain one or more reactive groups), that is forming a permanent attachment in the fibre and could not be removed by repeated treatment with boiling water under neutral conditions (Hao et al 2000, https://www.sciencedirect.com/topics/engineering/reactive-dye). Approximately 95% of reactive*

Dye class (chemistry)	**Principal applications** *(bold items are major usage)*	**Chemical class** *(bold chemical classes are most frequently used)*	**Examples (Colour index or CI name)**
(water-soluble; ***anionic****; polar; smaller molecule size)*	(textiles/cotton, nylon, silk, wool) and synthetic fibres (modified acrylics)	**azo** (including metallized azo), basic, formazan, oxazine, **phthalocyanine triphenylmethane**	Reactive Red 83, Reactive Red 83, Reactive Red 195 Reactive yellow 15 *(approximately 95% of reactive dyes are azo dyes; ESC 2013)*
Solvent *(water-insoluble (but solvent-soluble);* ***non-ionic****; nonpolar or little polar)*	Fats, gasoline, **inks,** lacquers, lubricants, **gasoline**, oils, **plastics**, stains, varnishes and **waxes**	**Anthraquinone**, **azo**, **phthalocyanine** and **triphenylmethane**	Sulphur Black 1, Solvent Blue 35 Solvent Red 26
Sulfur *(water insoluble)*	**Cotton**, leather, paper, polyamide fibres, **rayon**, silk and wood	Indeterminate structures.	Sulphur Black 1
Vat[10] *(water-insoluble but soluble with alkali;* ***non-ionic***	**Cotton**, **cellulosic fibres**, polyester-cotton, rayon and wool	**Anthraquinone** (including polycyclic quinines) and **indigoids**[11]	Vat Blue 4 (indanthrene), Vat Green 1, Vat Green 9

11.2. Chemistry, toxicity and ecotoxicity of synthetic and textile dyes

11.2.1. Colour chemistry

Most dyes are synthetic. Synthetic dyes are made from synthetic resources such as chemicals, petroleum by-products, and earth minerals (Moradi et al 2018). They are made as complex organic substances so that they can resist getting

dyes are ***azo dyes*** *(ESC 2013). This dye is popular due to their high wet fastness, brilliance and availability of a range of hues/colour (Hao et al 2000).*

[10] ***Vat*** *= The term vat dyes is derived from the large vessels used to apply the dye. A dye, such as indigo, that produces a fast colour by impregnating fibre with a reduced soluble form that is then oxidized to an insoluble form.*

[11] ***Indigoids*** *= related to or resembling indigo (a deep radish blue) especially in chemical structure and dyeing properties (https://www.merriam-webster.com/dictionary/indigoid\). The indigo dye is the most important dye in the manufacturing of* ***blue jeans*** *(Pereira and Alves 2012, p147).*

degraded upon contact with water, detergents or any other washing agents. Dye molecules can withstand degradation even when exposed to extreme heat sources, oxidising agents or strong light.

The one-ring, two-ring and three-ring aromatic hydrocarbons such as benzene, naphthalene and anthracene, respectively, are the basic building blocks for most **textile dyes** (Gregory 2007). Synthetic dyes are sometimes referred to as '**coal tar dyes**[12]', since they are manufactured from substances which, until recently, were only obtained from coal tar. As elaborated in Table 11.1, that some dyes are water-soluble; some are water-insoluble; some are anionic, cationic or non-ionic and some dyes fall under polar or non-polar categories (***note:*** dyes are soluble in water and/or an organic solvent, while pigments are insoluble in both types of liquid media *https://www.ncbi.nlm.nih.gov/books/NBK385442/*).

Dyes possess colour because they absorb light in the visible range of the spectrum (band)[13] at certain wavelength[14] (red: ~ 700-635 nm; orange: ~ 635-590 nm; yellow: ~ 590-560 nm; green: ~ 560-490 nm; blue: ~ 490-450 nm; violet: ~ 450-400 nm). In addition, dyes have a colour-bearing groups/ atom (responsible for colour) called **"chromophores**" (Pereira and Alves 2012, Khan and Malik 2014). Chromophore imparts colour by absorbing light in the visible region. The common chromophores include azo (–N=N–), carbonyl (=C=O), methine (-CH=), nitro (-NO_2), nitroso (N-OH, C=NH, -CH=N-, NO), quinoid and sulphur (C=S) (Pereira and Alves 2012, Khan and Malik 2014). They also possess "**auxochromes**[15]" (an electron-withdrawing or electron-donating substituent, which generate or intensify/deepen the colour of the chromophores, "colour helper") (Christie 2001, Pereira and Alves 2012). Some auxochromes are amine (–NH_3), carboxyl (–COOH), sulfonate (HSO_3) and hydroxyl (–OH) (Welham 2000, Khan and Malik 2014).

---

[12] ***Coal tar dyes*** *are artificial colouring agents made by combining various* ***aromatic hydrocarbons*** *like toluene, xylene, benzene, which are obtained from the distillation of bituminous coal. Coal tars are also made from petroleum distillates. Coal tar is a brown or black liquid of extremely high viscosity. These dyes are used in foods, over the counter and prescription* ***drugs***, ***textiles, cosmetics***, *and* ***personal care*** *products (hair dyes), shampoos, and deodorants. http://thetoxicfreefoundation.com/database/ingredient/coal-tar-dyes*

[13] ***Visible spectrum*** *is the observable region of the electromagnetic wave which is visible to human eyes. In the electromagnetic spectrum, the visible spectrum ranges from the infrared region to the UV region. We can detect the range of light spectrum from about 400 nanometres (violet) to about 700 nanometres (red) and after this range, the human eye is unable to see the rest of the electromagnetic waves (https://byjus.com/physics/wavelength-of-light/).*

[14] ***Wavelength*** *is defined as the property of a wave in which the distance between the identical points between the two successive waves are calculated. (https://byjus.com/physics/wavelength-of-light/).*

[15] ***Auxochrome*** *is a group of atoms attached to a* ***chromophore*** *which modifies the ability of that chromophore to absorb light.*

Dye molecules structure consists of an aromatic/ benzene ring or fused benzene ring which alternates with double and single bonds. The chemical classes (chromophores) of dyes more frequently employed on an industrial scale are **azo** (60-70% of colourants), **anthraquinone** (15% of colourant), arylmethane/phenylmethane, indigoid, **phthalocyanine** (2% colourants) and xanthenes derivatives (Pereira and Alves 2012, Przystaś et al 2012, *https://www.essentialchemicalindustry.org/materials-and-applications/colorants.html)* (see Fig. 11.2).

11.2.2. Persistence

Synthetic organic dyestuffs are not readily biodegradable (Anliker et al 1981, OECD 1988) and they are recalcitrant[16] in the biodegradation process (Puvaneswari et al 2006). Dyes persist for a long time because of their high stability to light or temperature, resistance to surfactants/detergents degradation and poor removal by wastewater treatments (Rajaguru et al 2001, Przystas et al 2012). The insoluble dyes (azo, disperse, solvent, sulfur and vat) have low biodegradability only 45-47% of insoluble dyes are known to be biodegradable; Clnar et al 2008) or difficult to degrade (Clnar et al 2008, Robert et al 2008, Rauf and Ashraf 2012, Hassan and Nemr 2017) and therefore, would persist in the environment for a long time. The half-life of widely used metal-complexed dyes (dyes composed of chromium, or cobalt, or copper or nickel, or copper) is reported to be 2 to 13 years (Copaciu et al 2013, Lellis et al 2019). Reactive dyes are most commonly used in textile industries to dye wool, cotton, silk, and acrylics (O'Neill et al 1999) and can remain in the environment for a long time (~50 years) due to their high stability (Pereira and Alves 2012).

[16] **Recalcitrant** = *uncooperative (resist biodegradation).*

Structures of chemical classes of dyes

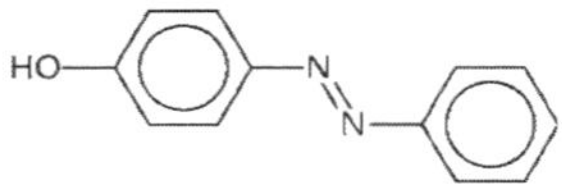

Azo ($C_{16}H_{11}N_2NaO_4S$)

*[(aromatic compounds (benzene or naphthalene) rings with one or more **nitrogen bonds** (–N=N– groups), constitute the largest class of synthetic dye (**60–70%** of all dyestuffs produced) (Carliell et al 1995); the three **metals** of importance in azo dyes are copper, chromium, and cobalt (Hunger 2003); (examples based on CI: Acid red 88, Acid red 97, Acid blue 113, Acid orange 7, Reactive red 120; Khan and Malik 2014)].*

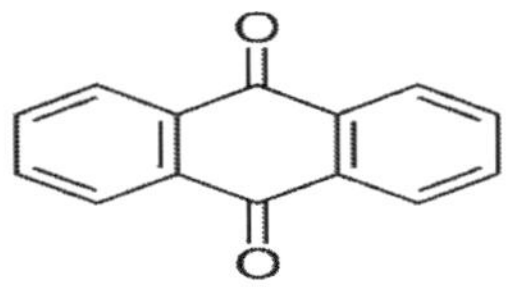

Anthraquinone ($C_{14}H_8O_2$)

[(anthraquinone dyes are based on 9,10-anthraquinone; constitute the second most important class of textile dyes, after azo dyes (Khan and Malik 2014) (examples based on CI.: Acid blue 25, Acid Blue 43, Acid green 25, Vat violet 1, Disperse red 60; https://en.wikipedia.org/wiki/Anthraquinone)].

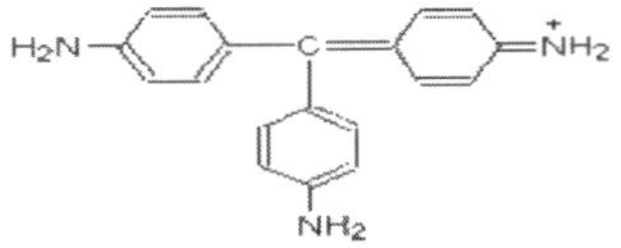

Triarylmethane[17]**/ triphenylmethane**[18]

($C_{19}H_{16}$)

*[arylmethane dyes are derived from **methane**, where some of the hydrogen atoms are replaced with aryl[19] ring; also referred to as phenylmethane dyes; examples based on CI: Basic green 4, Basic violet 1 (Hao et al 2000)*

[17] ***Arylmethane dyes*** *are derived from methane, diarylmethane dyes have two aryl rings and triarylmethane dyes contain three aryl rings (http://stainsfile.info/dyes/class/clsarylm.htm).*

[18] ***Malachite green (MG)*** *is one type of* ***triphenylmethane*** *dyes and is extensively used in both the textile industry and the fish farming industry due to its relatively low cost, ready availability, and high efficacy against fish microbial pathogens*

[19] ***Aryl*** *is a functional group or substituent derived from an aromatic ring, usually an aromatic hydrocarbon (Wang et al 2012).*

Indigoid[20] ($C_{16}H_{10}N_2O_2$)
[one of the oldest known class of organic dyes; indigo dye is an organic compound with a distinctive blue colour (example: Vat blue 1].

Phthalocyanine ($(C_8H_4N_2)_4H_2$)
[large, aromatic, macrocyclic[21]*, organic compound; colours range from reddish-blue to yellowish green; forms complexes with numerous* ***metals****; the majority are based on copper phthalocyanines; structurally related to the macrocyclic ring system porphyrin*[22]*; example (CI), Direct blue 86; Hunger 2003)].*

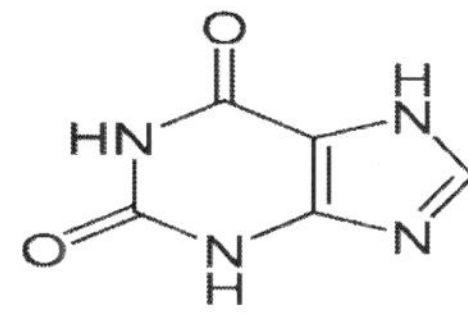

Xanthene[23] ($CH_2[C_6H_4]_2O$)
(two benzene rings joined by a methylene group and an oxygen atom; tend to be fluorescent[24] *yellow to pink to bluish red dyes].*

Fig. 11.2. Most frequently used chemical classes of dyes
*[(Azo, anthraquinonic, arylmethane, indigoid, phthalocyanine and xanthene) and their structures; (**C.I**= colour index; image structures are based on:https://en.wikipedia.org/wiki/Azo_dye#/media/File:Typical_azo_compound.svg; https://en.wikipedia.org/wiki/Anthraquinone#/media/File:Anthraquinone_acsv.svg; http://stainsfile.info/dyes/class/clsarylm.htm; https://en.wikipedia.org/wiki/Indigo_dye#/media/File:Indigo_skeletal.svg; https://en.wikipedia.org/wiki/Phthalocyanine#/media/File:Phthalocyanine.svg; https://en.wikipedia.org/wiki/Xanthene]*

11.2.3. Bioaccumulation

The log K_{ow} (octanol/ water partition coefficient) of the synthetic dyes varies. The most common synthetic dyes (anthraquinone, azo, indigo, phenylmethane, phthalocyanine, xanthene) have a log $K_{ow} > 3$ (Table 11.2). This means that these

[20] ***Indigoid*** *= of or relating to that group of vat dyes that have a molecular structure like that of indigo.*

[21] ***Macrocyclic*** *= molecules and ions containing twelve or more membered ring.*

[22] ***Porphyrins*** *are a group of heterocyclic macrocycle organic compounds; naturally occurring porphyrins are haemoglobin, chlorophyll a, and vitamin B12 (Hunger 2003).*

[23] ***Xanthene dyes*** *tend to be fluorescent, yellow to pink to bluish red, brilliant dyes.*

[24] ***Fluorescein*** *is an organic compound and dye. It is available as a dark orange/red powder slightly soluble in water and alcohol. It is widely used as a fluorescent tracer for many applications.*

dyes are supposed to bioaccumulate in fish and other aquatic organisms (the organic chemicals having a log K_{ow} values >3 are known to be hydrophobic and bioaccumulative) (Kibria et al 2010). The partition coefficient (log K_{ow}) is a useful indicator of the bioaccumulation tendency of dyestuffs (Anliker et al 1981).

The log K_{ow} is linearly correlated with log bioconcentration factors[25] (BCF) in aquatic organisms (Davies and Dobbs 1984, Khan and Malik 2014). For example, Anliker et al (1981) analysed 75 dyes for their partition coefficient (log K_{ow}) and BCF in fish where the researchers showed that **hydrophilic dyes** (water-soluble cationic and anionic dyes) had a lower BCF compared to hydrophobic dyes (non-ionic water-insoluble disperse dyes) (Fig. 11.3). Azo dyes (Solvent Red 1, log K_{ow} 4.6) are highly bioaccumulative with a BCF of 290,000 (Hou et al 1991, MSDS-Pubchem 2005d, Table 11.2). **The hydrophobic dyes**, i.e., disperse and solvent, and many aromatic amines (Fluorescent dyes) have high values of the partition coefficient (high log K_{ow}), and therefore, pose a high risk of bioaccumulation (Bafana et al 2011) (***note:*** most disperse dyes are azo compounds).

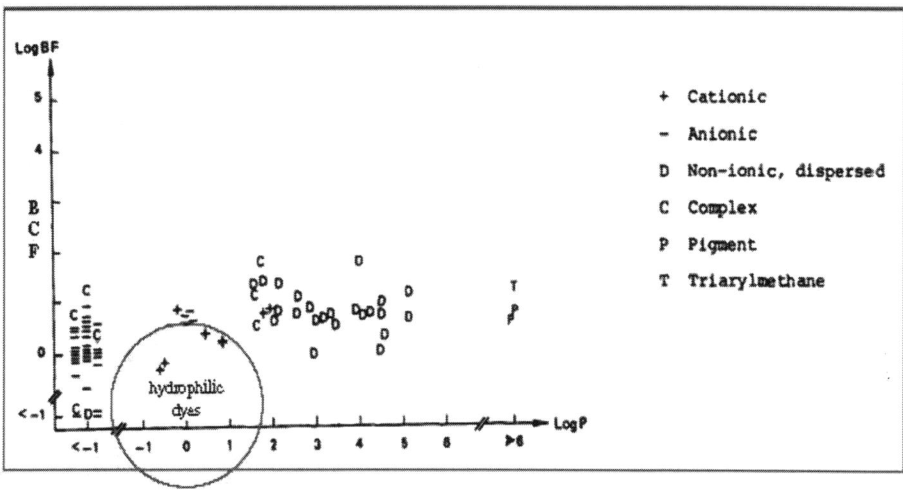

Fig. 11.3. The relationships between bioaccumulation factors (Y axis) in fish and partition coefficient (log K_{ow}) (X axis) of 72 dyestuffs (Anliker et al 1981, page 269).

[25]***The bioconcentration factor*** *(BCF) is defined as the ratio of the chemical accumulated in tissue to the concentration in water (concentration in fish ÷ concentration in water). Generally, high-bioconcentration factors tend to be associated with very hydrophobic compounds (having high log* K_{ow} *value). Conversely, low bioconcentration factors are associated with compounds having high-aqueous solubility/hydrophilic compounds (low log* K_{ow} *value) (Montgomery and Crompton 2017).*

Table 11. 2. Chemistry, toxicity and ecotoxicity of some common synthetic dyes.

[Ecotoxicity classification: extremely toxic: EC_{50}/LC_{50} < 1 mg/L; non-toxic: EC_{50}/LC_{50} > 100 mg/L; for definition of bioaccumulation, log K_{ow}, LD_{50}, EC_{50}, LC_{50} see footnotes given in chapter 7 or 9; ***BCF*** *= Bioconcentration Factor; ED = Endocrine disruptors;* ***na*** *= data is not available].*

Chemical classes of dyes	**Chemistry** Log K_{ow}	**Toxicity** (LD_{50}, carcinogenicity, BCF, ED)	**Ecotoxicity** (EC_{50}/LC_{50}: Algae, Daphnia and fish)
Anthraquinone	log K_{ow}: 3.4 (MSDS-Pubchem 2005a, IARC 2013)	**LD_{50}:** 5000 mg/kg (MSDS-Pubchem 2005a) **Carcinogenic:** Disperse Blue 1, Disperse orange 11, Disperse violet 1 (Hunger 2003, Gregory 2007) **BCF:** 12 (MSDS-Pubchem 2005a)	**Algae**, *Selenastrum capricornutum,* EC_{50}: 0.5 mg/L (disperse blue) for 96 h (Novotny et al 2006) (growth inhibition) ***Daphnia magna*** EC_{50}: > 0.048 mg/L for 48 h (ECHA 2006) **Fish**, *Lepomis macrochirus* (Bluegill) LC_{50} >5 mg/L/24 hr (Verschueren 2001, MSDS-Pubchem 2005a).
Arylmethane/ Phenylmethane	log K_{ow}: 5.3 (triphenylmethane) (MSDS-Pubchem 2005b)	**LD_{50}:** 2250 mg/kg (diphenylmethane) (MSDS-Pubchem 2005c) **Carcinogenic:** Basic violet 14, Acid violet 49, Basic red 9, Solvent yellow 34 (Hunger 2003, Gregory 2007) **BCF**: 870.96 (MSDS-Pubchem 2005c)	**Algae,** *Selenastrum capricornutum,* EC_{50}: 1 mg/L (growth inhibition) (di and triphenylmethane-Basic/methyl violet 1) (Little et al 1974, Hao et al 2000). **Daphnia:** na **Fish**, *Leuciscusidus melanotus* (Carp), LC_{50}: 7.5 mg/L for 48 h (Juhnke and Luedemann 1978, MSDS-Pubchem 2005c) (Diphenylmethane), **Fish**, fathead minnow, *Pimephales promelas* LC_{50}: 0.05 mg/L (di and triphenylmethane- Basic/methyl violet 1) (Little et al 1974, Hao et al 2000).
Azo	log K_{ow}: 4.6 (MSDS-Pubchem 2005d) (solvent red 1[26]) log K_{ow}: 4.5 (MSDS-	**LD_{50}**: 23160 mg/kg (Acid red 26) (MSDS-Pubchem 2005h) **LD_{50}:** 15200 mg/kg (Congo red) (MSDS-Pubchem 2005i)	**Algae,** *Selenastrum capricornutum,* EC_{50}: 4 mg/L (growth inhibition) (diazo-Acid blue 113) (Little et al 1974, Hao et al 2000). **Algae,** *Selenastrum capricornutum,* EC_{50}: 4.8 (congo red) for 96 h (Novotny et al 2006) (growth

[26] ***Solvent Red 1*** *= an azo dye*

Chemical classes of dyes	**Chemistry** Log K_{ow}	**Toxicity** (LD_{50}, carcinogenicity, BCF, ED)	**Ecotoxicity** (EC_{50}/LC_{50}: Algae, Daphnia and fish)
	Pubchem 2005g) (disperse orange 30) log K_{ow} = 3 (Malik et al 2014)	**Carcinogenic:** Acid dye, Acid red 26, Solvent yellow 1 and Solvent yellow 2 (Hunger 2003, Gregory 2007) **Endocrine disruptor:** Frog (*Silurana tropicalis*) (Sorian et al 2014), Rats (Kumar et al 2008, Suryavathi et al 2005) **BCF:** 290,000 (Hou et al 1991, MSDS-Pubchem 2005d) (solvent red 1)	inhibition) **Algae,** *Raphidocelis subcapitata*[27] EC_{50}: 0.239 mg/L (biomass effect) (mordant black 11) (Croce et al 2017) ***Daphnia similis*** EC_{50}: 0.1 mg/L (disperse red 1) (Vacchi et al 2013) ***Daphnia similis*** EC_{50}: 0.13 mg/L (disperse red 1) (Ferraz et al 2011) ***Daphnia magna,*** LC_{50}: 0.552 mg/L for 48 h (disperse orange 30) (Croce et al 2017) **Fish**, fathead minnow, *Pimephales promelas* LC_{50}: 10 mg/L (diazo- Acid blue 113) (Little et al 1974, Hao et al 2000) **Fish**, *Cyprinus carpio,* 13.4 mg/L for 96 h (MSDS-Millipore 2013) (Diisopropyl-azo-dicarboxylate).
Indigo	log K_{ow}: 3.72 (MSDS-Pubchem 2005f)	**LD_{50}:** 5000 mg/kg (Steingruber 2004, (MSDS-ThermoFisher 2011)	**Algae**, EC_{50}: 6.3 mg/L ***Daphnia magna*** EC_{50}: > 500 mg/L for 48h **Fish,** *Oryzias latipes* LC_{50}: > 1000 mg/L for 96h (MSDS-ThermoFisher 2011).
Phthalocyanine	log Kow: 6.6 (MSDS-ThermoFisher 2008)	**LD_{50}:** 10,000 mg/kg (OECD 1988)	**Plant**, rice, *Oryza sativa*, LC_{50}: >100 mg/L (OECD 1988) **Algae:** na **Daphnia:** na **Fish**, orange-red killifish, *Oryzias latipes,* LC_{50}: >100 mg/L for 48 h (OECD 1988, (MSDS-ThermoFisher 2008).

[27]***Raphidocelis subcapitata****, formerly known as Selenastrum capricornutum and Pseudokirchneriella subcapitata. This species is quite sensitive to the presence of toxic substances including metals and has a ubiquitous distribution, so is broadly used in ecotoxicology studies (https://en.wikipedia.org/wiki/Raphidocelis_subcapitata).*

Chemical classes of dyes	Chemistry Log K_{ow}	Toxicity (LD_{50}, carcinogenicity, BCF, ED)	Ecotoxicity (EC_{50}/LC_{50}: Algae, Daphnia and fish)
Xanthene	log K_{ow}: 4.23 (MSDS-Pubchem 2005e)	**LD_{50}:** 5000 mg/kg (Ash and Ash 2009)	**Algae:** na ***Daphnia pulex,*** LC_{50}: 0.423 mg/L for 48 h (Phloxine B[28]) (Walthall and Stark 1999) ***Daphnia pulex***, LC_{50}: 337 mg/L for 48 h (Fluorescein) (Walthall and Stark 1999) **Fish**, rainbow trout, *Salmo gairdneri* LC_{50}: 1372 mg/L for 96 h (Marking 1969) (Fluorescein).

11.2.4. Toxicity

LD_{50} values show that most dyes are **slightly non-toxic to non-toxic** having LD_{50} values between 2000 and 23,000 mg/kg (Table 11.2). Several common dyes are **carcinogens** including several azo (Acid Red 26, Solvent Yellow 2), anthraquinone (Disperse blue 1), benzidine, diphenylamine (Solvent Yellow 34) and triphenylamine (Acid Violet 49) (see Table 11.3). Congo Red (CR) is the most commonly used azo dyes worldwide. It is toxic to many organisms and is a suspected carcinogen as well as mutagen[29] (Xia et al 2013, Ismail et al 2019, *https://phys.org/news/2008-05-congo-red.html*). Azo dyes which are derived from benzidine[30] are carcinogens; exposure to them has classically been associated with urinary bladder cancer (Golka et al 2004). Dyes are also linked to **endocrine disruptions**[31] in organisms. Diazo dyes (Bismarck Brown Y) caused increased developmental defects in frog embryogenesis and early larval development in Western clawed frog, *Silurana tropicalis* (Sorian et al 2014). Rats that were exposed to textile effluents containing azo dyes experienced a reduction in several reproductive parameters such as lower sperm count and a decrease in reproductive protein levels (Kumar et al 2008, Suryavathi et al 2005).

---

[28] ***Phloxine B*** *(commonly known simply as phloxine) is a water-soluble red dye used for colouring drugs and cosmetics in the United States and colouring food in Japan.*

[29] ***Mutagen*** *= a change in the DNA of a cell.*

[30] ***Benzidine*** *has been widely used for the detection of blood and as a reagent in the manufacture of dyes. It is a confirmed human carcinogen (IARC group1) (https://pubchem.ncbi.nlm.nih.gov/compound/Benzidine).*

[31] ***Endocrine disruptors*** *(ED) are chemicals that can interfere with endocrine (or hormone) systems at certain doses. These disruptions can cause cancerous tumours, birth defects, and other developmental disorders (https://www.harathelabel.com.au/blogs/news/health-and-synthetic-dyes).*

Table 11.3. Dyes classified as a potential human carcinogens
[(Hunger 2003, Gregory 2007; *IARC = International Agency for Research on Cancer; na = data is not available*)].

Chemical class	Colour index number	IARC cancer group (IARC2019)
Anthraquinone	Disperse Blue 1	2B
Anthraquinone	Disperse Orange 11	2B
Azo	Acid dye	na
Azo	Acid red 26	na
Azo	Solvent yellow 1	na
Azo	Solvent yellow 2	na
Benzidine	Congo red	1
Diphenylamine	Solvent yellow 34	na
Ketonimine	Basic yellow 2	na
Triphenylamine	Basic violet 14	na
Triphenylmethane	Acid violet 49	na
Triphenylmethane	Basic red 9	na

11.2.5. Ecotoxicity

The ecotoxicity data (EC_{50}/ LC_{50} for algae, *Daphnia* and fish) for most synthetic dyes are not available or scarce. Based on data and information, anthraquinone, azo, indigo, phenylmethane, phthalocyanine, xanthene dyes are toxic to aquatic organisms (Table 11.2). Several dyes (**anthraquinone, di and triphenyl-methane, azo and xanthene**) are **extremely toxic** to aquatic organisms, with $EC_{50}/LC_{50} < 1.0$ mg/L: *Anthraquinone* [(*Daphnia magna* EC_{50}: > 0.048 mg/L for 48 h (ECHA 2006)]; *Di and triphenylmethane* [(Fish, fathead minnow, *Pimephales promelas* LC_{50}: 0.05 mg/L) (Little et al 1974, Hao et al 2000)]; *Azo* [(Algae, *Raphidocelis subcapitata,* EC_{50}: 0.239 mg/L (biomass effect) (mordant black 11) (Croce et al 2017); *Daphnia similis,* EC_{50}: 0.1 mg/L (Disperse red 1) (Vacchi et al 2013);] *Xanthene* [(*Daphnia pulex,* LC_{50}: 0.423 mg/L for 48 h

(Phloxine B) (Walthall and Stark 1999)]. The **indigo and phthalocyanine** dyes appear to be **non-toxic** having EC_{50}/ LC_{50} values is > 100 mg/L to aquatic organisms (see Table 11.2).

11.3. Impacts of synthetic dyes and textile effluents on social, economic and environmental sectors

Textile industries are known to consume a large volume of water (on an average of 200 L water per kg of fibre; Pereira and Alves 2012) and chemicals (as many as 2,000 different chemicals) for the wet processing of textiles (Khan and Malik 2014). During dyeing processes, the entire dye is not fixed to the fibre and therefore, high volumes of wastewaters containing dyes and other pollutants are produced which is released with effluents and ultimately finds its way into the wetland/lowland environment. As mentioned above, five major industries (textile, dyeing, paper and pulp, tannery and paint and dye manufacture) are responsible for the presence of dye effluents in the environment (waterways), of which, the textile industry alone accounts for 54% (Katheresan et al 2018).

11.3.1. Detection of dye-residues in the aquatic environments

Spills, accidental and often intentional discharge of dyes from industries, effluent (treated or untreated) from sewage treatment plants are some of the path through which dyes can enter the aquatic environments. Most of these dyes escape conventional wastewater treatment processes and remain in the environment, often reaching lakes, rivers, holding ponds and contaminating the waterways). A large variety of dyes can be found in real effluents. It has been estimated that 9% (O'Neill et al 1999) to 15% (Ismail et al 2019) of the total amount of dyestuffs produced in the world are discharged or lost with the textile wastewaters (O'Neill et al 1999, Ismail et al 2019). On the other hand, during the dyeing process, the loss of colourants to the environment can be 10 to 50% (reviewed by Chequer et al 2013) Dyes pose a potential environmental risk, particularly to aquatic ecosystems (refer to section 11.3.2) and they can be found at ppb (parts per billion or μg/L) levels in treated sewage effluent, rivers and sediments. Since **azo dyes** constitute the largest group of synthetic colourants used in industries, hence, they can be the most common synthetic dyes found in the environment (Zhao and Hardin 2007, Ali 2010). Only a few studies were carried out to assess the level of dyes in waterways. In a study conducted on the Yamaska River of Quebec, Canada, Maguire and Tkacz (1991) detected 15 dyes in water, sediment and suspended solid samples, and reported Disperse Blue 26 (azo dye), Disperse Blue 79 (azo dye) and Disperse Red 60 (anthraquinone dye) concentrations ranging from 3 μg/L to 17 μg/L. In Brazil, dispersed dyes have also been detected, such as Disperse Red 1 (azo dye), Disperse Blue 373 and Disperse Violet 93, in samples from the Ribeirão dos Cristais and Piracicaba rivers, with dye concentrations ranging from 0.002 μg/L to 3.5 μg/L (Carneiro et al 2010; Zocolo et al 2014).

Fig. 11.4. Examples of Bangladesh's rivers which are dying due to dyeing

*[**note:** Bangladesh's textile and apparel industry represents over 83% of the country's total exports and more than 45% of employment. Textile manufacturing in Bangladesh generating as much as 300 metric tons of wastewater per every ton of fabric, with a host of harmful chemicals. In Bangladesh's capital, Dhaka, the government has declared five rivers (the Balu, Turag, Bangshi, Buriganga, and Shitalakkhya, rivers as "biologically dead" due to dyestuff mixed effluent from the surrounding garment/ apparel factories with near to 0 mg/L dissolved oxygen (DO) concentration in it (it could be hypoxic: DO <2 mg/L or anoxic: DO: 0 mg/L conditions which is unsuitable for aquatic biota to survive). Bangladesh has grown into a cheap clothing manufacturing hub of the world. But poor regulation of the industry is taking a greater toll on its rivers and the health of the poor who live nearby or depends on rivers for agriculture, irrigation, fisheries, natural food, drinking, bathing, recreation and water transport (boating) etc. The government categorises textile dyeing industries as **"Red industries" (most polluting)** under Bangladesh Environment Conservation Act, 1995 and the Environment Conservation Rules, 1997 and made use of Effluent Treatment Plants (ETPs) mandatory for the factories; https://www.iamrenew.com/environment/bangladeshs-polluted-waters-rivers-dying-due-to-dyeing/#:~:text=In%20Bangladesh's%20capital%2C%20Dhaka%2C%20the, effluent%20from%20surrounding%20garment%20factories.&text=In%202021%2C%20textile%20industries%20in,million%20metric%20tons%20of%20fabrics].*

11.3.2. Impacts on water quality

The effluents discharged from textile factories are a mixture of dyes, other pollutants and metals. In fact, the dye effluents are generally high in **colour** (0.06-2500 Pt-CO (Platinum-Cobalt Scale), Table 11.4; Fig. 11.4), pH (4.15-12.8; Table 11.4), suspended solids (TSS) (15-8000 mg/L; Table 11.4), dissolved solids (TDS) (1120-23125 mg/L; Table 11.4), chemical oxygen demand (COD) (100-30000 mg/L; Table 11.4), biochemical oxygen demand (BOD) (9.8-6000 mg/L; Table 11.4), and temperature (21-62°C) (Table 11.4). A number of **heavy metals** have been detected in textile effluents including i. cadmium (Cd) (0.01-42 mg/L; Table 11.5), ii. cobalt (Co) (0.02-0.06 mg/L; Table 11.5), iii. chromium (Cr) (BDL-2.7 mg/L; Table 11.5, *note: BDL=below detection limit*), iv. copper (Cu) (0.09-<10 mg/L; Table 11.5), v. iron (Fe) (0.02-39 mg/L; Table 11.5), vi. manganese (Mn) (0.04-<10 mg/L; Table 11.5), vii. nickel (Ni) (0.013-<10 mg/L; Table 11.5), viii. lead (Pb) (BDL-2.07 mg/L Table 11.5) and ix. Zinc (Zn) (0.11-8.76 mg/L; Table 11.5). Heavy metals often used in dye fixatives and also in dyes (Mirghani et al 2008). According to Khan and Malik 2014, metals can be present in dyes for two reasons: first, metals are used as catalysts during the manufacture of some dyes and therefore, can be present as impurities; second, in some dyes, the metal is chelated[32] (bonds) with the dye molecule, forming an integral structural element (Khan and Malik 2014). Metal-complex dyes are used to improve light and wash fastness of dyed fabrics. The metals used most frequently in metal–complex dyes are Cu, Cr, Co, and Ni (Hao et al 2000, Gregory 2007). Cr (CrVi-hexavalent chromium) and Ni are carcinogenic (IARC group 1), and Co is possibly carcinogenic (IARC group 2B) (see Chapter 3, Table 3.7 for chemistry, toxicity and ecotoxicity of heavy metals).

11.3.3. Impacts on the aquatic environment and aquatic biota (fish, crab, algae)

The effluents discharged from textile factories (textile industries are the most users of synthetic dyes) are high in colour, pH, suspended solids, dissolved solids, biochemical oxygen demand and chemical oxygen demand, temperature and salts (Table 11.4). They are also with elevated levels of certain heavy metals (Cd, Co, Cr, Cu, Fe, Mn, Ni, Pb, Zn; Table 11.5) (David and Rajan 2014, Yaseen and Scholz 2019). **Colour** (Fig. 11.4) is the first contaminant. A very small amount of dyes (< 1 mg/L) in the water could cause a highly visible change in colour (Banat et al 1996). Water-soluble azo dyes at 10-50 mg/L concentration can cause waste streams to become highly coloured (Bae and Freeman 2007).

Coloured wastewater as effluents would affect the aesthetic merit and transparency of the receiving surface waterways (rivers, lakes, estuaries) and thus may impact recreational and water-based tourism industries. It can also affect the aquatic ecosystem by decreasing the passage of light penetration (i.e. reduce euphotic zone[33]) and photosynthetic activity causing oxygen deficiency in lakes, rivers and other bodies of water. Fish kills can occur in water bodies with low dissolved oxygen concentrations (hypoxic or anoxic conditions) resulting from the discharge of textile effluents. Textile effluents and textile dyes cause significant effects on fish, crab and algae and there is a risk of disruption of natural seafood production and primary production, some examples of which are listed below:

- Textile dyeing effluent caused an *abnormal behaviour* in **fish**, *Poecilia reticulata* (e.g. erratic swimming, hyper-excitation, rapid opercular movement and thick mucus, and *histopathological changes* in *P. reticulata* (e.g. enlargement of primary gill bar and detachment of secondary gill bar, disintegration of intestinal villi and infiltration of haemocytes into the lumen)) (Selvaraj et al 2015)

---

[32] ***Chelation*** *is a compound having a ring structure that usually contains a metal ion held by coordinate bonds (to combine with (a metal) so as to form a chelate ring) (https://www.merriam-webster.com/dictionary/chelating).*

[33] ***Euphotic zone*** *= The euphotic zone is the layer closer to the surface that receives enough light for photosynthesis to occur*

- Textile dye effluent reduced the nutritive value of freshwater female **crab,** *Spiralothelphusa hydrodroma* (loss of nutritive value of protein, carbohydrate and lipid in ovary, spermatheca, hepatopancreas, muscle, gills, brain, thoracic ganglia and eyestalk) (*S hydrodroma* is an important human food) (Sekar et al 2009).
- Dye effluent has strongly affected the rate of feeding, absorption and food conversion in Gangetic catla **fish** (*Labeo catla*) (Gita et al 2017) (*L. catla* is an important human food) (Sekar et al 2009).
- The textile effluents have caused an adverse effect on the ionic regulations in liver, kidney and muscle in proteinous edible freshwater **fish**, *Mastacembelus armatus* (a decrease of Na^+, Cl^- concentrations and an increase of K^+, Ca^{2+}, Mg^{2+} concentrations) (Karthieyan et al 2006) (*M. armatus* is an important human food).
- Disperse dyes (Disperse Red 60, Disperse Red 73 and Disperse Red 78) induced biochemical alterations in embryos of zebrafish (*Danio rerio*) at environmentally realistic concentrations (Meireles et al 2018).
- The exposure of textile dyes (Optilan Yellow, Drimarene Blue and Lanasyn Brown) inhibited the growth, pigmentation and elemental composition of the algal cells (freshwater green **algae**, *Chlorella vulgaris*) (Gita et al 2019). Therefore, the textile dyes could impact on primary production in the aquatic environment.
- Exposure of indigo dye effluent significantly reduced the growth, biomass production and the number of cells per coenobium (colony) in freshwater **microalga**, *Scenedesmus quadricauda* (Chia and Musa 2014).

Heavy metals (HMs) including As, Cd, Co, Cr, Cu, Fe, Hg, Mn, Ni, Pb and Zn present in the effluent (Table 11.5) may be toxic and could cause lethal and sub-lethal effects on aquatic organisms (fish) (Chapter 3 of this book has discussed in detail about the HMs effect on aquatic biota, therefore is not repeated further in this chapter). Textile dyes can accumulate in aquatic biota. For example, Malachite Green bioaccumulated in eggs, fry and adult muscle tissue of rainbow trout *Oncorhynchus mykiss* (Meinertz et al 1995). Belpaire et al (2015) also reported that textile dyes (malachite green, crystal violet, brilliant Green) were accumulated in wild European eel, *Anguilla* spp. in their muscle tissues sampled from the Belgian rivers, canals and lakes. Dye contaminated fish can be unfit for human consumption. Dyes are also linked to endocrine disruption to organisms, for example, diazo dyes (Bismarck Brown Y) caused increased developmental defects in frog embryogenesis and early larval development in the Western clawed frog (Sorian et al 2014). Moreover, the dye effluent can contaminate the surface and groundwater, therefore, such water may not be suitable for beneficial uses such as for irrigation and drinking (see Fig. 11.5. for possible impacts of textile effluents on the environment).

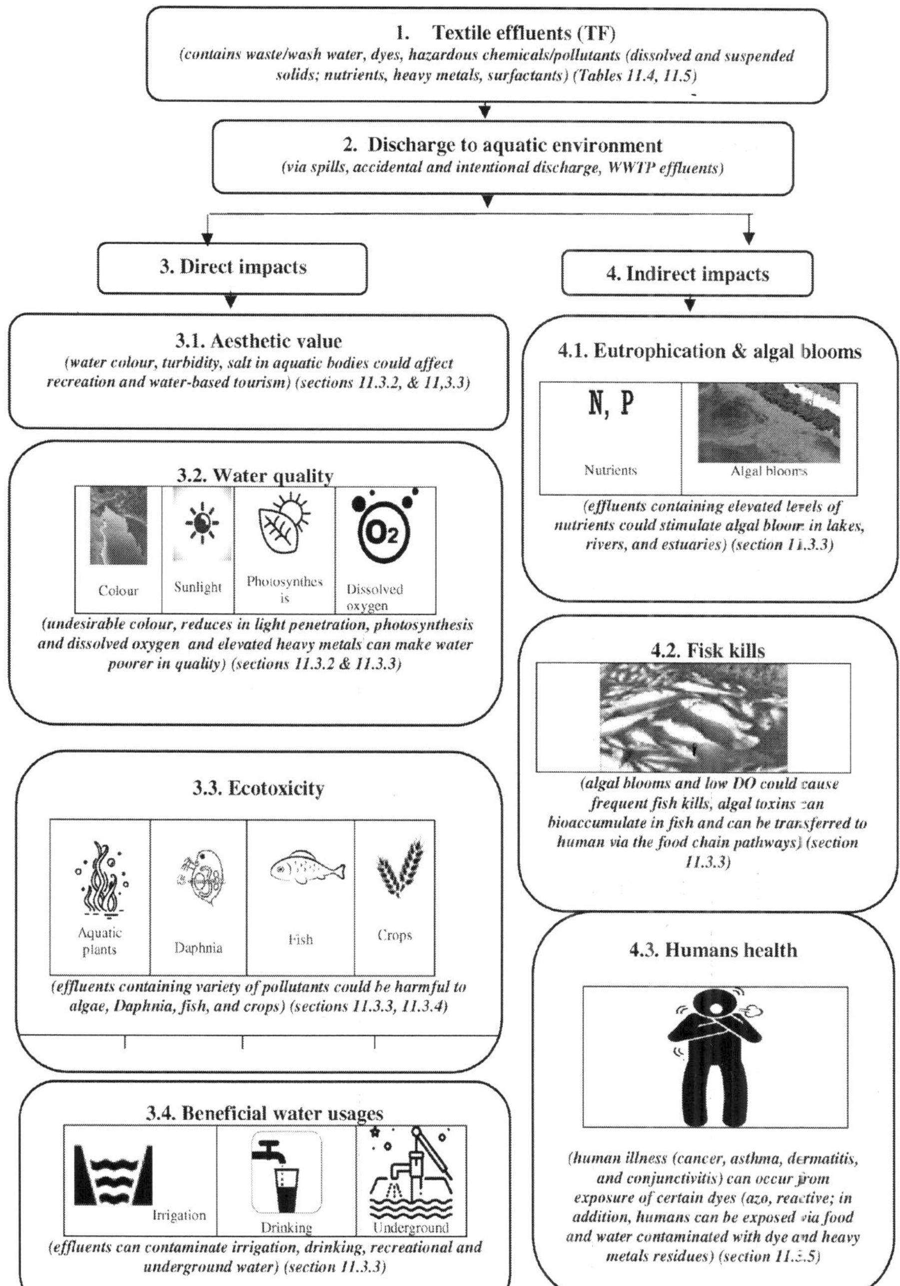

Fig. 11.5. Schematic representation of the possible direct and indirect impacts of textile effluents discharged into the environment.

11.3.4. Impacts on crops and vegetables

Irrigation of agricultural lands with effluents contaminated water can severely affect soil fertility and plant growth (Puvaneswari et al 2006). The soils contaminated by the azo dyes affected the germination, survival and growth of the plants of Broad bean (*Vicia faba*). The root-length, shoot-length and the number of leaves were reported to be severely affected with the increase of azo dye concentration in the soil (Rao et al 2019). Effluents from viscose (rayon) reduced the rate of seed germination and growth of maize crop plants (*Zea mays*) (Nirmalarani and Janardhanan 1988). The germination and growth of kidney bean (*Phaseolus aureus*) and lady's finger (*Abelmoschus esculentus*) crops inhibited and delayed when exposed to textile factory effluent (Ajmal and Khan 1985). Khan et al (2011) evaluated the impact of textile factory effluents at 0, 10, 25, 50 and 100% concentration on germination, chlorophyll contents, and root development in three leguminous crops, pea (*Pisum sativum*), lentil (*Lens esculentum*) and Gram (*Cicer arietinum*). The research exhibited a substantial reduction in total germination, root and shoot dry matter, number of root branches/plant, and chlorophyll contents when grown with a higher concentration (50 and 100% concentration) of textile effluents. The study reveals that the effect of textile effluent was crop-specific depending on the concentration of effluent and stage of growth.

11.3.5. Impacts on human's health

People can be exposed to dyes via contaminated water and food. In addition, workers who work at dye manufacturing factories or textile-producing factories can be exposed at their place of work via dermal absorption. The contamination of groundwater with dye-containing effluents can expose the general population via the oral route. Moreover, some synthetic dyes/textile dyes are not only highly toxic but also carcinogenic and mutagenic such as **azo** dyes (Gita et al. 2019, Lellis et al 2019) (see Table 11.3 for the list of dyes classified as potential carcinogens). Azo dyes can be absorbed after skin exposure, and such dermal exposure to Azo dyes can occur as an occupational hazard or from the use of cosmetic products (reviewed by Chequer et al 2011). Cationic dyes, along with benzidine (Azo dyes) were implicated in the high incidence of urinary bladder cancer[34] in the textile industry between 1930 and 1960 (Gregory 2007). Three Azo dyes that metabolised to benzidine are Direct Black 38, Direct Blue 6, and Direct Brown 95 (*https://www.ncbi.nlm.nih.gov/books/NBK304402/*). Some

[34] ***Bladder cancer*** *is when abnormal cells in the bladder grow and divide in an uncontrolled way. There are different types of bladder cancer: urothelial carcinoma is the most common form of bladder cancer (80-90%) (https://www.cancer.org.au/about-cancer/types-of-cancer/bladder-cancer.html)*

reactive dyes and disperse dyes are well- known skin sensitisers[35]. Breathing in respiratory sensitisers can cause occupational asthma *http://www.hse.gov.uk/textiles/dyes-dyeing.htm.* Some examples of reactive dyes which are classified as respiratory/ skin sensitisers are Reactive Yellow 25, Reactive Yellow 39, Reactive Orange 14, Reactive Red 65, Reactive Red 66, Reactive Orange 16, Reactive Violet 33, Reactive Blue 114; Gregory 2007). Reactive dyes can also cause problems in factory workers who manufacture the dyes as well as textile workers who handle the dyes in the dyeing process. Some reactive dyes caused contact dermatitis[36], allergic conjunctivitis[37], rhinitis[38], occupational asthma[39] or other allergic reactions to the workers who produce or handle reactive dyes (Hunger 2003). Certain disperse dyes have been implicated in causing allergic reactions, particularly when they are used for skin-tight, **close-fitting** clothes made from synthetic fibres (Gregory 2007).

11.4. Management of synthetic dyes and textile effluents

To reduce the risks and impacts of synthetic dyes/textile effluents on the environment and public health, the following approaches, measures and strategies can be implemented

11.4.1. Monitoring of effluent: The textile effluent quality at the receiving water bodies should be regularly monitored for water colour (Box 11.1), BOD, COD, TSS, TDS, temperature, nutrients (N, P) and heavy metals levels to safeguard aquatic biota, aquatic ecosystems, irrigation, and drinking water and human health and compared with the guideline/standard discharge limits (see Tables 11.6, 11.7). Methods to measure the colour of polluted water is briefly described in Box 11.1.

---

[35] ***Sensitisers*** *= They are substances which can cause sensitisation in the lungs (**respiratory** sensitisers) and/or skin (**skin** sensitisers) after exposure to them. A respiratory sensitiser is a substance which when inhaled can trigger an irreversible allergic reaction in the respiratory system. Once sensitised, individuals can have severe reactions to further exposure, even in small amounts. In the most extreme cases, individuals may develop asthma and/or dermatitis respectively https://www.ed.ac.uk/health-safety/guidance/hazardous-substances/sensitisers*

[36] ***Dermatitis*** *= a medical condition in which the **skin** becomes red, swollen, and sore, sometimes with small blisters, resulting from direct irritation of the skin (itchy) by an external agent or an allergic reaction to it. Examples of this condition are atopic dermatitis (eczema), dandruff (flakes of skin) and contact dermatitis (https://www.mayoclinic.org/diseases-conditions/dermatitis-eczema/symptoms-causes/syc-20352380).*

[37] ***Conjunctivitis,*** *also known as **pink eye**, is inflammation of the outermost layer of the white part of the eye and the inner surface of the eyelid. It makes the eye appear pink or reddish. Pain, burning, scratchiness, or itchiness may occur. The affected eye may have increased tears or be "stuck shut" in the morning*

[38] ***Rhinitis*** *= is irritation and inflammation of the mucous membrane inside the **nose**. Common symptoms are a stuffy nose, runny nose, sneezing, and post-nasal drip.*

[39] ***Asthma*** *= A condition in which a person's airways become inflamed, narrow and swell and produce extra mucus, which makes it difficult to breathe.*

Table 11.4. Physico-chemical characteristic of textile effluents*

[Yaseen and Scholz 2019; *https://www.bsr.org/reports/awqwg/BSR_AWQWG_Guidelines-Testing-Standards.pdf*; ***BOD*** *= Biological/biochemical oxygen demand (BOD);* ***COD*** *= Chemical oxygen demand;* ***EC*** *= Electrical conductivity;* ***na****= data is not available;* NO_3–N *= Nitrate–nitrogen;* Na_2CO_3 *= Sodium carbonate;* ***TDS*** *= Total dissolved solids;* ***TKN****= Total kjeldahl nitrogen;* ***TSS****= Total suspended solids;* ***Typical*** *= representative or combined samples; *see Table 11.6 for textile effluent discharge guidelines/ limits for physicochemical parameters].*

Parameters	1 Typical	2 Typical	3 Typical	4 Typical	5 Typical	5 Typical	7 Bangladesh	8 China	9 Greece	10 India	11 India	12 Malaysia	13 Mexico	14 Italy	15 Thailand	Remarks (ranges)
	Ghaly et al 2014	Hussein 2013	Upadhye and Joshi 2012	Kalra et al 2011	Al-Kdasi et al 2005, Turhan and Turgut 2009	Eswar-amoorthi et al 2008	Nahar et al 2018	Sun et al 2014	Avlonitis et al 2008	Bhuv-anes-wari et al 2016	Mane-kar et al 2014	Kehinde and Aziz 2014	Almazán-sánchez et al 2016	Tomei et al 2016	Nopkhu-ntod et al 202	
BOD[‡] (mg/L)	80-6000	80–6000	100–4000	100–4000	80–6000	300–500	na	na	na	1206-1750	368-458	80–6000	91.91	9.8	25	**BOD:** 9.8-6,000 mg/L
Chlorides (Cl^-) (mg/L)	na	200–6000	1000–6000	1000–6000	na	na	na	na	3000–6000	4005-4320	384-452	na	338.19	38.6	na	**Cl⁻:** 38.6-6,000 mg/L
Chlorine (Cl) (mg/L)	1000-6000	na	na	na	1000-1600	na	na	na	na	na	na	na	na	na	na	Cl: 1000-6,000 mg/L
COD[§]	150–	150–	150–	150–	150–	1000-	na	1114-	100	3880-	752-	150–	344	1017	750	**COD:** 100-

[‡] ***BOD (Biochemical oxygen demand)*** *is the amount of dissolved oxygen needed (i.e. demanded) by microorganisms (bacteria) to break down organic material present in each water sample/ wastewater at a certain temperature (20°C) over a specific time period (5 days). It is the most widely used parameter applied to wastewater (https://www.onsiteinstaller.com/online_exclusives/2017/08/the_installers_guide_to_bod5).*

[§]***COD (Chemical Oxygen Demand)*** *is the total measurement of all chemicals (organics & in-organics) in the water/wastewater.*

Parameters	1 Typical	2 Typical	3 Typical	4 Typical	5 Typical	5 Typical	7 Bangladesh	8 China	9 Greece	10 India	11 India	12 Malaysia	13 Mexico	14 Italy	15 Thailand	Remarks (ranges)
	Ghaly et al 2014	**Hussein 2013**	**Upadhye and Joshi 2012**	**Kalra et al 2011**	**Al-Kdasi et al 2005, Turhan and Turgut 2009**	**Eswar-amoorthi et al 2008**	**Nahar et al 2018**	**Sun et al 2014**	**Avlonitis et al 2008**	**Bhuv-anes-wari et al 2016**	**Mane-kar et al 2014**	**Kehinde and Aziz 2014**	**Almazán-sánchez et al 2016**	**Tomei et al 2016**	**Nopkhu-ntod et al 202**	
(mg/L)	12,000	10,000	10,000	10,000	12,000	1500		1350		4400	1120	30,000				30,000 mg/L
Colour (Pt-Co)**	50–2500	na	50–2500	50–2500	50–2500	50–2500	na	na	na		25-260	50–2500	330	0.66	420	**Colour**: 0.06-2500 Pt-Co
EC†† (µS/cm)	na	na	na	na	na	na	388.6-5405	na	1000	na	na	na	471.3	na	na	**EC:** 388-5405 µS/cm
Oil & grease (mg/L)	10-30	10-50	na	na	na	10-30	na	na	na	na	na	5-5.5	na	na	na	**Oil & grease**: 5-50 mg/L
pH	6–10	5.5–10.5	6–10	6–10	7–9	6–10	4.15-12.79	7.4-7.9	na	8.6-9.2	7.8-7.9	6.95–11.8	6.84	9	na	**pH:** 4.15-12.8
Phosphate (PO_4^{3-})	<10	na	na	na	na	na	na	na	na	72.8-86.8	BDL-3	na	287.08	3.2	na	**PO_4^{3-}:** 3.2-287 mg/L

****PT-Co (Platinum-Cobalt Scale)** *or* ***APHA-Hazen*** *Scale) is a colour scale that was introduced in 1892 by chemist Allen Hazen. The index was developed to evaluate pollution levels in wastewater. APHA colour, also referred to as the Hazen scale (https://en.wikipedia.org/wiki/APHA_color).*

††***EC (Electrical conductivity*** *is a measure of the* ***saltiness*** *of the water and is measured on a scale from 0 to 50,000 uS/cm. Electrical conductivity is measured in micro siemens (uS/cm) per centimetre (uS/cm).* ***Freshwater*** *is usually between 0 and 1,500 uS/cm and typical* ***seawater*** *have a conductivity value of about 50,000 uS/cm (https://mrccc.org.au/wp-content/uploads/2013/10/Water-Quality-Salinity-Standards.pdf).*

Parameters	1 Typical	2 Typical	3 Typical	4 Typical	5 Typical	5 Typical	7 Bangladesh	8 China	9 Greece	10 India	11 India	12 Malaysia	13 Mexico	14 Italy	15 Thailand	Remarks (ranges)
	Ghaly et al 2014	**Hussein 2013**	**Upadhye and Joshi 2012**	**Kalra et al 2011**	**Al-Kdasi et al 2005, Turhan and Turgut 2009**	**Eswar-amoorthi et al 2008**	**Nahar et al 2018**	**Sun et al 2014**	**Avlonitis et al 2008**	**Bhuv-anes-wari et al 2016**	**Mane-kar et al 2014**	**Kehinde and Aziz 2014**	**Almazán-sánchez et al 2016**	**Tomei et al 2016**	**Nopkhu-ntod et al 202**	
(mg/L)																
Sulphate (SO_4^{2-}) (mg/L)	600-1000	500-700	na	na	na	600-1000	na	na	na	2050-2250	320-380	na	227.06	4.5	na	**SO_4^{2-}:** 4.5-2250 mg/L
TDS (mg/L)	2900-3100	1500-6000	1800-6000	1800-6000	2900-3100	8000-12,000	1120-23,125	na	na	2070-2200	1670-2040	2900-3100	na	na	4800	**TDS:** 1120-23,125 mg/L
Tempera-ture (T) (°C)	35-45	33-45	35-45	35-45	na	35-45	na	na	na	28-29	43-46	21-62	28-30	na	na	**T:** 21-62°C
TKN (mg/L)	70-80	70-80	70-80	70-80	70-80	na	na	na	na	na	na	70-80	na	na	na	**TKN:** 70-80 mg/L
TSS‡‡ (mg/L)	15-8000	100-5000	100-5000	100-5000	15-8000	200-400	na	na	na	550-650	120-180	15-8000	na	na	90	**TSS:** 15-8,000 mg/L
Turbidity (NTU)	na	na	na	na	na	na	na	125-137	na	12.5-16.6	na	na	104.66	na	na	**Turbidity:** 12.5-137 NTU

‡‡***Total suspended solids (TSS)*** *are particles that are larger than* **2 microns** *found in the water column. Anything smaller than 2 microns (average filter size) is considered a* ***dissolved solid.*** *The high levels of total suspended solids will increase water temperatures and decrease dissolved oxygen (DO) levels (https://www.fondriest.com/environmental-measurements/parameters/water-quality/turbidity-total-suspended-solids-water-clarity/).*

Table 11.5. Heavy metals (HMs) in textile effluents* (mg/L)

*[**BDL**= below detection limit; **na** = data is not available; **Typical** = representative or combined samples; (*see Table 11.7 for textile effluent discharge guideline limits for HMs; Note: **typical sources of HMs in textile effluents**: As: fibres, incoming water, fugitive, treated timber; Cd: impurity in salt; Co: dyes, Cr- dyes, laboratory; Cu: dyes, incoming water, fibre; Hg: dye/commodity chemical impurities; Mn: permanganate strip; Ni: dyes; Pb: dyes, plumbing, shop; Zn: dyes, impurities in commodity, chemicals, incoming water* (Ismail et al 2019).

Heavy metals	1 Typical	2 Typical	3 Typical	4 Typical	5 Typical	6 Typical	7 Bangladesh	8 India	9 India	10 India	11 India	12 Malaysia	13 Nigeria	14 Pakistan	15 Spain	Remarks (HMs ranges)
	Ghaly et al 2014	Hussein 2013	Upadhye and Joshi 2012	Kalra et al 2011	Al-Kdasi et al 2005, Turhan and Turgut 2009	Eswar-amoorthi - et al 2008	Nahar et al 2018	Rastogi and Singh 2012	Bhard-waj et al 2014	Poorn-ima et al 2011	Manekar et al 2014	Kehinde and Aziz 2014	Odipe et al 2019	Imtia-zuddin et al 2012	Aouni et al 2012	
As	<10	na	na	na	na	<10	na	na	na	na	na	na	na	na	na	**As:** <10 mg/L
Cd	na	na	na	na	na	na	na	0.014-0.018	0.01	na	0.01-0.02	na	0.01-0.41	na	42	**Cd:** 0.01- 42 mg/L
Co	na	na	na	na	na	na	na	na	na	na	0.02-0.06	na	na	na	na	**Co:** 0.02- 0.06 mg/L
Cr	na	2-5	na	na	na	na	BDL ~2.15	0.01-0.09	0.11-0.21	0.04-2.7	0.03-0.07	na	na	1.05-1.86	na	**Cr:** BDL- 2.7 mg/L
Cu	<10	2-6	na	na	na	<10	0.09 ~21	0.10-0.21	0.17-0.28	na	0.03-0.11	na	na	0.07-5.14	na	**Cu**: 0.09- 21 mg/L
Fe	<10	na	na	na	na	<10	0.02--13.9	0.54-2.2	0.39-0.90	10.4-39	0.7-8.25	na	0.0025-0.96	1.08-3.11	na	**Fe**: 0.002- 39 mg/L

Heavy metals	1 Typical	2 Typical	3 Typical	4 Typical	5 Typical	6 Typical	7 Bangladesh	8 India	9 India	10 India	11 India	12 Malaysia	13 Nigeria	14 Pakistan	15 Spain	Remarks (HMs ranges)
	Ghaly et al 2014	Hussein 2013	Upadhye and Joshi 2012	Kalra et al 2011	Al-Kdasi et al 2005, Turhan and Turgut 2009	Eswar-amoorthi - et al 2008	Nahar et al 2018	Rastogi and Singh 2012	Bhard-waj et al 2014	Poorn-ima et al 2011	Manekar et al 2014	Kehinde and Aziz 2014	Odipe et al 2019	Imtia-zuddin et al 2012	Aouni et al 2012	
Hg	<10	na	na	na	na	<10	na	na	na	na	na	na	na	na	na	**Hg:** <10 - <10 mg/L
Mn	<10	na	na	na	na	<10	0.10~ 2.46	0.04-0.13	na	0.46-0.97	na	na	na	0.88-1.85	na	**Mn:** 0.04- <10 mg/L
Na	7000	400-2175	610-2175	610-2175	na	7000	na	na	na	1121-2015	na	na	na	na	1008	**Na**: 400- 7000 mg/L
Ni	<10	na	na	na	na	<10	na	0.78-1.55	0.11-0.22	0.013-0.98	0.07-0.3	na	na	0.06-1.53	na	**Ni:** 0.013- <10 mg/L
Pb	na	na	na	na	na	na	BDL ~ 0.44	0.51-1.26	0.02-0.10	1.01-2.07	0.1-0.13	na	0.0007-0.36	0.16-0.35	na	**Pb:** BDL- 2.07 mg/L
Zn	<10	3-6	na	na	na	<10	BDL ~ 8.76	0.29-1.24	0.11-0.51	0.19-2.5	0.01-0.3	na	0.51-1.07	2.36-6.03	na	**Zn**: BDL-0.11-8.76 mg/L

Box 11.1. Methods to measure the colour of polluted water.

The natural waters can be coloured due to the presence of *dissolved organic matter* (e.g. humic and fulvic acids originate from the soil and decaying **vegetable matter-tannins**) or *waste discharge from dyeing* operations (e.g. **textile** and **paper** manufacture); the *dissolution of* ***metals in pipes***; or badly *corroded* ***iron pipes*** (produce a brownish colour); or the corrosion of ***copper pipes*** (produce a blue-green colouration on sanitary ware and a faint blue colour in water in extreme cases); or the ***oxidation of iron*** (bore water having 'red water') or algal blooms (blue-green algae) (*file:///C:/Users/User1/Downloads/Colour_Facts_Sheet%20(3).pdf,* Kibria et al 2010). There are two recognised approaches to measure the colour of polluted water: ***Visual Comparison method*** *and* ***Spectrophotometric method.***

i. ***Visual Comparison methods*** – Here, the colour of a sample is determined by comparison to a standard colour. The standard is typically either a colour wheel or a set of calibrated concentrations of a known substance. Potassium chloroplatinate, K_2PtCl_6 and cobaltous chloride, $CoCl_26H_2O$ can be used to make the colour standards. **Filtration is not recommended** because it may remove some colour. Since colour is highly dependent on the pH of the sample, therefore, it is essential to specify the **pH** at which the colour was determined. For further information please refer to the Standard Methods for the Examination of Water and Wastewater, Section 2120B for Visual Comparison Method (AWWA 2014, Baird et al 2017).

ii. ***Spectrophotometric method*** – This method is applicable to **true colour** measurements. As the turbidity interferes with the analysis, therefore **filtration of sample is required**. A spectrophotometer is used to measure how much light is transmitted through the sample at several wavelengths of light. Potassium chloroplatinate and cobaltous chloride are the two reagents used to make the colour standards. Since colour is highly dependent on the pH of the sample, therefore, here also it is essential to specify the pH at which the colour was determined. However, pH adjustment is not required if the sample **pH** is between 4 and 10. Colour values obtained using a spectrophotometer are dependent on the **wavelength** used for the measurement (values ranging from 395 nm to 465 nm are generally used). For further information please refer to Standard Methods for the Examination of Water and Wastewater, Section 2120 C. Spectrophotometric-Single-Wavelength Method; Section 2120 D. Spectrophotometric-Multi-Wavelength Method; Section 2120 E. Tristimulus Spectrophotometric Method; and 2120 F. ADMI Weighted-Ordinate Spectrophotometric Method (AWWA 2014: Baird et al 2017, *http://arco-hvac.ir/wp-content/uploads/2018/04/AWWA_M12_Simplified_Procedures_for.pdf*).

11.4.2. Discharge of untreated effluents*:* The direct discharge of untreated textile effluents into rivers, canals, estuaries should be stopped, discouraged and be penalised via implementing "polluters to pay principle (P2P)", (i.e. the party responsible for producing pollution are responsible for paying for the damage done to the natural environment, natural assets and sensitive biodiversity and also pay for remedial measures).

11.4.3. Mandatory effluent treatment: The respective government should implement laws, acts and regulations that it is mandatory for all textile and apparel industries to treat effluent water before discharging to the environment. A wide range of technologies has been developed and now available for the removal of synthetic dyes from waters and wastewaters to decrease/ reduce their environmental impacts. Some of these technologies are highlighted in Box 11.2 (for more information, please consult Chequer et al 2011, Perreira and Alves 2012, Katheresan et al 2018):

Box 11.2. Various technologies for the removal of synthetic dyes.

i. Physical methods are straight forward and commonly used but could be very costly. It includes *membrane-filtration* (here, dye wastewater is passed through a membrane which separates dye particles from clean water); *coagulation and flocculation* (here, coagulation/flocculation inducing agents are added to dye wastewater to allow dye particles to clump together, clumps are then removed through filtration); *nano-filtration* and *ultra-filtration* (here, dye wastewater is passed through a thin-pored membrane which separates dye particles from clean water); *reverse osmosis* (it is a pressure-driven system where water is passed through an extremely thin membrane leaving contaminants on one side and water on the other side); and sorption techniques (activated carbon[40], biosorption by microorganisms (algae, yeasts, the bacterium, fungi to discolour dyes); chitin/ chitosan-based absorbents) (Perreira and Alves 2012).

ii. Biological methods are cheaper, commonly and extensively used. It includes *adsorption by microbial biomass* (here living organisms are allowed to adsorb dye molecules); *algal degradation* (here algae consumes dye particle for self-growth); *enzyme degradation* (extracted enzymes are used to degrade dye molecules); *fungal culture* (fungus are used to break down dye molecules while they consume dyes for their growth); *microbial cultures* (here, bacteria are mixed with chemicals or other bacteria to remove dye particles); *pure-* and *mixed-culture* (in this process, a mixture of algae, bacteria or fungus is grown with necessary chemicals to remove dye) (Perreira and Alves 2012, Katheresan et al 2018).

iii. Chemical methods are generally costly. It includes *oxidation* (here, oxidising agents are used to treat dye effluents where oxidising agents break down the complex dye molecules to carbon dioxide and water, usage of catalyst can further enhance the process and cost); *advanced oxidation process* (here, multiple

[40]***Activated carbon*** *is used to purify liquids and gases in a variety of applications, including municipal drinking water, food and beverage processing, odour removal, industrial pollution control. Activated carbon is produced from carbonaceous source materials, such as coconuts, nutshells, coal, peat and wood. The primary raw material used for activated carbon is any organic material with a high carbon content. Carbon adsorption helps in removing pollutants from air or water streams. Active carbon comes in two variations: Powder Activated Carbon (PAC) and Granular Activated Carbon (GAC). The GAC version is mostly used in water treatment (https://www.lenntech.com/products/activated-carbon.htm).*

oxidation process is done simultaneously to remove dye particles); *fenton reaction/ reagent* (here a mixture of catalyst and hydrogen peroxide is used to remove dye particles from wastewater); *ultraviolet irradiation* (here, UV light are used to decompose dye particles in wastewater); *ozonation (here* ozone is used to eliminate dye particles) (Katheresan et al 2018).

iv. ***Other methods*** - Various agricultural products can be used to remove dyes from aqueous solutions. These include cotton waste, rice husk, bark, sugar industry mud, palm-fruit bunch, Jack fruit peel, wood, orange peel, sugar cane dust, peat, and pith. Phytoremediation by plants is another cheaper alternative. Plants can accumulate organic pollutants by acting as filters. Phytoremediation involves growing plants in a contaminated matrix for a required growth period to remove contaminants from the matrix. The plants can be subsequently harvested, processed, and disposed of in an eco-friendly manner. *Lemna minor* (Duckweed) is an aquatic plant which often forms dense floating mats in eutrophic ditches and ponds and is a promising plant for the bioaccumulation of basic dyes from an aqueous solution (Reema et al 2011).

***11.4.4. Low-impact dyes*:** Research should be prioritised to develop dyes which do not contain toxic chemicals like HMs (heavy metals), do not require chemical mordants to fix them to the fibre, require less rinsing and have a high absorption rate in the fabric (~70%) (Choudhury 2018) to reduce impacts on the environment from the discharge of textile effluents. Low-impact dyes that meet the above criteria are classified by the Oeko-Tex[41] Standard 100 as eco-friendly. High absorption rates and a decreased use of rinse water create less wastewater and less pollution *(https://www.cattermoleconsulting.com/explaining-low-impact-dyes/).*

***11.4.5. Preventing exposure of dyes*:** In order to minimise the risk from reactive dyes (which can cause skin irritation skin sensitisation), exposure to dye dust should be avoided. This may be achieved by using liquid dyes, low dusting formulations and by using the appropriate personal protective equipment.

11.4.6. Alternative to azo dyes: Since benzidine (azo dyes) can cause cancer to humans, therefore, substitutes for these dyes should be developed and be used in textile industries.

***11.4.7. Awareness and education*:** Community should be educated about the threats and risks posed by dyes, dye contaminated water and food on ecological and public health risks. A number of dyes are carcinogenic (azo dyes). Reactive, vat and disperse dyes are well-known skin sensitisers. Using textile effluent contaminated water for irrigation can be harmful and could cause lethal and sub-lethal effects to fish and crops and vegetables (as discussed in sections 11.3.3. and 11.3.4).

---

[41]***Oeko-Tex*** *is a registered trademark, representing the product labels and company certifications issued and other services provided by the International Association for Research and Testing in the Field of Textile and Leather Ecology (which also calls itself Oeko-Tex in short).*

Table 11.6. Textile effluent guidelines/standards (traditional water quality) discharge limits for some selected countries *[**na**= data is not available].*

	Traditional water quality effluent discharge limits				
Parameters	**Australia** (SWGWQG 2010)	**Bangladesh** (Nahar et al 2018)	**China** (Ghaly et al 2014)	**India** (Ghaly et al 2014)	**USA (EPA)** (Ghaly et al 2014)
BOD (Biochemical oxygen demand)	≤ 30 mg/L	100 mg/L (Ghaly et al 2014)	60 mg/L	30 mg/L	50 mg/L
COD (Chemical Oxygen Demand)	30 mg/L	200 mg/L (Ghaly et al 2014)	200 mg/L	250 mg/L	80 mg/L
Colour (Global colour guidance values)	≤150 ADMI[42] units (Global Effluent Guidelines 2007). *[Offensive colour is not acceptable (visual observation); (test required using one of the following methods: (1) method producing results in ADMI units or (2) using spectrophotometry, measuring transmission at wavelengths 436 nm, 525 nm and 620 nm].*				
Colour (aesthetic colour guideline)	15 Hazen colour units[43] (HU) (NHMRC and NRMMC 2011). Up to 25 HU if the turbidity is low (NHMRC and NRMMC 2011).				
Colour	300 Pt-CO (Sztarr 2008); 150 Pt-CO	na	80 Pt-CO	na	7 (Lovibond[44])

[42] ***ADMI*** *stands for the American Dye Manufacturer's Institute. The ADMI Colour Index was developed to monitor the colour of wastewater effluent as an indicator of water quality. It has been estimated that annually up to 12% of the synthetic dyes used in textile manufacturing operations can be found in dye wastewater effluent. After being processed in treatment plants, 20% of those dye losses will enter the environment.*

https://support.hach.com/app/answers/answer_view/a_id/1000980/~/what-is-the-meaning-of-admi%3F-

https://support.hunterlab.com/hc/en-us/articles/201482339-What-scale-is-used-to-measure-the-color-of-waste-water-

[43]***Hazen colour*** *units is the colour scale used for measuring water quality can also be referred to as the Pt-Co (Platinum-Cobalt) / APHA (from American Public Health Association) / HU (Hazen Units) colour scale. It ranges from 0 (clean or distilled water) to 500 (very dark, polluted water). The colour of drinking water is often evaluated before it is treated to see if there are contaminants and then after it has been filtered / treated, to measure the effectiveness of the treatment and as a quality assurance measure – ahead of the water being sent out and the risk of health problems.*

[44]***Lovibond*** *also known as degrees lovibond or abbreviated as L is a scale for measuring colour originally introduced by Joseph Lovibond in the 1860's.*

	Traditional water quality effluent discharge limits				
Parameters	**Australia** (SWGWQG 2010)	**Bangladesh** (Nahar et al 2018)	**China** (Ghaly et al 2014)	**India** (Ghaly et al 2014)	**USA (EPA)** (Ghaly et al 2014)
	(SWGWQG 2010)				
Colour (trade waste) (practiced by different water authorities in Victoria, Australia)	Should not exceed 1,000 on the Platinum cobalt scale [Gippsland water, Victoria, Australia] (Sztarr 2008). The trade waste shall have a colour not exceeding 300 ADMI units. True Colour units [Goulburn Valley Water, Victoria, Australia] (Sztarr 2008). The limit for Colour shall be 300 ADMI units within the pH range of 6.0 to 10.0 relative to distilled water (the sample of such waste must be adjusted to a pH range of 7.0 to 8.0 [North East Region Water Authority, Victoria, Australia] (Sztarr 2008). The colour is not to be exceeded, at a pH of 8.0 of 300 platinum/cobalt units: [City of Devonport, Tasmania, Australia] (Sztarr 2008).				
DO (Dissolved oxygen)	Bangladesh: 4.5-8 mg/L: Canada: 6 mg/L; Hong Kong: > 4 mg/L (Ghaly et al 2014)				
Dye effluent discharge limit	International standards for dye effluent discharge into the environment				
EC (Electrical conductivity)		1200 µS/cm	na	na	na
International standards for dye effluent discharge into the environment	Biological oxygen demand: Below 30 mg/L. Chemical oxygen demand: Below 50 mg/L. Colour: Below 1 ppm. pH: Between 6 and 9. Suspended solids: Below 20 mg/L. Temperature: Below 42°C. Toxic pollutants: Not allowed to be released. (Mojsov et al 2016, Ballav et al 2018, Katheresan et al 2018).				
Nitrate	na	100 mg/L (Ghaly et al 2014)	na	10 mg/L	20 mg/L
Oil & grease	na	10 mg/L (Ghaly et al 2014)	10 mg/L	na	10 mg/L
pH	6-9	6.5-8 (Ghaly et al 2014)	6-9	5.5-9	6-9
Phosphate	na	na	1 mg/L	5 mg/L	5 mg/L

	Traditional water quality effluent discharge limits				
Parameters	**Australia** (SWGWQG 2010)	**Bangladesh** (Nahar et al 2018)	**China** (Ghaly et al 2014)	**India** (Ghaly et al 2014)	**USA (EPA)** (Ghaly et al 2014)
TDS (Total dissolved solids)	na	2100 mg/L	na	2100 mg/L	2000 mg/L
Temperature	≤ 37 ^{O} C	40-45 ^{O} C	na	50° C	40°C
TSS (Total suspended solids)	≤ 30 mg/L	100 mg/L (Ghaly et al 2014)	150 mg/L	100 mg/L	30 mg/L

Table 11.7. Heavy metals effluent discharge limits for some selected countries *[**na**= data is not available].*

	Heavy metals water quality effluent discharge limits *(see also chapters 2, 3 for guidelines of arsenic and heavy metals for other beneficial water usages)*				
	Australia (SWGWQG 2010)	**Bangladesh** (Nahar et al 2018)	**China** (Ghaly et al 2014)	**India** (Ghaly et al 2014)	**USA (EPA)** (Ghaly et al 2014)
Arsenic (As)	≤0.01 mg/L	na	na	na	na
Cadmium (Cd)	≤0.01 mg/L	0.5 mg/L	na	na	na
Cobalt (Co)	≤0.02 mg/L	na	na	na	na
Chromium (Cr)	≤ 0.10 mg/L	0.5 mg/L	na	0.10 mg/L	<0.10 mg/L
Copper (Cu)	≤0.25 mg/L	0.5 mg/L	1 mg/L	3 mg/L	< 1 mg/L
Iron (Fe)	na	2 mg/L	na	3 mg/L	20 mg/L
Mercury (Hg)	≤ 0.10 mg/L	na	na	0.01 µg/L	0.05 µg/L
Nickel (Ni)	≤0.2 mg/L	na	na	na	na

Lead (Pb)	≤0.10 mg/L	0.10 mg/L	na	na	na
Zinc (Zn)	≤ 1 mg/L	5 mg/L	5 mg/L	5 mg/L	<10 mg/L

11.5. Conclusion

Synthetic dyes can be considered as "**contaminants of emerging concern**" or emerging **contaminants**. Synthetic dyes are made from chemicals, petroleum by-products, and earth minerals. They include acid, azo, basic, direct, disperse, fluorescent, mordant, reactive, solvent and vat etc. Synthetic dyestuffs are not readily biodegradable but persist for a long time in the environment because of their high stability to light or temperature, resistance to surfactants/ detergents degradation and poor removal by wastewater treatment. Textile industries consume a large volume of water (200 L water per kg of fibre) and chemicals (as many as 2,000 different chemicals) for the wet processing of textiles. During dyeing processes, the entire dye is not fixed to the fibre and therefore, high volumes of wastewaters containing dyes and other pollutants are produced which is released along with the effluents [*note:* the textile industry alone accounts for 54% dye effluents in the environment, followed by dyeing (21%), paper and pulp (10%), tannery and paint (8%) and dye manufactures (7%)]. The effluents discharged from textile factories are high in colour, pH, suspended solids, dissolved solids, BOD (biochemical oxygen demand) and COD (chemical oxygen demand), temperature and salts and heavy metals (Cd, Co, Cr, Cu, Fe, Mn, Ni, Pb, Zn).

The discharge of dye residues into water bodies is causing ecological and human health risks worldwide (Fig. 11.5). A number of rivers in Bangladesh are dying due to dyeing (Fig. 11.4). Coloured wastewater as effluents could affect the aesthetic merit and transparency of the receiving surface waterways and thus may impact on recreational and water-based tourism industries. It can also affect the aquatic ecosystem via decreasing the passage of light penetration and photosynthetic activity causing oxygen deficiency in lakes, rivers and other bodies of water and thus could disrupt primary production in aquatic ecosystems. Several dyes (anthraquinone, di- and tri-phenylmethane, azo and xanthene) are extremely toxic to aquatic organisms (algae, *Daphnia* and fish) with EC_{50}/LC_{50} < 1 mg/L (section 11.2.5). Textile effluents and textile dyes containing heavy metals can cause significant effects on fish, crab, algae, crops and vegetables (section 11.3.3). Some textile effluents (Malachite Green- a benzidine-based dye) bioaccumulated in eggs, fry and adult muscle of fish rainbow trout and wild European eel. Therefore, there is a possibility of transfer of dye residues to humans via the food chain pathways. Dye contaminated fish can be unfit for human consumption.

Moreover, the dye effluent can contaminate the surface and groundwater, therefore, such water may not be suitable for beneficial uses such as for irrigation and drinking and recreation. Irrigation of agricultural lands with effluents contaminated water can severely affect soil fertility and plant growth (germination, survival and growth). Several common dyes are carcinogenic including several azo, anthraquinone, benzidine, diphenylamine and triphenyl-amine. Cr, Ni, Co generally found in textile effluents are carcinogenic and bioaccumulative to biota. The most frequently utilised textile dye in the world is Azo dye (60-70% of all dye used), the residue of which can remain in the environment for a long time (~50 years), is highly bioaccumulative and carcinogenic. To reduce the risks and impacts of synthetic dyes/textile effluents on the environment and public health, ongoing monitoring of textile effluents and comparing with guidelines/regulator discharge limits, preventing discharge of untreated effluents into waterways, enforcing mandatory textile effluent treatment before discharge in the environment, carryout research to innovate low impact dyes (dyes which donot contain toxic heavy metals, require less rinsing and have a high dye absorption rate in the fabric) and awareness and education of communities (on the threats and risks associated with dyes, dye contaminated water, and dye contaminated food) should be implemented.

11.6. References

Ajmal M and Khan AU 1985 Effect of a textile factory effluent on soil and crop plants. *Environmental Pollution* **37** 131–148

Ali H 20I0 Biodegradation of synthetic dyes-a review. Water Air Soil Pollution *DOI:.l 007/s 11270-010-0382-4*

Al-Kdasi A, Idris A, Saed K and Guan CT 2005 Treatment of textile wastewater by advanced oxidation processes — a review. *Global Nest Journal International* **6** 222–230

Almazán-sánchez PT, Linares-hernández I, Solache-Ríos MJ and Martínez- Miranda V 2016 Textile wastewater treatment using iron modified clay and copper-modified carbon in batch and column systems. *Water Air and Soil Pollution* 227 1–14

Anliker R, Clarke EA and Moser P 1981 Use of the partition coefficient as an indicator of bioaccumulation tendency of dyestuffs in fish. *Chemosphere* **10(3)** 263–274 *doi:10.1016/0045-6535(81)90026-6.*

Aouni A, Fersi C, Cuartas-Uribe B et al 2012 Reactive dyes rejection and textile effluent treatment study using ultrafiltration and nanofiltration processes. *Desalination* **297** 87–96

APHA (American Public Health Association) 1995 Visual ADMI colour method for the examination of water and wastewater. APHA standard method 120E APHA – American Public Health Association, Washington DC USA.

https://law.resource.org/pub/us/cfr/ibr/002/apha.method.2120.1992.pdf

Ash M and Ash I 2009 Handbook of preservatives. Synapse Information Resources, Inc.. Endicott, NY : Synapse Information Resources Retrieved from

https://app.knovel.com/hotlink/toc/id:kpHP000012/handbook-preservatives/ handbook-preservatives ISBN978-1-934764-39-8

Avlonitis SA, Poulios I, Sotiriou D, et al 2008 Simulated cotton dye effluents treatment and reuse by nanofiltration. *Desalination* **221** 259–267

AWWA 2014 Simplified procedures for water examinations - Manual of Water Supply Practices, M12 (6th Edition). Chapter 12. Colour. American Water Works Association (AWWA). Pages 117-122.

Bafana A, Devi SS and Chakrabarti T 2011 "Azo dyes: past, present and the future". *Environmental Reviews* **19** 350–371 *doi:10.1139/a11-018. ISSN 1181-8700.*

Baird, RB, Eaton, AD and Rice, EW 2017 Standard Methods for the Examination of Water and Wastewater 23rd Edition Part 2000 Physical and aggregate properties. 2120 Colour.

https://www.standardmethods.org/doi/abs/10.2105/SMWW.2882.017

Ballav N, Das R, Giri S et al 2018 L -Cysteine doped polypyrrole (Ppy@L-cyst): asuper adsorbent for the rapid removal of Hg^{+2} and efficient catalytic activity of he spent adsorbent for reuse, *Chemical Engineering Journal* **345** 621–630 *https://doi.org/ 10.1016/j.cej.2018.01.093.*

Banat I.M, Nigam P, Singh D and Marchant R 1996 Microbial decolorization of textile dye containing effluents: a review. *Bioresource Technology* **58(3)** 217-227

Bae J-S and Freeman HS 2007 Aquatic toxicity evaluation of new direct dyes to the *Daphnia magna. Dyes and Pigments* **73(1)** 81-85

Balapure K, Bhatt N and Madamwar D 2015 Mineralization of reactive azo dyes present in simulated textile waste water using down flow microaerophilic fixed film bioreactor. *Bioresource Technology* **175** 1-7 *http://dx.doi.org/10.1016/ j.biortech.2014.10.040.*

Belpaire C, Reyns T, Geeraerts C and Loco JV 2015 Toxic textile dyes accumulate in wild European eel *Anguilla anguilla Chemosphere* **138** 784–791

Bhardwa V, Kumar P and Singha G 2014 Toxicity of heavy metals pollutants in textile mills effluents. *International Journal of Scientific & Engineering Research* **5(7)** 664-666

Bhuvaneswari A, Asha B and Selvakumar D 2016 Start up and enhancement granulation in an anaerobic baffled reactor for the treatment of textile wastewater. *International Journal of Civil Engineering* **9** 645–652

Carliell CM, Barclay SJ, Naidoo N et al 1995 Microbial decolorisation of a reactive azo dye under anaerobic conditions. *Water South Africa* (Pretoria), **21** 61-66

Carneiro PA, Umbuzeiro GA, Oliveira DP and Zanoni MVB 2010 Assessment of water contamination caused by a mutagenic textile effluent/dyehouse effluent bearing disperse dyes, *Journal of Hazardous Materials* **174** 694–699 *https://doi.org/10.1016/j. jhazmat.2009.09.106.*

Chen CH, Chang CF and Liu SM 2010 Partial degradation mechanisms of malachite green and methyl violet B by Shewanella decolorationis NTOU1 under anaerobic conditions. *Journal of Hazardous Materials* **177(1-3)** 281-289 *http://dx.doi.org/10.1016/j.jhazmat.2009.12.030*

Chequer FMD, Dorta DJ and de Oliveira DP 2011 Azo dyes and their metabolites: Does the discharge of the Azo dye into water bodies represent human and ecological Risks? In Tech. In book: *Advances in Treating Textile Effluent* 27-48 *DOI: 10.5772/19872*

Chequer FMD, de Oliveira GAR, Anastácio Ferraz ERA et al 2013. Textile dyes: dyeing process and environmental impact. chapter 6. INTECH. *http://dx.doi.org/10.5772/53659*

Chia MA and Musa RI 2014 Effect of indigo dye effluent on the growth, biomass production and phenotypic plasticity of *Scenedesmus quadricauda* (Chlorococcales). *Anais da Academia Brasileira de Ciências (Annals of the Brazilian Academy Sciences)* **36** 419-428

Choudhury AKR 2018 Eco-friendly dyes and dyeing. *Advances Materials and Technologies for Environmental Applications* **2(1)** 145-146

Christie R 2001 Colour chemistry. The royal society of chemistry. Cambridge, United Kingdom

Çlnar OS, Yasar M, Kertmen K et al 2008 Effect of cycle time on biodegradation of azo dye in sequencing batch Reactor. *Process Safety and Environmental Protection* 86 455–460

Copaciu F, Opris O, Coman V et al 2013 Diffuse water pollution by anthraquinone and azo dyes in environment importantly alters foliage volatiles, carotenoids and physiology in wheat (*Triticum aestivum*). *Water, Air, and Soil Pollution* **224(3)** 1478

Cotillas S, Llanos J and Cañizares P et al 2018 Removal of procion Red MX-5B dye from wastewater by conductive-diamond electrochemical oxidation, *Electrochimca Acta* **263** 1–7 *https://doi.org/10.1016/j.electacta.2018.01.052*

Croce R, Cinà F, Lombardo A et al 2017 Aquatic toxicity of several textile dye formulations: Acute and chronic assays with *Daphnia magna* and *Raphidocelis subcapitata Ecotoxicology and Environmental Safety* **144** 79–87

Davies RP and Dobbs AJ 1984 The prediction of bioconcentration in fish. *Water Research* **18(10)** 1253-1262

David N S and Rajan MR 2014 Impact of dyeing industry effluent on groundwater quality by water quality index and correlation analysis. *Journal of Pollution Effects & Control* **2(2)** *DOI: 10.4172/2375-4397.1000126*

ECHA (European Chemical Agency) 2006 Anthraquinone EC number: 201-549-0 | CAS number: 84-65-1 *https://www.echa.europa.eu/mt/web/guest/registration-dossier/-/registered-dossier/2211/6/2/4*

ESC (Essential Chemical Industry) 2013 Colorants. The Essential Chemical Industry – online *https://www.essentialchemicalindustry.org/materials-and-applications/colorants.html*

Eswaramoorthi S, Dhanapal K and Chauhan DS 2008 Advances in textile wastewater treatment: the case for UV-Ozonation and membrane bioreactor for common effluent treatment plants in Tirupur, Tamil Nadu, India. Environmental Technology Awareness Series *https://www.scribd.com/document/2608600/Advances-in-Textile-Wastewater-Treatment*

Ferraz ERA, Umbuzeiro GA, De Almeida G et al 2011 Differential toxicity of disperse red 1 and disperse red 13 in the Ames test, HepG2 cytotoxicity assay, and *Daphnia* acute toxicity test. *Environmental Toxicology* **26** 489–497

Garg VK and Kaushik P 2007 Influence of textile mill wastewater irrigation on the growth of sorghum cultivars. *Applied Ecology and Environmental Research* **6(2)** 1–12

Ghaly AE, Ananthashankar R, Alhattab M and Ramakrishnan VV 2013 Production, Characterization and treatment of textile effluents: a critical review. *Journal* of *Chemical Engineering & Process Technology* **5(1)** 1000182 *http://dx.doi.org/10.4172/2157-7048.1000182*

Ghaly A, Ananthashankar R, Alhattab M and Ramakrishnan V 2014 Production, characterization and treatment of textile effluents: a critical review. *Journal of Chemical Engineering & Process Technology* **5** 1–18

http://dx.doi.org/10.4172/2157-7048.1000182

Gita S, Hussan A and Choudhury TG 2017 Impact of textile dyes waste on aquatic environments and its treatment. *Environment and Ecology* **35(3C)** 2349-2353

Gita S, Shukla, SP, Saharan N et al 2019 Toxic effects of selected textile dyes on elemental composition, photosynthetic pigments, protein content and growth of a freshwater chlorophycean Alga *Chlorella vulgaris. Bulletin of Environmental Contamination and Toxicology.* **102** 795–801 *https://doi.org/10.1007/s00128-019-02599-w*

Global Effluent Guidelines 2007 Environment, health and safety Handbook volume 2.2 April 2007, section V:

http://www.levistrauss.com/wp-content/uploads/2014/01/Global-Effluent-Guidelines.pdf

Golka K, Kopps S and Myslak ZW 2004 "Carcinogenicity of azo colorants: influence of solubility and bioavailability". *Toxicology Letters* **151(1)** 203–210. *doi:10.1016/j.toxlet.2003.11.016. PMID 15177655.*

Gregory P 2007 Toxicology of textile dyes. Chapter 3, pages 44-73. In book: Environmental aspects of textile dyeing. Woodhead Publishing Series in Textiles *DOI: 10.1533/9781845693091.44*

Gupta VK and Suhas 2009 Application of low-cost adsorbents for dye removal - A review. *Journal of the Environmental Management* **90** 2313-2342

Gupta VK, Kumar R, Nayak A et al 2013 Adsorptive removal of dyes from aqueous solution onto carbon nanotubes: A review. *Advances in Colloid and Interface Science* **193-194** 24-34

Hamprecht R and Westerkamp Aloys 2000 "Disperse Dyes". Ullmann's Encyclopaedia of Industrial Chemistry. Weinheim: Wiley-VCH. *doi:10.1002/14356007.a08_565*

Hao OJ, Kim H, and Chiang P-C 2000 Decolorization of wastewater. Critical Reviews in Environmental Science and Technology **30** 449–505 *doi.org/10.1080/10643380091184237*

Hassan MA and Nemr AL 2017 Health and environmental impacts of dyes: mini review. *American Journal of Environmental Science and Engineering* **1(3)** 64-67

Hou M, Baughman GL and Perenich TA 1991 Estimation of water solubility and octanol/water partition coefficient of hydrophobic dyes: Part II. reverse-phase high-performance liquid chromatography. *Dyes and Pigments* **16(4)** 261-271

Hunger K 2003 Editor Industrial Dyes Chemistry, Properties, Applications. WILEY-VCH Verlag GmbH & Co. KGaA, Weinheim. ISBN 3-527-30576-9. 660p

http://citeseerx.ist.psu.edu/viewdoc/download?doi=10.1.1.674.8611&rep=rep1&type=pdf

Hussein FH 2013 Chemical properties of treated textile dyeing wastewater. *Asian Journal of Chemistry* **25** 9393–9400 *http://www.asianjournalofchemistry.co.in/User/ViewFreeArticle.aspx?ArticleID=25_17_126*

IARC (International Agency for Research on Cancer) 2013 Some chemicals present in industrial and consumer products, food and drinking-water. Anthraquinone. *https://www.ncbi.nlm.nih.gov/books/NBK373184/*

IARC 2019 IARC monographs on the identification of carcinogenic hazards to humans. *https://monographs.iarc.fr/agents-classified-by-the-iarc/*

Imtiazuddin SM, Mumtaz M, Mallick KA 2012 Pollutants of wastewater characteristics in textile industries. *Journal of Basic and Applied Science* **8** 554–556

Ismail M, Akhtar K, Khan MI et al 2019 Pollution, toxicity and Carcinogenicity of organic dyes and their catalyticbio-remediation. *Current Pharmaceutical Design* **25** 3653-3671 *DOI: 10.2174/1381612825666191021142026*

Jamee R and Siddique R 2019 Biodegradation of Synthetic Dyes of textile effluent by microorganisms: *An environmentally and economically sustainable approach. European Journal of Microbiology and Immunology* **9(4)** 114–118

Juhnke I and Luedemann D 1978 Results of the investigation of 200 chemical compounds for acute fish toxicity with the golden orfe test *Zeitschrift for Wasser-Abwasser-Forschung* **11(5)** 161-164 (in German)

Karthieyan S, Jambulingam M, Sivakumar P et al 2006 Impact of textile effluents on freshwater fish *Mastacembelus Armatus* (Cuv. & Val) *E-Journal of Chemistry* **3(4)** 303-306, October 2006

Katheresan V, Kansedo J and Lau SY 2018 Efficiency of various recent wastewater dye removal methods: A review. *Journal of Environmental Chemical Engineering* **6** 4676–4697

Kalra SS, Mohan S, Sinha A and Singh G 2011 Advanced oxidation processes for treatment of textile and dye wastewater: a review. In: 2nd International conference on environmental science and development IPCBEE 4 IACSIT Press, *Singapore http://www.ipcbee.com/vol4/59-ICESD2011D30020.pdf*

Kehinde FO and Aziz HA 2014 Textile waste water and the advanced oxidative treatment process, an overview. *International Journal of innovative research in Science, Engineering and Technology* **3** 15310–15317

Khan MG, Daniel G, Konjit M et al 2011 Impact of textile wastewater on seed germination and some physiological parameters in Pea (*Pisum sativum* L.), Lentil (*Lens esculentum* L.) and Gram (*Cicer arietinum* L.). *Asian Journal of Plant Sciences* **10** 269-273

Khan S and Malik A 2014 Environmental and health effects of textile industry wastewater, chapter 4, pages 55-72. *In: Book. Environmental deterioration and human health. Natural and anthropogenic determinants* (Abdul Malik, Elisabeth Grohmann and Rais Akhtar Editors). ISBN 978-94-007-7890-0 (eBook). *DOI 10.1007/978-94-007-7890-0_4*, Springer Science+Business Media Dordrecht

Kibria G, Haroon AKY, Nugegoda D and Rose G 2010 *Climate change and chemicals: Environmental Biological aspects.* New India Publishing Agency (NIPA), New Delhi. 470 p. ISBN: 9789380235301

https://www.researchgate.net/publication/261216635_Climate_Change_and_Chemicals_Environmental_and_Biological_Aspects

Kim SY, Grady PL and Hersh SP 1983 Energy consumption and conservation in the fibre-producing and textile Industries. *Textile Progress* **13(3)** 1-14

Kumar V, Majumdar C and Roy P 2008 Effects of endocrine disrupting chemicals from leather industry effluents on male reproductive system." *Journal of Steroid Biochemistry and Molecular Biology* **111(3–5)** 208–216

Lellis B, Fávaro-Polonio CZ, Pamphile JA and Polonio JC 2019 Effects of textile dyes on health and the environment and bioremediation potential of living organisms. *Biotechnology Research and Innovation* **3(2)** 275-290

https://doi.org/10.1016/j.biori.2019.09.001

Little LW, Lamb JC, Chillingworth MA and Durkin WB 1974 Acute toxicity of selected commercial dyes to the fathead minnow and evaluation of biological treatment for reduction of toxicity. *Proc. 29th Ind. Waste Conf.*, pp 524–534, Purdue University, Lafayette, IN, 1974.

Maguire RJ and Tkacz RJ 1991 Occurrence of dyes in the Yamaska River, Quebec. *Water Pollution Research Journal of Canada* **26** 145–161

Manekar P, Patkar G, Aswale P et al 2014 Detoxifying of high strength textile effluent through chemical and biooxidation processes. *Bioresource Techno*logy**157** 44–51

Malik A, Elisabeth G and Rais A (Eds.) 2014 Environmental deterioration and human health. Natural and anthropogenic determinants. ISBN 978-94-007-7890-0. Springer

Marking LL 1969 Marking toxicity of rhodamine B and fluorescein sodium to fish and their compatibility with antimycin A. *The Progressive Fish – Culturist* **31(3)** 139-142 *https://echa.europa.eu/registration-dossier/-/registered-dossier/18690/6/2/2*

Meinertz JR, Stehly GR, Gingerich WH and Allen, JL 1995 Residues of [14C] malachite green in eggs and fry of rainbow trout (*Oncorhynchus mykiss*) (Walbaum) after treatments of eggs. *Journal of Fish Diseases* **18(3)** 239–247

Meireles G, Daam, MA, Sanches, ALM et al 2018 Red disperse dyes (DR 60, DR 73 and DR 78) at environmentally realistic concentrations impact biochemical profile of early life stages of zebrafish (*Danio rerio*). *Chemico-Biological Interactions* **292** 94–100

Mirghani MES, Kabbashi NA, Qudsieh IY and Elfaki FA 2008 A new method for the determination of toxic dye using FTIR spectroscopy.) *IIUM Engineering Journal* **9(2)** 27-38

Mojsov KD, Darko Andronikov, Janevski DA et al 2016 The application of enzymes for the removal of dyes from textile effluents. *Advanced Technologies* **5(1)** 81–86

Montgomery JH and Crompton TR 2017 Environmental chemicals desk reference. 1645 pages. Taylor & Francis Group. ISBN 9781498775731

Moradi R, Lashgari N, Kruger HG et al Metal-Free Synthetic Organic Dyes. Elsevier ISBN 9780128156476 MSDS-Millipore 2013 Diisopropyl-azo-dicarboxylate

https://us.vwr.com/assetsvc/asset/en_US/id/12557234/contents

MSDS Pubchem 2005a (updated 2020) Anthraquinone *https://pubchem.ncbi.nlm.nih.gov/compound/Anthraquinone#section=Environmental-Abiotic-Degradation*

MSDS-Pubchem 2005b (updated 2016) Triphenylmethane *https://pubchem.ncbi.nlm.nih.gov/compound/triphenylmethane*

MSDS-Pubchem 2005c (updated 2020) Diphenylmethane *https://pubchem.ncbi.nlm.nih.gov/compound/Diphenylmethane*

MSDS-Pubchem 2005d (updated 2020) Solvent red 1 *https://pubchem.ncbi.nlm.nih.gov/compound/Solvent-red-1*

MSDS-Pubchem 2005e (updated 2020) Xanthene *https://pubchem.ncbi.nlm.nih.gov/compound/Xanthene#section=ChemIDplus*

MSDS-Pubchem 2005f (updated 2020) Indigo *https://pubchem.ncbi.nlm.nih.gov/compound/Indigo*

MSDS-Pubchem 2005g (updated 2020) Disperse orange 30. 2-[N-(2-Cyanoethyl)-4-[(2,6-dichloro-4-nitrophenyl)azo]anilino]ethyl acetate

https://pubchem.ncbi.nlm.nih.gov/compound/Disperse-Orange-30

MSDS-Pubchem 2005h (updated 2020) Acid red 26 *https://pubchem.ncbi.nlm.nih.gov/compound/Acid-Red-26*

MSDS-Pubchem 2005i (updated 2020) Congo red *https://pubchem.ncbi.nlm.nih.gov/compound/Congo-red*

MSDS-ThermoFisher 2008 (revised 2018) Copper phthalocyanine

https://www.fishersci.com/store/msds?partNumber=AC269150250&productDescription=COPPER+PHTHALOCYANINE%2C+D+25GR&vendorId=VN00032119&countryCode=US&language=en

MSDS-ThermoFisher 2011 Indigo

https://www.fishersci.com/shop/msdsproxy?productName=AC212130250&productDescription=INDIGO%2C+SYNTHETIC+25GR&catNo=AC212130250&vendorId=VN00032119&storeId=10652

Nahar K, Chowdhury MAK, Chowdhury MAH et al 2018 Heavy metals in handloom-dyeing effluents and their biosorption by agricultural by products. *Environmental Science and Pollution Research* **25** 7954–7967 *https://doi.org/10.1007/s11356-017-1166-9*

NHMRC and NRMMC 2011 Australian Drinking Water Guidelines. Paper 6 National Water Quality Management Strategy. National Health and Medical Research Council, National Resource Management Ministerial Council, Commonwealth of Australia, Canberra. National Resource Management Ministerial Council, Commonwealth of Australia, Canberra.

Nirmalarani J and Janardhanan K 1988 Effect of South India Viscose factory effluent on seed germination seedling growth and chloroplast pigments content in five varieties of Maize (*Zea mays*) *Madras Agriculture Journal* **75** 41

Nopkhuntod S, Dararat S and Yimrattanabovorn J 2012 Removal of reactive dyes from wastewater by shale. *Songklanakarin Journal of Science and Technology* **34** 117–123

Novotny C, Dias N, Kapanen A et al 2006 Comparative use of bacterial, algal and protozoan tests to study toxicity of azo- and anthraquinone dyes. *Chemosphere* **63** 1436–1442

Odipe OE, Raimi MO, Suleiman F 2019 Assessment of heavy metals in effluent water discharges from textile industry and river water at close proximity: A comparison of two textile industries from Funtua and Zaria, North Western Nigeria. Madridge

Journal of Agricultural land Environmental Sciences **1(1)** 1-6 *doi: 10.18689/mjaes-1000101*

OECD (Organisation for Economic Co-operation and Development) 1988 Copper phthalocyanine CAS number 147-14-8

https://web.archive.org/web/20170516224832/http://www.inchem.org/documents/sids/sids/147148.pdf

O'Neill C, Hawkes FR, Hawkes DL et al 1999 Colour in textile effluents-sources, measurement, discharge consents and simulation: a review. *Journal of Chemical Technology Biotechnology* **74** 1009-1018

Pereira L and Alves M 2012 Chapter 4. Dyes — Environmental impact and remediation p.112-164. *In book:* A. Malik, E. Grohmann (eds.), *Environmental Protection Strategies for Sustainable Development,* Strategies for Sustainability *DOI 10.1007/978-94-007-1591-2_4*, © Springer Science+Business Media B.V. 2012

Poornima K, Vasudevan K and Vankateshwarlu M 2001. Evaluation of heavy metals in textile effluents in relation to soil and pond water *Electronic Journal of Environmental Sciences* **4** 79-84

Przystaś W, Zabłocka-Godlewska E and Grabińska-Sota E 2012 Biological removal of Azo and Triphenylmethane dyes and toxicity of process by-products. *Water Air and Soil Pollution* **223** 1581–1592

Puvaneswari N, Muthukrishnan J, and Gunasekaran P 2006 Toxicity assessment and microbial degradation of azo dyes. *Indian Journal of Experimental Biology Biol* **44** 618-626

Rao TVRK, Kumari C and Kumar A 2019 Effect of soil contamination by some azo dyes on the seed germination and plant growth of broad bean (*Vicia faba*). *Research Journal of Pharmaceutical, Biological and Chemical Sciences* **10(4)** 265 *https://doi.org/10.33887/rjpbcs/2019.10.4.37*

Rastogia A and Singh P 2012 Analysis of heavy metals in carpet and textile industry effluent of Bhadoi region, India. *Journal of Pharmaceutical Biomedical Sciences* **26(26)** 304-307

Rauf MA and Ashraf SS 2012 Survey of recent trends in biochemically assisted degradation of dyes. *Chemical Engineering Journal* **209** 520–530 *https://doi.org/10.1016/j.cej.2012.08.015.*

Rajaguru, P, Kalpana R, Hema A et al 2001 Genotoxicity of some sulfur dyes on tadpoles (*Rana hexadactyla*) measured using the comet assay. *Environmental and Molecular Mutagen* **38** 316–322

https://doi.org/10.1002/em.10027

Reema R M, Saravanan P, Kumar M et al 2011 'Accumulation of methylene blue dye by Growing *Lemna minor Separation Science and Technology* **46(6)** 1052-1058 *DOI: 10.1080/01496395.2010.528503*

Ribeiro AR and de Aragão Umbuzeiro G 2014 Effects of a textile azo dye on mortality, regeneration, and reproductive performance of the planarian, *Girardia tigrine*. *Environmental Sciences Europe* **26** article 22

Robert L, Joseph F and Alexander A 2008 Fisher's contact dermatitis in: Textiles and shoes. BC Decker Inc., Ontario, p.339-401

Rodríguez-Couto S, Osma JF and Toca-Herrera JL 2009 Removal of synthetic dyes by an eco-friendly strategy, *Engineering in Life Sciences* **9(2)** 116–123 *https://doi.org/10.1002/elsc.200800088*

Sekar P, Prasad SH and Raman MD 2009 Effect of textile dye industry effluent on the nutritive value of fresh water female crab *Spiralothelphusa hydrodroma* (Herbst). *Journal of Applied Sciences Research* **5** 2041-2048

Selvaraj D, Leena R, Kamal DC 2015 Toxicological and histopathological impacts of textile dyeing industry effluent on a selected teleost fish *Poecilia reticulata. Journal of Pharmacology and Toxicol*ogy **3** 26-30

Soriano JJ, Mathieu-Denoncourt J, Norman G et al 2014 Toxicity of the azo dyes Acid Red 97 and Bismarck Brown Y to Western clawed frog (*Silurana tropicalis*). *Environmental Science and Pollution Research* **21** 3582–3591 *DOI 10.1007/s11356-013-2323-4*

Steingruber E 2004 "Indigo and Indigo Colorants" Ullmann's Encyclopedia of Industrial Chemistry. Wiley- VCH, Weinheim. *doi: 10.1002/14356007.a14_149.pub2*

Sztarr Y 2008 Measuring colour in trade waste. 2nd Annual WIOA NSW Water Industry Engineers & Operators Conference Jockey Club – Newcastle 8 to 10 April 2008, pages 97-103.

http://www.wioa.org.au/conference_papers/08_nsw/documents/YolandaSztarr.pdf

Sun Y, Zheng H, Tan M et al 2014 Synthesis and characterization of composite flocculant PAFS–CPAM for the treatment of textile dye wastewater. *Journal of Applied Polymer Science* **31** 1–8

Suryavathi V, Sharma S, Sharma, S et al 2005 Acute toxicity of textile dye wastewaters (untreated and treated) of Sanganer on male reproductive systems of albino rats and mice. *Reproductive Toxicol*ogy **19(4)** 547–556

SWGWQG (Sustainable Water Group Water Quality Guidelines) 2010 Water quality guidelines. Sustainable Water Group. Business for Social Responsibility (BSR). Partnership of companies committed to managing water use and wastewater discharge in global textile supply chains.

https://www.bsr.org/reports/awqwg/BSR_AWQWG_Guidelines-Testing-Standards.pdf

Sztarr Y 2008 Measuring colour in trade waste 2nd Annual WIOA NSW Water Industry Engineers & Operators Conference Jockey Club – Newcastle 8 to 10 April 2008. pages 97-103.

http://www.wioa.org.au/conference_papers/08_Qld/documents/YolandaSztarr.pdf

Tomei MC, Pascual JS and Angelucci DM 2016 Analysing performance of real textile wastewater bio-decolourization under different reaction environments. *Journal of Cleaner Production* **129** 468–477

Turhan K and Turgut Z 2009 Decolorization of direct dye in textile wastewater by ozonization in a semi-batch bubble column reactor. *Desalination* **242** 256–263

Upadhye VB and Joshi SS 2012 Advances in wastewater treatment—a review. *International Journal of Chemical Science* **3** 264–268

Vacchi FI, Albuquerque AF, Vendemiatti JAS et al 2013 Chlorine disinfection of dye wastewater: Implications for a commercial azo dye mixture. *Science of the Total Environment* **44** 2302–309

Verdon E and Andersen WC 2017 "Certain Dyes as pharmacologically active substances in fish farming and Other aquaculture products" U.S. Food and Drug Administration papers 13. *http://digitalcommons.unl.edu/usfda/13*

Verschueren K 2001 Handbook of environmental data on organic chemicals. Volumes 1-2. 4th ed. John Wiley & Sons. New York, NY. 2001, p. 220

https://www.wiley.com/en-nl/Handbook+of+Environmental+Data+on+Organic+Chemicals,+4+Volume+Set,+5th+Edition-p-9780470171721

Walthall WK and Stark JD 1999 The acute and chronic toxicity of two xanthene dyes, fluorescein sodium salt and phloxine B, to *Daphnia pulex Environmental Pollution* **104** 207-215

Wang JA, Gao F, Liu Z et al 2012 Pathway and molecular mechanisms for malachite green biodegradation in Exiguobacterium sp. MG2 *PLOS ONE* **7(12)** e51808

Welham A 2000 The theory of dyeing (and the secret of life). *Journal of the Society of Dyers and Colourists* **116** 140–143

Yaseen DA and Scholz M 2019 Textile dye wastewater characteristics and constituents of synthetic effluents: a critical review. *International Journal of Environmental Science and Technology* **16** 1193–1226 *https://doi.org/10.1007/s13762-018-2130-z*

Yazdani MR 2018 Engineered adsorptive materials for water remediation. Development, characterization and application. Aalto University publication series. Doctoral dissertation 58/2018

file:///C:/Users/User1/Downloads/DoctoralDissertation_RozaYazdani2018.pdf

Yagub MT, Sen TK, Afroze S and Ang HM 2014 Dye and its removal from aqueous solution by adsorption: A review. *Adv. Colloid Interface Sci.* **209** 172-184

Vijaykumar MH, Vaishampayan PA, Shouche SY and Karegoudar TB 2007 Decolourization of naphthalene-containing sulfonated azo dyes by Kerstersia sp. strain VKY1. *Enzyme and Microbial Technology* **40** 204–211

Xia H, Chen L and Fang Y 2013 Highly efficient removal of Congo red from wastewater by Nano-Cao. *Separation Science and Technology* **48** 2681-2687

http://dx.doi.org/10.1080/01496395.2013.805340

Zhao X and Hardin I 2007 HPLC and spectrophotometric analysis of biodegradation of azo dyes by *Pleurotus ostreatus. Dyes Pigments* **73** 322-325

Zocolo GJ, dos Santos GP, Vendemiatti, J et al 2015 Using SPE-LC-ESI-MS/MS analysis to assess disperse dyes in environmental water samples. *Journal of Chromatographic Science* **53** 1257–1264 *https://doi. org/10.1093/chromsci/bmu221.*

Zollinger H 1987 Colour chemistry-synthesis, properties and applications of organic dyes and pigments. VCH, New York, 92-100

Web

https://p2infohouse.org/ref/11/10489/sectors71.html

https://shenglufashion.com/2019/08/16/wto-reports-world-textile-and-apparel-trade-in-2018/

https://www.essentialchemicalindustry.org/materials-and-applications/colorants.html; https://www.sciencedirect.com/topics/chemistry/metal-complex-dye

https://en.wikipedia.org/wiki/Category:Azo_dyes

https://en.wikipedia.org/wiki/Malachite_green

https://www.britannica.com/technology/direct-dye

https://www.sciencedirect.com/science/article/pii/B9781845696955500131

https://www.yourdictionary.com/mordant

https://organiclifestyle.com/articles/fibre-reactive-low-impact-dyes#mordants

https://www.sciencedirect.com/topics/engineering/reactive-dye

https://www.ncbi.nlm.nih.gov/books/NBK385442/

https://www.merriam-webster.com/dictionary/indigoid

http://thetoxicfreefoundation.com/database/ingredient/coal-tar-dyes

https://www.essentialchemicalindustry.org/materials-and-applications/colorants.html

https://byjus.com/physics/wavelength-of-light/

https://en.wikipedia.org/wiki/Anthraquinone

https://en.wikipedia.org/wiki/Azo_dye#/media/File:Typical_azo_compound.svg; https://en.wikipedia.org/wiki/Anthraquinone#/media/File:Anthraquinone_acsv.svg

http://stainsfile.info/dyes/class/clsarylm.htm

https://en.wikipedia.org/wiki/Indigo_dye#/media/File:Indigo_skeletal.svg; https://en.wikipedia.org/wiki/Phthalocyanine#/media/File:Phthalocyanine.svg

https://en.wikipedia.org/wiki/Xanthene

http://stainsfile.info/dyes/class/clsarylm.htm

https://en.wikipedia.org/wiki/Raphidocelis_subcapitata

https://pubchem.ncbi.nlm.nih.gov/compound/Benzidine

https://www.harathelabel.com.au/blogs/news/health-and-synthetic-dyes

https://www.merriam-webster.com/dictionary/chelating

https://www.iamrenew.com/environment/bangladeshs-polluted-waters-rivers-dying-due-to-dyeing/#:~:text=In%20Bangladesh's%20capital%2C%20Dhaka%2C%20the, effluent%20from%20surrounding%20garment%20factories.&text=In%202021%2C %20textile%20industries%20in,million%20metric%20tons%20of%20fabrics

https://www.bsr.org/reports/awqwg/BSR_AWQWG_Guidelines-Testing-Standards.pdf

https://www.onsiteinstaller.com/online_exclusives/2017/08/the_installers_guide_to_bod5

https://en.wikipedia.org/wiki/APHA_color

https://mrccc.org.au/wp-content/uploads/2013/10/Water-Quality-Salinity-Standards.pdf

https://www.fondriest.com/environmental-measurements/parameters/water-quality/turbidity-total-suspended-solids-water-clarity

https://www.ncbi.nlm.nih.gov/books/NBK304402/

http://www.hse.gov.uk/textiles/dyes-dyeing.htm

https://www.cancer.org.au/about-cancer/types-of-cancer/bladder-cancer.html

https://www.ed.ac.uk/health-safety/guidance/hazardous-substances/sensitisers

file:///C:/Users/User1/Downloads/Colour_Facts_Sheet%20(3).pdf

https://www.mayoclinic.org/diseases-conditions/dermatitis-eczema/symptoms-causes/syc-20352380 http://arco-hvac.ir/wp-content/uploads/2018/04/AWWA_M12_Simplified_Procedures_for.pdf

https://www.lenntech.com/products/activated-carbon.htm

https://www.cattermoleconsulting.com/explaining-low-impact-dyes/

https://support.hach.com/app/answers/answer_view/a_id/1000980/~/what-is-the-meaning-of-admi%3F-

https://support.hunterlab.com/hc/en-us/articles/201482339-What-scale-is-used-to-measure-the-color-of-waste-water-

https://en.wikipedia.org/wiki/Azorubine

https://pubchem.ncbi.nlm.nih.gov/compound/6321394

https://en.wikipedia.org/wiki/Fluorescein

https://en.wikipedia.org/wiki/Disperse_dye

https://phys.org/news/2008-05-congo-red.html

http://www.hse.gov.uk/textiles/dyes-dyeing.htm

■■■■

12 Sewage pollution Risks, Impacts and Management

Summary

Sewage water/wastewater is a cocktail of chemical pollutants/contaminants. These contaminants include biological micropollutants (bacteria, viruses), chemical micropollutants (heavy metals, hormones, household chemicals, oil compounds, personal care products, pesticides, pharmaceuticals, plastic additives, POPs, surfactants), nutrients (nitrogen and phosphorus) and solids. Several of these contaminants are persistent, toxic, bio-accumulative (PBT), as well as endocrine-disrupting and carcinogenic. Chemical micro-pollutants found in sewage water can have different impacts on aquatic biota, its environment and human health. This chapter reviews **i.** pathways of sewage water entering aquatic environments; **ii**. chemistry, toxicity and ecotoxicity of chemical micropollutants found in sewage water **iii.** impacts of chemical micropollutants found in sewage water on social, economic and environmental sectors including effluent concentrations compared to Predicted No-Effect Concentration (PNEC), **iv.** impacts of chemical micropollutants found in sewage water on aquatic biota including (algae, *Daphnia*, fish, crustaceans, birds), bioaccumulation, endocrine disruption, algal blooms and humans health; **v.** Linkages between sewage pollution and climate change; and **vii.** management of sewage pollution.

Contents

List of Tables

List of Figures

List of Box

12.1. Introduction

Sewage is wastewater from domestic, commercial and industrial sources. There are four types of wastewater: **i.** domestic wastewater (which carries disposed of water and wastes away from laundry, kitchen, bathroom and toilet and consists of *greywater* (bathroom, laundry, kitchen) and *blackwater* (urine and faeces); **ii.** industrial wastewater (which consists of chemicals, salts, acids); **iii**. agricultural wastewater (that consists of fertilisers and pesticides) and **iv.** urban wastewater (that consists of both domestic and industrial wastewaters including stormwater run-off). In sewage, the freshwater component is 99% and the remaining 1% is contaminants (EPA Vic 1995, Cunningham and Gharipour 2018, YVW 2020). These contaminants (Fig. 12.1) usually include the following:

- biological micropollutants/pathogenic microorganisms (bacteria, viruses);

- chemical micropollutants/ toxic substances (heavy metals, pesticides, pharmaceuticals, surfactants, oil, POPs etc);
- nutrients (nitrogen and phosphorus) and
- solids (in the form of suspended solids).

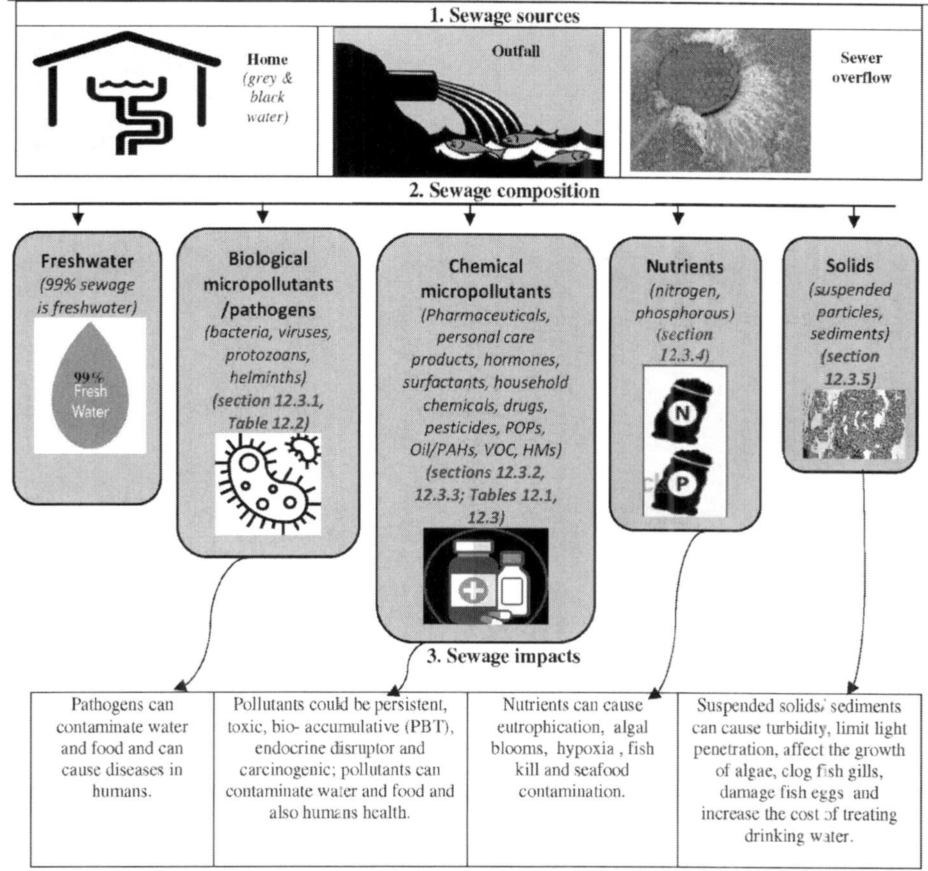

Fig. 12.1. Sewage sources, and composition and impacts of sewage water.

12.1.1. Pathways of sewage water entering aquatic environments

Sewage water/ wastewater can enter into the aquatic systems from surface run-off both from non-point and point sources (e.g. agricultural farms, urban areas, hospitals, health care facilities, and direct and indirect discharges from industries). Outfalls of wastewater treatment plants (WWTPs) and stormwater drain are other routes of entry of wastewater into the environment. Leaking of sewer pipes and flooding can cause entirely untreated sewage to enter rivers, lakes, reservoirs, coastal areas and other water sources, causing them to become polluted. Sewage spills can be another major route of entering wastewater into the environment. A sewage spill can occur due to various reasons including the following:

- blockages/ the build-up of sewer main pipes due to tree roots, fat, oil, grease and other foreign objects (e.g. improper disposal of items in toilets and sinks) (Fig. 12.2);
- break down of sewer main pipes; sewer line collapses caused by old and deteriorated pipes;
- heavy rainfall events causing the sewer overflows (e.g. tree branches, dirt and other debris can clog swear mains and drains);
- loss of electrical power in wastewater treatment plants (WWTP) or pump-stations; and
- aged and poorly maintained infrastructure.

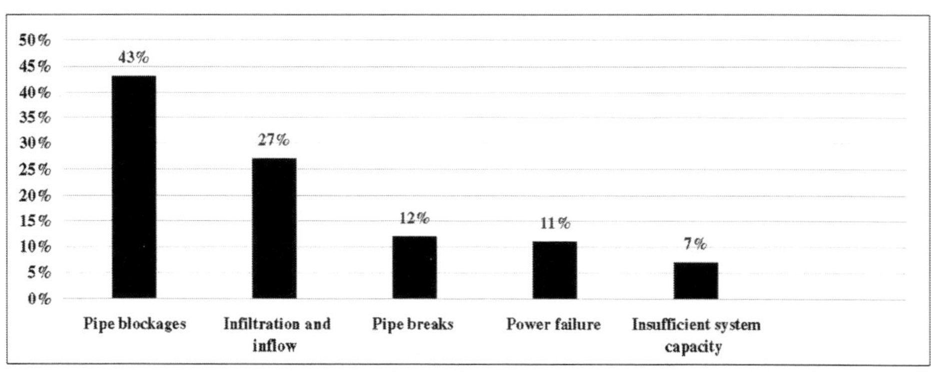

Fig. 12.2 Occurrence of sanitary sewer overflows by cause (USEPA 1996)

12.2. Chemistry, toxicity and ecotoxicity of chemical micropollutants[1] found in sewage water

Sewage water is a cocktail of chemical pollutants/ contaminants (stressors), the most common of which are **i.** ***Heavy metals*** (Cd, Cr, Cu, Pb, Hg, Ni, Zn), **ii.** ***Hormones*** (EE2, E2, E2), **iii.** ***Flame retardants*** (PBDE, PFOS, PFOA), **iv.** ***Plastic additives/ plasticisers*** (BPA, DEHP, DBP), **v.** ***Persistent organic pollutants (POPs)*** (Aldrin/dieldrin, DDT, endosulfan, endrin, heptachlor, PCBs), **vi.** ***Pesticides*** (atrazine, carbendazim, diazinon, diuron, glyphosate), **vii.**

---

[1] ***Micropollutants=*** *micropollutants are those substances that are generally present in water bodies in low concentrations (typically, in the range of ng–µg/L) but which can have detrimental effects on humans, the environment or drinking water supplies https://www.umweltbundesamt.de/sites/default/files/medien/1410/publikationen/180709_uba_pos_mikroverunreinigung_en_bf.pdf*

Pharmaceuticals (caffeine, carbamazepine, ciprofloxacin, diclofenac, erythromycin, ibuprofen, naproxen, ofloxacin, paracetamol, salicylic acid, sulfamethoxazole, trimethoprim), **viii. *Polycyclic aromatic hydrocarbons (PAHs)*** (acenaphthene, anthracene, benzo[b]fluoranthene, benzo[ghi]perylene, ethylbenzene, fluorene, indeno[1,2,3-c,d]pyrene, naphthalene, o-xylene, phenanthrene, toluene) and **ix. *Surfactants/ detergents*** (alkyl ethoxy sulfate, alkyl sulfate, linear alkyl benzene sulfonate, nonylphenol ethoxylates (NPEs)/ nonylphenol (NP)); **x. *Volatile organic compounds (VOCs)*** (ethylbenzene, toluene and O-Xylene). These lists of pollutants (i to x) are persistent, toxic, bioaccumulative (PBT). Some of them are also endocrine-disrupting and carcinogenic. Aquatic biota can be exposed to these pollutants via WWTP outfall or even surface water (lakes, rivers, estuaries) contaminated with raw sewage. Humans can be exposed via food and water contaminated with these pollutants. Table 12.1 provide some details about the chemistry, toxicity and ecotoxicity of pollutants with respect to their half-life, log K_{ow}, bioconcentration factors, LD_{50} (rat, birds), EC_{50}/LC_{50} (algae, Daphnia and fish). The following section is a summary of 12.1:

12.2.1. Persistent

[(very persistent: half-life in environmental/ biological samples >1 year)].

Several chemicals found in sewage water are very persistent (*half-life> 1 year*) including ***Heavy metals*** (Pb in sediment: up to 22.6 years, Hg in sediments: up to 2.1 to 180 years); ***Flame retardants*** (PBDEs in biological samples: up to 1-12 years, PFOS in biological samples: up to 2 3.4 years); ***Plastic additives/ plasticisers*** (DEHP in sediments: up to 11.5 months, nonylphenol: up to 60 years (sediments); ***POPs*** (aldrin/dieldrin in soil: up to 5 years, DDT in soil: up to 4-30 years, endrin in soil: up to 12 years, heptachlor in soil: up to 9-10 years, PCBs in soil: up to 10.9-11.2 years); ***Pesticides*** (carbendazim in soil: up to 2.08 years/25 months); ***Pharmaceuticals*** (caffeine in surface water: up to 10 years, carbamazepine in lake water: up to 15.6 years, ciprofloxacin in biosolids: up to 6.3 years, ibuprofen in lake water: up to 8 months, ofloxacin in biosolids: up to 3.65 years); ***Polycyclic aromatic hydrocarbons (PAHs)*** (very short half-life-few hours); ***Surfactants/ detergents*** (very short half-life-except nonylphenol); ***Volatile organic compounds (VOCs)*** (very short half-life- few hours) (see Table 12.1 for details).

12.2.2. Bioaccumulation (biota)

[(highly bioaccumulative: log K_{ow}> 4 or BCF> 500)]

Several chemicals found in sewage water are highly bioaccumulative (*log K_{ow}> 4 or BCF> 500*) including ***Heavy metals*** (BCF of Hg: up to 1,000,000-100,000,000, BCF of Cd: up to 10,000-1000,000), ***Hormones*** (log K_{ow} of EE2: >4), ***Flame retardants*** (log K_{ow} of PBDEs > 4), ***Plastic additives/ plasticisers*** (log K_{ow} of DEHP and DBP > 4); ***POPs*** (log K_{ow} of aldrin/dieldrin, DDT, endosulfan, endrin, heptachlor and PCBs > 4); ***Pharmaceuticals*** (log K_{ow} of diclofenac > 4.51); ***Polycyclic aromatic hydrocarbons (PAHs)*** (log K_{ow} of anthracene, benzo[a]pyrene, benzo[b]fluoranthene, benzo[ghi]perylene, fluoranthene, fluorene, indeno[1,2,3-c,d]pyrene, phenanthrene> 4.0); ***Surfactants/ detergents*** (log K_{ow} of nonylphenol > 4) (see Table 12.1 for details).

12.2.3. Toxicity (rat or birds)

[(extremely toxic: LD_{50} <50 mg/kg; very toxic: LD_{50} 50-500 mg/kg)].

12.2.3.1. LD_{50} of rat, birds

Several chemicals found in sewage water are extremely toxic (*LD_{50} <50 mg/kg*) to very toxic (*LD_{50} 50-500 mg/kg*) including ***Heavy metals*** (Cr LD_{50} rat: 52 mg/kg, Cd LD_{50} of rat: 472 mg/kg); ***Flame retardants*** (PFOS LD_{50} of rat: 154 mg/L, PFOA LD_{50} of rat: 450-680 mg/kg), ***Plastic additives/ plasticisers*** (DEHP LD_{50} of rat: 500-5000 mg/kg, NP LD_{50} of rat: 30-2000 mg/kg), ***POPs*** (aldrin/dieldrin LD_{50} of rat: 38 mg/kg, endosulfan: LD_{50} of rat: 44.9 mg/kg, endrin: LD_{50} rat of 17.8 mg/ kg); ***Pesticides*** (diazinon LD_{50} of birds: 2.7 mg/kg); ***Pharmaceuticals*** (diclofenac LD_{50} of birds: 0.1-0.2 mg/kg, naproxen LD_{50} of rat: 248 mg/kg); ***Polycyclic aromatic hydrocarbons (PAHs)*** (benzo[a]pyrene LD_{50} of rat: 50 mg/kg); ***Surfactants/ detergents*** (linear alkyl benzene sulfonate LD_{50} of rat: 401-1,900 mg/kg) (see Table 12.1 for details).

12.2.3.2. Carcinogenicity

[(IARC Group 1= carcinogenic to humans; IARC Group 2A= probably carcinogenic to humans or IARC Group 2B= possibly carcinogenic to humans)].

Several chemicals found in sewage water is carcinogenic (cancer-causing and grouped into 1, 2A and 2B by International Agency for Research on Cancer (IARC) including the following:

- IARC group 1: ***Heavy metals*** (Cd, Cr, Ni); ***Hormones*** (EE2, E2, E1); ***POPs*** (PCBs) ***Oil/petrochemical compounds*** (Benzo[a]pyrene); (see Table 12.1 for details)
- IARC group 2A*:* ***Heavy metals*** (Pb); ***Flame retardants*** (PBDE); ***POPs*** (aldrin/dieldrin; endosulfan); ***Pesticides*** (diazinon, glyphosate); (see Table 12.1 for details)
- IARC group 2B*:* ***Heavy metals*** (Hg); ***Flame retardants*** (PFOA), ***Plastic additives/ plasticisers*** (DEHP); ***POPs*** (heptachlor); ***Polycyclic aromatic hydrocarbons (PAHs)*** (benzo[b]fluoranthene, indeno[1,2,3-c,d]pyrene, naphthalene) ***Volatile organic compounds (VOCs)*** (ethylbenzene) (see Table 12.1 for details)

Table 12.1. Chemistry, toxicity, and ecotoxicity of selected chemicals found in sewage water

BCF: *Bioconcentration factors;* ***ED=*** *Endocrine disruptor;* ***IARC=*** *International Agency for Research on Cancer].*

[note: chapter 2 through chapter 11 provided in-depth details on the chemistry, toxicity and ecotoxicity of arsenic (chapter 2), heavy metals (chapter 3), microplastics (chapter 4), POPs (chapter 5), oil/petrochemicals (chapter 6), pesticides (chapter 7), algal toxins (chapter 8), pharmaceuticals (chapter 9), surfactants (chapter 10), and synthetic dyes (chapter 11). Therefore, these are not repeated in this chapter; for definitions of BCF, EC_{50}/LC_{50}, half-life, log K_{ow}, please consult Chapters 3, 6, 7 and 9].

Chemical group	**Persistence** *(half-life, d= days, mo= month; yr= year)*	**Bio-accumulation** *(log K_{ow}; BCF (bioconcentration factor)*	**Toxicity** *(LD_{50} of mammals, birds, bees; IARC cancer group (1, 2A, 2B); endocrine disruptor (ED))*	**Ecotoxicity** *(EC_{50}/LC_{50} of algae, Daphnia, fish)*
Heavy metals *(see **Chapter 3** of this book, which elaborately discussed risks and impacts of heavy metals)*				
Cadmium (Cd) *(heavy metals)*	half-life (biological): **10-30 yr** (WHO 2010)	**BCF: 10,000-1000,000;** highly bio-accumulative	LD_{50} rat: *225* mg/kg; **very toxic** to mammals; IARC **gr 1; ED**	EC_{50} algae: 0.0023 mg/L; **extremely** toxic to **algae**
Chromium (Cr) *(heavy metals)*	half-life (biological): **10.75 yr** (Petersen et al 2000)	BCF: 100-1,000; low-bioaccumulative	LD_{50} rat: **52** mg/kg; **extremely** toxic to mammals; IARC **gr 1,** CrVi	EC_{50} *Daphnia*: 0.162 mg/L; **extremely** toxic to ***Daphnia***
Copper (Cu) *(heavy metals)*	na	BCF: 100-26,000; bioaccumulative	LD_{50} rat: **472** mg/kg; **very** toxic to mammals	EC_{50} *Daphnia*: 0.003 mg/L; **extremely** toxic to ***Daphnia***
Lead (Pb) *(heavy metals)*	half-life (sediments): **22.6 yr** https://www.sciencedirect.com/topics/medicine-and-dentistry/lead-210	BCF: 499-1,700; low bioaccumulative	IARC **gr 2A**; **ED**	LC_{50} fish: 2.2 mg/L; toxic to aquatic life
Mercury (Hg) *(heavy metals)*	half-life (sediments): **2.1 y-180 yr** (Kudo et al 1975)	**BCF: 1,000,000-100,000,000** (MeHg); highly bio-accumulative	IARC **gr 2B-** (MeHg); **ED**	EC_{50} *Daphnia:* 0.0052 mg/L; **extremely** toxic to ***Daphnia***

Chemical group	**Persistence** *(half-life, d= days, mo= month; yr= year)*	**Bio-accumulation** *(log K_{ow}; BCF (bioconce-ntration factor)*	**Toxicity** *(LD_{50} of mammals, birds, bees; IARC cancer group (1, 2A, 2B); endocrine disruptor (ED))*	**Ecotoxicity** *(EC_{50}/LC_{50} of algae, Daphnia, fish)*
Nickel (Ni) *(heavy metals)*	half-life (environmental samples, Ni-63): **92 yr** (Beasley and Held 1969)	BCF: 370; low bioaccumulative	IARC **gr 1**	EC_{50} algae: 0.174 mg/L; **extremely** toxic to **algae**
Zinc (Zn) *(heavy metals)*	half-life: **243 d** https://en.wikipedia.org/wiki/Zinc	BCF: 50-1,130; low bioaccumulative	**ED**	EC_{50} algae: 0.106 mg/L; **extremely** toxic to **algae**
Hormones *(see **Chapter 9** of this book, which elaborately discussed risks and impacts of pharmaceuticals including hormones)*				
17α-ethinyl-estradiol (EE2) *(contraceptive/ hormone)*	half-life:>50 d	log K_{ow}: **4.15;** highly bio-accumulative	moderately toxic to mammals; IARC **gr 1**; **ED**	EC_{50} algae: 0.84 mg/L; **extremely** toxic to **algae**
17β-Estradiol (E2) *(natural hormone)*	half-life:>56 d	log K_{ow}: 3.94; moderately bioaccumulative	IARC **gr 1**; **ED**	EC_{50} fish: 0.00012 mg/L; **extremely** toxic to **fish**
Estrone (E1) *(natural hormone)*	na	log K_{ow}: 3.43; moderately bioaccumulative	IARC **gr 1**; **ED**	EC_{50} fish: 0.0000078 mg/L; **extremely** toxic to fish
Flame retardants *(based on chapter 5 of this book, which elaborately discussed risks and impacts of persistent organic pollutants (POPs) including flame retardants)*				
Polybrominated diphenyl ethers (PBDEs) *(flame retardants)*	half-life (humans): **1-12 yr** (Oulhote et al 2016)	log K_{ow}: **6.2-6.9;** highly bio-accumulative	LD_{50} rat: **500**-5000 mg/kg; **very toxic** to moderately toxic to mammals; IARC **gr 2A**, **ED**	LC_{50} fish : 0.027 mg/L; **extremely** toxic to **fish**
Perfluorooctane sulfonic acid (PFOS) *(flame retardants)*	half-life (biological sample): **3.4 y** (Li et al 2018)	log K_{ow}: **4.49;** highly bio-accumulative	LD_{50} rat: **154 mg/kg** (Lewis 2005)**;** **very** toxic to mammals**; ED**	LC_{50} fish: 7.8 mg/L; moderately toxic to aquatic life

Chemical group	Persistence *(half-life, d= days, mo= month; yr= year)*	Bio-accumulation *(log K_{ow}; BCF (bioconcentration factor)*	Toxicity *(LD_{50} of mammals, birds, bees; IARC cancer group (1, 2A, 2B); endocrine disruptor (ED))*	Ecotoxicity *(EC_{50}/LC_{50} of algae, Daphnia, fish)*
Perfluorooctanoic acid (PFOA) *(flame retardants)*	half-life (biological sample): **2.7 yr** (Li et al 2018)	log K_{ow}: **4.8-6;** highly bio-accumulative	LD_{50} rat: **450-680** mg/kg; **very toxic** to moderately toxic to mammals; IARC **gr 2B**, **ED**	EC_{50} *Daphnia*: 0.531 mg/L; **extremely** toxic to ***Daphnia***
Plastic additives/ plasticisers *(see **Chapter 4** of this book, which elaborately discussed risks and impacts of plastics)*				
Bisphenol A (BPA) *(Plastic additives/ plasticisers)*	half-life: 4.5 d (soil, water) (Abo et al 2016, *https://en.wikipedia.org/wiki/Bisphenol_A)*	log K_{ow}: 3.4; moderately bioaccumulative	moderately toxic to mammals; **ED**	EC_{50} algae: 2.5 mg/L; **very** toxic to algae
Di-(2-ethylhexyl) phthalate (DEHP) *(Plastic additives/ plasticisers)*	half-life: **347 d (11.5 mo)** (sediments) (NICNAS 2019)	log K_{ow}: **7.5**; highly bio-accumulative	LD_{50} rat: **500**-5000 mg/kg; **very toxic** to moderately toxic to mammals; IARC **gr 2B**; **ED**	EC_{50} fish: 0.32 mg/L; **extremely** toxic to fish
Di-n-butyl phthalate (DnBP)/ Dibutyl phthalate (DBP) *(Plastic additives/ plasticisers)*	half-life: 46 d (sediments) (NICNAS 2019)	(log K_{ow}: **4.45;** highly bio-accumulative	**ED**	EC_{50} algae: 0.4 mg/L; **extremely** toxic to **algae**
Nonylphenol (NP)	half-life: **60 yr** (sediments) (Shang et al 1999, Soares et al 2008)	(log K_{ow}: **4.30-4.6;** highly bio-accumulative	LD_{50} rat: **300**-2000 mg/kg; **very toxic** to moderately toxic to mammals; **ED**	EC_{50} *Daphnia*: 0.01 mg/L; **extremely** toxic to **Daphnia**
Persistent organic pollutants (POPs) *(see **Chapter 5** of this book, which elaborately discussed risks and impacts of POPs)*				
Aldrin/ dieldrin *(insecticide, POPs)*	half-life: **5 yr** (soil)	log K_{ow}: **4.32;** highly bio-accumulative	LD_{50} **rat**: **38** mg/kg, LD_{50} **birds**: **50** mg/kg; **extremely** toxic to mammals,	LC_{50} fish: 0.0037 mg/L; **extremely** toxic to **fish**

Chemical group	**Persistence** *(half-life, d= days, mo= month; yr= year)*	**Bio-accumulation** *(log K_{ow}; BCF (bioconcentration factor)*	**Toxicity** *(LD_{50} of mammals, birds, bees; IARC cancer group (1, 2A, 2B); endocrine disruptor (ED))*	**Ecotoxicity** *(EC_{50}/LC_{50} of algae, Daphnia, fish)*
			birds; IARC **gr 2A**; **ED**	
Dichloro-diphenyl trichloroethane (DDT) *(insecticide, POPs)*	half-life: **4 to 30 y** (soil)	log K_{ow}: **4.9-6.9;** highly bio-accumulative	moderately toxic to mammals, IARC **gr 2A**; **ED**	LC_{50} *Daphnia*: 0.0011 mg/L; **extremely** toxic to ***Daphnia***
Endosulfan *(insecticide, POPs)*	half-life: 5 to **8 mo** (soil)	log K_{ow}: **4.74-4.79;** highly bio-accumulative	LD_{50} **rat**: **44.9** mg/kg; **extremely** toxic to **mammals**, IARC **gr 2A; ED**	LC_{50} *Daphnia*: 0.0075 mg/L; **extremely** toxic to *Daphnia*
Endrin *(insecticide, POPs)*	half-life: **12 yr** (soil)	log K_{ow}: **5.6;** highly bio-accumulative	LD_{50} **rat**: **17.8** mg/kg; **extremely** toxic to mammals, **ED**	LC_{50} *Daphnia* 0.00004 mg/L**; extremely** toxic to *Daphnia*
Heptachlor *(insecticide, POPs)*	half-life: **9-10 mo** (soil)	log K_{ow}: **5.27;** highly bio-accumulative	moderately toxic to mammals, IARC **gr 2B; ED**	LC_{50} fish: 0.007 mg/L; **extremely** toxic to fish
Polychlorinated biphenyls *(PCBs, POPs)*	half-life: **10.9-11.2 yr** (soil) (Doick et al 2005)	log K_{ow}: **4.46-8.18;** highly bio-accumulative	low to non-toxic to mammals; IARC **gr 1**; **ED**	LC_{50} fish: 0.0015 mg/L; **extremely** toxic to fish
Pesticides *(see **Chapter 7** of this book, which elaborately discussed risks and impacts of pesticides)*				
Atrazine *(herbicide)*	half-life: **105 d** (soil)	log K_{ow}: 2.5	non-toxic to mammals**; ED**	LC_{50} fish: 0.0015 mg/L; **extremely** toxic to fish
Carbendazim *(fungicide)*	half-life: up to **25 mo** (soil)	log K_{ow}: 1.38	non-toxic to mammals**; ED**	LC_{50} *Daphnia*: 0.13 mg/L; **extremely** toxic to *Daphnia*
Diazinon *(insecticide)*	na	log K_{ow}: 3.3 moderately bioaccumulative	LD_{50} **birds: 2.7** mg/kg; **extremely** toxic to birds, IARC **gr 2A**; **ED**	LC_{50} *Daphnia*: 0.00096 mg/L; **extremely** toxic to *Daphnia*

Chemical group	**Persistence** *(half-life, d= days, mo= month; yr= year)*	**Bio-accumulation** *(log K_{ow}; BCF (bioconce-ntration factor)*	**Toxicity** *(LD_{50} of mammals, birds, bees; IARC cancer group (1, 2A, 2B); endocrine disruptor (ED))*	**Ecotoxicity** *(EC_{50}/LC_{50} of algae, Daphnia, fish)*
Diuron *(herbicide/ algicide)*	half-life: **4-8 mo** (soil)	log K_{ow}: 2.85;	non-toxic to mammals	LC_{50} fish: 5.6 mg/L; moderately toxic to *Daphnia*
Glyphosate *(herbicide)*	half-life: up to **174 d** (soil)	log K_{ow}: <-3.2	IARC **gr 2A**; **ED**	LC_{50} fish: 86 mg/L; moderately toxic to fish
Pharmaceuticals *(see **Chapter 9** of this book, which elaborately discussed risks and impacts of pharmaceuticals)*				
Caffeine *(stimulant)*	half-life: **10 yr** (surface water)	log K_{ow}: -0.07	moderately toxic to mammals	LC_{50} fish: 87 mg/L; moderately toxic to fish
Carbamazepine *(anticonvulsant)*	half-life: up to **15.6 yr** (lake)	log K_{ow}: 2.45	moderately toxic to mammals	LC_{50} *Daphnia*: 13.8 mg/L; moderately toxic to *Daphnia*
Ciprofloxacin *(antibiotic)*	half-life: up to **6.3 yr** (biosolids)	log K_{ow}: 0.4	moderately toxic to mammals	EC_{50} bacteria**: 0.006** mg/L; **extremely** toxic to **bacteria**
Diclofenac *(anti-inflammatory)*	half-life: up to 35 d	log K_{ow}: **4.51**; highly bio-accumulative	LD_{50} birds: **0.1-0.2** mg/kg; **extremely** toxic to birds	LC_{50} algae: 86 mg/L; moderately toxic to fish
Erythromycin *(antibiotic)*	half-life: up to 37.9 d	log K_{ow}: 3.06; moderately bioaccumulative	moderately toxic to mammals	EC_{50} bacteria**: 0.022** mg/L; LC_{50} algae: **0.02** mg/kg; **extremely** toxic to **bacteria and algae**
Ibuprofen *(anti-inflammatory)*	half-life: up to **247 d (8 mo)** (lake)	log K_{ow}: 3.97; moderately bioaccumulative	moderately toxic to mammals**; ED**	LC_{50} algae: **7.1** mg/L; **very** toxic to algae
Naproxen *(anti-inflammatory)*	half-life: 14 d	log K_{ow}: 3.18 moderately bioaccumulative	LD_{50} birds: **248** mg/kg; **very** toxic to mammals	LC_{50} algae: **7.1** mg/L; **very** toxic to **algae**
Ofloxacin *(antibiotic)*	half-life: up to **3.65 yr** (biosolids)	log K_{ow}: 0.35	moderately toxic to mammals	LC_{50} algae: **1.44** mg/L; **very** toxic to algae

Chemical group	**Persistence** *(half-life, d= days, mo= month; yr= year)*	**Bio-accumulation** *(log K_{ow}; BCF (bioconcentration factor)*	**Toxicity** *(LD_{50} of mammals, birds, bees; IARC cancer group (1, 2A, 2B); endocrine disruptor (ED))*	**Ecotoxicity** *(EC_{50}/LC_{50} of algae, Daphnia, fish)*
Paracetamol *(analgesic)*	half-life: **72 d**	log K_{ow}: 0.46	moderately toxic to mammals**; ED**	LC_{50} *Daphnia*: 9.2 mg/L; **very** toxic to ***Daphnia***
Salicylic acid *(anti-inflammatory)*	half-life: **82 yr** (aqueous)	log K_{ow}: 2.24	moderately toxic to mammals	LC_{50} fish: 37 mg/L; moderately toxic to fish
Sulfamethoxazole *(antibiotic)*	half-life: 35 d	log K_{ow}: 0.89	Non-toxic to mammals	LC_{50} *Daphnia*: 0.03 mg/L; **extremely** toxic to ***Daphnia***
Trimethoprim *(antibiotic)*	half-life: up to **100 d** (sediment)	log K_{ow}: 0.64	non-toxic to mammals; LD_{50} rat: > 5000 mg/kg	Sight to non-toxic to aquatic life LC_{50}> 100 mg/L)
Polycyclic aromatic hydrocarbons *(PAHs)* ***(petrochemicals)*** *(see chapter 6 of this book, which elaborately discussed risks and impacts of oil compounds)*				
Acenaphthene *(PAH)*	very short half-life (few hours)	log K_{ow}: 3.92 moderately bioaccumulative	non-toxic to mammals	EC_{50} algae: **0.51** mg/L; **extremely** toxic to **algae**
Anthracene *(PAH)*	very short half-life	log K_{ow}: **4.54; highly** bioaccumulative	non-toxic to mammals; **ED** (Kibria et al 2010, IEH 2012)	LC_{50} fish: **0.0028** mg/L; **extremely** toxic to fish
Benzo[a]-pyrene *(PAH)*	very short half-life	log K_{ow}: **6.04**; **highly** bioaccumulative	LD_{50} rat: **50>** mg/kg; **extremely** toxic; IARC **gr 1; ED** (Kibria et al 2010, IEH 2012)	LC_{50} *Daphnia*: 0.005 mg/L; **extremely** toxic to ***Daphnia***
Benzo[b]-fluoranthene *(PAH)*	very short half-life	log K_{ow}: **5.8**; **highly** bioaccumulative	IARC **gr 2B; ED** (Kibria et al 2010, IEH 2012)	LC_{50} *Daphnia*: 1.04 mg/L; **extremely** toxic to ***Daphnia*** (Wernersson and Dave 1997)
Benzo[ghi]-perylene *(PAH)*	very short half-life	log K_{ow}: 6.5; **highly** bioaccumulative	**ED** (Kibria et al 2010, IEH 2012)	na

Chemical group	**Persistence** *(half-life, d= days, mo= month; yr= year)*	**Bio-accumulation** *(log K_{ow}; BCF (bioconce-ntration factor)*	**Toxicity** *(LD_{50} of mammals, birds, bees; IARC cancer group (1, 2A, 2B); endocrine disruptor (ED))*	**Ecotoxicity** *(EC_{50}/LC_{50} of algae, Daphnia, fish)*
Fluoranthene *(PAH)*	very short half-life	log K_{ow}: **5.22; highly** bioaccumulative	moderately toxic to mammals; **ED** (Kibria et al 2010, IEH 2012)	LC_{50} fish: **0.0077** mg/L; **extremely** toxic to **fish**
Fluorene *(PAH)*	very short half-life	log K_{ow}: **4.18; highly** bioaccumulative	**ED** (Kibria et al 2010, IEH 2012)	LC_{50} *Daphnia*: 0.43 mg/L; **extremely** toxic to ***Daphnia***
Indeno[1,2,3-c,d]pyrene *(PAH)*	very short half-life	log K_{ow}: **7.0; highly** bioaccumulative	IARC **gr 2B; ED** (Kibria et al 2010, IEH 2012)	na
Naphthalene *(PAH)*	very short half-life		IARC **gr 2B; ED** (Kibria et al 2010, IEH 2012)	LC_{50} fish: **1.4** mg/L; **very** toxic to **fish**
Phenanthrene *(PAH)*	very short half-life	log K_{ow}: **4.46; highly** bioaccumulative	LD_{50} rat: 1800 mg/kg; moderately toxic to mammals; **ED** (Kibria et al 2010, IEH 2012)	LC_{50} *Daphnia*: 0.35 mg/L; **extremely** toxic to ***Daphnia***
Surfactants *(see chapter 10 of this book, which elaborately discussed risks and impacts of surfactants)*				
Alkyl ethoxy sulfate (AES) *(anionic)*	half-life: 2.4 d	log K_{ow}: 0.95-3.9	LD_{50} rat: 2000 mg/kg; moderately toxic to mammals	LC_{50} fish: **0.8** mg/L; **extremely** toxic to **fish**
Alkyl sulfate (AS) *(anionic)*	half-life: 160-460 h	log K_{ow}: 1.6-4.6	LD_{50} rat: 1400-8000 mg/kg; moderate to non-toxic to mammals	EC_{50} algae: 27 mg/L; toxic to algae
Linear alkyl benzene sulfonate (LAS) *(anionic)*	half-life: 3 d	log K_{ow}: 1.94-3.42	LD_{50} rat: 401-1900; moderate to non-toxic to mammals	LC_{50} fish: **4.1** mg/L; **very** toxic to **fish**
Nonylphenol ethoxylates	half-life: 15-20 d	log K_{ow}: **4.48; highly**	LD_{50} rat: 2000-5000; moderate to non-	EC_{50} algae: **0.025** mg/L;

Chemical group	**Persistence** *(half-life, d= days, mo= month; yr= year)*	**Bio-accumulation** *(log K_{ow}; BCF (bioconcentration factor)*	**Toxicity** *(LD_{50} of mammals, birds, bees; IARC cancer group (1, 2A, 2B); endocrine disruptor (ED))*	**Ecotoxicity** *(EC_{50}/LC_{50} of algae, Daphnia, fish)*
(NPEs)/ nonylphenol (NP) *(non-ionic)*		bioaccumulative	toxic to mammals**;** **ED**	**extremely** toxic to algae
Volatile organic compounds ***(petrochemicals)*** *(see chapter 6 of this book, which elaborately discussed risks and impacts of oil compounds)*				
Ethylbenzene *(BTEX)*	very short half-life	log K_{ow}: 3.27	LD_{50} rat: 3500 mg/kg; moderately toxic to mammals; IARC **gr 2B**	EC_{50} algae: **2.6** mg/L; **very** toxic to **algae**
Toluene *(BTEX)*	very short half-life	log K_{ow}: 2.73	non-toxic to mammals	LC_{50} fish: **5.5** mg/L; **very** toxic to **fish**
o-Xylene *(BTEX)*	very short half-life	log K_{ow}: 3.1	LD_{50} rat: 1800 mg/kg; moderately toxic to mammals	LC_{50} *Daphnia*: 1 mg/L; **extremely** toxic to ***Daphnia*** (Galassi et al 1988, MSDS-Pubchem-0-xylene 2005)

12.2.3.3. Endocrine disruption

Several chemicals found in sewage water are endocrine disruptor (ED) including the following:

Heavy metals (Cd, Pb, Hg, Zn); ***Hormones*** (EE2, E2, E1); ***Flame retardants*** (PBDE, PFOS, PFOA); ***Plastic additives/plasticisers*** (BPA, DEHP, DBP, NP); ***POPs*** (aldrin/dieldrin, DDT, endosulfan, endrin, heptachlor, PCBs); ***Pesticides*** (atrazine, carbendazim, diazinon, glyphosate); ***Pharmaceuticals*** (ibuprofen, paracetamol); ***Polycyclic aromatic hydrocarbons (PAHs)*** (anthracene, benzo[a]pyrene, benzo[b]fluoranthene, benzo[ghi]perylene, fluoranthene, fluorene, indeno[1,2,3-c,d]pyrene, naphthalene); ***Surfactants/detergents*** (nonylphenol) (see Table 12.1 for details).

12.2.4. Ecotoxicity (bacteria, algae, Daphnia, fish)

[(Extremely toxic: EC_{50}/LC_{50} < 1 mg/L; very toxic EC_{50}/LC_{50}: 1-10 mg/L)].

Heavy metals (Cd, Cr, Cu, Hg, Ni Pb: EC_{50}/LC_{50} of algae/*Daphnia*/fish <1 mg/L); ***Hormones*** (EE2, E2, E1: EC_{50} of algae/fish <1 mg/L); ***Flame retardants*** (PBDE, PFOS, PFOA: LC_{50} of fish <1 mg/L); ***Plastic additives/ plasticisers*** (DEHP, BBA, NP: EC_{50}/LC_{50} of algae/*Daphnia*/fish <1 mg/L); ***POPs*** (aldrin/

dieldrin; DDT, endosulfan; endrin, heptachlor, PCBs: LC_{50} fish< 1 mg/L), ***Pesticides*** (atrazine, carbendazim, diazinon: EC_{50}/LC_{50} of *Daphnia*/fish< 1 mg/L); ***Pharmaceuticals*** (ciprofloxacin, erythromycin, paracetamol, sulfamethoxazole: EC_{50}/LC_{50} of bacteria/algae/*Daphnia* <1 mg/L); ***Polycyclic aromatic hydrocarbons (PAHs)*** (acenaphthene, anthracene, benzo[a]pyrene, benzo[b]fluoranthene, fluoranthene, fluorene, phenanthrene: EC_{50}/LC_{50} of algae/ *Daphnia*/fish <1 mg/L); ***Surfactants/ detergents*** (alkyl ethoxy sulfate, nonylphenol: EC_{50}/LC_{50} of algae/ fish < 1 mg/L) ***Volatile organic compounds (VOCs)*** (o-xylene: LC_{50} *Daphnia*: 1 mg/L) (see Table 12.1 for details).

12.3. Impacts of biological and chemical micropollutants found in sewage water on social, economic and environmental sectors

12.3.1. Biological micropollutants

Sewage water contains disease-causing bacteria, viruses, protozoan parasites and helminths (Table 12.2). There is a possibility of transmission of waterborne diseases (such as gastroenteritis) to humans via pathogens in contaminated water or food. The human can also be exposed to these pathogens during primary (bathing, swimming) and secondary (boating, fishing) recreational activities. Sewage contaminated water may not be suitable for bathing or swimming as epidemiological studies have shown that there is a linear correlation between microbial water quality of bathing water and gastro-intestinal illnesses (Cabelli et al 1982). In fact, sewage leak generally forces the closure of beaches to swimming. Filter-feeding shell-fish organisms (mussels and oysters) accumulate pathogenic bacteria in their tissues. Filter feeders filter a great quantity of water. There is a prohibition of closing shell-fisheries areas during sewage spill (Owili 2003) worldwide. Consumption of pathogen contaminated shell-fish is a risk to human.

Table 12.2. Pathogens found in untreated sewage water/wastewater
(WHO 2006, UN Water 2015).
na*= information is not available*

Agent	Species	Disease
Bacteria	*Campylobacter jejune*	Gastroenteritis[2]
	Escherichia coli	Gastroenteritis
	E. coli O157:H7	Bloody diarrhoea, haemolytic uremic syndrome[3]

[2]***Gastroenteritis*** *= inflammation of the* ***stomach and intestines****, typically resulting from* ***bacterial toxins*** *or viral infection and causing vomiting and diarrhoea.*

Agent	Species	Disease
	Helicobacter pylori	Abdominal pain, peptic ulcers, gastric cancer
	Salmonella spp.	Salmonellosis[4], gastroenteritis, diarrhoea, fever
	Salmonella typhi	Typhoid fever
	Shigella spp.	Shigellosis, Dysentery
	Vibrio cholerae	Cholera[5]
Viruses	Adenovirus	Respiratory disease, eye infections
	Astrovirus	Gastroenteritis
	Calicivirus	Gastroenteritis
	Coronavirus	Gastroenteritis
	Coxsackie viruses	Herpangina, aseptic meningitis[6], respiratory illness, fever, paralysis, respiratory, heart and kidney disease
	Echovirus	Fever, rash, respiratory and heart disease, aseptic meningitis
	Enteroviruses	Gastroenteritis

[3]***Haemolytic–uremic syndrome*** *(HUS) is a group of* ***blood disorders*** *characterized by low red blood cells, acute kidney failure, and low platelets (https://en.wikipedia.org/wiki/Hemolytic%E2%80%93uremic_syndrome).*

[4]***Salmonellosis*** *is a type of gastroenteritis caused by* ***Salmonella bacteria****. In Australia, most Salmonella infections occur after eating* ***contaminated food*** *but also sometimes after contact with another person with the infection. Symptoms of Salmonella infection includes fever, headache, diarrhoea, abdominal pain, nausea and vomiting. Symptoms usually develop 6–72 hours after exposure to the bacteria, but sometimes up to 2 weeks (http://conditions.health.qld.gov.au/HealthCondition/condition/14/33/124/Salmonella-infection-salmonellosis).*

[5]***Cholera*** *is an infection of the small intestine by some strains of the bacterium* ***Vibrio cholerae*** *(Gram-negative, comma-shaped bacterium). The classic symptom is large amounts of watery diarrhea that lasts a few days. It is spread mostly by unsafe water and unsafe food that has been contaminated with human faeces containing the bacteria. Undercooked seafood is a common source. There are concerns that rising sea levels will increase rates of disease since cholera bacterium shows increased growth at water salinity between 5‰ and 25‰ (https://en.wikipedia.org/wiki/Cholera#:Borroto 1998).*

[6]***Meningitis*** *a serious disease in which the membranes surrounding the* ***brain and spinal cord*** *develop an infection and become swollen (= larger than normal), causing severe headache, a high temperature and sometimes death (https://www.oxfordlearnersdictionaries.com/definition/english/meningitis).*

Agent	Species	Disease
	Hepatitis A and E	Infectious hepatitis
	Norovirus	Gastroenteritis
	Parvovirus	Gastroenteritis
	Poliovirus	Paralysis, aseptic meningitis
	Rotavirus	Gastroenteritis
Protozoa	*Balantidium coli*	Balantidiasis (dysentery)
	Cryptosporidium parvum	Cryptosporidiosis[7]
	Cyclospora cayetanensis	Persistent diarrhoea
	Entamoeba histolytica	Amoebiasis[8] (amoebic dysentery)
	Giardia lamblia	Giardiasis[9]
Helminths	*Ascaris lumbricoides* (roundworm)	Ascariasis
	Ancylostoma duodenal and *Necator americanos*	Hookworm
	Clonorchis sinensis (liver fluke)	Clonorchiasis
	Fasciola (liver fluke)	Fascioliasis
	Fasciolopsis buski (intestinal fluke)	Fascioliasis
	Fasciolopsis buski (intestinal fluke)	Fascioloidiasis

[7]***Cryptosporidiosis*** *is a diarrhoeal disease caused by the* ***parasite, Cryptosporidium****, which infects the intestine. Cryptosporidium hominis causes the most infections in humans in Australia. The most common symptom of cryptosporidiosis are watery diarrhoea and stomach cramps (https://www.health.nsw.gov.au/Infectious/factsheets/Pages/cryptosporidiosis.aspx#.).*

[8] ***Amoebiasis*** *infection is most common in tropical areas with untreated water. It spreads through drinking or eating uncooked food, such as fruit, that may have been washed in contaminated local water.*

[9] ***Giardiasis*** *is an infection in the small intestine. It's caused by a microscopic parasite called Giardia lamblia.*

Agent	Species	Disease
	Opisthorchis viverrini	Opisthorchiasis
	Schistosoma (blood fluke)	Schistosomiasis (Bilharzia)
	Trichuris (whipworm)	Trichuriasis
	Taenia (tapeworm)	Taeniasis

12.3.2. Chemical micro-pollutants

12.3.2.1. Chemical concentrations in sewage/wastewater treatment plant (WWTP) effluent

Typical chemical micropollutants detected in sewage water/WWTP effluents include: **i.** heavy metals, **ii.** hormones, **iii.** household chemicals (cheating agents, corrosion inhibitors, flame retardants, perfluorinated compounds, plastic additives/plasticisers, sweeteners), **iv.** illicit drugs, **v.** persistent organic pollutants (POPs), **vi.** personal care products, **vii.** pesticides, **viii.** pharmaceuticals, **ix.** polycyclic aromatic hydrocarbons (PAHs), **x.** surfactants and **xi.** volatile organic compounds. Among all the types/ groups, pharmaceuticals were frequently detected, followed by household chemicals (flame retardants, plastic additives/ plasticisers), surfactants, PAHs, and pesticides. Table 12.3 provides an account of chemicals detected in sewage/waste in effluents, their concertation, as well as the Predicted No-Effect Concentration (**PNEC**) and Assessment factor (**AF**) of each detected pollutant. Chemicals which have exceeded the PNEC is considered as toxic/harmful to biota (algae, *Daphnia* and fish) and have been marked with an **asterisk (*)**. The following section is a summary of 12.3:

12.3.2.1.1. Heavy metals (09 types/groups detected)

i. Aluminium (Al), *ii. Cadmium (Cd), iii. Chromium (Cr), *iv. Copper (Cu), v. Iron (Fe), *vi. Lead (Pb), vii. Mercury (Hg), *viii. Nickel (Ni), and *ix. Zinc (Zn).

12.3.2.1.2. Hormones (07 types/groups detected)

*i. 17α-ethinylestradiol (EE2) (contraceptive), *ii. 17β-Estradiol (E2) (natural hormone), iii. Androsterone (natural hormone), iv. Estriol (E3) (natural hormone), *v. Estrone (E1) (natural hormone), vi. Progesterone (natural hormone), and vii. Testosterone (natural hormone).

12.3.2.1.3. Household chemicals (32 types/groups detected)

- ***Household chemicals (Chelating agents) (02 types/groups detected)***

*i. Ethylene-diamine-tetra-acetic acid (EDTA), and ii. Nitrilo-tri-acetic acid (NTA).

- ***Household chemicals (Corrosion inhibitors) (06 types/groups detected)***

i. 2-hydroxybenzothiazole (OHBT), ii. 2-methylthiobenzothiazole (MTBT), iii. Benzothiazole, iv. Benzothiazole-2-sulfonic acid (BTSA), v. Benzotriazole, and vi. Methylbenzotriazole.

- ***Household chemicals (Flame retardants- brominated) (04 types/groups detected)***

i. 2,2',4,4'-Tetrabromodiphenyl ether (BDE-47), ii. Decabromodiphenyl ether (BDE-209), iv. Pentabromodiphenyl ethers (BDE-99), and iv. Tetrabromobisphenol A (TBBA).

- ***Household chemicals (Flame retardants- chlorinated paraffins) (01 types/groups detected)***

i. Short-chain chlorinated paraffins (SCCPs, C10-13).

- ***Household chemicals (Flame retardants- phosphorous) (07 types/groups detected)***

i. Tri-iso-butyl phosphate (TiBP), ii. Tri-n-butyl phosphate (TnBP), iii. Triphenyl phosphate (TPP), iv. Tris(1,3-dichloro-2-propyl) phosphate (TDCP), v. Tris(2-chloro,1-methylethyl) phosphate (TCPP), vi. Tris(2-chloroethyl) phosphate (TCEP), and vii. Tris(butoxyethyl)-phosphate (TBEP).

- ***Household chemicals (Perfluorinated compounds) (02 types/groups detected)***

*i. Perfluorooctane sulfonic acid (PFOS), and ii. Perfluorooctanoic acid (PFOA).

- ***Household chemicals (Plastic additives/ plasticisers) (06 types/groups detected)***

*i. Bisphenol A (BPA), ii. Butyl benzyl phthalate (BBP), *iii. Di-(2-ethylhexyl) phthalate (DEHP), iv. Diethyl phthalate (DEP), v. Dimethyl phthalate (DMP), and *vi. Di-n-butyl phthalate (DnBP).

- ***Household chemicals (Sweeteners) (04 types/groups detected)***

i. Acesulfame, *ii. Cyclamate, iii. Saccharin, and *iv. Sucralose.

12.3.2.1.4. Illicit drugs (04 types/groups detected)

i. Amphetamine (AMP) (nervous system stimulant), ii. Cocaine (nervous system stimulant), iii. Ecstasy (MDMA) (psychedelic drug), and iv. THC-COOH (cannabis metabolite) (a psychoactive drug).

12.3.2.1.5. Persistent organic pollutants (POPs) (10 types/ groups detected)

i. Aldrin (insecticide), ii. Dichlorodiphenyltrichloroethane (DDT) (insecticide), *iii. Dieldrin (insecticide), *iv. Endosulfan (insecticide), v. Endrin (insecticide), *vi. Heptachlor (insecticide), vii. Hexachlorobenzene (fungicide), viii. Hexachlorocyclohexanes (HCHs) (insecticide), ix. Pentachloronitrobenzene (PCNB) (fungicide), and *x. Polychlorinated biphenyls (PCBs) (Σ 209 PCBs) (dielectric and coolant fluids).

12.3.2.1.6. Personal care products (16 types/groups detected)

i. 4-methylbenydlidene camphor (4-MBC) (UV Filters), ii. Benzophenone-3 (BP-3) (UV Filters), *iii. Chloroxylenol (biocides), *iv. Ethylhexyl methoxycinnamate (EHMC or OMC) (UV Filters), v. Ethyl-paraben (preservatives), vi. Galaxolide (HHCB) (fragrances), *vii. Iso E Super (OTNE) (fragrances), viii. Methyl-paraben (preservatives), ix. Musk ketone (fragrances), x. Musk xylene (fragrances), xi. N,N-diethyl-m-toluamide (DEET)/ (biocides), xii. Octocrylene (OC) (UV Filters), xiii. Octyl-tirazone (OT or ethylhexyl triazone) (UV Filters), xiv. Tonalide (AHTN) (fragrances), *xv. Triclocarban (biocides), and *xvi. Triclosan (biocides).

12.3.2.1.7. Pesticides (12 types/groups detected)

i. AMPA (Aminomethyl phosphonic acid) (degradation product of herbicide Glyphosate) (herbicide), *ii. Atrazine (herbicide), *iii. Carbendazim (fungicide), *iv. Diazinon (insecticide), *v. Diuron (herbicide/ algicide), *vi. Glyphosate (herbicide), *vii. Irgarol 1051 (algicide), viii. Isoproturon (herbicide), ix. MCPA (4-chloro-2-methylphenoxy) acetic acid) (herbicide), x. Mecoprop (herbicide), *xi. Terbuthylazine (herbicide), and xii. Terbutryn (herbicide).

12.3.2.1.8. Pharmaceuticals (46 types/groups detected)

i. Atenolol (*beta blocker*), ii. Azithromycin, (*antibiotic*), *iii. Caffeine (*stimulant*), *iv. Carbamazepine (*anticonvulsant*), *v. Ciprofloxacin (*antibiotic*), *vi. Clarithromycin (*antibiotic*), *vii. Clindamycin (*antibiotic*), *viii. Diatrizoic acid (*contrast medium*), ix. Diclofenac (*anti-inflammatory*), x. Diltiazem (*anti-hypertensive*), xi. Eprosartan (*anti-hypertensive*), *xii. Erythromycin (*antibiotic*), xiii. Fluconazole (*antifungal*), xiv. Furosemide (*diuretic*), xv. Gabapentin (*anticonvulsant*), xvi. Gemfibrozil (anti-cholesterol/lipid regulator), xvii. Hydrochlorothiazide (diuretic), *xviii. Ibuprofen (*Anti-inflammatory*), xix. Iohexol (*contrast medium*), xx. Iomeprol (*contrast medium*), xxi. Iopamidol (*contrast medium*), xxii. Ipromide (*contrast medium*), xxiii. Irbersartan (*anti-hypertensive*), xxiv. Ketoprofen (*anti-inflammatory*), xxv. Lisinopril (anti-hypertensive), xxvi. Losartan (*anti-hypertensive),* xxvii. Mefenamic acid (*anti-*

inflammatory), xxviii. Metformin (anti-diabetic), xxix. Metoprolol (beta blocker), xxx. Metronidazole (antibiotic), xxxi. Naproxen (anti-inflammatory), xxxii. Norfloxacin (antibiotic), xxxiii. Ofloxacin (antibiotic), xxxiv. Oxazepam (anxiolytic), xxxv. Paracetamol (analgesic), xxxv. Pravastatin (*anti-cholesterol/lipid regulator*), xxxvi. Primidone (anticonvulsant), *xxxvii. Propanolol (beta blocker), xxxviii. Ranitidine (antacid), xxxix. Salicylic acid (anti-inflammatory), xl. Simvastatin (anti-cholesterol/lipid regulator), xli. Sotalol (beta blocker), *xlii. Sulfamethoxazole (antibiotic), xliii. Tramadol (analgesic), xliv. Trimethoprim (antibiotic), xlv. Valsartan (anti-hypertensive), xvi. Venlafaxine (anti-depressant).

12.3.2.1.9. Polycyclic aromatic hydrocarbons (PAHs) (12 types/groups detected)

i. Acenaphthene, ii. Anthracene, iii. Benzo[a]pyrene, iv. Benzo[b]fluoranthene, v. Benzo[ghi]perylene, vi. Benzo[k]fluoranthene, vii. Fluoranthene, viii. Fluorene, ix. Indeno[1,2,3-cd]pyrene, x. Naphthalene, *xi. Phenanthrene, and xii. Pyrene.

12.3.2.1.10. Surfactants (12 types/groups detected)

i. 2,4,7,9-tetramethyl-5-decyne-4,7-diol (TMDD) (*non-ionic*), *ii. Alcohol ethoxylate (AE) (*non-ionic*), iii. Alkyl ethoxy sulfate (AES) (*anionic*), iv. Alkyl sulfate (AS) (*anionic*), v. Alkyldimethylbenzyl ammonium chlorides/ *Benzalkonium chloride (BACs) (*cationic*), *vi. Alkyltrimethyl ammonium chlorides (ATAC) (anionic), *vii. Dialkyldimethyl ammonium chlorides (DDAC) (*anionic*), *viii. Linear alkyl benzene sulfonate (LAS) (*anionic*),* ix. Nonylphenol ethoxylates (NPEs)/ nonylphenol (NP) (*non-ionic*), *x. Octylphenol ethoxylates (OPEs)/octylphenol (OP) (*non-ionic*), xi. Secondary alkane sulfonate (SAS) (*anionic*), and *xii. Soap (*anionic*).

12.3.2.1.11. Volatile organic compounds (06 types/groups detected)

i. 1,2,4-Trimethylbenzene, ii. Ethylbenzene, iii. m-Xylene+p-Xylene, iv. o-Xylene, v. Styrene, and vi. Toluene.

Table 12.3. Typical sewage/wastewater treatment plant (WWTP) chemical effluent concentration and PNEC (compiled by Margot et al 2015)

*[**note i**. chemicals which have exceeded the **PNEC** (Predicted No-Effect concentration) is marked with an * (asterisk) (in column four); PNEC is usually derived using **EC_{50}/ LC_{50}** values of the most sensitive species (algae, daphnia, fish); **note ii**. the assessment factor (**AF**) which represent the quality of the available ecotoxicity tests data are also included in column four; if **AF ≥ 1,000**, the PNEC is less reliable and if **AF = 10**, the PNEC is more reliable; **na**= data is not available)].*

Chemical group	Substance	Effluent concentration in sewage water/wastewater	Remarks *PNEC*[10], *AF*[11], *Toxic* (when effluent concentration > PNEC), *Persistence* (substances removed < 50% in conventional treatment) (Margot et al 2015)
1. Heavy metals	Aluminium (Al)	Total metals[12]: 68 µg/L; Dissolved metals: 20 µg/L (UK) (Gardner et al 2012)	PNEC: 0.4 µg/L AF: 10

[10]***Predicted no-effect concentration (PNEC)=*** *is the concentration of a substance in any environment below which, adverse effects will most likely not occur during the long-term or short-term exposure. In environmental risk assessment, PNECs is compared to actual or predicted environmental concentration (PEC) to determine if the risk of a substance is acceptable or not. PNEC is usually derived using EC_{50}/LC_{50} values of the most sensitive species (algae, Daphnia, fish) https://www.chemsafetypro.com/Topics/CRA/How_to_Calculate_Predicted_No-Effect_Concentration_(PNEC).html; Richman and Castensson 2008).*

[11]***Assessment factor (AFs)/Security factor*** *is used to calculate the PNEC (PNEC = [Concentration affecting the most sensitive species]/AF), which depends on the number and the quality of the available ecotoxicity tests. For **AF ≥ 1,000**, only acute ecotoxicity data are available, and thus the PNEC is less reliable whereas **AF of 10** means a long-term result (EC_{10} or NOECs) from three species representing different trophic levels are available, therefore PNEC is more reliable (Margot et al 2015). Species representing **trophic levels** are water (algae, Daphnia, fish), soil (earthworm, plants and microorganisms)); **AF 50** = two long-term results (EC_{10} or NOECs) from species representing two trophic levels are available; **AF 100** = one long-term results (EC_{10} or NOECs) from species representing two trophic levels are available; **AF 1,000** = at least one short term (EC_{50}/LC_{50}) from each of the trophic levels are available. (NOEC= No observed effect concentration is the highest tested concentration for which there are no statistically significant difference of effect ($p<0.05$) when compared to the control group in long-term ecotoxicity studies; EC_{10} is the effect concentration at which 10% effect (mortality, inhibition of growth, reproduction, etc) is observed compared to the control group. https://www.chemsafetypro.com/Topics/CRA/How_to_Calculate_Predicted_No-Effect_Concentration_(PNEC).html*

[12]***Total and dissolved metals=*** *Dissolved metals are **filtered prior to analysis** to remove the particulate matter. This is typically performed on groundwater. **Total metals** are performed on the sample as received and is generally tested on surface water, trade waste and potable water samples.*

Chemical group	Substance	Effluent concentration in sewage water/wastewater	Remarks **_PNEC_**[10], **_AF_**[11], **_Toxic_** *(when effluent concentration > PNEC)*, **_Persistence_** *(substances removed < 50% in conventional treatment) (Margot et al 2015)*
	*Cadmium (Cd)	Total metals: 0.05 µg/L Dissolved metals: 0.24 µg/L (UK) (Gardner et al 2012)	PNEC: 0.013 µg/L AF: 10 **Toxic**
	Chromium (Cr)	Total metals: 0.7 µg/L Dissolved metals: 0.6 µg/L (UK) (Gardner et al 2012)	PNEC: 2.4 µg/L AF: 10
	*Copper (Cu)	Total metals: 8.3 µg/L Dissolved metals: 5.6 µg/L (UK) (Gardner et al 2012)	PNEC: 0.05 µg/L AF: 10 **Toxic**
	Iron (Fe)	Total metals: 8.3 µg/L Dissolved metals: 5.6 µg/L (UK) (Gardner et al 2012)	PNEC: 56 µg/L AF: 10
	*Lead (Pb)	Total metals: 0.8 µg/L Dissolved metals: 0.4 µg/L (UK) (Gardner et al 2012)	PNEC: 0.021 µg/L AF: 10 **Toxic**
	Mercury (Hg)	Total metals: 0.016 µg/L Dissolved metals: 0.01 µg/L (UK) (Gardner et al 2012)	PNEC: 0.025 µg/L AF: 10 **Persistence**
	*Nickel (Ni)	Total metals: 4.9 µg/L Dissolved metals: 4.3 µg/L (UK) (Gardner et al 2012)	PNEC: 0.110 µg/L AF: 10 **Toxic, persistence**
	*Zinc (Zn)	Total metals: 30.9 µg/L Dissolved metals: 24.0 µg/L (UK) (Gardner et al 2012)	PNEC: 0.0085 µg/L AF: 10 **Toxic, persistence**
2. Hormones	*17α-ethinylestradiol (EE2) *(contraceptive)*	0.0005-0.002 µg/L (Lubliner et al 2010, Robert et al 2011, Gardner et al 2012)	PNEC: 0.000004 µg/L AF: 10 **Toxic**

Chemical group	Substance	Effluent concentration in sewage water/wastewater	Remarks **PNEC[10], AF [11], Toxic** *(when effluent concentration > PNEC),* ***Persistence*** *(substances removed < 50% in conventional treatment) (Margot et al 2015)*
	*17β-Estradiol (E2) *(natural hormone)*	0.013-0.04 μg/L (Lubliner et al 2010, Robert et al 2011, Gardner et al 2012)	PNEC: 0.00042 μg/L AF: 10 **Toxic**
	Androsterone *(natural hormone)*	0.0001 μg/L (Lubliner et al 2010)	PNEC: na AF: na
	Estriol (E3) *(natural hormone)*	0.01 μg/L (Lubliner et al 2010)	PNEC: 0.67 μg/L AF: 100
	*Estrone (E1) *(natural hormone)*	0.012-0.217 μg/L (Lubliner et al 2010, Robert et al 2011, Gardner et al 2012)	PNEC: 0.0036 μg/L AF: 10 **Toxic**
	Progesterone *(natural hormone)*	0.008 μg/L (Lubliner et al 2010)	PNEC: 1.0 μg/L AF: 100
	Testosterone *(natural hormone)*	0.012 μg/L (Lubliner et al 2010)	PNEC: 0.2 μg/L AF: 50
3. Household chemicals	***3.1. Chelating agents[13]***		
	*Ethylene-diamine-tetra-acetic acid (EDTA)	90 μg/L (Knepper 2003, Reemtsma et al 2006, Clara et al 2012)	PNEC: 39 μg/L AF: 10 **Toxic, persistence**
	Nitrilo-tri-acetic acid (NTA)	10 μg/L (Knepper et al 2003, Reemtsma et al 2006, Clara et al 2012)	PNEC: 156 μg/L AF: 10

[13]***Chelating agents*** *are chemical compounds that react with metal ions to form a stable, water-soluble complex. They are also known as chelants, chelators, or sequestering agents. Chelating agents have a ring-like centre which forms at least two bonds with the metal ion allowing it to be excreted. Chelating agents are usually organic compounds (a compound that contains carbon) https://www.drugs.com/drug-class/chelating-agents.html*

Chemical group	Substance	Effluent concentration in sewage water/wastewater	Remarks ***PNEC***[10], ***AF*** [11], ***Toxic*** *(when effluent concentration > PNEC),* ***Persistence*** *(substances removed < 50% in conventional treatment) (Margot et al 2015)*
	3.2. Corrosion inhibitors[14]		
	2-hydroxybe-nzothiazole (OHBT)	0.1 µg/L (Reemtsma et al 2006)	na
	2-methylthi-obenzothiazole (MTBT)	0.1 µg/L (Reemtsma et al 2006)	na **Persistence**
	Benzothiazole	0.3 µg/L (Reemtsma et al 2006)	PNEC: 30 µg/L AF: 50
	Benzothiazole-2-sulfonic acid (BTSA)	1.6 µg/L (Reemtsma et al 2006)	na **Persistence**
	Benzotriazole	6.6 µg/L (Loos et al 2013)	PNEC: 40 µg/L AF: 10 **Persistence**
	Methyl-benzotriazole	2.9 µg/L (Loos et al 2013)	PNEC: 40 µg/L AF: 10 **Persistence**
	3.3. Flame retardants[15] ***(brominated)***		
	2,2' ,4,4' -Tetrabro-modiphenyl ether (BDE-47)	0.0007-0.008 µg/L (Gardner et al 2012, Hope et al 2012)	PNEC: 0.5 µg/L AF: 10

[14]***Corrosion inhibitors*** *= a chemical compound that, when added to a liquid or gas, decreases the corrosion rate of a material, typically a metal or an alloy, that comes into contact with the fluid.*

[15]***Flame retardants*** *refer to a variety of substances that are added to combustible materials (plastics and textile) to prevent fires from starting or to slow the spread of fire and provide additional escape time. Flame retardants are used in four major areas: Electronics and Electrical Devices (TVs and computers); Building and Construction Materials (homes, offices and public buildings, including schools and hospitals); Furnishings and Transportation (aeroplanes, cars, trains). (https://www.chemicalsafetyfacts.org/flame-retardants/).*

Chemical group	Substance	Effluent concentration in sewage water/wastewater	Remarks ***PNEC***[10], ***AF*** [11], ***Toxic*** *(when effluent concentration > PNEC)*, ***Persistence*** *(substances removed < 50% in conventional treatment) (Margot et al 2015)*
	Decabromo-diphenyl ether (BDE-209)	0.002 µg/L (North 2004)	PNEC: 0.04 µg/L AF: 50
	Pentabromo-diphenyl ethers (BDE-99)	0.0006-0.009 µg/L (Gardner et al 2012, Hope et al 2012)	PNEC: 10 µg/L AF: 10,000
	Tetrabromobisphen ol A (TBBA	0.002 µg/L (Potvin et al 2012)	PNEC: 1.7 µg/L AF: 10
	3.4. Flame retardants (chlorinated paraffins)		
	Short chain chlorinated paraffins (SCCPs, C10-13)	0.06 µg/L (Zeng et al 2013)	PNEC: 0.5 µg/L AF: 10
	3.5. Flame retardants (phosphorous)		
	Tri-iso-butyl phosphate (TiBP)	0.16 µg/L (Meyer and Bester 2004, Loos et al 2013)	PNEC: 341 µg/L AF: 100
	Tri-n-butyl phosphate (TnBP)	0.3 µg/L (Meyer and Bester 2004, Loos et al 2013)	PNEC: 37 µg/L AF: 10
	Triphenyl phosphate (TPP)	0.05 µg/L (Meyer and Bester 2004, Loos et al 2013)	PNEC: 0.14 µg/L AF: 10
	Tris(1,3-dichloro-2-propyl) phos-phate (TDCP)	0.15 µg/L (Meyer and Bester 2004, Loos et al 2013)	PNEC: 10 µg/L AF: 50 **Persistence**
	Tris(2-chloro, 1-methylethyl) phosphate (TCPP)	1.5 µg/L (Meyer and Bester 2004, Loos et al 2013)	PNEC: 120 µg/L AF: 50 **Persistence**

Chemical group	Substance	Effluent concentration in sewage water/wastewater	Remarks ***PNEC*[10]*, AF* [11], *Toxic*** *(when effluent concentration > PNEC),* ***Persistence*** *(substances removed < 50% in conventional treatment) (Margot et al 2015)*
	Tris(2-chloroethyl) phosphate (TCEP)	0.35 µg/L (Meyer and Bester 2004, Loos et al 2013)	PNEC: 1440 µg/L AF: 50 **Persistence**
	Tris(butoxyethyl)-phosphate (TBEP)	0.44 µg/L (Meyer and Bester 2004, Loos et al 2013)	PNEC: 76 µg/L AF: 100
	***3.6. Perfluorinated compounds*[16]**		
	*Perfluorooctane sulfonic acid (PFOS)	0.012 µg/L (Guo et al 2010, Loos et al 2013)	PNEC: 0.01 µg/L AF: 10 **Toxic, persistence**
	Perfluorooctanoic acid (PFOA)	0.013 µg/L (Guo et al 2010, Loos et al 2013)	PNEC: 0.65 µg/L AF: 10 **Persistence**
	3.7. Plastic additives/ plasticizers		
	*Bisphenol A (BPA)	0.2 µg/L (Robert et al 2011, Gardner et al 2012)	PNEC: 0.175 µg/L AF: 10 **Toxic**
	Butyl benzyl phthalate (BBP)	0.18 µg/L (Bergé et al 2013)	PNEC: 1.81 µg/L AF: 10
	*Di-(2-ethylhexyl) phthalate (DEHP)	2.4 µg/L (Bergé et al 2013)	PNEC: 0.01 µg/L AF: 10 **Toxic**

[16]***Perfluorinated compounds*** *[PFCs] are organic substances in which all of the hydrogens of the hydrocarbon backbones are substituted with fluorine atoms. The fluorine-carbon bonds are extremely stabile conferring these substances with very high thermal and chemical stability. PFCs are persistent, and some of the substances bioaccumulate in the bioaccumulate in the biota (Stahl et al 2011).*

Chemical group	Substance	Effluent concentration in sewage water/wastewater	Remarks ***PNEC***[10], ***AF*** [11], ***Toxic*** *(when effluent concentration > PNEC),* ***Persistence*** *(substances removed < 50% in conventional treatment) (Margot et al 2015)*
	Diethyl phthalate (DEP)	0.34 µg/L (Bergé et al 2013)	PNEC: 102 µg/L AF: 10
	Dimethyl phthalate (DMP)	0.34 µg/L (Bergé et al 2013)	PNEC: 96 µg/L AF: 10
	*Di-n-butyl phthalate (DnBP)	0.57 µg/L (Bergé et al 2013)	PNEC: 0.34 µg/L AF: 10 **Toxic**
	3.7. Sweeteners[17]		
	Acesulfame	30 µg/L (Lange et al 2012, Kokotou and Thomaidis 2013)	PNEC: 2000 µg/L AF: 50 **Persistence**
	*Cyclamate	0.5 µg/L (Lange et al 2012, Kokotou and Thomaidis 2013)	PNEC: 0.001 µg/L AF:50
	Saccharin	0.5 µg/L (Lange et al 2012, Kokotou and Thomaidis 2013)	PNEC: 9983 µg/L AF: 50
	*Sucralose	10 µg/L (Lange et al 2012, Kokotou and Thomaidis 2013)	PNEC: 0.001 µg/L AF: na **Toxic**
4. Illicit drugs[18]	Amphetamine (AMP) *(nervous system*	0.002 µg/L (Kasprzyk-Hordern et al 2009, Bijlsma et al 2012)	PNEC: 4.9 µg/L AF: 1000

[17]***Sweetener*** *= any of various natural and artificial substances that provide a sweet taste in food and beverages. "diet drinks contain artificial sweeteners like saccharin"*

[18]***Illicit drugs*** *are substances that either* ***stimulate*** *(such as cocaine or amphetamines) or* ***inhibit*** *(such as heroin or sedative-hypnotics) the central nervous system or cause* ***hallucinogenic*** *effects (such as marijuana or LSD) to the effect that their use has been prohibited globally; https://www.sciencedirect.com/topics/medicine-and-dentistry/illicit-drug*

Chemical group	Substance	Effluent concentration in sewage water/wastewater	Remarks ***PNEC***[10], ***AF*** [11], ***Toxic*** *(when effluent concentration > PNEC),* ***Persistence*** *(substances removed < 50% in conventional treatment) (Margot et al 2015)*
	stimulant)		
	Cocaine *(nervous system stimulant)*	0.03 µg/L (Kasprzyk-Hordern et al 2009, Bijlsma et al 2012)	PNEC: na AF: na
	Ecstasy (MDMA) *(psychedelic drug)*	0.1 µg/L (Kasprzyk-Hordern et al 2009, Bijlsma et al 2012)	PNEC: na AF: na **Persistence**
	THC-COOH (cannabis metabolite) *(psychoactive drug)*	0.013 µg/L (Kasprzyk-Hordern et al 2009, Bijlsma et al 2012)	PNEC: na AF: na
5. Persistent organic pollutants (POPs)	Aldrin *(insecticide)*	0.001 µg/L (Katsoyiannis et al 2004)	PNEC: 0.036 µg/L AF: 10
	Dichlorodiphenyl-trich-loroethane (DDT) *(insecticide)*	0.001 µg/L (Katsoyiannis and Samara 2004)	PNEC: 0.001 µg/L AF: 10
	*Dieldrin *(insecticide)*	0.009 µg/L Katsoyiannis and Samara 2004)	PNEC: 0.001 µg/L AF: 10 **Toxic**
	*Endosulfan *(insecticide)*	0.0027 µg/L (Katsoyiannis and Samara 2004)	PNEC: 0.001 µg/L AF: 10 **Toxic**
	Endrin *(insecticide)*	0.0028 µg/L Katsoyiannis and Samara 2004)	PNEC: 0.034 µg/L AF: 10
	*Heptachlor *(insecticide)*	0.0064 µg/L (Katsoyiannis and Samara 2004)	PNEC: 0.046 µg/L AF: 10 **Toxic**
	Hexachloro-benzene	0.0017 µg/L (Katsoyiannis and Samara 2004)	PNEC: 0.004 µg/L

Chemical group	Substance	Effluent concentration in sewage water/wastewater	Remarks ***PNEC***[10]***, AF*** [11]***, Toxic*** *(when effluent concentration > PNEC),* ***Persistence*** *(substances removed < 50% in conventional treatment) (Margot et al 2015)*
	(fungicide)		AF: 10
	Hexachlorocyclohexanes (HCHs) *(insecticide)*	0.0062 µg/L Katsoyiannis and Samara 2004)	PNEC: 0.06 µg/L (AF:10)
	Pentachloronitrobenzene (PCNB) *(fungicide)*	0.014 µg/L (Katsoyiannis and Samara 2004)	PNEC: 1.3 µg/L AF: 10
	*Polychlorinated biphenyls (PCBs) *(Σ 209 PCBs) (dielectric and coolant fluids)*	0.02 µg/L (Hope et al 2012, Balasubramani et al 2014)	PNEC: 0.01 µg/L AF: 10 **Toxic**
6. Personal care products	4-methylbenydlidene camphor (4-MBC) *(UV Filters)*	0.8 µg/L (Balmer et al 2005, Kupper et al 2006)	PNEC: na (AF: na)
	Benzophenone-3 (BP-3) *(UV Filters)*	0.27 µg/L (Balmer et al 2005, Kupper et al 2006)	PNEC: 1.6 µg/L (AF: 50)
	*Chloroxylenol *(biocides)*	0.3 µg/L (Kasprzyk-Hordern et al 2009)	PNEC: 0.026 µg/L (AF: 5000) **Toxic**
	*Ethylhexyl methoxycinnamate (EHMC or OMC) *(UV Filters)*	0.036 µg/L (Balmer et al 2005, Kupper et al 2006)	PNEC: 0.027µg/L (AF: 1000) **Toxic (?)**
	Ethyl-paraben *(preservatives)*	0.0025 µg/L (González-Mariño et al 2011)	PNEC: 4210 µg/L (AF: 100)

Chemical group	Substance	Effluent concentration in sewage water/wastewater	Remarks ***PNEC***[10], ***AF*** [11], ***Toxic*** *(when effluent concentration > PNEC)*, ***Persistence*** *(substances removed < 50% in conventional treatment) (Margot et al 2015)*
	Galaxolide (HHCB) *(fragrances)*	0.085 µg/L (Bester et al 2008, Fatta-Kassinos et al 2010, Clara et al 2011)	PNEC: 6.8 µg/L AF: 10
	*Iso E Super (OTNE) *(fragrances)*	1.4 µg/L (Bester et al 2008, Fatta-Kassinos et al 2010, Clara et al 2011)	PNEC: 0.56 µg/L AF: 50 **Toxic**
	Methyl-paraben *(preservatives)*	0.019 µg/L (González-Mariño et al 2011)	PNEC: 4.0 µg/L AF: 50
	Musk ketone *(fragrances)*	0.017 µg/L (Bester et al 2008, Fatta-Kassinos et al 2010, Clara et al 2011)	PNEC: 0.0003 µg/L AF: 10,000
	Musk xylene *(fragrances)*	0.7 µg/L (Bester et al 2008, Fatta-Kassinos et al 2010, Clara et al 2011)	PNEC: 0.8 µg/L AF: 5,000
	N,N-diethyl-m-toluamide (DEET) *(biocides)*	0.017 µg/L (DGE 2013, Loos et al 2013)	PNEC: 0.01 µg/L AF: 10,000
	Octocrylene (OC) *(UV Filters)*	0.052 µg/L (Balmer et al 2005, Kupper et al 2006)	PNEC: 1000 µg/L AF: 100
	Octyl-tirazone (OT or ethylhexyl triazone) *(UV Filters)*	<0.034 µg/L (Kupper et al 2006)	PNEC: 800 µg/L AF: 100
	Tonalide (AHTN) *(fragrances)*	0.25 µg/L (Bester et al 2008, Fatta-Kassinos et al 2010, Clara et al 2011)	PNEC: 3.9 µg/L AF: 50
	*Triclocarban *(biocides)*	0.070 µg/L (Gardner et al 2012, Loos et al 2013, Lozano et al 2013)	PNEC: 0.0011 µg/L AF: 50

Chemical group	Substance	Effluent concentration in sewage water/wastewater	Remarks ***PNEC***[10]***, AF*** [11]***, Toxic*** *(when effluent concentration > PNEC),* ***Persistence*** *(substances removed < 50% in conventional treatment) (Margot et al 2015)*
			Toxic
	*Triclosan *(biocides)*	0.2 µg/L (Gardner et al 2012, Loos et al 2013, Lozano et al 2013)	PNEC: 0.0015 µg/L AF: 10 **Toxic**
7. Pesticides	AMPA (Aminomethyl phosphonic acid) (degradation product of herbicide Glyphosate) *(herbicide)*	4.0 µg/L (Gardner et al 2013, Gardner et al 2013)	PNEC: 11.0 µg/L AF: 100 **Persistence**
	*Atrazine *(herbicide)*	0.01 µg/L (Loos et al 2013, Margot et al 2013)	PNEC: 0.000001 µg/L AF: 10 **Toxic, persistence**
	*Carbendazim *(fungicide)*	0.1 µg/L (Oulton et al 2010, DGE 2013, Margot et al 2013)	PNEC: 0.000011 µg/L AF: 10 **Toxic, Persistence**
	*Diazinon *(insecticide)*	0.04 µg/L (Oulton et al 2010, DGE 2013, Margot et al 2013)	PNEC: 0.00005 µg/L AF: 10 **Toxic, Persistence**
	*Diuron *(herbicide/ algicide)*	0.07 µg/L (Singer et al 2010, DGE 2013, Loos et al 2013, Margot et al 2013)	PNEC: 0.000026 µg/L AF: 10 **Toxic, Persistence**
	*Glyphosate *(herbicide)*	0.85 µg/L (Gardner et al 2013, Gardner et al 2013)	PNEC: 0.2 µg/L AF: 10 **Toxic, Persistence**

Chemical group	Substance	Effluent concentration in sewage water/wastewater	Remarks ***PNEC***[10]***, AF*** [11]***, Toxic*** *(when effluent concentration > PNEC),* ***Persistence*** *(substances removed < 50% in conventional treatment) (Margot et al 2015)*
	*Irgarol 1051[19] *(algicide)*	0.007 µg/L (Singer et al 2010, DGE 2013, Margot et al 2013)	PNEC: 0.00000 45 µg/L AF: 10 **Toxic, Persistence**
	Isoproturon *(herbicide)*	0.025 µg/L (Singer et al 2010, DGE 2013, Loos et al 2013, Margot et al 2013)	PNEC: 0.174 µg/L AF: 10 **Persistence**
	MCPA (4-chloro-2-methylphenoxy) acetic acid) *(herbicide)*	0.3 µg/L (DGE 2013, Loos et al 2013)	PNEC: 1.4 µg/L AF: 10 **Persistence**
	Mecoprop *(herbicide)*	0.5 µg/L (Singer et al 2010, DGE 2013, Loos et al 2013, Margot et al 2013)	PNEC: 514 µg/L AF: 10 **Persistence**
	*Terbuthylazine *(herbicide)*	0.5 µg/L (Singer et al 2010, Margot et al 2013)	PNEC: 0.06 µg/L AF: 10 **Toxic, persistence**
	Terbutryn *(herbicide)*	0.02 µg/L (Singer et al 2010, Margot et al 2013)	PNEC: 0.2 µg/L AF: 10
8. Pharmaceuticals	Atenolol *(beta blocker)*	0.84-0.94 µg/L (Robert et al 2011, Kostich et al 2010)	PNEC: 64 µg/L AF: 50 **Persistence**
	Azithromycin *(antibiotic)*	0.22-0.41 µg/L (Lubliner et al 2010, DGE 2013)	PNEC: na AF: na **Persistence**

[19] ***Irgarol® 1051*** *is a highly specific and effective inhibitor of the photosynthesis of algae, thereby providing outstanding performance in their control. Combined with its very low water solubility, it is the ideal choice for use in antifouling coatings; https://www.basf.com/us/en/products/General-Business-Topics/dispersions/Products/irgarol/ irgarol_1051.html*

Chemical group	**Substance**	**Effluent concentration in sewage water/wastewater**	**Remarks** ***PNEC***[10]**,** ***AF*** [11]**,** ***Toxic*** *(when effluent concentration > PNEC),* ***Persistence*** *(substances removed < 50% in conventional treatment) (Margot et al 2015)*
	*Caffeine *(stimulant)*	0.18-0.82 μg/L (Lubliner et al 2010, Loos et al 2013, Margot et al 2013)	PNEC: 0.5 μg/L AF: 10 **Toxic**
	*Carbamazepine *(anticonvulsant)*	0.14-0.83 μg/L (Lubliner et al 2010, Robert et al 2011, Loos et al 2013, Kostich et al 2014)	PNEC: 0.089 μg/L AF: 10 **Toxic, persistence**
	*Ciprofloxacin *(antibiotic)*	0.0067-0.179 μg/L (Lubliner et al 2010, DGE 2013, Loos et al 2013, Kostich et al 2014)	PNEC: 0.0012 μg/L AF: 10 **Toxic**
	*Clarithromycin *(antibiotic)*	0.13-0.276 μg/L (Lubliner et al 2010, Robert et al 2011)	PNEC: 0.002 μg/L AF: 100 **Toxic, persistence**
	*Clindamycin *(antibiotic)*	0.05-0.07 μg/L (Loos et al 2013, Margot et al 2013, Kostich et al 2014)	PNEC: 0.0002 μg/L AF: 10,000 **Toxic (?), persistence**
	*Diatrizoic acid *(contrast medium)*	0.59-0.62 μg/L (Robert et al 201, Loos et al 2013)	PNEC: 0.5 μg/L AF: 10 **Toxic, Persistence**
	Diclofenac *(anti-inflammatory)*	0.26-0.65 μg/L (Robert et al 2011, Gardner et al 2012)	PNEC: 500 μg/L AF: 10
	Diltiazem *(anti-hypertensive)*	0.085-0.155 μg/L (Lubliner et al 2010, Kostich et al 2014)	PNEC: 1.64 μg/L AF: 5000
	Eprosartan *(anti-hypertensive)*	0.227-0.880 μg/L (Loos et al 2013, Margot et al 2013)	PNEC: na AF: **Persistence**

Chemical group	Substance	Effluent concentration in sewage water/wastewater	Remarks ***PNEC***[10], ***AF*** [11], ***Toxic*** *(when effluent concentration > PNEC),* ***Persistence*** *(substances removed < 50% in conventional treatment) (Margot et al 2015)*
	*Erythromycin *(antibiotic)*	0.0042-0.83 µg/L (Lubliner et al 2010, Robert et al 2011, Gardner et al 2012)	PNEC: 0.20 µg/L AF: 10 **Toxic, persistence**
	Fluconazole *(antifungal)*	0.108-0.110 µg/L (Margot et al 2013, Loos et al 2013)	PNEC: 0.006 µg/L AF: 5,000 **Toxic (?), persistence**
	*Furosemide *(diuretic)*	0.28 µg/L (Kostich et al 2014)	PNEC: 0.0006 µg/L AF: 1000 **Toxic?**
	Gabapentin *(anticonvulsant)*	1.9 µg/L (DGE 2013)	PNEC: 10 µg/L AF: 10,000 **Persistence**
	Gemfibrozil *(anti-cholesterol/ lipid regulator)*	0.14-1.4 µg/L (Lubliner et al 2010, Loos et al 2013, Kostich et al 2014)	PNEC: 312 µg/L AF: 10 **Persistence**
	Hydrochloro-thiazide *(diuretic)*	1.1 µg/L (Kostich et al 2014)	PNEC: 0.003 µg/L AF: 10,000 **Toxic (?), persistence**
	*Ibuprofen *(anti-inflammatory)*	0.081-0.46 µg/L (Lubinar et al 2010, Robert et al 2011, Gardner et al 2012, Loos et al 2013, Kostich et al 2014)	PNEC: 0.001 µg/L AF: 10 **Toxic**
	Iohexol *(contrast medium)*	0.015-0.19 µg/L (Margot et al 2013)	PNEC: 32000 µg/L AF: 100 **Persistence**
	Iomeprol *(contrast medium)*	0.38-10 µg/L (Loos et al 2013, Margot et al 2013)	PNEC: na AF: na **Persistence**

Chemical group	Substance	Effluent concentration in sewage water/wastewater	Remarks *PNEC*[10], *AF* [11], ***Toxic*** *(when effluent concentration > PNEC)*, ***Persistence*** *(substances removed < 50% in conventional treatment) (Margot et al 2015)*
	Iopamidol *(contrast medium)*	0.144-1.6 µg/L (DGE 2013, Loos et al 2013)	PNEC: na AF: na **Persistence**
	Ipromide *(contrast medium)*	2.4-2.7 µg/L (DGE 2013, Loos et al 2013)	PNEC: na AF: na **Persistence**
	Irbersartan *(anti-hypertensive)*	0.48-1.7 µg/L (Margot et al 2013, Loos et al 2013)	PNEC: µg/L AF: na
	Ketoprofen *(anti-inflammatory)*	0.086-0.19 µg/L (DGE 2013, Loos et al 2013)	PNEC: na AF: na **Persistence**
	Lisinopril *(anti-hypertensive)*	0.18 µg/L (Kostich et al 2014)	PNEC: na AF: na
	Losartan *(anti-hypertensive)*	0.51 µg/L (Margot et al 2013)	PNEC: 2860 µg/L AF: 50
	Mefenamic acid *(anti-inflammatory)*	0.87 µg/L (Robert et al 2011)	PNEC: 0.79 µg/L AF: 5000 **Persistence**
	Metformin *(anti-diabetic)*	10.35-**27.80** µg/L (Robert et al 2011, Lubinar et al 2014)	PNEC: 2200 µg/L AF: 50
	Metoprolol *(beta blocker)*	0.24-0.41 µg/L (DGE 2013, Kostich et al 2014)	PNEC: 146 µg/L AF: 50 **Persistence**
	Metronidazole *(antibiotic)*	0.68 µg/L (DGE 2013)	PNEC: 250 µg/L AF: 50 **Persistence**

Chemical group	**Substance**	**Effluent concentration in sewage water/wastewater**	**Remarks** ***PNEC***[10], ***AF*** [11], ***Toxic*** *(when effluent concentration > PNEC),* ***Persistence*** *(substances removed < 50% in conventional treatment) (Margot et al 2015)*
	Naproxen *(anti-inflammatory)*	0.027-0.46 µg/L (Lubinar et al 2010, Robert et al 2011, Loos et al 2013)	PNEC: 0.16 µg/L AF: 10
	Norfloxacin *(antibiotic)*	0.039-0.07 µg/L (Lubinar et al 2010, DGE 2013)	PNEC: 0.16 µg/L AF: 10
	Ofloxacin *(antibiotic)*	0.01-0.25 µg/L (Lubinar et al 2010, Gardner et al 2012, DGE 2013, Kostich et al 2014)	PNEC: µg/L AF: 10
	Oxazepam *(anxiolytic*[20]*)*	0.16-0.35 µg/L (Margot et al 2013, Loos et al 2013)	PNEC: 100,00 µg/L AF: 100 **Persistence**
	Paracetamol *(analgesic)*	0.008-0.079 µg/L (Margot et al 2013, Kostich et al 2014, Lubinar et al 2014)	PNEC: 0.18 µg/L AF: 10
	Pravastatin *(Anti-cholesterol/ lipid regulator)*	0.42 µg/L (Gardner et al 2012)	PNEC: na AF: na **Persistence**
	Primidone *(anticonvulsant)*	0.2 µg/L (DGE 2013)	PNEC: 15,000 µg/L AF: 100 **Persistence**
	*Propanolol *(Beta blocker)*	0.033-0.140 µg/L (DGE 2013, Kostich et al 2014, Gardner et al 2012)	PNEC: 0.1 µg/L AF: 10 **Toxic, Persistence**
	Ranitidine *(antacid)*	0.007-0.84 µg/L (Lubinar et al 2010, Loos et al 2013, Kostich et al 2014)	PNEC: 0.00025 µg/L AF: 1000 **Toxic (?)**
	Salicylic acid *(anti-inflammatory)*	0.078 µg/L (Oulton et al 2010, Verlicchi et al 2012)	PNEC: 3.2 µg/L AF: 10

20 Anxiolytic = a drug used to reduce anxiety.

Chemical group	Substance	Effluent concentration in sewage water/wastewater	Remarks ***PNEC[10], AF [11], Toxic*** *(when effluent concentration > PNEC),* ***Persistence*** *(substances removed < 50% in conventional treatment) (Margot et al 2015)*
	Simvastatin *(Anti-cholesterol/ lipid regulator)*	0.042-0.098 µg/L (Margot et al 2013, Kostich et al 2014)	PNEC: 62.5 µg/L AF: 100
	Sotalol *(beta blocker)*	0.435 µg/L (Robert et al 2011)	PNEC: na AF: na **Persistence**
	*Sulfameth-oxazole *(antibiotic)*	0.238-1.2 µg/L (Robert et al 2011, Loos et al 2013, Kostich et al 2014, Lubinar et al 2014)	PNEC: 0.013 µg/L AF: 15 **Toxic, persistence**
	Tramadol *(analgesic)*	0.256 µg/L (Loos et al 2013)	PNEC: na AF: na **Persistence**
	Trimethoprim *(antibiotic)*	0.15- 0.48 µg/L (Robert et al 2011, Loos et al 2013, Kostich et al 2014, Lubinar et al 2014)	PNEC: 100 µg/L AF: 10 **Persistence**
	Valsartan *(anti-hypertensive)*	1.6-2.1 µg/L (Margot et al 2013, Kostich et al 2014)	PNEC: na AF: na
	Venlafaxine *(anti-depressant)*	0.12-0.15 µg/L (Loos et al 2013, Margot et al 2013)	PNEC: 0.0057 µg/L AF: 5000
9. Polycyclic aromatic hydrocarbons (PAHs) *(oil/ petroch-emicals)*	Acenaphthene	0.023 µg/L (Fatone et al 2011, Hope et al 2012)	PNEC: 1.0 µg/L AF: 10
	Anthracene	0.002 µg/L (Fatone et al 201, Gardner et al 2012)	PNEC: 0.2 µg/L AF: 10
	Benzo[a]pyrene	0.0011 µg/L (Fatone et al 2011, Gardner et al 2012)	PNEC: 0.002 µg/L AF: 10

Chemical group	Substance	Effluent concentration in sewage water/wastewater	Remarks ***PNEC***[10], ***AF*** [11], ***Toxic*** *(when effluent concentration > PNEC),* ***Persistence*** *(substances removed < 50% in conventional treatment) (Margot et al 2015)*
	Benzo[b] fluoranthene	0.003 µg/L (Fatone et al 201, Gardner et al 2012)	PNEC: 0.054 µg/L AF:10,000
	Benzo[ghi] perylene	0.0011 µg/L (Fatone et al 2011, Gardner et al 2012)	PNEC: 0.00001 µg/L AF: 10,000 **Toxic (?)**
	Benzo[k] fluoranthene	0.002 µg/L (Fatone et al 2011, Gardner et al 2012)	PNEC: 0.0002 µg/L AF: 5000
	Fluoranthene	0.01 µg/L (Fatone et al 2011, Gardner et al 2012, Hope et al 2012)	PNEC: 0.06 µg/L AF: 10
	Fluorene	0.013 µg/L (Fatone et al 2011, Gardner et al 2012)	PNEC: 1.0 µg/L AF: 10
	Indeno[1,2,3-cd]pyrene	0.0014 µg/L (Fatone et al 201, Gardner et al 2012)	PNEC: na AF: na **Persistence**
	Naphthalene	0.049 µg/L (Fatone et al 2011)	PNEC: 37 µg/L AF: 10
	*Phenanthrene	0.012 µg/L (Fatone et al 2011)	PNEC: 0.00002 µg/L AF: 10 **Toxic**
	Pyrene	0.015 µg/L (Fatone et al 201, Gardner et al 2012)	PNEC: 0.12 µg/L AF: 10
Surfactants: ***anionic****= negatively charged;* ***cationic****= positively charged;* ***non-ionic****= uncharged; nonylphenol and octylphenol are the degradation products of NPEs and OPEs respectively*			
10. Surfactants/ detergents	2,4,7,9-tetramethyl-5-decyne-4,7-diol (TMDD) *(non-ionic)*	0.6 µg/L (Guedez and Püttmann 2011)	PNEC: 10 µg/L AF: 100

Chemical group	Substance	Effluent concentration in sewage water/wastewater	Remarks ***PNEC***[10], ***AF*** [11], ***Toxic*** *(when effluent concentration > PNEC),* ***Persistence*** *(substances removed < 50% in conventional treatment) (Margot et al 2015)*
	*Alcohol ethoxylate (AE) *(non-ionic)*	5.0 µg/L (Belanger et al 2006)	PNEC: 1.56 µg/L AF: 10,50 **Toxic**
	Alkyl ethoxy sulfate (AES) *(anionic)*	6.5 µg/L (Matthijs et al 1999)	PNEC: 18 µg/L AF: 10
	Alkyl sulfate (AS) *(anionic)*	5.7 µg/L (Matthijs et al 1999)	PNEC: 84 µg/L AF: 50
	*Alkyldimethylbenzyl ammonium chlorides or Benzalkonium chloride (BACs) *(cationic)*	0.5 µg/L (Kreuzinger et al 2007, Clara et al 2012)	PNEC: 0.12 µg/L AF: 10 **Toxic**
	*Alkyltrimethyl ammonium chlorides (ATAC) *(anionic)*	0.24 µg/L (Kreuzinger et al 2007, Clara et al 2012)	PNEC: 0.136 µg/L AF: 50 **Toxic**
	*Dialkyldimethyl ammonium chlorides (DDAC) *(anionic)*	1.2 µg/L (Kreuzinger et al 2007, Clara et al 2012)	PNEC: 0.91 µg/L AF: 10
	*Linear alkyl benzene sulfonate (LAS) *(anionic)*	50 µg/L (Feijtel et al 1996, Mungray and Kumar 2009)	PNEC: 21 µg/L AF: 10 **Toxic**
	*Nonylphenol ethoxylates (NPEs)/ nonylphenol (NP) *(non-ionic)*	6.0 µg/L (Bergé et al 2012)	PNEC: 0.5 µg/L (NPEs)/ 0.03 µg/L (NP) AF: 10 **Toxic**

Chemical group	Substance	Effluent concentration in sewage water/wastewater	Remarks ***PNEC*[10], *AF*[11], *Toxic*** *(when effluent concentration > PNEC),* ***Persistence*** *(substances removed < 50% in conventional treatment) (Margot et al 2015)*
	*Octylphenol ethoxylates (OPEs)/ octylphenol (OP) *(non-ionic)*	1.0 µg/L (Loyo-Rosales et al 2007)	PNEC: 21 µg/L (OPEs)/ 0.00021 µg/L (OP) AF: 10 **Toxic**
	Secondary alkane sulfonate (SAS) *(anionic)*	3 µg/L (Field et al 1995)	PNEC: 7.2 µg/L AF: 50
	*Soap *(anionic)*	150 µg/L (Matthijs et al 1999)	PNEC: 22-44 µg/L; AF: 10 **Toxic**
11. Volatile organic compounds *(oil/ petrochemicals)*	1,2,4-Trimethyl-benzene[21]	0.340 µg/L (Fatone et al 2011)	PNEC: 23.56 µg/L AF: 100
	Ethylbenzene	0.063 µg/L (Fatone et al 2011)	PNEC: 20 µg/L AF: 50
	m-Xylene+ p-Xylene	0.005 µg/L (Fatone et al 2011)	PNEC: 26 µg/L AF: 50
	o-Xylene	0.170 µg/L (Fatone et al 2011)	PNEC: 26 µg/L AF: 50
	Styrene	0.1 µg/L (Fatone et al 2011)	PNEC: 6.3 µg/L AF: 10
	Toluene	0.75 µg/L (Fatone et al 2011)	PNEC: 100 µg/L AF: 10

21 ***1,2,4-Trimethylbenzene*** *= an aromatic hydrocarbon, It occurs naturally in coal tar and petroleum.*

12.3.3. Impacts of micropollutants found in sewage/ wastewater on aquatic biota

Chemical micro-pollutants found in sewage water can have different impacts on aquatic biota as highlighted below:

12.3.3.1. Heavy metals

Heavy metals (Cd, Cr, Cu, Hg) can affect growth, survival and reproduction of aquatic plants and animals. Heavy metals are persistent, bioaccumulative and toxic (PBT). Metal pollution can harm aquatic organisms through lethal and sub-lethal effects and can reduce or eliminate species from an ecosystem through increased susceptibility to fish disease, mortality and decreased fecundity (Kibria 2016a). Chapter 3 discussed extensively on heavy metals and its impact on biota including freshwater and marine fish and therefore is not repeated further in this chapter.

12.3.3.2. Household chemicals (EDTA, flame retardants, perfluorinated compounds, plastic additives)

EDTA can cause eutrophication since the molecule contains approximately 10% of the nitrogen that could eventually be available to the aquatic microbiota (Sillanpää 1997, Oviedo and Rodríguez 2003). **Flame retardants** (PBDEs) caused alteration of thyroid hormones in fish, inhibiting sperm production, reduction in egg production, gonad development, and ovarian growth, swimming activity, fertilization and increase in larval mortality (Nugegoda and Kibria 2017). **Perfluorinated compounds (PFOS)**: PFOS has been detected in algae, birds, clams, crabs, fish, frog, mammals, oysters, shrimps, starfish, turtles and zooplankton (OECD 2002, Houde et al 2006, D'Hollander et al 2010). PFOS was found more toxic to Zebra fish (*Danio rerio*) and caused sub-lethal effects [pericardial edema (build-up of fluid), malformation like spinal curvature] (Ulhaq et al 2013). **Plastic additives** (bisphenol-A) and DEHP affected egg production in aquatic snails (Jobling et al 2004) and reproduction in daphnids respectively (OEHHA 2009). Plastic additive compounds (phthalates, bisphenol-A) fall within the category of PBT. Chapters 4 discussed extensively on plastic additives and its impact on marine life, seafood, honey, salt, sugar, beer, drinking water, tourism, shipping and therefore is not repeated further in this chapter.

12.3.3.3. Oil compound/volatile organic compounds (PAHs, BTEX)

Oil exposure to fish reduced larval survival, decreased swimming speeds, edema (buildup of fluid), necrosis (tissue death), neoplasia (tissue damage), arrhythmia (abnormal heartbeat), decreased reproductive success and immune capacity, reduced growth in fish (Incardona et al 2012, Hook et al 2016). Chapter 6 discussed extensively on oil compounds and its impact on biota (fish), seafood, mariculture, fisheries/fishing, tourism, wildlife, ecosystems (corals, mangroves, seagrasses) and therefore is not repeated further in this chapter.

12.3.3.4. Personal care products (biocides, UV filters, parabens, fragrances)

Biocides (triclocarban, triclosan) affected the growth of algae and freshwater crustaceans (Brausch and Rand 2011). UV filters (4-MBC[22] and EHMC[23]) have been found to accumulate in different organs of fish (trout, perch, roach), crustaceans and marine mussels, inhibited algal growth and acts as endocrine disruptors to aquatic organisms (Gago-Ferrero et al 2012). **UV filters** are toxic to phytoplankton, microalgae, protozoa, and crustaceans (Sánchez-Quiles and Tovar-Sánchez 2015). **Parabens** (methylparaben and propylparaben) are used as a preservative (preservative in food, cosmetics, pharmaceuticals). Most ingredients in cosmetic products (soaps, toothpastes, shampoos, hair dyes, make-up, deodorants, hair styling products, perfumes, sunscreen lotions, creams) may eventually end up in the aquatic environment through sewage treatment systems (DEPA 2013). The longer chain parabens can have potential estrogenic effects on fish as they induce the production of vitellogenin in rainbow trout after intraperitoneal injection or after exposure in water containing propylparaben (Bjerregaard et al 2003). Paraben occurrence was also reported in fish tissues from the Philippines and Spain (Haman et al 2015). **Synthetic fragrances** (HHCB[24]) affected the larval development in certain crustaceans (copepods) is affected by synthetic fragrance (HHCB) (Walters et al 2005, Peck and Hornbuckle 2006).

12.3.3.5. Pharmaceuticals (antibiotics, hormones)

Antibiotics inhibited the growth of microalgae and affected reproduction and hatching of juvenile crustaceans (water fleas) (Kümmerer 2009). Diclofenac (an *anti-inflammatory drug*/pharmaceutical) caused cell damage (kidney) in fish (trout) (Mehinto et al 2010). **Hormones:** Synthetic hormones 17α-ethinylestradiol (EE2) caused endocrine-disrupting effects in fish (low egg production and feminization of male fish) when exposed to a mixture of sewage effluents

---

[22]***4-MBC*** *= 4-Methylbenzylcathinone, or Benzedrone. Enzacamene is an organic camphor derivative that is used in the cosmetic industry for its ability to protect the skin against UV, specifically UV B radiation.*

[23]***EHMC*** *= Ethylhexyl methoxycinnamate (EHMC) is one of the most common UVB filters used worldwide.*

[24]***HHCB*** *= 1,3,4,6,7,8-Hexahydro-4,6,6,7,8,8,-hexamethylcyclopenta[g]-2-benzopyran. Also known as Galaxolide is a synthetic polycyclic musk used as an ingredient in a wide range of consumer products including perfumes, cosmetics, shampoos, lotions, detergents, fabric softeners, and cleaning agents.*

(Jobling and Tyler 2003, Jobling et al. 2004). EE2 was found toxic to fish and invertebrates at 1.0 ng/L or less (Länge et al 2001, Schulte-Oehlmann et al 2004). Several pharmaceuticals are PBT. Chapter 9 discussed extensively on pharmaceuticals and its impact on biota including bacteria, algae, *Daphnia* and molluscs, fish, frogs, birds, crops and therefore is not repeated further in this chapter.

12.3.3.6. Pesticides (herbicides, insecticides, fungicides)

Pesticides are PBT. It caused significant effects on non-target organisms including birds, earthworms, fish, frogs, pollinators (Kibria et al 2010, Kibria 2016b). Chapter 7 discussed extensively on pesticides and its impact on biota (birds, earthworms, fish, frogs, bees) and therefore, is not repeated further in this chapter.

12.3.3.7. Surfactants

Surfactants like nonylphenol (alkylphenol ethoxylates degradation product)) is extremely toxic to aquatic organisms (e.g. EC_{50} for algae is 0.02 mg/, LC_{50} for *Daphnia* and fathead minnow fish are 0.19 mg/L and 0.3 mg/L respectively). Surfactants can affect the growth, motility and photosynthetic ability of algae (Lewis 1990). Chapter 10 discussed extensively on surfactants and its impact on biota (fish), seafood, crops, vegetables and fruits and therefore is not repeated further in this chapter.

12.3.4. Impacts of nutrients on the aquatic environment

Sewage discharge leads to excess nutrients to enter aquatic environments. Nutrient enrichment can cause a facet of environmental problems including eutrophication, algal blooms in both fresh and marine environment. When algal bloom dies off and sinks to the bottom of the river bed or seafloor, bacteria feed on dead algae using up oxygen causing the water to become depleted of oxygen (creating hypoxic or anoxic environments) (Fig. 12.3). Hypoxia can cause mass mortality of fishes and benthos resulting in the loss of fisheries resources and impacting livelihoods of fishers. Water with freshwater algal blooms could be unsuitable for agriculture, drinking, livestock, irrigation, recreation and wildlife. Some of the algae such as *Microcysts* sp. produce toxins (Microcystins) which can accumulate in crops and seafood and can be transferred to humans via the food and water. Cyanobacterial blooms/ toxic blue-green algae (BGA) can affect the water quality by causing peculiar tastes, odours, discolouration and unsightly scum. In the coastal environment, nutrient enrichment due to sewage spill can cause proliferation of microalgae, dinoflagellates. Incidents of paralytic shell-fish poisoning (PSP) of wildlife and human are associated with dinoflagellates, *Alexandrium* spp., *Gymnodinium* spp. (Chapter 8 discussed extensively on nutrients, algal blooms, algal toxins and hypoxia, therefore, is not repeated further in this chapter.

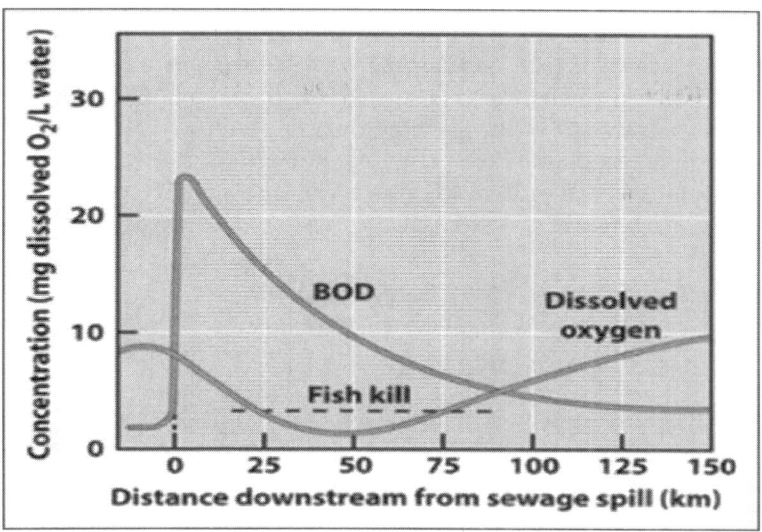

Fig. 12.3. Sewage spill and fish kills

When large scale sewage effluent enters a waterway, bacteria decompose sewage while using oxygen from the surrounding environment. It causes depletion of the dissolved oxygen of the water causing fish kill. Graph source: https://slideplayer.com/slide/6812917/ontang

12.3.5. Impacts of suspended solids/particles on the aquatic environment

Suspended solids/particles reduce the clarity of the water. It reduces light penetration affecting photosynthesis in aquatic environment/plants. It may clog/ coat fish gills and feeding apparatus in molluscs causing stress or mortality. It can smother fish eggs and benthic organisms (macroinvertebrates) and their habitats.

12.3.6. Impacts on humans health

Several micro-pollutants are toxic and bio-accumulative (can accumulate in crops, vegetables and seafood; e.g. heavy metals, pesticides, POPs, oil compounds) and can be transferred to human via the food chain. Some of the micro-pollutants are also endocrine disrupting (e.g. plastic additives, pesticides, nonyl-phenol) and carcinogenic to humans (e.g. heavy metals, pesticides, petrochemical compounds, POPs, nonylphenol) (Soares et al 2008, Kibria et al 2010). Sewage water contains disease-causing bacteria, viruses, protozoan parasites and helminths. There is a possibility of transmission of waterborne diseases to humans via pathogens in contaminated water or food.

12.4. Linkages between sewage pollution and climate change

Climate change could increase the incidence of sewer overflows due to increased rainfall intensity during storms, inundation/ flooding of wastewater treatment plants (due to sea-level rise) and corrosion of wastewater treatment equipment due to the rise in ambient and seasonal temperatures (Kibria et al 2016). Currently, the wastewater utility systems were not built to cope with the stresses of extreme rainfall or extreme temperatures projected in the future. For example, warmer temperatures and prolonged drought can change the corrosion patterns of sewage pipes and can lead to water/ sewer main pipes to break.

12.5. Management of sewage pollution

12.5.1. Industrial best practices

- Monitor **sewage overflow events** overtime regularly, isolate problematic areas and address the identified areas;
- Use **sewage spill hazard matrix** to calculate/ assess the risks associated with the spill as given in Table 12.4 (this spill hazard matrix is based on the i. volume of the sewage spill; ii. location where the sewage spill occurred (rural areas or residential areas or sensitive areas); iii. population density (high or low population); Table 12.4);
- **Treat wastewater** (to produce an effluent which will cause as little harm as possible when discharged to the surrounding environment, thereby preventing pollution compared to releasing untreated wastewater into the environment). As an example, sewage treatment at Werribee Treatment Plant, Victoria, Australia happens/ completes in several stages (Fig. 12.4): ***stage 1.*** wastewater (raw sewage) enters the aerobic ponds; ***stage 2.*** bacteria decompose/ digest the organic material and produce odour and methane gas (which is captured); ***stage 3.*** sludge (thick, soft, wet mud) containing the heavy metals and other chemicals are settled out; **stage 4.** wastewater moves into the aerobic pond; ***stage 5***. algae grow in the nutrient-rich water; ***stage 6***. bacteria and algae remove nitrogen from the nutrient-rich water; ***stage 7.*** zooplankton grow and feed on bacteria and algae; ***stage 8.*** treated wastewater, along with zooplankton, flows into the Port Phillip Bay, Victoria, Melbourne as effluents *(**note**: a number of hydrophilic pollutants (e.g. pharmaceuticals) have poor removal efficiency compared to hydrophobic pollutants; see Table 12.5);*

Table 12.4. Sewage spill hazard calculator* (kL = 1,000 litres).

[Sewage hazard calculator is based on: https://epa.tas.gov.au/Documents/Sewage%20Spill%20Notification%20Guidelines%20OCTFINALv2.pdf].*

LOCATION (of sewage spill)	**VOLUME (of sewage spill)**			
	Minor Spill (1 kL)	**Moderate Spill (1-10 kL)**	**Major Spill (10-50 kL)**	**Maximum Spill (> 50 kL)**
Low Risk Area *(spill in* ***rural*** *areas with* ***low population,*** *unlikely to contaminate natural waterways,* ***good dilution****)*	Not notifiable	Not notifiable	Notifiable	Notifiable
Medium Risk Area *(spill in* ***residential areas*** *with* ***medium population****, likely to harm or* ***contaminate*** *natural waterways, low dilution)*	Not notifiable	Notifiable	Notifiable	Critically Notifiable
High Risk Area *(spill in* ***sensitive areas*** *such as at playgrounds, schools, sporting facilities, public reserves, national parks, wetlands, recreation areas (lakes or coastal beach);* ***very likely to contaminate pristine*** *waterways or recreational water areas or* ***aquaculture/fishery facilities*** *(shell-fish fisheries)*	Notifiable	Notifiable	Critically Notifiable	Critically Notifiable

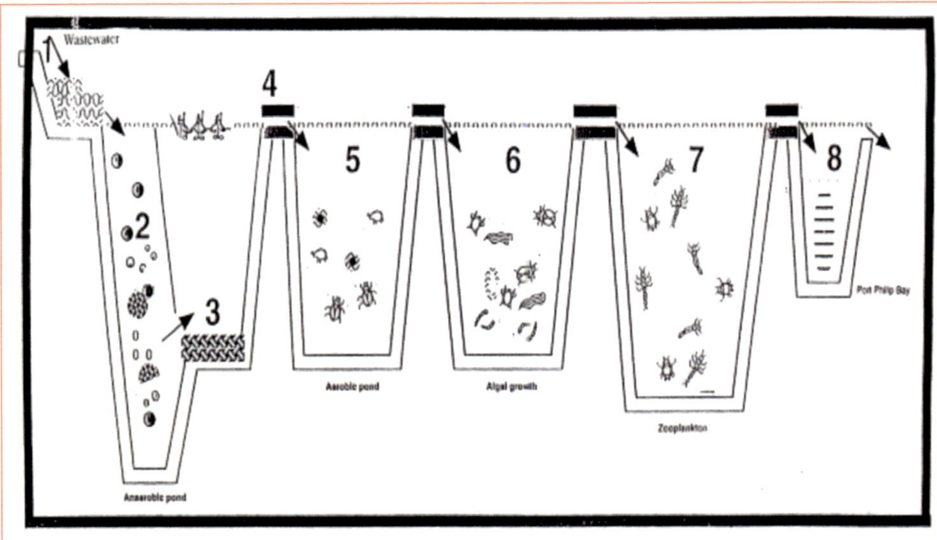

Fig. 12.4. Simplified sequence of zooplankton production at the Werribee Sewage Treatment Lagoons (imgae source; Kibria et al 1999, page 222).

Table 12.5. Pollutants removal efficiency of micropollutants found in sewage water (Margot et al 2015).

Category	Typical removal efficiency of micropollutants (in WWTP) (pollutants in bold are poor removal (< 50%)
Biocides *(Personal care products)*	Medium to high removal efficiency (**62-95%;** mainly removed by **sorption**) *[Chloroxylenol (95%) (Kasprzyk-Hordern et al 2009); N,N-diethyl-m-toluamide (DEET) (62%) (DGE 2013); Triclocarban (90% (Gardner et al 2013, Lozano et al 2013); Triclosan (90%) (Gardner et al 2013, Lozano et al 2013)].*
Chelating agents *(Household chemicals)*	Low to high removal efficiency (**5-95%;** mainly removed by **biodegradation)** *[**Ethylene-diamine-tetra-acetic acid (EDTA) (5%)** (Reemtsma et al 2006, Clara et al 2012, Margot et al 2015); Nitrilo-tri-acetic acid (NTA) (95%) (Reemtsma et al 2006, Clara et al 2012, Margot et al 2015)].*
Corrosion inhibitors *(Household chemicals)*	Mainly low removal efficiency (**15-30%;** mainly removed by **biodegradation)** *[2-hydroxybenzothiazole (OHBT) (60%) (Reemtsma et al 2006); **2-methylthiobenzothiazole (MTBT) (0%)** (Reemtsma et al 2006);*

Category	**Typical removal efficiency of micropollutants (in WWTP) (pollutants in bold are poor removal (< 50%)**
	Benzothiazole-2-sulfonic acid (BTSA) (15%) *(Reemtsma et al 2006, Margot et al 2013);* ***Benzotriazole (26%)*** *(DGE 2013, Margot et al 2013);* ***Methylbenzotriazole (30%)*** *(Margot et al 2013)].*
Flame retardants - brominated *(Household chemicals)*	High removal efficiency (**90-93%;** mainly removed by **sorption**) *[2,2',4,4'-Tetrabromodiphenyl ether (BDE-47) (91%) (Gardner et al 2013); Decabromodiphenyl ether (BDE-209) (90% (Nyholm et al 2013); Pentabromodiphenyl ethers (BDE-99) (93%) (Gardner et al 2013); Tetrabromobisphenol A (TBBA) (90%) (Potvin et al 2012)].*
Flame retardants - phosphorous *(Household chemicals*	Poor to high removal efficiency (**1-86%;** mainly removed by **biodegradation** and **sorption**) *[Tri-iso-butyl phosphate (TiBP) (86%) (Meyer and Bester 2004); Tri-n-butyl phosphate (TnBP) (60%) (Meyer and Bester 2004); Triphenyl phosphate (TPP) (65%) (Meyer and Bester 2004);* ***Tris(1,3-dichloro-2-propyl) phosphate (TDCP) (1%)*** *(Meyer and Bester 2004);* ***Tris(2-chloro,1-methylethyl) phosphate (TCPP) (1%)*** *(Meyer and Bester 2004);* ***Tris(2-chloroethyl) phosphate (TCEP) (1%)*** *(Meyer and Bester 2004); Tris(butoxyethyl)-phosphate (TBEP) (88%)].*
Fragrances *(Personal care products)*	Moderate to high removal (**54 to 85%;** mainly removed by **sorption**) *[Galaxolide (HHCB) (85%), Tonalide (AHTN) (85%) (Clara et al 2011, Kupper et al 2006), Iso E Super (OTNE) (65%),* ***Musk ketone (46%),*** *Musk xylene (54%) Bester et al 2008, Yang and Metcalfe 2006)].*
Heavy metals (Total)	High removal efficiency (**75 to 93%**) (except Ni; mainly removed by **sorption**) *[Al (93%) (Gardner et al 2013), Cd (85%) (Gardner et al 2013), Cr (73% (Ziolko et al 2011), Cu (86%)(Gardner et al 2013), Fe (78%)(Gardner et al 2013), Pb (89%)(Gardner et al 2013), Hg (79%)(Gardner et al 2013),* ***Ni*** *(31%) (Gardner et al 2013), Zn (75%) (Gardner et al 2013)].*
Hormones	High removal effcienecy (**76-99.9%;** mainly removed by **biodegradation)** *[Androsterone (99.9%) (Lubliner et al 2010), 17-ethinyl estradiol (EE2) (60%) (Gardner et al 2013, Margot et al 2013), 17-estradiol*

Category	Typical removal efficiency of micropollutants (in WWTP) (pollutants in bold are poor removal (< 50%)
	(E2) (90%) (Lubliner et al 2010, Margot et al 2013), Estrone (E1) (76%) (Lubliner et al 2010)].
Hydrophilic pollutants *(polar, highly soluble, hardly biodegradable, low log K_{ow})*	Poor to moderate removal efficiency: **< 50%** in WWTP *[Most pharmaceuticals, several herbicides, several household chemicals (chelating agents, corrosion inhibitors, sweeteners, EDTA, some phosphorus flame retardants, personal care products].*
Hydrophobic pollutants *(non-polar, high low K_{ow})*	**High removal** efficiency: **>70% in WWTP** *[Most heavy metals, PAHs, POPs, brominated flame retardants and several personal care products (surfactants, plastic additives, hormones)].*
Illicit drugs	Low to high removal efficiency (**15-98%;** mainly removed by **biodegradation)** [*Amphetamine (AMP) (98%) (Kasprzyk-Hordern et al 2009, Bijlsma et al 2012); Cocaine (79%) (Kasprzyk-Hordern et al 2009, Bijlsma et al 2012); **Ecstasy** (MDMA) **(15%)** (Kasprzyk-Hordern et al 2009, Bijlsma et al 2012); THC-COOH (cannabis metabolite) (98%)(Kasprzyk-Hor dern et al 2009, Bijlsma et al 2012)].*
Perfluorinated compounds *(Household chemicals)*	Poor to high removal efficiency (**1%;** mainly removed by **sorption**) *[**Perfluorooctane sulfonic acid (PFOS) (1%)** (Guo et al 2010); **Perfluorooctanoic acid (PFOA) (1%)** (Guo et al 2010)].*
Pesticides	Poor removal efficiency (**0-48%; hydrophilic** herbicides; removed by **biodegradation**) *[**AMPA** (Aminomethyl phosphonic acid) (degradation product of herbicide Glyphosate) **(0%)** (Gardner et al 2013); **Atrazine** (herbicide) **(23%)** (Singer et al 2010, Margot et al 2013); **Carbendazim** (fungicide) **(30%)** (Singer et al 2010, DGE 2013, Margot et al 2013)**; Diazinon** (insecticide) **(40%)** (Singer et al 2010, Margot et al 2013)**; Diuron** (herbicide/ algicide) **(33%)** (Singer et al 2010, DGE 2013, Margot et al 2013)**; Glyphosate** (herbicide) **(30%)** (Gardner et al 2013)**; Irgarol** 1051 (algicide) **(48%)** (Singer et al 2010, DGE 2013, Margot et al 2013)**; Isoproturon** (herbicide) **(42%)** (Singer et al 2010, DGE 2013, Margot et al 2013)**; MCPA** (4-chloro-2-methylphenoxy) acetic acid) (herbicide) **(28%)** (DGE 2013); **Mecoprop** (herbicide) **(25%)***

Category	Typical removal efficiency of micropollutants (in WWTP) (pollutants in bold are poor removal (< 50%)
	(Singer et al 2010, DGE 2013, Margot et al 2013); ***Terbuthylazine*** *(herbicide)* ***(24%)*** *(Singer et al 2010); Terbutryn (herbicide) (60%) (Singer et al 2010, Margot et al 2013)].*
Pharmaceuticals	High removal efficiency (>**70%;** mainly removed by both **biodegradation**) *[Caffeine (95%), (Oulton et al 2010, Margot et al 2013); Ibuprofen (80%) (Oulton et al 2010, DGE 2013, Verlicchi et al 2012); Paracetamol (100%) (Oulton et al 2010, DGE 2013, Verlicchi et al 2012); Salicylic acid (99%) (Oulton et al 2010, Verlicchi et al 2012)].*
Pharmaceuticals	Medium removal effcienecy (**30-70%;** mainly removed by both **biodegradation)** *[**Atenolol (41%)** (DGE 2013, Verlicchi et al 2012, Margot et al 2013); **Azithromycin (39%)** (DGE 2012, Verlicchi et al 2012, Margot et al 2013); **Bezafibrate (41%)** (DGE 2013, Margot et al 2013), Ciprofloxacin (65%) ((DGE 2013, Verlicchi et al 2012, Margot et al 2013); **Erythromycin (45%)** (Verlicchi et al 2012, Margot et al 2013); **Gemfibrozil (39%)** ((DGE 2013, Verlicchi et al 2012, Margot et al 2013); **Ketoprofen** (40%) (DGE 2013, Verlicchi et al 2012, Margot et al 2013); **Mefenamic acid (42%)** (DGE 2013, Verlicchi et al 2012, Margot et al 2013); **Naproxen** (40%) (DGE 2013, Margot et al 2013); Ranitidine (52%) (Verlicchi et al 2012); **Sulfamethoxazole (44%)** ((DGE 2013, Verlicchi et al 2012, Margot et al 2013); **Trimethoprim (35%)** (DGE 2013, Verlicchi et al 2012, Margot et al 2013)].*
Pharmaceuticals	Poor removal efficiency (**<30%;** mainly removed by both **biodegradation** and **sorption)** *[**Carbamazepine (16%)** (DGE 2013, Verlicchi et al 2012, Margot et al 2013); **Diclofenac (20%)** (DGE 2013, Verlicchi et al 2012, Margot et al 2013); **Fluconazole** (15%) (Margot et al 2013); **Metoprolol (25%)** (DGE 2013, Verlicchi et al 2012, Margot et al 2013); **Primidone (16%)** (Margot et al 2013, Oosterhuis et al 2013); **Propranolol (28%)** (DGE 2013, Verlicchi et al 2012, Margot et al 2013)].*
Plastic additives/ plasticisers *(Household chemicals)*	Moderate to high removal efficiency (**62-92%;** mainly removed by **biodegradation)** *[**Bisphenol A** (BPA) (80%) (Gardner et al 2013, Margot et al 2013); **Butyl benzyl phthalate** (BBP) (86%) (Bergé et al 2013); **Di-(2-ethylhexyl) phthalate** (DEHP) (69%) (Bergé et al 2013);*

Category	Typical removal efficiency of micropollutants (in WWTP) (pollutants in bold are poor removal (< 50%)
	Diethyl phthalate *(DEP) (92%) (Bergé et al 2013);* ***Dimethyl phthalate*** *(DMP) (62%) (Bergé et al 2013),* ***Di-n-butyl phthalate*** *(DnBP) (69%) (Bergé et al 2013)].*
Polycyclic aromatic hydrocarbons (PAHs)	High removal efficiency (**70-90%; **mainly removed by **sorption**) *[Acenaphthene (85%) (Fatone et al 2011); Anthracene (90%) (Fatone et al 2011); Benzo[a]pyrene (70%) (Fatone et al 2011); Benzo[b]fluoranthene (80%) (Fatone et al 2011); Benzo[ghi]perylene (80%) (Fatone et al 2011, Gardner et al 2013); Benzo[k]fluoranthene (90%) (Fatone et al 2011); Fluoranthene (80%) (Fatone et al 2011); Fluorene (90%) (Fatone et al 2011); Indeno[1,2,3-cd]pyrene (80%) (Fatone et al 2011, Gardner et al 2013); Naphthalene (60%)(Fatone et al 2011); Phenanthrene (80%) (Fatone et al 2011); Pyrene (70%) (Fatone et al 2011)].*
Poor removal pollutants	**Poor removal** efficiency (< **50**% in WWTP) ***[Chelating agents*** *(EDTA);* ***Corrosion inhibitors*** *(MTBT, BTSA, Methylbenzotriazole);* ***Fragrances*** *(Musk ketone);* ***Heavy metals*** *(Ni);* ***Illicit drugs*** *(Ecstasy);* ***Pesticides/ herbicides*** *(AMPA, Atrazine, Carbendazim, Diazinon, Diuron, Glyphosate, Irgarol 1051, Isoproturon, Mecoprop, Terbuthylazine);* ***Pharmaceuticals*** *(Carbamazepine, Diclofenac, Fluconazole, Metoprolol, Primidone, Propranolol);* ***Perfluorinated compounds*** *(PFOA, PFOS);* ***Phosphorous flame retardants*** *(TDCP, TCPP, TCEP);* ***Sweeteners*** *(Acesulfame, Sucralose)].*
POPs	High removal efficiency (**71-91%; hydrophobic** pollutants/pesticides mainly removed by **sorption**) *Aldrin (insecticide) (86%) (Katsoyiannis and Samara 2004); Dichlorodiphenyltrichloroethane (DDT) (insecticide) (83%) (Katsoyiannis and Samara 2004); Dieldrin (insecticide) (77%) (Katsoyiannis and Samara 2004); Endosulfan (insecticide) (84%) (Katsoyiannis and Samara 2004); Endrin (insecticide) (81%) (Katsoyiannis and Samara 2004); Heptachlor (insecticide) (91%) (Katsoyiannis and Samara 2004); Hexachlorobenzene (fungicide) (90%) (Katsoyiannis and Samara 2004); Hexachlorocyclohexanes (HCHs) (insecticide) (71%) (Katsoyiannis and Samara 2004); Pentachloronitrobenzene (PCNB) (fungicide) (75%) (Katsoyiannis and Samara 2004); Polychlorinated biphenyls (PCBs) (75%) (Blanchard et al 2004, Katsoyiannis and Samara 2004)].*

Category	**Typical removal efficiency of micropollutants (in WWTP) (pollutants in bold are poor removal (< 50%)**
Preservatives *(Personal care products)*	High removal efficiency (**95%;** mainly removed by **biodegradation**) *[Ethyl-paraben (95%) (Gonzalez-Marino et al 2011); Methyl-paraben (95%) (Gonzalez-Marino et al 2011)].*
Surfactants *(hydrophobic)*	High removal efficiency (**80-99%;** mainly removed by **biodegradation)** *[Alcohol ethoxylate (AE) (99%) (McAvoy et al 1998, McAvoy et al 2006); Alkyl ethoxy sulfate (AES) (99%) (Matthijs et al 1999, Scott and Jones 2000); Alkyl sulfate (AS) (99%)(Matthijs et al 1999, Scott and Jones 2000); Alkyldimethylbenzyl ammonium chlorides/Benzalkonium chloride (BACs) (95%) (Kreuzinger et al 2007, Clara et al 2012); Linear alkyl benzene sulfonate (LAS) (97%) (Feijtel et al 1996, Schowanek et al 2007, Mungray and Kumar 2009); Nonylphenol ethoxylates (NPEs) (90%); Octylphenol ethoxylates (OPEs) (80%), Soap (99%) (Scott and Jones 2000, Berna et al 2007)].*
Sweeteners *(Household chemicals)*	Low to high removal efficiency **(5-99%; mainly removed by** biodegradation) *[**Acesulfame (5%)** (Lange et al 2012, Kokotou and Thomaidis 2013); Cyclamate (95%) (Lange et al 2012, Kokotou and Thomaidis 2013); Saccharin (99%) (Lange et al 2012, Kokotou and Thomaidis 2013); **Sucralose (5%)** (Lange et al 2012, Kokotou and Thomaidis 2013)].*
UV filters *(Personal care products)*	High removal efficiency (**72-96%;** mainly removed by **biodegradation)** *[4-methylbenydlidene camphor (4-MBC) (72%) (Balmer et al 2005, Kupper et al 2006); Benzophenone-3 (BP-3) (90%) (Balmer et al 2005, Kupper et al 2006); Ethylhexyl methoxycinnamate (EHMC or OMC) (98%) (Balmer et al 2005, Kupper et al 2006); Octyl-tirazone (OT or ethylhexyl triazone) (96%) (Kupper et al 2006)].*
Volatile organic compounds (VOCs)	High removal efficiency (**70-98%;** mainly removed by **volatilisation);** *[1,2,4-Trimethylbenzene (82%)* (*Fatone et al 2011*); *Ethylbenzene (98%) (Fatone et al 2011); m-Xylene+p-Xylene (97%) (Fatone et al 2011); o-xylene (70%) (Fatone et al 2011); Styrene (96%) (Fatone et al 2011); Toluene (75%) (Fatone et al 2011)].*

- Maintain and upgrade **sewerage infrastructures** as required;
- Maintain and **monitor assets** to avoid persistent leakages, collapses, blockages and operational failures during dry and wet weather seasons;
- As an **emergency response**, the following measures may be taken when a sewage spill has already occurred **i.** erect a temporary weir or bund (to prevent sewage reaching downstream), **ii.** use sewerage pump or vacuum trucks (to remove pools of sewage or to pump-out overflowing pump wells); **iii.** call an environmental waste/ sewage clean-up company, septic tank company or city health department to clean up large sewage spills; **iv.** dilute the waterway or generate a water flow by using uncontaminated water (such as groundwater);
- If the spill occurred in a **sensitive waterway** having EPBC (Environment Protection and Biodiversity Conservation Act) embargo/ban (as per The Environment Protection and Biodiversity Conservation Act 1999, Australia), ensure the protection of **native fishes** or temporarily relocate native fish from the spilt waterway until the environment becomes safe;
- In some circumstances, **hydrated lime** (garden lime) may be spread (as a disinfectant) over the spilt area to absorb the sewage. This helps to neutralise wastewater so that it is less harmful to the environment and assists in reducing odours.
- **Sewage sludge** can be utilised for agricultural purposes (growing crops, vegetables) to recover the organic and inorganic nutrients (in this case, regular monitoring of residue levels in crops and vegetables would be required to safeguard the humans health);
- **Sewage sludge** can further be utilised for *land filling*, *incineration* (e.g. to use incineration ash as a source of phosphorus), *land farming, land reclamation* (e.g. to reclaim mine lands or landfill), *forestry* (e.g. forest fertilisation, silviculture for fibre crops), *resource recovery from biosolids* (e.g. a source of minerals and metals) and *energy* (e.g. bioenergy from digestion) and horticulture and landscaping fertilisers, erosion control, highway right-of way-vegetation) (Fijalkowski et al 2017);
- Sewage water can be **recycled** after removing solids and impurities, and can be used in sustainable landscape irrigation or to recharge groundwater aquifers or to meet commercial and industrial water needs. This will help to achieve sustainability and water conservation goals and prevents discharging the treated water to rivers and oceans. Recycled water can also be used for streamflow augmentation to benefit ecosystems and improve aesthetic uses (Kibria et al 2016);
- A **better management** of micropollutants include implementation of regulations (ban or restriction on the use of harmful compounds), proper

disposal of pollutants, use of chemicals with lower environmental impacts and improvement of the existing treatment technologies including the use of advanced treatments of hardly biodegradable/ hydrophilic/lower sorption micropollutants (such as several pharmaceuticals);

- Consider possible **climate change impacts** on sewerage assets such as increased rainfall intensity during storms, inundation/flooding of wastewater treatment plants (due to sea-level rise) and corrosion of wastewater treatment equipment due to rise in ambient and seasonal temperatures;
- **Monitor chemical pollutant** level in sewage effluents and compare with guideline threshold values to protect water quality and aquatic biota and human health (Table 12.6). In addition, it is worth to have a reconnaissance survey to assess concentrations of sterol compounds (as indicators of faecal contamination) using chemical markers (see Box 12.1).

12.5.2. Community best practices

- The community should be educated to minimise their water use when it's raining (e.g. delay washing clothes or run the dishwasher until the rain stops) to lessen the burden on the sanitary system;
- The community should be educated to refrain from flushing inappropriate items into the wastewater system (toilets) (such as fats, oils or grease, wet wipes, tampons, napkins, diapers, condoms, toys, syringes, newspapers and plastics etc.) since these substances clog both plumbing and the sewer system causing backup;
- The community should be educated to avoid planting trees and shrubs near household sewer lines since plant roots are one of the most common causes of the lateral sewage line problems.

Table 12.6. Water quality guidelines for the protection of aquatic ecosystems/ aquatic species for chemical micropollutants found in sewage water

(biocides, chelating agents, fragrances, heavy metals, hormones, nutrients, oil/petroleum compounds, parabens/preservatives, perfluorinated compounds, POPs, pesticides, plastic additives, surfactants, and UV filters);

*[**EQS**= Chronic environmental quality standards for inland surface waters for selected countries in Europe (Margot et al 2015); **na**= data is not available; **PNEC** = Predicted no-effect concentration, the concentration below which exposure to a substance is not expected to cause adverse effects].*

Type	Chemical name	EQS *(chronic environment quality standards) (Margot et al 2015)*	PNEC *(Predicted no-effect concentration) (Margot et al 2015)*	Environment Quality Indicators *(ANZECC & ARMCANZ 2000 SEPP 2018) 95% or 99% aquatic species protection trigger values (TV); **FW**= freshwater; **MW**= marine water; **SP**= species protection*
Biocides	DEEE +(*N,N*-diethyl-m-toluamide)	41 µg/L	0.8 µg/L	na
	Triclosan	na	0.0015 µg/L	na
Chelating agents	EDTA (ethylene-diamine-tetra-acetic acid)	220,00 µg/L	39 µg/L	na
Fragrances	Galaxolide (HHCB)	na	6.8 µg/L	na
	Musk ketone	na	10 µg/L	na
	Musk xylene	na	10 µg/L	na
	Tonalide (AHTN)	na	3.9 µg/L	na
Heavy metals	Cadmium	0.08-0.25 µg/L	0.013 µg/L	FW: 0.2 µg/L (95% SP)
	Chromium	3.4 µg/L	2.4 µg/L	FW: 1.0 µg/L (95% SP) MW: 4.4 µg/L (95% SP)
	Copper	1-28 µg/L	0.05 µg/L	FW: 1.4 µg/L (95% SP) MW: 1.3 µg/L (95% SP)

Type	**Chemical name**	**EQS** *(chronic environment quality standards)* *(Margot et al 2015)*	**PNEC** *(Predicted no-effect concentration)* *(Margot et al 2015)*	**Environment Quality Indicators** *(ANZECC & ARMCANZ 2000 SEPP 2018)* *95% or 99% aquatic species protection trigger values (TV);* ***FW****= freshwater;* ***MW****= marine water;* ***SP****= species protection*
	Iron	1.2 µg/L	56 µg/L	na
	Lead	1.2 µg/L	0.021 µg/L	FW: 3.4 µg/L (95% SP) MW: 4.4 µg/L (95% SP)
	Mercury	0.05 µg/L	0.025 µg/L	FW: 0.06 µg/L (95% SP) MW: 0.1 µg/L (95% SP) (inorganic)
	Nickel	4.0 µg/L	0.110 µg/L	FW: 11 µg/L (95% SP) MW: 7.0 µg/L (99% SP)
	Zinc	8.0-125 µg/L	0.008.5 ng/L	FW: 8 µg/L (95% SP) MW: 15 µg/L (99% SP)
Hormones	17 α- ethinyl-estradiol (EE2)	0.000037 µg/L	0.000004 µg/L	na
	17 α-estradiol (E2)	0.0004 µg/L	0.000042 µg/L	na
Nutrients	Total nitrogen	na	na	Unmodified streams: ≤150 to ≤900 µg/L. Slightly to moderately modified streams: ≤800 to ≤1100 µg/L. Highly modified streams: ≤1,200 to ≤1,3 00 µg/L
	Total phosphorous	na	na	Unmodified streams: ≤20 to ≤35 µg/L. Slightly to moderately modified streams: ≤50 to ≤60 µg/L. Highly modified streams: ≤110 µg/L

Type	Chemical name	EQS *(chronic environment quality standards)* *(Margot et al 2015)*	PNEC *(Predicted no-effect concentration)* *(Margot et al 2015)*	Environment Quality Indicators *(ANZECC & ARMCANZ 2000 SEPP 2018)* *95% or 99% aquatic species protection trigger values (TV);* ***FW****= freshwater;* ***MW****= marine water;* ***SP****= species protection*
Oil/ Petrochemical	0-Xylene (BTEX)	na	26 μg/L	FW: 350 μg/L (95% SP)
	Anthracene (PAH)	0.1 μg/L	0.2 μg/L	na
	Benzene (BTEX)	na	na	FW: 950 μg/L (95% SP) MW: 500 μg/L (99% SP)
	Benzo[*a*]pyrene (PAH)	0.00017 μg/L	0.002 μg/L	na
	Ethylbenzene (BTEX)	na	20 μg/L	50 μg/L (99% species)
	Fluoranthene (PAH)	0.0063 μg/L	0.06 μg/L	na
	Naphthalene (PAH)	2 μg/L	37 μg/L	FW: 16 μg/L (95% SP) MW: 50 μg/L (99% SP)
	Phenanthrene (PAH)	na	0.00002 μg/L	na
	Pyrene (PAH)	na	0.120 μg/L	na
	Toluene (PAH)	50 μg/L	100 μg/L	na
Parabens (preservatives)	Ethyl-paraben	na	21 μg/L	na
	Methyl-paraben	na	4 μg/L	na
Perfluorinated compounds	Perfluorooctane sulfonic acid (PFOS)	0.00065 μg/L	0.01 μg/L	na

Type	Chemical name	EQS *(chronic environment quality standards)* *(Margot et al 2015)*	PNEC *(Predicted no-effect concentration)* *(Margot et al 2015)*	Environment Quality Indicators *(ANZECC & ARMCANZ 2000 SEPP 2018)* *95% or 99% aquatic species protection trigger values (TV); **FW**= freshwater; **MW**= marine water; **SP**= species protection*
	Perfluorooctanoic acid (PFOA)	na	0.65 µg/L	na
Persistent organic pollutants (POPs)	Aldrin *(insecticide)*	0.01 µg/L	0.036 µg/L	na
	Dichloro-diphenyl trichloroethane (DDT) *(insecticide)*	0.025 µg/L	0.001 µg/L	FW: 11.0 µg/L (95% SP) MW: 7.0 µg/L (99% SP)
	Dieldrin *(insecticide)*	na	0.001 µg/L	na
	Endosulfan *(insecticide)*	0.005 µg/L	0.001 µg/L	na
	Endrin *(insecticide)*	na	0.0034 µg/L	FW: 0.01 µg/L (99% SP) MW: 0.004 µg/L (99% SP)
	Heptachlor *(insecticide)*	0.0000002 µg/L	0.046 µg/L	FW: 0.01 µg/L (95% SP)
	Hexachloro-benzene *(fungicide)*	0.01 µg/L	0.004 µg/L	na
	PCBs *(dielectric and coolant fluids)*	na	0.001 µg/L	FW: 0.3 µg/L (95% SP) PCBs (aroclor 1242) FW: 0.01 µg/L (95% SP; PCBs (aroclor 1254)
Pesticides	Aminomethyl phosphonic acid (AMPA)	1500 µg/L	11 µg/L	na

Type	Chemical name	EQS *(chronic environment quality standards)* *(Margot et al 2015)*	PNEC *(Predicted no-effect concentration)* *(Margot et al 2015)*	Environment Quality Indicators *(ANZECC & ARMCANZ 2000 SEPP 2018)* *95% or 99% aquatic species protection trigger values (TV); **FW**= freshwater; **MW**= marine water; **SP**= species protection*
	Atrazine	0.6 µg/L	0.000011 µg/L	FW: 3 µg/L (95% SP) protection (freshwater)
	Carbendazim	0.340 µg/L	0.000114 µg/L	na
	Diazinon	0.015 µg/L	0.000005 µg/L	FW: 0.01 µg/L (95% SP)
	Glyphosate	108 µg/L	0.2 µg/L	FW: 370 µg/L (95% SP)
	Irgarol 1051	0.0023 µg/L	0.0000045 ng/L	na
	Isoproturon	0.32 µg/L	0.174 µg/L	na
	MCPA	1.34 µg/L	1.4 µg/L	na
	Mecoprop	3.6 µg/L	514.7 µg/L	na
	Terbuthylazine	0.22 µg/L	0.06 µg/L	na
	Terbutryn	0.06 µg/L	0.2 µg/L	na
Pharma-ceuticals	Acetaminophen /Paracetamol	na	0.5 µg/L	na
	Bezafibrate	0.46 µg/L	0.00003 µg/L	na
	Caffeine	na	0.5 µg/L	na
	Carbamazepine	0.5 µg/L	0.089 µg/L	na
	Diclofenac	0.05 µg/L	0.5 µg/L	na
	Salicylic acid	na	0.5 µg/L	na

Type	**Chemical name**	**EQS** *(chronic environment quality standards)* *(Margot et al 2015)*	**PNEC** *(Predicted no-effect concentration)* *(Margot et al 2015)*	**Environment Quality Indicators** *(ANZECC & ARMCANZ 2000 SEPP 2018)* *95% or 99% aquatic species protection trigger values (TV); **FW**= freshwater; **MW**= marine water; **SP**= species protection*
	Sulfamethoxazole	0.6 µg/L	0.013 µg/L	na
Plastic additives	Bisphenol A (BPA)	1.5 µg/L	0.175 µg/L	na
	Butyl benzyl phthalate (BBP)	20 µg/L	1.81 µg/L	na
	Di-(2-ethylhexyl) phthalate (DEHP)	1.3 µg/L	0.01 µg/L	na
	Diethyl phthalate (DEP)	200 µg/L	102 µg/L	na
Surfactants	Alkyl ethoxy sulfate (AES)	na	18 µg/L	FW: 650 µg/L (95% SP)
	Linear alkylbenzene sulfonate (LAS)	na	21 µg/L	FW: 280 µg/L (95% SP)
	Nonylphenol ethoxylates (NPEs), Nonylphenol (NP)	0.013 µg/L (NP)	0.5 µg/L (NPEs) 0.03 µg/L (NP)	na
	Octylphenol ethoxylates (OPEs), Octylphenol (OP)	0.1 µg/L (OP)	21 µg/L (OPEs) 0.00021 µg/L (OP)	na
	Soap	na	22-44 µg/L	na

Type	Chemical name	EQS *(chronic environment quality standards)* *(Margot et al 2015)*	PNEC *(Predicted no-effect concentration)* *(Margot et al 2015)*	Environment Quality Indicators *(ANZECC & ARMCANZ 2000 SEPP 2018)* *95% or 99% aquatic species protection trigger values (TV); **FW**= freshwater; **MW**= marine water; **SP**= species protection*
UV Filters	Benzophenone-3 (BP-3)	na	1.6 µg/L	na
	Ethylhexyl methoxyc-innamate (EHMC or OMC)	na	0.027 µg/L	na

Box 12.1. Monitoring of microbial pathogens/ pollutants (faecal contaminants) using chemical markers (Ford et al 2010, Hussain et al 2010).

A reconnaissance survey can be made to assess concentrations of sterol[25] compounds (coprostanol[26] and cholestanol[27] as indicators of faecal contamination). Monitoring of contamination through assessing microbial (biological) indicators is limited by two factors, its culturability and organism characterisation leading to an

[25] ***Sterols**, also known as steroid alcohols, are a subgroup of the steroids and an important class of organic molecules. They are a type of lipid. Faecal sterols (also known as **5β-stanols**) are useful biomarkers (**biomarkers** = a naturally occurring molecule, gene, or characteristic by which a particular pathological or physiological process, disease, etc. can be identified) of animal faeces and have the important advantage of long-term preservation in soils and sediments due to their low solubility in water and their ability to bind to particulate organic matter. Their distribution in faecal material, called a **stanol fingerprint**, identifies a particular mammalian species on the basis of its diet (main sterol uptake). The most common techniques used for stanol fingerprinting are to identify whether faeces belong to **carnivores, omnivores or herbivores**. Mainly because of their respective diets, **cholesterol-derived 5β-stanols** are found in high proportions in the faeces of omnivores and carnivores (**coprostanol and epicoprostanol**), while 5β-stanols derived from β-sitosterol, a phytosterol (plant sterol), are found in high proportions in **herbivore faeces** (24-ethylcoprostanol and 24-ethylepicoprostanol) (Harrault et al 2019).*

[26] ***Coprostanol** (**5β-Coprostanol**) (**5β-cholestan**-3β-ol) is a 27-carbon **stanol** formed from the biohydrogenation of cholesterol (cholest-5en-3β-ol) in the gut of higher **animals and birds**. This compound has frequently been used as a biomarker for the presence of human faecal matter in the environment. The principal source of **5β-coprostanol** in the environment is from **human wastes**. Coprostanol has a low water solubility, and consequently a high octanol-water partition coefficient (log K_{ow} = 8.82) (https://en.wikipedia.org/wiki/Coprostanol).*

[27] ***Cholestanol** = A cholesterol derivative found in human faeces, gallstones, eggs, and other biological matter. (https://pubchem.ncbi.nlm.nih.gov/compound/Cholestanol#).*

inability to distinguish a particular source of contamination. On the other hand, chemical markers such as faecal sterols can provide advantages in determining the level and extent of contamination as they have environmental stability (low water solubility, log K_{ow} 6.8), ease of detection and can be used to discriminate between human and nonhuman contamination. A comprehensive 12 months survey was done within the Goulburn Murray Rural Water (G-MW), Australia to assess the microbial pathogen levels in waterways comprising irrigation channels, drains, and rivers during 2006-2007. One of the main objectives of the survey was to determine the concentrations of faecal sterols (especially coprostanol and cholestanol) in surface water of the G-MW water supply system to monitor and identify potential high input sites of faecal contamination (microbial pollution "hot spots").

A Before-After-Control for-Impact (BACI) approach was used in sampling, i.e. samples were taken from both upstream and downstream of a site. The selected pathogen monitoring sites were drains and channels close by a piggery, a slaughterhouse (Knackery), a dairy, and a poultry farm, and a town sewage treatment plant (Table 12.7). In addition to these locations, samples were also collected from three sites external to the water supply system (referred to here as "reference sites", which were assumed to have a minimal direct impact from human or intensive industry-derived wastewater). Results of this extensive and expensive research are given in Table 12.7 (to benefit the readers). The highlight of this monitoring is given below:

- very high concentrations of **coprostanol and cholestanol** (faecal indicator sterols) (in either surface water or bottom sediments) were recorded at four of the 17 sampling sites (sites 04, 05, 09, 11) (Table 12.7); these four sites are drain sites and impacted by **knackery and dairy**; coliforms and other pathogenic indicator counts were also previously found to be comparatively high at these four drain sites;
- coprostanol concentrations at each location varied from as low as 0 ng/L at the reference site to **11,327** ng/L in a surface water sample of a drain directly downstream of a dairy (site 09);
- the majority of the sites contained coprostanol in the range of **500 to 800 ng/L**;
- no faecal-associated sterol compounds were detected at the external reference sites (sites 01 and 02), these were assumed to be free from faecal contamination;
- sewage water discharge and/ or substantial water runoff maybe the principal factors contributing to faecal contamination of the supply drains and channels;
- the Australian guidelines for drinking water supplies state a threshold of **700 ng/L of coprostanol** which was detected at only **four drain sites** (sites 04, 05, 09, 11) in this study;
- coprostanol and cholestanol ratios may be employed to indicate that faecal contamination from human sources

Dr Golam Kibria, as the Senior Environmental Scientist with the G-MW (the lead author of this book) planned, lead and managed this expensive and extensive research during 2001-2012 periods. The research was carried out in collaboration with the researchers from the University of Melbourne, Australia.

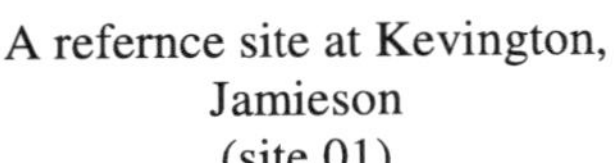

A refernce site at Kevington, Jamieson (site 01)

A channel ste close by a poultry farm, Kyabram (site 12)

A drain site close by a Knackery, Stanhope (site 04)

Photos of some selected sites (Photo courtesy: Dr Golam Kibria)

Table 12.7. Concentrations (ng/L) of coprostanol (Cop) and cholestanol (Chol) in surface water samples (channels, drains) impacted by knackery, piggery, dairy, poultry and sewage of Goulburn Murray waterways during October 2006 to October 2007 (data are based on Ford et al 2010, Hussain et al 2010).

*[**Chol** = cholestanol; **Cop** = coprostanol, **D**= the site was dry (no water); **d/s** = downstream; **nd** = not detected; **u/s** = upstream; the Australian guidelines for drinking water supplies threshold for coprostanol is 700 ng/L (NHMRC 2004; ***Bold** = data shown in columns and rows with an asterisk and bold indicates it exceeded Australian guidelines value of 700 ng/L for coprostanol].*

Site number and location		Nov 2006	Dec 2006	Jan 2007	Feb 2007	Mar 2007	Apr 2007	May 2007	Jun 2007	Jul 2007	Aug 2007	Sep 2007	Oct 2007
01. Reference site *(Kevington River, 17 km upstream of Jamieson)*	Cop	nd	-	-	-	nd	-	-	-	nd	-	-	nd
	Chol	nd	-	-	-	nd	-	-	-	nd	-	-	nd
02. Reference site *(Goulburn River, Goulburn weir, Nagambie)*	Cop	nd	nd	nd	nd	nd	nd	nd	nd	308	nd	nd	nd
	Chol	nd	400	395	nd	nd	nd	nd	nd	205	nd	nd	nd
03. Reference *(Natural wetlands Mooroopna)*	Cop	D	D	D	D	D	D	-	569	557	526	994	616
	Chol	D	D	D	D	D	D	-	420	389	351	664	406
04. Drains-Knackery *(Deakin drain 16, Stanhope) (u/s) (mean Cop= **1294** ng/L)*	Cop	***792**	***748**	D	D	D	531	519	***4,039**	***1,210**	***1,933**	***1,147**	***729**
	Chol	503	522	D	D	D	371	354	596	800	424	476	507

Site number and location		Nov 2006	Dec 2006	Jan 2007	Feb 2007	Mar 2007	Apr 2007	May 2007	Jun 2007	Jul 2007	Aug 2007	Sep 2007	Oct 2007
05. Drains Knackery *(Deakin drain 16, Stanhope) (d/s) (mean Cop=* ***852*** *ng/L)*	Cop	678	***822**	D	D	D	551	541	***2,024**	***1,257**	609	613	574
	Chol	460	262	D	D	D	420	400	479	817	449	407	407
06. Channels-Piggery *(Kyabram) (u/s) (mean Cop=* ***591*** *ng/L)*	Cop	600	623	nd	nd	557	nd	nd	***739**	524	524	nd	574
	Chol	402	402	395	nd	375	nd	nd	490	351	340	336	409
07. Channels-Piggery *(Kyabram) (d/s) (mean Cop=* ***573*** *ng/L)*	Cop	nd	622	nd	nd	nd	nd	nd	651	516	559	nd	517
	Chol	395	398	nd	nd	361	350	351	364	347	350	nd	341
08. Drains - Dairy *(drain 4, Girgarre) (u/s)*	Cop	D	D	D	D	D	D	D	D	910	D	-	-
	Chol	D	D	D	D	D	D	D	D	837	D	-	-
09. Drains-Dairy *(drain 4, Girgarre) (d/s) (mean Cop=* ***2249*** *ng/L)*	Cop	D	D	D	D	543	***11,327**	611	***804**	***2,120**	***1,153**	***833**	602
	Chol	D	D	D	D	390	837	341	647	1353	687	543	408
10. Drains-Dairy *(Mosquito drain 8, Girgarre) (u/s) (mean Cop=* ***589*** *ng/L)*	Cop	***723**	617	685	638	459	-	530	524	652	536	529	nd
	Chol	465	413	545	482	320	-	373	366	452	371	396	nd
11. Drain Dairy *(drain 4, Girgarre) (d/s) (mean Cop=* ***633*** *ng/L)*	Cop	657	***707**	633	627	433	-	***829**	***747**	567	676	534	554
	Chol	427	527	447	437	298	-	481	527	396	581	366	383
12. Channels-Poultry *(Kyabram) (u/s) (mean Cop=* ***542*** *ng/L)*	Cop	607	nd	nd	nd	644	nd	521	516	514	519	472	nd
	Chol	395	397	400	nd	392	356	341	344	341	343	313	340

Site number and location		Nov 2006	Dec 2006	Jan 2007	Feb 2007	Mar 2007	Apr 2007	May 2007	Jun 2007	Jul 2007	Aug 2007	Sep 2007	Oct 2007
13. Channels-Poultry *(Kyabram (d/s) (mean Cop= **524** ng/L)*	Cop	600	603	nd	610	543	nd	524	687	516	647	643	517
	Chol	403	420	415	428	386	379	380	757	360	553	457	347
14. Drains-Dairy *(Wyuna) (u/s) (mean Cop= **594** ng/L)*	Cop	647	620	nd	nd	nd	nd	-	D	D	D	nd	514
	Chol	395	435	400	nd	403	nd	-	D	D	D	291	350
15. Drain-Dairy *(Wyuna) (d/s) (mean Cop= **564** ng/L)*	Cop	605	600	620	605	564	516	D	533	520	514	nd	nd
	Chol	405	398	432	413	373	349	D	347	344	353	300	339
16. Channels-Sewage *(Mooroopna) (u/s) (mean Cop= **555** ng/L)*	Cop	***717**	603	nd	620	500	514	-	nd	517	513	nd	nd
	Chol	640	398	403	400	333	329	-	349	424	344	299	339
17. Channels-Sewage (Mooroopna) (d/s) *(mean Cop= **583** ng/L)*	Cop	nd	600	nd	607	549	514	nd	586	516	551	748	nd
	Chol	395	397	400	407	357	346	343	577	511	479	572	340

12.6. Conclusion

Sewage water/wastewater (e.g. blackwater, greywater) is a cocktail of chemical pollutants/contaminants. In sewage, the freshwater component is 99% and the remaining 1% is contaminants. These contaminants include: **i.** biological micropollutants (bacteria, viruses), **ii**. chemical micropollutants (heavy metals, hormones, household chemicals, oil compounds, personal care products, pesticides, pharmaceuticals, plastic additives, POPs, surfactants), **iii.** nutrients (nitrogen and phosphorus) and **iv**. solids. Several of these contaminants are persistent, toxic, bio-accumulative (PBT), as well as endocrine-disrupting and carcinogenic. Aquatic biota can be exposed to these pollutants via WWTP outfall or surface water contaminated with raw sewage. Humans can be exposed via food and water contaminated with these pollutants.

Several chemicals found in sewage effluents have **exceeded the PNEC** (Predicted No-Effect Concentration to algae, *Daphnia* and fish) (Table 12.3). Chemicals which have exceeded the PNEC are considered as toxic/ harmful to biota (algae, *Daphnia* and fish). The following pollutants have exceeded the PNEC limit:

- ***Heavy metals*** (Cd, Cu, Pb, Pb, Hg, Ni, Zn) (*out of 09 types detected)*;
- ***Hormones*** [17α-ethinyl-estradiol (EE2), 17β-Estradiol (E2), Estrone (E)] (*out of 07 types detected);*
- ***Household chemicals*** [ethylenediaminetetraacetic acid or EDTA), perfluorooctane sulfonic acid (PFOS) (flame retardant), bisphenol A (BPA), Di-(2-ethylhexyl) phthalate (DEHP), Di-n-butyl phthalate (DnBP), cyclamate (sweeteners), sucralose (sweeteners)] (*out of 32 types detected);*
- ***Pesticides*** [Atrazine (herbicide), Carbendazim (fungicide), Diazinon (insecticide), Diuron (herbicide/ algicide), Glyphosate (herbicide), Irgarol 1051 (algicide), Terbuthylazine (herbicide)] (out of 12 types detected);
- ***POPs*** [dieldrin, endosulfan, heptachlor, polychlorinated biphenyls (PCBs)] (*out of 10 types detected);*
- ***Personal care products*** [chloroxylenol (biocides), ethylhexyl methoxy-cinnamate ((UV Filters), iso E Super (fragrances), triclocarban (biocides), triclosan] (*out of 07 types detected);*
- ***PAHs/petrochemical compounds/ VOCs*** (phenanthrene) (*out of 18 types detected*);
- ***Pharmaceuticals*** (caffeine, carbamazepine, ciprofloxacin, clarithro-mycin, clindamycin, diatrizoic acid, erythromycin, ibuprofen, sulfamethoxazole) (*out of 46 types detected*); and
- ***Surfactants*** [alcohol ethoxylate (AE), alkyldimethylbenzyl ammonium chlorides/Benzalkonium chloride (BACs), alkyltrimethyl ammonium chlorides (ATAC), dialkyldimethyl ammonium chlorides (DDAC), linear alkyl benzene sulfonate (LAS), octylphenol ethoxylates (OPEs)/ octylphenol (OP), soap] (*out of 12 types detected*);

Several pollutants have poor removal efficiency in WWTP (<50%) **including chelating agents** (*EDTA*); **corrosion inhibitors** (*MTBT, BTSA, Methylbenzotriazole*); **fragrances** (*Musk ketone*); **heavy metals** (*Ni*); **illicit drugs** (*Ecstasy*); **pesticides/herbicides** (*AMPA, Atrazine, Carbendazim, Diazinon, Diuron, Glyphosate, Irgarol 1051, Isoproturon, Mecoprop, Terbuthylazine*); **pharmaceuticals** (*Carbamazepine, Diclofenac, Fluconazole, Metoprolol, Primidone, Propranolol*); **perfluorinated compounds** (*PFOA, PFOS*); **phosphorous flame retardants** (*TDCP, TCPP, TCEP*); **sweeteners** (*Acesulfame, Sucralose*). These pollutants are generally hydrophilic/ polar, highly soluble and hardly biodegradable. In contrast, **hydrophobic** pollutants have a high removal efficiency in WWTP (> 70%) by sorption or biodegradation (most heavy metals, POPs, brominated flame retardants, several personal care products as well as easily degradable pollutants) (such as surfactants, plastic additives, hormones, some pharmaceuticals and household chemicals) (Table 12.5).

Chemical micro-pollutants found in sewage water can have different impacts on aquatic biota through lethal and sub-lethal effects and can reduce or eliminate species from an ecosystem through increased susceptibility to disease, mortality and decreased fecundity. Sewage discharge can lead to excess nutrients to enter aquatic environments. Nutrient enrichment can cause a facet of environmental problems including eutrophication, algal blooms in both fresh and marine environment, hypoxia, fish kills and seafood contamination. Suspended solids/particles in sewage reduce the clarity of the water. It reduces light penetration affecting photosynthesis in aquatic plants. It may clog/ coat fish gills and feeding apparatus in molluscs causing stress or mortality. Pathogens (bacteria, viruses) can contaminate water and food and cause diseases in humans.

To reduce the risks and impacts of sewage pollution on aquatic biota and human health, a number of strategies and actions can be undertaken including use sewage spill hazard matrix to calculate/ assess the risks associated with the spill; maintaining and upgrading sewerage infrastructures to cope with climate change related-risks (e.g. overflows); frequent monitoring of assets to avoid persistent leakages, collapses, blockages and operational failures during dry and wet weather seasons; utilisation of sewage sludge where feasible (e.g. land reclamation, forestry), recycling of sewage water for landscape irrigation and monitoring chemical pollutants level in sewage effluents and comparing with guideline threshold values (Table 12.6) to protect water quality and aquatic biota and human health.

12.7. References

Abo R, Kummer N-A and Merkel BJ 2016 Optimized photodegradation of Bisphenol A in water using ZnO, TiO_2 and SnO_2 photocatalysts under UV radiation as a decontamination procedure. *Drinking Water Engineering and Science* **9(2)** 27–35 *doi:10.5194/dwes-2016-5*

ANZECC (Australian and New Zealand Environment and Conservation Council) & ARMCANZ (Agriculture and Resource Management Council of Australia and New Zealand) 2000 Australian and New Zealand guidelines for fresh and marine water quality, Volume 1, The Guidelines Chapters 1-7.

https://www.waterquality.gov.au/sites/default/files/documents/anzecc-armcanz-2000-guidelines-vol1.pdf; https://www.waterquality.gov.au/anz-guidelines/guideline-values/default/water-quality-toxicants/toxicants/detergents-2000

Balasubramani A, Howell NL and Rifai HS 2014 Polychlorinated biphenyls (PCBs) in industrial and municipal ffluents: concentrations, congener profiles, and partitioning on to particulates and organic carbon. *Science of the Total Environment* **473-474** 702–713

Balmer ME, Buser HR, Müller MD and Poiger T 2005 Occurrence of some organic UV filters in wastewater, in surface waters, and in fish from Swiss lakes. *Environmental Science and Technology* **39** 953–962

Beasley TM and Held EE 1969 Nickel-63 in marine and terrestrial biota, soil, and sediment. *Science* **164(3884)** 1161-1163 *DOI: 10.1126/science.164.3884.1161*

Belanger SE, Dorn PB, Toy R et al 2006 Aquatic risk assessment of alcohol ethoxylates in North America and Europe. *Ecotoxicology and Environmental Safety* **64** 85–99

Bergé A, Cladière M, Gasperi J et al 2012 Meta-analysis of environmental contamination by alkylphenols. *Environmental Science and Pollution Research* **19** 3798–3819

Bergé A, Cladiere M, Gasperi J et al 2013 Meta-analysis of environmental contamination by phthalates. *Environmental Science and Pollution Research* **20** 8057–8076

Berna JL, Cassant G, Hager CD et al 2007 Anaerobic biodegradation of surfactants: scientific review. *Tenside Surfactants Detergents* **44** 312–347

Bester K, Klasmeier J and Kupper T 2008 Emissions of OTNE (Iso-E-super) – mass flows in sewage treatment plants. *Chemosphere* **71** 2003–2010

Bijlsma L, Emke E, Hernández F and de Voogt P 2012 Investigation of drugs of abuse and relevant metabolites in Dutch sewage water by liquid chromatography coupled to high resolution mass spectrometry. *Chemosphere* **89** 1399–1406

Bjerregaard P, Andersen DN, Pedersen KL et al 2003 Estrogenic effect of propylparaben (propylhydroxybenzoate) in rainbow trout Oncorhynchus mykiss after exposure via food and water. *Comparative Biochemistry and Physiology Part C: Toxicology & Pharmacology* **136(4)** 309-317

Blanchard M, TeilMJ, Ollivon D et al 2004 Polycyclic aromatic hydrocarbons and polychlorobiphenyls in wastewaters and sewage sludges from the Paris area (France). *Environmental Research* **95** 184–197

Borroto RJ 1998 Global warming, rising sea level, and growing risk of cholera incidence: a review of the literature an evidence. *GeoJournal* **44(2)** 111-120

Brausch JM and Rand GM 2011 A review of personal care products in the aquatic environment: Environmental concentrations and toxicity. *Chemosphere* **82** 1518–1532

Cabelli VJ, Dufour AP, McCabe LJ and Levin MA 1982 Swimming-associated gastroenteritis and water quality. *Am J Epidemiol* **115** 606-616

Clara M, Gans O, Windhofer G et al 2011 Occurrence of polycyclic musks in wastewater and receiving waterbodies and fate during wastewater treatment. *Chemosphere* **82** 1116–1123

Clara M, Windhofer G, Weilgony P et al 2012 Identification of relevant micropollutants in Austrian municipal wastewater and their behaviour during wastewater treatment. *Chemosphere* **87** 1265–1272

Cunningham C and Gharipour M P 2018 Pipe dreams: urban wastewater treatment for biodiversity protection. *Urban Science* ***2(10)*** *18p doi:10.3390/urbansci2010010*

DEPA (Danish Environmental Protection Agency) 2013 Survey of parabens. Strandgade 29 1401 Copenhagen K, Denmark

https://www2.mst.dk/Udgiv/publications/2013/04/978-87-93026-02-5.pdf

DGE (Direction générale de l'environnement) 2013 Statement 2012 of the wastewater treatment in canton Vaud, Switzerland [Bilan 2012 del'epuration vaudoise]. 2013.

https://www.vd.ch/fileadmin/user_upload/themes/environnement/eau/fichiers_pdf/bilan_2012_epuration_Vaud.pdf

D'Hollander W, de Voogt P, De Coen W and Bervoets L 2010 Perfluorinated Substances in human food and other sources of human exposure. *In*: P. de Voogt (ed.), *Reviews of Environmental Contamination and Toxicology* **208** Perfluorinated alkylated substances: 179-215

Doick KJ, Klingelmann E, Burauel P et al 2005 Long-term fate of polychlorinated biphenyls and polycyclic aromatic hydrocarbons in an agricultural soil. *Environmental Science and Technology* **39(10)** 3663–3670

EPA Vic (Environment Protection Authority Victoria, Australia) 1995 Managing sewage discharges to inland waters. Publication 473 *https://www.epa.vic.gov.au/about-epa/publications/473*

Fatone F, Di Fabio S, Bolzonella D and Cecchi F 2011 Fate of aromatic hydrocarbons in Italian municipal wastewater systems: an overview of wastewater treatment using conventional activated-sludge processes (CASP) and membrane bioreactors (MBRs). *Water Research* **45** 93–104

Fatta-Kassinos D, Bester K and Kümmerer K (eds) 2010 Xenobiotics in the urban water cycle: mass flows, environmental processes, mitigation and treatment strategies. Series Environmental Pollution 16 Springer Nature. ISBN 978-90-481-3509-7

Feijtel T, Vits H, Murray-Smith R et al 1996 Fate of LAS in activated sludge wastewater treatment plants: a model verification study. *Chemosphere* **32** 1413–1426

Field JA, Field TM, Poiger T, Siegrist H and Giger W 1995 Fate of secondary alkane sulfonate surfactants during municipal wastewater treatment. *Water Research* **29** 1301–1307

Fijalkowski K, Rorat S, Grobelak A et al 2017 The presence of contaminations in sewage sludge. The current situation. *Journal of Environmental Management* **203** 1126-1136

Ford R, Hill J, Malik A, English J, Leo A and Kibria G 2008 Final report for the pilot pathogen study and extended pathogen monitoring survey in the central G-MW water supply system. Report prepared under a research collaboration agreement between Goulburn Murray Rural Water Authority and the University of Melbourne.G-MW Docs# 2489289; ISBN: 1-87635-607-3. *DOI: 10.13140/2.1.4942.1764.* *https://www.researchgate.net/publication/264084049_A_pilot_study_of_microbial_pathogen_monitoring_in_the_Central_Goulburn_Irrigation_Area_of_the_G-MW_water_supply_system_Australia_research_collaboration_between_GM_Water_Australia_and_the_University_of*

Galassi S, Mingazzini M, Viganò L et al 1988 Approaches to modeling toxic responses of aquatic organisms to aromatic hydrocarbons. *Ecotoxicology and Environmental Safety* **16(2)** 158-69 *doi: 10.1016/0147-6513(88)90030-9.*

Gardner M, Comber S, Scrimshaw MD et al 2012 The significance of hazardous chemicals in wastewater treatment works effluents. *Science of the Total Environment* **437** 363–372

Gardner M, Jones V, Comber S, et al 2013 Performance of UK wastewater treatment works with respect to trace contaminants. *Science of the Total Environment* **456–457** 359–369

Gago-Ferrero P, Díaz-Cruz MS and Barceló D. 2012 An overview of UV-absorbing compounds (organic UV filters) in aquatic biota. *Analytical and Bioanalytical Chemistry* **404(9)** 2597-2610 *doi: 10.1007/s00216- 012-6067-7.*

González-Mariño I, Quintana JB, Rodríguez I and Cela R 2011 Evaluation of the occurrence and biodegradation of parabens and halogenated by-products in wastewater by accurate-mass liquid chromatography-quadrupole-time-of-flight-mass spectrometry (LC-QTOF-MS). *Water Research* **45** 6770–6780

Guedez AA and Püttmann W 2011Occurrence and fate of TMDD in wastewater treatment plants in Germany. *Water Research* **45** 5313–5322

Guo R, Sim W-J, Lee E-S, Lee J-H and Oh J-E 2010 Evaluation of the fate of perfluoroalkyl compounds in wastewater treatment plants. *Water Research* **44** 3476–3486

Haman C, Dauchy X, Rosin C et al 2015 Occurrence, fate and behavior of parabens in aquatic environments: A review . *Water Research* **68** 1-1 1

Harrault L, Milek K, Jarde´ E et al 2019 Faecal biomarkers can distinguish specific mammalian species in modern and past environments. *PLoS ONE* **14(2)** e0211119 *https://doi.org/10.1371/journal. pone.0211119*

Hook S, Batley G, Holloway M et al 2016 Oil spill monitoring handbook. *In*: Sharon Hook, Graeme Batley, Michael Holloway, Paul Irving and Andrew Ross (eds.). CSIRO Publishing Locked Bag 10, Clayton South VIC 3169, Australia. 9781486306367 (epub)

https://publications.csiro.au/rpr/download?pid=csiro:EP158596&dsid=DS2

Hope BK, Pillsbury L and Boling B. 2012 A state-wide survey in Oregon (USA) of trace metals and organic chemicals in municipal effluent. *Science of the Total Environment* **417(418)** 263–272

Houde M, Martin JW, Letcher RJ, Solomon KR and Muir DC 2006 "Biological monitoring of polyfluoroalkyl substances: A review". *Environmental Science and Technology* **40(11)** 3463–73

Hussain MA, Ford R and Hill J 2010 Determination of fecal contamination indicator sterols in an Australian water supply system. *Environmental Monitoring and Assessment* **165** 147–157 *DOI 10.1007/s10661-009-0934-5*

IEH (Institute of Environment and Health) 2012 A review of latest endocrine disrupting chemicals research implications for drinking water. Final Report DWI:70/2/266

http://dwi.defra.gov.uk/research/completed-research/reports/DWI70_2_266.pdf

Incardona JP, Vines CA, Anulaciona BF, Baldwina DH et al 2012 Unexpectedly high mortality in Pacific herring embryos exposed to the 2007 Cosco Busan oil spill in San Francisco Bay. *Proceedings of the National Academy of Science of the USA* **109(2)** E51–E58. *doi: 10.1073/pnas.1108884109. Epub 2011 Dec 27*

Jobling S and Tyler CR 2003 Endocrine disruption in wild freshwater fish. *Pure Applied Chemistry* **75** 2219–223

Jobling S, Casey D, Rodgers-Gray T et al 2004 Comparative responses of molluscs and fish to environmental estrogens and an estrogenic effluent. *Aquatic Toxicology* **66** 207–222

Kasprzyk-Hordern B, Dinsdale RM and Guwy AJ 2009 The removal of pharmaceuticals, personal care products, endocrine disruptors and illicit drugs during wastewater treatment and its impact on the quality of receiving waters. *Water Research* **43** 363–380

Robert K, Rik E, Marion J et al 2011 Assessment of micropollutants from municipal wastewater: combination of exposure and ecotoxicological effect data for Switzerland. In Waste Water Evaluation & Management (ed) F S. S Einschlag, Rijeka, Croatia: *In Tech* 31–54 *DOI: 10.5772/16152*

Katsoyiannis A and Samara C 2004 Persistent organic pollutants (POPs) in the sewage treatment plant of Thessaloniki, northern Greece: occurrence and removal. *Water Research* **38** 2685–2698

Kibria G 2016a Trace metals/heavy metals and its impact on environment, biodiversity and human health -A short review. 5p. DOI: 10.13140/RG.2.1.3102.2568. *https://www.researchgate.net/publication/266618621_Traceheavy_Metals_and_Its_Impact_on_Environment_Biodiversity_a nd_Human_Health-_A_Short_Review*

Kibria G 2016b Pesticides and Its Impact on Environment, Biodiversity and Human Health- A Short Review. 5p. *DOI: 10.13140/RG.2.1.4487.4965/1*

https://www.researchgate.net/publication/266618490_Pesticides_and_Its_Impact_on_Environment_Biodiversity_and_Huma n_Health-_A_Short_Review

Kibria G, Nugegoda D, Fairclugh R, Lam P and Bradly A 1999 Utilization of wastewater-grown zooplankton: Nutritional quality of zooplankton and performance of silver perch *Bidyanus bidyanus* (Mitchell 1838) (Teraponidae) fed on wastewater-grown zooplankton. *Aquaculture Nutrition* **5** 221-227

Kibria G, Haroon AKY, Nugegoda D and Rose G 2010 Chemicals and Climate Change: Environmental and Biological Aspects. NIPA, India, 470 p *DOI: 10.13140/ 2.1.4384.0963* ISBN: 93-80235-30-1.

Kibria G, Haroon AKY and Nugegoda D 2016 Climate Change & Water Security. Impacts, Future Projections, Adaptations and Mitigations. ISBN: 978-93-85516-26-9.

DOI:10.13140/RG.2.1.1848.1528/2
https://www.researchgate.net/publication/303218799_Climate_Change_Water_Security

Knepper TP 2003 Synthetic chelating agents and compounds exhibiting complexing properties in the aquatic environment. *Trends in Analytical Chemistry* **22** 708–724

Kokotou MG and Thomaidis NS 2013 Determination of eight artificial sweeteners in wastewater by hydrophilic interaction liquid chromatography-tandem mass spectrometry. *Analytical Methods* **5** 3825–3833

Kostich MS, Batt AL and Lazorchak JM 2014 Concentrations of prioritized pharmaceuticals in effluents from 50 large wastewater treatment plants in the US and implications for risk estimation. *Environmental Pollution* **184** 354–359

Kreuzinger N, Fuerhacker M, Scharf S et al 2007 Methodological approach towards the environmental significance of uncharacterized substances—quaternary ammonium compounds as an example. *Desalination* **215** 209–222

Kudo A, Mortimer DC and Sanford Hart JS 1975 Factors influencing desorption of mercury from bed sediments. *Canadian Journal of Earth Sciences* **12(6)** 1036-1040 *https://doi.org/10.1139/e75-094*

Kümmerer K. 2009 Antibiotics in the aquatic environment – A review – Part I. *Chemosphere* **75** 417-434

Kupper T, Plagellat C, Brändli RC et al 2006 Fate and removal of polycyclic musks, UV filters and biocides during wastewater treatment. *Water Research* **40** 2603–2612

Länge R, Hutchinson TH, Croudace CP et al 2001 Effects of the synthetic estrogen 17 alpha Ethinylestradiol on the life cycle of the fathead minnow (*Pimephales promelas*). *Environmental Toxicology and Chemistry* **20** 1216-1227

Lange FT, Scheurer M and Brauch HJ 2012 Artificial sweeteners: a recently recognized class of emerging environmental contaminants—a review. *Analytical and Bioanalytical Chemistry* **403** 2503–2518

Lewis MA 1990 Chronic toxicities of surfactants and detergent builders to algae: a review and risk assessment. *Ecotoxicology and Environmental Safety* **20** 123–140

Lewis RJ Sr (ed) 2005 Sax's dangerous properties of industrial materials. 11th Edition. Wiley-Inter science, Wiley & Sons, Inc. Hoboken, NJ. 2004., p. 1902. *J. Am. Chem. Soc* ISBN 0-471-47662-5. *https://doi.org/10.1021/ja041002c*

Li Y, Fletcher T, Mucs D et al 2018 Half-lives of PFOS, PFHxS and PFOA after end of exposure to Contaminated drinking water. *Occupational and Environmental Medicine* **75** 46–51 *doi:10.1136/oemed-2017-104651*

Loos R, Carvalho R, Antonio DC et al 2013 EU-wide monitoring survey on emerging polar organic contaminants in wastewater treatment plant effluents. *Water Research* **47** 6475–6487

Loyo-Rosales JE, Rice CP and Torrents A 2007 Fate of octyl and nonylphenol ethoxylates and some carboxylated derivatives in three American wastewater treatment plants. *Environ Sci Technol* **41** 6815–6821

Lozano N, Rice CP, Ramirez M and Torrents A 2013 Fate of triclocarban, triclosan and methyl triclosan during wastewater and biosolids treatment processes. *Water Research* **47** 4519–4527

Lubliner B, Redding M and Ragsdale D 2010 Pharmaceuticals and personal care products in municipal wastewater and their removal by nutrient treatment technologies. *https://fortress.wa.gov/ecy/publications/publications/1003004.pdf.*

Margot J, Kienle C, Magnet A et al 2013 Treatment of micropollutants in municipal wastewater: ozone or powdered activated carbon? *Science of the Total Environment* **461–462** 480–498

Margot J, Rossi L, Barry DA and Holliger C 2015 A review of the fate of micropollutants in wastewater treatment plants. *WIREs Water doi: 10.1002/wat2.1090*

Matthijs E, Holt MS, Kiewiet A and Rijs GBJ 1999 Environmental monitoring for linear alkylbenzene sulfonate, alcohol ethoxylate, alcohol ethoxy sulfate, alcohol sulfate, and soap. *Environmental Toxicology and Chemistry* **18** 2634–2644

McAvoy DC, Dyer SD, Fendinger NJ et al 1998 Removal of alcohol ethoxylates, alkyl ethoxylate sulfates, and linear alkylbenzene sulfonates in wastewater treatment. *Environmental Toxicology and Chemistry* **17** 1705–1711

McAvoy DC, Eckhoff WS, Begley WM et al 2006 A comparison of alcohol ethoxylate environmental monitoring data using different analytical procedures. *Environmental Toxicology and Chemistry* **25** 1268–1274

Mehinto AC, Hill EM and Tyler CR. 2010 Uptake and biological effects of environmentally relevant concentrations of the nonsteroidal anti-inflammatory pharmaceutical diclofenac in rainbow trout (*Oncorhynchus mykiss*). *Environmental Science and Technology* **44** (**2**) 2176-2182

Meyer J and Bester K 2004 Organophosphate flame retardants and plasticisers in wastewater treatment plants. *Journal of Environmental Monitoring* **6** 599–605

MSDS-Pubchem-O xylene 2005 (revised 2020)

https://pubchem.ncbi.nlm.nih.gov/compound/2-xylene#section=Ecotoxicity-Values

Mungray AK and Kumar P 2009 Fate of linear alkylbenzene sulfonates in the environment: a review. *International Biodeterioration & Biodegradation* **63** 981–987

NHMRC (National Health and Medical Research Council) 2004 Guidelines for managing risks in recreational water. National Health and Medical Research Council, Commonwealth of Australia, 185.

NICNAS (National Industrial Chemicals Notification and Assessment Scheme, Australia) 2019 Phthalate esters: environment tier II assessment

https://www.nicnas.gov.au/chemical-information/imap-assessments/imap-assessments/tier-ii-environment-assessments/phthalates

North KD 2004 Tracking polybrominated diphenyl ether releases in a wastewater treatment plant effluent, Palo Alto, California. *Environmental Science and Technology* **38** 4484–4488

Nugegoda D and Kibria G 2017 Effects of environmental chemicals on fish thyroid function: Implications for fisheries and aquaculture in Australia. *General and Comparative Endocrinology* **244** 40-53 *doi: 10.1016/j.ygcen.2016.02.021. Epub 2016 Feb 2*

Nyholm JR, Grabic R, Arp HPH et al 2013 Environmental occurrence of emerging and legacy brominated flame retardants near suspected sources in Norway. *Science of the Total Environment* **443** 307–314

OECD (Organisation for Economic Co-operation and Development) 2002 Hazard assessment for perfluoro octane sulfonate (PFOS) and its salts. ENV/JM/RD (2002) 17/FINAL. *https://www.oecd.org/chemicalsafety/risk-assessment/2382880.pdf*

OEHHA 2009 Toxicological Profile for Di-(2-Ethylhexyl) Phthalate (DEHP). Report by Integrated Risk Assessment Branch, Office of Environmental Health Hazard Assessment, California Environmental Protection Agency. Scientific Reports | 7: 16999 | *DOI:10.1038/s41598-017-15462-x*

Oosterhuis M, Sacher F and ter Laak TL 2013 Prediction of concentration levels of metformin and other high consumption pharmaceuticals in wastewater and regional surface water based on sales data. *Science of the Total Environment* **442** 380–388

Oulhote Y, Chevrier J and Bouchard MF 2016 Exposure to Polybrominated Diphenyl Ethers (PBDEs) and Hypothyroidism in Canadian Women. *The Journal of Clinical Endocrinology & Metabolism* **101(2)** 590–598 *https://doi.org/10.1210/jc.2015-2659*

Oulton RL, Kohn T and Cwiertny DM 2010 Pharmaceuticals and personal care products in effluent matrices: a survey of transformation and removal during wastewater treatment and implications for wastewater management. *Journal of Environmental Moni*toring **12** 1956–1978

Oviedo C and Rodríguez J 2003 EDTA: the chelating agent under environmental scrutiny. Química Nova **26(6)** 901-905

http://www.scielo.br/pdf/qn/v26n6/a20v26n6.pdf; *http://dx.doi.org/10.1590/S0100-*40422003000600020

Owili MA 2003 Assessment of impact of sewage effluents on coastal water quality in Hafnarfjordur, Iceland. The United Nations University. Final project. *Corpus ID: 39510358 http://www.unuftp.is/static/fellows/document/monicaprf03.pdf*

Peck AM and Hornbuckle KC Synthetic Musk Fragrances in Urban and Rural Air of Iowa and the Great Lakes *Atmospheric Environment* **40(2)** 6101-6111

Petersen R, Thomsen JF, Jørgensen NK and Mikkelsen S 2000 Half-life of chromium in serum and urine in a former plasma cutter of stainless steel *Occupational and Environmental Medicine* **57** 140–142

Potvin CM, Long Z and Zhou H 2012 Removal of tetrabromobisphenol A by conventional activated sludge, submerged membrane and membrane aerated biofilm reactors. *Chemosphere* **89** 1183–1188

Reemtsma T, Weiss S, Mueller J et al 2006 Polar pollutants entry into the water cycle by municipal wastewater: a European perspective. *Environmental Science and Technology* **40** 5451–5458

Richman C and Castensson S 2008 Impact of waste pharmaceuticals: an environmental hazard or green wash. *The Pharmaceutical Journal* **280** 335-342

Sánchez-Quiles D and Tovar-Sánchez A 2015 Are sunscreens a new environmental risk associated with coastal tourism? *Environment International* **83** 158–170

Schowanek D, David H, Francaviglia R et al 2007 Probabilistic risk assessment for linear alkylbenzene sulfonate (LAS) in sewage sludge used on agricultural soil. Regulatory Toxicology and Pharmacology 49 245–259

Schulte-Oehlmann U, Oetken M, Bachmann J and Oehlmann J 2004 Effects of ethinylestradiol and methyltestosterone in prosobranch snails. *In*: Kümmer, K (ed.). *Pharmaceuticals in the Environment: Sources, Fate, Effects and Risks*. Second edition. Berlin, Germany: Springer. pp. 233-247

Scott MJ and Jones MN 2000 The biodegradation of surfactants in the environment. *Biochimica et Biophysica Acta* **1508** 235–251

SEPP (State Environment Protection Policy) 2018 SEPP Water Victoria Environment Protection Act 1970

http://www.gazette.vic.gov.au/gazette/Gazettes2018/GG2018S499.pdf

Shang DY, Macdonald RW and Ikonomou MG 1999 Persistence of nonylphenol ethoxylate surfactants and their primary degradation products in sediments from near a municipal outfall in the strait of Georgia, British Columbia, Canada. *Environ Sci Technol* **33** 1366–1372

Sillanpää M 1997 Environmental fate of EDTA and DTPA. *Reviews of Environmental Contamination and Toxicology* **132** 85-111 *DOI: 10.1007/978-1-4612-1964-4_3*

Singer H, Jaus S, Hanke I et al 2010 Determination of biocides and pesticides by on-line solid phase extraction coupled with mass spectrometry and their behaviour in wastewater and surface water. *Environmental Pollution* **158** 3054–3064

Soares A, Guieysse B, Jefferson B and Cartmell E 2008 "Nonylphenol in the environment: A critical review on occurrence, fate, toxicity and treatment in Waste-waters". *Environment International* **34(7)** 1033–1049 *doi:10.1016/j.envint. 2008.01.004. PMID 18282600*

Stahl T, Mattern D and Brunn H 2011 Toxicology of perfluorinated compounds. *Environmental Sciences Eur*ope **23** 38 *https://doi.org/10.1186/2190-4715-23-38*

Ulhaq M, Carlsson G, Örn S and Norrgren L 2013 Comparison of developmental toxicity of seven perfluoroalkyl acids to zebra fish embryos. *Environmental Toxicology & Pharmacol*ogy **36** 423-426

UN Water 2015 Wastewater Management. A UN-Water Analytical Brief *file:///C:/Users/User1/Downloads/UN-Water_Analytical_Brief_Wastewater_Management.pdf*

US EPA 1996. Sanitary sewer overflows. What are they and how can we reduce them? EPA 832-K-96-001 – Summer 1996 *https://www3.epa.gov/npdes/pubs/ssodesc.pdf*

Verlicchi P, Al Aukidy M and Zambello E 2012 Occurrence of pharmaceutical compounds in urban wastewater: removal, mass load and environmental risk after a secondary treatment — a review. *Science of the Total Environment* **429** 123–155

Walters A, Santillo D and Johnston P 2005 Environmental and human health concerns relating to synthetic musk compounds Greenpeace Research Laboratories, Department of Biological Sciences, University of Exeter, Exeter EX4 4PS, UK June 2005

http://www.greenpeace.to/publications/synthetic_musks_2005.pdf

Wernersson AS and Dave G 1997 Phototoxicity identification by solid phase extraction and photoinduced toxicity to *Daphnia magna Archives of Environmental Contamination and Toxicology* **32(3)** 268-273

WHO 2006 WHO guidelines for the safe use of wastewater, excreta, and greywater. World Health Organization, Geneva.

WHO 2010 Exposure of cadmium: A major public health concern.

https://www.who.int/ipcs/features/cadmium.pdf

Yang J-J and Metcalfe CD 2006 Fate of synthetic musks in a domestic wastewater treatment plant and in an agricultural field amended with biosolids. *Science of the Total Environment* **363** 149–165

YVW (Yarra Valley Water) 2020 Sewerage system. What is sewage?

https://www.yvw.com.au/about-us/student-resources/secondary/sewerage-system (accessed 25 June 2020)

Zeng L, Li H, Wang T, Gao Y, Xiao K, Du Y, Wang Y, Jiang G. 2013 Behavior, fate, and mass loading of short chain chlorinated paraffins in an advanced municipal sewage treatment plant. *Environmental Science*

Ziolko D, Martin OV, Scrimshaw MD and Lester JN 2011 An evaluation of metal removal during wastewater treatment: the potential to achieve more stringent final effluent standards. *Critical Reviews in Environmental Science and Technology* **41** 733–769

Web

https://www.sciencedirect.com/topics/medicine-and-dentistry/lead-210

https://en.wikipedia.org/wiki/Zinc

https://en.wikipedia.org/wiki/Bisphenol_A

https://www.chemsafetypro.com/Topics/CRA/How_to_Calculate_Predicted_No-Effect_Concentration_(PNEC).html;

https://slideplayer.com/slide/6812917/ontang

https://epa.tas.gov.au/Documents/Sewage%20Spill%20Notification%20Guidelines%20OCTFINALv2.pdf

■■■■

13 Climate Change, Pollutants and Ecotoxicity Risks, Impacts and Management

Summary

Both climate change and pollutants are a global problem and a major threat to social, economic and environmental sectors. The climate change stressors (temperature, salinity, ocean acidification (OA) and hypoxia)) can alter the environmental fate and transport of pollutants. It can also enhance the uptake and toxicity of pollutants to aquatic biota (algae, crustaceans, fish, and molluscs). This chapter reviews: **i**. effects of climate change stressors on pollutants release, mobilisation, and transportation in the environment; **ii**. interactive effects of stressors (climate change and pollutants) on aquatic biota (algae, crustaceans, fishes, frogs, and molluscs); **iii.** trends in the effect of climate change stressors on pollutant toxicity (increase or decrease) and **iv**. mitigation strategies and measures to reduce the risks and impacts of two stressors (climate change and pollutants).

Contents

List of Tables

13.1. Introduction

The UNFCC (United Nations Framework Convention on Climate Change) defines '*climate change*' as the change of climate which is attributed directly or indirectly to human activity that alters the composition of the global atmosphere and which is in addition to natural climate variability observed over a comparable time period (UNFCCC 1992). Climate change is caused by an increase in the concentration of greenhouse gases (GHGs) due to human activities. The main GHGs are carbon dioxide (CO_2), methane (CH_4), nitrous oxide (N_2O). Higher concentrations of GHGs will lead to increased heat-trapping, producing an enhanced greenhouse effect. The result is likely to cause a change in weather and climate of our planet (IPCC 2007a, Schmidhuber and Tubiello 2007). Such changes are **i**. the rise in temperatures; **ii.** increase in the frequency and intensity of extreme weather events (rainfall, floods, drought, storms/cyclones, bush/forest fires, snow/ice melts); **iii**. sea-level rise and alternation in salinity levels of water bodies; **iv.** ocean acidification (OA) (decrease in ocean pH), **v**. increase in eutrophication (and algal blooms) in water bodies; and **vi.** increase in hypoxia (decrease in dissolved oxygen level in aquatic environments, like lakes, rivers, estuaries and oceans (IPCC 2007a, b; Kibria et al 2016a, Kibria et al 2017, IPCC 2018) (Table 13.1). Temperature, extreme weather events, salinity, OA, eutrophication and hypoxia are referred to herein in this chapter as "**climate change stressors**". The climate change stressors can alter the environmental fate and transport of pollutants[1] (Noyes et al 2009, Bogdal and Scheringer 2011, Gilman and Jenssen 2011). It can also enhance the uptake and toxicity of pollutants to aquatic biota (algae, crustaceans, fish, and molluscs) (Schiedek et al 2007, Noyes et al 2009) (Table 13.1). Nonetheless, both climate change and pollutants are a global problem and a major threat to social, economic and environmental sectors. The interactive effects of these two stressors (climate

[1]**Pollutants**= *harmful chemicals, which are persistent, bioaccumulative and, toxic as well as endocrine-disrupting and carcinogenic (cancer-causing) such as POPs, pesticides, heavy metals.*

change and pollutants) would exacerbate the global pollution problems and increase threats and risks to aquatic ecosystems including aquatic biota.

13.2. Pollutants mobilisation and transportation induced by climate change

High temperature increases volatilisation of pollutants (POPs, pesticides (DDT), heavy metals) and increase release, mobilisation and degradation of pollutants (discussed in section 2.1). Extreme events (intense rainfall/precipitation and floods, bushfires/forest fires, snow/ice melts) increase mobilisation of both hydrophobic and hydrophilic pollutants (pesticides, heavy metals, nutrients, EDCs, PAHs, pharmaceuticals, PCDD/F, PCBs) from environmental reservoirs (soils, sediments, vegetation, permafrost, snow and ice etc) (discussed in sections 2.2.1-2.2.6). Table 13.2 provides details about the relationships between climate change stressors and pollutants mobilisation. The following is a summary of Table 13.2, which discusses the effects/influences of climate change stressors (temperature, extreme events, biotic transport, and other stressors) on environmental fate and transport of pollutants.

Table 13.1. Synopsis of interrelationships among climate change, pollutants and ecotoxicity

[Pollutants= POPs, DDT, heavy metals, PCBs, PAHs, dioxins, pharmaceuticals] [DO= dissolved oxygen; EE= intense rainfall, floods, forest fires, cyclones, snow melts]; ↑= Increase; ↓ = decrease].

Climate Change	**1. Temperature/ Global Warming (T°C)** *(rise in T°C)* ↑ **T⁰C**	**2. Extreme Events (EE)** *(increase in the frequency and intensity of rainfall, floods, bush/forest fires, storms/ cyclones, snow/ice melts)* ↑ **EE**	**3. Sea-Level Rise (SLR)** *(rise in sea level)* ↑ **SLR**	**4. Ocean Acidification (OA)** *(decrease in pH)* **pH** ↓	**5. Hypoxia** *(decrease in dissolved oxygen (DO))* **DO** ↓
Climate Change & Pollutants mobilisation	• **Increase in Temperatures:** Increase volatilisation of pollutants; increases release, mobilisation and degradation of pollutants (see section 13.2.1). • **Increase in Intense Rainfall/ precipitation and floods** : Increase mobilisation of pollutants from environmental reservoirs; increase wet deposits; increase leaching and degradation of pollutants (see section 13.2.2). • **Increase in Bushfires/forest fires:** Increase pollutants load into the environment; increase wash of pollutants from burnt slope into the waterways (post-bushfire rainfall events) (see section 13.2.2).				

Climate Change, Pollutants toxicity & Ecotoxicity[2]	• **Increase in Snow/ice melts:** Increase release and mobilisation hydrophilic and hydrophobic pollutants trapped in polar and alpine glaciers, sea ice & permafrost (see section 13.2.2). • **Increase in Temperatures:** Increase the toxicity of pollutants, increase oxygen consumption; increase bioaccumulation in biota; increase vitellogenin in male fish; decrease growth, reproduction and antioxidant defences in biota; and alteration of metallothionein levels in biota (see section 13.3.1). • **Increase in Salinity:** Increase or decrease the toxicity of pollutants (see section 13.3.2). • **Increase in Ocean acidification (low pH):** Increase or decrease the toxicity of pollutants; enhance bioaccumulation of metals; affects reproduction; affects clam shells (thinner shells); alter the metal binding proteins level (metallothionein and ferritin); increase metal mobilisation (see section 13.3.3). • **Increase in Hypoxia (low DO):** Increase in toxicity of pollutants (see section 13.3.4).

13.2.1. Temperature

High temperature increases volatilisation of pollutants (POPs, pesticides (DDT), heavy metals (Hg)) and increase release, mobilisation and degradation of pollutants from environmental reservoirs (soils, sediments, vegetation, permafrost, snow and ice, polar and alpine glaciers) into the air, soil and water (Noyes et al 2009, Bogdal and Scheringer 2011). For instance, the higher temperatures have already enhanced: **i**. transportation of pollutants (PCBs) from temperate areas to the Arctic (Lamon et al 2009), **ii.** the atmospheric concentrations of the POPs[3] (hexachlorobenze and PCBs) in the Great Lakes Region (USA) (Ma et al 2004); **iii**. POPs and other pesticides from soils to the atmosphere (Noyes et al 2009); **iv**. the degradation of POPs (Bloomfield et al

[2]***Ecotoxicity is*** *referred to here as the toxic effects caused by chemical pollutants/stressors (single pollutant or a mixture of pollutants) on aquatic ecosystems and aquatic biota including algae, crustaceans, fish and molluscs.*

[3]***POPs*** *= Persistent organic pollutants (POPs). POPs are chemicals of global concern due to their potential for long-range transport, persistence in the environment, ability to bio-magnify and bio-accumulate in ecosystems, as well as their significant negative effects on human health and the environment. They include pesticides, industrial chemicals, and by-products. POPs recognised in Stockholm Convention are Aldrin, Alpha hexachlorocyclohexane, Beta hexachlorocyclohexane, Chlordane, Chlordecone, DDT, Dieldrin, Dioxins (PCDD), Endrin, Endosulfan, PCDF (Furans), Heptachlor, Hexabromobiphenyl, Heptabromodiphenyl ether, Hexabromocyclododecane, Hexachlorobenzene, Hexabromodiphenyl ether, Lindane, Mirex, Pentabromodiphenylether, Pentachlorobenzene, Polychlorinated biphenyls (PCBs), Perfluorooctane sulfonic acid, its salts and perfluorooctane sulfonyl fluoride (PFOS), Tetrabromodiphenyl ether, and Toxaphene) (https://www.environment.gov.au/protection/ chemicals-management/pops; https://en.wikipedia.org/wiki/Persistent_organic_pollutant).*

2006, Noyes et al 2009), and pesticides (DDT into DDE) (Schenker et al 2007); and **v.** photodegradation rate of several phenyl-urea pesticides/ herbicides (linuron, chlorotoluron, diuron, and isoproturon) (Benitez et al 2006) in soils, sediments and surface waters. The dissipation of atrazine was found to be 2-3 times higher at warmer temperatures (Queensland) compared to colder (Tasmania) temperatures in Australia (Kookana et al 2010). In fact, climate change alters the fate of pollutants (pollutants concentrations) via *solvent switching* (pollutants partitioning into different physical phases such as solid, liquid, gas) and *solvent depletion* (pollutants bioaccumulation, biomagnification, trophic structure alterations, hydrological processes, organic carbon cycling, freezing, snow melting mechanisms) (Macdonald et al 2003a,b, Noyes et al 2009). Further, the warmer climates are projected to increase insect pests and plant diseases (Kibria et al 2010), diseases in livestock and aquaculture (Kibria et al 2013), and in humans (malaria, and other vector such as dengue fever) (Bogdal and Scheringer 2011), water and food-borne diseases (OECD 2019)). This will require increased use of pesticides and pharmaceuticals hence, there is a possibility of more emissions of chemical pollutants in the environment (see Table 13.2 for details).

13.2.2. Extreme events

Climate change is projected to increase the frequency and intensity of extreme events such as heavy rainfall/ precipitation, floods, droughts, hot days, heat waves, cyclones/ hurricanes/ storms, bush fires/ forest fires (IPCC 2007a,b). Global warming (warmer oceans) is likely to intensify cyclone activity and heighten storm surges. This would cause greater surges to move further inland, threatening larger areas (Wheeler 2011). The increase in the frequency of high-water events would cause flooding of low coastal zones and potential inundation of thousands of kilometres. Extreme weather events increase mobilisation of both hydrophobic and hydrophilic pollutants (pesticides, heavy metals, nutrients, EDCs, PAHs, pharmaceuticals, PCDD/F, PCBs) from environmental reservoirs into the environment (sections 13.2.2.1 to 13.2.2.6).

13.2.2.1. Floods: Floods would cause increased mobilisation of pollutants from environmental reservoirs (soils, sediments, vegetation, permafrost, snow and ice, glaciers) (IPCC 2007a,b, Bogdal and Scheringer 2011, UNEP/AMAP 2011) to the environment. Some examples are highlighted below: **i**. floods contributed significantly in re-emission and redistribution of POPs formerly stored in sediment and agricultural soils (Bogdal and Scheringer 2011); **ii.** floods transported and remobilised DDT, PCDD/F, benzenehexachloride/ lindane into sediments of the River Mulde in Germany (Wilken et al 1994); **iii.** floods mobilised and contaminated arable soil with DDT, PCBs, and PAHs in Bohemia, (Pulkrabova et al 2008); and **iv**. floods transported very high levels of PCDDs/PCDFFs in the soil from the rivers Elbe and Mulde in Central Europe (Umlauf et al 2005) (see Table 13.2 for details).

13.2.2.2. Rainfall/ precipitation: In areas, where it is projected to increase in run-off/ rainfall/ precipitation due to climate change, in those areas there may be an increased discharge of pollutants (pesticides, heavy metals, metalloids, nutrients, EDCs, pharmaceuticals, dioxins) into lakes, rivers, estuaries (Kibria et al 2010, Kibria 2014). The regions which had more intense precipitation, there the depositions of POPs (hexachlorocyclohexane, toxaphene) (Macdonald et al 2003a,b) and pesticides from air to the surface water was enhanced as wet depositions (Burgoa and Wauchope 1995, Bogdal and Scheringer 2011). Increased precipitation also enhanced: **i**. the degradation of pesticides (Van den Berg et al 1999); **ii**. the distribution of POPs (hexachlorocyclohexane and toxaphene) to aquatic systems (Macdonald et al 2003a); and **iii**. nutrient pollution loads (nitrate, ammonia nitrogen, and phosphate) and affected the seawater quality and eutrophication in the Shenzhen Bay, China (Xu et al 2019). According to a recent research carried out by Jarsjö et al 2020 reveals that climate change (intense rainfall) would cause mobilisation of arsenic (As) and lead (Pb) in those regions where average groundwater levels are expected to increase in the future. Moreover, the leaching and potential movement of atrazine to both surface water and groundwaters is likely to be high in areas where there is expected to be higher rainfalls (Kookana et al 2010).

13.2.2.3. Drought: Research carried out in Australia (Goulburn Murray water catchment, Victoria) as part of Global pollution monitoring programme found that climate variability (dry vs wet year) can influence the transport of heavy metals into waterways. Elevated levels of copper (Cu) were found during the dry year (drought) while elevated levels of zinc (Zn) were found during the wet year (floods) showing climate variability affects outflows from the catchment, thereby affecting the inputs, transport and availability of individual metals. Copper fungicides are typically insoluble in water and it may be mainly transported to the waters by wind during dry years. On the other hand, zinc salts (in fertiliser) are usually more water soluble and therefore more zinc may be washed to the waters during wet years (Kibria et al 2012).

13.2.2.4. Storms/ hurricanes: The concentrations of *POPs/ organochlorine* compounds (aldrin, heptachlor and methoxychlor, DDT, DDE), *heavy metals* (arsenic, iron, lead, mercury) and *semivolatile organic compounds* (benzidine, benz[a]anthracene, benzo[b]-flouranthene, benzo[a]pyrene, N-nitroso-dimethylamine and N-nitroso-di-n-propylamine) elevated several folds in soils, sediments and run-off waters shortly after hurricane Katrina in the USA in August 2005 (Presley et al 2005). Similarly, the concentrations of pesticides (atrazine, carbaryl, diquat dibromide, fipronil, fluoranthene, imidacloprid) were found increased with increasing storm intensity in water bodies in Volusia County, Florida and Portland, Oregon in The USA under different storm intensities (Chiovarou and Siewicki 2007) (see Table 13.2 for details).

13.2.2.5. Bushfire/ forest fires: Bushfire/ forest fires are projected to increase in many countries due to climate change (e.g. Australia, Europe, the USA) (Kibria 2017). Bush fires/ wildfires/ forest fires could increase pollutants loads via: **i.**

mercury (Hg) load into the atmosphere as gaseous Hg as well as into aquatic environments as run-off as ash Hg (Ku et al 2018; *https://phys.org/news/2018-12-wildfire-ash-mercury.html*); **ii.** the releases of most toxic chemicals such as dioxins into the atmosphere and waterways (Kibria et al 2010, Kibria 2017); and **iii**. the post-bushfire rainfall washout of residues including nutrients, heavy metals, pesticides, and POPs from burnt slope into the nearby streams and rivers (Armitage et al 2011, Kallenborn et al 2012, Kibria 2017) (see Table 13.2 for details).

***13.2.2.6. Melting/ thawing of snows/ice*:** Melting/ thawing of snows/ ice (due to climate change) would lead to release and mobilisation of significant amounts of POPs, Hg, organochlorine pesticides (e.g. DDT), fungicide hexachlorobenzene (HCB), heavy metals (manganese, zinc) and other hydrophilic and hydrophobic chemicals trapped in polar and alpine glaciers, sea ice and as well as in permafrost[4] (Macdonald et al 2003a,b, Geisz et al 2008, Bogdal et al 2009, Noyes et al 2009, Kibria et al 2010, Bogdal and Scheringer 2011, Kallenborn et al 2012, Zaharescu et al 2016). Snowmelt/ melting of permafrost increased: **i**. in metal concentrations (calcium, manganese and zinc) and deteriorated water quality in mountain streams in the USA (Todd et al 2012) and **ii.** enhanced wet deposition of POPs to aquatic and terrestrial ecosystems (Macdonald et al 2003a, Meyer and Wania 2008). The melting of sea ice (sea-ice retreat due to rising temperatures) of marine waters of the Western coast of Spitsbergen (Norway) during the winters (2004–2010) caused the release of fungicide hexachlorobenzene (HCB) into the air (Ma et al 2011, Kallenborn et al 2012) (see Table 13.2 for details).

13.2.3. Biotic transport of pollutants

Climate-induced fluctuations in the migratory patterns of fish could play an important role in altering the local and global transport of POPs (Burek et al 2008). For example, the rise in temperatures would cause fish to move towards north and south poles to escape the heat or higher temperatures (Perry et al 2005, Last et al 2011). Thus, fish exposed to hydrophobic/ bioaccumulative pollutants in one location can transport substantial quantities of pollutants to other locations (see Table 13.2 for details).

13.2.4. Wind speeds, ocean currents, sea-level rise, salinity and soil erosion

Wind speeds (Bogdal and Scheringer 2011), ocean currents (Macdonald, et al 2003b, Yamashita et al 2008, Bogdal and Scheringer 2011), sea-level rise (Kallenborn et al 2012), salinity (Bogdal and Scheringer 2011) and soil erosion

[4]***Permafrost*** *is any ground that remains completely frozen—32°F (0°C) or colder—for at least two years straight. These permanently frozen grounds are most common in regions with high mountains and in Earth's higher latitudes—near the North and South Poles*

(ACIA 2005) could also increase pollutants transport and loads. **Higher wind speeds** promoted the efficient intercontinental transfer of the PCBs promoting their dispersal in the global atmosphere (Kallenborn et al 2012) (see Table 13.2 for details).

Table 13.2. Effects of climate change stressors on pollutants release, mobilisation, and transportation in the environment.

*[**Albedo**= is a measure of how much light that hits a surface is reflected without being absorbed; **Ammonia nitrogen**= NH_3-N; **Ca=** calcium; **EDC**= endocrine disrupting chemicals; **EE**= extreme events (intense rainfall/precipitation, floods, drought, bush fires/forest fires cyclones/storms/hurricanes, snow/ice melts); **POP**= persistent organic pollutants; **DDT**= dichlorodiphenyltrichloroethane; **DDE**= dichlorodiphenyldichloroethylene; **Dioxins**= POPs including PCDD, PCDF, PCBs; **HCB**=Hexachlorobenzene; **Hg** = mercury; **Mn** = manganese; **Nitrate**= NO_3-N; **OC**= organochlorine; **PAH**= polycyclic aromatic hydrocarbons; **PCB**= polychlorinated biphenyl; **PCDD**= polychlorinated dibenzodioxins; **PCDF**= Polychlorinated dibenzofurans; **Phosphate**= PO_4^{3}; **SO_4**= sulfate; **Zn** = zinc].*

Climate change stressors *(arranged alphabetically)*	**Pollutants**	**Examples of research carried out and related to pollutants mobilisation, transportation and enhancement induced by climate change stressors**	**References**
Biotic transport of pollutants	POPs	Rise in temperatures cause fish to move towards north and south poles to escape the heat or higher temperatures (Perry et al 2005, Last et al 2011). Thus, fish exposed to hydrophobic/ bioaccumulative pollutants in one location can transport substantial quantities of pollutants to other locations.	Burek et al 2008, Perry et al 2005, Last et al 2011
Bush fires/ forest fires	Dioxins	Bushfire/forest fires are projected to increase in many countries (Australia, the USA) and during a bush fire, the combustion process release the most toxic chemicals such as dioxins into the atmosphere and waterways	Kibria et al 2010, Kibria 2017
Bush fires/ forest fires	Hg	Wildfires/ forest fires could increase substantial amounts of Hg load into the atmosphere (as gaseous Hg) as well as into aquatic environments as run-off (as ash Hg) *https://phys.org/news/2018-12-wildfire-ash-mercury.html*	Ku et al 2018
Drought	Cu, Zn	Climate variability (dry vs wet year) can influence the heavy metals inputs, transport into waterways. Elevated levels of Cu were found during the dry year (drought) while elevated levels of Zn were found during the wet year (floods) in Australia.	Kibria et al 2012

Climate change stressors *(arranged alphabetically)*	**Pollutants**	**Examples of research carried out and related to pollutants mobilisation, transportation and enhancement induced by climate change stressors**	**References**
Floods	Dioxins *(PCDDs/ PCDFs)*	A monitoring study conducted in Central Europe following the huge flooding of the rivers Elbe and Mulde in 2002 found very high levels of PCDDs/PCDFFs in soil in periodically flooded pasture land riverside of the dikes.	Umlauf et al 2005
Floods	POPs	Climate change-related flooding events may significantly contribute to re-emission and redistribution of POPs formerly stored in sediments and agricultural soils.	IPCC 2007 a, b
Floods	POPs	Floods can contribute significantly in re-emission and redistribution of POPs formerly stored in sediment and agricultural soils.	Bogdal and Scheringer 2011
Floods	POPs	Floods and flood related erosion would cause increased mobilisation of pollutants (POPs) from environmental reservoirs (soils, vegetation, permafrost, snow and ice, glaciers) to the environment.	Bogdal and Scheringer 2011, UNEP/AMAP 2011
Floods	POPs (*DDT, lindane, PCDD/F*)	Flooding events transported and remobilised contaminants (DDT, PCDD/F, benzenehexa-chloride/ lindane) into sediments of the river Mulde and its tributaries and adjacent areas in Germany.	Wilken et al 1994
Floods	POPs (*DDT, PCBs, PAHs*)	A 4-year survey soon after catastrophic floods that occurred in the western part of Bohemia in August 2002 documented those floods mobilised and contaminated arable soil with DDT, PCBs, and PAHs.	Pulkrabova et al 2008
Hurricanes	POPs, heavy metals, volatile organic compound	The concentrations of *POPs/organochlorine compounds* (aldrin, heptachlor and methoxychlor, DDT, DDE), *heavy metals* (arsenic, iron, lead, mercury) and *semivolatile organic compounds* (benzidine, benz[a]anthracene, benzo[b]-flouranthene, benzo[a]pyrene, N-nitroso-dimethylamine and N-nitroso-di-n-propylamine) elevated several folds in soils, sediments and runoff waters shortly after hurricane Katrina.	Persley et al 2005

Climate change stressors *(arranged alphabetically)*	**Pollutants**	**Examples of research carried out and related to pollutants mobilisation, transportation and enhancement induced by climate change stressors**	**References**
Ocean current	POPs	Changes in the ocean current patterns would change in POP mobility and distribution.	Macdonald, et al 2003b, UNEP 2010, Kallenborn et al 2012
Ocean current	POPs *(per-fluorinated acids)*	Ocean currents around the world have been speeding up over the past two decades as the planet warms (particularly in the global tropical oceans) (Hu et al 2020). Ocean currents will affect the transport of POPs, in particular water-soluble per-fluorinated acids (Yamashita et al 2008, Bogdal and Scheringer 2011).	Yamashita et al 2008, Bogdal and Scheringer 2011, Hu et al 2020
Rainfall/ precipitation	Nutrients *(nitrate, ammonia nitrogen, phosphate)*	Rainfall-runoff pollution loads accounted for 60–80% of the total nutrient pollution loads (nitrate, ammonia (NH^3) nitrogen, and phosphate) and affected the seawater quality and eutrophication in the estuary area and the inner part of the Shenzhen Bay, China.	Xu et al 2019
Rainfall/ precipitation	As, Pb	Climate change would cause mobilisation of As and Pb in those regions where average groundwater levels are expected to increase in the future.	Jarsjö et al 2020
Rainfall/ precipitation	Pesticides	Increases in the intensity and frequency of rain and storm events promoted the wet deposition of pesticides to terrestrial and aquatic systems.	Burgoa and Wauchope 1995, Chiovarou and Siewicki, 2007, Dabrowski et al 2002
Rainfall/ precipitation	Pesticides	More frequent and intense rainfall events may increase erosion of pesticide-rich soil particles from fields to drains and surface waters.	Bloomfield et al 2006
Rainfall/ precipitation	Pesticides	More frequent and intense rainfall events may lead to more by-pass flow increasing the likelihood of rapid pesticide movement to drains, surface waters and vertically below the soil layer.	Bloomfield et al 2006
Rainfall/ precipitation	Pesticides	Decreases in precipitation would result in reduced washout of currently used pesticides resulting in	Armitage et al 2011

Climate change stressors *(arranged alphabetically)*	Pollutants	Examples of research carried out and related to pollutants mobilisation, transportation and enhancement induced by climate change stressors	References
		their greater potential for atmospheric transport to higher latitudes.	
Rainfall/ precipitation	Pesticides	Increased precipitation resulted in elevated soil moisture which enhanced the degradation of pesticides.	Van den Berg et al 1999
Rainfall/ precipitation	Pesticides, POPs	Precipitation increases will enhance the potential for pesticide and POPs run-off into the aquatic system, whereas decreases in precipitation, will ameliorate chemical runoff but may increase persistence.	Noyes et al 2009
Rainfall/ precipitation	Plastics	High run-off due to high rainfall can have high plastic inputs in waterways compared to dry season. The abundance and distribution of plastic debris in Hong Kong beaches were found significantly higher in the wet season than in the dry season.	Cheung et al 2016
Rainfall/ precipitation	POPs	In regions projected to have more precipitation, depositions of POPs from air to surface will increase as wet depositions.	Bogdal and Scheringer 2011
Rainfall/ precipitation	POPs	Increased precipitation enhanced wet deposition of POPs to aquatic and terrestrial ecosystems	Macdonald et al 2003a, Meyer and Wania 2008
Rainfall/ precipitation	POPs *(hexachloro-cyclohexane, toxaphene)*	Increases in precipitation enhanced the distribution of POPs (hexachlorocyclohexane (HCH)) and toxaphene)) to aquatic systems.	Macdonald et al 2003a
Rainfall/ precipitation	POPs, pesticides	An increase in precipitation will increase the wet deposition of PM (particulate matter), POPs, and pesticides.	Noyes et al 2009
Rainfall/ precipitation	Atrazine	The leaching and potential movement of atrazine to both surface water and groundwaters is likely to be high in areas where there is expected to be higher rainfalls.	Kookana et al 2010
Salinity	Lubricating oil in plastics	The sorption of lubricating oil in PE and PS increases with increasing concentration of salinity.	Hu et al 2017

Climate change stressors *(arranged alphabetically)*	**Pollutants**	**Examples of research carried out and related to pollutants mobilisation, transportation and enhancement induced by climate change stressors**	**References**
Salinity	POPs	Alteration in salinity levels due to climate change would change the solubility of POPs in water (POPs are less soluble in water if salinity is higher).	Bogdal and Scheringer 2011
Salinity	Triclosan in plastics	An increase of salinity enhanced the sorption of triclosan in polyethylene plastics.	Wu et al 2016
Sea level rise	POPs	Sea level rise would increase erosion and increase the release of POPs from secondary source.	Kallenborn et al 2012
Snow/ ice melts (sea ice)	Pesticides (HCB)	The melting of sea ice (sea-ice retreat due to rising temperatures) of marine waters of the western coast of Spitsbergen, Norway during the winters (2004–2010) caused the release of fungicide hexachlorobenzene (HCB) into the air.	Ma et al 2011, Kallenborn et al 2012
Snow/ ice melts	Heavy metals	Reduced snow covers enhanced the mobilisation of naturally occurring heavy/trace metals in high altitude environments in the USA.	Zaharescu et al 2016
Snow/ ice melts	Hg, POPs	Melting/thawing of snows and permafrost would lead to release of very significant amounts of Hg, POPs trapped in ice in polar and alpine glaciers and as well as to a loss of permafrost.	Kibria et al 2010, Bogdal and Scheringer 2011
Snow/ ice melts	POPs	Melting snow, ice, and glaciers will release and remobilise POPs sequestered in the frozen matrices.	Noyes et al 2009
Snow/ ice melts	POPs	Ice/ snow melt enhances wet deposition of POPs to aquatic and terrestrial ecosystems. Water-soluble organic contaminants can be discharged in greatly elevated concentrations at an early stage of melting, while the bulk of the hydrophobic chemicals attached to particles is often released at the end of the melt period.	Macdonald et al 2003a, Meyer and Wania 2008
Snow/ ice melts	POPs	Snow and snowmelt may increase contaminant levels. For example, contaminants can be readily adsorbed to snow surfaces and transported to the	Macdonald et al 2003a, Meyer and

Climate change stressors *(arranged alphabetically)*	**Pollutants**	**Examples of research carried out and related to pollutants mobilisation, transportation and enhancement induced by climate change stressors**	**References**
		ground, whereas, snow melts (due to rise in temperature) increases the concentration of a contaminant in meltwater.	Wania, 2008
Snow/ ice melts (Alpine)	POPs	Melting alpine glaciers would be a new source for POP released in Europe in the future.	Bogdal et al 2009
Snow/ ice melts (permafrost)	Heavy metals *(Ca, Mn, Zn)*	Todd et al (2012) reported that the climate change-driven increasing temperatures and changes in snowmelt/melting of permafrost timing was associated with the increases in metal concentrations (Ca, Mn, Zn) and deterioration of water quality in different naturally acidic mountain streams in the USA (based on 30-year stream water chemistry data set).	Todd et al 2012
Snow/ ice melts (sea ice)	POPs	The predicted rapid loss of sea ice in the Arctic and polar environments would influence POPs composition, sources, pathways and distribution processes on a global scale.	Kallenborn et al 2012
Snow/ ice melts (sea ice)	POPs	Reduced sea ice coverage in the polar regions would increase POP evaporation from open surfaces (sea–land).	Macdonald et al 2003b, UNEP 2010, Kallenborn et al 2012
Snow/ ice melts (sea ice)	POPs/ DDT	Increased melting of glacial ice during the summer melting periods would release considerable amounts of POPs (Bogdal et al 2009) including currently used pesticides. There is evidence that previously deposited stocks of DDT in glacial ice are now being released into the Antarctic marine waters with glacier melting (Geisz et al 2008).	Geisz et al 2008, Bogdal et al 2009, Kallenborn et al 2012
Soil erosion	POPs	Climate change will exacerbate soil erosion which will increase in mobilisation of POPs present in soils and transfer to rivers, lakes and oceans.	ACIA 2005
Soil moisture	Pesticides, POPs	Enhanced soil moisture will increase the hydrolytic degradation of pesticides but may not affect POPs since they are relatively resistant to hydrolysis.	Noyes et al 2009

Climate change stressors *(arranged alphabetically)*	**Pollutants**	**Examples of research carried out and related to pollutants mobilisation, transportation and enhancement induced by climate change stressors**	**References**
Storms	Pesticides *(six pesticides)*	Concentrations of pesticides (atrazine, carbaryl, diquat dibromide, fipronil, fluoranthene, and imidacloprid) were found increased with increasing storm intensity in water bodies in Volusia County, Florida and Portland, Oregon in the USA under different storm intensities.	Chiovarou and Siewicki 2007
Stormwater	Plastics	Lightweight plastic products and particles (e.g., bags, films, clothing fibres, pellets and plastic bottles) are more easily transported greater distances by extreme weather events such as stormwater.	USEPA 2016
Temperature	Atrazine *(herbicide)*	The rate of dissipation of atrazine was found to be 2-3 times higher at warm subtropical Queensland sites in Australia compared to the colder Tasmanian site in Australia (Kookana et al 2010). Hence, climatic conditions (soil temperature) appears to be one of the key factors that can affect atrazine dissipation.	Kookana et al 2010
Temperature	DDT *(OC insecticide)*	The warmer climates would intensify the propagation and spread of malaria and other vector-borne diseases in tropical countries requiring more DDT use to control malaria (Bogdal and Scheringer 2011). This will result in more DDT residues in the environment ***(note:** DDT is still may be in use in those countries who are not signatory to Stockholm Convention).*	Bogdal and Scheringer 2011
Temperature	OPs	Significant temperature rises in the Arctic ocean due to change in albedo[5] properties would increase POP evaporation from open surfaces (sea–land).	Kallenborn et al 2012
Temperature	PAHs in plastics	The greater adsorption of PAHs was found at 21°C compared to 10°C in plastic (polyethene).	Crawford and Quinn 2017

[5] ***Albedo** is a measure of how much light that hits a surface is reflected without being absorbed. Something that appears white reflects most of the light that hits it and has a high albedo, while something that looks dark absorbs most of the light that hits it, indicating a low albedo (https://climate.ncsu.edu/edu/Albedo).*

Climate change stressors *(arranged alphabetically)*	**Pollutants**	**Examples of research carried out and related to pollutants mobilisation, transportation and enhancement induced by climate change stressors**	**References**
Temperature	PCBs *(POPs)*	Higher temperatures have increased primary and secondary volatilisation emissions of PCBs and enhanced transport of PCBs from temperate regions to the Arctic.	Lamon et al 2009
Temperature	PCBs *(POPs)*	Chemicals with properties similar to PCBs may experience enhanced mobility due to increased volatilisation relating to the rise in temperature, and increased rain fall in Europe.	Paul et al 2012
Temperature	Pesticides	Water temperature increases will increase the hydrolysis of pesticides to less or more bioactive degradates.	Lamon et al 2009
Temperature	Pesticides	Higher temperatures will increase the volatilisation of pesticides and degradation of pesticide in the soil and surface waters.	Bloomfield et al 2006
Temperature	Pesticides	The warmer climates and climate extreme events could be more favourable to the proliferation of insect pests and plant diseases requiring increased uses of pesticides and thereby, more emissions of pesticides in the environment in the future.	Kibria et al 2010
Temperature	Pesticides	Global warming (rise in temperatures) could lead to enhanced volatilisation of pesticides relative to soil and water.	Noyes et al 2009
Temperature	Pesticides	Global warming (rise in temperatures) could lead to enhanced volatilisation and microbial degradation of pesticides relative to soil and water.	Noyes et al 2009
Temperature	Pesticides *(urea herbicides)*	Increased water temperature increased the photodegradation rate of several phenyl-urea herbicides (linuron, chlorotoluron, diuron, and isoproturon). The sequence of photodecomposition rates was: linuron>chlorotoluron>diuron>isoproturon.	Benitez et al 2006
Temperature	POPs	An increase in temperature will increase the volatilisation of pollutants (POPs) into the air from environmental reservoirs (surface water, soils, vegetation, permafrost, snow and ice).	Bogdal and Scheringer 2011

Climate change stressors *(arranged alphabetically)*	**Pollutants**	**Examples of research carried out and related to pollutants mobilisation, transportation and enhancement induced by climate change stressors**	**References**
Temperature	POPs	The loss of permafrost associated with rising temperatures will re-release pollutants (POPs) from frozen soils making them available for atmospheric partitioning or run-off to aquatic systems.	Macdonald et al 2003a
Temperature	POPs	Water temperature increases will increase the solubility of POPs resulting in higher levels in water.	Noyes et al 2009
Temperature	POPs (DDE)	Higher temperatures will increase the volatilisation of POPS and degradation of POPs in reservoirs (soils, vegetation and seawater) (Bogdal and Scheringer 2011). The degradation of POPs may include the formation of transformation products such as DDE from DDT (Schenker et al 2007).	Schenker et al 2007, Bogdal and Scheringer 2011
Temperature	POPs, hexachloro-benzene, PCBs	Elevated air temperatures increased the volatility and atmospheric concentrations of the POPs, hexachlorobenzene and PCBs, in the Great Lakes Region, the USA.	Ma et al 2004
Wind speeds	PCBs	Higher wind speeds promoted the efficient intercontinental transfer of the PCBs promoting their dispersal in the global atmosphere.	Kallenborn et al 2012
Wind speeds	POPs	Modified wind fields and higher wind speeds would promote atmospheric transport of airborne particles and associated POPs and increased deposition of POPs.	Bogdal and Scheringer 2011
Wind speeds	POPs	Higher wind speeds could lead to faster and more effective transport of POPs. As a consequence, higher levels of POPs from the primary and secondary sources can move to regions downwind (in the direction in which the wind blows).	Bogdal and Scheringer 2011

13.3. Interactive effects of stressors (climate change and pollutants) on aquatic biota (algae, crustaceans, fish, and molluscs)

The bioavailability and toxicity of pollutants are likely to increase in response to climate change (rising temperatures/global warming, alteration of salinity due to SLR, reduced pH due to OA, and increased hypoxia due to eutrophication/algal blooms) (Tables 3 to 6). The interactive effects of climate change stressors (temperature, salinity, OA/low pH and hypoxia) and pollutants (e.g. POPs, DDT, pesticides, heavy metals) may cause significant impact on aquatic biota via increasing toxicity, reducing growth and survival and enhancing bioaccumulation of pollutants (sections 13.3.1 to 13.3.4):

13.3.1. Interactive effects of [Temperature + Pollutants]

Over the 20th century, the earth's surface temperature has increased by 0.76°C, of which most of the warming occurred between 1976 and 2000. Scientists believe it is likely for a further rise in earth's temperature by another 1.8°C (range 1.1°C-2.9°C) to 4.0°C (range 2.4°C-6.4°C) by the end of 21st century (IPCC 2007a, b; Schmidhuber and Tubiello 2007). The temperature increases in rivers and sea-surface (due to climate change) are projected to cause substantial impacts on fisheries, aquaculture, and seafood security across the globe (negative, positive and mixed) (Kibria et al 2017). At higher temperatures, the metabolism of aquatic organisms is increased and oxygen concentration is reduced in water (oxygen solubility in water is inversely related with temperature) and therefore, the rate of water inflow into the animal (fish, crustaceans, molluscs) must increase to extract more oxygen, which can increase the entrance of dissolved chemical pollutant(s) into their body (Kennedy and Walsh 1997, Noyes et al 2009, Patra et al 2009, Marques et al 2010, Anawar 2013, Kibria et al 2016b). In fact, an underlying mechanism of this interactive toxicity is that temperature alters the toxicokinetics[6] of chemical pollutants in exposed biota (Noyes et al 2009, Gilman and Jenseen 2011, DeLorenzo 2015). The interactive effects of these two stressors (climate change-related increase in temperature and pollutants) would exacerbate such impacts on aquatic biota as discussed below:

13.3.1.1. Increase in toxicity of pollutants with increasing temperature to aquatic biota

Some selected examples of the increase in toxicity pollutants due to the increase in temperature to aquatic biota are listed below:

Algae *(phytoplankton):* The toxicity of Ametryn, atrazine, diuron, irgarol (herbicides) exposed to phytoplankton, *Dunaliella tertiolecta* increased with increased temperature (DeLorenzo et al 2013);

Crustaceans *(crabs, grass shrimps, water flea):* the toxicity of the following pollutants (cadmium, carbaryl, chlorothalonil, corexit, cypermethrin,

---

[6]***Toxicokinetics***= *is the movement and fate, also referred to as the disposition, of toxicants. It includes the processes of absorption, distribution, and elimination (including biotransformation and excretion) of toxicants in an organism (https://www.sciencedirect.com/topics/pharmacology-toxicology-and-pharmaceutical-science/toxicokinetics).*

deltamethrin, fenvalerate, methyl parathion, trimethyltin, scourge[7] (resmethrin and piperonyl butoxide; a synthetic pyrethroid insecticide), terbufos, and trichlorfon) increased with increased temperature to crustaceans such as: **i.** Cd exposed to fiddler crab, *Uca pugilato* (O'Hara 1973); **ii.** carbaryl (carbamate insecticide) exposed to grass shrimp, *Palaemonetes pugio* (Johnson and DeLorenzo 2011); **iii**. chlorothalonil (fungicide) exposed to grass shrimp, *P. pugio* (DeLorenzo et al 2009); **iv**. Corexit (oil dispersants) exposed to grass shrimp, *P. pugio* (National Research Council 2005); **v**. cypermethrin, deltamethrin and fenvalerate exposed to water fleas (*Daphnia magna*) (Ratushnyak et al 2005); **vi**. methyl parathion exposed to crab, *Chasmagnathus granulate* (Monserrat and Bianchini 1995); **vii**. trimethyltin (organotin compound) exposed to fiddler crab, *U. pugilato* (Wright and Roosenburg 1982); **viii.** scourge (synthetic pyrethroid) exposed to grass shrimp, *P. pugio* (DeLorenzo et al 2009); and **ix**. terbufos and trichlorfon (organophosphate pesticides (OP)) exposed to grass shrimp, *P. pugio* (Breken-Folse et al 1994).

***Fish** (catfish, eel, mosquito fish, sheepshead minnows, silver perch, rainbowfish, rainbow trout, and western carp, gudgeon):* the toxicity of the following pollutants (atrazine, chlorpyrifos, dieldrin, enodsulfan, terbufos and trichlorfon) increased with increased temperature to a number fish species, such as: **i.** atrazine (herbicide) exposed to catfish, *Ictalurus punctatus* (Gaunt and Barker 2000); **ii.** chlorpyrifos and endosulfan exposed to Australian warm water fish species (silver perch, *Bidyanus bidyanus)* and a cold-water rainbow trout, *Oncorhynchus mykiss*) (Patra et al 2015); **iii**. dieldrin (OC insecticide) exposed to the freshwater darter, *Etheostoma nigrum* (Silbergeld 1973); **iv**. endosulfan exposed to eel, *Anguilla anguilla* (Ferrando et al 1987); **v.** endosulfan exposed to mosquito fish, *Gambusia affinis* (Nowak and Sunderam 1991); **vi.** terbufos and trichlorfon (OP insecticides) exposed to sheepshead minnows, *Cyprinodon variegatus* (Breken-Folse et al 1994).

***Frogs** (green frog, leopard frogs):* the toxicity of the following pollutants (carbaryl, esfenvalerate) increased with increased temperature to frogs, such as: **i.** carbaryl (carbamate insecticide) exposed to green frog *Rana clamitans* (Boone and Bridges 19 99); **ii**. esfenvalerate (pyrethroid) exposed to larval leopard frogs (*Rana* spp.) (Materna et al 1995).

***Molluscs** (clam, scallop):* the toxicity of the following pollutants (2,4-D, corexit, dicamba, mecoprop) increased with increased temperature to molluscs, such as: **i**. Corexit 9500 (oil dispersants) exposed to scallop, *Argopecte irradians* (National Research Council 2005); **ii.** 2,4-D, dicamba, and mecoprop (herbicides) to clam, *Mya arenaria* (Greco et al 2011).

---

[7]***Scourge** used primarily to control adult mosquitoes. It contains resmethrin and piperonyl butoxide as active ingredients. Resmethrin is a man-made pyrethroid insecticide.*

13.3.1.2. Decrease in toxicity of pollutants with increasing temperature to aquatic biota

Some selected examples of decrease in toxicity **of** pollutants due to increase in temperature to aquatic biota are listed below:

Crustaceans** (crab, grass shrimp, white-leg shrimp)**: the toxicity of the following pollutants (chlorpyrifos, DDT, dimethoate, nitrophenol) decreased with increased temperature; **i.** nitrophenol exposed to grass shrimp, *P. pugio* (Breken-Folse et al 1994); **ii.** DDT exposed to crab, *Callinectes sapidus* (Koenig et al 1976). The increased biotransformation rates of DDT at a higher temperature may have caused to yield less DDT toxic effects on crab (Koenig et al 1976); and iii. dimethoate, chlorpyrifos exposed to post-larvae, and juveniles of white-leg shrimp, *Litopenaeus vannamei*, the research shows that an increase in temperature from 15 to 25°C decreased the acute toxicity of both dimethoate, chlorpyrifos (Pawara et al 2020). It may demonstrate that the toxicity effect on white-leg shrimp depends on its life stage and type of organic phosphates used;

***Fish** (sheepshead minnows):* the toxicity of nitrophenol exposed to fish sheep-shead minnows, *Cyprinodo variegatus* decreased with increased temperature (Breken-Folse et al 1994).

13.3.1.3. Other effects to aquatic biota

Some other interactive effects (temperature + pollutants) observed and documented in aquatic biota are listed below:

Algae: **i.** Higher temperature caused decreased growth rate, cell densities, and starch content in algae, *Dunaliella tertiolecta* exposed to herbicides (ametryn, atrazine, diuron, irgarol) (DeLorenzo et al 2013).

***Crustaceans** (crabs, crayfish, daphnia):* **i.** Higher temperatures affected reproduction rates of *Daphnia pulex* interactively exposed to pyrimethanil (fungicide) (Scherer et al 2013); **ii.** A ten-fold increase in acute lethality/toxicity was observed at elevated temperature with methyl parathion exposed to crabs (*Chasmagnathu granulata*) (at a higher temperature it may favour enzymatic activation of the organophosphate over degradation and excretion) (Monserrat and Bianchini 1995); **iii.** A 3–4-degree Celsius changes in temperature caused three-fold increases in metal-induced oxygen consumption in crayfish (Khan et al 2006); **iv.** Less DDT toxic effects was observed to crab, *Callinectes sapidus* at a higher temperature (at the higher temperature, there could be increased biotransformation rates of DDT which may have caused to yield less DDT toxic effects to crab) (Harwood et al 2009) (see Table 13.3 for details).

***Fish** (eel, fathead minnows, flatfish, trout):* **i.** increased vitellogenin[8] (VTG) (egg yolk precursor protein) was found in male fathead minnows, *Pimephales promelas* exposed to a mixture of estrogenic chemicals (Brian et al 2008); **ii.** Increased metal uptake in the Mediterranean flatfish, *Solea senegalensis* and altered metallothionein[9] (metal-binding protein) levels and metal partitioning into liver and kidney was observed (Siscar et al 2014); **iii.** Increased mortality of eel, *Anguilla anguilla* was observed exposed to endosulfan (Ferrando et al 1987); **iv.** Increased mortality of juvenile rainbow trout (*Oncorhynchus mykiss*) was observed exposed to endosulfan (Capkin et al 2006); **v.** Both rainbow trout (*Oncorhynchus mykiss*) and mosquito fish (*Gambusia holbrooki*) showed a linear increase in endosulfan toxicity at increasing temperatures (Nowak and Sunderam 1991, Patra et al 2015). This may be related to higher degradation and metabolisation observed at higher temperatures (Ghadiri and Rose 2001). In addition, endosulfan can cause damage to fish gills as well as liver and kidney of fish (Capkin et al 2006); **vi.** Decreased yolk utilisation was observed in fish, *Fundulus heteroclitus* exposed to Number 2 fuel oil (Bunker A oil) (Linden et al 1980) (see Table 13.3 for details).

Frogs: **i**. Carbaryl exposure resulted in lower survival of frog, *Rana clamitans* (tadpoles) at higher temperatures (Boone and Bridges 1999); **ii**. Endosulfan exposure affected feeding rate and predator avoidance of the Australian frog (*Limnodynastes peronii*) at a higher temperature (Broomhall 2004); and **iii.** Esfenvalerate (synthetic pyrethroid) exposure increased frog's (*Rana* spp.) mortality at elevated temperature (Materna et al 1995) (see Table 13.3 for details).

***Molluscs** (clam, Mediterranean mussel):* **i.** A decrease in detoxification enzyme activity and an increase in bioaccumulation of benzo[a] pyrene (PAH) was observed to the Mediterranean mussel *Mytilus galloprovincialis* at a higher temperature (Kamel et al 2012); **ii.** 2,4-D, dicamba, and mecoprop (herbicides) inhibited the antioxidant defences to clam, *Mya Arenaria* at a higher temperature (Greco et al 2011) (see Table 13.3 for further details).

---

[8]***Vitellogenin** is a precursor (that produces another compound) protein of egg yolk normally in the blood or haemolymph only of females that is used as a biomarker in vertebrates of exposure to environmental estrogens which stimulate elevated levels in males as well as females.*

[9]***Metallothioneins** (MTs) are a family of small, highly conserved, cysteine-rich (amino-acid) metal-binding proteins that are important for zinc and copper homeostasis, protection against oxidative stress, and buffering against toxic heavy metals (https://www.sciencedirect.com/topics/neuroscience/metallothionein#).*

Table 13.3. Trends in the effect of climate change stressors on pollutant toxicity and interactive effects of temperature + pollutants on ecosystem biota (*algae, crustaceans, fish, frogs and molluscs*)
[↑= increase in toxicity with increasing in temperature; ↓decrease in toxicity with increasing in temperature; OC= organochlorine; OP= organophosphate; PAH= polycyclic aromatic hydrocarbon; POPs= persistent organic pollutants].

Climate change stressor effects on biota *(temperature /global warming)*	**Pollutants**	**Ecotoxicity** *[examples of research carried out and related to interactive effects of two stressors (temperature and pollutants) on algae, crustaceans, fish, and molluscs; ↑= increase in toxicity with increasing in temperature; ↓decrease in toxicity with increasing in temperature].*	**References**
Algae (phytoplankton, *Dunaliella tertiolecta)*	Ametryn, atrazine, diuron, irgarol *(herbicides)*	↑ Toxicity of herbicide to phytoplankton, *D. tertiolecta* increased with increased temperature; toxicity trend: 35°C> 25°C) (caused decreases in growth rate, cell densities, and starch content in algae).	DeLorenzo et al 2013
Biofilms[10]	Diuron, Imazalil, Prochloraz, Simazine and Chlorpyrifos *(Pesticides)*	↓ A research conducted a manipulative experiment using 24 artificial streams to examine the individual and combined effects of i. warming (18.5° C, 20.2°C), ii. hydrological stress and iii. chemical stress caused by pesticide exposure on river biofilms. Hydrological stress caused strong negative impacts on river biofilms, whereas effects of warming and pesticide exposure were less intense.	Romero et al 2019
Crustaceans (grass shrimp, *Palaemonetes pugio*)	4-nitrophenols *(used to make drugs, fungicides, dyes)*	↓ Increased temperature (from 17° C to 27°C) decreased the toxicity of nitrophenol to grass shrimp, *P. pugio*.	Breken-Folse et al 1994
Crustaceans *(grass shrimp, Palaemonetes pugio)*	Corexit *(oil dispersants)*	↑ The toxicity of Corexit to grass shrimp increased with increasing temperatures (from 17°C to 27°C.	National Research Council 2005
Crustaceans (Water flea, *Daphnia pulex)*	Pyrimethanil *(fungicide)*	↑ A study assessed the effects of fungicide pyrimethanil (at an environmental relevant concentration) and increasing temperature (15°C, 20° C, 25°C) on daphnids, *D. pulex*. The results show that multiple stress factors	Scherer et al 2013

[10]***Biofilms**= microorganisms such as algae, cyanobacteria.*

Climate change stressor effects on biota *(temperature /global warming)*	**Pollutants**	**Ecotoxicity** *[examples of research carried out and related to interactive effects of two stressors (temperature and pollutants) on algae, crustaceans, fish, and molluscs; ↑= increase in toxicity with increasing in temperature; ↓decrease in toxicity with increasing in temperature].*	**References**
		(pyrimethanil pollution + temperature) affected reproduction rates of *D. pulex* interactively. Exposure of pyrimethanil reduced daphnid reproduction, which may be connected to temperature. For example, at high temperatures, a rapid and increased uptake of pyrimethanil may have occurred (due to higher metabolic rates at higher temperatures).	
Crustaceans (blue crabs, *Callinectes sapidus)*	DDT *(OC insecticide)*	↓ Increased temperature decreased the toxicity of DDT body burdens to crab, *C. sapidus* (Koenig et al 1976). At higher temperature, there is an increased biotransformation rates of DDT which may have caused to yield less DDT toxic effects on crab (Harwood et al 2009).	Koenig et al 1976, Harwood et al 2009
Crustaceans (crabs, *Chasmagnathus granulata)*	Methyl parathion *(OP insecticide)*	↑ A ten-fold increase in acute lethality/toxicity with crabs (*C. granulata*) were observed with temperature change from 12°C to 30°C exposed to methyl parathion. It may be that the higher temperature favours enzymatic activation of the organophosphate over degradation and excretion.	Monserrat and Bianchini 1995
Crustaceans (crayfish *Orconectes immunis)*	Cd, Cu, Pb, Zn *(heavy metals)*	↑ The toxicity of Cd, Cu, Pb and Zn to the juvenile crayfish, *O. immunis* was significantly increased under elevated temperature. A 3–4-degree Celsius change in temperature caused three-fold increases in metal-induced oxygen consumption.	Khan et al 2006
Crustaceans (fiddler crab, *Uca pugilator)*	Cd *(heavy metal)*	↑ Cd toxicity increased with increasing temperature in the fiddler crab, *U. pugilator*. Cd was least toxic at 10°C and most toxic at 30°C.	O'Hara 1973
Crustaceans (fiddler crab, *Uca pugilator)*	Trimethyltin *(organotin compound)*	↑ An increased metal toxicity with increasing temperature was reported with larval crab, *U. pugilator* exposed to trimethyltin. Trimethyltin was most toxic at 28°C and least toxic at 10°C.	Wright and Roosenburg 1982

Climate change stressor effects on biota *(temperature /global warming)*	**Pollutants**	**Ecotoxicity** *[examples of research carried out and related to interactive effects of two stressors (temperature and pollutants) on algae, crustaceans, fish, and molluscs; ↑= increase in toxicity with increasing in temperature; ↓decrease in toxicity with increasing in temperature].*	**References**
Crustaceans (grass shrimp, *Palaemonetes pugio*)	Bifenthrin *(pyrethroid insecticide)*	↑ The toxicity of bifenthrin increased at the higher temperature/ climate change conditions (30°C) to grass shrimp, *P. pugio* compared to 20°C (standard conditions).	Al-Yassein 2019
Crustaceans (grass shrimp, *Palaemonetes pugio*)	Carbaryl *(carbamate insecticide)*	↑ Increased temperature increased toxicity of carbaryl to grass shrimp, *P. pugio* (toxicity trend: 35°C>25°C).	Johnson and DeLorenzo 2011
Crustaceans (grass shrimp, *Palaemonetes pugio*)	Chlorothalonil *(fungicide)*	↑ Toxicity of chlorothalonil to grass shrimp, *P. pugio* increased with increased temperature; toxicity trend: 35°C >25°C).	DeLorenzo et al 2009
Crustaceans (grass shrimp, *Palaemonetes pugio*)	Ibuprofen *(nonsteroidal anti-inflammatory drug)*	↑ The toxicity of ibuprofen increased at the higher temperature/ climate change conditions (30°C) to grass shrimp, *P. pugio* compared to 20°C (standard conditions).	Al-Yassein 2019
Crustaceans (grass shrimp, *Palaemonetes pugio*)	PBDE-47 *(brominated flame retardant)*	↑ The toxicity of PBDE-47 increased at the higher temperature/climate change conditions (30°C) to grass shrimp, *P. pugio* compared to 20°C (standard conditions).	Al-Yassein 2019
Crustaceans (grass shrimp, *Palaemonetes pugio*)	Scourge[(R)] (resmethrin and piperonyl butoxide) *(synthetic pyrethroid insecticide)*	↑ Toxicity of scourge to grass shrimp, *P. pugio* increased with increased temperature; toxicity trend: 35°C >25°C).	DeLorenzo et al 2009
Crustaceans (grass shrimp, *Palaemonetes pugio*)	Terbufos, trichlorfon *(OP insecticides)*	↑ Increased temperature (from 17°C to 27°C) increased the toxicity of the organophosphate pesticides (terbufos and trichlorfon) to grass shrimp, *P. pugio*.	Breken-Folse et al 1994
Crustaceans (grass shrimp, *Palaemonetes pugio*)	Triclosan *(antibacterial/ antibiotic)*	↑ The toxicity of tricloson increased at the higher temperature/ climate change conditions (30°C) to grass shrimp, *P. pugio* compared to 20°C (standard conditions).	Al-Yassein 2019

Climate change stressor effects on biota *(temperature /global warming)*	**Pollutants**	**Ecotoxicity** *[examples of research carried out and related to interactive effects of two stressors (temperature and pollutants) on algae, crustaceans, fish, and molluscs; ↑= increase in toxicity with increasing in temperature; ↓decrease in toxicity with increasing in temperature].*	**References**
Crustaceans (water fleas (*Daphnia magna*)	Fenvalerate, cypermethrin, deltamethrin *(pyrethroid insecticides)*	↑ The toxicity of pyrethroids (fenvalerate, cypermethrin and deltamethrin) increased at the higher temperatures (28°C compared to 23°C) in water fleas (*D. magna*).	Ratushnyak et al 2005
Crustaceans (White-leg shrimp, *Litopenaeus vannamei)*	Dimethoate, chlorpyrifos *(OP insecticide)*	↓ An experiment was carried out to assess the acute toxicity of two organophosphate pesticides (dimethoate and chlorpyrifos) on post-larvae and juveniles of *L.vannamei* at different ambient temperatures (15°C, 25° C and 34°C) regimes. Acute toxicity indices showed that lower temperature (15°C) induced significantly higher pesticide toxicity (dimethoate and chlorpyrifos) in both life stages (post-larvae, and juveniles) of *L. vannamei* compared to higher temperatures (25° C and 34 °C)	Pawara et al 2020
Fish (fathead minnows, *Pimephales promelas*)	17β-estradiol, 17α-ethinylestradiol, 4-tertnonylphenol, 4-tertoctylphenol, bisphenol-A) *(estrogenic chemicals)*	↑ A transient (short time) increase of the vitellogenin (VTG) (yolk precursor protein) was observed in male fathead minnows, *P. promelas* at higher temperatures (30°C vs. 20°C) upon exposure to a mixture of estrogenic chemicals. The temperature-dependent increase in VTG was observed only during the first 24 h of exposure, demonstrating that the effects of elevated temperature were more pronounced early in the exposure period.	Brian et al 2008
Fish (sheepshead minnows, *Cyprinodon variegatus*)	4-nitrophenol (insecticide)	↓ Increased temperature (from 17°C to 27° C) decreased the toxicity of nitrophenol to sheepshead minnows, *C. variegatus.*	Breken-Folse et al 1994
Fish *(flat fish, Solea senegalensis)*	Ag, Cd, Cu and Mn, Co, Fe, Hg, Se and Zn *(heavy metals)*	↑ Increasing temperature (from 15°C to 20°C) increased metal uptake in the Mediterranean fish, *S. senegalensis* and altered metallothionein (metal binding protein) levels and metal partitioning into liver and kidney.	Siscar et al 2014

Climate change stressor effects on biota *(temperature /global warming)*	**Pollutants**	**Ecotoxicity** *[examples of research carried out and related to interactive effects of two stressors (temperature and pollutants) on algae, crustaceans, fish, and molluscs; ↑= increase in toxicity with increasing in temperature; ↓ decrease in toxicity with increasing in temperature].*	**References**
Fish (catfish, *Ictalurus punctatus*)	Atrazine *(herbicide)*	↑ The toxicity of atrazine to catfish, *I. punctatus* increased with increasing temperatures.	Gaunt and Barker 2000
Fish (silver perch, *Bidyanus bidyanus,.* rainbow trout, *Oncorhynchus mykiss*)	Endosulfan *(OC insecticides)*	↑ Warm water species (silver perch) exhibited maximum toxicity with endosulfan at around 30°C and 35 °C than at 15°C, 20°C and 25°C. Endosulfan toxicity with cold water rainbow trout fish increased linearly with increasing temperatures.	Patra et al 2015
Fish (silver perch, *Bidyanus bidyanus*, rainbowfish, *Melanotaenia duboulayi*, western carp gudgeon, *Hypseleotris klunzingeri.* rainbow trout, *Oncorhynchus mykiss*)	Chlorpyrifos *(OP insecticide)*	↑ Chlorpyrifos were found toxic to all species (warm water and cold-water species) with increasing temperature.	Patra et al 2015
Fish (Freshwater darter, *Etheos-toma nigrum*)	Dieldrin *(OC insecticide)*	↑ The toxicity of dieldrin to the freshwater darter, *E. nigrum* increased with increasing temperatures.	Silbergeld 1973
Fish (eels, *Anguilla anguilla)*	Endosulfan *(OC insecticide)*	↑ Endosulfan, exposed to eel, *A. anguilla* exhibited a positive temperature correlation. The percentages of mortality were higher at 29°C than at 22°C.	Ferrando et al 1987
Fish (rainbow trout, *Oncorhynchus mykiss*)	Endosulfan *(OC insecticide)*	↑ The juvenile rainbow trout (*O. mykiss*) when exposed to the endosulfan caused increased mortality with increased temperature (as water temperature increased from 13°C to 16°C).	Capkin et al 2006

Climate change stressor effects on biota *(temperature /global warming)*	**Pollutants**	**Ecotoxicity** *[examples of research carried out and related to interactive effects of two stressors (temperature and pollutants) on algae, crustaceans, fish, and molluscs; ↑= increase in toxicity with increasing in temperature; ↓decrease in toxicity with increasing in temperature].*	**References**
Fish (rainbow trout, *Oncorhynchus mykiss*)	PCBs *(POPs)*	↑ The concentration of hydroxylated PCB (OH–PCBs) in the trout plasma was found increased with increasing temperature. The rising temperature (8°C, 12°C, and 16°C) enhanced biotransformation of PCBs to the toxicologically active hydroxylated PCB (OH-PCBs)[11] metabolites in rainbow trout, *O. mykiss*.	Buckman et al 2007
Fish (sheepshead minnows, *Cyprinodon variegatus*)	Terbufos, trichlorfon *(OP insecticides)*	↑ Increased temperature (from 17°C to 27° C) increased the toxicity of the OP insecticides (terbufos and trichlorfon) to sheepshead minnows, *C. variegatus*.	Breken-Folse et al 1994
Fish (mosquitofish, *Gambusia affinis*)	Endosulfan *(OC insecticide)*	↑ Increasing toxicity of endosulfan increased with increasing temperature in mosquito fish, ***Gambusia affinis*** at 25°C and 30° C and but decreased at 35° C (Nowak and Sunderam 1991). Reduced toxicity of endosulfan at the highest temperature (35° C) was possibly a result of faster degradation by hydrolysis, dissipation by volatilisation, or detoxification and metabolism (Shivaramaiah et al 2005).	Nowak and Sunderam 1991, Shivaramaiah et al 2005
Fish (mummichog/ killifish *Fundulus heteroclitus*)	Number 2 fuel oil *(Bunker A oil, light crude oil-diesel oil)*	↑ Oil exposure decreased yolk utilisation in fish, *F. heteroclitus* and the effect was intensified under increased temperature (from 20 to 30°C).	Linden et al 1980
Frogs (Green frog *Rana clamitans*)	Carbaryl *(carbamate insecticide)*	↑ The toxicity of the insecticide carbaryl increased with temperature increases from 17°C to 27°C in the green frog (*R.*	Boone and Bridges 1999

[11]***Hydroxylated polychlorinated biphenyls (OH-PCBs)*** *are produced in the environment by the oxidation of PCBs through a variety of mechanisms, including metabolic transformation in living organisms and abiotic reactions with hydroxyl radicals (OH) (https://www.ncbi.nlm.nih.gov/pmc/articles/PMC3812322/#:.) Hydroxyl radicals are highly reactive (easily becoming hydroxy groups) and consequently short-lived.*

Climate change stressor effects on biota *(temperature /global warming)*	**Pollutants**	**Ecotoxicity** *[examples of research carried out and related to interactive effects of two stressors (temperature and pollutants) on algae, crustaceans, fish, and molluscs; ↑= increase in toxicity with increasing in temperature; ↓decrease in toxicity with increasing in temperature].*	**References**
		clamitans). Increased temperature resulted in lower survival of frogs (tadpoles).	
Frogs (Australian frog *Limnodynastes peronii*	Endosulfan *(OC insecticide)*	↑ In an experiment, the eggs of the Australian frog (*L. peronii*) were reared under a high (20°C) and low temperature (14°C) regime and exposed to the insecticide endosulfan. There was a negative effect of endosulfan on feeding rate and predator avoidance that was proportionally worse for the tadpoles reared at a higher temperature.	Broomhall 2004
Frogs (leopard frogs, *Rana* spp.)	Esfenvalerate *(synthetic pyrethroid insecticide)*	↑ Esfenvalerate toxicity increased at elevated temperatures exposed to larval leopard frogs (*Rana* spp.). Temperature also influenced frog's mortality with significantly greater mortality at 22°C than at 18°C.	Materna et al 1995
Insects (bollworm (*Earias vitelli*)	Endosulfan *(OC insecticide)*	↑ Higher temperature and higher humidity increased toxicity of endosulfan to bollworm, *E. vitelli*.	Satpute et al 2007
Molluscs (mussel *Mytilus gallopro-vincialis)*	Benzo[a] pyrene *(PAH)*	↑ Interactions between temperature and benzo[a] pyrene (BAP) were observed with the mussel *Mytilus galloprovincialis*. Increasing temperature was correlated with a decrease in detoxification enzyme activity and an increase in BAP accumulation and toxicity.	Kamel et al 2012
Molluscs (scallop *Argopecten irradians)*	Corexit 9500 *(oil dispersant)*	↑ Increasing temperature has been shown to increase oil dispersant toxicity. The dispersant Corexit 9500 was more toxic to the scallop *A. irradians* at higher temperature (20°C compared to 2°C).	National Research Council 2005

Climate change stressor effects on biota *(temperature /global warming)*	**Pollutants**	**Ecotoxicity** *[examples of research carried out and related to interactive effects of two stressors (temperature and pollutants) on algae, crustaceans, fish, and molluscs; ↑= increase in toxicity with increasing in temperature; ↓decrease in toxicity with increasing in temperature].*	**References**
Molluscs (estuarine soft-shell clam, *Mya arenaria)*	2,4-D, dicamba, and mecoprop *(herbicides)*	↑ Increased temperature increased toxicity of herbicides to clam, *M. arenaria* (toxicity trend: $7^0C>18^0C$) (inhibited the antioxidant[12] defences as measured by malondialdehyde[13] content, cytochrome C oxidase[14] and superoxide dismutase[15] activity)	Greco et al 2011

13.3.2. Interactive effects of [Salinity + Pollutants]

Sea level rise (SLR) is the average increase in the level of the world's oceans. Global warming or increases in temperatures cause the oceans to warm and expand in volume inducing a rise in the sea levels. Furthermore, warmer climate facilitates melting of glaciers, ice caps and ice sheets causing the further addition of water to the oceans (Kibria and Haroon 2016). The latest IPCC report predicts a sea-level rise of 0.18-0.38 m (low GHG- B1 scenario) and 0.2 to 0.59 m (high GHG -A1F1 scenario) at the end of this century (Solomon et al 2007). Rising sea level is one of the most catastrophic consequences of global warming/climate change and a major threat to coastal habitats, coastal aquaculture and fisheries worldwide (Kibria et al 2017). SLR will alter environmental salinity (Noyes et al 2009, Gilman and Jenssen 2011) or make salinity changes (Schiedek et al 2007). Salinity can alter the toxicity of pollutants. The solubility and bioavailability of toxicants (POPs, pesticides) could be affected by salinity (salting-out effect= substances which are less soluble at high salt concentration). Organic compounds

[12]**.Antioxidants** *are compounds that inhibit oxidation. Antioxidants are compounds that may help delay or even prevent cell damage (https://www.medicalnewstoday.com/articles/325873).*

[13]***Malondialdehyde*** *is an organic compound with the formula $CH_2(CHO)_2$, and a by-product of* ***lipid metabolism*** *in the body. Malondialdehyde is a highly reactive compound, and one of the many reactive electrophile species that cause toxic stress in cells and form covalent protein adducts, called advanced lipoxidation end products (ALE) (https://pubchem.ncbi.nlm.nih.gov/ compound/Malonaldehyde).*

[14]*The* ***enzyme cytochrome c oxidase*** *(COC) is a large transmembrane* ***protein*** *complex. CCO is composed of 13 protein subunits (https://www.sciencedirect.com/topics/agricultural-and-biological-sciences/cytochrome-c-oxidase).*

[15]***Superoxide dismutases*** *(SOD) are a group of* ***enzymes*** *that catalyse the dismutation of superoxide radicals (O_2-) to molecular oxygen (O_2) and hydrogen peroxide (H_2O_2), providing cellular defence against reactive oxygen species (note:* ***dismutation*** *reaction between two identical molecules in which one is reduced and the other oxidized; disproportionation) (https://www.sciencedirect.com/topics/neuroscience/superoxide-dismutase).*

are known to be less soluble and more bioavailable in saltwater than in freshwater due to the "salting out" effect whereby the water molecules are strongly bound by salts making them unavailable for dissolution of organic chemicals (Noyes et al 2009, Schwarzenbach et al 2003). Moreover, the osmoregulatory abilities of fish, crustaceans and molluscs can be altered/ decreased by salinity (Gilman and Jenssen 2011). The interactive effects of two stressors (climate change-related rise in sea level/salinity and pollutants) would exacerbate such impacts on aquatic biota as discussed below:

13.3.2.1. Increase in toxicity with increasing salinity to aquatic biota

Some selected examples of the increase in toxicity of pollutants due to increase in salinity to aquatic biota are listed below:

Algae *(phytoplankton, pondweed):* the toxicity of the following herbicides (ametryn, atrazine, diuron, irgarol) increased with increasing salinity: **i.** ametryn, atrazine, diuron, irgarol exposed to phytoplankton, *Dunaliella tertiolecta*; **ii.** Atrazine exposed to pondweed, *Stuckenia pectinate* (Hall et al 1997) (see Table 13.4 for details).

Crustaceans *(brine shrimp, grass shrimp):* the toxicity of the following pollutants (2,4-dinitrophenol[16]**,** aldicarb, bifenthrin, chlorothalonil, dimethoate, ibuprofen, PBDE-47, tebufenozide) increased with increasing salinity: **i.** 2,4-dinitrophenol exposed to grass shrimp, *Palaemonetes pugio*; **ii.** aldicarb, dimethoate, and tebufenozide exposed to brine shrimp (Song and Brown 1998, DeLorenzo 2015); **iii.** bifenthrin exposed to grass shrimp, *P. pugio* (Al-Yassein 2019); **iv.** chlorothalonil exposed to grass shrimp, *P. pugio* (DeLorenzo et al 2009); **v.** ibuprofen exposed to the grass shrimp, *P. pugio* (Al-Yassein 2019); **vi.** PBDE-47 exposed to grass shrimp, *P. pugio* (Al-Yassein 2019) (see Table 13.4 for details).

Fish *(stickleback):* the toxicity of the following pollutants (aldrin, atrazine, dieldrin, DDT, methoxychlor, guthion, malathion, CoRal and toxaphene) increased with increasing salinity: **i.** aldrin, dieldrin, DDT, methoxychlor, guthion, malathion, CoRal and toxaphene exposed to fish, three-spined stickleback, *Gasterosteus aculeatus* (Katz 1961); **ii.** atrazine exposed to fish, *Cyprinodon variegatus* (Hall et al 1994) (see Table 13.4 for details).

Molluscs *(oyster):* the toxicity of the following pollutants (diesel oil) increased with increasing salinity; **i.** diesel oil exposed to oyster (*Crassostrea gigas)* (Zanette et al 2011).

---

[16]**2,4-Dinitrophenol** *(2,4-DNP or DNP) is an organic compound. DNP is used as an antiseptic and as a non-selective bioaccumulating pesticide. DNP is particularly useful as an herbicide alongside other closely related dinitrophenol herbicides like 2,4-dinitro-o-cresol (DNOC), dinoseb and dinoterb.(https://en.wikipedia.org/wiki/2,4-Dinitrophenol).*

13.3.2.2. Decrease in toxicity with increasing salinity to aquatic biota

Some selected examples of decrease in toxicity of pollutants due to increase in salinity to aquatic biota are listed below:

Crustaceans *(copepod, fiddler crab, grass shrimp, white-leg shrimp)*: the toxicity of the following pollutants (4-nitrophenol[17], atrazine, cadmium, chlorpyrifos, DDT, dimethoate, scourage, trimethyltin) decreased with increasing salinity to crustaceans: **i.** 4-nitrophenol exposed to grass shrimp, *P. pugio* (Breken-Folse et al 1994); **ii**. Cd exposed to fiddler crab, *Uca pugilator* (O'Hara 1973); **iii.** chlorpyrifos and dimethoate exposed to post-larvae, and juveniles of white-leg shrimp, *Litopenaeus vannamei* (Pawara et al 2020); **iv.** scourge exposed to adult grass shrimp, *P. pugio* (DeLorenzo et al 2009); **v.** trimethyltin exposed to larval fiddler crab, *U. pugilator (*Wright and Roosenburg 1982); **vi.** atrazine exposed to copepod, *Eurytemora affinis* (Hall et al 1994); and **vii**. chlorpyrifos, DDT exposed to copepod, *Microarthridion littorale* (Staton et al 2002) (see Table 13.4 for details).

Fish *(three-spined Stickleback, grass shrimp, gulf killifish, sheepshead min nows):* the toxicity of the following pollutants (2,4-dinitrophenol, 4-nitrophenol, Corexit, endrin, chlordane, and lindane) decreased with increasing salinity to fish: i. 2,4-dinitrophenol exposed to sheepshead minnows, *C. variegatus* (Breken-Folse et al 1994); ii. 4-nitrophenol exposed to sheepshead minnows, *C. variegatus* (Breken-Folse et al 1994); iii. Cd exposed to grass shrimp, *C. variegatus* (Hall et al 1995); iv. Corexit exposed to gulf killifish *Fundulus grandis (*Kuhl et al 2013); and v. endrin, chlordane, and lindane exposed to three-spined Stickleback, *Gasterosteus aculeatus exposed (*Katz 1961) (see Table 13.4 for details).

Molluscs *(bivalve):* the toxicity of the following pollutants (diesel oil) decreased with increasing salinity to molluscs: i. diesel oil exposed to marine bivalve, *Mytilus edulis* (Tedengren and Kautsky 1987).

13.3.2.3. Other effects to aquatic biota

Some other interactive effects (salinity + pollutants) observed and documented in aquatic biota are listed below:

Algae: **i.** increased salinity caused decreased growth rate, cell densities, and starch content in algae, *Dunaliella tertiolecta* exposed to herbicides (ametryn, atrazine, diuron, irgarol) (DeLorenzo et al 2013) (see Table 13.4 for details);

---

[17] **4-Nitrophenol** *(also called p-nitrophenol or 4-hydroxynitrobenzene) is a phenolic compound that has a nitro group at the opposite position of the hydroxyl group on the benzene ring. 4-Nitrophenol is used as the precursor for the preparation of phenetidine and acetophenetidine, indicators, and raw materials for fungicides. Since 1998 DNP has been withdrawn from agricultural use (https://en.wikipedia.org/wiki/4-Nitrophenol).*

Molluscs: **i.** decreased salinity increased metal bioavailability and bioaccumulation in blue mussel, *M. edulis* (Zanette et al 2011); **ii**. low salinity increased diesel oil toxicity and affected respiration and ammonia excretion to marine bivalve *Mytilus edulis* (Tedengren and Kautsky 1987) (see Table 13.4 for further details).

Table 13.4. Trends in the effect of climate change stressors on pollutant toxicity and interactive effects of salinity + pollutants on ecosystem biota (*algae, crustaceans, fish, and molluscs].*

[OC= organochlorine insecticide; OP= organophosphate insecticide; PAH= Polycyclic aromatic hydrocarbons; ↑= increase in toxicity with increasing salinity; ↓decrease in toxicity with increasing salinity].

Climate change stressor effects on biota *(salinity)*	**Pollutants**	**Ecotoxicity** *[examples of research carried out and related to interactive effects of two stressors (salinity and pollutants) on algae, crustaceans, fish, and molluscs; ↑ = increase in toxicity with increasing salinity; ↓decrease in toxicity with increasing salinity].*	**References**
Algae (sago pondweed *Stuckenia pectinate)*	Atrazine *(herbicide)*	↑ Atrazine was more toxic at the higher salinity (12 ppt[18]) to pondweed, *S. tuckenia* (based on plant dry weight) (toxicity trend: 12 ppt>6 ppt>1 ppt).	Hall et al 1997
Algae (phytoplankton, *Dunaliella tertiolecta)*	Ametryn, atrazine, diuron, irgarol *(herbicides)*	↑ Elevated salinity and herbicides exposure increased toxicity to algae, *D. tertiolecta)* (caused significant decreases in growth rates, cell densities and starch content) (toxicity trend: 40 ppt> 20 ppt).	DeLorenzo et al 2013
Crustaceans (grass shrimp, *Palaemonetes pugio*)	2,4-dinitrophenol *(a pesticide used as wood preservative)*	↑ Toxicity of 2,4-dinitrophenol increased to grass shrimp, *P. pugio* as salinity increased.	Breken-Folse et al 1994
Crustaceans (grass shrimp, *Palaemonetes pugio*)	4-nitrophenol *(insecticide)*	↓ Decreased 4-nitrophenol toxicity to grass shrimp, *P. pugio* with increased salinity.	Breken-Folse et al 1994
Crustaceans (brine shrimp (*Artemia* spp.)	Aldicarb dimethoate, tebufenozide	↑ Increased salinity increased toxicity of aldicarb, dimethoate, and tebufenozide to brine shrimp	Song and Brown 1998, DeLorenzo

18 ppt *= 'parts per thousand' or ppt (the amount of salt in 1,000 g of water); most of the ocean has a salinity of between 34 ppt and 36 ppt.*

Climate change stressor effects on biota *(salinity)*	**Pollutants**	**Ecotoxicity** *[examples of research carried out and related to interactive effects of two stressors (salinity and pollutants) on algae, crustaceans, fish, and molluscs; ↑ = increase in toxicity with increasing salinity; ↓ decrease in toxicity with increasing salinity].*	**References**
	(pesticides / insecticides)	(*Artemia* spp.).	2015
Crustaceans (grass shrimp, *Palaemonetes pugio*)	Bifenthrin *(pyrethroid insecticide)*	↑ The toxicity of bifenthrin increased at the higher salinity (35 ppt) to grass shrimp, *P. pugio* compared to 20 ppt.	Al-Yassein 2019
Crustaceans (fiddler crab, *Uca pugilator)*	Cadmium *(heavy metal)*	↓ Cd toxicity decreased with increasing salinity in the fiddler crab, *U. pugilator* (i.e. Cd was most toxic at 10 ppt and least toxic at 30 ppt).	O'Hara 1973
Crustaceans (grass shrimp, *Palaemonetes pugio*)	Chlorothalonil *(fungicide)*	↑ Toxicity of the chlorothalonil to grass shrimp. *P. pugio* increased with increased salinity (toxicity trend: 30 ppt>20 ppt).	DeLorenzo et al 2009
Crustaceans (white-leg shrimp, *Litopenaeus vannamei)*	Dimethoate, chlorpyrifos *(OP insecticide)*	↓ Acute toxicity indices showed that lower salinities (5 and 15 ppt) induced significantly higher pesticide toxicity in both life stages (post-larvae, and juveniles) of white-leg shrimp, *L. vannamei* compared to 25 ppt.	Pawara et al 2020
Crustaceans (grass shrimp, *Palaemonetes pugio*)	Ibuprofen *(non-steroidal anti-inflammatory drug)*	↑ The toxicity of ibuprofen increased at the higher salinity (35 ppt) to grass shrimp, *P. pugio* compared to 20 ppt.	Al-Yassein 2019
Crustaceans (grass shrimp, *Palaemonetes pugio*)	PBDE-47 *(brominated flame retardant)*	↑ The toxicity of PBDE-47 increased at the higher salinity (35 ppt) to grass shrimp, *P. pugio* compared to 20 ppt.	Al-Yassein 2019
Crustaceans (grass shrimp, *Palaemonetes pugio*)	Scourge *(resmethrin and piperonyl butoxide; synthetic pyrethroids)*	↓ Toxicity of the Scourge to adult grass shrimp, *P. pugio* reduced with increased salinity (toxicity trend: 20 ppt> 30 ppt).	DeLorenzo et al 2009

Climate change stressor effects on biota *(salinity)*	**Pollutants**	**Ecotoxicity** *[examples of research carried out and related to interactive effects of two stressors (salinity and pollutants) on algae, crustaceans, fish, and molluscs; ↑ = increase in toxicity with increasing salinity; ↓ decrease in toxicity with increasing salinity].*	**References**
Crustaceans (grass shrimp, *Palaemonetes pugio*)	Triclosan *(antibacterial/ antibiotic)*	↑ The toxicity of tricloson increased at higher salinity (35 ppt) to grass shrimp, *P. pugio* compared to 20 ppt.	Al-Yassein 2019
Crustaceans (fiddler crab, *U. pugilator, zoeae stage)*	Trimethyltin *(organotin compound)*	↓ Trimethyltin toxicity increased with decreasing salinity to larval fiddler crab, *U. pugilator* (trimethyltin was most toxic at 10 ppt and least toxic at 30 ppt).	Wright and Roosenburg 1982
Crustaceans (copepod nauplii, *Eurytemora affinis*)	Atrazine *(herbicide)*	↓ Decreased atrazine toxicity to copepod, *E. affinis* with increasing salinity (toxicity trend: 5 ppt> 15 ppt> 25 ppt).	Hall et al 1994
Crustaceans (harpacticoid copepod, *Microarthridion littorale)*	Chlorpyrifos, DDT *(insecticides)*	↓ Decreased pesticide (chlorpyrifos, DDT) toxicity to copepod, *M. littorale* with increasing salinity (measured as % survival) (toxicity trend: 3 ppt >35 ppt> 12 ppt).	Staton et al 2002
Fish (sheepshead minnows, *C. variegatus*)	2,4-dinitrophenol *(pesticide used as wood preservative)*	↓ Toxicity of 2,4-dinitrophenol decreased to sheepshead minnows, *C. variegatus*) as salinity decreased.	Breken-Folse et al 1994
Fish (sheepshead minnows, *Cyprinodon variegatus*)	4-nitrophenol *(insecticide)*	↓ Toxicity decreased as salinity increased for 4-nitrophenol to sheepshead minnows, *C. variegatus*.	Breken-Folse et al 1994
Fish (three-spined stickleback, *Gasterosteus aculeatus*)	Aldrin (OC), dieldrin (OC), DDT (OC), methoxychlor (OC), guthion (OP), malathion (OP), CoRal (OP) and toxaphene (OC) *(pesticides)*	↑ Increased salinity increased pesticide toxicity to the fish, *G. aculeatus* for aldrin, dieldrin, DDT, methoxychlor, guthion, malathion, CoRal and toxaphene.	Katz 1961

Climate change stressor effects on biota *(salinity)*	**Pollutants**	**Ecotoxicity** *[examples of research carried out and related to interactive effects of two stressors (salinity and pollutants) on algae, crustaceans, fish, and molluscs; ↑ = increase in toxicity with increasing salinity; ↓ decrease in toxicity with increasing salinity].*	**References**
Fish (estuarine sheepshead minnows, *Cyprinodon variegatus)*	Atrazine *(herbicide)*	↑ Increased atrazine toxicity to fish, *C. variegatus* with increasing salinity (toxicity trend: 25 ppt> 15 ppt>5 ppt).	Hall et al 1994
Fish (sheepshead minnows, *C. variegatus*)	Cadmium *(heavy metal)*	↓ Cd toxicity increased with decreasing salinity to *C. variegatus* (5 ppt> 15 ppt>25 ppt)	Hall et al 1995
Fish (Gulf killifish *Fundulus grandis)*	Corexit *(oil dispersant)*	↓ Toxicity of the Corexit to the gulf killifish *F. grandis* was significantly greater at decreasing salinity (trend of toxicity: 4 ppt> 12 ppt> 18 ppt). It is likely that reduced biodegradation of toxic components of Corexit and toxic crude oil components in low-saline environments (4 ppt) caused the highest toxicity to fish.	Kuhl et al 2013
Fish (three-spined Stickleback, *Gasterosteus aculeatus*)	Endrin (OC), chlordane (OC), and lindane (OC) *(pesticides)*	↓ Increased salinity decreased toxicity of endrin, chlordane, and lindane to fish, *G. aculeatus.*	Katz 1961
Fish (three-spined Stickleback, *Gasterosteus aculeatus*)	Sevin (carbamate) and heptachlor (OC) *(pesticides)*	The salinity had no effect on sevin (carbamate) and heptachlor to fish, *G. aculeatus.*	Katz 1961
Molluscs (blue mussel *Mytilus edulis*)	Cadmium (Cd) *(heavy metal)*	↓ Decreased salinity (15 ppt) increased metal bioavailability and bioaccumulation (in soft tissues) in the blue mussel *M. edulis.* At 35 ppt. there was lower bioaccumulation.	Fischer 1986

Climate change stressor effects on biota *(salinity)*	**Pollutants**	**Ecotoxicity** *[examples of research carried out and related to interactive effects of two stressors (salinity and pollutants) on algae, crustaceans, fish, and molluscs; ↑ = increase in toxicity with increasing salinity; ↓ decrease in toxicity with increasing salinity].*	**References**
Molluscs (oyster, *Crassostrea gigas)*	Diesel oil *(PAH)*	↑ Exposure to diesel oil (1 ppm) at 35 ppt resulted in 53% oyster (*C. gigas.)* mortality, whereas no mortality was seen at any of the lower salinities (9 ppt, 15ppt, or 25 ppt). At 25 ppt salinity, diesel oil exposure caused a significant increase in GST (glutathione S-transferase activity[19]) (but not at the other salinity levels).	Zanette et al 2011
Molluscs (bivalve *Mytilus edulis*)	Diesel oil *(PAH)*	↓ Lower salinity (6.3 ppt) increased toxicity of diesel oil to the marine bivalve, *M. edulis*. At lower salinity, diesel oil caused negative effects on respiration and ammonia (NH_4-N) excretion to bivalve, *M. edulis*. Toxicity trend: 6.3 ppt > 28 ppt.	Tedengren and Kautsky, 1987

13.3.3. Interactive effects of [Ocean acidification + Pollutants]

The ocean absorbs approximately ~30% of atmospheric carbon dioxide (CO_2) resulting from human activities including fossil fuel burning, industries, cement manufacturing, deforestation and land-use changes. CO_2 dissolves in water and produces weak carbonic acid ($H_2CO_3^-$) which causes a decrease in ocean pH (due to an increase in hydrogen ion concentration or H+). This is called 'Ocean acidification' (Roessig et al 2005, Meehl et al 2007, Turley et al 2010). The higher absorption of CO_2 has already acidified the surface layers of the ocean causing an overall decrease of 0.1 pH units since the pre-industrial period, which is equivalent to a 30% increase in hydrogen ion concentration or acidity. The surface ocean pH is projected to decrease by 0.3-0.4 pH units by 2100 relative to pre-industrial conditions, equivalent to a 150% increase in acidity (H+) and 50% decrease in CO_3^{2-} (Meehl et al 2007, Wittmann and Pörtner 2013). OA can alter the toxicity of pollutants such as metals through speciation in seawater (Pascal et al 2010) as well as releasing metals bound in sediments (Ardelan et al 2009) as a free toxic form. Moreover, OA can be associated with metal mobilisation since

[19] ***Glutathione transferases*** *(GSTs.) GST are enzymes* ***found*** *mainly in the cytosol (intracellular fluid). It plays a critical role in cellular detoxification against xenobiotics and noxious compounds as well as against oxidative stress (https://www.nature.com/articles/s41389-017-0025-3).*

the free ion activity of metals increases with acidity (Lopez et al 2010). In addition, the metal toxicity due to OA/ hypercapnia (hypercapnia= elevated CO_2) could be metal-specific (Ivanina et al 2013). OA is projected to adversely affect many organisms that use calcium carbonate for their skeletons and shell including krill, pteropod, molluscs, corals, echinoderms, and fish (Kibria et al 2017). The interactive effects of two stressors (climate change-related rise in OA and pollutants) would exacerbate such impacts on aquatic biota as discussed below:

13.3.3.1. Increase in toxicity of pollutants with increasing OA/ low pH

Some selected examples of the increase in toxicity pollutants due to increase in OA (low pH) to aquatic biota are listed below:

***Algae** (kelp):* increased toxicity in two kelp species, *M. pyrifera* and *U. pinnatifid* exposed to Cu (Leal et al 2018) (see Table 13.5 for details).

***Crustaceans** (amphipod):* **i.** increased the toxicity of metals/ metalloid (As, Cd, Cr, Cu, Fe, Mn, Ni Pb, Zn) exposed to amphipods *(C. volutator)* (Roberts et al 2013); **ii.** increased Cu and Cd toxicity to meiobenthic copepod, *Amphiascoides atopus* (Pascal et al 2010) (see Table 13.5 for details).

***Molluscs** (bivalves, clams, cuttlefish, blue mussels):* **i.** enhanced metal accumulation (As, Cr, Cu, Hg, Ni, Pb) in clams, *Ruditapes philippinarum* (Lopez et al 2010); **ii.** enhanced bioaccumulation of Cd (in the gills, mantle and adductor muscle) in three bivalve species, *M. edulis, T. granosa, and M. meretrix* (Shi et al 2016); **iii.** suppressed immune-related functions in haemocytes of clams, *M. mercenaria* and oysters *C. virginica* exposed to Cd (Ivanina et al 2014); **iv.** reduced the burying activity and the predator avoidance of bivalves exposed to metals (Cd, Cu and Pb) (Ivanina and Sokolova 2015); **v.** enhanced the Cd and Cu accumulation in mantle tissues in the hard clam *M. mercenaria* and oysters, *C. virginica.* The higher Cd and Cu burdens can be associated with elevated mRNA expression of metal binding proteins (metallothionein and ferritin) (Götze et al 2014); **vi.** enhanced bioaccumulation of Cu and Pb in blue mussels, *Mytilus edulis* (Han et al 2014); **vii.** increased toxicity of resmethrin to calm larvae, *M. mercenaria* (Garcia et al 2014); **viii.** enhanced Ag accumulation in the eggs of cuttlefish, *Sepia officinalis* (Lacoue-Labarthe et al 2009); **ix.** enhanced the uptake of both Ag and Zn in the embryo of squid, *Loligo vulgaris* (Lacoue-Labarthe et al 2011); **x.** increased copper uptake in the hard clam, *M. mercenaria* (Ivanina et al 2013) (see Table 13.5 for details).

Polychaetes: suppressed fertilisation, sperm motility (movement) and larval survival in early life history stages of the polychaete, *Arenicola marina* exposed to Cu (Campbell et al 2014) (see Table 13.5 for details).

13.3.3.2. Decrease in toxicity of pollutants with increasing OA/ low pH

Some selected examples of the decrease in toxicity of pollutants due to increase in OA (low pH) to aquatic biota are listed below:

***Crustaceans** (copepod):* **i.** significantly reduced Hg accumulation in copepod, *Tigriopus japonicus.* The results may suggest that OA can potentially mitigate the Hg toxicity to some important life traits in marine copepods (Li et al 2017) (see Table 13.5 for details).

***Fish** (cod):* i. significantly decreased the toxicity of copper to cod *Gadus morhua.* via possibly limiting the copper-induced osmoregulatory disturbance (Larsen et al 1997) (see Table 13.5 for details).

***Molluscs** (clam, cuttlefish)* **i.** decreased Cd uptake/ accumulation in the eggs of cuttlefish *Sepia officinalis* (Lacoue-Labarthe et al 2009); and **ii**. decreased cadmium uptake in the hard clam, *M. mercenaria* (see Table 13.5 for details).

13.3.3.3. Other effects

Some other interactive effects (OA/l ow pH+ pollutants) observed and documented are listed below:

Crustaceans: **i.** suppressed the reproduction of the nauplii[20] of marine copepods *Tisbe battagliai* (Fitzer et al 2013); ii. OA/low pH caused DNA damage in amphipods, *C. volutator* (Roberts et al 2013) (see Table 13.5 for details).

***Molluscs** (clam):* **i.** OA/ hypercapnia caused thinner shells to clam larvae with irregular edges compared to normocapnia (pH 9) (Garcia et al 2014); **ii**. the metal toxicity due to OA/hypercapnia is metal specific since increased Cu uptake and decrease Cd uptake were found in the hard clam, *M. mercenaria* (Ivanina et al 2013) (see Table 13.5 for further details).

---

[20] ***nauplii**= larval stage of many crustaceans.*

Table 13.5. Trends in the effect of climate change stressors on pollutant toxicity and interactive effects of [OA*/ low pH /elevated CO_2] + pollutants on ecosystem biota (*algae, crustaceans, fish, and molluscs].*

*[*__note:__ ocean acidification (OA), low pH, elevated pCO_2 and hypercapnia are synonyms and interchangeably used; __As__= arsenic; __Cr__= chromium; __Cu__=, copper; __Hg__= mercury; __hypercapnia__ = elevated CO_2; __Ni__= nickel (Ni); __Pb__= lead; __pCO_2__ = partial pressure; __μatm__= micro atmospheres ↑= increase in toxicity with increasing OA/low pH; ↓ decrease in toxicity with increasing OA/low pH].*

Climate change stressor effects on biota (Low pH/elevated CO_2/Ocean Acidification)	Pollutants	Ecotoxicity [examples of research carried out and related to interactive effects of two stressors ([OA/ low pH/ elevated CO_2] and pollutants) on algae, crustaceans, fish, and molluscs; ↑ = increase in toxicity with increasing OA; ↓ decrease in toxicity with increasing OA].	References
Algae/ kelps (*Macrocystis pyrifera* and *Undaria pinnatifida*)	Copper (Cu) (heavy metal)	↑ Low pH/ OA (pH 7.65) increased the toxicity of Cu to two kelp species, M. pyrifera and *U. pinnatifida*. Meiospore[21] and germling growth[22] for both species declined under OA.	Leal et al 2018
Crustaceans (marine copepods *Tisbe battagliai*)	Copper (Cu) (heavy metal)	↑ The interactive effects of two stressors (low pH/OA: pH <7.75 and Cu) suppressed the reproduction of the nauplii[23] of marine copepods, *Tisbe battagliai*. The decline in naupliar production was greater with copper at decreasing pH than for decreasing pH alone.	Fitzer et al 2013
Crustaceans (marine copepod *Tigriopus japonicus*)	Mercury (Hg) (heavy metal)	↓ OA (elevated CO_2 = pCO_2 1,000 μatm) significantly reduced Hg accumulation in copepod, *Tigriopus japonicus*. The results suggest that OA can potentially mitigate the Hg toxicity to some important life traits in marine copepods. The combined OA and Hg exposure decreased the total-Hg concentrations by 52, 73, 75, and 83%, respectively in copepods.	Li et al 2017
Crustaceans (Amphipod,	Mixture of trace metals/	↑ Elevated OA (pCO_2=1,140 μatm) increased the toxicity of metals/ metalloid	Roberts et al 2013

[21] ***Meiospores**= haploid spores resulting from meiosis.*

[22] ***Germling growth=** germination of an algal spore/immature gametophyte.*

[23]***Nauplii**= larval stage of many crustaceans.*

Climate change stressor effects on biota (Low pH/elevated CO_2/Ocean Acidification)	**Pollutants**	**Ecotoxicity** [examples of research carried out and related to interactive effects of two stressors ([OA/ low pH/ elevated CO_2] and pollutants) on algae, crustaceans, fish, and molluscs; ↑ = increase in toxicity with increasing OA; ↓ decrease in toxicity with increasing OA].	**References**
Corophium volutator)	metalloid in sediments (As, Cd, Cr, Cu, Fe, Mn, Ni, Pb, Zn)	(As, Cd, Cr, Cu Fe, Mn, Ni Pb, Zn) to amphipods (C. volutator) causing a 2.7-fold increase in DNA damage.	
Crustaceans (meiobenthic harpacticoid copepod, Amphiascoides atopus)	Cd, Cu (heavy metal)	↑ Increased OA/ elevated CO_2/ decreased pH resulted in increased Cu and Cd toxicity in the meiobenthic copepod, *Amphiascoides atopus*. OA can alter the toxicity of waterborne metals by influencing their speciation in seawater (Pascal et al 2010). Moreover, OA can release metal bounds in sediments (Ardelan et al 2009) rendering them to be more toxic in a free form.	Ardelan et al 2009, Pascal et al 2010
Fish (cod Gadus morhua)	Copper (Cu) (heavy metal)	↓ OA /hypercapnia (elevated CO_2) (~9,900 µatm $_{P}CO_2$) significantly decreased the toxicity of copper to cod *Gadus morhua*. Hypercapnia significantly decreased the toxicity of copper, by limiting the copper induced osmoregulatory disturbance.	Larsen et al 1997
Molluscs (juveniles of the clam *Ruditapes philippinarum*)	Arsenic (As), Chromium (Cr), Copper (Cu), Lead (Pb), Nickel (Ni), Mercury (Hg), (heavy metals)	↑ OA/ low pH enhanced/increased metal accumulation (As, Cr, Cu, Hg, Ni, Pb) in clams *Ruditapes philippinarum*. Acidification can be associated with metal mobilisation since the free ion activity of metals increases with acidity. The research may demonstrate that OA can increase metal accumulation in the biota associated with an originally non-toxic environment.	Lopez et al 2010
Molluscs (three marine bivalve species, *Mytilus edulis*, *Tegillarca granosa*, and *Meretrix*	Cadmium (Cd) (heavy metal)	↑ OA/ low pH (pH 7.4-7.8) enhanced bioaccumulation of Cd (in the gills, mantle and adductor muscle) in three bivalve species, *M. edulis*, *T. granosa*, and *M. meretrix*. OA may have increased in the concentration of Cd in the acidified seawater, further, OA may damage of	Shi et al 2016

Climate change stressor effects on biota (Low pH/elevated CO_2/Ocean Acidification)	**Pollutants**	**Ecotoxicity** [examples of research carried out and related to interactive effects of two stressors ([OA/ low pH/ elevated CO_2] and pollutants) on algae, crustaceans, fish, and molluscs; ↑ = increase in toxicity with increasing OA; ↓ decrease in toxicity with increasing OA].	**References**
meretrix)		epithelia of bivalves (resulting in easier Cd penetration) and OA may have hampered Cd exclusion.	
Molluscs (hard shell clam *Mercenaria mercenaria* and *Eastern oyster Crassostrea virginica*)	Cadmium (Cd) (heavy metal)	↑ OA/ elevated $_{P}CO_2$ (Elevated $_{P}CO_2$ (~800 µatm and 2,000 µatm) combined with Cd exposure suppressed immune-related functions in haemocytes of clams *M. mercenaria* and oysters *C. virginica*.	Ivanina et al 2014
Molluscs (marine bivalves)	Cadmium (Cd), Copper (Cu), Lead (Pb) (heavy metals)	↑ OA/ low pH accompanied by metals (Cd, Cu and Pb) reduced the burying activity and slowed the rate of the protective valve closing of marine bivalves and the predator avoidance.	Ivanina and Sokolova 2015
Molluscs (*Crassostrea virginica* (eastern oyster), *Mercenaria mercenaria* (hard shell clam)	Copper (Cu) and Cadmium (Cd) (heavy metals)	↑ OA/ elevated $_{P}CO_2$ ($_{P}CO_2$ ~800 and ~1500 µatm) enhanced the Cd and Cu accumulation in mantle tissues in the hard clam *M. mercenaria* and *oysters C. virginica*. Higher Cd and Cu burdens were associated with elevated mRNA expression of metal binding proteins (metallothionein and ferritin).	Götze et al 2014
Molluscs (Blue mussels, *Mytilus edulis*)	Copper (Cu), Lead (Pb) (heavy metals)	↑ OA/ low pH (pH 6.2) enhanced bioaccumulation of Cu and Pb in blue mussels, *Mytilus edulis*.	Han et al 2014
Molluscs (bivalve, clam larvae, *Mercenaria mercenaria*	Resmethrin (Pyrethroid insecticide)	↑ OA/ **hypercapnia** (elevated CO_2/ low pH -pH 7) increased the toxicity of resmethrin to clam larvae, *Mercenaria mercenaria*. OA/ hypercapnia caused thinner shells to larvae with irregular edges compared to **normocapnia** (pH 9) (effect of bicarbonate HCO_3).	Garcia et al 2014
Molluscs (cuttlefish, *Sepia officinalis*)	Silver (Ag) (heavy metals)	↑ OA/ low pH (pH 7.6, pH 8.5) enhanced Ag accumulation in the eggs of cuttlefish *Sepia officinalis*. ↓ OA/ low pH (pH 7.6,	Lacoue-Labarthe et al 2009

Climate change stressor effects on biota (Low pH/elevated CO_2/Ocean Acidification)	**Pollutants**	**Ecotoxicity** [examples of research carried out and related to interactive effects of two stressors ([OA/ low pH/ elevated CO_2] and pollutants) on algae, crustaceans, fish, and molluscs; ↑ = increase in toxicity with increasing OA; ↓ decrease in toxicity with increasing OA].	**References**
		pH 8.5) decreased Cd uptake/ accumulation in the eggs of cuttlefish *Sepia officinalis*.	
Molluscs (Squid, *Loligo vulgaris*)	Silver (Ag), Zinc (Zn) (heavy metals)	↑ OA/ low pH (pH 7.6, pH 8.5) enhanced the uptake of both Ag and Zn in the embryo of squid, *Loligo vulgaris*.	Lacoue-Labarthe et al 2011
Molluscs (hard clam, *Mercenaria mercenaria*)	Cadmium, copper (heavy metals)	↑ OA/ hypercapnia increased copper uptake in the hard clam, *M. mercenaria*; ↓ OA/hypercapnia decreased cadmium uptake in the hard clam, *M. mercenaria*. The study shows that hypercapnia modifications to metal-toxicity were metal specific.	Ivanina et al 2013
Molluscs (hard clam, *Mercenaria mercenaria*)	Cadmium, copper (heavy metals)	↑ OA/ hypercapnia increased copper uptake in the hard clam, *M. mercenaria*. ↓ OA/hypercapnia decreased cadmium uptake in the hard clam, *M. mercenaria*. The study shows that hypercapnia modifications to metal-toxicity were metal specific.	Ivanina et al 2013
Polychaete (*Arenicola marina*)	Copper (Cu) (heavy metal)	↑ OA/ low pH (7.47, 7.77) and accompanied by Cu negatively affected/ suppressed fertilisation success, sperm motility (movement) and larval survival in early life history stages of the polychaete Arenicola marina (compared to single OA or copper exposures). The results may demonstrate that OA will increase the toxicity of copper to marine organisms.	Campbell et al 2014

13.3.4. Interactive effects of [Hypoxia + Pollutants]

Hypoxia (DO <2 mg/L) are associated with excess nutrient inputs from both point- and non-point sources (Diaz and Rosenberg 2008; *https://gulfhypoxia.net/about-hypoxia/*). When algae/algal blooms die and sink to the bottom, they are decomposed by bacteria, using the available dissolved oxygen (Diaz and Rosenberg 2008). This causes the hypoxic environment (Diaz and Rosenberg 2008; *https://oceanservice.noaa.gov/podcast/feb18/nop13-*

hypoxia.html). Hypoxia is one of the most significant stressors affecting aquatic systems worldwide, and its severity and prevalence are projected to rise due to increases in nutrient input (eutrophication) and climate change (Gewin 2010). Areas of hypoxic and anoxic (no oxygen) conditions or "dead zones" are generally projected to increase with increasing seawater temperatures (IPCC 2014). Coastal hypoxia is a major threat to coastal ecosystems and coastal biodiversity across the globe, resulting in so-called ''dead zones'' and triggering both lethal (mortality) and sub-lethal stresses (reduced growth and reproduction) and causing damage/loss of fisheries resources (Renaud 1986, Rabalais et al 2001). Hypoxia can cause forced migration of biota, reduction of suitable habitat, increased vulnerability to predation and disruption of life cycle stages (Rabalais et al 2001, Service 2004). The interactive effects of two stressors (climate change-related rise in hypoxia and pollutants) would exacerbate such impacts on aquatic biota as discussed below:

13.3.4.1. Increase in toxicity of pollutants with increasing hypoxia/ low DO

***Fish** (catfish, zebrafish):* **i.** increased toxicity of atrazine to catfish (*I. punctatus*) (Gaunt and Barker 2000); **ii**. increased toxicity of Cu to zebrafish, *Danio rerio* (Fitzgerald et al 2016); **iii** damaged electrical stability, rhythm of the heartbeat and myocardial dysfunction (heart attack) in fish, *Brycon amazonicus amazonicus and Hoplias malabaricus.* (see Table 13.6 for further details).

***Molluscs** (clams, oyster):* **i**. anoxia (no oxygen) enhanced Cd uptake in oyster, *C. virginica* (Ivanina et al 2010); **ii.**. increased the toxicity of resmethrin to larval clams, *M. mercenaria;* (Garcia et al 2014); iv. suppressed immunity of exposed oysters (*C. virginica*). (Barnett et al 2020) (see Table 13.6 for further details).

Table 13.6. Trends in the effect of climate change stressors on pollutant toxicity and interactive effects of [hypoxia/ low DO] + pollutants) on aquatic biota (*algae/ crustaceans/ fish/ molluscs].*

[DO= dissolved oxygen; ↑= increase in toxicity with increasing hypoxia; ↓ = decrease in toxicity with increasing hypoxia].

Climate change stressor effects on biota *(Hypoxia/ DO <2 mg/L)*	**Pollutants**	**Ecotoxicity** *(examples of research carried out and related to interactive effects of two stressors ([hypoxia/ low DO] and pollutants) on fish, and molluscs; ↑= increase in toxicity with increasing hypoxia; ↓= decrease in toxicity with increasing hypoxia].*	**References**
Fish (catfish, *Ictalurus punctatus)*	Atrazine (herbicide)	↑ **Hypoxia** (decreased DO; DO= < 2 mg/L) increased toxicity of atrazine to catfish (*I. punctatus*) .	Gaunt and Barker 2000

Climate change stressor effects on biota *(Hypoxia/ DO <2 mg/L)*	**Pollutants**	**Ecotoxicity** *(examples of research carried out and related to interactive effects of two stressors ([hypoxia/ low DO] and pollutants) on fish, and molluscs; ↑= increase in toxicity with increasing hypoxia; ↓= decrease in toxicity with increasing hypoxia].*	**References**
Fish (zebrafish, *Danio rerio*)	Copper (Cu) *(heavy metal)*	↑ Hypoxia increased toxicity of Cu to zebrafish, *D. rerio* in the late development stage (hatched larvae).	Fitzgerald et al 2016
Fish (Neotropical freshwater fish species, matrinxã, *Brycon amazonicus and* traíra, *Hoplias malabaricus)*	Mercury (Hg) *(heavy metal)*	↑ The interactive effects two stressors (hypoxia + Hg) caused damage in electrical stability, the rhythm of the heartbeat and myocardial dysfunction (heart attack) in fish, *B. amazonicus and H malabaricus.* As a result, swimming ability, foraging capacity, and growth and/ or reproductive performance of fish can be impaired.	Monteiro et al 2020
Molluscs (Oyster, *Crassostrea virginica)*	Cadmium (Cd) *(heavy metal)*	↑ **Anoxia** (DO~ 0 mg/L) enhanced Cd uptake (2.7-fold higher tissue Cd burdens) in the oyster, *C. virginica.* Oysters spent more time actively ventilating in hypoxia than normoxia (**normoxia** = DO> 6 mg/L).	Ivanina et al 2010
Molluscs (bivalve, clam larvae, *Mercenaria mercenaria)*	Resmethrin *(pyrethroid insecticide)*	↑ Hypoxia (DO= 2 mg/L) increased the toxicity of resmethrin to larval clams, *M. mercenaria.*	Garcia et al 2014
Molluscs (Eastern oyster, *Crassostrea virginica*)	Tributyltin (TBT) *(organotin compound)*	↑ Hypoxia in combination with Tributyltin (TBT) (hypoxia + TBT treatment) suppressed immunity of exposed oysters (*C. virginica*). It also caused a decrease in haemocytes.	Barnett et al 2020

13.4. Pollutants, climate change and ecotoxicity covered in other chapters

Chapters 1 to 12 of this book discussed briefly the linkages between specific pollutants and climate change and the toxic effects of pollutants on aquatic biota (ecotoxicity) and therefore, is not repeated further in this chapter. Table 13.7 identified (in **bold** in column 3) aspects related to pollutants, climate change and ecotoxicity along with other main aspects covered in each chapter.

Table 13.7. Showing the main aspects covered in previous chapters (chapters 1 to 12) including climate change, pollutants and ecotoxicity

Chapter	Pollutants *(examples of pollutants covered)*	Main aspects covered in each chapter
1	**Pollution and pollutants**	• Pollution types and the world's most polluted places • The Stockholm Convention on Persistent Organic Pollutants • Harmful pollutants and priority pollutants • Linkages between pollution and poverty • **Linkages between pollution and climate change** • Linkages between pollution and sustainable development goals (SDGs) • Control and management of pollutants
2	**Arsenic** *[As_2O_3, As_2O_5]*	• Groundwater arsenic contamination areas in the world • Concentration of arsenic in ground waters of the world • Chemistry, toxicity and **ecotoxicity of arsenic** • LD_{50}/ EC_{50}/ LC_{50} of different arsenic species • Arsenic contamination of drinking water; rice and vegetables; cows' milk and dung; freshwater and marine fish) • Facts about arsenobetaine in marine organisms • Arsenic effects on human health & social life • Managing arsenic problems
3	**Heavy metals** *[Cd, Cu, Cr, Hg, Ni, Pb, Zn]*	• Major industrial sources of heavy metals • Bioavailability, bioaccumulation and bioconcentration of heavy metals • Heavy metal's transportation- grasshopper effect • Chemistry, toxicity and **ecotoxicity of heavy metals** • Heavy metals residues in Australian waterways • Cadmium, copper, lead, mercury, and zinc contamination of food (rice, crops and vegetables, freshwater and marine fish) • Elevated levels of mercury in child–bearing women in 25 Countries • Hidden heavy metals in face makeup • **Linkages among heavy metals load, bioaccumulation and climate change** • Effects of on human health • Management and remediation of heavy metals
4	**Plastic (including plastic additive**	• Common plastics, their application and recycling potential • Chemical additives in plastic materials

Chapter	Pollutants *(examples of pollutants covered)*	Main aspects covered in each chapter
	compounds) *[BBP, DEHP, DBP, BPA, PBDEs, nonylphenol and heavy metals]*	• Toxicity and **ecotoxicity of plastic additive compounds** • Plastic residues in global waterways and plastic pollution "Hot spots" in the world • Contaminants/pollutants adsorbed in various types of plastic polymers • Plastics in seafood, honey, sugar, beer, salt and drinking water • Impacts on marine life, aesthetic value, recreation and tourism, shipping and the local economy • Impacts on human health • **Linkages between plastic pollution and climate change** • Management and mitigation measures
5	**Persistent Organic Pollutants (POPs)** *[PBDE; PCB; PCDD; PCDF; PFOA; PFOS]*	• POP chemicals (Dioxins, Furans and PCBs, PBDEs, PFOS, PFOA • Currently controlled under the Stockholm Convention • Characteristics and sources of POPs • Chemistry, toxicity and **ecotoxicity of POPs** • Impacts on the environmental matrices, aquatic biota, and food • PCDD, PCDF and PCBs in fish and human breast milk (Global survey) • PFOS and PFOA in algae, birds, clams, crabs, fish, frogs, mammals, oysters, shrimps, starfish, turtles and zooplankton • Impacts on humans health • **Linkages between POPs and climate change** • Management of POPs
6	**Oil/petroleum products** *[PAHs (benzo[a]pyrene, naphthalene); BTEX (ethylbenzene, toluene)*	• Oil groups and their characteristics • The fate of oils in the environment (weathering processes) • Toxicity and **ecotoxicity of oils** • Impacts on fish toxicity, seafood security, mariculture, fisheries and fishing, tourism, wildlife, ecosystems (corals, seagrasses, mangroves), humans health, salt production, agriculture and human's health • Management of oil spill and oil pollution • Recovery and restoration of ecosystems after oil spills
7	**Pesticides** *[atrazine, azoxystrobin chlorpyrifos, DDT, diuron, endosulfan, glyphosate,*	• Classification of pesticides and the history of pesticides used • Chemistry, toxicity and **ecotoxicity of pesticides** • List of pesticides considered as an endocrine disruptor

Chapter	Pollutants *(examples of pollutants covered)*	Main aspects covered in each chapter
	imidacloprid]	• Pesticide residues in food (fruits, vegetables, grain, seafood), water (surface, groundwater and sediments), human (women) breast milk • Impacts on non-target organisms (birds, earthworms, fish, frogs, bees) • Impacts on human's health • **Linkages between pesticides and climate change** • Management of pesticides risk
8	**Nutrients** *[nitrogen , phosphorous; toxins (anatoxin, microcystins, ciguatoxins, PSP]*	• Connection among eutrophication, algal blooms and hypoxia • Chemistry, toxicity and **ecotoxicity of algal toxins** • **Ecotoxicity of hypoxia** • Impacts of algal toxins on water quality; crops/ edible plants; bioaccumulation in seafood organisms; fish kill; livestock; wildlife; recreation and humans • Impacts of hypoxia on marine life; corals; fish reproduction; habitat/ ecosystem compression; tourism; fisheries, Seafood security and livelihoods; ocean acidification and chemicals release and production • **Linkages among algal blooms, hypoxia and climate change** • Management of nutrient pollution, algal blooms, algal toxins and hypoxia
9	**Pharmaceuticals** *[acetylsalicylic acid/ aspirin; carbamazepine; ciprofloxacin; diclofenac; ibuprofen; 17α-ethinylestradiol; paracetamol; triclosan]*	• Classification of pharmaceuticals (human and veterinary) • Chemistry, toxicity and **ecotoxicity of pharmaceuticals** • Environmental risk assessment of pharmaceuticals based on i. PEC/PNEC (Risk quotient (RQ), ii. PBT; iii. bioaccumulation (log K_{ow}) and **ecotoxicity** • Worldwide distribution of estrogens/ hormones (E1, E2, E3, EE2) and other pharmaceuticals in surface water and drinking water • Impacts on aquatic biota (bacteria, algae, daphnia, molluscs, fish), wildlife (frogs, birds) and crops • Impacts on human health • **Linkages between pharmaceuticals use and climate change** • Management of pharmaceuticals
10	**Surfactants** *[linear alkyl benzene sulfonate (LAS); nonylphenol]*	• Classification and industrial applications of surfactants. • Chemistry, toxicity and **ecotoxicity of surfactants**

Chapter	Pollutants *(examples of pollutants covered)*	Main aspects covered in each chapter
		• Impacts on environmental compartments, water quality and aquatic biodiversity, foam pollution, algal blooms, seafood, crops, vegetables, fruits • Impacts on human health • Management of surfactants
11	**Synthetic and textile dyes** *[acid, azo, basic, direct, disperse, reactive, solvent and vat]*	• Classification of dyes • Chemistry, toxicity and **ecotoxicity of synthetic and textile dyes** • Impacts on water quality, the aquatic environment and aquatic biota (fish, crab, algae), crops and vegetables • Impacts on human's health • Management of synthetic dyes and textile effluents
12	**Sewage pollutants** *[Heavy metals; hormones, flame retardants; plastic additives/ plasticisers; POPs; pesticides; pharmaceuticals, PAHs; surfactants; Volatile organic compounds (BTEX)]*	• Chemistry, toxicity and **ecotoxicity of micropollutants in sewage water** • Pathogens found in untreated sewage water/wastewater • Chemical concentrations in sewage/ wastewater treatment plant (WWTP) effluent • Typical sewage/wastewater treatment plant (WWTP) chemical effluent concentration and PNEC • Impacts of micropollutants found in sewage/wastewater on aquatic biota • Impacts on human's health • **Linkages between sewage pollution and climate change** • Management of sewage pollution

13.5. Management of climate change and pollutants risks

In order to reduce the risks and impacts of two stressors (climate change and pollutants) we would be in need to implement a number of strategies, actions and measures, some of which are highlighted below:

Reducing greenhouse gas emissions (CO_2, CH_4, N_2O) that causes climate change via: **i.** use and production of renewable energy (solar, wind, nuclear, hydro, biofuels); **ii.** supporting low-carbon development (LCD) pathways; **iii.** encouraging the use of energy-efficient appliances; promoting climate-smart agriculture, livestock, fisheries and aquaculture; **iv.** reduce, reuse and recycle (3Rs) waste including plastic and food waste; **v.** afforestation, reforestation and restoration of forests (as carbon sinks); **vi.** awareness, education, capacity building on climate change risks, mitigation and resilient adaptation; **vii.** implement community behaviour and life-style change programmes on LCD; and **viii.** identify and include climate change risk calculator in all future development programmes.

Table 13.8. Toxicity matrix of some selected inorganic pollutants (heavy metals/ metalloid) induced by climate change stressors (temperatures, salinity, ocean acidification and hypoxia)

[↑= increase in toxicity of pollutants with increasing climate stressors; ↓= decrease in toxicity of pollutants with increasing climate stressors;Symbols: A=algae, C crustaceans; F= fish; M =molluscs; - = data is not available; number shown with a superscript are references listed underneath of this table].

Climate Stressors	Arsenic (As)	Cadmium (Cd)	Copper (Cu)	Chromium (Cr)	Iron (Fe)	Nickel (Ni)	Lead (Pb)	Mercury (Hg)	Silver (Ag)	Zinc (Zn)
Temperature	-	↑ [$C^{1,2}$, F^{3}]	↑ [C^{1}, F^{3}]	-	↑ [F^{3}]	-	↑ [C^{1}]	↑ [F^{3}]	↑ [F^{3}]	↑ [C^{1}]
Salinity	-	↓ [C^{4}, $F^{5,6}$]	-	-	-	-	-	-	-	-
Ocean acidification	↑ [C^{10}, M^{13}]	↑ [$C^{10,11}$, $M^{14,15,16}$] ↓ [M_{22}]	↑ [A^{7}, $C^{8,10,11}$, $M^{13,16,17,18,20}$] ↓ [F^{12}]	↑ [$C^{10,}$ M^{13}]	↑ [C^{10}]	↑ [C^{10}, M^{13}]	↑ [C^{10}, $M^{13,18,20}$]	↑ [M^{13}] ↓ [$C^{9,}$ F^{12}]	↑ [M^{19},^{25}]	↑ C^{10}, M^{19}
Hypoxia	↑ [F^{21}]	↑ [M^{24}]	↑ [F^{22}]	-	-	-	-	↑ [M^{23}]		-

[1]Khan et al 2006; [2] O'Hara 1973; [3]Siscar et al 2014; [4]O'Hara 1973; [5]Hall et al 1995; [6]Fischer 1986; [7]Leal et al 2018; [8]Fitzer et al 2013; [9] Li et al 2017; [10]Roberts et al 2013;[11] Pascal et al 2010;[12]Larsen et al 1997; [13] Lopez et al 2010; [14] Shi et al 2016; [15]Ivanina ct al 2014; [16]Ivanina and Sokolova 2015;[17] Götze et al 2014;[18] Han et al 2014; [19]Lacoue-Labarthe et al 2011; [20] Ivanina et al 2013; [21] Gaunt and Barker 2000;[22] Fitzgerald et al 2016;[23] Monteiro et al 2020; [24] Ivanina et al 2010; [25] Lacoue-Labarthe et al 2009

Table 13.9. Toxicity matrix of some selected organic pollutants (POPs, pesticides, pharmaceuticals and oil compounds) induced by climate change stressors (temperatures, salinity, ocean acidification and hypoxia)

[↑= increase in toxicity of pollutants with increasing climate stressors; ↓= decrease in toxicity of pollutants with increasing climate stressors; Symbols: A=algae, C crustaceans; F= fish; M =molluscs; - = data is not available; number shown with a superscript are references listed underneath of this table].

	Atrazine	Benzo [a] pyrene	Chlor-pyrifos	Corexit	DDT	Dieldrin	Diuron	Endo-sulfan	17α-Ethinyl estra-diol	Ibupro-fen	PBDE	PCB	Trime-thyltin/ Tributyltin	Triclosan
Tempe-rature	↑ [A[1], F[8]]	↑ M[15]	↑ F[9] ↓ [A[25], C[6]]	↑ [C[2],M[16] ↓ F[22]	↑ [F[21]] ↓ [C[3,20]]	↑ [F[10]]	↑ [A[1]]	↑ [F[9,11,12,14]]	↑ [F[7]]	↑ [C[5]]	↑ [C[5]]	↑ [F[13]]	↑ [C[4]]	↑ [C[5]]
Salinity	↑ [A[17,18], F[19]] ↓ [C[19]]	-	↓ [C[6,20]]	-	↓ [C[20]]	↑ [F[10,21]]	-	-	-	↑ [C[5]]	↑ [C[5]]	-	↑ [C[4]]	↑ [C[5]]
Ocean acidifi-cation	-	-	-	-	-	-	-	-	-	-	-	-	-	-
Hypoxia	↑ [F[23]]	-	-	-	-	-	-	-	-	-	-	-	↑ [M[24]]	-

[1]DeLorenzo et al 2013; [2] National Research Council 2005; [3] Koenig et al 1976,[4] Wright and Roosenburg; [5] Al-Yassein 2019; [6] Pawara et al 2020; [7]Brian et al 2008; [8]Gaunt and Barker 2000; [9]Patra et al 2015;[10]Silbergeld 1973; [11]Ferrando et al 1987; [12]Capkin et al 2006; [13]Buckman et al 2007; [14]Nowak and Sunderam 1991; [15]Kamel et al 2012;[16]National Research Council 2005; [17]Hall et al 1997;[18]DeLorenzo et al 2013; [19]Hall et al 1994; [20] Staton et al 2002;[21]Katz 1961; [22] Kuhl et al 2013; [23] Gaunt and Barker 2000;[24] Barnett et al 2020; [25]Romero et al 2019

Reducing pollutants loads into the environment that causes significant harmful impacts on aquatic biota via **i.** reduce and restrict the production, use and sale of POPs listed under the Stockholm Convention (eliminate and restrict); **ii.** encourage and enforce acts, regulations and good practices for the management of pesticides, heavy metals and POPs residues; **iii.** effectively restrict the use of banned chemicals and pesticides; **iv**. contain/isolate pollutants (via litter traps, retention basins/artificial wetlands, or landfill); **v**. continue research where there are a baseline data gap related to the interactive effects of two stressors (climate change and pollutants) on aquatic biota; **vi.** awareness, education, capacity building on pollutants risks, **vii.** adopt polluters pay principle; and **viii.** regular monitoring programmes should be in place to monitor harmful, high-risk priority pollutants like dioxins, furans, PCBs, PBDEs, PFAS, pesticides, heavy metals, and pharmaceuticals (antibiotics) level in all the environmental matrices possible (water, soil, sediments, biota, food, human breast milk and compare with guidelines threshold values (e.g. water quality guidelines, maximum residue limits (MRL).

13.6. Conclusion

Both climate change and pollutants are a global problem and a major threat to social, economic and environmental sectors. The interactive effects of the two stressors (climate change and pollutants) would exacerbate global pollution problems and can pose significant threats and risks to aquatic ecosystems via increasing pollutant loads and by causing deleterious effects on aquatic biota (algae, crustaceans, fish, and molluscs). Evidence gathered in this review (sections 13.3.1-13.3.4) found that the synergetic and additive effects of the two stressors impaired growth and reproduction, increased vitellogenin (egg yolk precursor protein), altered metal-binding proteins (metallothionein) levels and enhanced bioaccumulation in aquatic biota (see Tables 13.8, 13.9). High temperature increased volatilisation of pollutants (POPs, DDT, pesticides, heavy metals) and increased release, mobilisation and degradation of pollutants. Extreme weather events increased mobilisation of both hydrophobic and hydrophilic pollutants (pesticides, heavy metals, nutrients, EDCs, PAHs, pharmaceuticals, PCDD/F, PCBs) from environmental reservoirs (soils, sediments, vegetation, permafrost, snow and ice, polar and alpine glaciers) into the environment.

The toxicity of several pollutants increased with increasing temperatures, salinity, ocean acidification and hypoxia (climate change stressors) to aquatic biota are summarised and listed below:

- **↑ Increase toxicity of pollutants with increasing <u>temperatures</u> to aquatic biota:** *2,4-D, ametryn, atrazine, bifenthrin, cadmium, carbaryl, chlorothalonil, corexit, cypermethrin, dicamba, deltamethrin, dieldrin, diuron, enodsulfan, esfenvalerate, ibuprofen, irgarol, mecoprop, methyl* *parathion, PBDE-47, scourge/ resmethrin, terbufos, trichlorfon, terbufos, trichlorfon and triclosan.*

- ↑ **Increase toxicity of pollutants with increasing salinity to aquatic biota:** *2,4-dinitrophenol, aldicarb, aldrin, ametryn, atrazine, bifenthrin, chlorothalonil, CoRal, DDT, dieldrin, diesel oil, dimethoate, diuron, guthion, ibuprofen, irgarol, malathion, methoxychlor, PBDE-47, tebufenozide, and toxaphene.*
- ↑_**Increase toxicity** of **pollutants with increasing ocean acidification/ low pH to aquatic biota:** *As, Ag, Cd, Cr, Cu, Fe, Hg, Mn, Ni Pb, Zn), and pyrethroid insecticide (resmethrin).*
- ↑ **Increase toxicity of pollutants with increasing hypoxia/** decrease in DO **to aquatic biota:** *Atrazine, cadmium, copper and mercury, resmethrin and tributyltin.*

This review paper also reveals that the toxicity of a few pollutants actually decreased with increasing temperatures, salinity, ocean acidification to aquatic biota are summarised and listed below:

- ↓ **Decrease toxicity of pollutants with increasing temperature to aquatic biota:** *chlorpyrifos, DDT, dimethoate, and nitrophenol*
- ↓ **Decrease toxicity of pollutants with increasing salinity to aquatic biota:** *2,4-dinitrophenol, 4 - nitrophenol, atrazine, cadmium, chlorpyrifos, chlordane, Corexit, DDT, endrin, dimethoate, lindane, scourage, and trimethyltin.*
- ↓ **Decrease toxicity of pollutants with increasing ocean acidification/ low pH to aquatic biota:** *cadmium, copper and mercury.*

13.5. References

ACIA 2005 Arctic climate impact assessment. Cambridge University Press. *https://www.amap.no/documents/doc/arctic-arctic-climate-impact-assessment/796*

Al-Yassein RN 2019 The Effects of climate change on the ecotoxicology of contaminants of emerging concern: flame retardants, contemporary use pesticides and pharmaceuticals personal care products on the estuarine grass shrimp, *Palaemontes pugio (Doctoral dissertation). Retrieved from https://scholarcommons.sc.edu/etd/5472*

Anawar HM 2013 Impact of climate change on acid mine drainage generation and contaminant transport in water ecosystems of semiarid and arid mining areas. *Physics and Chemistry of the Earth* **58–60** 13–21

Ardelan M, Steinnes E, Lierhagen S and Linde SO 2009 Effect of experimental CO_2 leakage on solubility and transport of seven trace metals in seawater and sediment. *Science of the Total Environment* **407** 6255–6266

Armitage JM, Quinn C and Wania F 2011 Global climate change and contaminants – an overview of opportunities and priorities for modelling the potential implications for long-term human exposure to organic compounds in the Arctic, *Journal of Environmental Monitoring* **13(6)** 1432–1546

Barnett AF, Gledhill JH, Griffitt RJ et al 2020 Combined and independent effects of hypoxia and tributyltin on mRNA expression and physiology of the Eastern oyster (*Crassostrea virginica*). *Scientific Reports* **10** 10605 *https://doi.org/10.1038/s41598-020-67650-x*

Benitez FJ, Real FJ, Acero JL and Garcia C 2006 Photochemical oxidation processes for the elimination of phenyl-urea herbicides in waters. *Journal of Hazardous Materials* **138(2)** 278–287

Bloomfield, J. P., R. J. Williams, D. C. Gooddy, J. N. Cape and P. Guha 2006. Impacts of climate change on the fate and behaviour of pesticides in surface and groundwater—a UK perspective. *Science of the Total Environment* **369** 163–177

Bogdal C, Schmid, P, Zennegg, M et al 2009 Blast from the past: melting glaciers as a relevant source for persistent organic pollutants. *Environmental Science & Technology* **43** 8173–8177

Bogdal C and Scheringer M 2011 Environmental fate and long-range transport of POPs. Chapter 3, pages 15-20. In: UNEP/AMAP. Climate change and POPs; Predicting the Impacts. Report of the UNEP/AMAP Expert group, Secretariat of the Stockholm Convention, Geneva. 62 pp. *file:///C:/Users/User1/Downloads/UNEP-POPS-GMP-REP-ClimateChangePredictImpacts-5.En%20(1).pdf*

Boone MD and Bridges CM 1999 The effect of temperature on the potency of carbaryl for survival of tadpoles of the green frog (*Rana clamitans*). *Environmental Toxicology and Chemistry* **18(7)** 1482–1484

Brecken-Folse JA, Mayer FL, Pedigo LE and Marking LL 1994 Acute toxicity of 4-nitrophenol, 2,4-dinitrophenol, terbufos and trichlorfon to grass shrimp (*Palaemonetes* spp.) and sheepshead minnows *Cyprinodon variegatus* as affected by salinity and temperature. *Environmental Toxicology and Chemistry* **13** 67–77

Brian JV, Harris CA, Runnalls TJ et al 2008 Evidence of temperature-dependent effects on the estrogenic response of fish: implications with regard to climate change. *Science of the Total Environment* **397** 72–81

Broomhall SD 2004 Egg temperature modifies predator avoidance and the effects of the insecticide endosulfan on tadpoles of an Australian frog. *Journal of Applied Ecology* **41(1)** 105–113

Buckman AH, Brown SB, Small J et al 2007 Role of temperature and enzyme induction in the biotransformation of polychlorinated biphenyls and bioformation of hydroxylated polychlorinated biphenyls by rainbow trout (*Oncorhynchus mykiss*). *Environmental Science and Technology* 41 3856–3863

Burek KA, Gulland FMD and Ohara TM 2008 Effects of climate change on Arctic marine mammal health. *Ecological Applications* **18(2)** S126-134

Burgoa B and Wauchope RD 1995 Pesticides in run-off and surface waters. *Environmental behaviour of agrochemicals.* John Wiley and Sons Ltd

Campbell AL, Mangan S, Ellis RP and Lewis C 2014 Ocean acidification increases copper toxicity to the early life-history stages of the polychaete *Arenicola marina* in artificial seawater. *Environmental Science & Technology* **48** 9745–9753

Capkin E, Altinok I and Karahan S 2006 Water quality and fish size affect toxicity of endosulfan, an organochlorine pesticide, to rainbow trout. *Chemosphere* **64(10)** 1793–800

Cheung PK, Cheung LTO, and Fok L 2016 Seasonal variation in the abundance of marine plastic debris in the estuary of a subtropical macro-scale drainage basin in South China. *Science of the Total Environment* **562** 658–665

Chiovarou ED and Siewicki TC 2007 Comparison of storm intensity and application timing on modelled transport and fate of six contaminants. *Science of the Total Environment* **389(1)** 87-100

Crawford CB and Quinn B 2017 Microplastic pollutants. Chapter 6. The interactions of microplastics and chemical pollutants. Persistent organic pollutants (POPs). Page 131-156. Elsevier Inc.

Dabrowski JM, Peall SKC, Reinecke AJ et al 2002 Runoff-related pesticide input into the Lourens River, South Africa: basic data for exposure assessment and risk mitigation at the catchment scale. *Water Air & Soil Pollution* **135** 265–268

DeLorenzo ME, Wallace SC, Danese LE and Baird TD 2009 Temperature and salinity effects on the toxicity of common pesticides to the grass shrimp *Palaemonetes pugio*. *Journal of Environmental Science and Health* B **44** 455–460 *DOI: 10.1080/03601230902935121*

DeLorenzo ME, Danese L and Baird TD 2013 Influence of increasing temperature and salinity on herbicide toxicity in estuarine phytoplankton. *Environmental Toxicology* **28(7)** 359-371

DeLorenzo ME 2015 Influence of increasing temperature and salinity on herbicide toxicity in estuarine phytoplankton. *Environmental Toxicology* **28(7)** 359-371 *Health* B **44** 455–460 *DOI: 10.1080/03601230902935121*

Diaz RJ and Rosenberg R 2008 Spreading Dead Zones and Consequences for Marine Ecosystems. *Science* **321** 926-929. *DOI: 10.1126/science.1156401*

Ferrando MD, Andreu-Moliner E, Almar MM and Cebrian C, Nunez A 1987 Acute toxicity of organochlorine pesticides the European eel, *Anguilla anguilla*: The dependency exposure time and temperature. *B Environ Contam Tox* **39** 365–369

Fischer H 1986 Influence of temperature, salinity, and oxygen on the cadmium balance of mussels *Mytilus edulis*. *Marine Ecology Progress Series* **32** 265–278

Fitzer SC, Caldwell GS, Clare AS et al 2013 Response of copepods to elevated $_{P}CO_2$ and environmental copper as co-stressors: A multigenerational study. *PLoS ONE* **8** e71257 *https://doi.org/10.1371/journal.pone.0071257*

Fitzgerald JA, Jameson HM, Dewar VH Fowler et al 2016 Hypoxia suppressed copper toxicity during early development in zebrafish embryos in a process mediated by the activation of the HIF signalling pathway. *Environmental Science and Technology* **50** 4502–4512 *DOI: 10.1021/acs.est.6b01472*

Garcıa RN, Chung KW, Key PB et al 2014 Interactive effects of mosquito control insecticide toxicity, hypoxia, and increased carbon dioxide on larval and juvenile Eastern oysters and hard clams. *Archives of Environmental Contamination and Toxicology* **66** 450–462

Gaunt P and Barker SA 2000 Matrix solid phase dispersion extraction of triazines from catfish tissues; examination of the effects of temperature and dissolved oxygen on the toxicity of atrazine. *International Journal of Environment and Pollution* **13(1-6)** 284-312 *DOI: 10.1504/IJEP.2000.002320*

Geisz HN, Dickhut RM, Cochran WA et al 2008 Melting glaciers: a probable source of DDT to the Antarctic marine ecosystem. *Environmental Science and Technology* **42** 3958–3962

Gewin V 2010 Dead in the water *Nature* **466** 812–814

Ghadiri H and Rose CW 2001 Degradation of endosulfan in a clay soil from cotton farms of western Queensland. *Journal of Environmental Management* **62** 155– 169

Gilman A and Jenssen BM I 2011 Impacts of climate change on toxicological and ecotoxicological effects of POPs exposure. Chapter 5, pages 29-41. In: UNEP/ AMAP. Climate change and POPs; Predicting the Impacts. Report of the UNEP/ AMAP Expert group, Secretariat of the Stockholm Convention, Geneva. 62 pp. *file:///C:/Users/User1/Downloads/UNEP-POPS-GMP-REP-ClimateChangePredictImpacts-5.En%20(1).pdf*

Götze S, Matoo OB, Beniash E et al 2014 Interactive effects of CO_2 and trace metals on the proteasome activity and cellular stress response of marine bivalves *Crassostrea virginica* and *Mercenaria mercenaria. Aquatic Toxicology* **149** 65–82 *DOI: 10.1016/j.aquatox.2014.01.027*

Greco L, Pellerin J, Capri E et al 2011 Physiological effects of temperature and a herbicide mixture on the soft-shell clam *Mya arenaria* (Mollusca, Bivalvia). *Environmental Toxicology and Chemistry* **30** 132–141

Han ZX, Wu DD, Wu J et al 2014 Effects of ocean acidification on toxicity of heavy metals in the bivalve *Mytilus edulis* L. *Synthesis and Reactivity in inorganic Metal-Organic and Nano-Metal Chemistry* **44** 133–139

Hall L, Ziegenfuss M, Anderson R et al 1994 Influence of salinity on atrazine toxicity to a Chesapeake Bay copepod *Eurytemora affinis* and fish *Cyprinodon variegatus. Estuaries* **17** 181–186

Hall Jr LW, Ziegenfuss MC, Anderson RD and Lewis BL 1995 The effect of salinity on the acute toxicity of total and free cadmium to a Chesapeake Bay copepod and fish. *Marine Pollution Bulletin.* **30** 376–384

Hall Jr LW, Anderson DM and Ailstock MS, 1997 Chronic toxicity of atrazine to sago pondweed at a range of salinities: Implications for criteria development and ecological risk. *Archives of Environmental Contamination and Toxicology* **33** 261–267

Harwood AD, You J and Lydy MJ 2009 Temperature as a toxicity identification evaluation tool for pyrethroid insecticides: Toxicokinetic confirmation. *Environmental Toxicology and Chemistry* **28** 1051–1058

Hu J.-Q, Yang S.-Z, Guo L et al 2017 Microscopic investigation on the adsorption of lubrication oil on microplastics. *Journal of Molecular Liquids* **227** 351-355

Hu S, Sprintall J, Guan C et al 2020 Deep-reaching acceleration of global mean ocean circulation over the past two decades. *Science Advances* **6** eaax7727

IPCC (Intergovernmental Panel on Climate Change) 2007a. Summary for Policymakers. In: Climate Change 2007: The Physical Science Basis. Contribution of Working Group I to the Fourth Assessment Report of the Intergovernmental Panel on Climate Change. In: Solomon, S., D. Qin, M. Manning, Z. Chen, M. Marquis, K.B. Averyt, M. Tignor and H.L. Miller (eds.). Cambridge University Press, Cambridge, United Kingdom and New York, NY, USA, 996 p.

IPCC (Intergovernmental Panel on Climate Change) 2007b Climate change 2007. Impacts, adaptation and vulnerability. Contribution of working group II to the fourth assessment report of the Intergovernmental panel on climate change. Cambridge University Press.

IPCC (Intergovernmental Panel on Climate Change) 2014 Climate Change 2014 Impacts, Adaptation, and Vulnerability Part A: Global and Sectoral Aspects Contribution of Working Group II to the Fifth Assessment Report of the Intergovernmental Panel on Climate Change. Cambridge/New York: Cambridge University Press.

IPCC 2018. Summary for Policymakers - Special Report. IPCC Press Release 2018/24/PR (2018).

Ivanina AV, Sokolov EP and Sokolova IM 2010 Effects of cadmium on anaerobic energy metabolism and mRNA expression during air exposure and recovery of an intertidal mollusk *Crassostrea virginica. Aquatic Toxicology* **99** 330–342

Ivanina AV, Beniash E, Etzkorn M et al 2013 Short-term acute hypercapnia affects cellular responses to trace metals in the hard clams *Mercenaria mercenaria. Aquatic Toxicology* **140/141** 123–133

Ivanina AV, Hawkins C and Sokolova IM 2014 Immunomodulation by the interactive effects of cadmium and hypercapnia in marine bivalves *Crassostrea virginica* and *Mercenaria mercenaria. Fish & Shellfish Immunology* **37** 299–312

Ivanina AV and Sokolova IM 2015 Interactive effects of metal pollution and ocean acidification on physiology of marine organisms. *Current Zoology* **61(4)** 653–668

Jarsjö J, Andersson-Sköld Y, Fröberg M et al 2020 Projecting impacts of climate change on metal mobilization at contaminated sites: Controls by the groundwater level. *Science of the Total Environment* **712** 135560

Johnson SE and DeLorenzo ME 2011 Effects of increased temperature on pesticide toxicity in the grass shrimp *Palaemonetes pugio*. NOAA Office of Education, Science and Education Symposium, 2–4 August 2011, Silver Spring, MD

Kamel N, Attig H, Dagnino A, Boussetta H and Banni M 2012 Increased temperatures affect oxidative stress markers and detoxification response to benzo[a]pyrene exposure in mussel *Mytilus galloprovincialis. Archives of Environmental Contamination and Toxicology* **63** 534–543

Katz M 1961. Acute toxicity of some organic insecticides to three species of salmonids and to the three-spine stickleback. *Transactions of the American Fisheries Society* **90** 264–268

Kennedy CJ and Walsh PJ 1997 Effects of temperature on xenobiotic metabolism. In: Wood, C.M., McDonald, D.G. (Eds.), Global Warming – Implications for Freshwater and Marine Fish. Cambridge University Press, pp. 303–324.

Khan MA, Ahmed SA, Catalin B et al 2006 Effect of temperature on heavy metal toxicity to juvenile crayfish *Orconectes immunis* (Hagen). *Environmental Toxiclogy* **21** 513–520

Kallenborn R, Halsall C, Dellong M and Carlsson P 2012 The influence of climate change on the global distribution and fate processes of anthropogenic persistent organic pollutants. *Journal of Environmental Monitoring* **14** 2854–2869

Kibria G, Haroon AKY, Nugegoda D and Rose G 2010 Climate change and chemicals: Environmental Biological aspects. New India Publishing Agency. New Delhi. 470 p. ISBN: 9789380235301. *DOI: 10.13140/2.1.4384.0963*

Kibria G, Lau, TC and Wu R 2012 Innovative 'Artificial Mussels' technology for assessing spatial and temporal distribution of metals in Goulburn–Murray catchments waterways, Victoria, Australia: Effects of climate variability (dry vs. wet years). *Environment International* **50** 38-48

Kibria G, Haroon AKY and Nugegoda D 2013 Climate change and Agricultural food production: Impacts, vulnerabilities and remedies. New India Publishing Agency (NIPA), New Delhi, India, 300 p. ISBN: 9789381450512. *DOI: 10.13140/2.1.3245.4081*

Kibria G 2014 Climate Change and Chemical Contaminants. *DOI: 10.13140/RG.2.1.2733.6169*

https://www.researchgate.net/publication/261216780_Climate_Change_and_Chemical_Contaminants

Kibria G and Haroon AKY 2016 Climate change impacts on wetlands of Bangladesh, its biodiversity and ecology, and actions and programs to reduce risks. Chapter 10. Pages: 189-204. In: Prusty, A.K, Chandra, R. and Azeez, P.A. (eds.), Wetland Science: Perspectives from South Asia. Springer.

Kibria G, Haroon AKY and Nugegoda D 2016a Climate Change and Water security. Impacts, future projections, adaptations and mitigations. New India Publishing Agency (NIPA), New Delhi, India, 330 p, ISBN: 978-93-85516-26-9. *DOI: 10.13140/RG.2.1.1848.1528/2*

https://www.researchgate.net/publication/303218799_Climate_Change_Water_Security

Kibria G, Hossain MM, Mallick D, Lau TC and Wu R 2016b Trace/heavy metal pollution monitoring in estuary and coastal area of Bay of Bengal, Bangladesh and implicated impacts. *Marine Pollution Bulletin* **105** 393–402

http://dx.doi.org/10.1016/j.marpolbul.2016.02.021

Kibria G 2017 Bush fire/Forest fire/Wildfire Risk Increases Globally but Why? *DOI: 10.13140/RG.2.2.35126.60481*

Kibria G, Haroon AKY, Nugegoda D 2017 Climate change impacts on tropical and temperate fisheries, aquaculture, and seafood security and implications - A review. *Livestock Research for Rural Development* 29 (1)

https://www.lrrd.org/lrrd29/1/kibr29012.htm

Koenig C, Livingston R and Cripe C 1976 Blue crab mortality: Interaction of temperature and DDT residues. *Archives of Environmental Contamination and Toxicology* **4** 119–128

Kookana R, Holzb G, Chris Barnes C et al 2010 Impact of climatic and soil conditions on environmental fate of atrazine used under plantation forestry in Australia. *Journal of Environmental Management* **91(12)** 2649-2656

Ku P, Tsz-Ki Tsui M, Nie X, Chen H et al 2018 Origin, reactivity, and bioavailability of Mercury in wildfire ash. *Environmental Science and Technology* **52(24)** 14149-14157 *https://pubs.acs.org/doi/abs/10.1021/acs.est.8b03729*

Kuhl AJ, Nyman JA, Kaller MD and Green CC 2013 Dispersant and salinity effects on weathering and acute toxicity of Sweet Louisiana crude oil. *Environmental Toxicology and Chemistry* **32** 2611–2620

Lacoue-Labarthe T, Martin S, Oberhänsli F et al 2009 Effects of increased pCO_2 and temperature on trace element (Ag, Cd and Zn) bioaccumulation in the eggs of the common cuttlefish *Sepia officinalis*. *Biogeosciences* **6** 1–13

Lacoue-Labarthe T, Réveillac E, Oberhänsli F et al 2011 Effects of ocean acidification on trace element accumulation in the early-life stages of squid *Loligo vulgaris*. *Aquatic Toxicology* **105** 166–176

Lamon L, Waldow HV, McLeod M et al 2009 Modeling the global levels and distribution of polychlorinated biphenyls in air under a climate change scenario. *Environment Science and Technology* **43** 5818-5824

Larsen BK, Poertner HO and Jensen FB 1997 Extra- and intracellular acid- base balance and ionic regulation in cod *Gadus morhua* during combined and isolated exposures to hypercapnia and copper. *Marine Biology* **128** 337–346

Last PR, White WT, Gledhill DC et al 2011 Long-term shifts in abundance and distribution of a temperate fish fauna: a response to climate change and fishing practices. *Global Ecology and Biogeography* **20** 58–72

Leal PP, Hurd CL, Sander SG et al 2018 Copper pollution exacerbates the effects of ocean acidification and warming on kelp microscopic early life stages *Scientific Reports* **8** 14763

Li Y, Wang W-X and Wang M 2017 Alleviation of mercury toxicity to a marine copepod under multigenerational exposure by ocean acidification. *Scientific Reports* **7** 324 *DOI:10.1038/s41598-017-00423-1*

Linden O, Laughlin R, Sharp JR and Neff JM 1980 The combined effect of salinity, temperature and oil on the growth pattern of embryos of the killifish *Fundulus heteroclitus* Walbaum. *Mar. Environ. Res.* **3** 129–144

Lopez IR, Kalman J, Vale C and Blasco J 2010 Influence of sediment acidification on the bioaccumulation of metals in *Ruditapes philippinarum*. *Environmental Science and Pollution Research* **17** 1519–1528

Ma J, Hung H, and Blanchard P 2004 How do climate fluctuations affect persistent organic pollutant distribution in North America? Evidence from a decade of air monitoring. *Environmental Science and Technology* **38(9)** 2538–2543

Ma JM, Hung HL, Tian C and Kallenborn R 2011 Revolatilization of persistent organic pollutants in the Arctic induced by climate change. *Nature Climate Change* **1** 255–260

Macdonald RW, Mackay D, Li YF and Hickie B 2003a How will global climate change affect risks from long- range transport of persistent organic pollutants? *Human and Ecological Risk Assessment* **9(3)** 643–660

Macdonald RW, Harner T, Fyfe J et al 2003b AMAP Assessment 2002, The Influence of Global Change on contaminant pathways to, within, and from the Arctic. Arctic Monitoring and Assessment Programme (AMAP), report, 2003, Oslo, Norway, vol. xii, p. 65.

Marques A, Nunes ML, Moore SK and Strom MS 2010 Climate change and seafood safety: human health implications. *Food Research International* **43** 1766–1779

Materna EJ, Rabeni CF and Lapoint TW 1995 Effects of the synthetic pyrethroid insecticide, esfenvalerate, on larval leopard frogs (*Rana* spp.). *Environmental Toxicology and Chemistry* **14(4)** 613–622

Meehl G A, Stocker T F, Collins W et al 2007 Global climate projections. Climate Change 2007: The Physical Science Basis. Contribution of Working Group I to the Fourth Assessment Report of the Intergovernmental Panel on Climate Change. In: S. Solomon, D. Qin, M. Manning, Z. Chen, M. Marquis, K. B. Averyt, M. Tignor and H. L. Miller, (eds.), Cambridge University Press, Cambridge. p. 747-845

Meyer T and Wania F 2008 Organic contaminant amplification during snowmelt. *Water Research* **42** 1847–1865

Monserrat J and Bianchini A 1995 Effects of temperature and salinity on the toxicity of a commercial formulation of methyl parathion to *Chasmagnathus granulata* (Decapoda, Grapsidae). *Brazilian Journal of Medical and Biological Research* **28(1)** 74–78

Monteiro DA, Taylor EW, McKenzie DJ et al 2020 Interactive effects of mercury exposure and hypoxia on ECG patterns in two Neotropical freshwater fish species: *Matrinxã, Brycon amazonicus* and traíra, *Hoplias malabaricus Ecotoxicology* **29(4)** 375-388 *doi: 10.1007/s10646-020-02186-4. Epub 2020 Mar 12.*

National Research Council 2005 Oil Spill Dispersants: Efficacy and Effects. Washington: The National Academies Press.

Nowak B and Sunderam RIM 1991 Toxicity and bioaccumulation of endosulfan to mosquitofish, *Gambusia affinis* (Baird & Girard). *Internationale Vereinigung f€ur Theoretische und Angewandte Limnologie*: Verhandlungen **24** 2327–2329

Noyes PD, McElwee MK, Hilary D. Miller HD et al 2009 The toxicology of climate change: Environmental contaminants in a warming world. *Environment International* **35** 971–986

OECD 2019. Pharmaceutical residues in freshwater. Hazards and Policy responses. *https://www.oecd.org/environment/resources/pharmaceutical-residues-in-freshwater-policy-highlights.pdf*

O'Hara J 1973 The influence of temperature and salinity on the toxicity of cadmium to the fiddler crab *Uca pugilator. Fisheries Bulletin* **71** 149–153

Pascal P-Y, Fleeger JW, Galvez F and Carman KR 2010 The toxicological interaction between ocean acidity and metals in coastal meiobenthic copepods. *Marine Pollution Bulletin* **60** 2201–2208

Patra RW, Chapman JC, Lim RP et al 2009 Effects of temperature on ventilatory behaviour of fish exposed to sublethal concentrations of endosulfan and chlorpyrifos. *Environmental Toxicology and Chemistry* **28** 2182–2190

Patra RW, Chapman JC, Lim RP et al 2015 Interactions between temperature and contaminant toxicity to freshwater fish. *Environmental Toxicology and Chemistry* **34(8)** 1809–1817

Paul AG, Hammen VC, Hickler T et al 2012 Potential implications of future climate and land-cover changes for the fate and distribution of persistent organic pollutants in Europe. *Global Ecology and Biogeography* **21** 64–74

Pawara AP, Sanayea SV, Soorambail Shyama S et al 2020 Effects of salinity and temperature on the acute toxicity of the pesticides, dimethoate and chlorpyrifos in post-larvae and juveniles of the white leg shrimp. *Aquaculture Reports* **16** 100240

Perry AL, Low PJ, Ellis JR and Reynolds JD 2005 Climate change and distribution shifts in marine fishes. *Science* **30** 1912-1915

Presley SM, Rainwater TR, Austin GP et al 2005 Assessment of pathogens and toxicants in New Orleans, LA following hurricane Katrina. *Environmental Science and Technology* **40** 468-474

Pulkrabova´ J, Suchanova M, Tomaniova M et al 2008 Organic pollutants in areas impacted by flooding in 2002: A 4-Year Survey. Bulletin of *Environmental Contamination and Toxicology* **81(3)** 299-304 *DOI: 10.1007/s00128-008-9486-6*

Rabalais NN, Harper Jr., DE, Turner RE 2001 Responses of nekton and demersal and benthic fauna to decreasing oxygen concentrations. *In*: Rabalais NN and Turner RE (eds.), *Coastal and estuarine studies: coastal hypoxia consequences for living resources and ecosystems.* American Geophysical Union, Washington D C, 115–128

Ratushnyak AA, Andreeva MG, Trushin MV 2005 Effects of type II pyrethroids on *Daphnia magna*: dose and temperature dependences. *Riv Biol-Biol Forum* **98(2)** 349–357

Renaud ML 1986 Hypoxia in Louisiana coastal waters during 1983: Implications for fisheries. *Fishery Bulletin* **84** 9–26

Roberts DA, Birchenough SR, Lewis C et al 2013 Ocean acidification increases the toxicity of contaminated sediments. *Global Change Biology* **19** 340–351 *DOI: 10.1111/gcb.12048*

Roessig J, Woodley C M, Cech J and Hansen L J 2005 Effects of global climate change on marine and estuarine fishes and fisheries. *Reviews in Fish Biology & Fisheries* **14** 251-275

Romero, F, Acuña C, Carme Font C et al 2019 Effects of multiple stressors on river biofilms depend on the time scale. *Scientific Reports* **9** 15810 *https://doi.org/10.1038/s41598-019-52320-4*

Satpute NS, Deshmukh SD, Rao NGV et al 2007 Temperature-dependent variation in toxicity of insecticides against *Earias vitella* (Lepidoptera: Noctuidae). *Journal of Economic Entomology* **100(2)** 357-360. *doi: 10.1603/0022-0493(2007)100 [357:tvitoi]2.0.co;2.*

Scherer C, Seelanda, Oehlmanna O and Ruth Müller R 2013 Interactive effects of xenobiotic, abiotic and biotic stressors on *Daphnia pulex*—Results from a multiple stressor experiment with a fractional multifactorial design. *Aquatic Toxicology* **138–139** 105–115

Schenker U, Scheringer M and Hungerbühler K 2007 Including degradation products of persistent organic pollutants in a global multi-media box model. *Environmental Science and Pollution Research* **14(3)** 145–152

Schiedek D, Sundelin B, Readman JW and Macdonald RW 2007 Interactions between climate change and contaminants. *Marine Pollution Bulletin* **54** 1845–1856

Schmidhuber J and Tubiello F N 2007 Global food security under climate change. *PNAS.* **104 (50)** 19703-19708

Schwarzenbach RP, Gschwend PM and Imboden DM. 2003 Solubility and activity coefficient in water. Environmental organic chemistry. 2nd Edition. Hoboken, NJ: John Wiley & Sons, Inc

Service RF 2004 Oceanography. New dead zone off Oregon coast hints at sea change in currents. *Science* **305** 1099–1099

Shi W, Zhao X and Han Y et al 2016 Ocean acidification increases cadmium accumulation in marine bivalves: a potential threat to seafood safety. *Scientific Reports* **6** 20197 *doi: 10.1038/srep20197*

Shivaramaiah HM, Sanchez-Bayo J Al-Rifai J and Kennedy IR. 2005 The fate of endosulfan in water. *Journal of Environmental Science and Health* B **40** 711–720

Silbergeld EK 1973 Dieldrin. Effects of chronic sublethal exposure on adaptation to thermal stress in fresh-water fish. *Environmental Science & Technology* **7(9)** 846-849

Siscar R, Torreblanca A, del Ramo J and Solé M, 2014 Modulation of metallothionein and metal partitioning in liver and kidney of *Solea senegalensis* after long-term acclimation to two environmental temperatures. *Environmental Research* **132** 197–205

Solomon S, Qin D, Manning M et al 2007 Technical Summary. In: Climate Change 2007: The Physical Science Basis. Contribution of Working Group I to the Fourth Assessment Report of the Intergovernmental Panel on Climate Change. In: Solomon, S., D. Qin, M. Manning, Z. Chen, M. Marquis, K.B. Averyt, M. Tignor and H.L. Miller (eds.). Cambridge University Press, Cambridge, United Kingdom and New York, NY, USA. 91 p

Song MY and Brown JJ 1998 Osmotic effects as a factor modifying insecticide toxicity on *Aedes* and *Artemia. Ecotoxicology and Environmental Safety* **41** 195–202

Staton JL, Schizas NV, Klosterhaus SL et al 2002 Effect of salinity variation and pesticide exposure on an estuarine harpacticoid copepod *Microarthridion littorale* (Poppe), in the southeastern US. *Journal of Experimental Marine Biology and Ecology* **278** 101-110

Todd AS, Manning AH, Verplanck PL et al 2012 Climate-change-driven deterioration of water quality in a mineralized watershed. *Environmental Science and Technology* **46** 9324–93

Tedengren M and Kautsky N 1987 Comparative stress response to diesel oil and salinity changes of the blue mussel *Mytilus edulis* from the Baltic and North seas. *Ophelia* **28** 1–9

Turley C M , Eby M, Ridgwell A J et al 2010 The societal challenge of ocean acidification. *Marine Pollution Bulletin* **60** 787-792

Umlauf G, Bidoglio G, and Christoph E 2005 The situation of PCDD/Fs and Dioxin-like PCBs after the flooding of river Elbe and Mulde in 2002. *Acta Hydrochimica et Hydrobiologica.* **33(5)** 543–554

UNEP (United Nations Environment programme) 2010 Climate Change and POPs; predicting the impacts, Report of the UNEP/AMAP expert group, United Nations Environmental Programme, report, p. 65

UNEP (United Nations Environment programme)/ AMAP (Arctic Monitoring and Assessment Programme) 2011 Climate change and POPs; Predicting the Impacts. Report of the UNEP/AMAP expert group, Secretariat of the Stockholm Convention, Geneva. 62 pp.

https://www.amap.no/documents/doc/climate-change-and-pops-predicting-the-impacts/753

UNFCC 1992 United Nations Framework Convention on Climate Change (UNFCC). FCCC/INFORMAL/84 GE.05-62220 (E) 20070. 24p *https://unfccc.int/resource/docs/convkp/conveng.pdf*

Van den Berg F, Kubiak R, Benjey WG et al 1999 Emission of pesticides into the air. *Water, Air & Soil Pollution* **115** 195–218

Wheeler D 2011 Quantifying Vulnerability to Climate Change: Implications for Adaptation Assistance. Washington, DC: Center for Global Development. *www.cgdev.org/content/publications/detail/142475*

Wilken M, Walkow F. Jager E and Zeschmar-Lahl B 1994 Flooding area and sediment contamination of the river Mulde (Germany) with PCDD/F and other organic pollutants. *Chemosphere* **29(9-11)** 2237-2252

Wittmann A C and Pörtner H O 2013 Sensitivities of extant animal taxa to ocean acidification. *Nature Climate Change* 3 995–1001

Wright D and Roosenburg W 1982 Trimethyltin toxicity to larval *Uca pugilator*: Effects of temperature and salinity. *Archives of Environmental Contamination and Toxicology* **11** 491–495

Wu C, Zhang K, Huang X and Liu J 2016 Sorption of pharmaceuticals and personal care products to polyethylene debris. *Environmental Science and Pollution Research* **23(9)** 8819-8826

Yamashita N, Taniyasu S, Petrick G, Wei Si, Gamo T, Lam PKS, Kannan K 2008 Perfluorinated acids as novel chemical tracers of global circulation of ocean waters. *Chemosphere* **70(7)** 1247-1255 *doi: 10.1016/j.chemosphere.2007.07.079. Epub 2007 Sep 12.*

Zaharescu DG, Hooda PS, Carmen I. Burghelea CI et al 2016 Climate change enhances the mobilisation of naturally occurring metals in high altitude environments. *Science of the Total Environment* **560–561** 73–81

http://dx.doi.org/10.1016/j.scitotenv.2016.04.002

Zanette J, de Almeida EA, da Silva AZ et al 2011 Salinity influences glutathione S-transferase activity and lipid peroxidation responses in the *Crassostrea gigas* oyster exposed to diesel oil. *Science of the Total Environment* 409 1976–1983

Xu H, Zhang Y, Zhu X and Zheng M 2019 Effects of rainfall-runoff pollution on eutrophication in coastal zone: a case study in Shenzhen Bay, southern China. *Hydrology Research* **50.4** 1062- 1075 *doi: 10.2166/nh.2019.012*

Web

https://www.environment.gov.au/protection/chemicals-management/pops; https://en.wikipedia.org/wiki/Persistent_organic_pollutant

https://phys.org/news/2018-12-wildfire-ash-mercury.html

https://climate.ncsu.edu/edu/Albedo

https://www.sciencedirect.com/topics/pharmacology-toxicology-and-pharmaceutical-science/toxicokinetics)

https://www.sciencedirect.com/topics/neuroscience/metallothionein#

https://www.ncbi.nlm.nih.gov/pmc/articles/PMC3812322/#:

https://www.medicalnewstoday.com/articles/325873

https://pubchem.ncbi.nlm.nih.gov/compound/Malonaldehyde

https://www.sciencedirect.com/topics/agricultural-and-biological-sciences/cytochrome-c-oxidase

https://www.sciencedirect.com/topics/neuroscience/superoxide-dismutase

https://www.nature.com/articles/s41389-017-0025-3

https://gulfhypoxia.net/about-hypoxia/

https://oceanservice.noaa.gov/podcast/feb18/nop13-hypoxia.html

■■■■

Appendix - 1 Index

B

C

G

H

M

■■■■

Appendix - 2
Other books of the authors of the similar series

2010

Golam Kibria, A K Yousuf Haroon, Dayanthi Nugegoda and Gavin Rose 2010. *Chemicals and Climate change: Environmental and biological aspects.* New India Publishing Agency (NIPA), New Delhi, India. 472 pages. ISBN 9789-38-0235-301.

2013

Golam Kibria, A K Yousuf Haroon and Dayanthi Nugegoda 2013. *Climate Change and Food Production: Impact, Vulnerabilities and Remedies*. New India Publishing Agency (NIPA), New Delhi, India. 312 pages. ISBN 978-93-81450-51-2.

2016

Golam Kibria, A K Yousuf Haroon and Dayanthi Nugegoda 2016. *Climate change and Water Security: Impacts, Future Scenarios, Adaptations & Mitigations.* New India Publishing Agency (NIPA), New Delhi, India. 314 pages. ISBN: 978-93-85516-26-9.

2021

Golam Kibria, A K Yousuf Haroon, Gavin Rose, Md. M. Maruf Hossain and Dayanthi Nugegoda 2021. *Pollution Risks, Impacts & Management. Social, Economic and Environmental Perspectives.* Scientific Publishers, Jodhpur - 342 001, Rajasthan, India. 833 pages. ISBN: 9789-3891-84969

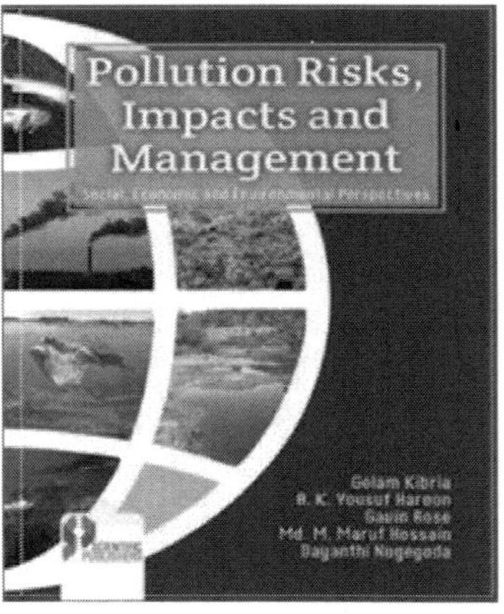